# Grapes & wines

# Grapes & wines

a comprehensive guide to varieties and flavours–
the key to enjoying modern wine

OZ CLARKE
&
MARGARET RAND

PAINTINGS BY LIZZIE RICHES
PRINCIPAL PHOTOGRAPHER MICK ROCK

WEBSTERS
LITTLE, BROWN AND COMPANY
BOSTON NEW YORK LONDON

A LITTLE, BROWN/WEBSTERS BOOK
This edition first published in 2001 by
Little, Brown and Company (UK)
Brettenham House
Lancaster Place
London WC2E 7EN

Created and designed by
Websters International Publishers Limited
Axe and Bottle Court
70 Newcomen Street
London SE1 1YT
*www.websters.co.uk*
*www.ozclarke.com*



A CIP catalogue for this book is
available from the British Library

ISBN 0-316-85726-2

**Colour separations** by P. T. Repro Multi Warna, Indonesia
**Printed and bound** by Star Standard Industries (PTE), Singapore

**Paintings** Lizzie Riches
Lizzie Riches is represented by The Portal Gallery, London
**Photography** Cephas Picture Library/principal photographer Mick Rock

**Editorial Director & Chief Editor** Fiona Holman
**Art Director & Art Editor** Nigel O'Gorman
**Editorial Assistant** Jo Turner
**Other editorial assistance** Kim Parsons, Maggie Ramsay, Helen Spence, Phillip Williamson
**Maps** Andrew Thompson
**Bottle photography** Stephen Marwood
**Desktop publishing** Keith Bambury
**Indexer** Angie Hipkin
**Managing Editor** Anne Lawrance
**Production** Kâren Connell

CORPORATE SALES
Companies, institutions and other organizations wishing to make bulk purchases of this or any other Oz Clarke title published by Little, Brown should contact the Special Sales Department on +44(0)20-7911 8089

***Page 1:*** *Cabernet Sauvignon at budbreak in early spring.*
***Page 2:*** *Bunches of Merlot grapes ready for harvest in the Sonoma Valley, California.*
***Page 3:*** *Zinfandel has been adopted by the Californians as their very own classic grape.*
***Page 4:*** *Tinto Fino (better known as Tempranillo) grapes on a 60-year-old vine at Dominio de Pingus, Ribera del Duero, Spain.*

# CONTENTS


**All about Grapes 6–31**

Introduction 6
The Story of the Vine 8
Vines Today 10
The Global Picture 12
Where Grapes Grow 14
In the Vineyard 18
In the Winery 28

**A-Z of Grape Varieties 32-295**

Key to the Classic Grape Paintings 296
Which Grapes Make Which Wines 298
Glossary of Technical Terms 304
Index of Grape Names and Their Synonyms 308
General Index 311
Bibliography 320
Acknowledgments 320


## SPECIAL FEATURES
### Classic Grapes


Cabernet Sauvignon 46 
Chardonnay 62 
Chenin Blanc 74 
Garnacha Tinta/ Grenache Noir 92 
Gewürztraminer 102 
Merlot 126 
Muscat 144 
Nebbiolo 154 
Pinot Noir 174 
Riesling 190 
Sangiovese 208 
Sauvignon Blanc 218 
Sémillon 230 
Syrah/Shiraz 244 
Tempranillo 256 
Viognier 274 
Zinfandel 286 

### Major Grapes


Albariño 36
Barbera 40
Cabernet Franc 44
Carmenère 60
Dolcetto 86
Malbec 118
Malvasia 120
Marsanne 124
Mourvèdre 140
Pinot Blanc 170
Pinot Gris 172
Pinotage 186
Roussanne 204
Silvaner 242
Touriga Nacional 268

# Introduction

So what is a grape, then? Well, it's juice and flesh, obviously. It's skin, obviously. It's pips, and I suppose it could be the stalks as well. And then what? And then everything, that's what.

If we have any interest in wine and in flavours, we have to be interested in the grape variety itself. If we have any interest in how a wine matures and changes with age, we have to know about the potential of the particular grape. If we care about the style of a wine, whether it should be sweet or dry, fizzy or fortified or still, each different grape variety's peculiar talents will be of prime importance. Do we like the flavour of oak barrel aging in our wine? Some grapes take to oak, again some don't – it's vital to know which ones love the kiss of oak and which ones loathe its hot embrace. And are we fascinated by how completely different wines taste when they come from different countries and from different regions within those countries? Without the consistent character of each different grape variety to use as a measuring point, mere comparison of place would be meaningless. However far we delve into all the things that influence the flavours of our wine, it all comes back to the grape.

**Above:** *Chardonnay and Pinot Noir are just about two of the most famous classic grape varieties in the world. Both hail from Burgundy and are responsible for two of the world's most magical wine styles – Pinot Noir for haunting sensuous reds and Chardonnay for stunningly rich, honeyed, nutty, buttery whites. Yet isn't it strange that neither red or white Burgundy trumpets the grape names on the label.* **Facing page:** *Double magnums stored away in the vintage bottle cellar at Château Canon-la-Gaffelière in St-Émilion, Bordeaux.*

And if that's the case, it seemed to me it was high time we had a book which took a really long hard look at the world of grapes, the wines they create and the flavours they contribute. A book which recognized that grape varieties are the most important factor in the flavour of a wine.

I mean, think about it. I give you a glass of pale golden green wine. It's got a wonderful pungent scent of gooseberry and passionfruit and lime. You taste it and the acidity crackles against your teeth, the exhilarating attack of citrus fruit scours your palate clean and makes your mouth drool with desire for food. Who made the wine? No idea. Where does it come from? It could be the Loire Valley in France. But it could also be South Africa or Chile, it could be Spain or northern Italy. And it could certainly be New Zealand. So. The four corners of the earth, really. But the grape variety? When the wine smells and tastes like that, you *know* it is Sauvignon Blanc. The unique, brilliantly recognizable character of the wine is down to the grape variety – Sauvignon Blanc – above all else. It is refined by the relative talent of the men and women who grow the grape and vinify the wine. It is modified or intensified by the local conditions under which it grows. But the core of the flavour comes from the grape.

Now, Sauvignon Blanc is a very dramatic grape. But so is Viognier with its powerful scent of apricot and may blossom. So is Gewürztraminer with its explosion of musky rose petals and lychees. So is Muscat with its overpowering aroma of hothouse grapes. Riesling is more subtle, but the unmistakeable balance of high acidity with floral notes and citrus fruit is unique to the grape. Chardonnay's nutty, oatmealy ripeness is created with the help of oak barrel aging, but no other grape achieves quite that taste, however similarly you treat it, wherever it is grown.

Red wine grapes are frequently less outspoken, and just at present the obsession with over-zealous use of new oak to age wines is spoiling the thrilling individuality of many grapes' flavours – but good varieties still shine through. Tannic sturdiness and blackcurrant fruit mark out Cabernet Sauvignon in a way no other grape can replicate. The ethereal scent and strawberry/cherry fruit of Pinot Noir, the damson fruit and violet perfume of Malbec, the sour cherry and herbal rasp of Sangiovese, the brilliant chocolate and smoky black plum blast of Shiraz – all of these experiences and many more are above all else due to the particular characteristics of the grape variety.

Obviously, every wine book has to talk about grape varieties to a greater or lesser extent. But it is remarkable how, over the years, grapes seem to have been relegated to a subordinate role when they are so evidently of such massive importance. Well, one of the reasons has to be that until the advent of modern 'New World' winemaking techniques that allow the winemaker to pinpoint the potential flavour of the grape and then maximize it, I suspect that few people – winemakers, wine writers and wine drinkers all – actually had much idea of what a grape variety was supposed to taste like. It was easier to say that a wine's particular taste derived from where it was grown, that it tasted of what the French call 'terroir'. Indeed, the wines often did have a minerally or earthy flavour which probably did emanate from the vineyard and from old-fashioned winemaking, rather than the actual grape itself. That's why, until recently, many experts and critics were obsessed with the minutiae of a wine's birthplace rather than its chief component – the juice of the grape itself.

But when the New World producers brashly barged their way through into our wine consciousness, everything changed. The Australians and Californians, New Zealanders, South Africans and Chileans didn't have much of a story to tell when it came to the traditions and historical importance of their vineyards – many of these selfsame vineyards had only just been planted. The one story they could tell, and the one their ultra-modern winemaking allowed them to tell, was that of the grape itself and the flavour it imparted to the wine.

And that's the story I'm hoping to tell here. The advent of labelling most wines first and foremost by their grape variety – 'varietal labelling' this is called – has done a brilliant job in simplifying the world of wine so that we can all make informed choices about what we like and what we don't like. But it has done more. As we get to know more about wine we crave variety. Varietal labelling makes it so easy to extend our knowledge and experience. Of course we can depend upon favourite countries and regions, just as we can depend upon favourite producers and vintages, but the grape variety blazes the way: everything else follows.

And so it does in this book. We look at the varieties of the world in enormous detail, much of it unpublished before. We look at the history, the places, the people, the wine styles and the flavours. The whole world of wine with the wine grape at its heart, is what this book is about.

While planning the project I realized I simply couldn't handle it all by myself. I knew I needed someone to work with who was a first class wine intellectual who could shoulder the bulk of the research, make the best of the brilliant, up-to-date and minutely detailed data that we would obtain from every important vineyard area worldwide and then put masses of often quite indigestible source material into a highly readable form. I also needed someone who, when I was flying off on my frequent flights of fancy, could rap me firmly on the knuckles and say 'Come on Oz. Back to earth. This is a serious book of research as much as a joyous celebration of grape varieties in all their multi-flavoured glory'. Well, she didn't quite say that; she'd have edited out most of the final sentence for a start. But the person who shared in the creation of this book with me is Margaret Rand. And without her talent and determination we'd never have got it done.

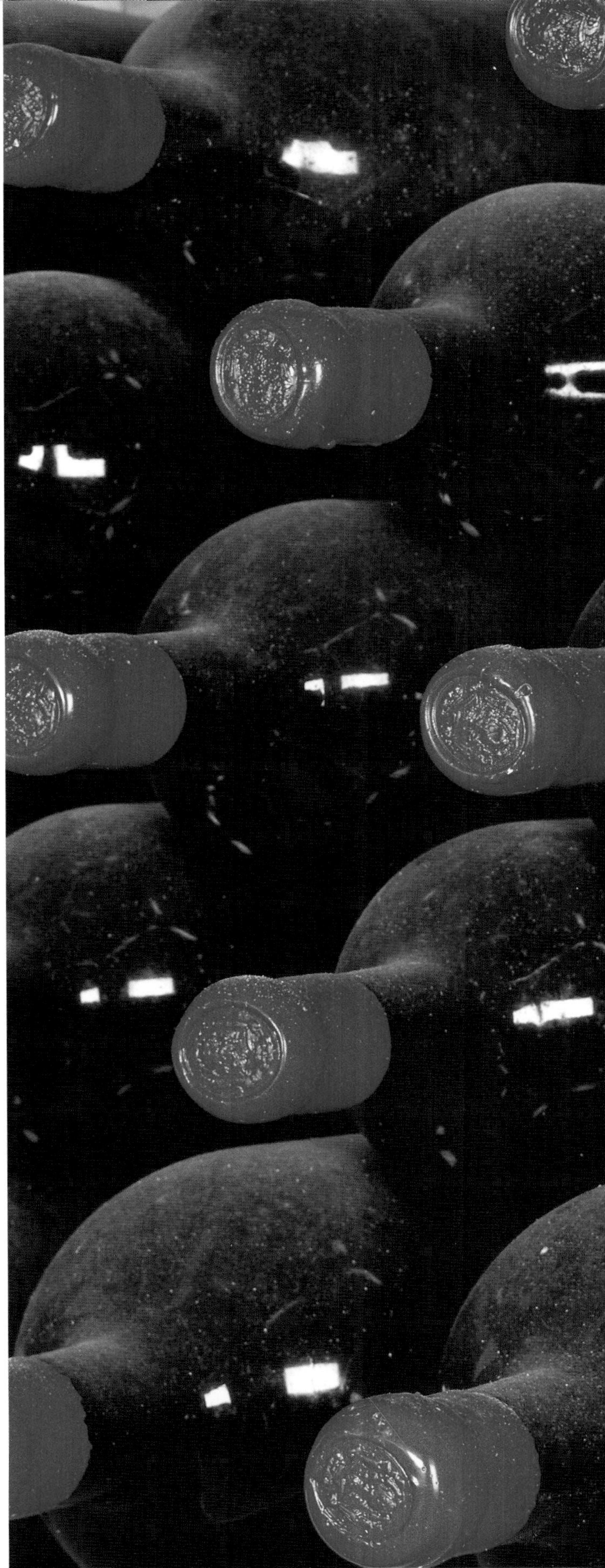

# THE STORY OF THE VINE

I'M NOT SURE ANYONE ACTUALLY planted the first vines, anymore than I'm sure that anyone intentionally made the first wine – which might even have been simply a mistake in the storage of dried eating grapes that then fermented by accident. Who knows? But vines grow wild. They always did. They still do. And a long time ago, people began eating their fruit.

Wine has been made for millennia: an early piece of evidence, in the form of a wine-stained amphora, dates from 3500 BC. But man was probably taking advantage of wild vines long before that; the European wine vine, *Vitis vinifera*, is believed to have made its first appearance in the area that is now Georgia and Armenia. Georgian archeologists believe they can date findings of cultivated grape pips to 7000–8000 BC. But just how and when the change was made from picking wild grapes to taming the vine is not known. It probably happened separately in a number of different places; it probably happened over a long period of time. However, by about 3000 BC winemaking was well advanced in Egypt and connoisseurs were making the same distinctions between different qualities as they do now.

The difference between wild vines and cultivated ones is that wild ones have male plants and female plants, whereas cultivated ones are hermaphrodite. One can speculate that as vines were deliberately planted it was the chance hermaphrodite ones that fruited best and thus, over a period of many, many years, came to dominate. In archaeological sites the pips of wild vines can be distinguished from those of cultivated vines by their different shape; but wild vines are no means extinct.

The *Vitis* genus is both diverse and widespread, and when the first settlers landed on the east coast of what is now the United States, the vines they found growing there were wild. Wild vines still grow there, as they do in Asia – and even in Europe, if you count odd outcrops of American vines, imported in the late 19th century as rootstocks after phylloxera destroyed most European vines, and allowed to fruit and seed themselves. European wild vines are probably now extinct, though their genes continue in their domesticated descendents.

One insatiably curious Californian wine producer, Warren Winiarski of Stag's Leap Wine Cellars in the Napa Valley, has travelled to Pakistan to take cuttings of wild vines from lands around the Karakoram Highway, the road that parallels the old Silk Road, which was one of the great trade routes of the ancient world. Wine made from such grapes would be as close as we can get to what our remote ancestors might have drunk – though we'd hopefully make it better.

## Vitis vinifera and wine

All vines belong to the genus *Vitis*, which in turn belongs to the Vitaceae (formerly Ampelidaceae) family (see chart opposite). The *Vitis* genus includes around 60 species and is generally divided into two sections: Euvites, which contains nearly all the American, Asian and European vine species, including the European wine vine, *Vitis vinifera*; and Muscadiniae, which is sometimes considered a separate genus.

This book focuses on *Vitis vinifera*; other species of vine, like *Vitis labrusca*, *Vitis riparia* or *Vitis berlandieri*, are important to wine largely because they provide, either directly or by way of crossings, the rootstocks onto which *Vitis vinifera* vines are grafted. (See page 18.)

There are also hybrids: vines whose parents are of different vine species, as opposed to crosses, which have both parents of the same species.

*This copy of a wall painting from the tomb of Kha'emwese at Thebes, c.1450 BC, is both a technical aide-mémoire, ensuring that the departed will be well supplied with wine in the next world, and perhaps an expression of pleasure in all the different stages of the cultivation of the vine and the making of wine. For us it is also invaluable documentary evidence of how grapes were grown and wine made in an early but extremely sophisticated civilization. The viti- and vinicultural techniques shown here are far from primitive, even if the wines that resulted might not be to our modern taste.*

The usual object of breeding hybrids is to combine some genetic advantage of a *labrusca* or *rupestris* vine with the better wine flavours obtained from *vinifera*: non-*vinifera* varieties give pungently scented wine often described as 'foxy', though the aroma is in fact more reminiscent of may blossom or nail varnish. Maybe that's why winemaking was slow to catch on among the early settlers in North America. But these vines may have resistance to disease (particularly phylloxera) or cold: hybrids of American vines with *vinifera* are widely planted in many more northerly states in the USA because *vinifera* vines find it difficult to withstand the winters. Seyval Blanc, a so-called French hybrid (French hybrids are a group of hybrid vines bred in France in the late 19th and early 20th centuries in an attempt to find a solution to phylloxera), is much planted in England.

Hybrids are becoming increasingly sophisticated. There are some now being bred at research stations in Germany that are said to produce wine indistinguishable from the wine of certain *vinifera* varieties. These latest breedings involve not simply a crossing of an American vine with a *vinifera* vine, but numerous back-crossings to stabilize the European character. There is one, for example, that is said to be very close in flavour to Pinot Blanc, but which has no Pinot Blanc genes; another is said to be indistinguishable from Riesling; this does have some Riesling somewhere in its ancestry. The best red hybrids resemble Merlot, with good colour, low acidity, and blackberry-cherry fruit.

At the moment hybrids are not permitted for the production of quality wine in the European Union. But then, European wine bureaucracy is notoriously conservative and inward-looking. Luckily, the rest of the world doesn't have such problems, and I've tasted fascinating wines made from grape varieties that haven't even been named yet in places like Australia and Argentina and at massive yields of between 20 and 30 tons per acre. You could ask, does the world of wine actually need any new flavours, especially at such high yields? Don't we already have a fantastic variety of grapes, grown just about everywhere the sun shines, giving us a fantastic variety of flavours to savour? Yes, sure. But what if some hybrid produced the best tasting wine the world has ever seen? Well, I want to taste it. Don't you?

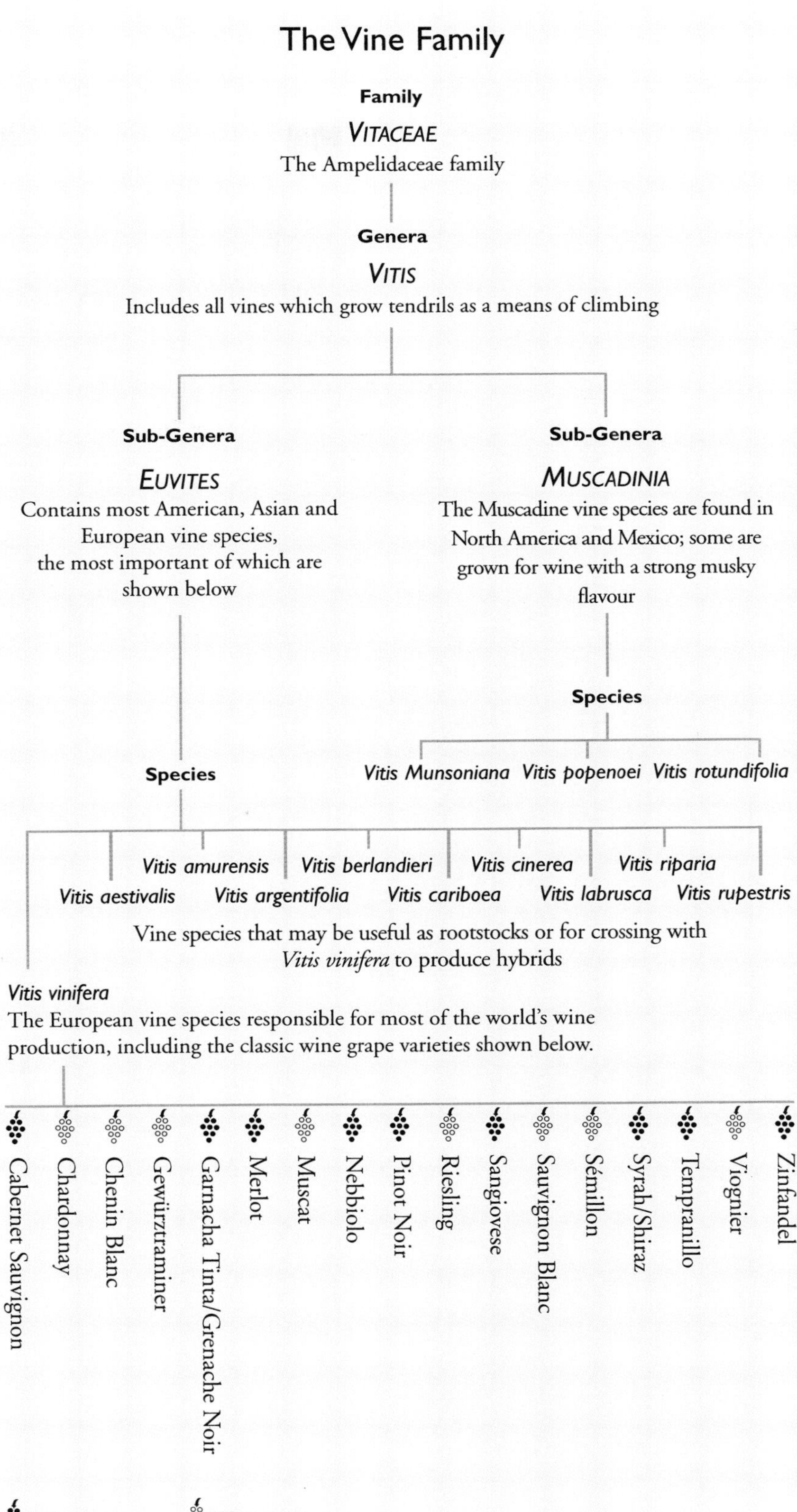

# VINES TODAY

THEY SAY THERE ARE SOME 8000 VARIETIES of vine – that includes wild varieties, those used for table grapes, those used for raisins, as well as wine grape vines – well, that's 8000 that we know about. New vines are still being described, and we've lost a fair few varieties over the years as well. So, why so many varieties?

*To you, me and most growers – and indeed to every bureaucrat in charge of enforcing wine laws – Pinot Noir **(left)** and Pinot Gris **(right)** are different varieties. Try putting Pinot Noir on your label when you used Pinot Gris, and you won't get far. Yet to a geneticist they are the same, because they have the same DNA.*

The vine, *Vitis* – whether *vinifera* or another type – has been on the planet a long, long time. It has had ample time to cross-breed. It also mutates constantly and unpredictably, although some families of vines, notably the Pinot family, are far more prone to mutation than others. *Vitis vinifera* can outdo most fruits in this respect: its heterozygous nature (meaning that its already complex gene pattern can be readily rearranged) enables it to adapt itself with surprising ease to new circumstances.

Which leads us to another question: at what point is a variation on an existing vine considered to be a separate vine variety? This is a complicated matter, and a grower's answer is different to the answer of an ampelographer (or vine geneticist).

To a grower, a vine is a different variety if it looks different. The points of difference that matter are the shape and appearance of the leaves – the depth of the indentations, the appearance of the veins – the hairiness of the shoot tips, the shape and tightness or looseness of the bunches, the colour of the berries, the appearance of the canes, buds, flowers and seeds. Thus, to a grower, Pinot Noir and Pinot Gris are separate varieties.

To vine geneticists like Dr Carole Meredith of the University of California at Davis (UCD), however, Pinot Noir and Pinot Gris are merely colour forms of the same variety. They have the same DNA, therefore they are the same vine. 'In genetic terms,' she says, 'each variety is descended from a single unique original seedling. All vines of that variety can be traced back to that original seedling and have been derived from it by cuttings or buds.'

Pinot Noir is notorious for its variation: some strains grow upright and produce large, loose bunches; others do the opposite. They can be so different in appearance, and the wine can be so different in quality and flavour, that as consumers we might be justified in wondering if all these different strains should really all bear the same name. Yes, says Meredith, they should. Their DNA is the same. To produce a new variety you have to produce a seedling. You can do this by crossing two separate varieties, but even a seedling of the same variety (*Vitis vinifera* vines, remember, are hermaphrodite) is unlikely to be true to its parent. (The propensity of *Vitis vinifera* to produce seedlings that are different to its parent is the reason why vine propagation is done with cuttings.) Vines may mutate so that they look different and even behave differently, but they are still the same variety, and always will be; and this ability to mutate is one reason why the same variety can succeed in different climates and circumstances. And it is the basis of clonal selection.

### What is a clone?

Good question. 'Clone' is a word that is going to crop up a lot in this book. If you take a cutting of a plant – it might be a vine, it might be a pelargonium – you have a clone. The second plant will be genetically identical to the first. Any properly observant grower will be aware of which vines in his vineyard are the best. They may be the healthiest, or they may give grapes that ripen a little earlier or a little later, or have particularly good flavours. Whatever their positive attributes, by propagating cuttings from those vines rather than others, that grower is practising a form of clonal selection.

### Massal versus clonal selection

This 'home-made' selection is known as massal selection, which is a rather odd term for what wine growers in traditional areas have been doing for centuries. The alternative, clonal selection proper, is done in the laboratory: it involves taking cuttings from vines that have been observed to be especially good in some way, but the object is not only to reproduce their good qualities, but also to eradicate their virus diseases.

Some virus diseases may eventually kill the vine; others may make it chronically weak and underproducing. There is no easy way of curing virus diseases once a vineyard is infected: the viruses live on pieces of root in the soil and are spread by nematode worms. By growing cuttings at high temperatures (around 38°C/100°F) and then taking a cutting of just the very point of the growing tip it is possible, since the spread of viruses through the vine is always just a fraction behind the growth of the vine, to have a piece of vine that is free of viruses. From that you grow your new vine. (There is some debate about how much clonal variation is itself the result of varying viral incidence.)

Clones take a long time – often around 15 years – to be made commercially available to growers: the plants must be selected over several generations to ensure that they stay true to type. (Mutations may become more stable later on – or less stable.) Their resistance to disease and poor weather, and their suitability to different soil types must be tested; and, most importantly of all, the quality of their wine must be judged. Many clones produced in the 1970s produced high yields but poor quality wine; much current clonal research is aimed at putting right those mistakes.

### GM vines: a future prospect?

Look. Genetically modified vines exist. Are they good? Are they bad? That depends on your views on GM in general. But they exist, so we're going to talk about them. It's true that because wine grapes are not a commodity in the sense of maize or soya, large agro-industrial companies have (thankfully) less interest in them. Once planted, vines generally remain in their vineyards for 30 years or more, unlike other crops which must be replanted from seed every year. With maize or soya there is the double profitability of every year selling seed genetically modified to resist a particular weedkiller, and then selling the weedkiller to go with it; with vines the position is different. What research there is is aimed at virus and fungus resistance: Riesling vines genetically modified for resistance to fungus infection are already being grown at the Geilweilerhof Institute for Grapevine Breeding in Germany, though for experimental purposes only so far. The genetic complexity of *Vitis vinifera* makes genetic modification further off than is the case with some other plants: there are several genes responsible for mildew resistance alone, and not all have yet been isolated. Some *Vitis* species, like *Vitis riparia* and *Vitis amurensis*, are genetically resistant to winter cold; others have genes that protect them against phylloxera or nematodes.

Vine breeding programmes are also aimed at exploiting these natural advantages to produce vines that are both disease-resistant and high quality, but though some modern crossings are undoubtedly fine – Scheurebe is a smashing vine, and Pinotage, a South African invention of 1926, is a fascinating original – most are not highly regarded and many have fallen by the wayside. Müller-Thurgau continues to be a widely grown vine (it dates from 1882), but most of Germany's producers would place it somewhere below diet cola in their ranking of boons of the modern world.

Genetically modified yeasts already exist, though they're used mostly by the brewing and distilling industries so far. The aim of such yeasts is more efficient conversion of sugar to alcohol, and perhaps some built-in anti-microbial action. But winemaking, thank goodness, demands more from its yeasts, and when winemakers choose yeasts (which may be either wild or selected and bred in laboratories) more subtle questions of flavour development play a large part in their choice.

### Vine confusion: a question of identity

The identification of vines is a fascinating mixture of science and happenstance. Throughout human history people have traded and migrated; and they've taken their vines with them. With those vines have gone their local names; but new local names might later have been appended. So the same vine may be known by many different names in many different countries or regions. Conversely, the same name might be given to many different vines, even in the same region. Perplexed? Well, me too. But the result, for the ampelographer, is a rich melting pot of confusion, out of which a few hard facts are gradually being pulled.

In the early 1990s, for example, it was realized that many of the Sauvignon Blanc vines planted in Chile were in fact the less aromatic Sauvignonasse, or Sauvignon Vert; a few years later ampelographers identified a large proportion of Chilean Merlot as Carmenère, a vine which had been much planted in Bordeaux before phylloxera. Indeed, it was a star performer, but now plays no part there. Is it any good? Chile suggests it is. In Italy, an official difference between Pinot Blanc and Chardonnay was made only in 1978; the close resemblance between the two is reflected in one of the traditional French names for Chardonnay: Pinot Chardonnay. And there are dozens more examples. If such confusion can arise over classic varieties, how many lesser varieties are languishing, misnamed and misunderstood, in the world's vineyards?

*This is how infant vines are brought into the world. Each is a cutting of a parent vine, and so genetically identical to its parent, and indeed to its siblings. Nevertheless, when sold to different growers and planted in different soils, climates and conditions, each will behave slightly differently. Planting laboratory-produced clones is a way of weeding out poor quality strains of a vine, but it need not produce identikit wines.*

# The Global Picture

First of all, let's get our bearings. This map shows where wine is grown: it gives the bare facts of latitude. It shows that wine is grown between about 32° and 51° north, and about 28° and 44° south. If you're thinking of making wine, that's where you should stick your pin. After that, though, things get a little more complicated.

Northern hemisphere vineyards, as you can see from the figures above, tend to be at higher latitudes than those in the southern hemisphere. Partly this is for the very simple reason that most land masses in the southern hemisphere don't extend all that far south: only the curving tail of South America goes past the 50° mark. That South Africa doesn't extend any further south is certainly a great source of frustration to some of the producers there, who have difficulties in finding sites cool enough for some varieties – Pinot Noir, for example.

But the question of latitude is not just a simple one of temperature. The northern hemisphere is generally warmer than the southern at the same latitude, both in the growing season and over the course of the whole year. This is partly because of the greater land mass, and partly (as far as western Europe is concerned) because of the warming effect of the Gulf Stream. So comparing a northern latitude with a southern one can be misleading. Because it is possible (just) to ripen grapes in southern England does not mean that the same grapes would ripen at the same latitude in Chile or Argentina. They wouldn't.

The effect of latitude on wine flavour is enormous. Lower latitudes have early springs and late autumns, and thus provide the long ripening season that winemakers the world over covet: grapes that ripen slowly on the vine have time to develop plenty of flavour and aroma as well as just sugar. But on the other hand, low latitudes tend to mean high temperatures, so that the grapes quickly reach high sugar levels, and lose acidity, which is the opposite of what growers want.

By contrast, higher latitudes mean shorter, cooler summers, but longer hours of daylight. More hours of daylight, and sunshine that is less fierce but lasts longer during the day, means more efficient photosynthesis, better retention of acidity and better development of flavour and aroma. Grapes ripen relatively faster at higher latitudes – in Bordeaux they reckon to have about 100 days

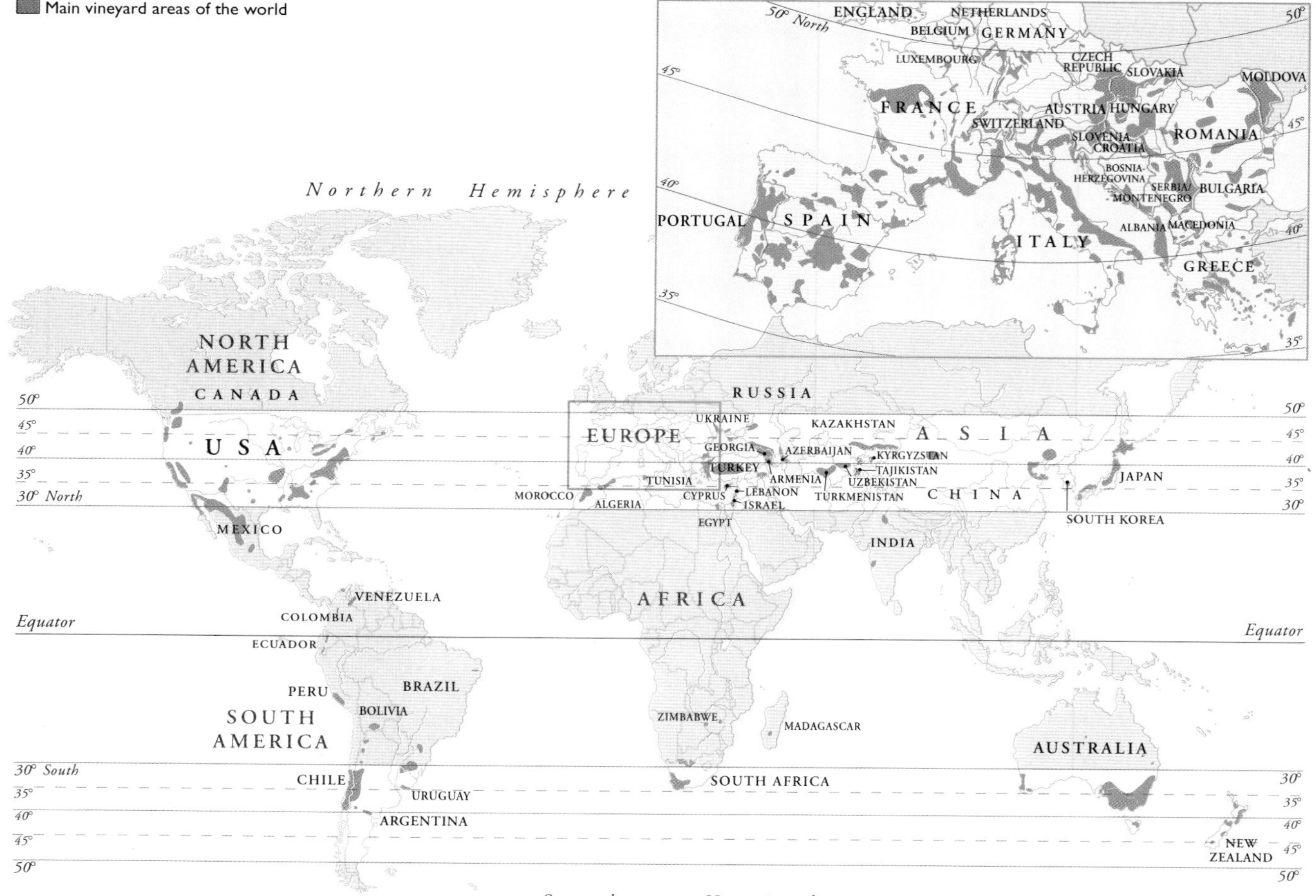

between flowering and picking, whereas on California's Central Coast the figure is more like 150 – but if the temperature is too cool there is nevertheless a risk that the grapes won't ripen fully before autumn sets in.

Ideally, therefore, high latitudes need to be tempered with warm climates, and vice versa. We'll be looking at climate more closely on page 14 onwards. The third element in the equation – and the element that balances the other two – is the grape variety.

### What grows where

Contrary to what one might expect, the total amount of land under vine has decreased in the past 50 years. In 1951 there were 8,845,130ha of vineyard in the world; in 1998 (the most recent year for which figures exist), the area was 7,799,000ha. Vineyard area rose from 1951 to reach a peak in the late 1970s of 10,213,000ha; at this stage government-sponsored vine-pull schemes in Europe and the USSR started reducing the total. The aim in the EC was to drain the wine lake; the aim in the USSR was to reduce alcohol abuse. By 1998 both these initiatives were effectively over; the European one, at least, had succeeded in its aim.

But that figure disguises more local changes. In Africa, more land in the South (3000ha) was planted between 1997 and 1998, while Algerian vineyards were being uprooted, albeit at a slower rate than in recent years. In the same 12-month period the USA planted an extra 9000ha, and Chile an extra 12,000ha; China planted 22,000ha, bringing its total to 149,000ha, though the vast majority of these vines produce table grapes and raisins. Europe's vineyard area fell by 47,800ha in the same period: 16,000ha went in Azerbaidjan, 14,000ha in Moldova and 13,600ha in the Ukraine.

There was, however, a moderate rise of wine growing areas in the 15 countries of the European Union, but this only amounted to 2100ha. Europe's vineyard still covers 5,041,000ha, which makes it far and away the biggest grape-producing area in the world. To put that in perspective, Australia had around 122,915ha of vineyards in 1999; in France, Languedoc-Roussillon alone had 250,000 ha.

### The move from white to red

*The world is wild for red wine. Producers can't plant red grape vines fast enough: in Australia new plantings of Chardonnay have virtually ended apart from premium sites – although between 1996 and 1999 production of Chardonnay increased by 55%. Red wine production in Australia is set to double in the next few years, while white production will increase by only 5%. In Italy red and white production was in perfect balance in 1995; two years later red production was up to 65%.*

Planting figures for red grape varieties
BETWEEN 1990 AND 1999

| | South Africa | California (USA) | Australia | Chile |
|---|---|---|---|---|
| (1990) | 15.7% | 46.4% | 32.25% | 62.85% |
| (1999) | 26.1% | 54.4% | 55.26% | 65.83% |

*Vineyard latitudes in the Southern Hemisphere don't get much higher than this: Central Otago, in New Zealand's South Island, lies at 44° 42'S, and according to climatic statistics it should be impossible to grow grapes here. In fact, though, the summer's extremely long sunshine hours, and the almost complete lack of summer cloud cover combine with favourable local mesoclimates to produce conditions suitable for both Chardonnay and Pinot Noir.*

# Where grapes grow

If climate and latitude alone determined where vines grew, life would be so much simpler. You would simply look at the sunshine, the temperature, the rainfall and the chance of frost, and work out what to plant where. Hey presto: great wine, every time. Unfortunately it's not quite that simple. Weather is only one of many factors that determine the quality and style of wine. When considering where grapes grow it's worth remembering that European vineyards were seldom planted after close analysis of the weather or the soil. If by chance they turned out to be great vineyards, then people tried to work out why. It's been a long process of discovery, and it's not over yet.

Winemakers have been studying Burgundy's Côte d'Or for years, but we still don't know precisely what it is about this little stretch of French vineyard that produces such marvellous Pinot Noir and Chardonnay. And if people can't agree on that, it's not surprising that they also can't agree on which of the attributes of the Côte d'Or you should try to imitate if you want to make great Pinot elsewhere, or how closely you should imitate them. Should you find somewhere that mimics the Burgundian climate? Or is the climate in fact a disadvantage? If you think that, then you'll seek somewhere warmer and drier. Should you be trying to copy the soil? And if so, should you be looking at its structure, its mineral content, or what?

Your views on these matters have historically depended on where you live. Growers setting out to make great wine in the New World have traditionally focused on climate as the main determining factor in wine quality. Those in the Old World have usually quoted what their fathers and grandfathers had quoted before them: terroir.

*If it wasn't for the water, there wouldn't be any wine here. The Finger Lakes in New York State have a relatively small surface area but are extremely deep and seldom freeze in winter, so act as a very efficient heat store. Vineyards on the surrounding hillsides are protected against extreme winter temperatures and late spring frosts, and enjoy milder autumns than areas further away.*

## Terroir

This is still a relatively poorly understood concept – but it is important to realize that 'terroir' does not mean 'soil'. The terroir of a vineyard is the sum of all its parts: its geology, its climate, its topology, its water-holding ability and the amount of sun it receives. So the soil, both topsoil and subsoil, is important, as are the mineral components of the soil. How fertile or infertile it is, and its depth and structure, which affects how well or poorly drained it is, are also factors. Altitude, steepness of slope and exposure to the sun, all matter, as does the mesoclimate, or climate particular to that vineyard. From the French point of view (and it is most of all a French concept) it is the terroir that makes each vineyard different. It underpins the Appellation Contrôlée system not least because, as the underlying factor behind wine quality and style, it should show in the wine no matter who the winemaker is or what he or she does to the wine. Winemakers come and go; the terroir remains.

However, good viticulture and winemaking can permit the expression of the terroir while bad viticulture and winemaking can mask it. And since good vineyard practice can mean installing drainage where necessary, and since good winemaking in northern Europe does not exclude chaptalization (the addition of sugar before or during fermentation to increase a wine's alcoholic strength; see page 29), growers are in practice not absolute slaves to what their terroir dictates. More and more French winemakers, I notice, are amending their definition of terroir to include the effects of man's hand on the land. Yet it is still possible to find some who attribute to terroir aspects of wine flavour that other nations might attribute to other causes. I have heard Alsace growers assert that in youth their wines have the flavour of the grape, but that later the flavour of the terroir comes to dominate. Other people might call the latter flavour that of maturity and bottle age. And the more blinkered members of the wine trade can still sometimes be heard extolling bottles that are badly made and reeking of sulphides as exhibiting 'terroir'.

## Climate

It's not surprising that so many winegrowers pay more attention to climate than to any other factor: its effects are obvious and undeniable. Even the most dedicated *terroiriste* is likely to blame the weather rather than the terroir when his vines are hit by spring frost (although those with vineyards sited outside frost pockets may well smugly praise their terroir when they escape). Rain or high winds during flowering; drought in late summer, causing photosynthesis, and thus the ripening process, to stop; summer rain that provides the humidity in which rot can attack the grapes; rain at harvest which dilutes the juice; hail at any time: the climatic hazards faced by the winegrower are almost endless. In Europe, growers are largely stuck with existing appellation boundaries and the weather conditions that go with them. In the New World you can plant a vineyard where you like – which in practice means, since the perfect vineyard site does not exist, that up to a point you choose your problems rather than inherit them.

Nevertheless, the weather during the growing season does not make much difference to the quality of the vintage until *véraison* (the onset of ripening, when the hard, green immature fruit softens and takes on the colour of the ripe grape) takes place. Spring frosts or excessive water stress may reduce quantity, but as far as quality is concerned it is the last couple of months before the harvest that really matter. The summer of 2000 in Bordeaux, for example, was notable for cold and rain until *véraison*; at that point the skies miraculously cleared, the sun shone and the wines were hailed as the best for years.

### Cool or warm

There is no absolute definition of what constitutes a cool, warm or hot climate for viticulture: what feels cool to a grower in California's Napa Valley might seem warm to a grower in the Saar Valley of Germany. A cool climate is usually understood to mean one where only early-ripening grape varieties will ripen: grapes like Pinot Noir, Riesling, Chardonnay or Gewürztraminer. An intermediate climate will ripen later-ripening varieties like Merlot, Cabernet Sauvignon and Syrah. In warm climates you get very late-ripening grapes like Mourvèdre, Grenache and Touriga Nacional, but also earlier-ripening varieties such as Muscat if the aim is sweet fortified wines. Carneros in California, New Zealand's South Island, Burgundy and Germany come under the heading of cool climate; Bordeaux, Tuscany, California's Napa Valley, Chile's Maipo and Australia's Coonawarra are intermediate. The south of France, the Douro Valley in Portugal, and McLaren Vale in Australia are warm.

Hot climates grow grapes mostly for raisins or table grapes, though wine is made nevertheless in some tropical climates in South America and elsewhere.

### Mediterranean or Continental

A Mediterranean climate has the sort of mild winters and long, warm summers that prevail around the Mediterranean. A Continental climate is more extreme, with colder winters and hotter, sometimes shorter, summers. More inland parts of Europe, such as Germany, Austria and Hungary, have a Continental climate. Regions near the sea, like Bordeaux, have a maritime climate: the proximity of a large body of water tempers extremes of heat and cold, but also brings the danger of storms and rain.

Having an expanse of water near a vineyard can in fact be vital in marginal climates (a marginal climate is one where a given grape variety will only just ripen: marginal climates often produce the most complex, elegant wines – in years when there's enough sun, that is. The downside of the equation is raw, green wine when the sun doesn't shine). A large river or lake, by acting as a heat store, can raise the average temperature by a vital degree or two. It's one of the reasons why Riesling will ripen on the south-east- to south-west-facing banks of the Mosel, but not further away from the river's edge.

### Defining ripeness

What exactly is ripeness? On the face of it it seems a silly question. We all know the difference between a ripe plum and an unripe one. But take the analogy of bananas. Some people consider a banana to be ripe when the skin is yellow all over. To others, a ripe banana has a skin speckled with black. In wine, a grape that is considered ripe in Champagne would be thought unripe in the Napa Valley. The high acid/low sugar 'ripeness' that is crucial for making sparkling wines of balance and vivacity in Champagne makes painfully acidic table wines. Napa's sun-soaked grapes make full-bodied table wines but would make fat, flabby fizz.

High sugar levels do not, on their own, constitute ripeness. A ripe grape has brown seeds, not green, and the stalk is lignified to the first joint of the bunch. When you taste it (and all good wine producers judge ripeness by taste first, and by analysis second) a red grape should have no vegetal green bean flavour, and its skin tannins should be soft and velvety. In other words the aim is physiological ripeness, and in warmer climates this occurs at higher sugar levels than in cooler climates. This is why ripe grapes in Australia's Barossa Valley will tend to produce more alcoholic wines than grapes of similar physiological ripeness in Bordeaux. The battle in warm regions is to attain full physiological ripeness without head-banging levels of alcohol: get the equation wrong and you end up with 14 per cent alcohol and unripe flavours.

Even so, a lot of growers pick their grapes too early – they're often nervous because autumn rains are on the way – and in cooler areas a bad summer or autumn may mean the grapes simply can't ripen. In white grapes this means high levels of sharp-tasting malic acid; leaving the grapes on the vine for longer would mean lower levels of malic acid, which

### THE WINKLER AND AMERINE HEAT SUMMATION SCALE

It is only right to give a proper account of the Winkler and Amerine Degree Day System, since it governed so much of what was planted and where in California from the 1940s on up until relatively recently.

Albert Julius Winkler (1894–1989) was a scientist at the University of California at Davis (UCD), the leading US viticultural research institute as well as the leading US college for aspiring winegrowers and makers. He and Maynard Amerine (1911–98), also of UCD, devised a system of classifying the regions of California by heat summations. They assumed a growing season that extended from 1 April to 31 October, and worked out the number of 'degree days' in different regions. This was done by calculating the amount by which the temperature exceeded 50°F/10°C, which is the lowest temperature at which the vine will grow. They divided California into five different regions, with Region I, the coolest, having fewer than 2500 degree days, to Region V, with more than 4000. Specific grape varieties were then recommended for each region.

When applied to California the system is broadly reliable, since it so happens that in California temperature is a pretty good guide to overall climate. Elsewhere it works less well: applied to Australia, for example, it would deny the possibility of making good Semillon in New South Wales's Hunter Valley – yet Hunter Semillon is one of the classic wine styles of Australia.

*In a very hot country you may have to go very, very high up to find a suitable climate for growing vines. In Catamarca in Argentina Michel Torino has planted vines at around 2400 metres: luckily the Andes happen to be there to provide such sites.*

falls as the grapes ripen, and higher levels of the riper-tasting tartaric acid.

In red grapes there is the additional factor of tannin. The current buzz word among red wine makers is 'texture': what is required if a red wine is to be properly fashionable is tannin that is silky and velvety in the mouth, with none of the harsh greenness of unripe tannins. Very ripe tannins can be produced by leaving the grapes on the vine for as long as possible – and for that you need a relatively cool climate that will hold the sugar level back and prevent the grapes shrivelling to raisins while you wait for the tannins to ripen. Producing red grapes with perfectly ripe, thick skins – the source of much tannin – is thus a preoccupation of growers.

### Climate and photosynthesis

Vines produce sugar by photosynthesis, the process in which sunlight is used to combine water with atmospheric carbon dioxide. The sugar thus made in the leaves is transported to the grapes. At low temperatures photosynthesis, and thus the ripening process, stops. But it also stops at high temperatures, and in drought conditions. In very hot, dry spells, if no irrigation is available, grapes may shrivel, but they won't ripen.

Photosynthesis is one of those flashpoints where the difference in attitude between the Old World and the New can be discerned. All growers accept its fundamental importance to ripening. But where Old World growers might attribute all wine flavours to the terroir, a New World grower might well focus on photosynthesis as the source of all flavour compounds. From this viewpoint, climate, the pruning and training of the vine (termed 'canopy management' – we'll be talking more about it on page 24) and all aspects of viticulture are important only in so far as they affect photosynthesis.

### The heights and the slopes

It is a relief to come to a topic about which there is relatively little disagreement. One of the few incontrovertible statements one can make about vineyards is that higher altitudes are cooler. In California they reckon that the temperature cools by 2.2°C (36°F) for every 329m (1080ft) of altitude. This may not be a conveniently round figure, but it shows that if you want the longer summer that goes with lower latitude, you can get a cooler climate – and thus get the possibility of very long hang time, with all the benefits to ripeness and flavour that that implies – by heading up into the mountains. It's worth remembering that 19th-century vineyard pioneers in California chose the hillsides, not the valley floors.

Go too high, of course, and you may find that your grapes won't ripen properly. On the Côte d'Or and in Hungary's Tokaj region it is the middle of the slopes that produce the best, ripest grapes. In Portugal's Douro Valley the highest vineyards may ripen two to three weeks after those nearer the river, and may indeed be too cool for port; but they come into their own for unfortified table wine, which needs less sugar but more acidity.

But slopes have another advantage, too: they get more direct sunshine. In cool climates that means better ripeness. They also tend to have poorer, thinner, better drained, less fertile soil, which acts as a natural brake on the vigour of the vine; this in turn helps to control the amount of grapes the vine produces, and gives more concentrated flavours.

Planting on valley floors is certainly desirable if you want to produce large quantities of inexpensive wine. Most of Chile's vines are planted on the flat land of the Central Valley, and for a producer to start planting on the slopes – even on relatively gentle ones – is like taking his feet off the bottom of the swimming pool for the first time. It took one quality-oriented company (Valdivieso) four years of anguished discussions to take the plunge and start planting on slopes.

But this is not to suggest that only sloping vineyards are good. The gravelly Médoc looks remarkably flat to the untutored eye, and the stony vineyards of Marlborough in New Zealand, which are about as flat as land can be, nevertheless produce excellent Sauvignon Blanc. The floor of the Napa Valley is also hardly devoid of good wine.

### Back to the soil

When a leading California producer, talking about Chardonnay, says, 'We know that limestone has important effects, but we don't know why. We can't demonstrate the difference organoleptically', you know that he's a long way from the traditional New World belief that soil was just the stuff that held the vine upright. Growers in Germany's Mosel Valley will point to the particular smoky flavour that Riesling attains on their slate soil. One leading New Zealand grower reckons that his clay soil gives his wines body and texture, that the soil's high calcium content gives 'nerve' and longevity, and that silica increases the aroma.

Are they all imagining it? Can wine actually gain particular flavours and qualities from the soil? (The French term *goût de terroir* embraces more than just the soil: see Terroir, page 14.) Leading Australian viticulturist Richard Smart belongs to the other camp: he believes that the effects of the soil on quality, never mind on flavour, are only indirect, in that the depth and distribution of the roots, and thus the vine's efficiency in finding water and nutrients, are affected by soil structure. Actual flavour from the soil? No. Not according to Smart.

The physical structure of soil should, however, be differentiated from its chemical attributes. Most authorities agree that the most important aspect of soil is its water-holding capacity, and how easily the vine can access that water.

What vines like best is well-drained soils that nevertheless have an adequate, but not too high, supply of water. The chalk of Champagne is an example: chalk drains freely yet maintains a high water table. Clay, by contrast, drains poorly, yet can be so dense that much of its water content may be inaccessible to the vine's roots.

Wet soils are cold soils; well-drained soils are warmer. Well-drained soils are thus likely to encourage budbreak earlier in the season, and aid ripening; they can be suitable for later-ripening varieties in marginal climates because they speed them up. A cold soil might be better suited to an earlier ripening variety. In Bordeaux, for example, Merlot does better on the clay soils of Pomerol and St-Émilion, where the later-ripening Cabernet Sauvignon would not ripen. Cabernet needs the warmer, freely draining gravel soil of the Médoc. Dark soils, too, are warmer; light-coloured soils are colder. And very stony soils, such as those found in parts of Châteauneuf-du-Pape, conduct and hold the heat well, again aiding ripening.

Some aspects of the chemical and mineral content of soil are more clearly understood and can be related to other crops. There is certainly a direct relationship between excess nitrogen and excess growth: vines getting too much nitrogen produce enormous quantities of leaves that in turn shade the grapes and hinder ripening. Too much potassium in the soil can reduce the acidity of wine. (Potassium-based fertilizers, very popular in the 1970s, are now at last out of fashion.) High levels of organic matter mean very fertile soils, and these, too, can mean problems of excess vigour in the vine. But the vine must have some organic matter in the soil. Very poor, infertile soils can indeed encourage the vine to put down very deep roots in search of nutrients, but if it doesn't find those nutrients anywhere it won't flourish.

## COMMON VINEYARD SOILS

**What you see at the surface of the vineyard may be material deposited by rivers, glaciers or erosion, or it may be the topsoil: the weathered bedrock combined with organic matter. If this layer is too fertile the vine will spread its roots sideways, rather than sending them deep – sometimes as much as 30 metres – into the subsoil to establish a stable foundation. The subsoil, or pure weathered bedrock, may be relatively uniform or may consist of several strata.**

**Alluvial:** soil deposited by rivers. Alluvial soils are very fertile, and contain sand, silt and gravel. **Argillaceous:** includes clays, marls, shales and others. **Calcareous:** contains calcium and magnesium carbonates. These are cool, water-retaining soils; usually alkaline, they give grapes with high acidity. **Chalk:** a type of limestone. It is cool and alkaline, and combines good drainage with sufficient moisture retention. **Clay:** cold, acid, poorly drained soil. Solid clay is difficult for vines, but clay mixed with other soils can be excellent for vine growing. **Clayey-loam:** even more fertile than loam on its own. Can drain poorly in wet weather. **Ferruginous clay:** clay with an admixture of iron. **Gneiss:** a type of granite. **Granite:** warm, mineral-rich soil that tends to produce grapes low in acidity. **Gravel:** pebbly, well-drained soil, generally infertile. It is acid, so produces grapes with low acidity. Gravel over limestone gives wines with more acidity than gravel over clay. **Limestone:** carbonate-rich soil. There are various sorts of limestone, of which chalk is one. Different limestones have different water retention. Limestone soils generally give grapes with high acidity. The Kimmeridgian soil of Chablis is a calcareous clay containing Kimmeridgian limestone. **Loam:** warm, fertile soil – generally too fertile, in fact, for fine wine. Loam contains clay, silt and sand. **Marl:** calcareous clay. Marl is cold soil that holds back ripening and gives wines with high acidity. **Sand:** fine-grained, warm soil that drains freely. **Sandstone:** rock with sand-sized particles bound together by minerals or forced together by pressure. **Sandy-loam:** loam with a large admixture of sand. Sandy loam is warm and well drained. **Schist:** warm, splintery, crystalline rock derived from shale, sandstone or granite. **Shale:** warm, reasonably fertile soil. **Slate:** warm, well-drained soil. Slate soils like those in the Mosel can consist almost entirely of thinly split fragments of rock.

*Large stones like these in Martinborough, New Zealand, can act as a heat store, soaking up heat during the day to advance ripening in what is a cool, marginal climate.*

*The deep beds of gravel here in Pessac-Léognan are warm and provide good drainage, which is vital in Bordeaux's damp climate, and can make all the difference between good wine and mediocre.*

*Again, good drainage and warmth are the key advantages of the slate of the Middle Mosel. There is so little actual soil here that one wonders how vines live at all.*

# IN THE VINEYARD

GREAT WINE IS MADE in the vineyard. That is an article of faith with every good winemaker these days. High-tech winemaking has achieved all it can (and it can achieve a lot); if you want to raise quality another notch, you must turn your attention to clones, rootstocks, canopy management, density of planting and yields. And, almost as a reaction to the necessity of technology in the winery, more and more growers are looking at untechnological, even apparently irrational practices like biodynamism. After the high-tech revolution of the 1980s and '90s, we now have the post-technological revolution. Back to basics. And there's nowhere more basic than the vineyard.

Thirty years ago it would have been impossible to predict the current obsession with earth and roots, with worms and microbiology, with leaves and sunlight and water stress. Then, at the beginning of the revolution that brought technology into the winery, people dreamed of stainless steel, of big shiny vats with temperature control, of new presses and filters and heat exchangers. And these exceedingly expensive pieces of equipment did indeed change the face of wine: they made possible the current world of clean, tasty, inexpensive bottles.

The idea was that these wines would be produced from equally efficient vineyards; a grower's budget therefore went on chemical fertilizers, herbicides, pesticides, anti-rot sprays and, where necessary and legally permissible, irrigation. Virus-free clones provided higher yields than ever before. Oh brave new world.

But highly efficient, high-tech vineyards, while they produce healthy, sound grapes, don't necessarily produce wines of much complexity or interest. Winemakers, being a competitive bunch, always want to do better. And so now we have another revolution: one in which the pesticides and fertilizers are being used far more sparingly, and in many cases being abandoned altogether. Holes are being dug in vineyards so that root distribution can be examined. Clones are being constantly reselected for quality, not for quantity of yield. And more attention is at last being paid to that necessity of modern viticulture, rootstocks and their ability to influence quality.

*Newly grafted Pinot Noir vines. The red wax is for protection in the early weeks. The choice of rootstock is crucial to the quality of the future wine.*

### Phylloxera: the coming of the plague

In most parts of the world it is necessary to graft vines on to other rootstocks. It is the only protection there is against *Phylloxera vastatrix*, the tiny aphid that at one stage of its complicated life cycle feeds on vine roots and kills the vine. Phylloxera invaded from the east coast of the United States in the second half of the 19th century, and vines throughout Europe were wiped out. It found its way to Australia in 1877 (though parts of Australia are still phylloxera-free), California in 1873 and New Zealand and South Africa by 1885.

This is how New Zealand winemaker David Hohnen described the first appearance of phylloxera on a four-year-old vineyard next to Cloudy Bay winery in 1989: 'In the full flush of summer growth, leaves yellowed on stunted shoots and the vines struggled to survive. Not all were struck. The diabolical contagion feared by all vignerons appeared as patches of jaundice in an otherwise dense green canopy.' The final infected vines at Cloudy Bay were pulled out in 1998, and all are now grafted.

The emergence of Biotype B in California attacked vines which had been grafted on to AXR1 rootstocks which were only semi-resistant to phylloxera, and widespread replanting took place in the 1990s.

Can phylloxera ever disappear? Probably not, but some authorities believe that if all vines in an area are grafted, infestation may eventually drop below a critical mass.

*A familiar sight in 1990s California wine country: burning phylloxera-affected vines that had been planted on insufficiently resistant rootstocks.*

### A question of graft

The vines used for phylloxera-resistant rootstocks are bred from native American species like *Vitis riparia*, which have natural immunity. There are dozens of different rootstocks now in common use, but it takes some years to determine by means of field trials just which rootstocks are most suitable in which circumstances and on which site. Some, like 333 EM and 41 B, are particularly tolerant to lime in soil, and are thus suitable for chalky soils like those found in Champagne. Others, like 110 R, are tolerant of drought; SO 4, Ramsey, Dog Ridge, 1613 C and others are resistant to nematodes, but SO 4 gives high yields and roots only shallowly; Rupestris St George, 99 R and 110 R are very vigorous, and produce huge canopies in their scions, which can be a problem; 101-14 and Riparia Gloire, by contrast, have very low vigour and so can be used to curb the vigour of vines in very fertile soils.

I mentioned nematodes just now. These are tiny worm-like organisms that live in the soil and spread viruses – and they in themselves can be a good enough reason for grafting vines on to resistant rootstock, even if there's no phylloxera problem in your region. Even in relatively disease-free Chile, where nematodes are not generally a big problem, they can flourish in the sandy parts of Casablanca, and nematode-resistant rootstocks are being found to be necessary if vines are to flourish there.

It was only really in the latter part of the 20th century that producers began to realize the importance of having the correct rootstock. At Château Angélus in St-Émilion, for example, the current generation say that when they took over in the 1980s, only about 30 per cent of the vines were on the right rootstocks. The choice of rootstock can affect the berry size, the leaf area and the leaf-to-berry ratio, all of which have quality implications for the wine.

### To clone or not to clone

While clones of some varieties like Pinot Noir, as we've seen on pages 10–11, can vary widely in quality, in vine varieties where there is less variation the choice of clone probably only accounts for between 1 and 5 per cent of the quality of the wine. So your choice of clone if you are growing Riesling is less important than if you are growing the genetically more variable Pinot Noir. And in both cases good viticulture is more important.

Some top producers are dubious of the value of clonally-selected vines. Jacques Seysses of leading Burgundy estate Domaine Dujac believes that in the wrong hands they are a bad thing: because they give higher yields it is necessary to put far more work into the vineyard, and thin the crop rigorously. It is an irony familiar to all producers in Europe that clonally-selected vines give yields that are higher than the legal appellation contrôlée maxima (see page 23).

Does having a variety of clones in a vineyard give greater complexity to the wine? Yes, say some growers; no, say others. In any case, many New World countries have only a limited range of clones officially available. The Mendoza clone of Chardonnay, for example, which is still found quite widely in the New World – actually it is a massal selection rather than a laboratory clone – produces high yields of loosely-structured wine. But it is only quite recently that sufficient quantities of better clones have become easily available.

This paucity has given rise to the most elusive sort of clone of all, the so-called Samsonite clone: vine cuttings brought in illegally, tucked into luggage, to avoid the

### OLD VINES

In the case of vines, age and beauty generally (though not always) go hand in hand. European vines are not permitted to produce appellation contrôlée (or the equivalent) wine until their fourth year, but that is still infancy in vine terms. After about 20 years vines become less vigorous and produce smaller crops of more concentrated fruit, and at a certain age – usually about 30 years – the producer has to decide if he values the increased quality more than the drop in yield. Most vines are uprooted at this point, and replaced.

Wine from old vines thus has some extra cachet. The only problem is defining how old is old. Some wines which blazon the words 'old vines' or 'vieilles vignes' on the label may be made from vines of only about 35 years, which is not even middle-aged, far less old.

So where can you find really old vines? Australia has some of the oldest. There is plenty of Shiraz and Grenache there, especially in the Barossa, that was planted a century ago: d'Arenberg can even boast some vines of 120 years. Chile has many vines of 60 years or more, and California has old Zinfandel and Carignane vines that are coming up to their centenary. Madera County in particular has large plantings of old Carignane left over from a previous fashion. Lighter wines made from different grapes came into vogue, and nobody ever bothered to uproot the old vines.

In Piedmont, Italy, Marcarina has some centenarian Dolcetto on its own roots in the cru of La Morra, and in Campania there is some Syrah and Aglianico, around the same age, which has grown to the size of small apple trees. In Germany's Mosel Valley, Ernst Loosen has some Riesling vines of 100 years old in the Ürziger Würzgarten vineyard.

Old vines certainly produce wine of greater intensity, but don't forget that many of the great 1961 Pomerols and St-Émilions came from vines that had been replanted after the devastating winter of 1956.

### PRE-PHYLLOXERA VINES

Technically these are vines that date from before phylloxera took hold at the end of the 19th century – which would make them very old indeed. But the phrase is sometimes used to refer to ungrafted vines – and these are less rare than one might suppose. In Europe some of the most famous are at Quinta do Noval in the Douro Valley in Portugal, which produces its rare Nacional port from half-a-dozen small terraces of ungrafted vines. The vines all around are grafted, and attempts to extend ungrafted vines to other parts of the vineyard have always failed. For some reason phylloxera never affected those particular parts of the vineyard – nobody knows why.

Elsewhere in Europe the Middle Mosel is still free of phylloxera and there are many vines planted on their own roots.

In the New World ungrafted vines are remarkably common. They are the rule in Chile, where phylloxera has never struck; there are also many in Australia, especially in South Australia, and some in New Zealand.

*Gnarled Shiraz vines over a century old growing in Clare Valley, South Australia. Vines like these survived because fashion passed them by, yet now they make some of the world's most sought-after red wines. Old-vine Grenache and Mourvèdre are in almost as much demand now as old-vine Shiraz.*

*Spraying the vines in early spring in Marlborough, New Zealand. Chemicals are used in a far more restrained way nowadays, but are still essential for a healthy crop in most vineyards.*

lengthy quarantine imposed by many countries in an effort to curb the spread of vine diseases. Selfish? Irresponsible? Offenders – and many top producers are offenders – say they get their cuttings checked out by laboratories, and much faster than would be the case if they brought them in openly.

The debate over the quality implications of clonal versus massal selection will continue. To quote Aubert de Vilaine of the Domaine de la Romanée-Conti, Burgundy: 'There are advantages to clonal selection and advantages to massal selection. A great wine needs a population: it is a mistake to have only athletes. But they must still be the sort of people you want.'

### In sickness and in health

The vine is subject to so many ailments and pests that sometimes one wonders how it manages to survive at all. One Oregon vineyard lists its particular local enemies thus: 'Crown gall, weevils, thrips, birds, bears, other critters.' Another vineyard, this time in California, knows when the grapes are ripe because whole tribes of raccoons come and eat them. In Germany wild boar can be a problem; in Australia it's kangaroos. Deer, rabbits and birds are universal predators on defenceless grapes. That is, of course, assuming that the grapes succeed in surviving downy mildew (which likes warm, wet weather), powdery mildew (which prefers dry weather), early bunch stem necrosis and grey rot, and that the vine manages to avoid black goo, Pierce's disease, fan leaf virus, leaf roll virus and eutypa, to name only a few.

Vine diseases can be bacterial, fungal, viral or phytoplasma; some are controllable, some preventable, some deadly. Some of the fungal diseases, including downy mildew, powdery mildew, anthracnose and grey rot can be controlled with the use of fungicide sprays, of which Bordeaux mixture, a solution of lime, copper sulphate and water, is the oldest.

*Mustard being grown as a cover crop between rows of 100-year-old Zinfandel vines in Sonoma County, California. Mustard can help to control virus-spreading nematodes in the soil.*

Virus diseases like fan leaf and leaf roll are spread either by cuttings or by nematodes in the soil. Even vines that are virus-free when planted tend to become infected – though they certainly stay healthier longer than young vines that have no pretensions to being virus-free. (As one Burgundian puts it, 'You'll stay in better shape if you go to Africa healthy than if you go there unhealthy.') But then even the term virus-free is only relative: of about 20 identified viruses only six are really dangerous, and it is those of which virus-free vines are free. When it comes to the others, vines must take their chance.

Pierce's disease, on the other hand, is a bacterial disease which kills vines fast. It is one of the main targets of national quarantine regulations, and is spread by an insect called a sharpshooter which has sadly little respect for national boundaries; the disease is so far confined to certain parts of the US – including Napa – and Central and South America, but while the blue-green sharpshooter is relatively manageable and home-loving and seldom strays more than 15m from streams, the larger and more mobile glassy-winged sharpshooter, which originates in the Southern states of the US, is beginning to encroach into Oregon, Napa and Sonoma, and is capable of spreading Pierce's disease very fast and very efficiently. Sharpshooters are found in Europe as well. European growers are not looking forward to an encounter with Pierce's disease.

Other insects, this time leaf hoppers, spread flavescence dorée, one of a group of phytoplasma diseases known as grapevine yellows. This will kill young vines and weaken old ones, and is potentially even more destructive than phylloxera, partly because of the speed with which it spreads, and partly because there is no treatment. Northern Italy suffered greatly from it in 1995; it is found in France, Germany, Switzerland, Australia, New York State and elsewhere.

The only route open to growers is to control the leaf hopper population; this can be done by spraying insecticide. But here one comes up against one of the main tenets of modern viticulture: that biodiversity is a good thing, that sprays of all kinds should be reduced as much as possible, and that it is desirable, where possible, to move in the direction of organic viticulture.

## Going organic

It is simply not feasible for all wine producers to become fully organic – even if there were any international agreement on what constitutes organic viticulture, which so far there is not. Warm, dry climates are best suited to thoroughgoing organic production, and regions like the South of France, Chile and California have a built-in advantage in not suffering from the humidity that can lead to mildew, rot and other ailments. In cooler, wetter regions like much of Germany, for a grower to go organic may mean losing an estimated 30 per cent of his crop in many years. Total loss is not unknown.

The compromise position is known as integrated management, or *lutte raisonnée* in French. Again, it is not closely defined, but means using chemicals as little as possible and only when necessary. It is such a broad church that it manages to accommodate virtually all producers who are not organic or biodynamic – at least, they all say they practise integrated management, though sometimes one wonders if all they are seeking is the economic advantage that comes with not being over-lavish with chemicals.

Those who seek real integrated management may cease to use herbicides, or other methods of weed control like ploughing or hoeing between the rows of vines, and instead encourage cover crops, which may be cut during the season or ploughed in. Some such crops, like mustard, can help to control nematodes and so are obviously beneficial; others, like many clovers, can unfortunately encourage leaf hoppers. The increase in ground cover in vineyards and the resulting increase in the leaf hopper population is certainly a factor in the spread of grapevine yellows. Biodiversity works both ways.

When you start to interfere in an ecosystem, even though you act with the best possible intentions, you cannot be sure of the effects of your actions. If, when planting a vineyard in virgin land, you spray against a pest, you may find you have destroyed a predator of that pest. Organic and semi-organic producers prefer to introduce predators to control pests – in Chile, a coleopter called Ambrysellus can be used against red spider; ladybirds can be used against aphids. (This is, of course, just another form of interference, but so far the results seem to be more beneficial to the vine.) Sexual confusion (the use of pheromones to control populations of certain insects) can be used against grapeworm and other pests.

All this is far more work than simply loading up a tractor with a container of chemical spray. It requires far closer study of your vineyards, and far more time spent on seeking solutions, but more and more serious growers seem to be prepared to make the commitment. No doubt many more would go fully organic but for the devastating reality of rot and mildew. Bordeaux mixture is permitted under organic rules, but presumably only for pragmatic and historic reasons, since it eventually builds up in the soil and can cause copper toxicity.

Full-blown organic viticulture forbids the use of any industrially synthesized compound, though the details vary from organization to organization. Fertilizers must

### A YEAR IN THE VINEYARD

**Work in the vineyard follows a basic pattern the world over, but the timing of each operation is dependent on both the prevailing climate and the weather conditions of the year.**

**Winter:** the leaves fall, the sap descends and the vine becomes dormant. Pruning can be done at any time during the winter.

**Spring:** the sap starts to rise, and the first signs of growth appear: the young vine must be protected against frost from now until early summer. The buds are also vulnerable to pests and diseases: sprays help to control these enemies of the vine, though organic and biodynamic growers use other methods. Ploughing and hoeing aerates the soil and clears weeds; fertilizer may be applied. As the ground warms up, new vines can be planted. Once the vines begin to shoot, the new growth needs to be tied to the wires, otherwise the foliage would shade the fruit and prevent it from ripening; the final trellising (page 24–25) takes shape.

**Summer:** eight weeks after budbreak, the vine flowers for about 10 days and then the fruit sets; cold or wet weather at this time can cause poor fruit set and thus reduce the quantity of the harvest. Summer pruning, or leaf plucking, may be necessary to allow more sunlight to reach the fruit. At *véraison* (the point at which the fruit changes colour) a green harvest may be done to reduce the size of the crop, and superfluous clusters removed. Netting or bird scarers may be used to protect against bird damage.

**Autumn:** picking usually begins in September/October in the northern hemisphere, or February/March in the southern.

*Budbreak on a Cabernet Sauvignon vine; this normally occurs 20–30 days after the sap starts to rise in the spring.*

*Young Pinot Noir clusters begin to appear just three weeks after flowering. Rain and cold may yet prevent the young berries from developing evenly.*

*Cabernet Sauvignon changing colour at véraison; this marks the beginning of ripening and occurs about 100 days after flowering.*

## BIODYNAMISM: NEW AGE HIPPIEDOM OR SOUND SCIENCE?

Biodynamism is the most extreme position taken by growers who follow the path from conventional viticulture to integrated management to organic. It is based on the theories of Rudolf Steiner (1861–1925), and emphasizes the importance of working with the movements of the planets and cosmic forces to achieve health and balance in the soil and in the vine.

It is different in kind as well as in degree from organic viticulture. Biodynamism uses natural herbal, mineral and organic preparations in homeopathic quantities: key applications are of horn dung, horn silica and dung compost. The first is made by burying a cow's horn filled with dung over the winter; it benefits the roots. Horn silica again involves a cow's horn, but this time filled with powdered silicum and buried over the summer; it aids photosynthesis. Dung compost is self-explanatory but not simple, since ideally every vineyard should have its own recipe, to reflect its unique needs; it helps the soil.

These preparations are mixed with water and 'dynamized' by stirring them first one way and then the other: horn manure should be stirred into warm water if possible, and stirred briskly for precisely one hour. This should be done by hand, not by machine: one of the leading exponents of biodynamism, Loire producer Nicolas Joly, compares the patterns created in the water to the patterns of Celtic art. 'The air is unceasingly taken into the spirals and intensifies all sorts of exchanges,' he says. The solution should then be applied at sunrise to the leaves of the vine, in quantities of just 40-50 litres of water per hectare.

The notion that briskly stirring a liquid introduces air is not strange to the many winemakers who stir the lees of maturing wine: they know that creating a spiral gives a different result to moving a stirrer backwards and forwards. The suggestion that this somehow relates to some lost Celtic wisdom might, however, upset those who have problems accepting the odder elements of biodynamism. Joly also stresses that manure must be buried in a cow's horn rather than a bull's horn because the former accentuates the 'primordial feminine', and when choosing manure, you should take into account not just the diet of the animal (for choice, one third leaves, one third roots and one third hay) but also its temperament: horses, he says, are dominated by heat.

What is also controversial is the stress on the influence of the stars and planets on the growth and well-being of the vine. Different times of the day and different phases of the moon are held to be suitable for different activities and different remedies: growers turning to biodynamism must be prepared to be out in their vines at three or four in the morning if required. It is an extremely dedicated, expensive and time-consuming way of growing grapes. Perhaps the oddest thing of all is that it seems to work. Indeed, the argument has moved from whether biodynamism works to why it works, and one grower is now working with scientists to try and discover the reason. Biodynamism seems to produce better wines with purer flavours and better balance. You can taste the difference; with organic viticulture this is not always the case.

Is this because biodynamism means paying intensely detailed attention to your vines and soil – in other words, would that degree of care pay off whether you were biodynamic or not? What you put on the soil must make a difference: Dominique Lafon of Domaine des Comtes Lafon in Burgundy found that after he went biodynamic the acid balance returned to his wines after just three years. 'You can feel in the wine that the vine is healthy,' he says. Soil expert Claude Bourguignon finds that microbial life in the top 30cm of soil is much the same in an organic vineyard as in a biodynamic one. But, he says, if you go deeper than this, the microbial life in a biodynamic one is far greater. Reactions between the roots of the vine, which can descend 30m into the subsoil, and the microfauna around them, generate the oxygen and minerals that are vital for a strong, healthy plant. Deep-rooting vines give thicker, stronger grapeskins, and thus more resistance to disease, more flavour, and longer maturing wine.

For many biodynamic growers in France, one of the attractions of the system is that it enables their wines to reflect their terroir: the application of standard commercial preparations, they fear, has a 'dumbing-down' effect on the individuality of each vineyard.

therefore be natural – compost and manure – and the addition of these is also normal in integrated management.

Do organic methods produce better wines? One feels they should; yet too many wines with organic accreditation are poor quality. The range of quality is in fact exactly the same as is found in non-organic wines: from disappointing to very good. Perhaps you can be a careful vine grower without being a good winemaker as well; whatever the reason, an organic logo on a label is not in itself a guarantee of a good wine.

### The optimum crop

Optimum, you notice, not biggest, smallest or even best. Winemakers, like everyone else, have to cut their coats according to their cloth, and it is sad but true that some possible improvements in quality are not initiated, particularly in less fashionable appellations, because the improved wine would not fetch a substantially higher price, even though the investment might be substantial. On the other hand, among fashionable appellations and fashionable producers the sky seems to be the limit, be it in the lengths they will go to to improve their wines, the amount of money they will invest, or the prices they can charge. Life is no fairer to wine producers than to anyone else.

### Yields

The rule used to be so simple: it was that in wine, quantity and quality did not go together. Higher yields equalled lower quality, and vice versa.

The experience of the New World, where vines are generally irrigated, has changed the picture somewhat. Work on canopy management, especially in New Zealand, has revealed a great deal about the production of good quality wines at relatively high yields in fertile soil.

What has become clear is that you don't make high quality wine merely as a result of having low yields. In practice, most (though not all) of the world's great wines come from low-yielding, low-vigour vines – but while low yields are associated with high quality, they do not in themselves confer it.

It's a question of getting the right amount of sunshine on to the leaves and grapes, and thus encouraging optimum ripening. Low-yielding, low-vigour vines have small, open, leaf canopies, and thus good leaf and fruit exposure to the sun. (This may be one reason why old vines, with their less vigorous leaf growth, give such good grapes.) Shade means less sugar, less flavour, often a streak of unripeness and, in red wines, less colour.

If you halve the yield you may double the price of the wine, but not necessarily the quality: there is an optimum yield, which is the point at which the vineyard is in balance. That point is different for every grape variety in every vineyard in every vintage; not surprisingly, relatively few vineyards are in perfect balance. One Bordeaux proprietor points to only one château there that he knows to be in perfect balance – and it's not his.

Above and below that optimum yield the law of diminishing returns sets in. Reduce your yields too harshly and there will not be the increase in quality to justify the reduction in quantity. Increase your yields too generously and eventually quality will fall faster than quantity rises as the wine becomes increasingly dilute. To take the example of Riesling grown on the steep slopes of the Mosel Valley: there the turning point for quality is said to be between 120 and 150 hectolitres per hectare; after that quality drops. Even the lower of those figures is far more than a good Mosel grower will admit to; but then asking a wine grower about his yields is a little like a doctor asking a patient how much he drinks. The doctor makes a mental adjustment of the figure he is given, and perhaps we should do the same.

Yield is expressed in various ways in vineyards around the world. Some countries favour tonnes of fruit per hectare; others prefer tons per acre. Many European countries express yields as hectolitres of juice per hectare of vines. It is difficult to convert hectolitres per hectare to tonnes per hectare with precision because, of course, much depends on how much juice is extracted from the grapes; in fact more efficient extraction of juice is one reason (though by no means the only one, or even the main one) why vineyard yields have been rising in recent years. Roughly speaking, one metric tonne of grapes will produce between 550 and 750 litres of juice, depending on how heavily they are pressed. White grapes may yield less than red, because very gentle pressing is essential for many varieties if harsh flavours are to be avoided. Riesling is an example of this: its thick skins, if pressed hard, will yield coarse-tasting phenolic compounds. But as a rough guide, one tonne per hectare is more or less equal to seven hectolitres per hectare. One ton per acre is more or less equal to 17.5hl/ha. (See Measurements page 307.)

*The direction of the rows of vines can make a big difference to how your grapes ripen: you might want to protect the vines from the prevailing wind or you might want to let the wind in, for greater ventilation. These vines are in Marlborough, New Zealand. Most regions show little consensus.*

European wine laws, though not usually those outside Europe, impose limits on yields for appellation wines. France, however, also has something called the *plafond limite de classement*, which allows the yield in any given year to be raised by as much as 20 per cent or even more over the basic yield. This is why yields in Bordeaux, even for leading appellations, may be as high as 55 or 60hl/ha, when the official base yield is just 45hl/ha.

Many, many vineyards around the world give yields that are too high for optimum quality. In Alsace, for example, between 1945, when the appellation was set up, and 1975, when the first Grand Cru vineyards were established, the basic permitted yield was increased two-and-a-half times. In Germany, whole lakes of poor quality Müller-Thurgau are produced at 300hl/ha.

## Density of planting

What is crucial, however, is not the yield per hectare or acre but the yield per vine. This is based on the number of bunches per vine, and the weight of the bunches, both of which are a direct result of viticulture – planting density, pruning, pest control and other vineyard management policies – and, of course, the weather.

Planting density can vary hugely, even within countries and regions. In Spain's Penedès, it varies from 800 to 2000 vines per hectare; in Australia and New Zealand it can be as low as 1000 per hectare in some vineyards or as high as about 4000/ha in others, with the odd experimental planting of 9000/ha; in California the traditional figure is about 1125 per hectare. Chile's usual figure is about 3000 per hectare, though one company is trying 25,000 per hectare in the South. In Burgundy's Côte d'Or, there are usually 10,000 vines per hectare, though the Domaine de la Romanée-Conti is experimenting with 16,000 per hectare; in Bordeaux's Médoc, there are usually 10,000 per hectare while in Entre-Deux-Mers this can go down to 2700 per hectare.

Generally growers are moving towards higher densities as part of the search for quality, but the choice of density ideally depends on the soil. In fertile, irrigated soils where the vines are vigorous, each vine will need enough room to spread its canopy out: a vigorous canopy cramped into a small space means shaded leaves and fruit. On less fertile soils quality can improve if the vines are more crowded and yet produce no more wine per hectare. But densities have changed over time in Europe as well: Ancient Roman viticulture used to have some 50,000 vines per hectare of vineyard.

It is often the method of vineyard management that, rightly or wrongly, determines the density: 19th-century European viticulture, with a vine density of 10,000–15,000 vines per hectare of vineyard, was designed around the horse. In Chianti, until quite recently, the usual density was 2700 vines per hectare, and was designed around the tractor.

## Canopy management: letting in the sun

Canopy management embraces pretty well everything one might do to a shoot or a leaf, from tying it in a particular place on a trellis to positioning it in the sun to cutting it off. It thus covers all aspects of pruning and training vines. The purpose of it is to get vines into that coveted balance in which they will produce the optimum yield of optimum quality grapes.

It takes about eight leaves to mature a cluster of grapes. But those eight leaves must have sunlight; if they are shaded by other leaves the vine will produce too little sugar and the crop won't ripen. (Dense canopies also encourage humidity and therefore rot and other ailments.) The answer is not necessarily to prune the vine hard: if you do that to a vigorous vine it will simply produce an even denser canopy, with more shading effect and consequent green flavour. Leaf plucking, or removal of excess leaves in summer, can help this.

If a vine is in good balance the weight of prunings cut off in winter will be between one-tenth and one-fifth the weight of the crop of grapes. But growers have to bear in mind that changes in viticultural methods won't show any benefits for a couple of years: with vines you're always working for the next vintage but one.

Different trellising systems are being tried out for many grape varieties, all over the world. For example, in New Zealand a combination of adequate sunshine, yet cool climate and soils of often high potential vigour can give problems of underripeness. In such conditions the vines grow fast and furiously and the canopy ends up shading the fruit: the result is green, harsh, underripe flavours. The problem is exacerbated by irrigation and by the application of fertilizers. Elaborate systems like Scott Henry (see right) have been successful in producing riper grapes.

*Geneva double curtain trellising in action in Argentina. Systems like this are being tried around the world, though are not suitable for all situations.*

Canopy management can be adapted to any need: in California's Napa Valley, when Stag's Leap Wine Cellars wanted to change from fruit-driven Sauvignon Blanc to a more herbaceous style of wine, it began by encouraging greater vigour and greater shading for greener flavours.

## Pruning and training

The traditional forms of trellising used in France, Germany and some other parts of Europe have the general name of Vertical Shoot Position, or VSP. This term embraces everything from gobelet to the pergolas used in parts of Italy, and includes the guyot and cordon methods.

The shoots are trained upwards as they grow, and the fruit positions itself towards the bottom of the canopy. Europe's low potential vigour soils produce relatively open canopies with this system, and leaf plucking and correct pruning to keep the vine in balance are generally enough to produce canopies of the required openness. In high vigour sites, on the other hand, VSP systems may lead to excess shading.

Newer systems are masterpieces of architecture. Canopies are divided vertically or horizontally, or both: some shoots may be trained upwards while others are, with difficulty, trained downwards.

Most systems are either head or cordon trained, and either spur or cane pruned. In head training the spurs or canes that produce the fruit-bearing shoots are positioned close together at the top of the trunk. In cordon training there are one or two long arms from which the fruit-bearing shoots will grow. These are part of the vine's permanent structure. In spur pruning all the canes are cut back to one or two buds each. In cane pruning, one or more canes from a previous year's growth are left behind at pruning, though cut to the required length. These canes bear the buds that will shoot to produce fruit.

## Green harvest

If your crop of grapes is too heavy – and many are – you can go through the vineyard during the summer and cut off the excess clusters. The idea is that the remaining clusters will have more concentrated flavours. It is not an ideal solution, and does nothing to

***Gobelet*** *A head trained, spur pruned system traditional to Mediterranean countries. The vine forms a low bush which shades the fruit from excessive heat.*

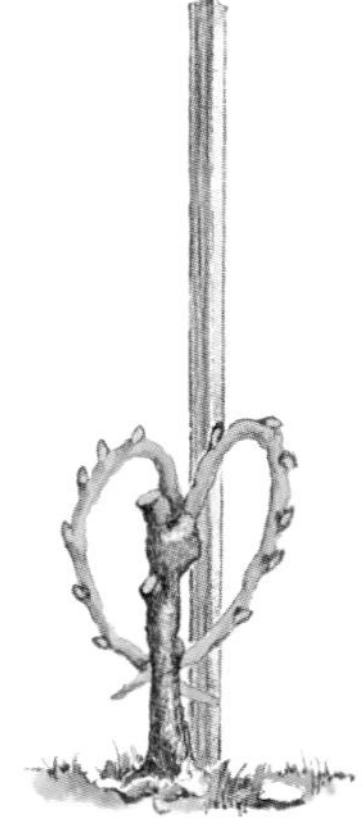

***Mosel arch***
*A head trained, cane pruned system used on the Mosel. Each vine has its own supporting post, with wires to which the foliage is tied.*

get the vine into balance. The ideal is to have the vine producing the right quantity in the first place, but this end is extremely hard to achieve – and since frost, disease and hail can all reduce the size of the crop, many growers like to have a few extra clusters on each vine as an insurance policy.

### Irrigation

It used to be said that vines had to struggle to produce great wine. The trouble is, vines that have to struggle too much produce wine that is very far from great; and the main cause of stress in vines is lack of water. Drought, as we've seen, simply causes the vine to shut up shop: photosynthesis, and thus the ripening process, ceases. A little water stress can be a good thing, but not too much.

Managing irrigation properly is one of the greatest skills a grower must learn. Traditionally it was banned in European countries, but it is creeping in through the back door: it is used in the Wachau in Austria, in many parts of Spain and in parts of Italy. There is a case for permitting irrigation on the hot, dry terraces of the Douro Valley in Portugal – not every year, but just when it is needed to

*An alternative approach to canopy management is minimal pruning, seen here in the Napa: all manipulation of the canopy is jettisoned, and pruning is stopped. The vines look alarmingly shaggy, but those who have tried it maintain that over several years the vine finds its own balance.*

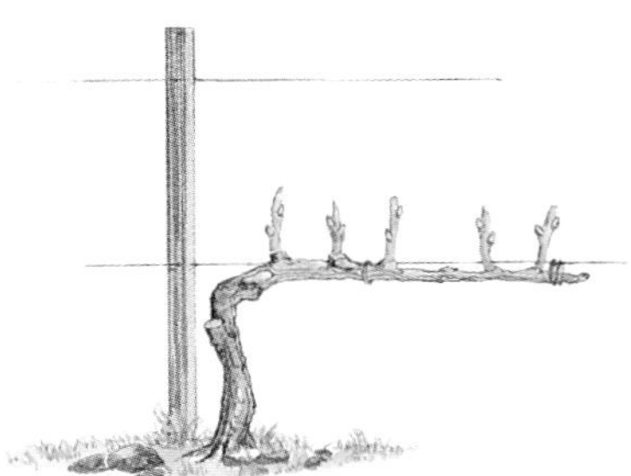

***Cordon de royat*** *A cordon trained, spur pruned vine. This is the system used in Champagne. The spurs are the short pieces of cane left by the pruner.*

***Lyre*** *A cordon trained system in which the canopy is horizontally divided.*

***Scott Henry*** *In this system half the shoots are trained up and half are trained down. In high-vigour sites a larger crop can be ripened, with higher sugar levels and lower acidity.*

***Double Guyot*** *An example of a head trained, cane pruned vine. At the end of the season the canes will be cut off and new ones selected for the next year.*

***Geneva double curtain*** *A cordon trained system with the canopy horizontally divided and all the shoots hanging down. In high-vigour sites vines can ripen more fruit this way, though grapes may get too much sun, leading to phenolic, oily-tasting wine.*

nudge the grapes to full physiological ripeness. It has even been discussed in Germany, and in hot summers with no rain it could have benefits for quality. Provided, of course, that it is not overdone.

In much of the New World rainfall patterns are different from those in Europe. If you get no rain at all between spring and autumn, as is often the case, then obviously you have to irrigate to be able to grow vines. Burgundy, for example, gets 60 per cent of its annual rainfall during the growing season: Napa gets just 15 per cent. But trying to use irrigation to replicate European rainfall patterns in the New World doesn't necessarily work either: that's what leading Bordeaux winemakers Bruno Prats, formerly of Château Cos d'Estournel, and Paul Pontallier, winemaker at Château Margaux, originally tried to do at Domaine Paul Bruno, their property in Chile. They had thought that imitating a wet spring, followed by water stress up to the harvest, would produce better quality; it didn't. 'It was a huge lesson in modesty,' they say.

Growers see irrigation as one of the more effective tools for adjusting what nature has given them: 'a way of imitating great terroir', as one puts it. But there are no easy answers about how best to do it.

Over-generous irrigation, for example, simply increases the amount of wine you make, by effectively diluting it with water. Likewise, if you irrigate after *véraison* (the time when the grape changes colour, from green to golden or black) you are simply inflating a balloon with water. What may be cleverer from the quality point of view is to introduce deficit watering just after flowering. This means giving the plant only enough water to grow sufficient leaves and shoots to ripen a crop; it causes mild stress which causes vegetative growth to slow and then stop, but not enough to stop photosynthesis and the ripening process. It also seems to affect cell division inside the grapes, so you get smaller berries. This means a higher proportion of skins and pips to flesh, and thus more colour, flavour and intensity.

The usual method of irrigation is drip irrigation. Thin rubber tubes run along the rows of vines, delivering carefully controlled quantities of water, drop by drop, to each vine. Ideally, constant monitoring of the humidity of the soil enables the quantity of water delivered to be correctly regulated. In Chile and Argentina, however, some vineyards still rely on the more old-fashioned ditch irrigation, in which ditches dug at intervals through the rows are periodically flooded. Obviously this is far less controllable, and most modern estates have abandoned it. Nevertheless, there is a school of thought which suggests that a large amount of water given to the vines in one go may encourage deeper rooting. Certainly one of the problems of drip irrigation is that it encourages root growth near the surface of the soil, when what all growers want is deep roots that search out nutrients at lower levels.

*Drip irrigation of young vines in New South Wales, Australia. Each vine has its own controlled water supply, and soil humidity can be regulated. Nutrients can also be delivered to the vine this way.*

*Flood irrigation in Argentina. This is cheaper and far less controllable: it cannot, for example, be regulated to induce the sort of mild water stress that is desirable for top quality.*

### Finally, picking

The grape harvest may have an image of jolly pickers tucking into rustic banquets washed down with litre bottles of last year's wine, but the reality is more likely to be one of sunburn and backache – or cold fingers and rheumatism, if you happen to be picking in November or December at the northernmost reaches of the northern hemisphere. (In Germany, Canada and the East Coast of the USA, picking may even take place on a freezing day in January, if you are aiming to make Icewine. For this, the grapes are picked and pressed while still frozen; very concentrated, sweet juice runs from the press, and the water content stays behind in the form of ice.)

And in many regions, the grape harvest takes the form of one man on a mechanical harvester, perhaps picking in the cool of the night before the morning sun warms the grapes too much for good winemaking.

Some producers distrust mechanical harvesters but there is no universally best method of picking, and how you pick your grapes must be governed by the dictates of ripeness, the weather and the type of wine you want to make. Plus, of course, your budget.

Picking at optimum ripeness is, whatever the climate, the most difficult to achieve. A cool spell in late summer may hold the grapes back, only for a hot week to accelerate them to the brink of overripeness. In those circumstances you need to act fast, since overripe grapes will produce soft, flabby wines which you will have to acidify (if you're allowed to). Or the grapes may be ripening so dreadfully slowly that you fear they will never get there at all, and perhaps rain is forecast. Do you then pick them before they are properly ripe, hoping to chaptalize – add sugar – (if you're allowed to) to compensate for the lack of sugar, or do you hold your breath and wait? One thing you should not do is pick in the rain: it is equivalent to simply pouring water into the juice.

When you have to pick fast, mechanical harvesters come into their own. They can work round the clock, and in hot climates picking at cool night-time temperatures is a quality plus. They work by repeatedly hitting the vines so that the fruit, either in the form of loose berries or whole clusters, falls on to conveyor belts. Inevitably there is some risk of bruising the fruit, and if the grapes have to travel a long distance to the winery there may be a risk of excessive skin contact with juice from broken grapes, but the effects of oxidation can be minimized with sulphur dioxide. (See In the Winery, pages 28–31.)

Machines are crucial where labour is scarce or expensive, but they cannot select the best grapes in the vineyard the way that well-trained pickers can. If you want to leave imperfect clusters behind on the vine, you have to have two-legged pickers (preferably with long experience). In Sauternes, and for Germany's Beerenauslesen and Trockenbeerenauslesen, where it is essential to pick selected grapes at particular levels of noble rot, you have to have pickers making several passages or *tries* through the vineyard. Like Sauternes, Monbazillac also relies on rigorous selection in the vineyard if the grapes are to achieve the super-ripeness needed for sweet wines. Nevertheless, mechanical harvesters were permitted for a period here until they were phased out from 1994. This is just one example of how a region's laws can favour the lazy.

Other things being equal, however, there are no quality differences implied in machine versus hand picking. Machines are far gentler than they used to be, and not all human pickers are good at following instructions.

## BOTRYTIS CINEREA: A BETTER KIND OF ROT

Not all rot on grapes is undesirable. *Botrytis cinerea*, when it affects ripe white grapes, produces luscious sweet wines such as Sauternes, Tokaji or Beerenauslese. (It destroys the colour of red grapes although very rare red Beerenauslese is made.) Ironically it is the same fungus that, in unripe grapes, causes grey rot or botrytis bunch rot, which is the biggest headache for any grower coping with a damp climate.

Grapes affected by noble rot turn golden, then pinky purple, then go brown as they dehydrate. Eventually they shrivel and become covered in a grey mould. The fungus metabolizes both sugar and acids, generally reducing the sugar content of the grape by one third, tartaric acid by five-sixths and malic acid by one third. But the loss of water, which is partly the effect of the fungus and partly the effect of evaporation as the skins are punctured by the fungus, results in great concentration of what is left.

*Here botrytis is attacking Chenin Blanc grapes; some berries are completely discoloured and mouldy, while others are still unaffected and green.*

But that is not the whole story. The chemical make-up of the grape juice is changed by the fungus, and glycerol, acetic acid and enzymes such as laccase are formed. The result is that wines made from botrytis-affected grapes have a flavour utterly different to that of other sweet wines, and combine it with such longevity that sometimes they seem almost immortal.

*Machine harvesting of Chardonnay in Narbonne, Languedoc, France. Picking at night like this means that the grapes are cooler when they arrive at the winery, and oxidation can be kept to a minimum.*

# In the Winery

Winemakers have got into the habit of saying 'wine is made in the vineyard'. Does this mean that as soon as the grapes arrive at the winery the die is cast? If the style and quality of the wine is inherent in the grape, is there nothing the winemaker can do to improve upon the raw material? Can he only destroy, not create? Well, obviously not. The winemaker still has choices. He can stress flavours or suppress them; he can coax a wine to precocious maturity or hold it in a state of arrested development. But if he does nothing but sit on his butt reading the newspaper, the best grapes in the world will swiftly turn into vinegar. So come on, you winemakers. Don't be self-deprecating. That's a pretty important job you're doing in there among your barrels and vats.

The first thing a decent winemaker must have is a vision of the style and flavour he wants to achieve in his wine. Indeed, the vines should have been cultivated with that vision in mind. In the winery he may favour minimal intervention, or else every technique under the sun, including the use of selected yeasts, must concentration, oak chips or new oak barrels, pre-fermentation cold maceration and goodness knows what. There is no New World-Old World pattern to winemaking any more. In Australia, for example, Chateau Reynella still uses a traditional 19th-century basket press for its wines. In Bordeaux, must concentration of reds, using space-age technology, is commonplace.

### Yeast: wild or cultured?

The winemaker may allow the native yeasts, present in the cellar and on the grape skins, to ferment the must to wine, or he may prefer the greater predictability of cultured, selected strains of yeast. If he chooses the latter, he may further choose a yeast that emphasizes particular aromas and flavours or one that is neutral in its effect on aroma.

The native yeast population in any cellar or vineyard will contain many different strains, and the balance of strains will vary from year to year. Winemakers who rely on wild yeasts welcome these differences, and the subtle year-to-year variation they give to the wines. The wine may also gain greater complexity from this varied population, though the relative unpredictability of wild yeasts means that there is a small risk of the fermentation sticking before it is complete, and refusing to restart. This can result in bacterial spoilage, and at best means that the wine will be sweet with unfermented grape sugar (residual sugar) where none was intended. If cultured yeast is used, the wild yeasts must be killed by the addition of sulphur to the must: sulphur is used throughout the winemaking process as an all-purpose anti-oxidant and antiseptic.

### Fermentation temperature

Controlling the temperature of fermentation was the 20th century's single biggest advance

## How wine is made

| Rosé wine | Red wine | White wine | Sweet wine |
|---|---|---|---|
| Rosé wine is made either by fermenting the juice of red grapes with the skins for a brief period, until the desired degree of colour is obtained, or by blending red and white wine (the latter method is illegal in the EU except in Champagne). | Red grapes are usually crushed and may be wholly or partly destemmed.<br><br>The juice is fermented with the skins and (if wanted) the stems. Chaptalization or acidification may be done at this stage.<br><br>During fermentation the skins and stems rise to form a layer, or cap, on top of the fermenting must or juice. This must be kept broken up and in contact with the must if colour and tannin are to be extracted. The usual method is to pump juice over the cap (remontage). Alternatively, the cap may be punched down (pigeage) either manually or mechanically.<br><br>The skins and juice, and the stems if present, are left to macerate after the end of fermentation, to extract more colour and tannin. The wine is run off the skins into wooden barrels or into vats made of stainless steel, cement or fibreglass.<br><br>The skins are pressed to extract all the liquid; this is called press wine. It will be matured separately and blended in later. | White grapes are destemmed and crushed.<br><br>The press separates juice and skins.<br><br>The juice is fermented in wood, stainless steel, cement, or fibreglass. Chaptalization or acidification may be done at this stage.<br><br>The new wine is run off into wooden barrels or into vats made of stainless steel, cement or fibreglass. | Sweet wine may be made by stopping the fermentation at the desired balance of alcohol and residual sugar, either by adding sulphur to kill the yeasts, or by centrifuging the wine to remove the yeasts. If the must is very high in sugar the fermentation may stop of its own accord before all the sugar has been turned into alcohol. |

The wine may undergo malolactic fermentation. This changes the sharp-tasting malic acid to the rounder-tasting lactic acid.

While the wine is maturing in vat or barrel the fine lees (the dead yeasts left from fermentation) may be stirred periodically (bâtonnage).

Alternatively the wine may be racked off its lees by being run from one barrel or vat to another.

All the vats or barrels are tasted, and the final blend put together.

The wine may be fined and/or filtered and/or cold stabilized to remove impurities.

**The wine is bottled.**

in winemaking. Wines fermented at cool temperatures retain their aromas, freshness and varietal character; those fermented too hot lose these desirable attributes and taste tired and stewed.

However, just because cool is good, colder is not always better. The temperature has to be above 10°C/50°F for yeasts to work effectively, and wines fermented at very low temperatures can be boringly neutral. Between 12° and 20°C is normal for white wine, and temperatures at the lower end of this scale favour the development of tropical fruit flavours, because the esters that are formed are those also found in tropical fruits; reds are fermented hotter, at between 25° and 30°C, to extract colour and tannin.

### Chaptalization and acidification

It takes approximately 18 grams of sugar in 1 litre of must to produce one degree of alcohol. If the must is insufficiently high in sugar, more can be added in the form of beet or cane sugar either before or during fermentation. The process is known as chaptalization, and is legal (and normal) in northern Europe and cool climates elsewhere, though illegal in the warmer climates of southern Europe. In southern Europe, however, 'enrichment' by the addition of concentrated grape must is often permitted, and has the same effect of raising alcohol levels. There are legal limits on how much the alcohol level may be raised by chaptalization or enrichment.

Acidification is the warm climate counterpart of chaptalization: adding acidity to compensate for its lack in the must. It may be added before fermentation or afterwards, but if you want it to blend subtly, adding before fermentation is better. Tartaric is the best, least intrusive acid though malic and citric can also be used.

### Must concentration

This can be thought of as a way of fattening wines for market. It involves removing water from must, by evaporating water at low temperature in a vacuum. The effect is to concentrate all the flavours in the wine. A process called reverse osmosis can also be used to remove water from wine, thus concentrating it. Must concentration is currently popular in Bordeaux as a way of producing the rich, fat wines that fashion demands. However, the jury is still out on the long-term development of such wines. It may be that it makes absolutely no difference to flavour and development in the long run, in which case it is useful only as a way of producing the sort of lush flavours that win press plaudits at the all-important moment when the young wines go on sale, in the spring following the vintage. Or it may be that must concentration handled badly can unbalance the wines because, remember, when you concentrate the must, you concentrate all its component parts, good and bad. More fruit might be fine. But more tannin? More acid?

*A sorting table, like this one at Robert Mondavi's Carneros winery in California, is common at leading estates. The first selection of grapes is done by the pickers in the vineyard, but when the grapes arrive at the winery they must be checked again, and any unhealthy or underripe clusters removed.*

### Maceration of red wines

Red wines are usually left to macerate with the skins after the fermentation is finished. The aim is to extract colour, tannins and other phenolic compounds, all of which are necessary to red wine, though just how much colour and tannin is desirable or possible depends on the style of wine required and the grape variety being used.

The approach to tannin, in particular, has changed greatly in recent years. Nobody these days wants harsh tannins in wine. Red wines are drunk much younger than they used to be, and having to cellar a wine merely because it is undrinkable young is not an attractive option. Yet of course fine reds do improve and develop with age, and so they still need tannin just as much as ever. But the tannins must be ripe, and not as obtrusive as of old. Techniques for handling tannins are one of the most crucial parts of modern red winemaking, and they start in the vineyard, with growers paying great attention to the need for ripe skins and pips, not just high sugar levels, in the grapes.

Different compounds are extracted at different temperatures and at different concentrations of alcohol. Tannin, for example, is extracted faster if there is alcohol present – so if you want more tannin you macerate the wine with the skins after fermentation.

If you want colour but not excessive tannin then you may want to do a 'cold soak': macerate the juice with the skins, at low temperature, before fermentation; a warm soak, at over 30°C, for 24 hours before fermentation is also being tried by some winemakers. White wines may be given skin contact too, before fermentation, to extract the flavours and aromas present in the skins. But overdoing this can also give harsh flavours. As always with wine, a little may be good, but more is not necessarily better.

### Carbonic maceration

If you want light fruity reds, bright in colour and low in tannin, this is the process to use. It is most usually associated with Beaujolais – a wine which fits the above description perfectly – but the process is used in many parts of the world, on a wide variety of grapes, to produce soft, easy-to-drink and enjoyably fragrant reds.

It involves putting whole clusters of grapes, uncrushed and unbroken, into a closed fermentation vat, and smothering the lot in carbon dioxide. With oxygen excluded, an intracellular fermentation takes place within each berry which, apart from producing some alcohol, also produces aromatic compounds and glycerol, and reduces malic acid levels. This sort of maceration may continue for between one and three weeks; usually the grapes are then pressed and a normal fermentation completes the transformation of sugar into alcohol.

In practice events are never quite this clear-cut. The weight of all those grapes tends to crush the ones at the bottom, so that the juice runs; it is only the grapes at the top of the vat that undergo a pure intracellular fermentation. Nevertheless, the wines have a characteristic aroma and flavour of bananas or bubblegum that can be very attractive. Such wines are intended for drinking young.

Even if it's not actual carbonic maceration, similar principles apply to reds made by whole cluster fermentation anywhere in the world. Even if the vat isn't doused in carbon dioxide, this gas will be given off by the fermentation of the juice at the bottom of the vat, and will have the effect of excluding some or all oxygen at the top. If any winemaker uses whole cluster fermentation, he's doing it to get bright, forward fruit flavours into the wine.

Carbonic maceration is not used for white wines, which acquire pretty strange flavours if treated this way.

### Malolactic fermentation

This secondary, bacterial fermentation changes the sharp-tasting malic acid in wine to the riper-tasting lactic acid and is almost universal in red wine. It does not reduce the total acidity in the wine, but makes it taste less aggressive, and adds to the wine's apparent weight in the mouth. In white wines it gives a creamy, buttery flavour, but reduces the up-front fruitiness. A winemaker can decide to put all, part, or none of his wine through malolactic fermentation, depending on the flavour he wants.

### The effect of oak

Wine may be stored in a variety of materials, including stainless steel, concrete, fibreglass and wood. (Wood usually, but not always, means oak; but in Savennières in the Loire, for example, chestnut and acacia are traditional.) In most of these the wine, if it is kept at low temperature and not allowed to oxidize, will not change or develop, and will retain its youth and freshness. Indeed, one Sauternes estate, Château Gilette, keeps its wine in concrete vats for a couple of decades before bottling it. At bottling the wine tastes almost as young as it did when first made.

But aging in oak barrels is different. Firstly, there is a small exchange of gases through the pores of the oak; this slow oxygenation softens the astringency of the young wine and reduces the fresh primary aromas. Old oak barrels, which impart no flavour of their own to wine, are sometimes used to achieve this.

### Composition of the grape

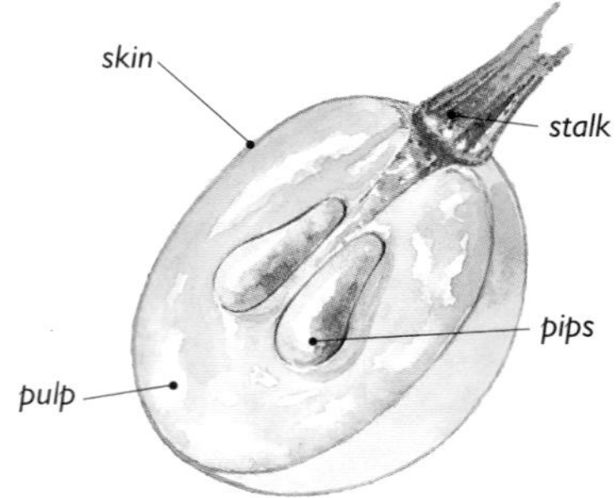


*The juice, or squeezed pulp, of the grape contributes water, sugar, acids and flavour compounds, but it is not the only ingredient of wine. Colour, tannins and flavour compounds also come from the skins, and, depending on the type of wine, some tannin may also come from the stalks. The pips contain bitter oils, and modern winemakers avoid crushing them.*

But new oak barrels also give particular flavours to wine, most obviously vanillin. This has a delicious affinity with the flavour of some grapes, notably Cabernet Sauvignon and Chardonnay, and these are commonly aged in small new oak barrels (usually 225-litre barriques, often known as *barriques Bordelais* simply because they are the size traditionally used in Bordeaux) for maximum oak flavour. If overused, however, new oak can dominate the flavour of the wine and make all wines taste alike.

French oak generally imparts more subtle flavours than American oak, though American oak treated in the same way as French (air dried rather than kiln dried; split rather than sawn) can be more subtle.(French oak barrels are also approximately twice the price of their American equivalents.) Oak from Minnesota and Wisconsin is much used, though some winemakers find it too tannic; Oregon oak is now beginning to be used, though only on a small scale so far. French oak may come from a variety of different regions: Limousin oak is wide-grained and tannic and often used for brandy; Tronçais, Allier and Nevers is tight-grained and much used for wine; Vosges, too, is tight-grained. Oak from the countries of the former Yugoslavia, often called Yugoslavian or Slavonian oak, has long been popular with Italian producers, who favour it for their large old casks. German oak, which can give a spicy note, may also be used, and Russian oak is beginning to make an appearance.

But it is not just the oak that imparts different characters to the wine: the choice of cooper is equally important. Winemakers experiment with different coopers just as much as with different oaks.

And then there is the question of toast. All barrels are bent into shape over a fire, and the fire inevitably toasts the inside of the barrel. Barrels given a heavy toast impart a spicy, toasty, roasted coffee flavour to wines, but give less oak flavour as the toast acts as a barrier between wine and wood tannins. Barrels with a medium toast will create wine with more tannin and vanilla; barrels with a light toast will impart more tannin again.

Obviously, the smaller the barrel the greater the effect of the wood on the wine. The most common size is the 225-litre Bordeaux barrique. The Burgundy *pièce* is slightly bigger, at 228 litres, and while most regions have their speciality sizes, few producers in Chablis still use the traditional *feuillette* of 132 litres. Traditional German

*The red wine fermentation chai at Domaine de Chevalier in Pessac-Léognan, Bordeaux. The post-fermentation maceration takes place in steel vats like these at most estates; the wine is often run into barriques for the malolactic fermentation.*

*Oaked wine has not necessarily seen the inside of a barrel. Other and cheaper forms of oak (**above**) include granular virgin oak, oak beans or cubes, and oak chips, with two different toast levels. These can be dunked into fermenting wine to give some new oak flavour. The traditional way is by aging the wine in new oak barrels, which are shaped and toasted over a fire, like these (**left**) at the coopers Tonnellerie Lasserre in the Bordeaux region. The precise degree of toast will have been specified in advance by the buyer of the barrel.*

barrels are the Mosel's 1000-litre Fuder and the Rhine's 1200-litre Stück. Italian *botti* come in various sizes, as do the pipes for aging port. In Australia hogsheads are commonly 300 litres, but nobody is bound by tradition these days and winemakers choose their barrel sizes according to the effect they want the wood to have on their wine style.

New oak barriques have another effect as well: they help to fix the colour of red wines, and polymerize the tannins. (In other words the molecular chains of the tannins are lengthened so that the tannins taste softer.) The sooner the wine goes into barrique, the better: the malolactic fermentation is often done in barrique for this reason.

Oak chips have the same effect of fixing colour and polymerizing tannins; in addition, if they are added during racking they reduce the necessity of adding sulphur dioxide, since they actively protect the wine against oxidation. They also give a touch of oak flavour, but they can't imitate the oxidative effects of proper barrel aging. Oak chips are generally illegal in Europe – which is not the same thing as saying they are not used.

### BLENDS VS. VARIETALS

It is a fact of modern wine that while consumers like varietal wines (those made from a single grape variety), winemakers often find blends more interesting – the reason being that a good blend is more than the sum of its parts, while only a few grape varieties are intrinsically interesting enough on their own to produce complex, profound wines. Many authorities believe that no matter how good a varietal wine, 5 per cent of another grape will always make it better.

Red Bordeaux (and usually white Bordeaux as well) is a blend; Burgundy, red or white, is a varietal wine. Chianti was traditionally a blend, though nowadays may be a varietal; Barolo is by law a varietal. Blends are a useful way of hedging your bets in regions where not every grape variety can be relied upon to ripen perfectly every year: Bordeaux is one example, Champagne another. Other blends may evolve because one or other grape is in short supply, or because in some regions it was usual to add white grapes to red wine to soften the astringency of the red: Viognier in Côte-Rôtie is an example. These days, better handling of tannins in red wines, and greater ripeness from leaving the grapes on the vine longer, solves that problem.

### Co-pigmentation

A red herring, or a rediscovery of old wisdom? It's a fancy name for what happens in the old practice of fermenting a few white grapes with the red. It still happens in Côte-Rôtie, in the Rhône, where a few per cent of Viognier may be added to Syrah – but it is not the same as blending white wine with red. In Côte-Rôtie the red and white grapes must be fermented together, and this could be crucial. It seems that if a red fermentation has what are called 'co-factors' added before fermentation, then these co-factors may produce extra colour. Paradoxically, these co-factors may be found in certain white grapes. Yalumba of Australia, which is busy researching co-pigmentation, thinks that quercetin, flavonoids and procyanidins from the skins and pips of Viognier are just such co-factors – and that if you add Viognier to a Shiraz ferment you get more colour than you would from Shiraz on its own. The colour also seems to be more stable over time, mouthfeel and texture are enhanced, and flavours and aromas are more vibrant. Does this apply only to Viognier and Shiraz? More research is needed. And while it is true that white grapes have often been fermented with red purely out of convenience, it is also possible that our ancestors had observed that the wines were better that way.

# A–Z OF GRAPE VARIETIES

The A–Z section from page 34 to page 295 covers entries on grape varieties from all over the world.

The symbols after each heading indicate whether the grape is dark-skinned, usually for making red or rosé wine, or light-skinned, usually for making white wine. If you cannot find the grape variety you want in the A–Z, it may be described under another synonym, so try the Index of Grape Names and Their Synonyms on page 308.

The A–Z section contains special features on 17 of the world's great classic grape varieties, along with information on geography, including a world map locating major plantings, history, viticulture and vinification, finding the grape around the world and enjoying the grape. A Consumer Information box with Best producers and Recommended wines and Maturity charts with advice on when to drink various wines complete each special feature.

In addition to the classic grapes, the A–Z contains two-page features on a further 15 major varieties. Best producers have been given for all the classic and major grapes and, where appropriate, also for other minor A–Z entries.

*Immaculate-looking vines, here in the Quintessa Vineyard, near Rutherford in the Napa Valley, California with Mount St Helena looming in the distance. Most of the classic French grape varieties are grown in California and recent replantings have done much to match varieties to the most suitable locations. This vineyard grows the red Bordeaux varieties of Cabernet Sauvignon, Cabernet Franc and Merlot and the end result is a rich complex and concentrated wine.*

## ABOURIOU

Tannic, low-acid grape of south-west France, found in a few lesser wines. It's on the way out, and won't be much missed.

## AGIORGITIKO

If a grape is named after St George, English-speaking drinkers might expect to find it in England, whose patron saint St George has been since the 13th century. But this patron of armies is also venerated by Greeks, and his eponymous grape makes some of the best reds in Greece.

Its cultivation is centred on the Peloponnese, where it is the sole grape in the gutsy, spicy, ripe, plum-tasting wines of Nemea. Here it is cultivated at between 250 and 800m, with the longest-living wines coming from the slopes of the high plateau of Asprokambos; elsewhere the grape's low acidity can limit its wines' longevity. But it has good fruit and colour, blends well with Cabernet Sauvignon, and on its own can make attractive rosés. It is also known as Mavro Nemeas. Best producers: (Greece) Antonopoulos Vineyards, Boutari, Skouras.

## AGLIANICO

This potentially high-quality Italian grape arrived from Greece, from where it was brought by the Phoenicians: 'Aglianico' is a corruption of the Italian word for Hellenic, 'Ellenico'. Whether it is identical to any variety currently grown in Greece is not clear. In any case, it seems not to have moved far from where it would have landed, in southern Italy: Campania and Basilicata are where it is found. It is concentrated in the provinces of Benevento and Avellino in the former, and in Potenza and Matera in the latter. Calabria, Puglia and the island of Procida, off the Neapolitan coast, also grow some Aglianico.

Its most prestigious (though nevertheless underrated) wine is Aglianico del Vulture, Basilicata's sole DOC wine. The production zone is centred on Mount Vulture, and vines are grown at between 450 to 600m; soils are volcanic, though more recently planted areas have more clay. Quality is generally good, and rising, and Taurasi in Campania also produces examples worth buying.

The weighty, concentrated, berried, sometimes smoky flavours of the wine, and its good acidity in warm climates, are attracting interest from the warmer parts of Australia, where the grape is being considered for planting in the Riverland, as well as in the cooler McLaren Vale and Margaret River. It is early

*Aglianico vines on the slopes of Monte Vulture, Basilicata in southern Italy. Aglianico is suddenly one of the most fashionable grapes of a newly fashionable region: look for it on a label near you soon. Alternatively, look for it on an Australian label.*

budding and rather late ripening, and favours a dry, sunny climate. Best producers: (Italy) Antonio Caggiano, D'Angelo, D'Antiche Terre, Di Majo Norante, Feudi di San Gregorio, Galardi, Mastroberardino, Montesole, Montevetrano, Odoardi, Paternoster, Orazio Rillo, Sasso, Giovanni Struzziero, Terredora di Paolo, Villa Matilde.

## AIDANI

This Greek grape grows on Santorini and other islands: it is attractively floral-scented and generally used for blending. There is also a black Aidani Mavro, some of which is used for the local sweet dried-grape wines. Best producers (white): Koutsouyanopoulos, Markezinis.

## AIRÉN

Airén is the major white grape of the vast La Mancha region in central Spain, and the low density of planting there – usually between 1200-1600 vines per hectare – means that the region has long held the title of largest area planted to one variety in the world. But that is changing as fashion turns towards red grapes.

Prices for red grapes are three times or more those for Airén, and that differential is especially painful given the huge investments made in state-of-the-art equipment for Airén. (As one consultant winemaker puts it, 'the tartaric acid bills are very high in La Mancha.') Reds can earn it back; Airén can't. A lot has been uprooted, and a lot more will be.

Ironically, the wine has never been so good. The grapes' thick skins and reasonably generous yields (even with irrigation yields seldom reach the legal limit of 85 hectolitres per hectare, but at such low density that is still quite a lot of grapes per vine) make it ideal for this region of extremes of heat and cold. The full weight of modern technology gets thrown at it, and the wines are faultlessly made, utterly clean and fresh – but usually have no character whatever. That's why they have been unable to command any price premium. You can't dislike them, but there's no real reason to buy them unless they're cheap.

A large part of La Mancha's production of Airén is distilled, and much eventually finds its way into port as fortifying spirit. Some is blended with red wine to make light-coloured reds at low prices – La Mancha is one of the few DO (or equivalent) regions able to do this under EU law. And *aficionados* of traditional methods will be glad to know that some is still made in the old yellow, oxidized way for the local market. Best producers: (Spain) Ayuso, Vinícola de Castilla, Rodriguez y Berger.

## ALBANA

An all-too-often unexciting Italian grape, given ideas way above its station by the granting of DOCG status to Albana di Romagna in 1987. The reasons usually given for this promotion are political and pragmatic: the authorities wanted a white DOCG, and the other possibilities were out of the question. The best Albana is exotically aromatic, with honeyed, soft fruit; too many examples, though, lack aroma and settle for being merely correctly made. Albana can be dry, off-dry, sweet or *passito*, and the latter two, though accounting for a very small proportion of production, can be the most interesting. Raisined grapes or, sometimes, nobly rotten ones are used, and there is some experimentation with barrique aging.

Plantings are concentrated in Emilia-Romagna in central Italy. The vine yields generously and is quite fussy about its growing conditions: it needs good rainfall but is susceptible to grey rot in damp conditions. The most commonly planted clone, Albana Gentile di Bertinoro, has thick skins which offer some protection against rot, and give quite deep-coloured wines: some authorities consider it a separate variety because of its smaller bunches and lower propensity to produce masses of vegetation. Its synonyms include Greco and Greco di Ancona, but it is not related to Greco di Tufo. Best producers:

(Italy) Celli, Umberto Cesari, Leone Conti, Stefano Ferrucci, Fattoria Paradiso, Tre Monti, Uccellina, Zerbina.

## ALBARIÑO

See pages 36-37.

## ALBAROLA

Italian grape found in Liguria and used both for somewhat neutral wine and for the table. In Cinqueterre and La Spezia, its centres as a wine grape, it is blended with other grapes.

## ALBILLO

White Spanish grape that gives wine low in acidity. It's found pretty widely across the country, but especially around Madrid. Best producers: (Spain) Dehesa de los Canonigos.

## ALEATICO

Sweet, scented red Aleatico is, at its best, an appealing dessert wine. It may be a dark mutant of Muscat Blanc à Petits Grains (see pages 144–153) and shares the same heady aroma of roses. It is found in Italy in Lazio and Puglia and on the island of Elba and on the French island of Corsica, and there have been recent plantings in southern Tuscany. It is also found in Kazakhstan and Uzbekistan, and to a very small extent in Chile. It has a rare white form, Aleatico Bianco, which is less fertile than standard white Muscat. Best producers: (Italy) Avignonesi, Francesco Candido, Le Pupille.

FRANCESCO CANDIDO

*A gloriously rose-scented sweet red made from Aleatico grapes. It's an almost impossible wine to match with food; drink it after dinner, and savour it.*

## ALFROCHEIRO PRETO

Interesting Portuguese grape found in Dão and Bairrada, and further south in the Alentejo and Terras do Sado. It was introduced to the Dão after phylloxera, but its history before that is obscure. It is susceptible to rot, but gives wines with good colour and blackberry and spice flavours and soft tannins. Its name in the Dão is Pé de Rato, or 'mouse paw' – either because of the Portuguese passion for naming grapes for animal references (see Esgaña Cão, Periquita, Rabo de Ovelha) or perhaps because of unusual local food and wine matches. Best producers: (Portugal) Caves Aliança, Quinta dos Roques, Sogrape.

## ALICANTE BOUSCHET

A heat-loving Teinturier grape, Alicante Bouschet is a crossing of Petit Bouschet and Grenache, bred by Henri Bouschet in 1866. Petit Bouschet was itself a crossing of Teinturier du Cher with Aramon (see page 38), produced earlier by Henri's father. The idea was to have a deep-coloured grape to blend with the prolific but light-coloured Aramon, but one that was of better quality than Teinturier du Cher. Alicante Bouschet (of which there are umpteen versions, crossed at different times; some are worse than others) spread rapidly through the south of France in the years after 1885, and was planted, too, in the South-West (including Bordeaux), in Burgundy and in the Loire Valley. In Spain a version of the grape, Alicante-Henri-Bouschet, is known as Garnacha Tintorera (see Garnacha, pages 92–101).

It is now in decline everywhere in France, and extinct in many places. (Its blending partner Aramon, which has one-fifteenth of the colour of Alicante, is also in decline.) It is early ripening and quite high-yielding, though at very high yields of 200 hectolitres per hectare or more, which it can attain on flat, fertile land, the alcohol level drops to 10 per cent or less.

In Portugal's Alentejo region it can produce wine of good colour, tannin and fruit, and is also grown with varying results in southern and central Italy, Israel, North Africa and former Yugoslavia. In Chile it is usually blended with other varieties like Cabernet Sauvignon, though it can make good concentrated, rough-fruited varietal wines. In California it supplied home winemakers during Prohibition, but its acreage has fallen drastically since. A few serious winemakers in Napa and Sonoma produce some, but lack of structure is its main failing. Best producers: (California) Papagni, St Francis, Topolos at Russian River; (Portugal) Quinta da Abrigada, Quinta do Carmo, Herdade de Mouchão, J P Ramos.

## ALIGOTÉ

Burgundy's second white grape variety comes a long way after Chardonnay both in reputation and in area planted (500ha as opposed to 12,000ha for Chardonnay). On the Côte d'Or, where it was once interplanted with Chardonnay for the sake of its acidity, it is now confined to the hilltop vineyards and to the plain; there are under 250ha in the Chablis region and a few hectares scattered in other departments. Bourgogne Aligoté is traditionally mixed with cassis to make kir, but in a good year, and particularly from the village of Bouzeron at the northern end of the Côte Chalonnaise, where it now has its own appellation, the wine is well worth drinking on its own for the sake of its fresh, buttermilk flavour. Occasional examples can age well, but generally it is for drinking young. It does not take well to oak aging. There is also a patch of vines at Die east of the Rhône Valley.

In Eastern Europe, where Romania, Bulgaria, Russia and the CIS countries grow it in large quantities, it is grown on flat land and its Burgundian hillside yields of 50-70hl/ha can triple, with a consequent drop in quality. There is also some in Chile and in California. Best producers: (Burgundy) Coche-Dury, Jayer-Gilles, Denis Mortet, Ponsot, Daniel Rion, A et P de Villaine.

## ALTESSE

See Roussette, page 206.

## ALVARINHO

See Albariño, pages 36–37.

## AMIGNE

There are only 20ha of this vine planted in the world, and 16 of them are around the Swiss town of Vétroz in the Valais. Its wines are often slightly off-dry and may be late-harvested: at their best they have concentration, length and individuality. Their flavour has been described as resembling that of brown bread, but they tend to lack acidity – as does brown bread (of course). Best producers: (Switzerland) Bon Père Germanier, Caves Imesch.

## ANCELLOTTA

A lesser part of the Lambrusco blend (see page 116), Ancellotta is found also in many other parts of Italy. It is blended with Sangiovese and many other varieties for the sake of its deep colour and good ripeness. Best producer: (Italy) Mariana Mantovana.

## ANSONICA

See Inzolia, page 115.

## ANTÃO VAZ

A white grape much grown in the hot Alentejo, Estremadura and Terras do Sado regions of southern Portugal. It withstands heat, but produces pretty bland wines. Best producers: (Portugal) José Maria da Fonseca, J P Ramos.

# Albariño

In terms of aroma, Albariño can be up there with Viognier and Gewürztraminer: exotic, suggestive, with scents that seem far beyond anything that could possibly come from a mere grape.

Yet Albariño is different, in that its wine is generally light to the point of being barely ripe in Portugal's Vinho Verde region: certainly an Australian winemaker, used to super-ripe grapes, would not be impressed with the meagre level of ripeness attained by Alvarinho (its Portuguese name) in Vinho Verde. High yields are the chief reason: old-style pergolas are commonly used in Portugal producing hefty crops of grapes that rarely exceed 8.5 per cent potential alcohol. Indeed, not all the Alvarinho in Vinho Verde is grown in vineyards. Many vines are still grown up poplar trees round the margins of fields, in a manner that would have been familar two millennia ago. Even when trained on wires, as is the modern way, canopies have to be large to allow for the vine's vigour in a wet, humid climate. Thirty to 40 buds per vine is normal and at that level of cropping the wines can reach 12–12.5 per cent alcohol.

There are pergolas north of the border in Spanish Galicia, too, where the size of the vineyard has increased eight times since the Rías Baixas DO was awarded in 1988. It's hardly surprising: Albariño is by far Spain's most fashionable white grape variety. According to some reports consumption of Albariño in Galicia alone is about three times the production of the area, and I have to say that some of the expensive and disappointingly neutral examples from Galicia that I've seen recently rather support this theory.

There is some experimentation with barrique fermentation and aging and, indeed, in Portugal some planting of Alvarinho further south, in Dão. Here the vines are being pruned and trained with a much smaller canopy, and lower crops. Alvarinho withstands heat well: it will be interesting to see if, when treated like this, it becomes even more Viognier-like, in weight and texture as well as perfume.

**The taste of Albariño**

Imagine Viognier, but lighter – all apricots and white peach, but with less weight and richness, and more acidity. The grape has thick skins and a high proportion of skins and pips to flesh, which means lots of aroma, but a slight tendency to bitterness. Greater ripeness might eradicate the latter, but would reduce the wine's ethereal lightness.

**Pazo de Señorans**

*A classic modern Albariño, full-flavoured but remarkably citrus, and intended for early drinking. Vintages vary greatly from year to year in this part of Spain: in the best years the wine has good balance and acidity.*

*Agro de Bazán's Granbazán Ambar from Rías Baixas in Spain's Galicia region is an unusually long-lived example of Albariño, and will age well for two or three years. In the winery it is given skin contact before fermentation, and not released until a year old: most Albariños are released for sale before Christmas of the year of the harvest.*

**José Maria da Fonseca**

*Alvarinho and Loureiro are blended with Moscatel de Setúbal to give freshness and acidity to J M da Fonseca's Quinta de Camarate Branco from the new Terras do Sado region bordering the Atlantic coast.*

**Left:** *Lagar de Fornelos vineyards in the Rías Baixas DO, Galicia in north-west Spain. Albariño's thick skins help protect it against rot in the wet climates of Galicia and of Vinho Verde, across the border in northern Portugal. All that water, plus the vine's natural vigour, means that whether it is trained on pergolas or wires it must be allowed to have a big, spreading canopy. If the canopy is cramped and dense, rot and underripeness become big problems.*
**Above:** *There is one theory that Albariño is related to Riesling, and another that it is the same grape as south-west France's Petit Manseng. Madeira also has an Alvarinho Liláz grape which is said to be different to the Alvarinho of northern Portugal.*

## CONSUMER INFORMATION

### Synonyms & local names

Albariño is the grape's Spanish name and Alvarinho the Portuguese name. Cainho Branco is a Portuguese synonym. The Spanish grape called Albarín Blanco is probably identical.

### Best producers

**SPAIN/Galicia/Rías Baixas** Agro de Bazán, Aldea de Abaixo, Domínguez Borrajo, Quinta de Couselo, Granja Fillaboa, Adegas Galegas, Lagar de Fornelos, Lusco do Miño, Marqués de Vizhoja, Gerardo Méndez Lázaro, Morgadío, Pazo de Barrantes, Pazo de Señorans, Pazo de Villarei, Robaliño, Bodegas Salnesur, Santiago Ruiz (Bodegas Lan), Terras Gauda, Valdamor, Valdumia, Bodegas de Vilariño-Cambados; **Ribeira Sacra** Adegas Moure.
**PORTUGAL/Vinho Verde** Quinta de Alderiz, Quinta d'Além, Quinta da Aveleda, Quinta da Baguinha, Casa de Compostela, António Esteves Ferreira, Felgueiras co-op, Quinta da Franqueira, Muros de Melgaço, Quintas de Melgaço, Moncão co-op, Quinta do Monte, Manuel Rodrigues de Oliveira, Doña Paterna, Manoel Salvador Pereira, Ponte de Lima co-op, Casa de Sezim, Sogrape; **rest of Portugal** D F J Vinhos, José Maria da Fonseca.

## RECOMMENDED WINES TO TRY

### Ten Spanish varietal wines

**Agro de Bazán** *Granbazán Ambar Rías Baixas*
**Adegas Galegas** *Pedro de Soutomaior Rías Baixas* and *Moure Abadia da Cova Ribeira Sacra Albariño*
**Lusco do Miño** *Rías Baixas Lusco Albariño*
**Marqués de Vizhoja** *Torre la Moreira Rías Baixas*
**Gerardo Méndez Lázaro** *Rías Baixas do Ferreiro Cepas Velhas*
**Pazo de Barrantes** *Rías Baixas*
**Pazo de Señorans** *Rías Baixas Selección de Añada*
**Bodegas Salnesur** *Rías Baixas Condes de Albarei Carballo Galego*
**Bodegas de Vilariño-Cambados** *Martín Códax Rías Baixas Gallaecia*

### Ten Portuguese Alvarinho wines

**Quinta de Alderiz** *Vinho Verde Alvarinho*
**Quinta da Aveleda** *Aveleda Vinho Verde Alvarinho*
**António Esteves Ferreira** *Soalheiro Vinho Verde Alvarinho*
**Muros de Melgaço** *Vinho Verde Alvarinho*
**Quintas de Melgaço** *Vinho Verde Alvarinho*
**Monção co-op** *Deu la Deu Vinho Verde Alvarinho*
**Manuel Rodrigues de Oliveira** *Encosta dos Castelos Vinho Verde Alvarinho*
**Doña Paterna** *Vinho Verde Alvarinho*
**Manoel Salvador Pereira** *Dom Salvador Vinho Verde Alvarinho*
**Sogrape** *Morgadio da Torre Vinho Verde Alvarinho*

### Seven blended wines

**Aldea de Abaixo** *Señorío da Torre Rías Baixas*
**Quinta de Couselo** *Rías Baixas*
**D F J Vinhos** *Grand'Arte Alvarinho/Chardonnay*
**José Maria da Fonseca** *Quinta de Camarate Branco Seco Terras do Sado*
**Adegas Galegas** *Rías Baixas Veigadares*
**Santiago Ruiz** *Rías Baixas*
**Terras Gauda** *Rías Baixas*

## ARAGONEZ

See Tempranillo, pages 256–265.

## ARAMON

This highly productive variety covered the plains of the south of France from the mid-19th century, when it was planted for its resistance to oidium, until the 1960s when the only slightly better quality Carignan began to take its place. At low yields and in good sites it can give concentrated, earthy, spicy, somewhat rustic wines; at high yields it always needed the extra colour provided by its traditional blending partner, Alicante Bouschet. Today it is rarely taken seriously except by the Mas de Daumas Gassac estate north of Montpellier at Aniane, where it produces herby, rich-textured wine.

## ARBOIS

A grape of the Loire Valley, now in decline. The wine is relatively low in acidity and supple, but of unremarkable quality.

## ARINTO

Any grape that can retain its acidity in the baking summer temperatures of southern Portugal is bound to be popular; and acidity is Arinto's *raison d'être*. Further plus points are that it can age well in bottle and when made well has considerable finesse and appealing lemony, peachy fruit. Some is now being fermented and aged in new French or Portuguese oak, and provided the oak is handled with a light touch this seems to add complexity to the wine.

It forms 75 per cent of the blend in Bucelas, the other quarter being taken by Esgana Cão, and is on the increase in Bairrada, Alentejo and Ribatejo. In Vinho Verde it is called Pedernã: here again its high acidity is valued. Its other names – Arinto Cachudo, Arinto Miudo, Arinto do Dão, Arinto Galego and others – may be subvarieties, or may just be synonyms for the same vine. Best producers: (Portugal) Alcântara Agricola, Quinta do Avelar, Quinta do Boavista, Esporão, J M da Fonseca, Quinta da Murta, Luis Pato, Quinta dos Pesos, Quinta da Romeira, Quinta do Valdoeiro.

## ARNEIS

This elegantly and exotically perfumed Piedmontese grape has only found popularity as a varietal wine relatively recently. Its traditional role was as a softener for Nebbiolo in Barolo and elsewhere, and a few rows would be planted alongside the Nebbiolo for this purpose. Few dedicated vineyards existed until a couple of producers began to take a serious interest in the vine in the 1970s and 1980s. The first example I had was from the great Barbaresco producer, Bruno Giacosa.

VIETTI

*Vietti was one of the companies which rescued Arneis from extinction in the 1970s, and its wine is still one of the best: herby, nutty and dry.*

Without that change in fortune it might well have become extinct. Its problems for the grower include low acidity, a tendency to oxidation, susceptibility to powdery mildew, low yields and a somewhat temperamental nature, but better viticultural practices, and planting on the chalky, sandy soils of the Roero gives good acidity and structure, and improved clones can help with the mildew problem. Blending in some wine grown on sandy clay soil will add perfume. Typically the wine has a quite powerful aroma of almonds and peaches and occasionally of hops.

Plantings are on the rise in the Roero, encouraged by high prices; the DOC wine is called Roero Arneis. Some producers vinify and/or age the wine in oak; there are also *passito* versions. It is also supposed to be added in small quantities to Nebbiolo-based red Roero, though this practice is on the decline. It is being assessed for planting in Australia. Best producers: (Italy) Almondo, Araldica, Brovia, Carretta, Ceretto, Cascina Chicco, Correggia, Deltetto, Giacosa, Malabaila, Malvirà, Montaribaldi, Angelo Negro, Castello di Neive, Prunotto, Vietti, Gianni Voerzio.

## ARNSBURGER

This German crossing of two Riesling clones is grown on Madeira for table wines. It produces generous yields of slightly floral wines.

## ARRUFIAC

A variety grown in the deep south-west of France for alcoholic, perfumed, somewhat heavy wine. It is found in the Pacherenc du Vic-Bilh appellation in the Vic-Bilh hills in north-east Béarn. It is usually blended with Gros and Petit Manseng, and Courbu, and is beginning to receive a little more attention from producers than heretofore.

## ASSARIO BRANCO

A name that seems to belong to more than one Portuguese vine. There is an Assario Branco in the Dão region that has a range of synonyms including Arinto Galego, Boal Cachudo, Malvazia Fina and Arinto; there is another one in the Alentejo, which might or might not be the same as Palomino. Best producer: (Portugal) Casa de Santar.

## ASSYRTICO

Steely, minerally fruit, high acidity and good length are the keynotes of this high quality Greek grape. It is the main vine on the island of Santorini, where the vineyards are ungrafted and many vines are 70 years old. It is also grown in Attica, Halkidiki and Drama on the Greek mainland. It may be blended with other grapes, including the less acidic Savatiano and the fatter, creamier Malagousia and also with Aidani in sweet dried-grape wines. Best producers: (Greece) Domaine Carras, Koutsouyanopoulos, Markezinis.

## ATHIRI

A Greek grape grown both for wine and for the table. It makes decent quality wine with a delicate aroma, and is often blended with other grapes, particularly Assyrtico. Best producer: (Greece) Domaine Carras.

## AUBUN

This vine of Mediterranean France is now in retreat, replaced by grapes of better colour and better quality. James Busby took some cuttings to Australia in the early 1830s, and it can still occasionally be found there. A few plantings also exist in California.

## AUXERROIS

The Cahors name for Cot, or Malbec. The variety was once grown in 30 different *départements* of France, but has greatly declined. See Malbec, pages 118–119.

## AVESSO

Portuguese grape grown mostly in the south-east of the Vinho Verde region. Yields are high and the grapes are large; alcohol levels are quite high, and acidity is lower than with most Vinho Verde grapes, so the wines feel relatively full and weighty. It is said to be the

*Harvesting Baga grapes from 70-year-old vines in Bairrada, its principal home in Portugal. Baga can be an austere, tannic grape and needs careful handling in the winery to bring out its fruit. A proportion of Touriga Nacional in the blend can also work wonders.*

same vine as Spain's Jaén Blanco. Best producers: (Portugal) Caves do Casalinho, Quinta de Covela, Paco de Teixeró.

## AZAL BRANCO

Late-ripening, high-acid Portuguese grape grown for Vinho Verde in northern Portugal. There is also a Tinto version, used for red Vinho Verde. Best producers: (Portugal) Quinta do Outeiro de Baixo, Caso do Valle.

## AZAL TINTO

Northern Portuguese vine grown for red Vinho Verde, especially around the town of Amarante. The wines have a high level of malic acid.

## BACCHUS

This heavily-scented German vine is the result of crossing a Silvaner x Riesling with Müller-Thurgau. It reaches high sugar levels but lacks acidity, and only when it is fully ripe does its exotic character really emerge. But at low yields or at Beerenauslesen or Trockenbeerenauslesen levels it can be overpowering: with Bacchus, more is not necessarily better.

Low yields, though, are seldom a problem in the Rheinhessen, its main field of operations, where it is blended into everyday QbA wines. There is also some planted in Mosel-Saar-Ruwer and Franken, and it is grown in England, where it can be aromatic in ripe years, slightly catty in cooler ones. This is in fact about the closest that England can come to the herbal pungency of Sauvignon Blanc. It should not be confused with the *Labrusca riparia* vine of the same name, found in the past in the US state of Virginia. Best producers: (England) Barkham Manor, Chapel Down, Three Choirs; (Germany) Juliusspital, Markgraf von Baden/Schloss Salem, Klaus Zimmerling.

## BACO

Baco Noir is a French hybrid widely planted in the eastern USA and in Canada for soft, fruity, smoky-tasting red wines. It is a crossing of Folle Blanche with a *Vitis riparia* vine, and was produced in 1894; in the years that followed it was planted quite widely in many parts of France. It has now been largely uprooted from French vineyards, as has its stablemate Baco 22A or Baco Blanc. Best producers: (Canada) Stonechurch Vineyards.

Baco Blanc, also a French hybrid, was bred in 1898 by the same eponymous nurseryman, and was a crossing of Folle Blanche with Noah, an American hybrid. Until the late 1970s Baco Blanc was the main grape for Armagnac in south-west France but is in the process of being phased out in favour of Ugni Blanc. It could also be found at one time in New Zealand.

## BAGA

Baga's bugbears are tannins of the most aggressive sort, coupled with high acidity. This Portuguese vine forms most or all of the blend in Bairrada, for many years probably Portugal's top red table wine region, and is also found in nearby Dão and further south in Ribatejo. Traditionally the tannin problem was exacerbated by fermenting the stalks with the must, and softened (up to a point) by leaving the wines in bottle for some years – as many as 10, 15 or even 20.

These days vinification techniques are aimed at controlling the astringency and producing rounder, softer wines from the start, in which Baga's rich, berried fruit is to the fore. Other varieties, especially Touriga Nacional (see pages 268–269), may be blended in.

Baga lasts very well in bottle, even up to 20 years. It attains greater depth with age, but it is never a wine of enormous finesse. It is fairly high yielding, giving up to 8 tonnes per hectare in Dão, and more than 12 tonnes per hectare for Vinho Regional wines.

To make a serious wine it needs to be grown on very well exposed slopes, but even so, one of the most go-ahead producers in the region, Sogrape, attributes the improvement in quality in Dão in the 1990s to the reduction of Baga and the inclusion of other grapes in the blend. Since the grape is thought to have been introduced to the Dão from Bairrada after phylloxera, and ousted Touriga Nacional in the process, perhaps it is not its fault if it does not perform so well there. Elsewhere varietal Baga, which manages to balance the slightly dry tannins with good fruit, can be very attractive and occasionally shows quite piercing blackcurrant fruit.

Its synonyms include Bago de Louro, Poerininha, Tinta Bairrada, Tinta de Baga and Tinta Fina. Best producers: (Portugal) Caves Aliança, Quinta das Bágeiras, Buçaco Palace Hotel, Quinta do Carvalhinho, D F J Vinhos, Gonçalves Faria, Quinta de Foz de Arouce, Caves Messias, Luís Pato, Caves Primavera, Quinta da Rigodeira, Casa de Saima, Caves São João, Sogrape, Caves Velhas.

## BARBAROSSA

An obscure Italian variety found in Emilia-Romagna and on the French island of Corsica. In Provence it becomes known as Barbaroux and is occasionally used as part of the blend for Côtes de Provence. It is not the same grape as the equally obscure Barbarossa found across the Italian border in Liguria. Best producer: (Italy) Fattoria Paradiso.

# Barbera

BARBERA IS BEST KNOWN outside Italy for being Piedmont's second-best red grape, after Nebbiolo, but inside Italy it has a wider market. It grows virtually all over the country, popping up in the most unlikely blends. The wine can be young and fruity, or dark, serious (sometimes a little self-consciously serious) and barrique-aged: new oak barriques seem to have a more natural affinity for Barbera's straightforward cherryish, sappy flavours than for Nebbiolo's exotic perfumes. But while Barbera gains complexity with oak it also loses some varietal character; and there are many very good examples, rich and sumptuous, which have never seen the inside of a barrique.

It is believed to have originated in the Monferrato hills, in central Piedmont, and here it is still planted on the best sites; further west, in Barolo and Barbaresco, it has to concede these to Nebbiolo. Overall about half the Piedmontese vineyard is given over to Barbera, and it is from Piedmont that the finest, most concentrated examples come. Barbera d'Alba has the most complexity and power, along with a deep colour; Barbera d'Asti is brighter in colour, with elegance and finesse. It is in Piedmont, too, that there is a traditionalist-versus-modernist debate similar to that over Nebbiolo, although Barbera has more conclusively swung towards modernism and use of the barrique.

The grape's high acidity makes it ideal for warm climates, and its low tannin and good colour are very much in the currently fashionable mould. Indeed, the New World is now showing quite a bit of interest in all things Italian, and Barbera is an obvious candidate for anywhere warm. But in fact it has been more than just a bit player in the Americas for some time. In the pre-Cabernet days in California it was an important variety – especially since many wineries were of Italian origin – and it is making a bit of a comeback now. The same is true in Argentina, where it came over with the tides of Italian immigration, and, despite being largely used for everyday wines, has shown that its fruity, low acid style can be a success in the modern world. In both countries yields will have to be kept down if we are to see how good New World Barbera could be.

Certainly it is making this transcontinental leap more convincingly than Nebbiolo, and Australia, too, is starting to produce some decent examples.

Elsewhere in Europe, there is a little in Slovenia, Greece, Romania and Israel. But watch this space.

### The taste of Barbera

Barbera can be young and cherry-fresh; or weighty, moreish and with a sour-cherry twist at the end; or again barrique-aged, plummy and rounder, with a touch of spice. This is the most serious style, with vibrant aromas and lots of body; but Barbera never has quite such exuberance as Cabernet Sauvignon, for example. It has a danger of becoming a little raisiny if overripe, but the acidity rarely fails, whatever the style, and we should see more and more of this adaptable grape in the future.

*Braida's Bricco dell'Uccellone was the wine that, back in the early 1980s, showed the way forward for Barbera. To make a single-vineyard, barrique-aged version was pioneering at that time; other Piedmontese producers took note. Originally the wine was sold as Vino da Tavola; now it has the DOC of Barbera d'Asti.*

IL PODERE DELL'OLIVOS

*When Jim Clendenen makes wine at Au Bon Climat in California, the wines are French in inspiration, so when he wanted to experiment with Italian grapes he started this label. The Barbera is dark, chewy and tannic.*

ELIO ALTARE

*Altare's Larigi is the epitome of the modern Piedmontese approach to Barbera: the wine is barrique aged and has plenty of bright, intense fruit.*

***Left:*** *New oak barriques in the cellar of Angelo Gaja in Barbaresco, Piedmont. The wine takes up polysaccharides from the oak, and these increase its richness and reduce its astringency. New wood also gives particular sorts of tannins – hydrolysable – to the wine which act as antioxidants.* ***Above:*** *Barbera grapes. All the best Piedmontese Barberas are made from yields of less than 45hl/ha; from yields above that level the wines can be very attractive, but they will not be great. As for where Barbera comes from: there is no mention of it in Piedmont before the end of the 18th century, and DNA analysis suggests that it could be related to Mourvèdre.*

## CONSUMER INFORMATION

**Synonyms & local names**
Local names include Barbera d'Asti, Barbera Dolce, Barbera Fine, Barbera Forte, Barbera Grossa, Barbera Riccia and Barbera Vera.

**Best producers**
**ITALY/Piedmont/Barbera d'Alba** Gianfranco Alessandria, Elio Altare, Azelia, Enzo Boglietti, Brovia, Domenico Clerico, Elvio Cogno, Aldo Conterno, Giacomo Conterno, Conterno-Fantino, Corino, Matteo Correggia, Elio Grasso, Giuseppe Mascarello, Mauro Molino, Monfalletto-Cordero di Montezemolo, Andrea Oberto, Armando Parusso, Pelissero, Ferdinando Principiano, Prunotto, Albino Rocca, Bruno Rocca, Luciano Sandrone, Paolo Scavino, Seghesio, Vajra, Mauro Veglio, Vietti, Gianni Voerzio, Roberto Voerzio; **Barbera d'Asti** Araldica Vini Piemontesi, La Barbatella, Pietro Barbero, Bava, Bertelli, Braida, Cascina Castlèt, Chiarlo, Garetto, Giuseppe Contratto, Coppo, Roberto Ferraris, Franco M Martinetti, Il Mongetto, Prunotto, Scarpa, F & M Scrimaglio, La Spinetta, La Tenaglia, Terre da Vino, Trinchero, Viarengo, Vietti, Cantina di Vinchio e Vaglio Serra, Viticoltori dell'Acquese; **Langhe** Elio Altare, Bongiovanni, Fratelli Cigliuti, Clerico, Luigi Einaudi, Gaja, Fiorenzo Nada, Marchesi di Gresy; **Monferrato** Giulio Accornero e Figli, La Barbatella, Rocche dei Manzoni; **Emilia-Romagna** Montesissa, La Stoppa, La Tosa.
**AUSTRALIA** Brown Brothers, Garry Crittenden, Montrose.
**USA/California** Bonny Doon, Il Podere dell'Olivos, Renwood, Youngs.
**ARGENTINA** Nieto Senetine, Norton.

### RECOMMENDED WINES TO TRY

**Ten Barbera d'Alba wines**
Elvio Cogno *Barbera d'Alba Bricco del Merlo*
Aldo Conterno *Barbera d'Alba Conca Tre Pile*
Elio Grasso *Barbera d'Alba Vigna Martina*
Armando Parusso *Barbera d'Alba Ornati*
Ferdinando Principiano *Barbera d'Alba Pian Romualdo*
Albino Rocca *Barbera d'Alba Gèpin*
Paolo Scavino *Barbera d'Alba Affinato in Carati*
Vietti *Barbera d'Alba Scarrone Vigna Vecchia*
Gianni Voerzio *Barbera d'Alba Ciabot della Luna*
Roberto Voerzio *Barbera d'Alba Riserva Pozzo dell'Annunciata*

**Ten Barbera d'Asti wines**
La Barbatella *Barbera d'Asti Superiore Vigna dell'Angelo*
Bava *Barbera d'Asti Superiore Stradivario*
Bertelli *Barbera d'Asti San Antonio Vieilles Vignes*
Braida *Barbera d'Asti Bricco dell'Uccellone* and *Barbera d'Asti Ai Suma*
Coppo *Barbera d'Asti Pomorosso*
Franco M Martinetti *Barbera d'Asti Montruc*
Pietro Barbero *Barbera d'Asti La Vignassa*
Prunotto *Barbera d'Asti Costamiole*
Scarpa *Barbera d'Asti La Bogliona*

**Five other Italian Barbera wines**
Giulio Accornero e Figli *Barbera del Monferrato Superiore Bricco Battista*
Elio Altare *Langhe Larigi*
Gaja *Langhe Rosso Sitorey*
La Stoppa *Colli Piacentini Barbera della Stoppa*
La Tosa *Colli Piacentini Gutturnio Vignamorello*

**Five New World Barbera wines**
Bonny Doon *Ca' del Solo Barbera (California)*
Garry Crittenden *King Valley I Barbera (Australia)*
Il Podere dell' Olivos *Santa Maria Valley Barbera (California)*
Norton *Barbera (Argentina)*
Renwood *Amador County Barbera (California)*

### BAROQUE

A grape of south-west France, possibly a crossing of Folle Blanche and Sauvignon Blanc. The aroma recalls the latter to some extent. The wine is quite alcoholic, but the vine is no longer much planted. Also spelt 'Barroque'.

### BASTARDO

Portuguese name for the grape known in France's Jura region as Trousseau (see page 270). In Portugal it is one of the permitted grapes for Port, though it is not of high enough quality to be one of the recommended five varieties. It is also found further south, especially in Dão. It is valued as a blending grape – it gives good alcohol and substance to a blend – but it yields poorly and so is being grown less and less. There is apparently an old Portuguese saying to the effect that to plant Bastardo is an excellent way for a grower to become poor. It does however have potential for fair quality if grown and made well. It is grown to a small extent over the border in Galicia under the names of Merenzao or María Ardona; some Australian 'Touriga' is in fact Bastardo. There is a white grape called Bastardo grown in Portugal and the Canaries, which may just be a white version of red Bastardo. Best producers: (Portugal) Caves Aliança, Caves São João, Quinta do Giesta.

### BICAL

A quite robust, ageworthy grape used for sparkling and still wines in the Portuguese regions of Bairrada and Dão; in the latter it is known as Borrado das Moscas, or 'Fly Droppings', allegedly because of its speckled skin (see right). The wine combines high acidity with high alcohol and can become honeyed after some years in bottle. Bical is said to be aromatic, though many modern varietal examples show little sign of this. Skin contact before fermentation seems, however, to produce a slightly soapy, flowery aroma. Its high acidity means that it is also used for sparkling wine in Bairrada. Best producers: (Portugal) Caves Aliança, Luís Pato, Caves São João, Sogrape.

### BLACK MUSCAT

A grape found to a limited extent in California and in our hothouse in Kent when I was a kid. It is the same as Muscat of Hamburg (see pages 144–153), and makes the most wonderfully rose-scented wines, both fortified and unfortified, though I'm afraid I just ate the gorgeous juicy taste bombs straight off the vine. Best producers: (California) Quady, Philip Togni.

### BLANC DE MORGEX

A speciality (albeit on a small scale) of Italy's Valle d'Aosta, making table wine, DOC wine and sparkling. The style is lean, clean and fresh, but not exciting. Best producer: (Italy) Cave du Vin Blanc de Morgex et de la Salle.

### BLANC FUMÉ

An occasional French synonym for Sauvignon Blanc, used in the Pouilly-sur-Loire area in the central Loire. A wine labelled Fumé Blanc is not necessarily from the Sauvignon grape, although this was a name coined by the Californian producer, Robert Mondavi for his oak-aged version of Sauvignon Blanc. The term Fumé Blanc has since passed into the New World wine lexicon, though without any clear definition of what it means. It usually, though not always, denotes an oak-aged Sauvignon Blanc or Sémillon.

### BLAUBURGER

An Austrian variety produced in the 1920s by crossing Portugieser and Blaufränkisch. The wines are generally straightforward, low in acidity, and light, though a few growers make more concentrated, blackberry-scented wines. Maverick Burgenland producer Willi Opitz makes a sweet wine from it. Best producer: (Austria) Willi Opitz.

### BLAUBURGUNDER

The German and Austrian synonym for Pinot Noir. Spätburgunder is a more common synonym in Germany. Best producers: (Austria) Albert Gesellmann, Fritz Wieninger.

### BLAUER PORTUGIESER

A lightweight, high-yielding table and wine grape found in Austria and Germany, where it's known as Portugieser (see page 188). Most authorities think it originated in Austria, though there is a view that it arrived there from Portugal. At best it makes everyday wines, easily giving crops of 160hl/ha. The colour is pale, and acidity is low. In Austria its synonyms are Vöslauer and Badener. Grown widely in Niederösterreich, it is the country's third most planted black grape but is gradually declining.

### BLAUER SPÄTBURGUNDER

A German synonym for Pinot Noir (see pages 174–185). Plain Spätburgunder is a more common name.

### BLAUFRÄNKISCH

A potentially good quality grape found mainly in Austria, Germany (where it is called both Lemberger and Limberger, see page 116) and points east – points west, too, since in the USA's Washington State it is grown under the name of Lemberger and makes tiny amounts of really lovely blackberryish red.

The best Austrian examples are intense and zesty, with flavours of red cherries and redcurrants. Tannin and acidity are high, and while some new oak aging can be a good thing, the wines are too often over-oaked. The vine produces 100hl/ha or more with ease, but the wines get thin and weedy if made from overcropped grapes. In a blend low-yield Blaufränkisch brings structure and acidity. The vine needs warmth, and flourishes in the Neusiedlersee and southern Burgenland regions of Austria. In Hungary it is called Kékfrankos; in the Czech Republic and Vojvodina, Frankovka; and in north-east Italy, Franconia. Best producers: (Austria) Feiler-Artinger, Albert Gesellmann, Gernot Heinrich, Kollwentz, Nittnaus, Ernst Triebaumer.

### BOAL

The name of Boal or Bual has been given indiscriminately to many Portuguese vines: there is Boal Bagudo, Boal Cachudo, Boal Carrasquenho, Boal Branco, Boal de Alicante, Boal Espinho and Boal Bonifacio, or Vital, and a raft of others.

Boal Branco is the version recommended for Madeira, but even though the law now states that any wine labelled Boal should be made from this grape plantings have not increased, and are still minute. Lack of demand for expensive Madeira is one reason: the traditional, noble Madeira varieties, of which Boal is one, already cost more for the shipping houses to buy than Tinta Negra Mole, the 'one-size-fits-all' workhorse grape of the island, and there is a limit to what the small Madeira market will bear. The other reason is the unwillingness of the growers to plant what have become unfamiliar grapes. Boal, in any case, is a poor yielder, being subject to poor fruit set if there are high winds at flowering time. Best producers: (Portugal) Barros e Sousa, Blandy, Cossart Gordon, Henriques & Henriques, Leacock, Rutherford & Miles.

### BOBAL

A dark-coloured, robust grape used for bulk wine and grape concentrate in south-east Spain. In Utiel-Requena replanting with

Tempranillo is recommended, but Bobal still accounts for 84 per cent of the vineyard there. Some winemakers are now turning to it because of the greatly increased price of Tempranillo grapes. If the wine is made well it has attractive black cherry fruit. Best producers: (Spain) Augusto Egli, Gandia.

## BOGAZKERE

Turkish grape giving decent, deep-coloured, alcoholic wine.

## BOMBINO BIANCO

High-yielding Bombino Bianco is found in Emilia-Romagna, the Marche, Lazio and in southern Italy. It also goes under the name Pagadebit due to its propensity to pay a grape grower's debts by virtue of its reliable cropping. Another synonym is Straccia Cambiale. In Abruzzo it is known as Trebbiano d'Abruzzo, although it is not related to true Trebbiano (see page 270).

A great deal has long been shipped from southern Italy northwards to Germany for blending with strongly aromatic German varieties to produce EU table wine or cheap Sekt. It also makes decent raisins. But the variety deserves better than that. In the north of Italy its wines can be distinctly tasty, though seldom very aromatic; as Trebbiano d'Abruzzo it can display so much more depth than true Trebbiano that one wonders why it was ever lumbered with such an unflattering name. But a lot depends on how it is cultivated and handled. Best producers: (Italy) Giovanni d'Alfonso del Sordo, Fattoria Paradiso, Aldo Pugliese.

## BONARDA

If you relish confusion, you'll love this one. Bonarda is not one but three Italian grapes, and all are probably different from the grape called Bonarda in Argentina.

Italy first: the Bonarda found in Piedmont is, in theory, blended into Gattinara and Ghemme, along with Nebbiolo (see pages 154–163) and Croatina (see page 85). This is Bonarda Piemontese – proper Bonarda, if there is such a thing. If you move over to Lombardy and Emilia-Romagna Bonarda is called Uva Rara but in Oltrepò Pavese in Lombardy and in Colli Piacentini in Emilia-Romagna, however, you find that Croatina there is called Bonarda di Gattinara or Bonarda di Cavaglia. In fact, there are two grapes called Bonarda here: Bonarda Grossa and Bonarda Piccola, or Bonarda Pignola. Both are subvarieties of Croatina.

In the Novara and Vercelli hills in Piedmont there are said to be two clones: Bonarda di Gattinara, from Vercelli, and Bonarda Novarese, from Novara. The most important Italian Bonarda quantitively is the one more properly called Croatina from Oltrepò Pavese and Colli Piacentini, which makes soft and simple reds of deep colour and a certain plummy richness.

In Argentina (where there is a great deal of Bonarda) it is difficult to pin down precisely what Bonarda is. It may be the same as California's Charbono, which in turn may have something to do with Dolcetto. It is very late ripening, but in warm spots can outclass Malbec. It does, however, need to be allowed to ripen thoroughly, or else it produces the same sort of green, vegetal flavours found in underripe Malbec. The consultant Dr Alberto Antonini for one feels that old-vine Bonarda is currently streets ahead of anything being made in Argentina from more famous grapes such as Sangiovese, Nebbiolo or Barbera. Best producers: (Italy) La Fracce, Mazzolino, Vercesi del Castellazzo; (Argentina) La Agricola, Anubis.

## BORRADO DAS MOSCAS

A Portuguese grape which goes under this name (it translates as 'Fly Droppings') in Dão. In Bairrada it is called Bical (see left). It has both high acidity and high alcohol. It may, indeed, taste of fly droppings as well, but, mercifully, I can't verify this. Best producers: (Portugal) Quinta das Mais, Quinta de Saes, Quinta dos Roques.

## BOUCHET

Cabernet Franc (see pages 44–45) traditionally goes by this name in the Right Bank regions of Bordeaux, in the appellations of St-Émilion, Pomerol, Fronsac and others.

## BOURBOULENC

One of five grapes used in the blend for white Châteauneuf-du-Pape in the southern Rhône, Bourboulenc has an incisive quality and a modicum of citrus perfume which makes it popular throughout the southern Rhône and Languedoc, as well as in much of Provence. Except in the La Clape sub-zone of Coteaux du Languedoc, it is always a minority partner in the blend. If picked too early it tastes lean and neutral, but when ripe (and it is a late ripener) it has good richness and depth, as well as citrus acidity and angelica freshness. Best producers: (France) de Caraguilhes, de Lastours, de Pech-Redon, de la Rivière Haute, la Rouquette-sur-Mer.

## BOUVIER

A vine that was discovered rather than deliberately bred – in 1900 by Clotar Bouvier in Nether Styria in Austria (now Slovenia). It reaches high sugar levels but has low acidity, and is used for sweet wines of generally unremarkable quality in Austria's Burgenland region, though I've had one or two good fat examples. The best are blended with some other more acidic variety, often Welschriesling. Best producers: (Austria) Alois Kracher, Lenz Moser.

## BRACHETTO

One of Italy's more unusual grapes (which is saying something), Brachetto makes every style from dry and still to its more usual type, which is sweet and sparkling. The colour is light red and the flavour reminiscent of wild strawberries of the most aromatic sort. It is found in Piedmont, and has its own DOC in Acqui. As a sweet red aromatic sparkler it is a refreshing oddity, but as a still *passito* wine it has more character and can last for many years in bottle. It may or may not be the same as the French grape Braquet which is found in the wines of Bellet, near Nice. Best producers: (Italy) Viticoltori dell'Acquese, Banfi Strevi, Bertolotto, Braida, Contero, Matteo Correggia, Piero Gatti, Domenico Ivaldi, Giovanni Ivaldi, Giuseppe Marenco, Scarpa.

## BROWN MUSCAT

The name given to the dark-skinned version of Muscat Blanc used in Rutherglen and Glenrowan in North-East Victoria, Australia for fortified sweet wines. Best producers: (Australia) All Saints, Campbells, Chambers, McWilliams, Morris, Seppelt.

## BRUNELLO

The name given to the Sangiovese grape by the producers of the Montalcino zone in Tuscany. It was long thought to be a separate clone, but now appears not to be. See Sangiovese, pages 208–217.

## BUAL

Boal, the name given to several different vines in Portugal, has long had its name Anglicized to Bual in English-speaking countries. See Boal (left).

## BUKETTRAUBE

There is some of this light, acidic, rather ordinary vine in South Africa. It is an authorized grape in Alsace, but you'd have to look hard to find any.

# Cabernet Franc

Cabernet Franc is, in fact, the original Cabernet grape: the far more famous Cabernet Sauvignon is Franc's offspring. Yet nowadays Franc is the minor member of the family and is often dismissed as a barely necessary seasoning component in the red wines of Bordeaux. This isn't fair. Sure, Cabernet Sauvignon is deeper, darker, richer, more tannic – but Franc has a delightfully mouthwatering perfume and a smooth, soothing texture that can tame the aggression and power of Cabernet Sauvignon. In Bordeaux it also ripens more easily and in difficult years produces much sweeter, more balanced fruit than Cabernet Sauvignon. Indeed, the cool soils of St-Émilion and Pomerol hardly ever ripen Cabernet Sauvignon, whereas Franc thrives there. It also thrives in the cooler soils of the Loire Valley, and though some misguided growers do mix it with Cabernet Sauvignon, Franc by itself is generally a far better drink, and, from Chinon and Bourgueil, can be one of France's most lovely red wines.

It probably originated in Bordeaux and was taken from there to the Loire by Cardinal Richelieu, who sent it to his abbey of St-Nicolas-de-Bourgueil in the 17th century, where it was planted by the Abbé Breton. It is certainly treated with far more respect in the Loire. It rarely gets the warmest spots of soil in Bordeaux, but in the Loire it is regarded as highly soil sensitive: the wines from sandy chalk soils are weightier than those from chalk or gravel, and the differences between appellations may be less marked than the differences within appellations. For instance, St-Nicolas-de-Bourgueil is not appreciably different to Bourgueil. Both have gravel terraces and both have tuffeau slopes and it is the difference between these, not the appellations, that makes the difference. When not overoaked, the wine here displays thrilling texture and flavour as well as great longevity.

In the northern Italy region of Friuli Cabernet Franc has often been confused with Carmenère and is often over-cropped, to the detriment of ripeness and weight. Elsewhere in Italy, and across the world, it is increasingly planted as a blending partner for Cabernet Sauvignon, perhaps with Merlot as well, by producers who prefer the subtlety and complexity of a blend. In parts of Canada, New York State and Washington State it can be more successful than Cabernet Sauvignon, and occasional good varietal versions are appearing from Australia, Chile and California.

**The taste of Cabernet Franc**

At its best, Cabernet Franc has an unmistakeable and ridiculously appetizing flavour of raspberries, also pebbles washed clean by pure spring water and a refreshing tang of blackcurrant leaves. This is the kind of flavour that gets your taste buds going from Chinon and Bourgueil in France's Loire Valley. Northern Italy can often achieve something similar, and New World examples, rare but good, generally emphasize the raspberry, sometimes to the point of jamminess, consequently losing a bit of the leafiness.

*For much of his career Charles Joguet has been the leading winemaker in the Chinon appellation in the western Touraine, making wines of great richness of flavour – wines that in fact make one think of Bordeaux rather than of the Loire Valley. Clos de la Dioterie, from very old vines, is the best of the lot.*

Château Cheval-Blanc

*There is some 60% Cabernet Franc in the blend of Cheval-Blanc, and that's a lot for the St-Émilion appellation. But there is an unusual amount of extremely suitable gravelly soil in the vineyard.*

La Jota

*A deep, dense, concentrated example from California's Howell Mountain, a region known for the structure, power and richness of its red wines.*

***Left:*** *Sorting machine-harvested Cabernet Franc to remove leaves and unhealthy grapes at Château de Targé in the Saumur-Champigny appellation in the Loire Valley. Cabernet Sauvignon never does as well here as its parent – the Loire is too far north and that bit cooler which suits Cabernet Franc, but makes Cabernet Sauvignon struggle in all but the best years.* ***Above:*** *Cabernet Franc grapes. The vine is very prone to mutation, but it lacks Cabernet Sauvignon's intensity and richness which is why Cabernet Sauvignon is a far more popular 'improver' grape worldwide.*

## CONSUMER INFORMATION

### Synonyms & local names

There are many French alternatives: the most important are Bouchet, sometimes found in St-Émilion, Pomerol and Fronsac on Bordeaux's Right Bank, and Breton in the Loire Valley. In Italy Cabernet Franc wine is often labelled simply as Cabernet. Bordo and Cabernet Frank are Italian synonyms.

### Best producers

**FRANCE/Bordeaux** Ausone, de Beauregard, Belair, Canon, Canon-la-Gaffelière, Cheval-Blanc, Clos des Jacobins, Clos l'Église, la Conseillante, Corbin-Michotte, Dassault, l'Évangile, Figeac, la Gaffelière, Lafleur, Larmande, Soutard, Tertre-Daugay, Tour-Figeac, Trottevielle, Vieux-Château-Certan; **Loire Valley** Philippe Alliet, Bernard Baudry, de Bonnevaux, Bouvet-Ladubay, Caslot-Galbrun, Clos des Marronniers, Clos Rougeard, Max Cognard, de la Coudraye, Couly-Dutheil, Daheuiller, Pierre-Jacques Druet, Filliatreau, Ch. du Hureau, Pierre Jamet, Charles Joguet, Langlois-Château, Logis de la Bouchardière, de Nerleux, Ogereau, Ch. Pierre-Bise, Olga Raffault, Richou, des Rochelles, des Roches Neuves, de la Sansonnière, Joël Taluau, de Targé, de Tigné, du Val Brun, de Villeneuve.
**ITALY** Ca' del Bosco, Marco Felluga, Gasparini, Franz Haaz, Pojer & Sandri, Quintarelli, Ronco dei Roseti, Ronco del Gelso, Russiz Superiore, San Leonardo, Schiopetto.
**USA/California** Havens, La Jota, Justin, Pride Mountain, Viader; **Washington State** Chinook; **New York State** Hargrave, Paumanok.
**CANADA** Chateau des Charmes, Pelee Island, Thirty Bench.
**AUSTRALIA** Chatsfield, Clonakilla, Frankland Estate, Grosset.
**NEW ZEALAND** Esk Valley, Providence.
**CHILE** Santa Rita, Valdivieso.
**SOUTH AFRICA** Bellingham, Warwick.

## RECOMMENDED WINES TO TRY

### Bordeaux reds with a significant percentage of Cabernet Franc

See Best producers left.

### Ten Loire Valley wines

Domaine Phillipe Alliet *Chinon Vieilles Vignes*
Bernard Baudry *Chinon les Grézeaux*
Clos Rougeard *Saumur-Champigny le Bourg*
Pierre-Jacques Druet *Bourgueil Cuvée Beauvais*
Filliatreau *Saumur-Champigny Vieilles Vignes*
Ch. du Hureau *Saumur-Champigny Cuvée Lisgathe*
Charles Joguet *Chinon Clos de la Dioterie*
Domaine des Rochelles/Jean-Yves Lebreton *Anjou-Villages Brissac*
Domaine des Roches Neuves *Saumur-Champigny Cuvée Marginale*
Ch. de Villeneuve *Saumur-Champigny le Grand Clos*

### Five Italian wines containing Cabernet Franc

Ca' del Bosco *Maurizio Zanella*
Marco Felluga *Carantan*
Pojer & Sandri *Trentino Rosso Faye*
Quintarelli *Alzero*
Russiz Superiore *Collio Cabernet Franc*

### Five New World wines with a significant percentage of Cabernet Franc

Esk Valley *Hawkes Bay Reserve The Terraces (New Zealand)*
Frankland Estate *Great Southern Olmo's Reward (Australia)*
Grosset *Clare Valley Gaia (Australia)*
Santa Rita *Maipo Valley Triple C (Chile)*
Viader *Napa Valley Estate Wine (California)*

# Cabernet Sauvignon

King Cab they call it. King Cab the colonizer, the conqueror. Cab the corrupter of other cultures, laying waste other grape varieties and other wine styles round the world with the brutal power of its broadsword, from Tuscany to Bulgaria, from Chile to Spain.

Yet at the same time Cabernet Sauvignon is the consumer's friend. It was the first grape to give such upfront flavours to red wine, flavours that were so easy to recognize and admire, that they turned on generations of drinkers who'd never come near a bottle of red before.

Cabernet is both these things. It has been the most insidious of colonizers, infiltrating almost by stealth – and welcomed by consumers , to whom it offers a lifeline and a recognizable name. Virtually every winemaking country where red vines will ripen has some Cabernet Sauvignon somewhere. Why?

*Aristocratic and magnificent, Cabernet Sauvignon is represented here by the sunburst, the emblem of France's king Louis XIV, also known as Le Roi Soleil or the Sun King. His brilliant court at the Palace of Versailles was filled with images of Louis' glory. The painting captures Cabernet Sauvignon's self-importance and regal position in the world of wine.*

Partly because it tastes recognizably the same, wherever it grows. That's its appeal to consumers: it's as good as a brand name. Its appeal to producers has been different: they know they can sell anything labelled Cabernet Sauvignon. But there's something else, as well. It's a very obliging grape to grow and vinify. It doesn't want to give any trouble, and it will grow almost anywhere that's reasonably warm. If you're a grower in an underrated region struggling to find the right way to grow your local vines but also trying to find a way to modernize your traditional wine styles, a judicious addition of Cabernet Sauvignon can be just what you need.

Of course there's a danger here. Once Cabernet Sauvignon is in a region it tends to stay there – and its powerful personality means that it will hijack any wine to which it is added. That's why many people see it as a ruthless detroyer of diversity. Yet they have to admit that it makes some of the most wonderful wines in the world. The blackcurrant and cigar-box scented wines of Pauillac are, for many wine lovers, the greatest creations of Bordeaux. In California, the tiny output of boutique wineries – dense, impenetrable Cabernet Sauvignons – are bid for with a frenzy that produces prices too insane to contemplate for the normal drinker.

The great wines of Pauillac, based on Cabernet Sauvignon, are an absolute delight, but their classic flavour is relatively simple, like so many great recipes in the kitchen. Blackcurrant fruit, seasoned by the closely related scents of cedar wood, pencil shavings and cigar box. That's the formula, as simple, as perfect as bacon and eggs or apple pie and cream. And consequently, when the modern wine world was expanding like crazy in the 1970s and 1980s, Cabernet offered a classic style that seemed easy to understand, and, those pioneers thought, easy to replicate. As it happens, the great Bordeaux reds have proved very difficult to replicate, but all efforts to do so have brought forth many exciting, and excitingly different, interpretations of Cabernet Sauvignon from around the globe. And the similarities of fruit and texture – sturdiness of tannin, a dark ripeness of black cherry or blackcurrant fruit, and a distinct propensity to develop cedar and cigar-box perfume with age – are ultimately of more importance than the differences.

Perhaps Cabernet Sauvignon does lack the perfumed subtlety of Pinot Noir, perhaps it doesn't possess the heady sensual onslaught of Shiraz or the easygoing plumpness of Merlot; certainly it doesn't demand the concentration and effort required by Nebbiolo or Sangiovese – but it is always itself. Wherever you plant it, however little money you have to invest in grand wineries and golden toasted heaps of new oak barrels, you can still make a recognizable, enjoyable Cabernet. Prince or pauper, peasant or plutocrat, Cabernet Sauvignon will express itself reliably and recognizably for them all.

**Cabernet Sauvignon: from Grape to Glass**

*Geography and History page 48; Viticulture and Vinification page 50; Cabernet Sauvignon around the World page 52; Enjoying Cabernet Sauvignon page 56*

# Geography and History

CABERNET SAUVIGNON GETS EVERYWHERE. Everywhere the sun shines, everywhere a grape will ripen. Everywhere someone decides they want to make a 'serious' red wine, they'll be planting Cabernet Sauvignon. And though it technically needs a fair amount of heat to ripen, that doesn't stop the optimists. Even England has produced the odd – and I mean odd – bottle from vines grown in plastic tunnels. Germany has approved the vine for cultivation as far north as the Mosel Valley, where in order to have the faintest hope of ripening it would need the best and hottest sites – those currently allocated to Riesling. It's a mad world.

But the point is, Cabernet Sauvignon is massively saleable, both as a wine type and as a name on the label. If you're a producer, everyone who drinks red wine in your

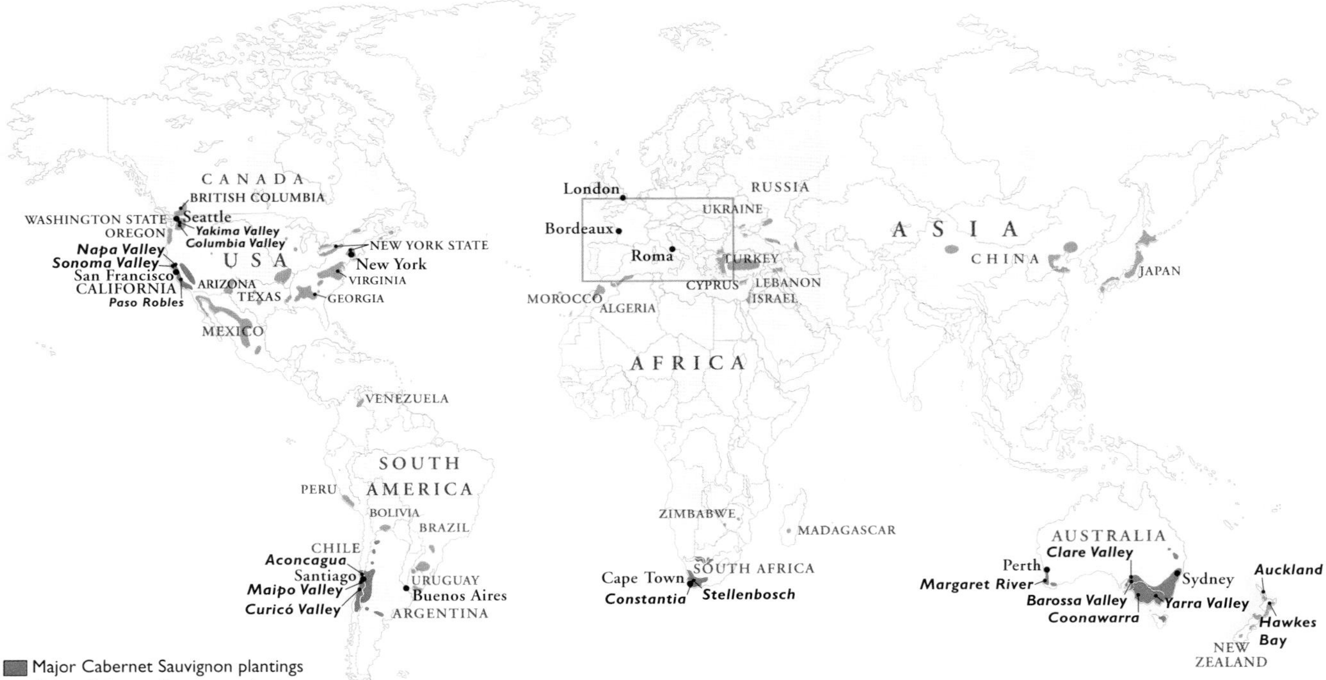

Major Cabernet Sauvignon plantings
Other Cabernet Sauvignon plantings

Major planting figures for Cabernet Sauvignon
AREA PLANTED (HECTARES)

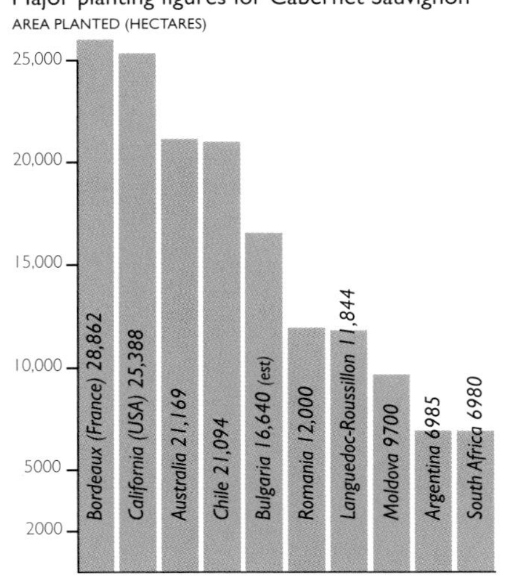

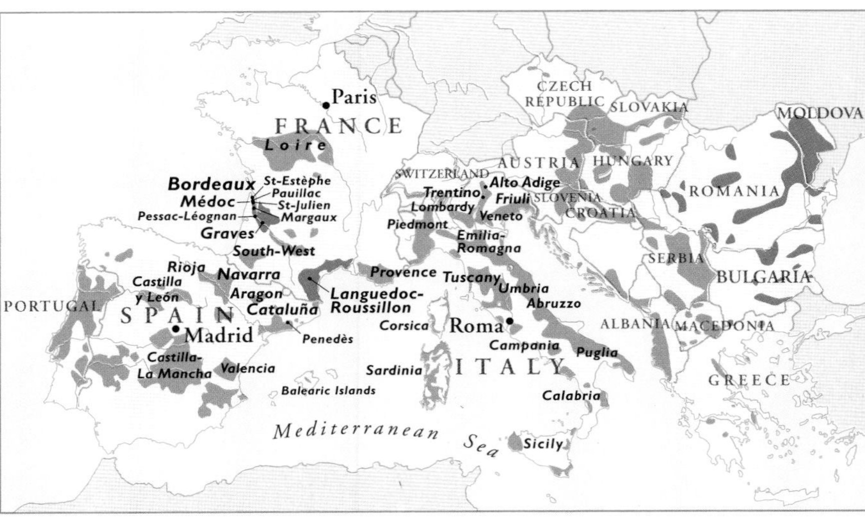

longed-for export markets has heard of Cabernet Sauvignon, so if you're making wine in a country or a region with little international reputation, and you want to have at least half a chance of selling in markets like the UK or USA, Cabernet Sauvignon is a red grape that will open doors for you. Your local vine varieties may be more interesting to the curious, but Cabernet Sauvignon is safe. It will sell. And if your local varieties are frankly uninspiring, there's nothing like a dollop or two of Cabernet slipped into the vat to make a far more appealing final blend.

Is Cabernet Sauvignon driving out other varieties? In the end, probably not. The biggest enemy of any wine region is not interloper grape varieties; it is the inability to sell wines internationally for a decent profit. In Portugal, winemakers have largely opted to rejuvenate their industry without the aid of Cabernet Sauvignon, but they have had European Union funds to call on. Bulgaria's vinous success in the 1980s was heavily based on Cabernet, and still is, but other local varieties are far from extinct. Cabernet is ubiquitous, but not all-embracing.

### Historical background

The myths surrounding the origins of Cabernet Sauvignon have always been disappointingly unimaginative compared to those of, say, Syrah. For example, because the word Sauvignon is a bit like 'sauvage', perhaps it was originally a wild grape? But isn't that true of every other *Vitis vinifera* variety?

Or: in the 18th century Cabernet Sauvignon used to be known in Bordeaux as Petite Vidure; perhaps it took its name from the hardness of its wood (*vigne dure* = vidure)? Is there then a link with Carmenère, another historic Bordeaux variety, which used to be known as Grande Vidure? One of Cabernet/Vidure's synonyms was Bidure; perhaps it was descended from the vine Pliny the Elder named as Biturica after the tribe, the Bituriges, who founded Bordeaux?

The truth, recently revealed by DNA fingerprinting by John Bowers and Carole Meredith at the University of California at Davis, is that Cabernet Sauvignon was a chance crossing of Cabernet Franc and Sauvignon Blanc. The clue, ironically enough, was there in the name all along – and in the flavour. With the benefit of hindsight, how did we miss the simple fact that it tastes like both its parents? (Think of the grassiness of unripe Cabernet Sauvignon, and the grassiness of Sauvignon Blanc.) Come to that, if it was a chance crossing, how come it's taken its parents' names?

We may never know when the crossing of the two occurred, nor when the name Petite Vidure fell by the wayside, but when the great wine estates of the Médoc were planted in the 18th century by members of Bordeaux's newly rich *noblesse de la robe*, Cabernet Sauvignon was established enough to form a major part of the blend for red Bordeaux. Its popularity in the Médoc is said to have been the work of Baron Hector de Brane, owner of Château Mouton in Pauillac until 1830, and his neighbour Armand d'Armailhacq.

*Entrance to the new chai at Château Lafite-Rothschild in Pauillac. The 1980s saw a building boom in Bordeaux, with cranes and diggers moving into what seemed like every second property. It was funded by steeply rising prices and a series of good vintages, plus increased demand for fine wine all over the world. But luckily for the Bordelais, the poorer vintages of the 1990s did not see a return to hard times.*

*A glimpse of history – bottles dating back to 1848 in the cellar at Château Margaux. Red Bordeaux is one of the world's great classic wine styles, largely because of its phenomenal ability to age. Amazingly these bottles could still taste good.*

*Traditional fining of red wines uses egg whites – six per barrique here at Château Léoville-Barton. In Spain and Portugal the yolks are used for sticky yellow cakes. I've never worked out what they do with the leftover yolks in Bordeaux.*

# Viticulture and Vinification

THE FACT THAT JUST ABOUT every wine country in the world grows Cabernet Sauvignon is inclined to make us take it for granted. If it grows everywhere, the logic goes, it must be easy to grow, making few demands on the knowledge or skill of the grower. It is true that it is less fussy about climate and soil than many varieties, is relatively disease-resistant and succeeds in producing wine that is recognizably Cabernet no matter where it is planted. But there are only a few places in the world where varietal Cabernet by itself is as good as or better than a blend. As one Australian winemaker memorably puts it: 'Bordeaux's greatest coup was in convincing the rest of the world that great red Bordeaux is pure Cabernet.'

## Climate

Distinguishing between the importance of climate and that of soil is always difficult, but seems more so with Cabernet Sauvignon precisely because at most quality levels it reflects its soil less than some grapes. In Bordeaux, soil type has traditionally determined what is planted where, but it is ultimately the temperature of the soil that is crucial; in Australia and California, more emphasis is given to climatic factors. But to quote Patrick Campbell of Laurel Glen Winery in California's Sonoma Valley: 'Cabernet Sauvignon at the top level should speak of a site. That may not be possible or even necessary at lower levels, but at the top of the pyramid, Cabernet must be from somewhere.' In other words, top Cabernet must taste not just of its variety, but of its 'place' as well.

Cabernet Sauvignon needs warmth to ripen. It needs a warmer climate than Pinot Noir, or it will turn out green and sappy, with a flavour of green bell peppers; too much warmth, however, turns it soft and jammy, with a flavour of cooked blackcurrants. Pyrazines, the odour compounds that give Cabernet Sauvignon the green, herbaceous part of its flavour profile, are destroyed by sunlight as the grapes ripen; the detection threshhold on the palate is 2ng/l. In warmer climates levels can fall from around 30ng/l at *véraison* to 1ng/l at picking.

Cutting-edge winery architecture at Viña Almaviva, a joint venture between Concha y Toro and Baron Philippe de Rothschild of Bordeaux in Chile's Maipo Valley. Outside investment and know-how – and a swanky name – equal premium prices. Luckily, in this instance, they make good wine as well – and quite a few such grandiose schemes don't deliver on wine quality.

Excessive green, vegetal flavours have been the bane of many regions which hoped their cool climates would give them Cabernets of Médoc-like elegance. In California these flavours are sometimes known as 'Monterey veggies'. The Monterey region is both cool and windy, and since vines shut down in high winds, eliminating excess vegetal flavours in Monterey Cabernet is very difficult, even with leaf removal. It is becoming clear that Russian River, too, is on the cool side for Cabernet.

The minty flavour sometimes found in Cabernet Sauvignon, particularly in Coonawarra in Australia, and in Washington State, is most likely a result of a marginally cool climate. But soil could also be a factor. It crops up occasionally in Pauillac, for example, but not in Margaux.

Could the minty flavour derive from the proximity of eucalyptus trees? Scientists say no; many of the more 'folklorique' of proprietors say yes. I'd say yes and no. Eucalyptus trees are especially common in Australia and the Napa and Sonoma Valleys in California. A famous Napa Valley producer used to leave a couple of wagons of grapes under the eucalyptus trees each harvest so that the eucalyptus gum dripped onto the grapes. He made very minty Cabernets and Pinot Noirs. He believed the sticky gum added something. But then you taste wines with a eucalyptus scent grown miles away from the nearest gum tree, and you wonder. It could equally well derive from the clone, or from excess shade. And I have yet to find a producer claiming that the proximity of a main road to his vineyard gives his wines a flavour of car exhaust, or that sheep in the next paddock make his wines taste pungently farmyardy.

## Soil

The fame of Cabernet Sauvignon was originally based on the gravel soils of Bordeaux's Médoc and Graves: they produced the wines that made the rest of the world want the vine. Cabernet Sauvignon likes gravel simply because it is warm. It drains well, warms up quickly in spring, and holds the heat well. All these factors suit this late budding, late ripening variety because they help to coax the grapes to ripeness in the marginal climate of somewhere like the Médoc. That is not to say that there are no spots in Bordeaux's St-Émilion, or even Pomerol, where Cabernet might thrive. But take gravel away from Bordeaux and you wouldn't have much Cabernet: the more clayey or limestone soils of St-Émilion and Pomerol are generally too cool. Nevertheless, in the 1960s until the mid-1970s the bureaucrats made it obligatory to plant Cabernet Sauvignon in St-Émilion – and, what was more, plant high-yielding clones on vigorous SO4 rootstocks. The results, says Stephan von Neipperg of Château Canon-la-Gaffelière, were 'fine if you wanted to make cola.' Rum and Cabernet, anyone?

Elsewhere, where the climate is warmer, gravel *per se* seems to be less crucial than soil that is well-drained and of poor potential vigour. In Coonawarra, South Australia, there is terra rossa over limestone, and in the Rutherford and Oakville areas of California's Napa Valley the vine thrives on alluvial soil.

## Yields

The 1970s clones that were developed for high yields – and which were often responsible for giving green, thin, herbaceous flavours to the wine – are increasingly being uprooted and replaced with newer, better ones. However, these new virus-free clones inevitably give higher yields.

In Bordeaux the Classed Growth châteaux regularly achieve 60hl/ha – the legal maximum is 50hl/ha, but there is also the *plafond limite de classement*, a legal dodge that enables appellation contrôlée regions to increase their yields in prolific years. Most years are now prolific enough to qualify. To keep yields down to an acceptable level good producers frequently do a green harvest – chopping off excess clusters at *véraison* – and when the wine (which is usually chaptalized) is made they then select more or less stringently for the *grand vin*. Surely the wines would be better if yields were lower in the first place? That so many châteaux are experimenting with must concentration techniques indicates a lack of balance somewhere in the equation.

## At the winery

Cabernet Sauvignon's high ratio of pip to pulp – almost 1:12, compared to Sémillon's 1:25 – and its high phenolic content mean that it can withstand both fairly high temperatures at

Coonawarra soil: terra rossa soil over limestone. It's the drainage of this soil that makes it so good for Cabernet in what is one of the coolest and wettest vineyard areas of Australia.

fermentation, and long maceration. Fermentation temperatures of up to 30°C are usual, and in Bordeaux, a maceration of three weeks was traditional because the cellar staff used to take the opportunity to shut the doors and go hunting. Where softer, earlier drinking wines are the aim, the maceration may only be a few days. In Australia and other New World countries, carbonic maceration has sometimes been used to produce soft, juicy wines.

## Cabernet Sauvignon and oak

The wine has a startling affinity with new oak, blending its blackcurrant flavours brilliantly with the vanilla and spice of the barrels. Indeed, it is the success of Cabernet in new oak that has made the 225-litre barrique Bordelais effectively the standard size wine barrel throughout the world.

In Australia, California and elsewhere, American oak, which gives a more assertively vanilla flavour, may also be used; but winemakers' sensitivity to the risk of over-oaking is increasingly leading them to use a mix of the two, or a mix of new and used barrels. In addition, more American oak is being processed like French oak, which gives it more subtle flavours, and different types of American oak are being identified: Oregon oak, for example, is seen as being more powerful in flavour than that from Missouri, Pennsylvania or Virginia.

## THE BORDEAUX BLEND

Cabernet Sauvignon is almost never bottled as a varietal wine in Bordeaux: it usually lacks enough flesh in the middle palate, and needs its somewhat lean profile filled out with the fatter Merlot and the perfumed, fruity Cabernet Franc. This is the classic blend for red Bordeaux. Not that there's a standard recipe, even in Bordeaux: each château has its own balance of vines, depending on its soil and climate, and its *grand vin* may or may not reflect that balance exactly.

With such a fickle climate, including the threats of severe frosts and then heavy rains at vintage time, much depends in Bordeaux on the year. The reason that the region evolved its particular mix of grapes was that not every vine could be relied upon to ripen every year. Having several varieties in your vineyards means that if one variety is hit by late frost, another may survive to give you a crop at harvest time.

Such pragmatism may not be necessary in warmer sites like the northern part of the Napa Valley, or Western Australia's Margaret River. Here varietal Cabernet can be very successful – though even so it is not uncommon for a touch of Merlot to be added to tweak the final flavour. In cooler climates like New Zealand, Cabernet blends are usually more successful than varietal Cabernet Sauvignons.

In Bordeaux Petit Verdot and Malbec may also be added, though Carmenère, which was important in Bordeaux before phylloxera, is now hardly grown there. Malbec is is grown patchily – there's some in Fronsac – and the lighter soil of Margaux means that a small percentage of Petit Verdot is often added there for its dark colour and violet perfume. Petit Verdot is also grown in corners of Australia, California, New York State, New Zealand and Spain for blending with Cabernet.

Further back in history the classic Bordeaux blend included Syrah, which might have been grown either in Bordeaux or in the Rhône Valley. This is echoed today in the classic Australian red blend of Cabernet and Shiraz; but Cabernet may be blended in other countries with almost every imaginable red grape. Tuscany, for instance, has made a notable success of blending it with Sangiovese, while the regions of Cataluña and Navarra in northern Spain make very good Cabernet-Tempranillo blends.

# Cabernet Sauvignon around the World

NOW THAT BORDEAUX HAS STARTED to make wines of New World richness, and the New World is showing that its wines have structure and longevity, it is becoming increasingly difficult to ascribe particular flavour profiles to particular places. Styles can be defined as much by individual winemakers and individual sites as by regions.

### Bordeaux

Cabernet Sauvignon is not the most widely planted grape in Bordeaux: that honour goes to Merlot. Back in the 19th century, Cabernet's popularity was increasing rapidly: growers loved its resistance to rot (its thick skins and loose clusters help here) as well as its tannic structure, acidity and good flavours. But the 1852 oidium epidemic in Bordeaux revealed its great susceptibility to that disease. Growers turned to Merlot instead, and confined Cabernet to the gravel outcrops of the Médoc and Graves.

But where there is less gravel – in the northern part of the Médoc, for example – Cabernet can be too austere for fun, never mind for fashion, and needs plenty of Merlot to fatten it up. In St-Émilion it is a minority grape, with Cabernet Franc and Merlot taking over; and in Pomerol's clay it is hardly found at all.

Even in its most favoured spots, Cabernet Sauvignon will not produce sensational wines every year. On average, each ten-year period produces three top-class years, three poor years and four that are somewhere in between.

There's no doubt that most of the longest-lived red Bordeauxs have a high proportion of Cabernet Sauvignon in their blend, though it's worth remembering that even in the Médoc and Graves, where Cabernet Sauvignon is likely to be the biggest single grape variety in the blend at the top châteaux, it may still be in a minority against the various other varieties in the blend.

Its style in the Médoc varies from the mineral austerity of St-Estèphe through violet-scented intensity in Margaux, classic lead pencils and blackcurrant in Pauillac, cedar and cigar boxes in St-Julien, softer and rich in Moulis and somewhat muddy in Listrac to minerally again in Pessac-Léognan. Lesser regions like the southern Graves produce good blackcurrant flavours, without the intensity of the best sites.

### Other French Cabernets

Cabernet Sauvignon is grown in the South-West for lookalike red Bordeaux wines, and it produces its customary blackcurrant fruit, but in a lighter style than the best of Bordeaux can offer. In the Midi it is used either straight or as a blender, to improve the flavour and fruit of tougher, less immediately appealing grapes like Carignan. As a varietal it has made a home in the Pays d'Oc, where styles, often reflecting the training or birthplace of the winemaker, can be more like those of Australia. There's a little in the Loire, but it has trouble ripening here; Cabernet Franc is far more successful.

### Italy: the super-Tuscan phenomenon

Cabernet Sauvignon in Italy is no longer the dangerous interloper it was once perceived to be. Winemakers have worked with it, learnt about it – and gone on to learn more about their native grape varieties. Cabernet on its own is no longer the first choice for anyone wanting to make serious wines.

The grape was present in Italy long before Italy was a single country. It arrived in Piedmont in 1820 and is still grown there, even, it is said, being introduced into Barolo in significant proportions. Such a thing would of course be illegal for DOC Barolo, but is believed to improve the colour and make the wine fruitier – both of which aims can be hard to attain with Nebbiolo.

Legal blends of Cabernet and Nebbiolo take the DOCs of Langhe or Monferrato; Cabernet and Barbera also blend well, with or without the addition of Nebbiolo, though the addition of two high tannin grapes to the high-acid Barbera can require the use of some new wood to add some sweet spice.

Piedmontese varietal Cabernets range from very good to excellent, but seem to need the best vineyard sites.

Cabernet's history in Tuscany has been still more controversial (apart from in Carmignano, where it has been part of the DOC blend since 1975), partly because of the arrival in the mid-1970s of the so-called super-Tuscans – top-class wines which deliberately went outside the DOC system, often in order to add Cabernet, or by making Cabernet as a varietal.

Cabernet in Tuscany has a beautiful deep blackcurrant and black cherry sweetness, and retains its acidity even when the alcohol

**CHÂTEAU RAUZAN-SÉGLA**
*The Wertheimer Group, owner of Chanel, bought this underperforming château in 1993 and has spent more than FF40 million on improvements.*

**CHÂTEAU LÉOVILLE-LAS-CASES**
*A super-expensive second growth Médoc château where the aim is to produce super-concentrated wine that may even outprice the first growths.*

**MARCHESI ANTINORI**
*Antinori's Solaia was one of the prototype super-Tuscans, and one of the earliest Tuscan attempts at Cabernet Sauvignon. It remains one of the world's best.*

reaches 14 per cent, as it can. The marriage of Cabernet and Sangiovese has also proved superbly fruitful; it is up to the winemaker to find a balance between the assertive Cabernet and the less dramatic Sangiovese.

Cabernet Sauvignon has long been conspicuous in Lombardy, where it is often blended with Merlot. Bordeaux-style blends can also be found in Emilia-Romagna, the Veneto, Friuli-Venezia Giulia and, to a lesser extent, the Alto Adige and Trentino. Teroldego and the Valpolicella grapes are other blending partners in their respective regions. Cabernet Sauvignon usually suffers from insufficient ripeness in the Alto Adige and Trentino – the dreaded green bean flavour again – particularly when it is trained on high pergolas and overproduces, as it usually is and does. Better clones, Guyot-trained, can help this problem, but so can growing the earlier-ripening Cabernet Franc instead of Cabernet Sauvignon.

In the South, Cabernet is blended with every conceivable red grape: Gaglioppo in Calabria; Merlot and Aglianico in Campania; Nero d'Avola in Sicily; Cannonau and Carignano in Sardinia.

## Spain

Nearly every region of Spain has some Cabernet Sauvignon planted, though often only as an experiment. But Cabernet experiments have a habit of turning out well. It has already shown it can produce good varietals in Penedès, and it was introduced to Rioja in the mid-19th century by the Marqués de Riscal. There are currently some 70ha planted there, and there are moves to get it added to the list of approved varieties. It is already being used at many bodegas. Is it a cuckoo in the nest? It seems to have less pronounced tannins in Rioja than in most places, and in cooler spots, like Haro, it doesn't ripen well, so I don't think we need to worry too much.

## Other Europe

The bargain red wine of the 1980s, Bulgarian Cabernet Sauvignon, has lost popularity to fruitier, softer versions from Australia and Chile. Cabernet is widely grown in Hungary, Moldova, Romania, the former Yugoslavia and throughout the former Eastern Bloc, and individual examples can be attractive, particularly where Western winemaking techniques are available. It is grown on a small scale in Austria, but seldom ripens well there, and some growers are replacing it with Merlot. It is successful in Greece and Israel, and forms part of the blend at Chateau Musar in Lebanon.

### Leap of faith

*Back in 1976 there was a tasting in Paris of top French wines against the best of their counterparts from California. It was organized by wine merchant Steven Spurrier, and its effect was electric. It is known as the Judgement of Paris tasting – and it awarded top place among the reds to Stag's Leap Wine Cellars 1973 Cabernet Sauvignon.*

*Warren Winiarski (left) of Stag's Leap planted his Cabernet vines in 1970, after years of exploring the Napa Valley and noting where the vegetation changed, where there was frost damage, and what the growing conditions were – and this was an unfashionable attitude at the time in California. The emphasis then was on grape variety, not place.*

*Winiarski's first career was teaching political theory at the University of Chicago. His fascination with wine began when a friend brought a bottle of wine to lunch – it came from the East Coast and was made from hybrid grapes. Eventually Winiarski and his wife decided to make wine themselves, and drove across the desert to California.*

*Winiarski's Paris triumph made the cover of* Time Magazine*; there is now a bottle of Stag's Leap 1973 in the Smithsonian Institution.*

STAG'S LEAP WINE CELLARS

*The Fay Vineyard is named after Nathan Fay, who made Winiarski realize that Stag's Leap was the Cabernet site he'd been searching for.*

## USA: California

The appearance of a new biotype of phylloxera in California in the 1980s, and the subsequent replanting of many vineyards, did not bring about the reduction in the amount of Cabernet Sauvignon, and the increase in other varieties that some people hoped for. Quite the reverse, in fact: Cabernet's acreage more than doubled between 1988 and 1998. In the Napa Valley north of Yountville the vineyards are almost solidly Cabernet Sauvignon now, with some Cabernet Franc and Merlot: there is very little Chardonnay or Sauvignon Blanc still grown in this part. The smaller hillside regions like Mount Veeder, Howell Mountain, Diamond Mountain and Spring Mountain, with their slow-aging, tightly structured styles of wine, are more Cabernet-dominated than before.

Sonoma's leading Cabernet region, Alexander Valley, has been replanted with better clones that are less likely to give green, herbaceous flavours, and more Merlot and Cabernet Franc have been planted. But for every producer experimenting with percentages of Merlot, Cabernet Franc, Malbec or Petit Verdot in his wine there is likely to be another increasing the proportion of Cabernet Sauvignon to emphasize the strong character that the grape has here.

Growers have generally learnt to make better balanced wines. Dry Creek Valley, Sonoma Mountain and Sonoma Valley are up and coming regions, and Mendocino County is showing great promise. Further south, Paso Robles in San Luis Obispo County, Santa Cruz Mountain and Monterey's Carmel Valley have only small amounts of Cabernet, but are making some stylish wines. Of the more established regions, Stags Leap District makes supple, well-structured, black cherry wines, and warmer Oakville and Rutherford are more blackcurrants and plums, richer and with firmer, dusty tannins.

The main stylistic difference in California is between hillside and valley floor wines. Hillsides, with their thinner soil, give lower yields (1–2 tons per acre, compared to 4–8 on the valley floors). The berries are smaller, and flavours are more austere and intense, chewier and less opulent, in a slow-maturing, Bordeaux style. But there are exceptions, notably the valley-floor-grown Opus One, with its Bordeaux-style restraint.

Oak is being used with a lighter hand than in the past, though we are not about to return to the ill-fated 'food wine' styles of the early 1980s, with their lean, attenuated fruit. California Cabernets are balanced at a higher level of alcohol than red Bordeaux: generally 14 per cent and over, to Bordeaux's 12.5–13 per cent. At lower alcohol levels California Cabernet can lack flavour.

That the best wines can age is not in doubt. The top 1978s still have another ten years in them, and while the 1980s saw swings of style that make generalizations difficult, the 1990s were kind to Cabernet. A succession of excellent vintages has produced wines that, from the top producers, will last.

## USA: Washington State

Cabernet Sauvignon could have greater long-term potential here even than Merlot – particularly now that better vineyard management is beginning to reduce its unripe green flavours. But Merlot has proved faster at showing its worth, and Cabernet Sauvignon is in second place, with 1416 acres (573ha) planted in 1999. It needs the hottest sites: the Yakima Valley is generally on the cool side, and warmer parts of the Columbia Valley are more suitable. Its great plus, as far as the growers are concerned, is its resistance to winter cold.

Cabernet's trademark here is its bright fruit, and it can make early-drinking styles – though there are an increasing number of growers who are producing impressively dark, brooding reds.

## Rest of North America

There are small quantities of Cabernet Sauvignon planted in Oregon (465 acres/188ha in 1999), mostly in the Umpqua and Rogue Valleys; other states, including Texas and Arizona, also grow it. Its toughness in the face of cold winters makes it attractive to Canadian growers, though the resulting wines are generally austere.

## Australia

Coonawarra led the way here in the 1970s with bright, often minty fruit and fine structure, though Cabernet's reputation at the top end is now based as much on the black-fruited, dustily herby and tightly structured wines of Margaret River, the balance and elegance of Yarra Valley, the sweet, focused fruit of Clare Valley and the full, heavy wines of the Barossa – an area that has proved itself better suited to Shiraz. At the

*The owners of Screaming Eagle in the Napa Valley (above) aim to make California's greatest wine, bar none. With just 200 cases of intensely concentrated Cabernet made each year, few people will ever have the chance to judge if they're succeeding. Not surprisingly, the wine has cult status.*

less expensive end there are abundant ripe-fruited examples blended from different regions and even between states.

In the 1980s Australia, like California, went through a phase of favouring lower alcohol (around 11–12 per cent), lighter wines. In the early 1990s the wines became much better balanced and the fruit riper; regional differences also began to be better defined. This heyday of Cabernet coincided with the very good (in most regions) vintages of 1990 and 1991.

Now Shiraz has taken over the title of Most Desirable Red Grape: in 1998 100,000 tonnes of Cabernet were harvested, to 150,000 tonnes of Shiraz. Lovers of Australian Cabernet should not mourn, however: plantings are still rising, and the best wines are easier to obtain. The top wines are less extracted than they were, and less likely to be over-oaked.

### New Zealand

Hawkes Bay is the key region here. The ripest Cabernet Sauvignons come from this climatically diverse North Island region, though often still retain a green flavour as a reminder of its relatively cool climate. This is exacerbated by high yields: the fertile alluvial soil of much of the flat land means vigorous vines. Better-adapted rootstocks and lower yields can help produce riper fruit that emphasizes cassis flavours; canopy management has already made big improvements to ripeness; and development of warm gravel beds in areas like Gimblett Road are all producing better Cabernet results. Waiheke Island, in Auckland, also has some impressive wines to its credit. Even so, blends with Merlot are nearly always more interesting than pure Cabernet Sauvignon.

### Chile

Climate is the main factor in determining what gets planted where in Chile – though once producers start planting on the slopes, soil becomes more important. Terroir is the new buzz-word: Maipo gives broad, blackcurranty, generous Cabernets with a distinctive dusty note, while Aconcagua wines are more structured, more closed in but still sweet and ripe at heart. Warmer Curico gives richer, softer Cabernets, and Colchagua gives fast-developing wines with soft tannins, sweet fruit and less acidity. High yields may mean that these differences are not always apparent in the bottle, but when Chile gets the balance between yield and ripeness right, her Cabernet Sauvignons can be some of the most irresistible in the world.

*Kim Goldwater of Goldwater Estate (above) on Waiheke Island, off New Zealand's North Island, believes that Cabernet Sauvignon is the most difficult grape of all to grow. There are very few places in the world, he says, where it does really well. I'd say that's a bit open to question – although Waiheke Island is certainly one of the few places Cabernet has a chance in New Zealand.*

### Rest of South America

Not surprisingly, Cabernet Sauvignon is found in just about every wine-producing country in Central and South America. In Mexico, the wines can be earthy and four-square; in Uruguay, they can have nicely balanced blackberry fruit. In Argentina it is typically blended with Malbec, especially at the top end. These premium wines, full of tobacco and leathery fruit at their best, can have considerable aging potential. Simpler examples tend to have sweeter, lighter fruit and are made for early drinking.

### South Africa

Cabernet from new clones, coming on stream in the mid-1990s, has ripe, sweet fruit in place of the high acidity and unripe herbaceous notes of the old, virused clones. Later picking, and better winemaking which avoids volatility, are also crucial. Location is starting to be a major factor in style: Constantia Cabernet has minty, herbal flavours to Stellenbosch's structure and weight. A blend of the two might be about right. The west coast above Cape Town is proving to be an interesting new region.

**SHAFER**

*Napa is one of the world's classic Cabernet regions, and this is Shafer's top Cabernet, with the capacity to age for 20 years or more.*

**PETALUMA**

*Petaluma wines are marked by their purity of fruit. This wine is predominantly Cabernet, with some softening Merlot blended in.*

**TE MATA**

*One of New Zealand's top reds, this Hawkes Bay blend adds Merlot and Cabernet Franc to a backbone of Cabernet Sauvignon in true Bordeaux style. Getting the Cabernet to optimum ripeness still needs good viticulture, in the form of yield control and canopy management.*

**MONTES**

*This supple, balanced blend of 80% Cabernet Sauvignon with Merlot and Cabernet Franc is one of Chile's new super-expensive super-reds.*

# Enjoying Cabernet Sauvignon

CABERNET SAUVIGNON MAKES WINES that can shine at a century old, yet it can also make delightful wine that is unbeatable a mere six months after vintage. Bordeaux is where Cabernet Sauvignon first showed what it could do, in terms of both flavour and longevity. Questions are sometimes raised about the aging ability of the currently fashionable richer, fleshier style of red Bordeaux. The 1982 vintage, the first in this style, has aged unpredictably, with some leading wines fading surprisingly quickly. 1990 has also come through extremely fast. However, two vintages don't make a rulebook, and it still seems fair to say that top red Bordeaux needs at least ten years to come round, and in a good vintage should last two or three decades longer. This does not accord with traditional practice in France, which is to drink them within a few years of the vintage. Although top Italian and Spanish Cabernets could easily take ten years of aging, most will also be consumed at only a few years old.

Cabernet changes so much in bottle that it would be a shame to forget the pleasures of mature wine. But it would also be a shame to forgo the pleasures of young Cabernet. The top Cabernets of Australia need around ten years, but most Aussie Cabs are excellent at five years old, and many peak at only two to three years old. Top Californian vintages like 1978 can last for two, perhaps three decades, but since so much California Cab is drunk on release without any further aging, it isn't surprising to find that, below the top level, most are ready at two to three years old. South American Cabernets – especially Chilean examples – are bursting with flavour at only a couple of years old though there is no reason to think they won't age. South African examples, though softer and riper than they used to be, still often need six to eight years. New Zealand Cabernets are usually ready quite young but do age well, with sweet blackcurrant fruit, even if they rarely lose their streak of leafy greenness.

## The taste of Cabernet Sauvignon

There is no mistaking the blackcurrant scent of Cabernet. Young wines taste of black cherry and plum; mature wines add the classic nose of pencil shavings, cedar and cigar boxes.

At lower ripeness levels Cabernet exhibits a telltale greenness, a green bell pepper nose that at its worst is raw and vegetal. It is a flavour winemakers try to avoid. More pleasant tastes and smells are those of tobacco, mint and eucalyptus, and fruits like blackberry and black cherry; blackcurrant is generally present, even in less than super-ripe examples, though less so in Bordeaux and Napa. Overripe Cabernet goes jammy-tasting – stewed blackcurrants at worst. Some growers like to pick a mixture of slightly underripe, perfectly ripe and slightly overripe grapes, believing that the combination of all these flavours gives extra complexity. Others say no, the ideal is to pick everything at optimum ripeness: that way the grape tastes most like itself.

New World examples, particularly at the less expensive end, show sweeter fruit than Bordeaux of equivalent quality: they are juicier and more forward and, at lower price levels, more attractive. Basic Bordeaux Rouge, regardless of how much Cabernet Sauvignon it contains, is more likely to emphasize the austerity of the style over the fruit.

*Opus One is a joint venture between a Bordeaux château (Mouton-Rothschild) and a California producer (Robert Mondavi). The first vintage was 1979, and at $50 was more expensive than any other California wine. It's been overtaken in price since. Château Margaux has never embarked on any joint venture, being unwilling either to risk putting its own wine in the shade, or to make something inferior. However, Margaux winemaker Paul Pontallier does have a Chilean joint venture: Domaine Paul Bruno – with ex-Cos d'Estournel owner Bruno Prats.*

### Matching Cabernet Sauvignon and food

All over the world Cabernet Sauvignon makes full-flavoured, reliable reds: the ideal food wine. Classic combinations include *cru classé* Pauillac with roast milk-fed lamb; super-Tuscans with *bistecca alla Fiorentina*; softer, riper New World Cabernet Sauvignons with roast turkey or goose. Cabernet Sauvignon seems to have a particular affinity for lamb but it partners all plain roast or grilled meats and game well and would be an excellent choice for many sauced meat dishes such as steak and kidney pie, beef stews, rabbit stew and any substantial dishes made with mushrooms.

## CONSUMER INFORMATION

### Synonyms & local names

France's many historical synonyms include Petite Vidure and Bidure.

### Best producers

**FRANCE/Bordeaux** Cos d'Estournel, Ducru-Beaucaillou, Grand-Puy-Lacoste, Gruaud-Larose, Haut-Brion, Lafite-Rothschild, Latour, Léoville-Barton, Léoville-Las-Cases, Lynch-Bages, Ch. Margaux, Mouton-Rothschild, Pichon-Longueville, Pichon-Longueville-Comtesse de Lalande, Rauzan-Ségla; **Midi** Mas de Daumas Gassac.
**ITALY** Antinori, Castello Banfi, Ca' del Bosco, Col d'Orcia, Gaja, Lageder, Maculan, Montevetrano, Ornellaia, Poliziano, Le Pupille, Querciabella, Castello dei Rampolla, San Leonardo, Sassicaia, Tasca d'Almerita, Terriccio, Tua Rita.
**SPAIN** Marqués de Griñón, Torres.
**PORTUGAL** Esporão.
**AUSTRALIA** Cape Mentelle, Cullen, Giaconda, Howard Park, Leasingham, Lindemans, Moss Wood, Mount Mary, Penfolds, Petaluma, Taylors, Wynns, Yarra Yering.
**NEW ZEALAND** Alpha Domus, Esk Valley, Goldwater, Matua Valley, Stonyridge, Te Mata, Villa Maria.
**USA/California** Araujo, Beringer, Bryant Family, Caymus, Dalla Valle, Diamond Creek, Dominus, Dunn, Grace Family, Harlan, La Jota, Laurel Glen, Peter Michael, Mondavi, Newton, Phelps, Ridge, Screaming Eagle, Shafer, Silver Oak, Spottswoode, Stag's Leap Wine Cellars; **Washington State** Andrew Will, Leonetti.
**SOUTH AMERICA** Almaviva, Carmen, Catena, Errázuriz, Mondavi/Chadwick, Montes (Alpha 'M'), Santa Rita, Tarapacá, Michel Torino, Valdivieso, Viña Casablanca, Weinert.
**SOUTH AFRICA** Boekenhoutskloof, Jordan, Kanonkop, Meerlust, Saxenburg, Thelema.

## RECOMMENDED WINES TO TRY

### Ten more Bordeaux Classed Growths

Domaine de Chevalier *Pessac-Léognan*
Ch. Ferrière *Margaux*
Ch. Haut-Bailly *Pessac-Léognan*
Ch. Lafon-Rochet *St-Estèphe*
Ch. Lagrange *St-Julien*
Ch. La Lagune *Haut-Médoc*
Ch. Léoville-Poyferré *St-Julien*
La Mission-Haut-Brion *Pessac-Léognan*
Ch. Montrose *St-Estèphe*
Ch Palmer *Margaux*

### Ten other good Bordeaux wines

Ch. d'Angludet *Margaux*
Ch. Chasse-Spleen *Moulis*
Ch. La Gurgue *Margaux*
Ch. Labégorce-Zédé *Margaux*
Ch. Maucaillou *Moulis*
Ch. Monbrison *Margaux*
Ch. Pibran *Pauillac*
Ch. Potensac *Médoc*
Ch. Poujeaux *Moulis*
Ch. Sociando-Mallet *Haut-Médoc*

### Ten New World lookalikes

Araujo *Eisele Vineyard Cabernet Sauvignon (California)*
Catena *Alta Cabernet Sauvignon (Argentina)*
Marqués de Griñon *Eméritus (Spain)*
Mondavi/Chadwick *Seña (Chile)*
Penfolds *Bin 707 Cabernet Sauvignon (Australia)*
Ridge *Montebello Cabernet Sauvignon (California)*
Stonyridge *Larose (New Zealand)*
Thelema *Cabernet Sauvignon (South Africa)*
Andrew Will *Klipsun Cabernet Sauvignon (Washington)*
Wynns *Coonawarra John Riddoch Cabernet Sauvignon (Australia)*

### Ten Italian Cabernet Sauvignons

Antinori *Solaia*
Castello Banfi *Tavernelle*
Col d'Orcia *Olmaia*
Montevetrano
Ornellaia
Poliziano *Le Stanze*
Le Pupille *Saffredi*
Castello dei Rampolla *Sammarco*
San Leonardo
Tasca d'Almerita *Cabernet Sauvignon*

### Ten good-value Cabernet Sauvignons

Beringer *Knights Valley Cabernet Sauvignon (California)*
Esporão *Alentejo Cabernet Sauvignon (Portugal)*
Jordan *Cabernet Sauvignon (South Africa)*
Leasingham *Classic Clare Cabernet Sauvignon (Australia)*
Lindemans *Bin 45 Cabernet Sauvignon (Australia)*
Baron Philippe de Rothschild *Mapa Cabernet Sauvignon (Chile)*
Taylors *Clare Valley Cabernet (Australia)*
Michel Torino *Don David Cabernet Sauvignon (Argentina)*
Valdivieso *Reserve Cabernet Sauvignon (Chile)*
Wynns *Coonawarra Cabernet Sauvignon (Australia)*

*Cabernet Sauvignon's small berries can make reds soft and fruity enough to enjoy at one to two years old, as well as the majority of the world's genuine long-distance wines.*

### Maturity charts

Cabernet Sauvignon is potentially one of the longest lasting of red grapes, but much depends on the producer.

**1996** Médoc Super-Second Cru Classé

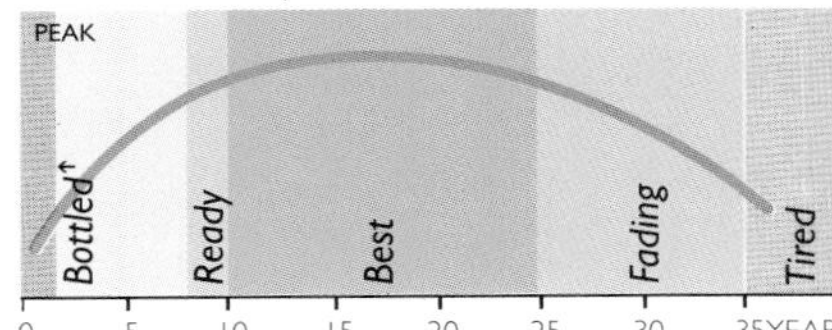


*1996 was a very good vintage in much of the Médoc: Margaux was a little more mixed in quality than St-Julien, St-Estèphe or Pauillac.*

**1995** Top Napa Cabernet Sauvignon

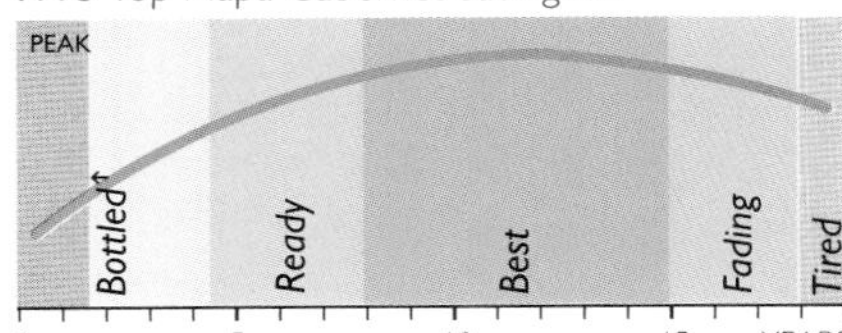


*1995 was an outstanding vintage for Napa Cabernet. The wines have concentration and, at the top level, a tight structure that will ensure long life.*

**1998** Coonawarra Cabernet Sauvignon

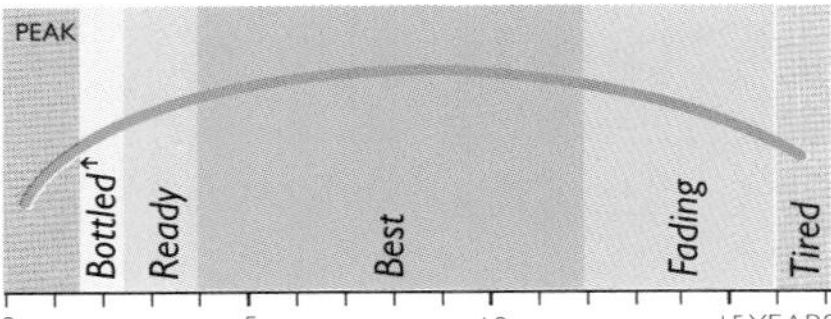


*A rich, ripe year for Coonawarra Cabernet, and considerably better than 1997. The wines have fine balance and opulent fruit.*

### CALADOC

A recent French crossing of Grenache and Malbec with good colour and body, enticing loganberry fruit and the capacity to age in bottle. It is attracting attention in South America and Portugal, among others.

### CALABRESE NERO

A synonym for Nero d'Avola, see page 164.

### CAMARATE

Portuguese grape or grapes. There is Camarate in the Douro, Ribatejo, Dão and elsewhere. It's also a synonym for white Fernão Pires (see page 89).

### CANAIOLO

A perfumed red grape that seems to have been the main constituent of Chianti until the late 19th century, and thereafter was used for softening the astringency of Sangiovese. It is no longer a required part of the Chianti blend, and has been in decline since the onset of phylloxera, when it proved a tricky vine to graft. The available clones have also been generally poor, though there are moves to remedy this. Some Tuscan growers still treasure their Canaiolo, and blend it with Sangiovese. There is some in Lazio, the Marche and Sardinia, and a white version, Canaiolo Bianco, in Umbria. The Orvieto name for this latter is Drupeggio.

### CANNONAU

The Sardinian name for Garnacha/Grenache (see pages 92–101). In Sardinia it gives powerful, often slightly earthy, toffeeish table wines and some exciting, rich fortified wines.

### CARIGNAN

A hot climate vine that probably did more than another other grape to fill Europe's wine lake in the late 20th century, Carignan is now in decline, but not fast enough. It still covers large tracts of Languedoc-Roussillon, producing yields of up to 200hl/ha with ease. The wine has quite dark colour, loads of tannin and acidity, plenty of astringency, and generally gives very little pleasure. However, it has shown itself to be eminently suitable for vinifying by the Beaujolais method of carbonic maceration when the colour deepens, the astringency softens and a rustic but attractive fruit and perfume appear out of nowhere. Blend this with Grenache or Syrah and good wine can result.

Only in exceptional sites, with first-class exposure and good drainage, and with very good winemaking can it produce fine wine on its own. However, such wines do exist, both in the south of France and elsewhere. When Carignan makes a serious wine yields are low and the vines are usually old – perhaps as much as 50 years. Paul Draper of Ridge Vineyards in California makes some excellent Carignan from vines planted in 1880. Here, as almost always, it is blended with other varieties. It can also make attractive herb-streaked rosés.

Outside Languedoc-Roussillon it is found in the southern Rhône, but it cannot travel too far because, being both late budding and late ripening, it needs a warm climate to ripen. It is also susceptible to rot and both kinds of mildew. It is not found in south-west France for this reason. There is also a white version, Carignan Blanc, found in Languedoc-Roussillon.

In Italy it is found as Carignano, especially in Sardinia, but also in Lazio. In Spain it is called Cariñena, and in fact originated in Aragon. Today, however, it plays only a small part in the wine named after it, and covers just 6 per cent of the Cariñena vineyard. Instead it is mostly found in Cataluña, especially in Ampurdán-Costa Brava, Priorat and Tarragona. Under its synonym of Mazuelo it may be a small part of the Rioja blend, and is valued for the very acidity that is not relished in the Languedoc. Its colour and tannin are also useful in Rioja.

California can, as Ridge Vineyards demonstrates, produce some very good Carignane, as it is called here, but it can also produce high-yielding Carignane quite as dreadful as any other. There are some 8000 acres (3100ha) of Carignane in California, and nearly all the vines are old, simply because no-one has got round to replanting with something else. Best producers: (France) Aupilhac, Clos Centeilles, la Dournie, Mont Tauch co-op, de Pech-Redon, Rabiéga, la Voulte-Gasparets; (Italy/Sardinia) Argiolas, Mauritania, Santadi co-op; (California) Cline Cellars, Ridge.

### CARIGNANO

The Italian name for Carignan (see above).

### CARIÑENA

The Spanish name for Carignan (see above).

### CARMENÈRE

See pages 60–61.

### CASTELÃO

This is the principal name for the southern Portuguese grape more widely known by its nickname of Periquita or 'little parrot', a name it acquired from a vineyard, Cova de Periquita, where it was planted by José Maria da Fonseca in the early 1850s. Its other names are João de Santarém, Mortágua de Vide Branca, Tinta Merousa (in parts of the Douro), Bartolomeu (in Alenquer), Bastardo Espanhol (in Madeira) and Trincadeira (in Estremadura and Bairrada), though the latter is also sometimes used in the Douro as a synonym for Tinta Amarela.

It mostly makes appealing, upfront wines, quite low in acidity and high in alcohol, generally on the light side and with good raspberryish fruit. The Setúbal peninsula seems to be one of its best regions, though a little added acidity may be necessary for balance and here, particularly in Palmela, ripe, rich reds appear. It is prone to overripeness in Alentejo but can still produce some big juicy reds. Much is blended with other grapes, though it also appears as a varietal under several of its names.

Castelão Nacional, alias Castelão Português, is an unrelated, red-fleshed Teinturier grape found in southern Portugal. Best producers: (Portugal) Quinta da Abrigada, Quinta do Casal Branco, J M da Fonseca, J P Vinhos, Pegos Claros, Casa Santos Lima, Sogrape.

### CATARRATTO

Widely planted Sicilian white grape that can attain good quality if yields are controlled, but only sometimes does so. The name covers several related varieties. In 1990 Catarratto was the second most planted vine in Italy, after Sangiovese, with 65,000ha. It used to be grown for Marsala, but now is mostly either distilled or turned into grape concentrate. It features in several DOCs in Sicily, and if well made the wine can be crisp and vaguely interesting. Best producers: (Italy) Calatrasi, Rapitalà Adelkam, Spadafora.

### CATAWBA

On the eastern US seabord New York State grows Catawba for pink wines in various styles. Its skins are dark pink in colour, and it needs help from thermovinification to produce anything that could be described as red. The wine is decidedly 'foxy'-tasting. It was first spotted growing beside the river Catawba in North Carolina in 1801, and is perhaps a *labruscana,* a crossing of *Vitis labrusca* and *Vitis vinifera.* Its synonyms include Mammoth Catawba and Francher Kello White. Best producers: (New York State) Conestoga, Tucquan, Naylor Wine Cellars, Mount Hope.

### CENCIBEL

The name given to Tempranillo in central and southern Spain. See pages 256–265.

*Sunset over snow-covered vineyards and Lac Léman at La Tour-de-Marsens, Lavaux in the Swiss canton of Vaud. Sometimes called Dorin in the Vaud, Chasselas produces 99 per cent of the canton's white wine and suffers from a certain* folie de grandeur*; elsewhere in Switzerland it may be called Fendant.*

## CERCEAL

A name given to several Portuguese grapes, among them the Madeira vine more commonly known as Sercial (see page 240). Other Cerceals are found in the Dão, Bairrada and Ribatejo regions.

## CESANESE

Old, interesting but relatively rare vine found in Lazio, near Rome. Best producers: (Italy) Casale della Ioria, Casale Marchese, Villa Simone.

## CÉSAR

César seems to have been grown in northern Burgundy for nearly 2000 years, but it is not permitted for AC wine in the Côte d'Or, and it is only found in the Yonne where it is, in any case, practically extinct. It gives small quantities of dark-coloured, rather astringent wine. One or two examples sampled from Chile weren't much better.

## CHAMBOURCIN

A French hybrid, one of the best in existence, that produces wines of an intensely purple colour and a pronounced flavour of black cherries and plums, sometimes with a touch of spice or game. The wine is best drunk when young and fresh. It has only been planted since the 1960s and is found to a small extent in Australia's New South Wales, and in France's Pays Nantais at the western end of the Loire Valley, and in south-west France, though only for table wine. In Australia Chambourcin is sometimes blended with Shiraz for the sake of its colour; however, varietal versions are now beginning to appear on the market, bursting with meaty, black cherry fruit.

## CHARBONO

A now rare Californian vine that may be the same as Italy's Dolcetto (see pages 86–87) and/or the Bonarda of Argentina (see page 43). Dolcetto might also be the same as the Charbonneau of France's Savoie region, better known as Douce Noire. But neither it nor Charbono in California are very widely planted. The odd Californian example is strong and smokily rich. Best producers: (California) Duxoup, Fife, Parducci.

## CHARDONNAY

See pages 62–73.

## CHASAN

A recent crossing of Palomino (see page 165) and Chardonnay which, in mixing the neutrality of the former with the flavour of the latter, manages to produce a lightweight, neutral imitation of Chardonnay. It is grown in the south of France, where the local name for Palomino is Listan. For some reason it is recommended for planting in every French region except Alsace.

## CHASSELAS

Switzerland's favourite grape variety reaches peaks of quality there that it attains nowhere else, although it is widely planted throughout the world, from Chile to the Ukraine. As befits what seems to be a very old vine, it comes in many variations. Its origins are unknown: perhaps it was cultivated around Byzantium, or perhaps it came from Egypt. The ampelographer Galet inclines to the more probable theory that it originated in Switzerland and spread out along various river valleys, taking different names as it went. Fendant and Perlan are two alternative Swiss names.

In Switzerland its main interest for the drinker is its ability to reflect its terroir: on granite soil it tastes flowery with good acidity, on chalk it is fruity and honeyed, on the deeper, more clay soils of Epesses it has more weight and character, and in Dézaley, also in the Vaud, it is minerally. But 'weight' is always a relative term with Chasselas: it is, at best, a lightweight, neutral wine, even though it has more acidity in Switzerland than elsewhere.

It is grown in Germany as Gutedel (see page 114), and in Austria, where there is relatively little planted, as Wälscher and Moster. In France it is being replaced both in Alsace and in Pouilly-sur-Loire, where it was widely planted before phylloxera and produced table grapes for the markets of Paris. But the coming of the railways meant that the Midi could get its earlier ripening grapes to Paris faster, and Pouilly lost its market. It began then to make wine from its Chasselas, and the appellation of Pouilly-Fumé was so called to distinguish the superior Sauvignon Blanc wine from the Chasselas-based Pouilly-sur-Loire. The best French Chasselas comes from Crépy, in Savoie, where the wines are like the Swiss versions, only even lighter – if that's possible.

Not all its worldwide plantings go into the wine vat: it is much grown as a table grape. Most of its Romanian crop, for example, is destined for the table. Hungary, Moldova, the Ukraine, north and south Italy and North Africa all have some Chasselas, as does Chile. But the grape known in California as Golden Chasselas is most likely to be Palomino (see page 165). Best producers: (France) Serge Dagueneau, Kientzler, Pfaffenheim co-op, de Ripaille, Guy Saget, Schoffit; (Switzerland) Henri Badoux, Louis Bovard, Les Frères Dubois, Robert Gilliard, Caves Imesch.

# Carmenère

WHAT DID YOU GET if, in the 19th century, you took cuttings of Merlot from Bordeaux and planted them in Chile? A field mix of Merlot and Carmenère, that's what. And (according to most estimates) between 60 and 90 per cent Carmenère, which perhaps says something about the relative unimportance of Merlot in pre-phylloxera red Bordeaux.

The two vines look very similar, the only difference being that Carmenère's young leaves are red underneath, while Merlot's are white, and that the central lobe of the Merlot leaf is longer. In Bordeaux Carmenère used to be considered to be just as good as Cabernet Sauvignon, but unlike the latter it proved an irregular yielder when grafted and it was phased out in the 20th century. It ripens some three weeks after Merlot, which makes a field blend (which is what most Chilean 'Merlot' is) inconvenient. If you pick when the Merlot is ripe you get an aggressive green pepper flavour from the Carmenère; if you pick when the Carmenère is ripe you get jammy, overripe Merlot. The difference between the two varieties was officially recognized in Chile in 1996, and it has been possible to label wines as Carmenère since 1998. New vineyards are planted with Merlot and Carmenère separated, but it will be a long time before the older mixed vineyards are superseded.

Chilean growers are, however, learning how to grow Carmenère. It dislikes irrigation or rain between winter and harvest time: water at this time exacerbates the green pepper flavour, as do poor soils which cause the vine to need more water. Because it gets high sugar levels before the tannins are ripe it needs a long growing season, but in too hot a site the alcohol goes too high and the balance disappears. Even so, it is rapidly proving itself to be a really interesting grape with an unusual savoury quality to its taste.

There are a few vines – literally a handful – still in Bordeaux and rumours of a few in California. And there may be quite a lot in northern Italy. Elsewhere? Who knows. Lots of cuttings were taken from Bordeaux in the 19th century, and planted in many different countries. We may still turn up some Carmenère in surprising places.

**The taste of Carmenère**

Carmenère's low acidity gives it really sweet-tasting fruit, which makes it even more important to keep the green peppers under control: when ripe it has blackberry, black plum and spice flavours, rich, round tannins, and a marvellous savoury array of flavours – coffee, grilled meat, celery and soy sauce. This sweet/savoury flavour plus a full mouth-massaging texture make Carmenère a real original whose character often improves the palate of both Cabernet Sauvignon and Merlot.

*Carmenère from the MontGras winery. Even recently planted vineyards, such as the ones belonging to MontGras in Chile's Colchagua Valley, are a mixture of Merlot and Carmenère. To make a varietal Carmenère it is necessary to get an ampelographer to identify each vine, and then pick them separately. But of course the law does allow a small admixture of another grape type in a varietal wine.*

CASA SILVA

*Rich, ripe, balanced fruit from a company that aims at subtlety and complexity in its wine. The first vintage in which Chilean producers were permitted to label a wine as Carmenère was 1998.*

TERRANOBLE

*Terranoble is a firm believer in new oak barrels for aging Carmenère, considering that without oak the wine can be too soft. It does have lower acidity than Merlot, so can benefit from a bit of added structure, particularly with all the soft lush ripeness it gets in Chile.*

**Left:** *The Carmenère vines here at Caliterra's Arboledas estate in the Colchagua Valley were planted in 1997 – which means that they probably really are Carmenère. The first vintage was in 2000.*
**Above:** *Carmenère's late-ripening habit, dislike of water during the growing season and tendency to give green flavours in cool years mean that it is more subject to vintage variation than most other varieties in Chile. In addition, the vines seem to need to be mature to give good flavours: many winemakers agree that vines of under eight years old are likely to give vegetal-tasting wines without the admixture of fruit richness that marks out the grape's eventual personality. But, whatever its faults, it adds immeasurably to the character of what is still called 'Merlot' in Chile.*

## CONSUMER INFORMATION

### Synonyms & local names

Grande Vidure is the best known of several historic Bordeaux synonyms. This name was occasionally used in Chile, when the grape was first identified there, but may no longer be used for wines shipped to the EU.

### Best producers

**CHILE** Almaviva, Apaltagua, Bisquertt, Caliterra, Carmen, Casa Donoso, Casa Lapostolle, Casa Silva, Concha y Toro, Curicó co-op (Los Robles), De Martino, Luis Felipe Edwards, Gracia, Los Robles, Mondavi/ Chadwick, MontGras, Santa Rita, Terranoble, Veramonte.

## RECOMMENDED WINES TO TRY

### Ten Chilean Carmenère wines

Bisquertt *Casa La Joya Gran Reserva Carmenère*
Caliterra *Arboleda Carmenère*
Carmen *Grande Vidure Reserve*
Casa Silva *Carmenère Reserva*
Concha y Toro *Terrunyo Carmenère*
De Martino *Reserva de Familia Carmenère*
Luis Felipe Edwards *Carmenère*
Gracia *Carmenère Reserva Especial Callejero*
Los Robles *Riserva Privada Carmenère*
MontGras *Carmenère Reserva*

### Ten top Chilean reds containing Carmenère

Almaviva
Apaltagua *Carmenère*
Casa Donoso *1810 Cabernet/Carmenère*
Casa Lapostolle *Clos Apalta*
Casa Silva *Gran Reserva (tinto)*
Mondavi/Chadwick *Seña*
Santa Rita *Syrah/Cabernet/Carmenère and Triple C*
Terranoble *Carmenère Gran Reserva*
Veramonte *Primus*

MONDAVI/CHADWICK
*There is just 16% Carmenère in this pricy, much-hyped, Cabernet-based wine from this joint venture between Californian Robert Mondavi and the Chilean company, Errázuriz. But it's an interesting reflection on how climate affects wine styles: a Cabernet-Carmenère blend in Bordeaux would be lean and mean. Here in Chile it's rich and spicy.*

E·R

# CHARDONNAY

IF I HAD TO NAME THE GRAPE I've drunk the most of over the years, I would have to say Chardonnay. It's the biggest name in white wine: it signifies white wine even to those drinkers who buy anything with the word Chardonnay on the label without being entirely clear that Chardonnay is a grape. But I'm not talking about what I drink most of now – I'm such a promiscuous toper that I'm always relentlessly seeking new grape experiences. But that's because there are now so many experiences to be had. And that's a new phenomenon, caused by the remarkable wine revolution we've witnessed over the last 15 to 20 years. A wine revolution based on easy-going, utterly pleasant, fruity, sometimes creamy, sometimes spicy wines that you couldn't help but like. A revolution led by one grape above all others – Chardonnay.

*There seem to be more flavours associated with Chardonnay than with any other grape and it also has a wonderful affinity with new oak barrels. So here carved in the fresh new oak are many of these flavours, including cloves, hazelnuts, warm brioche and a host of different fruits.*

I can see how those early Californian and Australian trailblazers, back in the late 1960s and 1970s, chose Chardonnay to plant in their vineyards because one of the few European white wines with any reputation was white Burgundy – and the grape they used for white Burgundy was Chardonnay. They couldn't take the Burgundy vineyards home, but they could try to study how the great wines were made – and they could plant the same grape – Chardonnay. Which they did, with awesome determination.

The rest of the world was watching, and it saw the rapidity with which Chardonnay took off. It saw how the drinking public swooned over this new experience – white wine that was soft, dryish, definitely fruity, often with a whiff of vanilla. And it sold like hot cakes to an entirely new generation of wine drinkers. Every single wine producer in every country who was just beginning to wonder if he or she could possibly attract the world's wine buyers to even consider taking a slurp of their grog was hit by the same blinding explosion of light at the same time. And that starburst of light was a glittering sign in the brains of the winemakers of the world that flashed the single word – CHARDONNAY – with all the brilliance of a Broadway billboard full of 1000-watt bulbs.

And that's why I've drunk more Chardonnay than any other wine. Every time I try to discover a new region or country – what's the wine I can't avoid? Chardonnay. The south of France saw Chardonnay as their great white hope. Spain caught on, Italy even more so, Bulgaria, Hungary, Romania, Greece, India – yes, India, and it's not too bad. Did I mention Moldova, Slovenia, Israel, China, England? I should have done. They've all got Chardonnay too.

But I haven't even started on the New World. California, Australia, Chile, Argentina, New Zealand, South Africa – everywhere in the New World anyone has ever thought of planting a vine, Chardonnay is pretty well top of the list. Suitable conditions or not – it doesn't matter. Chardonnay can make serviceable wine where it's too cold and where it's too hot. And every time you head off to try to sell your wine you know that one of the first questions asked will be – do you have a Chardonnay? And you'll smugly answer – yes.

So, if I've drunk all this Chardonnay, what's it been like? Well, it's almost all been drinkable. Is that it? Drinkable? But hang on. I'm talking here about me rushing in to try the first offerings of country after country over the years, and usually thinking – these guys have got a long way to go – but if there's one wine I can safely take in to dinner, the fledgling Chardonnay is it. As the countries have grown in confidence and ability, other varieties may overtake Chardonnay, lesser-known varieties may produce more thrilling, individualistic flavours, but Chardonnay's remarkable reliability, the fact that there seems to be a formula available worldwide nowadays showing even the most humble outfit how to make decent Chardonnay, means that Chardonnay has almost become a generic for dry white with a bit of flavour. Is that what you want? Sure. 'I'll have a glass of Chardonnay.' And that'll do nicely, sir.

**Chardonnay: from Grape to Glass**

*Geography and History page 64; Viticulture and Vinification page 66; Chardonnay around the World page 68; Enjoying Chardonnay page 72*

# Geography and History

CHARDONNAY IS A GLOBAL BRAND, every bit as ubiquitous as Nike or Levi's. The only parts of the wine world where Chardonnay cannot be found are parts of France, notably Bordeaux, where wine laws prohibit planting, despite the fact that it might make pretty decent grog. Elsewhere, until a few years ago, it seemed that every wine area was either an actual or a potential Chardonnay site.

But then the world switched to red wine. And just in time, I'd say. You heard far too many tales of old red wine vines being grubbed up and replaced with Chardonnay. Ancient Negroamaro and Primitivo in Italy's far South ripped out to be replaced by high-yield cash-cow Chardonnay. Old Grenache and Mataro, Barbera and Shiraz being dragged by their ancient roots to the bonfire

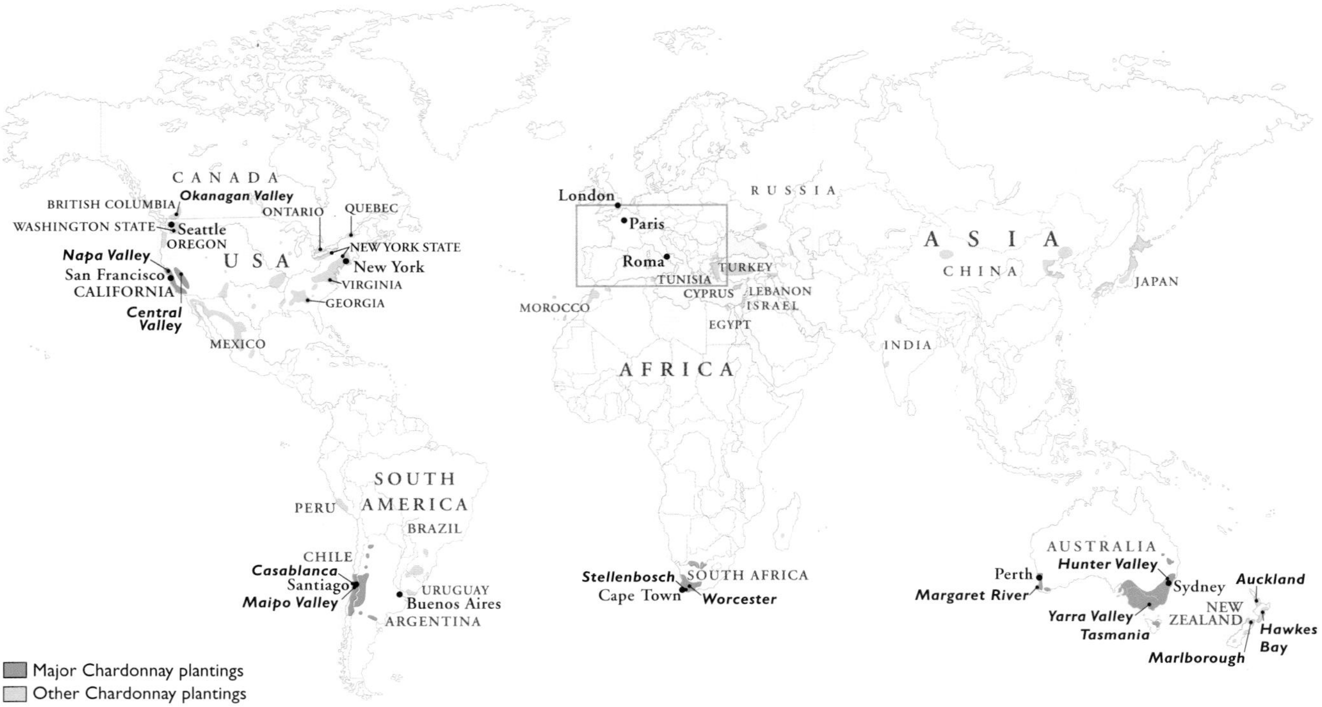

CANADA
BRITISH COLUMBIA
Okanagan Valley
ONTARIO
QUEBEC
WASHINGTON STATE
Seattle
OREGON
USA
NEW YORK STATE
New York
VIRGINIA
GEORGIA
Napa Valley
San Francisco
CALIFORNIA
Central Valley
MEXICO
London
Paris
Roma
RUSSIA
TURKEY
TUNISIA
CYPRUS
LEBANON
ISRAEL
EGYPT
MOROCCO
AFRICA
ASIA
CHINA
JAPAN
INDIA
SOUTH AMERICA
PERU
BRAZIL
CHILE
Casablanca
Santiago
Maipo Valley
URUGUAY
Buenos Aires
ARGENTINA
Stellenbosch
SOUTH AFRICA
Cape Town
Worcester
AUSTRALIA
Hunter Valley
Perth
Margaret River
Sydney
Yarra Valley
Tasmania
Auckland
NEW ZEALAND
Hawkes Bay
Marlborough
Major Chardonnay plantings
Other Chardonnay plantings

Major planting figures for Chardonnay

AREA PLANTED (HECTARES)
45,000
40,000
15,000
10,000
5000
2000
California (USA) 41,509
Australia 16,855
Burgundy (France) 12,996
Italy 11,800
Champagne (France) 8552
Languedoc-Roussillon (France) 7982
Chile 6705
South Africa 5938
Moldova 5400
Argentina 3373

ENGLAND
GERMANY
LUXEMBOURG
Champagne
Paris
Chablis
Alsace
Württemberg
Baden
CZECH REPUBLIC
SLOVAKIA
MOLDOVA
Loire
FRANCE
Burgundy
Côte d'Or
Mâconnais
SWITZERLAND
Valais
AUSTRIA
HUNGARY
Alto Adige
Trentino
SLOVENIA
CROATIA
ROMANIA
Murfatlar
Lombardy
Emilia-Romagna
SERBIA
BULGARIA
Rioja
Navarra
Somontano
Aragon
Languedoc-Roussillon
PORTUGAL
SPAIN
Cataluña
Ribatejo
Madrid
Penedès
Roma
ITALY
MACEDONIA
Macedonia
Castilla-La Mancha
Valencia
Extremadura
GREECE
Peloponnese
Mediterranean Sea

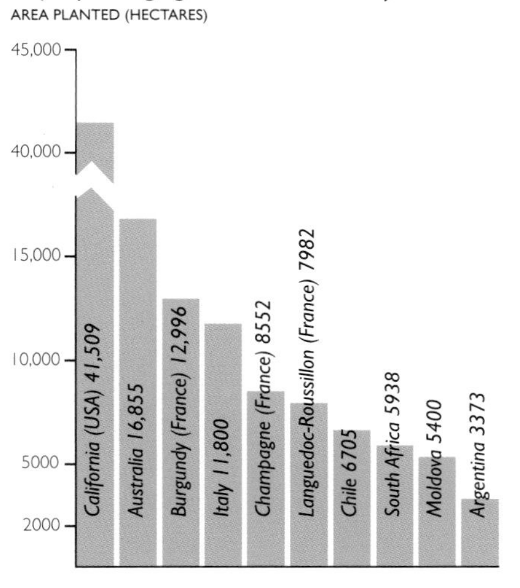

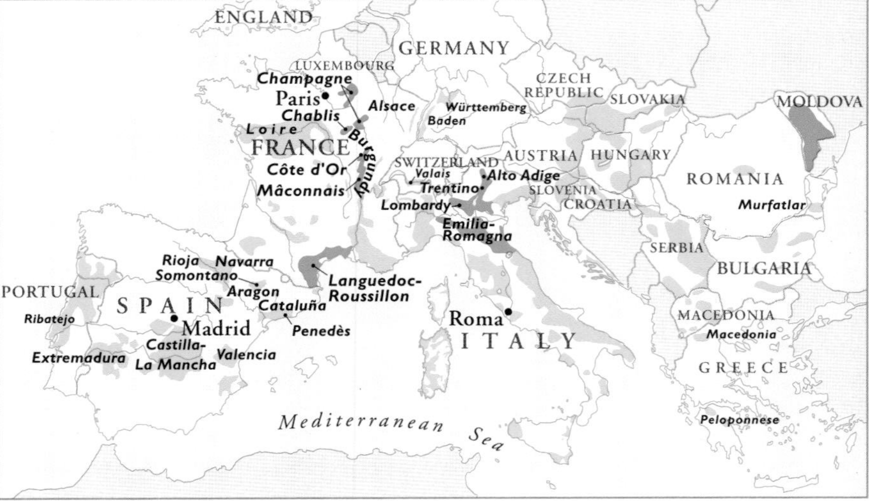

so that short-term Chardonnay could be milked briefly for profit, then forgotten. Chardonnay was too successful.

Look at the map. It's everywhere. From coolest to hottest, from England to India. Chardonnay is the world's most adaptable grape and can do pretty well any job you ask of it. With a little sweetness left in and a whiff of spice from oak chips, it has become the new Liebfraumilch for so-called 'entry level' drinkers. Yet it also makes some of the greatest white wines of all – the Grands Crus of Burgundy.

How can Chardonnay achieve so much at both the bottom and the top level? Well, remarkably for such a great grape, it is relatively neutral in itself. It provides a vehicle for every possible sort of terroir to express itself and a canvas for every possible winemaker's trick or talent. We need Chardonnay. No other white grape is as versatile for the consumer, or as easy-going for the producer. Mass-produced, it is a cheap commodity, but in the hands of a top winemaker it gives flavours that can still surprise and fascinate.

## Historical background

Now that DNA fingerprinting has revealed that Chardonnay is the offspring of Pinot Noir and Gouais Blanc we can reject all other, more exotic theories: that Chardonnay originated in the Middle East, or that it was brought to France from Cyprus. Chardonnay may have travelled the world in the last few decades, but for the first centuries of its existence it seems to have stayed very much at home. Gouais Blanc was planted in north-eastern France in the Middle Ages; so was Pinot Noir. The latter was much admired, while the former was not admired at all. Few misalliances can have had such spectacular results.

Confusion between Chardonnay and Pinot Blanc, a white mutation of Pinot Noir, is longstanding – it is only quite recently that French winemakers have stopped referring to their grape as Pinot Chardonnay, having eventually accepted assurances that the two are not related. Well, now – thanks to the work of John Bowers and Carole Meredith at the University of California at Davis, and Jean-Michel Boursiquot and Patrice This of the University of Montpellier, we know that they are related. And what's more, they look extremely similar. The leaf shape is almost the same, but where Chardonnay has naked veins on top of the leaf, Pinot Blanc does not. The hairs on the shoot tips are slightly different.

There is a third vine that looks confusingly like both Chardonnay and Pinot Blanc: Auxerrois. Some French nurseries can be a little cavalier about correct identification, at least, it seems, when they are selling to foreigners: the 'Chardonnay' vines that were sold to Baden in the mid-1980s turned out to be a mixture of Chardonnay and Auxerrois; the same happened when South Africa bought 'Chardonnay' in the 1980s. One German vine expert even recalls visiting Chablis a few years ago and realizing that by no means all the vines he saw there were Chardonnay.

*The seven Grands Crus of Chablis are situated on the appellation's warmest, south-west-facing slopes just above the town of Chablis. The differences between them are subtle but real, and easily disguised by too much new oak in the winery. Les Clos (above) produces the nuttiest, most honeyed wine of all.*

*Anne-Claude Leflaive, of Domaine Leflaive, is a leader of Burgundy's biodynamic pack. 'The vines are more resistant to disease this way,' she says.*

*Le Montrachet: for most people the wine from here is the pinnacle of white Burgundy. The slope is gentle – just 3 per cent – and the soil thin, marly limestone.*

# Viticulture and Vinification

THERE COULD NOT HAVE BEEN a white grape better suited than Chardonnay to launching the New World into the big league of winemaking. It cheerfully adapts to most soils and most climates, with its only problem in warm spots being a tendency to overripen rapidly and lose acidity.

In the decades when the winemaker was king, and the technical jiggery-pokery of the winery the focus of all attention, Chardonnay responded brilliantly, strutting its stuff with all the aplomb of a supermodel parading Versace in the morning and Armani after lunch. It is all too easy to make all Chardonnays taste alike – and, in fact, in blind tastings it is increasingly difficult to tell apart the top wines of California, Australia and Burgundy.

Yet Chardonnay can express terroir so well and react so well to sensitive winemaking that the one hope offered by an excess of Chardonnay in a red wine-obsessed world is that it may encourage more individuality, instead of just production-line lookalikes.

## Climate

With Chardonnay, terroir does make a difference even though it settles down happily in a far greater range of terroirs than most other varieties; it reflects its terroir in the wine.

In cool climates it produces leaner, steelier flavours and reasonably keen acidity, though no other region so far has been able to mimic the mineral steeliness of Chablis. Chablis, Champagne and Tasmania are among the few regions where Chardonnay can taste green and unripe, though good viticulture and the grape's early-ripening nature can combine to defeat this in all but cold and wet years. But it buds early, as well, so is prone to frost damage in the spring: frost is the bane of most Burgundians' lives. Even if your vineyards are frost-free, cool, wet weather during flowering can produce the uneven fruit set to which Chardonnay is also particularly prone. Late pruning, which delays flowering by up to a fortnight, can push flowering into a period of warmer, drier weather.

At the other end of the temperature scale it turns broad, heavy and flabby, with insufficient acidity to balance its melony fruit. Ideally, Chardonnay likes a long, slow ripening period in which it can develop flavour. In warm climates the temptation can be to pick before the acidity plummets; but more acidity in such cases can mean less flavour.

## Soil

Limestone, chalk and clay are important for Chardonnay, though it will produce good flavours on a huge variety of soils. In France, key Chardonnay regions juggle these three soil types in different proportions: solid chalk in Champagne; limestone with clay on the Côte d'Or. The soil of the heart of Chablis is Kimmeridgian chalky marl and marly limestone. The scattered areas of Petit Chablis, the lowliest Chablis appellation, spill on to Portlandian limestone, which gives less finesse.

On shallow, limy soils Chardonnay produces its tightest, most minerally wines; clay gives more weight and depth. In Meursault Perrières, for example, where the topsoil is barely 30cm deep over the limestone, the wine is restrained, powerful and slow maturing; in Meursault Charmes, where the topsoil is nearly 2m deep, the wine is rounder, richer and more seductive.

In Walker Bay, South Africa, Anthony Hamilton Russell of Hamilton Russell Vineyards believes that his low-vigour, stony, shale-derived soils, with their high proportion of clay, compensates for a climate that is warmer than Burgundy to produce relatively tight, restrained, minerally Chardonnay. Fruit grown on sandstone in the same area produces far richer and broader wine.

In the warm Hunter Valley in New South Wales, by contrast, Chardonnay is grown on the same light sandy soil that is favoured for Semillon: drainage is all important in a region where rain can make heavy, poorly drained soil impassable.

*Jean-François Coche-Dury in his cellar in Meursault in the Côte de Beaune. He has holdings all over Meursault as well as just 0.3ha of Corton-Charlemagne: holdings of top Côte d'Or vineyards are mostly small and precious. Coche-Dury wines are some of Burgundy's greatest.*

### Yields

Chardonnay's ability to ripen and attain good sugar levels even in unpromising sites makes it less yield-sensitive than, say, Pinot Noir. Go over about 80hl/ha in the low-vigour soils of France, however, and you're risking serious loss of quality – and watery, dilute wine. On the hills of the Midi 50–60hl/ha is probably about right for vin de pays. For really top Burgundian Chardonnay yields of about 30–45hl/ha are probably necessary. In Champagne yields may well exceed 100hl/ha – finesse, not concentration, is the aim here. On high-vigour soils, yields of 60–100hl/ha may well be necessary for vine balance. New Zealand manages to combine high yields with some of the most intense Chardonnay flavours in the New World; so it is possible

### Clones and sub-varieties

In Burgundy, there is little quality difference between massal selection and clones. The clones, of course, start off virus-free, which massal selected cuttings do not; many growers prefer clones for this reason. Each has its disadvantages: plant virus-free clones and your yields will be too high; plant virused cuttings and your vines run a serious risk of dying in ten or 15 years.

Commercially available clones worldwide have improved hugely in the past 15 years, and choosing the right clone can make all the difference between producing a wine that sells for a low price, and one that attracts a much higher one. New Dijon clones, better adapted to cooler climates, are helping produce richer flavours in Oregon. New Zealand produces remarkably powerful flavours despite generally poor clones, and Chile is also short of decent clones. Chardonnay Musqué, with its surprisingly Muscat-like aroma, is found in the Mâconnais, particularly in Clessé; there is also a Chardonnay Rose, with intriguing pink berries. Clone 166 also gives a Muscat-like aroma; this used to be widely planted in South Africa but has now been almost entirely replaced.

### At the winery

This somewhat neutral grape variety, once it gets into the winery, can be subjected to as many different treatments as the face of an aging socialite. All are directed towards influencing, refining or even adding flavour. Techniques usually thought of as Burgundian – barrel fermentation, malolactic fermentation and *bâtonnage*, or lees stirring – add, or bring out, buttery, creamy flavours in the wine; New World techniques – maceration with the skins, cold fermentation, ultra-hygiene – give tropical, fruit-forward flavours. From the 1980s onwards each camp has increasingly adopted the techniques of the other, which has led to cleaner, fresher white Burgundy and more elegant, restrained and subtle New World Chardonnays. Present a winemaker blind with a line-up of top Chardonnays from around the world, and it is increasingly likely that a Burgundian, an Italian, an Australian and a Californian will all mistake each other's wines for their own.

In Burgundy, and increasingly elsewhere for top wines, wild yeasts are allowed to have their way with the must, though producers might resort to cultured yeast if the fermentation is too slow in starting. The more mature the grapes, the longer the fermentation may last – though few continue as long as one of Domaine des Comtes Lafon's 1963 Chardonnays, which persisted for four or five years; the wine was eventually bottled in 1968.

Fermentation in barrel requires the right proportion of sediment to juice and, of course, clean sediment. Too many solids in the juice, says Dominique Lafon, give 'really weird wines' with green, bitter flavours. Fining the juice before fermentation, which might be necessary with machine-picked grapes, can, however, make the juice difficult to ferment.

In Burgundy and Champagne chaptalization is standard, because there simply isn't enough sugar ripeness in the grapes. In the warmer Mâconnais, chaptalization is also virtually de rigueur – but for different reasons. Because their grapes tend to lack acidity, Mâconnais growers often pick a few days early to capture the rapidly fading acid in the grape. But the trouble is an early-picked grape lacks interesting flavour development. So you pick tasteless grapes early for their acid, then add sugar to make a reasonably balanced wine. (Have you never wondered why so many Mâcon whites are so dull?) The Mâconnais really is an area where a bit of re-education in both winery and vineyard would be welcome.

In warmer climates acidification – the addition of tartaric, malic or citric acid, or any combination of those to the must – is equally common.

### THE OAK AND AROMA QUESTIONS

Chardonnay is not by any standards an aromatic grape. Pick one off a vine and it tastes nondescript, unless it has a touch of deliciously apricot-flavoured botrytis. Yet pour a glass of the blowzier sort of Chardonnay (a style rapidly going out of fashion) and you'll get a blast of pineapple and mango, butterscotch and toast, all held together by oak and willpower rather than by any intrinsic structure. Where does it all come from?

Early New World Chardonnays were deliberately made to be opposite in style to white Burgundy, the original classic style of Chardonnay – nutty, oatmealy and subtle. Australian techniques were based on those developed to get flavour out of the even more neutral Sultana grape, so aromatic yeasts were used for wines ripened to high sugar levels in warm climates; skin contact was employed to extract more flavour; fermentation was in stainless steel, and at the low temperatures that produce tropical fruit flavours, and after fermentation the wine was aged in forests of new oak barrels, which were probably made of assertively vanilla-ish American oak. The result was an instantly-likeable fruit cocktail; not bad for such a neutral grape.

Nowadays a mixture of aromatic yeasts and the neutral Prise de Mousse yeast – the one used in Champagne and for other sparkling wines – is more common, and wild yeasts are increasingly used for top wines. Skin contact is used by most winemakers from time to time, but many also avoid it completely by whole-cluster pressing; in large wine factories skin contact happens willy-nilly if they pick and crush the grapes faster than they can press them. The crushed grapes have to wait in tanks until the presses are ready for them.

Cooler climates are producing wines with better acidity, and early bottled, unoaked Chardonnay is enjoying a revival. But when oak is used it is handled more sensitively, and tighter grained French oak, and a mixture of new and old barrels, are more usual – though the quantity of oak used in many wineries could fall still further, and the wine (and perhaps the forests of France) would benefit.

# Chardonnay around the World

THE DIFFERENCE BETWEEN CHARDONNAY STYLES is increasingly not one of regions, but one of climates and techniques – although it is true that nobody outside Burgundy has yet produced a convincing imitation of the sublime flavours of Montrachet. But watch this space. The wine world is full of ambitious producers busting a gut to do so.

## Burgundy

Great white Burgundy is the epitome of Chardonnay. It is the wine that persuaded the rest of the world to plant Chardonnay, and to try and copy the flavours of the Côte d'Or. Yet the absence of the grape's name from most Burgundy labels means that it seemed to appear from nowhere and take over the world in one mighty bound.

It is planted widely on the Côte de Beaune; relatively little on the Côte de Nuits. Montrachet is smoky and immensely concentrated; Meursault buttery and oatmealy; Puligny-Montrachet structured, savoury and tight; Chassagne-Montrachet nuttier, and Corton-Charlemagne rich yet minerally. Further north in Chablis, where the geology is quite different, the wine takes on a flinty, austere mineral character, especially if it is made and aged without oak.

Further south in Burgundy the flavours are often more rustic, generally less complex. Here the grape name is more likely to appear on the label – although strictly speaking appellation contrôlée wines are not supposed to announce their grape variety. It is hard to see the point of this rule, and perhaps it will be quietly dropped sooner or later.

## Champagne

Chardonnay does not reach full ripeness here, even on the best, east-facing exposures of the Côte des Blancs. It's not the hours of sunshine that hold Chardonnay back – Champagne has as many of those as Alsace – it's the temperature. The west wind sweeps in across these low hills, keeping the annual mean temperature to around 10.5°C – just half a degree above the absolute minimum needed to ripen grapes. Winemakers here look for creamy, nutty, flowery characters in the Chardonnay; for vinosity, elegance and aroma. Mention the word 'fruit' to them and they say 'Oh yes, of course'; but one feels that it comes a good way down the list of priorities.

Because Champagne is usually a blend of regions and more often than not a blend of grapes, the different villages and districts are valued for the qualities they bring to the party. The Côte des Blancs villages of Cramant, Oger, Mesnil and Vertus all add the desired attributes of elegance and aroma, while Chardonnay from the eastern end of the Montagne de Reims is leaner and zesty; that from the Côte de Sézanne to the south is richer and creamier.

## Other French Chardonnays

Chardonnay has spread relentlessly from its bases in Burgundy and Champagne into the Jura, Savoie, the Ardèche, the Loire and even Alsace. These wines are all light, and generally fairly lean. In the Midi, where the grape is frequently given the Australian treatment, the wines are bigger and more tropical. In Languedoc it is best grown on the hills, and produces better quality than in Roussillon, where it is three or four degrees hotter on average; the higher altitudes of the Limoux or Pic St-Loup vineyards may well be capable of producing better quality still. The precipitous fall in prices of bulk Chardonnay from the Midi in 1999 and 2000 and the continued over-stocking since is likely to deter further planting in the foreseeable future – which is a pity. The feeling was that the Midi could produce superb semi-New World Chardonnays. But a continued lack of ambition and an obsession with high yields rather than flavour has caused the results to be very disappointing. Only a few growers in the Limoux region have regularly shown what can be achieved.

## Rest of Europe

In Italy, Tuscan Chardonnay is becoming increasingly refined as the vines grow older. Over-oaking can still be a problem, as can over-generous yields in the north. There may well be some confusion, either intentional or unintentional, with Pinot Blanc: Nicolas Belfrage, in *Barolo to Valpolicella,* quotes an unnamed Alto Adige producer as saying 'the best Chardonnay in Alto Adige is that made with Pinot Bianco'.

In Lombardy much goes to the sparkling wine industry. Some of the best still wines come from Piedmont, where the cooler climate gives the wines greater elegance. All over Italy it is blended with every conceivable white grape: Cortese, Favorita, Erbaluce, Ribolla, Albana, Trebbiano, Vermentino,

RENÉ ET VINCENT DAUVISSAT

*Wines from this domaine are object lessons in how to use oak with subtlety. The barrels are* feuillettes, *smaller than the 132-litre size traditional to Chablis.*

JEAN-NOËL GAGNARD

*Jean-Noël Gagnard's Bâtard-Montrachet is rare and in great demand. This slow-developing wine takes around ten years to fully reveal its powerful personality.*

BILLECART-SALMON

*This family-owned Champagne house makes wines of great elegance and delicacy. The name dates from 1818, when a M. Billecart married a Mlle Salmon.*

*Vineyards owned by the Chalone winery in the remote Gavilan Mountains above Soledad in Monterey County, California. In the 1970s Chalone was one of the first Californian producers to seek out limestone soil and the white wines have always possessed wonderful depth and balance.*

Procanico, Incrocio Manzoni, Verdeca, Grecanico, Catarratto, Nuragus, Viognier – and even Nebbiolo vinified off the skins.

In Spain, Chardonnay has shown itself more able to produce serious white wines than any native Spanish variety – with the exception of Verdejo, and even Verdejo often gets a bit of help from Sauvignon Blanc. Penedès, Navarra, Somontano and Costers del Segre are Chardonnay's strongholds; it is grown experimentally in Rioja, and there are moves to get it approved there. In Penedès it may be blended into sparkling Cava, generally with good results.

There is only a little Chardonnay in Portugal, but it is made (often by Australian winemakers) in a rich, nutty New World style. Austria's is somewhat leaner, and over-oaking can be a problem. It is found throughout Eastern Europe, with light, attractive examples coming from Hungary, often under Australian tutelage, and more varied quality from Bulgaria's large plantings. Bulgarian Chardonnay used to have a resinous dustiness and an unsettling perfume; that's what we were saying a decade and more ago. At last there are signs that the Bulgarians are beginning to get Chardonnay more or less right. Slovenian examples can be quite good; Swiss ones are light and attractive. There are isolated but good examples from Israel and Greece.

## USA: California

Californian production of Chardonnay will probably be almost 45 million cases a year by 2001; that's getting on for double 1996's figure of 22.5 million cases. Much of it is in Napa and Sonoma counties, with a lot of new plantings in Monterey, but an awful lot is also in the torrid Central Valley. Post-phylloxera replanting has concentrated it where it should be, in the cooler spots: Carneros gives brisk acidity; and Russian River a bit more substance and flinty fruit. Monterey Chardonnay is reminiscent of mango or guava; Santa Maria and Santa Barbara are richer and more tropical again. Alexander Valley Chardonnay is creamy and silky. Carneros, Russian River and Anderson Valley are all key areas for sparkling wine.

Styles are evolving in more than one direction. On the one hand there is movement towards more subtle, Burgundian wines, using indigenous yeasts, fermentation in small oak barrels, some lees stirring, and bottling without fining or filtration. On the other hand the traditional Californian style still finds favour here: there is no malolactic fermentation, and the wines are barrel aged but not barrel fermented; these wines can age well without developing any great extra complexity. Run-of-the-mill California Chardonnay often has some residual sugar which, together with high alcohol levels and low acid levels, can leave it tasting distinctly off-dry and, frankly, not that refreshing.

How far can you go in modifying styles in the Napa, where the intrinsic style is big and alcoholic? It is possible to remove some alcohol by using reverse osmosis, osmotic distillation or with a contraption called a Spinning Cone. Experiments suggest that reverse osmosis is the best method, and that if you gradually reduce the alcohol from

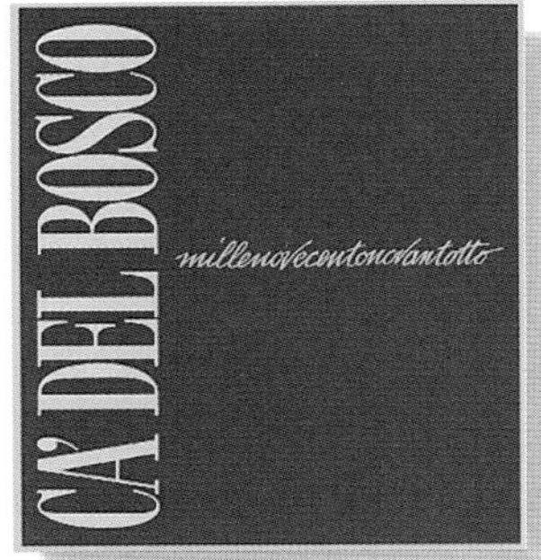


**Ca' del Bosco**

*Winemaker Maurizio Zanella's aim with this Chardonnay is to equal Burgundy. So it's barrel fermented, and is a deep, smoky, buttery delight.*

**Newton**

*Newton introduced the concept of unfiltered Chardonnay to California. It's an intense, refined wine of tremendous complexity.*

**Roederer Estate**

*Roederer uses grapes from the cool, damp Anderson Valley in northern California to produce a fine, sparkling wine in the style of Champagne.*

14 per cent down to 12 per cent you will hit two or three different levels of alcohol where the wine is in balance. Choosing one of these allows a winemaker to fine-tune a wine in the cellar. Others prefer to do it earlier on in the process, in the vineyard, with a combination of clever choice of site, canopy management and control of yields.

### Rest of North America

Washington Chardonnay is not dissimilar to California Chardonnay, generally with an emphasis on fruit flavour rather than creamy texture, but the top producers are beginning to produce good, savoury, nutty, Burgundian styles. Oregon's wines are becoming richer as it plants better, Dijon, clones. In New York State Chardonnay flourishes in all four regions – Lake Erie, Hudson River, Long Island and Finger Lakes – helped by its resistance to winter cold, and the best examples are very good. In Canada, too, it seems at home, producing surprisingly rich wines in Ontario, and lighter and less oaky ones in Quebec and British Columbia.

### Australia

Chardonnay production increased here by 55 per cent between 1996 and 1999, and it became the country's leading premium variety. But there its advance stopped: prices have fallen, and new plantings now are of red varieties, especially Shiraz, Tempranillo and Italian reds.

Nevertheless, Australian Chardonnay is here to stay. Styles are constantly evolving, with more refined, cooler climate wines at the top end, and more complexity lower down the scale, derived from Burgundian techniques like lees stirring. Over-oaking is much less of an affliction than it was, and there is less charring on the oak, but many wineries could reduce still further the amount of new oak they use.

The Chardonnays of today are a remarkable contrast with those of the 1980s, when deep golden-coloured, broad, fat, oily wines full of rich butterscotch flavours were the rule. Now colours are paler, structure is better and the fruit flavours are those of nectarine or white peach instead of melon. Warmer regions routinely add acidity, and this is apt to show on the palate, but winemaking generally is more hands-off than it once was.

Chardonnay displays regional variations less than some grapes, but Hunter Chardonnay is buttery, viscous and opulent: the style of Chardonnay most often thought of as Australian, in other words. More elegant wines of varying styles but pretty uniformly high quality come from the Yarra Valley. There is considerable complexity in the wines of the Eden Valley and Adelaide Hills and a grapefruit and white peach note in those of Padthaway. Margaret River produces wines of outstanding concentration and complexity, and the wines of Tasmania are delicate and citrus-flavoured when the fruit is ripe. The Riverland makes everything from inexpensive bulk wines to surprisingly high-quality examples. The coolest regions, including Tasmania, Geelong and the Macedon Ranges, have climates comparable to that of Champagne, and produce sparkling wine of great finesse and style.

**Australian Chardonnay: a new classic**

*Chardonnay vineyards at Leeuwin Estate in Margaret River, Western Australia (right). Leeuwin Estate makes complex Chardonnays that approach top Burgundies in their structure and ageability – and few people could have dreamt, 30 years ago, that Australia was capable of such quality. The grape is no newcomer to Australia. It was first planted in the 19th century, but winemakers didn't fall in love with it until after Murray Tyrrell produced a commercial version in the Hunter Valley in 1971, having acquired his vines by hopping over the fence into a neighbouring Penfolds vineyard to remove a few thousand prunings. The first reactions to the wine were, ironically, that Australians would never drink white wine with oak. Wine show judges gave it six marks out of 20, and as Murray's son Bruce Tyrrell says: 'even the spit bucket gets eight.'*

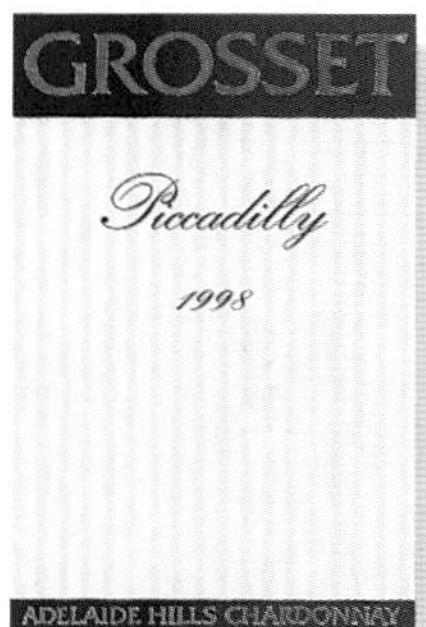


GROSSET

*This is the direction of top Australian Chardonnay today: cool climate fruit from regions like Adelaide Hills in South Australia and subtle aging in oak barrels.*

## New Zealand

New Zealand makes some of the most intense, powerful and balanced Chardonnays in the New World, marrying ripeness and depth with good acidity. Indeed, as producers appear to have lost their confidence in making really snappy Sauvignon, they seem to have discovered how to make deep, rich Chardonnay in its place. The most powerful come from Hawkes Bay, though Marlborough and Canterbury can produce impressive results. Wairarapa, Nelson and even Auckland also do well, and as better clones come on line, New Zealand should improve even more.

## Chile and South America

As winemaker Ignacio Recabarren puts it, 'in terms of quality, winery improvements took Chardonnay from under the table to on top of the table. To get to the roof we must work on viticulture. We are currently about halfway to the roof.'

One of the battles is getting better clones into the vineyards; the other major factors here are climate and yields. The latter are generally high to very high; the former, in this long thin country, are infinitely adjustable. The Casablanca Valley is the coolest climate spot so far being widely planted: plantings started here in about 1990, and Chardonnay covers 70–75 per cent of the vineyards. There's a frost risk in spring here; Recabarren characterizes it as as cool as Mâcon, but warmer than Marlborough. However, Casablanca is closer to the Equator than comparable regions elsewhere, which means that the grapes can be left on the vine for longer and still benefit from warmth from the sun. The other side of the coin is that leaving the grapes on the vine longer means lower acidity.

High yields do not in themselves imply lower quality: the balance of the vine is the crucial point. Casablanca's yields are high for two reasons, one of which is specific to the valley and tied to its frost risk; the other is common to the whole of Chile, and indeed to other Southern Hemisphere countries.

Casablanca's frost risk is very real. Usually it would be combated by heaters in the vineyards, by air propellers, or by water sprinklers. But heaters produce pollution, and are likely to be prohibited because of the valley's proximity to polluted Santiago. Sprinklers, however, use four times as much water as drip irrigation, and water is scarce here. Propellers need an inversion layer, which is not always there. So Casablanca growers are moving to a system of leaving extra shoots on the vines at pruning to allow for frost damage. However, these shoots must be thinned after the danger of frost has passed, and before flowering if the vines are spur pruned; before the berries are pea-sized if the vines are cane pruned. Not to do so will mean too high a crop in that year.

The second reason for high yields is that more generally, Chile's level of sunlight induction on the vine's half-formed buds is higher than that in Europe. Dijon clones therefore naturally give higher crops here than they would back home in Burgundy.

The climate, even of Casablanca, varies according to where you are: the lowest spots, by the town of Casablanca, are the coolest, and while you might pick Chardonnay in Maipo, in the Central Valley, in the second week of February, picking in Casablanca does not start until a month or six weeks later.

Styles in Chile are evolving from simple, fruity and oaky, and are gaining more interest and complexity at the top end.

So far Argentinian Chardonnay is lagging behind Chilean, partly because of overproduction, and partly because the vines grown there were usually brought in to make sparkling wines, and ripeness and concentration of flavour were not priorities. New clones and ambitious wine producers will make an enormous difference, as will the development of high-altitude, relatively cool climate regions like Tupungato.

## South Africa

Chardonnay covers 5.7 per cent of the vineyard here, and plantings are concentrated in the Robertson, Stellenbosch and Worcester regions. A typical premium South African Chardonnay is all or part barrel-fermented, with vanilla and tropical fruit flavours, and around 13.5 per cent alcohol. There is no real national style yet, and the differing permutations of soil, climate, clone, yield and producer can give anything from bulk-produced, dilute wines to much more tight, complex ones. It is also increasingly used for sparkling wine. Climates are seldom very cool here, and even the coolest parts, like Walker Bay, are warmer than Burgundy, though perhaps cooler than Adelaide Hills.

Poor clones have, until recently, hindered attempts to make classy Cape Chardonnay, but the will is there and good enough examples are appearing all round the country to make pundits who said the Cape couldn't grow Chardonnay look rather foolish.

GLENORA

*This barrel-fermented Chardonnay hails from New York State's Finger Lakes region. Steely acidity marries brilliantly with rich, toasty oak flavours.*

SERESIN

*Seresin's first vintage was 1996 but the company, established by film producer Michael Seresin, already makes one of New Zealand's classiest Chardonnays.*

CELLIER LE BRUN

*This Champagne-method specialist was one of the pioneers of New Zealand's Marlborough region, and still makes outstanding, toasty, creamy Blanc de Blancs.*

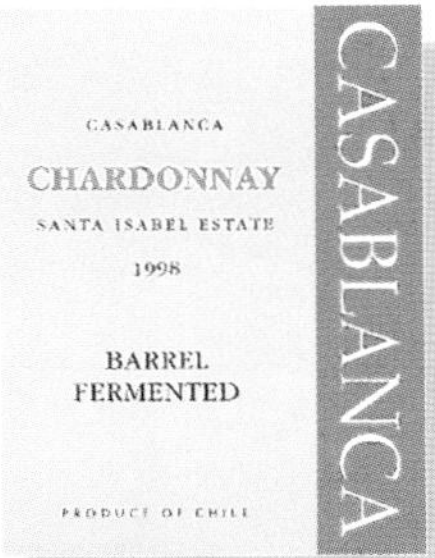


VIÑA CASABLANCA

*Chile is only just developing the concept of single-estate wines. Santa Isabel is a superb property in the cool Casablanca Valley.*

# Enjoying Chardonnay

TO GET THE MOST OUT OF a great Chardonnay – great, not merely good – you have to give it bottle age. True, Corton-Charlemagne can be utterly seductive from the barrel, but such wines are so rare and so expensive that drinking them straightaway can seem wickedly frivolous. Keeping wine like that for several years somehow both amortizes the expense and prolongs the pleasure. Am I showing my puritan streak? No. I'm showing that I'm such a hedonist that I can enthusiastically embrace the waiting period because I know that the eventual pleasure will be massive.

In a top year, from a leading producer, such Grand Cru white Burgundies can last up to 30 years, and really should not be drunk before they are eight or ten. But much depends on the style of the producer and how concentrated the wine is to begin with. Premiers Crus are less long-lived – 20 years might be the upper limit – and village wines should be drunk earlier again, within eight to ten years. Bourgogne Blanc and Côte Chalonnaise are best drunk within five years; the best Pouilly-Fuissé can be allowed up to eight. Most Mâcon should be drunk immediately. Basic Chablis needs two or three years, and Grand Cru needs at least eight and will last much longer.

There are as many rules for Chardonnays elsewhere as there are producers. Concentrated, balanced, cool climate wines from areas as diverse as Carneros, Russian River, Yarra Valley and Margaret River may easily last a decade or more. Others may tire after only a few years. Much depends on the philosophy and skill of the producer. But simple, warm climate Chardonnays do conform to an unbreakable rule: drink them early. Leave them even a few months too long and they will start to fade and flatten out and lose their brief, bright-eyed burst of fragrance and fruit.

### The taste of Chardonnay

The taste of Chardonnay is hard to pin down: unoaked wines from high-yielding vineyards may taste of not very much, while a minerally, concentrated bottle from the Côte d'Or or Margaret River may display a greater range of flavours than almost any other white grape. Some of these flavours may derive at least in part from vinification techniques; some come from the terroir, some from just the climate, some from the clone. Barely ripe Chardonnay tastes of green apples; riper, cool climate examples where the grapes had a long hang time have flavours of pears and acacia, lemons and grapefruit, nuts and biscuits, butter, honey and popcorn. There may also be a minerally, flinty or smoky streak to the fruit, some toast from the oak and, of course, balancing acidity which may seem tight and piercing in youth. White Burgundy, and occasionally other cool climate wines, may have a vegetal, slightly cabbagey tinge.

Warmer climate Chardonnays get tropical, with mango, cream, banana, pineapple, melon and peach, plus butterscotch and more butter, honey and toast. There may be some spice, too, or boiled sweets. Everything but the kitchen sink, really.

Champagne is often just crisp and creamy or flowery in youth, and develops fresh bread flavours and winey depth with age.

*Kistler is dedicated to making great Chardonnay of Burgundian-style complexity in California's Sonoma Valley. Back in Burgundy one of the ways the Domaine des Comtes Lafon achieves that is with wild yeasts: the yeasts in the cellar change year by year, says Dominique Lafon, and that's good. 'I don't want the same product every year. Different yeasts may dominate in different years.'*

### Matching Chardonnay and food

With its broad spectrum of flavours and styles – from steely, cool-climate austerity to tropical lusciousness – there is a Chardonnay for almost every occasion, and most are superb with roast chicken or other white meat. The relatively lean end of the spectrum is one of the best choices for simple fish dishes. Top Burgundy and the really full, rich New World blockbusters need rich fish and seafood dishes. Oaky Chardonnays are good with tricky-to-match smoked fish, and work well with garlicky dips such as guacamole, as well as with spicy, coconutty, South-East Asian food; they are also good all-rounders with, say, festive turkey.

## CONSUMER INFORMATION

### Synonyms & local names

The world's favourite white wine grape variety is rarely called by any other name and most of the old European synonyms have fallen into disuse; Austria, especially in Steiermark, still calls it Morillon.

### Best producers

**FRANCE/Chablis** René et Vincent Dauvissat, Droin, Laroche, Michel, Raveneau; **Côte d'Or** G Amiot, J-M Boillot, Bonneau du Martray, Michel Bouzereau, Carillon, Coche-Dury, Marc Colin, Drouhin, J-N Gagnard, Gagnard-Delagrange, Vincent Girardin, Jadot, F Jobard, Lafon, Lamy-Pillot, Dom. Leflaive, Dom. Leroy, Matrot, Bernard Morey, Niellon, Ramonet, Roulot, Sauzet, Verget; **Mâconnais** Barraud, Ferret, Ch. Fuissé, Guffens-Heynen, Merlin, Rijckaert, Robert-Denogent, Thévenet, Valette.
**AUSTRIA** Velich.
**GERMANY** Johner, Rebholz.
**ITALY** Antinori, Banfi, Bellavista, Ca' del Bosco, Colterenzio co-op, D'Alessandro/Manzano, Gaja, Isole e Olena, Lageder, Planeta, Ruffino, Castello della Sala, Tasca d'Almerita, Vie di Romans.
**SPAIN** Enate, Manuel Manzaneque, Torres.
**AUSTRALIA** Cullen, Evans & Tate, Giaconda, Green Point/Domaine Chandon, Grosset, Howard Park, Leeuwin, Lenswood, Nepenthe, Penfolds, Petaluma, Pierro, Plantagenet, Rosemount, Shaw & Smith, Tyrrell's, Vasse Felix.
**NEW ZEALAND** Cloudy Bay, Kim Crawford, Deutz, Hunter's, Kumeu River, Matua Valley, Montana, Neudorf, Seresin, Sileni, Te Mata, Vavasour, Villa Maria.
**USA/California** Arrowood, Au Bon Climat, Beringer, Calera, Chalone, Ferrari-Carano, Kistler, Landmark, Marcassin, Matanzas Creek, Merryvale, Peter Michael, Pahlmeyer, Roederer Estate, Saintsbury, Steele, Robert Talbott; **Washington State** Chateau Ste Michelle.
**SOUTH AFRICA** Glen Carlou, Hamilton Russell, Meerlust, Morgenhof, Mulderbosch, Rustenberg, Springfield, Thelema, Vergelegen.
**SOUTH AMERICA** Casa Lapostolle, Catena, Errázuriz, Montes, Viña Casablanca.

## RECOMMENDED WINES TO TRY

### Ten classic white Burgundies

**Bonneau du Martray** *Corton-Charlemagne*
**Coche-Dury** *Meursault les Perrières*
**René et Vincent Dauvissat** *Chablis la Forêt*
**Joseph Drouhin** *Beaune Clos des Mouches*
**Girardin** *Chassagne-Montrachet les Caillerets*
**Louis Jadot** *Chevalier-Montrachet les Demoiselles*
**Domaine Leflaive** *Le Montrachet*
**Raveneau** *Chablis les Clos*
**Robert-Denogent** *Pouilly-Fuissé les Carrons*
**Verget** *Bâtard-Montrachet*

### Five Burgundy-style lookalikes

**Antinori** *Castello della Salla Cervaro (Italy)*
**Grosset** *Piccadilly Chardonnay (Australia)*
**Hamilton Russell** *Chardonnay (South Africa)*
**Penfolds** *Yattarna Chardonnay (Australia)*
**Velich** *Tiglat Chardonnay (Austria)*

### Ten ripe, toasty New World wines

**Au Bon Climat** *Le Bouge d'àcôté Chardonnay (California)*
**Chateau Ste Michelle** *Cold Creek Vineyard Chardonnay (Washington)*
**Kistler** *Dutton Ranch Chardonnay (California)*
**Kumeu River** *Chardonnay (New Zealand)*
**Leeuwin Estate** *Art Series Chardonnay (Australia)*
**Marcassin** *Lorenzo Vineyard Chardonnay (California)*
**Montes** *Alpha Chardonnay (Chile)*
**Petaluma** *Chardonnay (Australia)*
**Pierro** *Chardonnay (Australia)*
**Springfield** *Méthode Ancienne Chardonnay (South Africa)*

### Five Champagnes (Blanc de Blancs)

**Billecart-Salmon** *Blanc de Blancs Vintage*
**Deutz** *Blanc de Blancs Vintage*
**Jacquesson** *Blanc de Blancs Grand Cru Vintage*
**Krug** *Clos de Mesnil Vintage*
**Ruinart** *Blanc de Blancs Vintage*

### Five sparkling wines

**Bellavista** *Franciacorta Gran Cuvée Satèn Non-vintage (Italy)*
**Deutz** *Marlborough Cuvée Blanc de Blancs Vintage (New Zealand)*
**Green Point/Domaine Chandon** *Blanc de Blancs Vintage (Australia)*
**Nyetimber** *Première Cuvée Blanc de Blancs Vintage (England)*
**Roederer Estate** *L'Ermitage Vintage (California)*

### Five unoaked/lightly oaked wines

**Nepenthe** *Unwooded (Australia)*
**Plantagenet** *Omrah Unoaked (Australia)*
**Vavasour** *Dashwood (New Zealand)*
**Vie di Romans** *Friuli Isonzo Ciampagnis Vieris (Italy)*
**Viña Casablanca** *Santa Isabel (Chile)*

*Chardonnay looks so similar to Pinot Blanc that in Italy they were only officially differentiated in 1978; in France in 1872. Small quantities of Pinot Blanc can still be found on the otherwise Chardonnay-dominated white vineyards of the Côte d'Or.*

### Maturity charts

Most Chardonnay should be drunk early: only top wines are designed to improve with years of bottle age.

**1996** Chablis Grand Cru

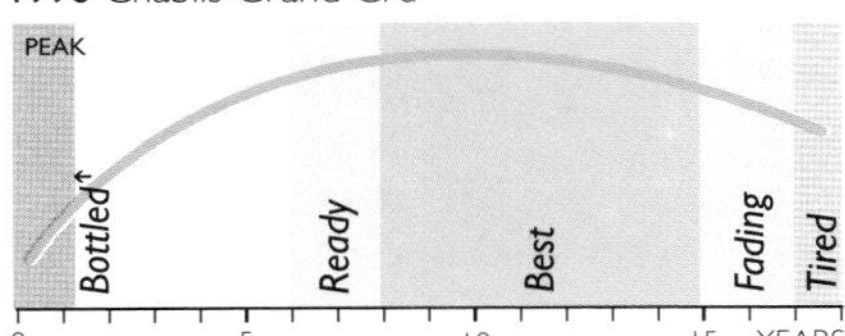


*1996 was an excellent year for white Burgundy generally, and for Chablis in particular. The wines have beautiful balance and racy, ripe acidity.*

**1997** Côte de Beaune White (Premier Cru)

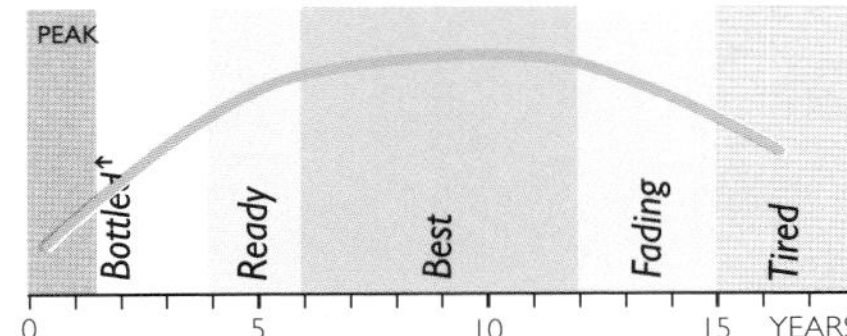


*A warm vintage that produced early-maturing, often soft wines. Some producers acidified. The wines should not be cellared for too long.*

**2000** Adelaide Hills Chardonnay

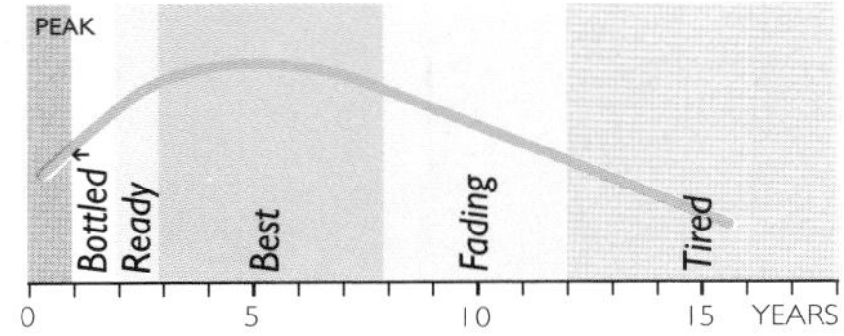


*A difficult vintage in which good viticulture paid off. Some growers were caught out and produced poor quality; others made successful, balanced wines.*

# Chenin Blanc

It hasn't always been easy being a Loire Valley grape. For years on end nobody elsewhere took much notice of you. If they wanted Sauvignon Blanc they went to New Zealand. If they wanted Chenin Blanc – well, on the whole they didn't want Chenin Blanc.

Which in a sort of way was to Chenin's advantage. Its producers made no effort to change the style of their wines when the world was mad for fat, oaky-buttery, tropically-fruited whites; they just went on making lean, minerally wines that at their best took years – decades, even – to mature to a bone-dry, honeyed depth. Unfashionable? You couldn't get more unfashionable than Chenin Blanc. Well, Aussie Shiraz was unfashionable, once. In the long run it's not always a bad thing. It means that when – if – the spotlight does turn on you again, you can attack the business of making great wine unencumbered by tradition because it was a tradition nobody wanted. The 'young guns' can do it all their way.

*Floating on air and water, the château of Chenonceau stretches across the river Cher in Touraine, with the river Loire beyond. Chenin Blanc was first planted here in the heart of the Loire Valley in the 15th century and Anjou and Touraine are still where it produces its most exciting wines, whether sweet or dry, sparkling or still.*

But there has to be a switch there to turn the spotlight back on. In Australia Shiraz had the world-class Penfolds Grange to blaze a trail. The Chenin Blanc had to look to another area altogether – Bordeaux and its Sauternes region. During the 1980s Sauternes had a series of splendid vintages for rich dessert wines, so by the 1990s prices had risen dramatically and there was an enthusiastic band of well-heeled consumers who'd got the sweet wine bug and who would pay. The only other French region with a classic sweet wine tradition was the Loire Valley, and during the 1990s, the Loire was luckier than Bordeaux with its weather. Spurred on by promises of the mighty dollar and with a brood of young winemakers who were much more aware of the modern wine world than their parents, the misty side valleys of the Loire, and in particular Bonnezeaux, Quarts de Chaume and Coteaux du Layon south of Angers, produced a series of gorgeous, rich, and startlingly original sweet wines from noble-rotted Chenin Blanc grapes. Suddenly Chenin had a standard-bearer for quality.

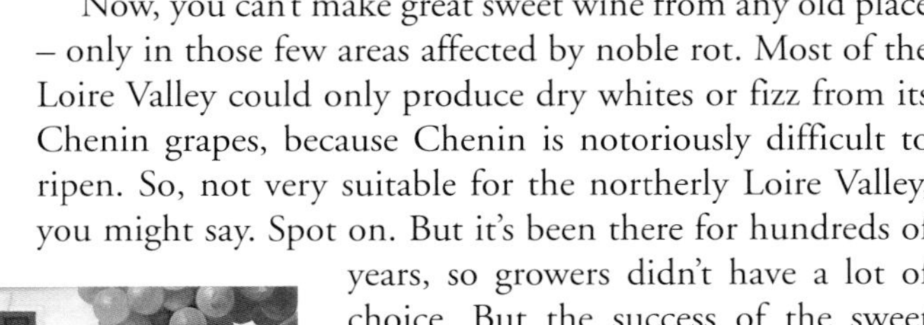

Now, you can't make great sweet wine from any old place – only in those few areas affected by noble rot. Most of the Loire Valley could only produce dry whites or fizz from its Chenin grapes, because Chenin is notoriously difficult to ripen. So, not very suitable for the northerly Loire Valley, you might say. Spot on. But it's been there for hundreds of years, so growers didn't have a lot of choice. But the success of the sweet whites at last spurred on the other producers to say 'If the sweet white guys can do it why not us?' It took a bit of an adjustment to the mind set: like it or not, New World attitudes about stainless steel, cleanliness and maximizing extraction of fruit flavours had to be adopted. And in the vineyards they realized they simply had to find a way to ripen the grapes, however unwilling Chenin was to perform in the chilly fields of Anjou and Touraine. Lower yields, better growing methods and that old-fashioned virtue – the courage to wait out poor autumn weather until your grapes tasted ripe and ready to pick – all of these have transformed Chenin in the Loire.

And now, suddenly, as winedrinkers begin to look for alternatives to Chardonnay and Sauvignon Blanc, Chenin Loire is showing it can provide bone dry, minerally, angelica and honey-scented wines of real interest. And in so doing, the Loire is at last providing a lead for other parts of the world with plantings of Chenin. South Africa has Chenin Blanc in nearly a quarter of all its vineyards; South America and California have a good deal, and even places like Australia and New Zealand have some. Already New Zealand, Australia and South Africa have shown how beautiful its greengage and angelica fruit can be, how well it responds to lower yields, how warmly it takes to barrel fermentation. I don't want them to start making Loire lookalikes, even if they could; nor do I want the growers of Savennières to start pretending they live in Stellenbosch. But I do want Chenin Blanc to come out of its shell once and for all and show the world what a good and original grape it is.

**Chenin Blanc: from Grape to Glass**

*Geography and History page 76; Viticulture and Vinification page 78; Chenin Blanc around the World page 80; Enjoying Chenin Blanc page 82*

# Geography and History

A GLANCE AT THIS MAP MIGHT make one think that Chenin Blanc is one of the world's favourite grapes. It covers nearly a quarter of the vineyard in South Africa; it flourishes in some of California's warmest spots; Argentina produces it in abundance. It can be found in Canada, New Zealand, Australia, Brazil, Uruguay and Mexico. Yet, with a few exceptions, none of these plantings produce wines that are interesting enough to warrant international attention – or to entitle it to so much space in this book.

The reason Chenin Blanc can be regarded as a classic white grape variety – indeed, one of the world's finest – is one small region of France. Anjou-Touraine, where the vineyards centre on the river Loire and its tributaries, is the source of classic Chenin. This is where the vine regarded as

Major Chenin Blanc plantings
Other Chenin Blanc plantings

Major planting figures for Chenin Blanc

AREA PLANTED (HECTARES)

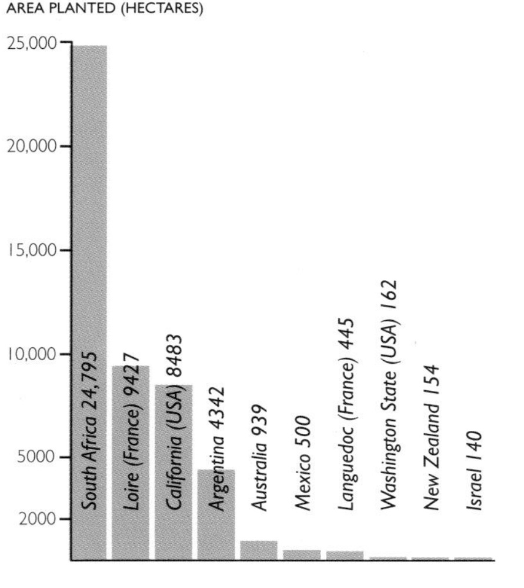


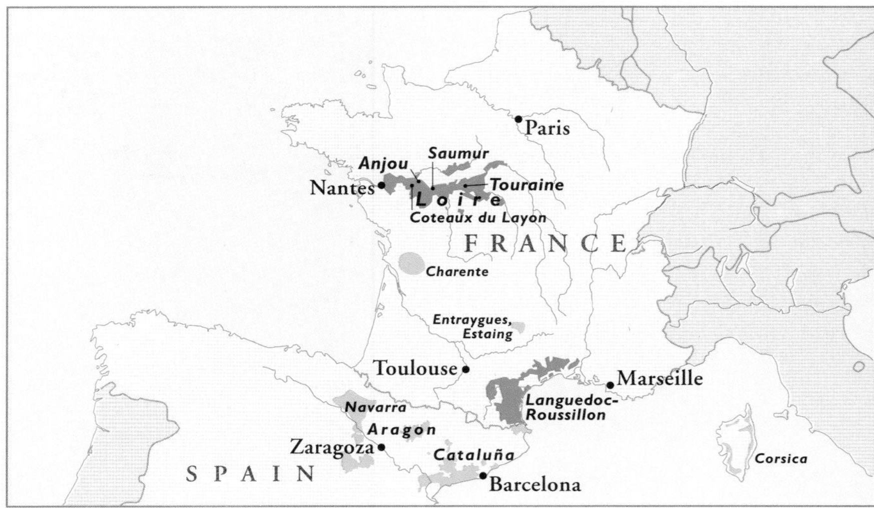

a workhorse grape in most of its homes produces startlingly intense, concentrated and long-lived wines, which can be sweet, medium or dry.

Why only here? Especially when Anjou-Touraine is so cool, and Chenin is so ponderously late-ripening. Growers there measure other grape varieties' ripening cycles by how much sooner than Chenin their fruit is ready to pick. Yet this is where the human element enters in. Chenin is a vigorous vine and will produce almost as many grapes as you want it to. Overproduce in Anjou and the result is some of the nastiest acid white produced in the whole of France. But keep the yields right down, in Savennières, or Saumur, or Vouvray, and the result is unique, steely wine with a flavour of greengages and angelica, that ages sublimely for a decade or more. As for sweet wine, Chenin is prone to noble rot, and in the side valleys of the Loire's tributaries, autumn mists and warm sunny days produce small crops of noble rotted Chenin that is both piercingly fresh yet succulent when young and which will age for generations to a deep amber brilliance of honey and barleysugar and quince.

## Historical background

Chenin Blanc has been growing in the Loire Valley for over a thousand years, and probably originated there: its first mention seems to be in 845 at the abbey of Glanfeuil on the left bank of the Loire. In 1445 it was planted in Touraine by Thomas Bohler, lord of Chenonceaux, and by his brother-in-law Denis Briçonnet, abbot of Cormery at Mont-Chenin, and the vine seems to have retained the name of this locality.

South Africa now has the world's highest plantings of Chenin – it first travelled there in 1652 with Jan van Riebeeck, the Dutch East India Company's first Commander at the Cape and the country's first winemaker: no doubt Chenin's ability to retain acidity even in warm climates, and its obligingly high yields, endeared it to the early settlers.

It seems to be genetically fairly stable and less given to mutation than other very old *Vitis vinifera* vines: genetic variation is not regarded as a problem in the Loire Valley, and most growers approve of the quality of the commercially available clones, even if they disapprove of the higher yields they usually produce.

One vine, sometimes referred to by Loire growers as a sub-variety of Chenin, is however not Chenin at all. It's actually Verdelho. A little can sometimes be found interplanted in the Chenin vineyards of the Loire: it is strictly forbidden by the appellation rules, but it's been there a long, long time and because it ripens two weeks before Chenin, it can be useful in a cool year. Its association with these vineyards is hardly new: back in 1928, in his book *Le Vigneron Angevin*, Dr Maisonneuve referred to Verdelho as 'an interesting *cépage*...which at first glance has a certain resemblance to Chenin Blanc'.

*Château de la Roche-aux-Moines viewed from above the Coulée-de-Serrant vineyard at Savennières, with the river Loire on the left. The terroir of Savennières is far from homogeneous, but all the vineyards share a steep slope down to the Loire, and hot, dry soil.*

*The Moulin de la Montagne amid Chenin Blanc vineyards at Bonnezeaux. It's windier here than in the Coteaux du Layon vineyards nearer the river but, in spite of this, noble rot arrives virtually every year.*

*Chenin Blanc grapes being harvested in Vouvray. The comparative rarity of noble rot in this appellation means more dry wines, and medium-sweet wines that keep the varietal flavours of Chenin.*

# Viticulture and Vinification

LOIRE GROWERS OF CHENIN BLANC both love the vine and despair of it. 'It's an ungrateful vine,' says Florent Baumard of Domaine Baumard in Coteaux du Layon: you do everything for it, you control its yield, you cosset it and fuss over it, 'and then at the moment of harvest, it rots, or stops ripening.' Perhaps a warning should be sent to South Africa, telling Chenin aficionados just what they are in for if they fall in love with the vine. When it is ripe, and properly balanced, it is wonderful, but unripe Chenin is one of the nastiest wines possible, and high-yielding Chenin is dull and bland. It is by no means an all-purpose vine, although that is what it has been used for in the Cape, where only recently have a few exciting low-yield examples emerged from a sea of forgettable dross.

## Climate

France's Loire Valley is a marginal growing region for Chenin Blanc, which of course is one reason why it produces wines of such great finesse there. Go even a little way further north and the wines become not so much refined as acidic: Jasnières is the most northerly appellation for Chenin in France, and the wines here are distinctly thinner, or *pointu*, as the the French say. Further south, the vine is grown in small quantities in the Languedoc, a local enologist having recommended it as being suitable for the region.

In the Loire, the traditional problem has always been one of unripeness. In cool, wet years, Chenin there has a flavour of green apples and hard, unyielding acidity, raw and unpleasant. But times have changed: growers point to global warming as the cause of their generally improved climate, which sometimes enables them to pick ten days or so earlier than in the past and still have ripe grapes.

It is climate, too, far more than soil, that determines whether Loire Chenin is to be sweet or dry. Vouvray and Montlouis have much the same climate and make excellent fizz as well as bone dry, medium and sweet wines. Coteaux du Layon is less continental and more influenced by the Atlantic Ocean, and so gets more botrytis than Vouvray and Montlouis. Savennières is windy and less foggy, and thus less subject to botrytis; nevertheless, according to Mme de Jessey of Domaine du Closel, while at the beginning of August there is more sugar in the grapes in Coteaux du Layon than in Savennières, by 1 October the situation is reversed, and Savennières has the higher sugar.

The picture is complex, to say the least, and is complicated by the abundance of rivers, large and small, in the region, which means that humidity varies from one vineyard to the next. The climate also varies hugely from year to year.

## Soil

If it is climate that determines whether Loire Chenin is to be dry or sweet, it is the soil that gives the wine its style. Chenin reflects its soil as clearly as Riesling or Pinot Noir. Sandy soils give light wines which are attractive young, and which mature relatively early. Clay gives weightier, richer wines, and is conducive to the development of botrytis: Coteaux du Layon is rich in clay, with a chalk subsoil. Limestone gives wines with fine, scything acidity, wines from silex are vivacious, tingling on the tongue. Calcareous clay – or *argilo-calcaire* – produces perhaps the most rounded wines, with both acidity and weight: Vouvray has plenty of this soil, as well as some silex in the best sites. Savennières is one of the world's driest wines, and has dark bluish schist, fairly friable on top but very hard underneath, hot and well drained; there is schist and quartz, too, at Bonnezeaux.

## Yields

Chenin Blanc is vigorous enough to produce wine like water, if that is what you require of it. Even on the poorest soils of the Loire, it will give 80hl/ha unless checked; on richer soils it can do twice that. In South Africa Chenin is generally left to overproduce to its heart's delight. Though good producers claim to keep their crop down to 50hl/ha,

*Nicolas Joly of the Savennières estate, Coulée-de-Serrant, here seen with one of his Chenin Blanc vines, is a passionate, almost mystical believer in biodynamism, the system based on the early 20th-century teachings of Rudolf Steiner. His wines are some of the most individual Chenins in the world.*

greedy producers will produce at least three times that amount – and counting. Chenin at this level is simply not interesting: most of the problems of South African Chenin can be related to overcropping. The rest are probably caused by planting Chenin in regions too warm for it to give interesting wines.

For dry wines, 40–50hl/ha is a generally accepted yield if quality is the aim. (The legal maximum in Savennières, for example, is 50hl/ha.) For sweet wines, Bonnezeaux's legal maximum is 25hl/ha, and most growers get 20–25hl/ha; the same applies in Coteaux du Layon. At Domaine Font Caude at Montpeyroux in the Languedoc, Alain Chabanon grows it for sweet wine on poor *argilo-calcaire* soil, with yields of between 5 and 11hl/ha.

To keep yields to these levels a green harvest – i.e. a thinning of the crop in midsummer – may be necessary, as well as a choice of *riparia* or *rupestris* rootstocks to reduce vigour. Many estates also sow grass between the rows to reduce vigour.

Old vines, being less vigorous, also play their part. The average age of the vines at Château de Fesles in Bonnezeaux is 35 years, and at Domaine du Closel in Savennières there are some vines approaching their centenary. (Old vines give more complex botrytis flavours, too: botrytis on young vines gives floral flavours, but not much depth. With old vines, it's the opposite.) Curiously, phylloxera seems no longer to be a problem at Closel. Here, Mme de Jessey has some vines still on their own roots, propagated by the age-old technique of layering; Château d'Épiré, also in Savennières, practises the same method on some vines. So far neither domaine has found any sign of the dreaded louse returning.

*Gaston Huet (above), the grand old man of Vouvray, has passed on the running of his famous family estate, Le Haut Lieu, to his son-in-law Noël Pinguet, who has introduced biodynamic methods.*

### At the winery

One major factor distinguishing New World Chenin wines from Loire versions is the fermentation temperature. In the Loire temperatures of between 16 and 20°C are normal: that's quite warm for a white wine these days. But what the Loire growers actively want to avoid is the tropical fruit aromas and flavours produced by fermentation at colder temperatures: it is these flavours that mark New World Chenins.

Fermentation in the Loire may take place in stainless steel, glass fibre, cement, old wood or new wood, though the flavour of new oak is unwanted. Usually a proportion of the barrels (a fifth to a third) is renewed each year. In Savennières chestnut or acacia barrels are traditional: neither gives a vanilla flavour to the wine, though chestnut gives a slightly buttery taste, and the flavour of acacia is stronger. In the New World the flavour of new oak is considered more welcome, adding toastiness and spice.

However, Loire winemaking has changed greatly in the last decade. The producers acknowledge that what consumers want is less acidity and more fruit flavour in their wines. Clean Chenin grapes have shown themselves to be very amenable to skin contact – leaving the lightly crushed grapes to macerate in their own juice before fermentation starts. This brings out the greengage and angelica flavours. To reduce acidity some producers (by no means all) let the malolactic fermentation take place, thereby reducing acidity and fattening the texture of the wine, and many age the wine on the fine lees, stirring the lees and wine together by *bâtonnage*. Even dry wines may be made with a few grams of residual sugar.

Bottling times vary: Philippe Foreau in Vouvray seeks wines (both sweet and dry) with high acidity and freshness; he therefore avoids the malolactic fermentation and bottles in the spring following the vintage; at Château de Fesles in Bonnezeaux the wine normally spends up to 24 months in wood before bottling.

### TRI, TRI AND TRI AGAIN

One of the keys to making great Chenin in the Loire Valley is successive selected pickings, or *tries*. Over a period of a month to six weeks in the late autumn pickers must be sent through the vineyards picking perhaps whole clusters, but more likely individual grapes, at precisely the right degree of ripeness. If grey rot, rather than the desired noble rot, is invading the vineyards then affected grapes must be removed in separate pickings. And the decision on whether to pick today or wait for greater ripeness, whether to make a dry wine, a demi-sec or a sweet one, is made on a day-to-day basis and may change as the vintage progresses.

The Loire Valley's convoluted map of soil and climate means that Chenin Blanc never ripens all at the same time. Grapes from warm schist soils ripen before cooler clay ones; noble rot appears more regularly where the river mists creep up the hills at night, to be dispelled by sunshine during the day. In addition, most producers have the option of making wines at different levels of sweetness. Every year, in other words, they are trying to hit several moving targets.

The degree of the grapes' ripeness and the lateness of the season between them will make the grower decide how much dry wine to make. In a cool year in Vouvray a lot will go to make sparkling wine; in great years almost no dry wine at all is made. Philippe Foreau of Clos Naudin says that if your grapes aren't ripe enough for *triage*, then they're not ripe enough to make sweet wine. But here again years vary: in some years grapes of 12.5° potential alcohol will be ripe, in other years not.

Four to six pickings are normal for Vouvray, perhaps three for Savennières. The purpose of each picking depends on the year: growers taste their grapes every day. The first picking might be for dry wine; others might be for sweet. If, in a hot, dry year there is no botrytis the grapes may be picked *passerillé*, or shrivelled; botrytis may arrive later, or it may not.

Château de Fesles in Bonnezeaux reckons to pick one or two grapes per cluster on the first picking, the majority of the grapes in the second and third pickings, and the remaining one or two per cluster thereafter. The third and fourth pickings, they say, are generally the finest.

# Chenin Blanc around the World

WINE STYLES IN THE LOIRE are moving towards less acidity; elsewhere in the world the move towards quality is only just beginning. We don't know yet quite how versatile Chenin will prove to be, but examples from South Africa, Australia, New Zealand and the Americas suggest we could have a rising star on our hands.

## France: Loire Valley

The year that saw the beginning of today's revival of Chenin quality in the Loire Valley was 1985. In that year, when botrytis was plentiful, growers rediscovered the art of selection in the vineyard. France was in the middle of a golden decade, the sun shone on a regular basis, money was being made and a whole battalion of growers suddenly decided to put quality first. Thank goodness they did. Loire Chenin had been on the slide since the 1960s, and we were in real danger of losing one of France's unique wine styles.

Good growers aim for the greatest possible richness in their grapes; climate, along with vintage variations, soil and the exposure of the slope will determine whether the result will be sweet or dry, or in between. In Anjou and Saumur there is more chance of botrytis: in Bonnezeaux and Quarts de Chaume, the tiny sub-regions of Coteaux du Layon, botrytis can be relied upon to appear to some degree practically every year. Further up the river in Touraine, Vouvray and Montlouis get botrytis only about four years in ten. In non-botrytis years the sweet wines in these regions are made from overripe and shrivelled – *passerillé* – grapes, and this goes part of the way to explaining the difference in flavour between sweet Vouvray and Bonnezeaux or Quarts de Chaume.

Sweet Vouvray is much more likely to have the characteristic varietal flavour of Chenin: a flavour of apples, greengage and minerals, that matures to flavours of honey and acacia and quince. The sweet, botrytized wines of Anjou-Saumur are more likely to have more typically botrytized flavours of peach, barleysugar and marzipan. (*Botrytis cinerea* tends to destroy varietal character: see page 27.)

Soil makes a difference too: the calcareous clay and silex of Vouvray gives different styles to the schist of Savennières and Bonnezeaux. Early bottling, in the spring after the vintage, and the avoidance of the malolactic fermentation (both more likely in Vouvray) preserve freshness; sweet wines given long aging in wood before bottling are likely to have more rounded, Sauternes-like characters.

The sweet wines are seldom high in alcohol: 12–13 per cent is normal for Vouvray Moelleux, with the grapes having been picked at a potential alcohol of 18–20 per cent. If the alcohol level rises any higher, the wine runs the risk of becoming unbalanced and heady.

Among the dry wines, Anjou Blanc Sec and Saumur Blanc are the simplest and earliest maturing (and up to 20 per cent of Chardonnay or Sauvignon may be added to the blend for Saumur); Savennières and Vouvray Sec are the most complex and longest-lasting. The maximum alcohol in Savennières is 13.5 per cent: if in a very ripe year the wine has more than that then a special dispensation must be sought from the appellation contrôlée authorities.

## Other French Chenins

There is a little Chenin Blanc planted in Limoux in the western Languedoc, where it is blended with Mauzac and Chardonnay for sparkling wine, and elsewhere in the region, for example further east at Montpeyroux in the Coteaux du Languedoc appellation.

## South Africa

With nearly a quarter of the total vineyard area, Chenin Blanc is still by far the most widely grown variety in South Africa, even though the amount planted declined throughout the 1990s. Its traditional local name, Steen, is gradually going out of fashion as interested producers, led by the Chenin Blanc Producers' Association, set up in the late 1990s, begin to take the grape more seriously. Most is planted in the regions of Paarl and Worcester and in Swartland, around Malmesbury, and the majority – 89.8 per cent in 1998 – is crushed by the co-operatives. In the past much of the harvest ended up being distilled, but more recently it has started finding its way into bottle, though the flood of pale, fruitless – and cheap – Chenins that are being exported to hit a low price point are doing the reputations of South Africa and Chenin Blanc no good.

The first aims of the Chenin Blanc Producers' Association are to identify the best vineyards, many of which are very old, and to try and discover which soils and climates suit the vine best.

PATRICK BAUDOIN

*This Grains Nobles cuvée from Patrick Baudoin's estate is from 1996, an excellent year. The wine is very ripe and forward, with concentrated, honeyed flavours.*

DOMAINE DU CLOS NAUDIN

*Now run by Philippe Foreau, Clos Naudin has vines in some of the best Vouvray sites: the wines combine raciness with good weight, and age extremely well.*

DOMAINE DU CLOSEL

*Mme Michèle de Jessey of Domaine du Closel devotes the park around her château to hay for her cows, which in turn provide manure for her vines.*

*There is far more Chenin Blanc grown in South Africa (above) than in the Loire Valley, but it is only very recently that growers have started to take it seriously. Research into the best sites is only just beginning, but in all the Cape's 25,000ha of Chenin there must be some interesting spots. Surely. But as well as identifying these, South Africa has to evolve a style. It can't, and doesn't want to, make Vouvray lookalikes.*

Whatever styles emerge, they are unlikely to be copies of the Loire. South African Chenins are drinkable earlier and are not intended to be long-lived – yet. And anyway, South African growers do not necessarily admire the style of Loire Chenin. What the French see as finesse and subtlety the South Africans see as raw fruit and acidity. While South Africans are now putting some emphasis on bottle aging, they are still more likely to seek flavours of guava and banana, pear and pineapple than the sort of mineral tightness that takes ten years in bottle to open out.

### California

The majority of California's Chenin is planted in the Central Valley, and is treated as a bulk wine suitable for inexpensive blends. It lacks the tight acid structure of Loire Chenin, and when made with residual sugar, as it often is, the result is loose-knit and bland. A great deal has been uprooted: the total acreage is down to 21,410 acres/8665ha (1998) from over 40,000 acres/16,188ha in 1986. However, occasional good examples do appear. Clarksburg in the Sacramento Delta produces some stylish fruit, and there are odd good Chenins from elsewhere.

### Rest of North America

In Oregon, Chenin's 1990 figure of 44 acres/18ha has declined until it has vanished off the statistics. There is still some in Washington State, though here, too, it is in decline. Terminal? More than likely.

### Australia

Australia's 939ha of Chenin Blanc are widely scattered across the country. The vine generally produces wines with soft fruit-salady flavours; critic James Halliday describes the taste as 'tutti-frutti'. However, the Swan Valley and Margaret River in Western Australia have produced some much tauter, more impressive examples.

### New Zealand

The occasional good sweet wine here is an exception to the general rule that New Zealand's Chenin is a useful high-acid blending partner for Müller-Thurgau in inexpensive blends. Every vintage in New Zealand seems to produce one or two Chenins of such balance and beauty – all greengage and angelica, flecked with honey, sharpened up with lemon acidity – that you cry: why doesn't New Zealand produce more? Simple reason. You can sell Chardonnay and Sauvignon Blanc for a lot more money. So Chenin remains a very marginal variety, despite being very suitable for New Zealand's climate. In 1999 there were 154ha, mostly in the North Island.

### Rest of the world

Chenin is widely planted in South America and Mexico; there is also some in Israel. It is usually vinified carelessly, with cavalier use of sulphur, so it's a bit difficult to know whether it could be any good or not. I'd say it could be, with better winemaking.

**CLOS ROUGEARD**

*The oak-fermented Saumur Blanc from Clos Rougeard comes from old vines that give wines of great structure and richness that are not dominated by the oak.*

**NEDERBURG**

*The nobly-rotten Edelkeur from Nederburg in South Africa is one of the best examples of dessert Chenin Blanc from the Cape.*

**MILLTON VINEYARD**

*A weighty, quite complex, off-dry Chenin from New Zealand produced by biodynamic methods. These can, it seems, succeed even in warm, humid Gisborne.*

# Enjoying Chenin Blanc

SWEET LOIRE CHENIN BLANC IS one of the most long-lived of all wines. In Vouvray, the richly botrytized years of 1921, 1945, 1947, 1955, 1959 and 1976 are all drinking well now; a sweet Vouvray reaches maturity at around 12 years old, and will last for a century. Coteaux du Layon from Moulin Touchais, a family company that makes a speciality of older wines, seems to be at its best with 20 to 30 years in bottle, though a decade in bottle would be a more usual rule of thumb for those who want to drink top quality Coteaux du Layon at perfect maturity. As so often happens, it is often the years with the highest acidity that last the longest; you can measure their progress in generations rather than decades.

Demi-sec wines also need about ten years to really shine; again, this applies to wines from very good producers. Lighter, less concentrated wines can be drunk earlier. Good demi-sec Chenins will stay at their peak for 30 years or so.

The finest dry wines are hardly faster to mature. Good dry Vouvray needs a decade or so; even sparkling Vouvray can last a few years, though sparkling Saumur (like the fresh, fruity, but bone dry still Saumur) should be drunk young.

Savennières reaches its peak after ten years or so, and lasts for many, many years. When the producers of Savennières formed themselves into an association in the 1950s, one of their members volunteered to supply the wine for their first formal dinner, saying that 'I have enough of the 1851 available to go with the main course.'

However, Chenin does go through a closed phase. It is delicious for a few months after bottling, but then may retire into itself for seven or eight years before it begins to emerge with the beguiling flavours of maturity.

## The taste of Chenin Blanc

No other grape can imitate the flavours of Loire Chenin Blanc. In extreme youth the taste of crisp green apples is mixed with greengage, angelica and something earthy, chalky, minerally; the dry wine can taste pretty hard in youth, particularly once it begins to close up in bottle. But even in this least generous phase you can peer through the acid carapace and get a brief glimpse of complexities to come. And trust your instincts. They do come.

In most cases acidity is a little less aggressive than it used to be but that's still comparative. It may be softer acidity, made less assertive by the malolactic fermentation; it may just be the riper acidity of more mature grapes from lower yielding vines. *Bâtonnage* or lees stirring also helps to clothe the acidity in more creamy weight. So young Chenin is less unapproachable than it was. But this mineral thing is still there, and it doesn't go, in non-botrytized wines, even when maturity brings a flood of acacia and honey, brioche, quince and greengage.

Botrytized Chenin wines have less of the Chenin green apples, and more of the peach and pineapple, barleysugar, marzipan, quince and cream that comes with botrytis. They are generally less weighty and alcoholic than Sauternes, and they never seem to lose that piercing Chenin acidity, however sweet they are.

New World versions of Chenin are more tropical in flavour – all bananas and guava and pineapple – and less tight in structure, with less minerality.

*In Vouvray, 1997 was a year of plentiful botrytis, so Huet's Le Haut Lieu Moelleux will have extra layers of complexity. It will need at least ten years in bottle to mature. Pierre Soulez' Château de Chamboureau comes from La Roche-aux-Moines, a 33-ha sub-appellation within Savennières, where the exposure to the sun is especially good.*

### Matching Chenin Blanc and food

Chenin Blanc makes wines ranging from averagely quaffable dry whites to the great sweet whites of the Loire Valley. The lighter wines can be good as apéritifs or with salads as well as with light fish or chicken dishes. The medium-sweet versions usually retain enough of their acidity to counteract the richness of pâté and creamy chicken and meat dishes such as the Loire speciality of pork with prunes. The sweet wines are good with most puddings and superb with those made with slightly tart fruit. They are also marvellous with fresh fruit, foie gras or blue cheese.

## CONSUMER INFORMATION

### Synonyms & local names

Traditionally called Pineau or Pineau de la Loire in the Loire; Steen has long been its traditional name in South Africa but Chenin Blanc is becoming more usual as the grape achieves higher status; called Pinot Blanco in parts of South America and in Mexico.

### Best producers

**LOIRE VALLEY/Bonnezeaux** de Fesles, Godineau, des Grandes Vignes, Laffourcade, Petits Quarts, Petit Val, René Renou, de la Sansonnière, de Terrebrune, la Varière; **Coteaux de l'Aubance** Jean-Yves Lebreton/des Rochelles, V & V Lebreton/de Montgilet, C Papin/de Haute Perche, Richou; **Coteaux du Layon** Pierre Aguilas, Patrick Baudoin, Baumard, de la Bergerie, du Breuil, Cady, Pascal Cailleau, Philippe Delesvaux, des Forges, Fresne, des Grandes Vignes, Jolivet, Maurières, Ogereau, de Passavant, Petit Val, Pierre-Bise, Jo Pithon, Joseph Renou, de la Roulerie, des Sablonettes, du Sauveroy, Soucherie, Yves Soulez, Touche Noire, la Varière; **Montlouis** Berger, Chidaine, Deletang, Dominique Moyer, Taille aux Loups; **Quarts de Chaume** de Baumard, Bellerive, Laffourcade, de Maurières, Pierre-Bise, Jo Pithon, Joseph Renou/du Petit Metris, Rochais/de Plaisance, Pierre Soulez, la Varière; **Saumur** Clos Rougeard, du Hureau; **Savennières** des Baumard, Clos de Coulaine, Clos de la Coulée-de-Serrant, Clos des Maurières, Closel, d'Épiré, des Forges, Laffourcade, Laroche/aux Moines, de Plaisance, Pierre Soulez, Yves Soulez, Pierre-Yves Tijou; **Vouvray** des Aubuisières, Bourillon-Dorléans, Marc Brédif, Champalou, Clos Baudoin, Clos Naudin, de la Fontainerie, Gaudrelle, Huet, François Pinon.
**SOUTH AFRICA** Flagstone, Kanu, Mulderbosch, Nederburg, Spice Route.
**NEW ZEALAND** Collards, Esk Valley, Millton Vineyard.
**USA/California** Chalone, Chappellet, Dry Creek Vineyards, Husch, Pine Ridge; **Washington State** Hogue Cellars, Kiona, Paul Thomas.

## RECOMMENDED WINES TO TRY

### Ten sweet Loire wines

**Patrick Baudouin** *Coteaux du Layon Sélection des Grains Nobles*
**Clos Naudin** *Vouvray Moelleux Réserve*
**Domaine Philippe Delesvaux** *Coteaux du Layon Cuvée Anthologie*
**Ch. de Fesles** *Bonnezeaux*
**Jo Pithon** *Coteaux du Layon St-Aubin Clos des Bois*
**René Renou** *Bonnezeaux Les Melleresses*
**Domaine Richou** *Coteaux de l'Aubance Cuvée les Trois Demoiselles*
**Domaine de la Sansonnière** *Bonnezeaux*
**Ch. Soucherie** *Coteaux du Layon la Tour*
**Domaine de la Taille aux Loups** *Montlouis Cuvée des Loups*

### Ten medium-sweet Loire wines

**Domaine des Aubuisières** *Vouvray les Girardières Demi-sec*
**Didier Champalou** *Vouvray*
**Clos Naudin** *Vouvray Demi-sec*
**Huet** *Vouvray Clos du Bourg Demi-sec*
**Domaine des Liards** *Montlouis Vieilles Vignes Demi-sec*
**Gaston Pavy** *Touraine Azay-le-Rideau*
**Ch. Pierre-Bise** *Anjou le Haut de la Garde*
**Domaine Richou** *Coteaux de l'Aubance Sélection*
**Taille aux Loups** *Montlouis Demi-sec*
**Ch. Gaudrelle** *Vouvray Réserve Spéciale*

### Ten classic dry Loire wines

**Dom. des Aubuisières** *Vouvray le Marigny Sec*
**Dom. des Baumard** *Savennières Clos de la Bergerie*
**Bourillon-Dorléans** *Vouvray Coulée d'Argent Sec*
**Clos de la Coulée-de-Serrant** *Savennières Coulée-de-Serrant*
**Clos Rougeard** *Saumur Blanc*
**Dom. du Closel** *Savennières Clos du Papillon*
**Domaine Deletang** *Montlouis les Batisses Sec*
**Huet** *Vouvray le Haut Lieu Sec*
**Ch. du Hureau** *Saumur Blanc*
**Joël Gigou** *Jasnières Cuvée Clos St-Jacques*

### Five sparkling wines

**Dom. de l'Aigle** *Crémant de Limoux non-vintage*
**Clos Naudin** *Vouvray Pétillant Vintage*
**Gratien & Meyer** *Crémant de Loire Brut non-vintage*
**Huet** *Vouvray Mousseux Vintage*
**Langlois-Château** *Crémant de Loire Quadrille Brut non-vintage*

### Five New World wines

**Chalone Vineyard** *Chenin Blanc (California)*
**Chappellet** *Napa Valley Old Vine (California)*
**Kanu** *Chenin Blanc Wooded (South Africa)*
**Millton** *Te Arai Chenin Blanc (New Zealand)*
**Mulderbosch** *Steen-op-Hout (South Africa)*

*When Botrytis cinerea attacks white grapes like this Chenin Blanc the berries first go mauve-brown in colour and then begin to shrivel. At that point good growers will send the pickers through the vineyards for a first tri, with instructions to pick only the shrivelled berries.*

### Maturity charts

Chenin Blanc from the Loire Valley can be immensely long-lived. It can also be somewhat unfriendly in youth.

**1997** Savennières (dry)

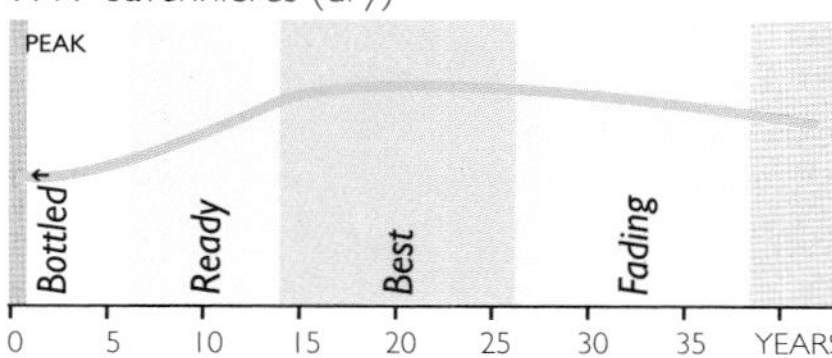


*An excellent year in Savennières. Many wines are being made to be approachable earlier than they used to be; this chart applies to traditional styles.*

**1997** Quarts de Chaume

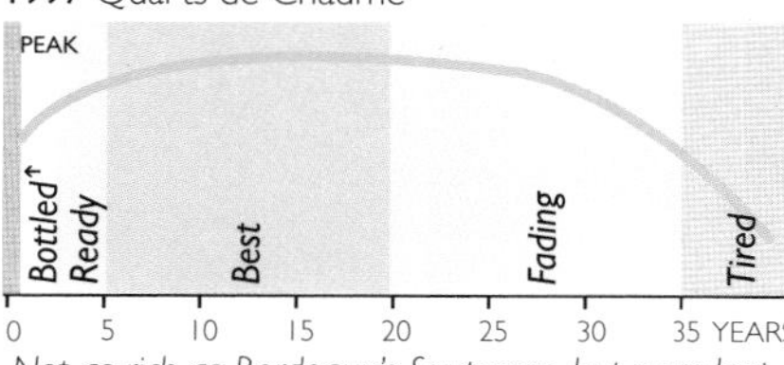


*Not as rich as Bordeaux's Sauternes, but succulent and sweet with a thrilling fruity acid that keeps them fresh and exciting for two decades at least.*

**1999** South Africa old vine

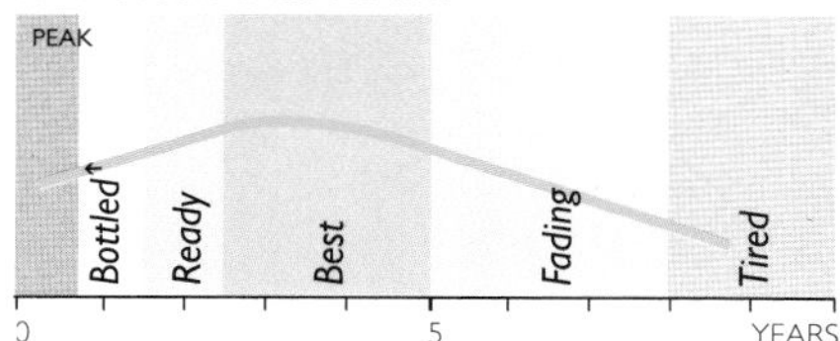


*Chenin dominates South Africa's vineyards and the wine is generally nondescript. However, old bush vine Chenin fermented in barrel can be lovely, toasty stuff.*

## CHIAVENNESCA

The name for Nebbiolo in the Valtellina region of Lombardy, northern Italy. See pages 154–163.

## CIENNA

A new red crossing produced in Australia and currently being grown experimentally by Yalumba at Wrattonbully, near Coonawarra in South Australia. Its parents are Cabernet Sauvignon and the Spanish Sumoll and first results are quite tasty.

## CINSAUT

This vine's reputation for high quantity and poor to middling quality is only partly deserved. It can certainly give high yields, but at low yields it can produce characterful, lush, sweetly rich wines which play a part in such star names as Lebanon's Chateau Musar. Here yields are as low as 25hl/ha. The Languedoc, too, can produce Cinsaut with finesse, providing that yields are kept down and the wine is given a long maceration with the skins. It is aromatic in youth, with soft, supple fruit, and is often used to calm down the tougher Carignan. But if allowed to yield heavily its quality falls rapidly. It is classed as a *cépage améliorateur* or improving variety, but lacks the prestige of other grapes used in a similar way such as Syrah or Mourvèdre. On its own it can make attractive rosé.

It buds relatively late and is susceptible to mildew and oidium, though in Algeria it has proved more resistant than Aramon to drought and drying winds like the Sirocco. It has long been a popular grape in North Africa, and is the main variety in Corsica.

In South Africa its popularity has been eclipsed by its offspring, Pinotage (see pages 186–187), the result of crossing Cinsaut (known locally as Hermitage) with Pinot Noir (see pages 174–185). It now accounts for just over 4 per cent of the South African vineyard area, and is used for fortified as well as table wines. As a table grape it can be found in France under the name of Oeillade: the berries are apparently judged too small to be of interest to export markets. Best producers: (France) de l' Amarine, de Caraguilhes, Clos Centeilles, Mas de Daumas Gassac, Mas Jullien, de Pech-Redon, Val d'Orbieu, Vignerons Catalans.

CHATEAU MUSAR

*This blend of Cabernet Sauvignon, Cinsaut and Syrah is not only Lebanon's most famous wine, but a world-class classic.*

## CLAIRETTE

Once the great standby of southern French white blends, Clairette seems to have had its day. It doesn't fit the modern idiom: it is too unstructured, too quick to oxidize, too high in alcohol and too low in acidity. And that is quite apart from the unfortunate fact that it happens to be a white grape in an era made for red wine.

It still plays a part, albeit a shrinking one, in many wines, however. On its own it makes Clairette de Bellegarde and Clairette du Languedoc, the second of which can be dry, sweet or *rancio*. It appears with Muscat Blanc à Petits Grains in the Clairette de Die appellation, and pops up in Châteauneuf-du-Pape, Côtes du Rhône, Côtes de Provence, Cassis, Bellet, Palette and many other southern appellations, often along with higher-alcohol grapes like Ugni Blanc. In the right terroir and at yields below 50 hl/ha it can make interesting wine: fat and perhaps a little heavy, and with 14 per cent alcohol, but with a certain musky attraction. Modern winemaking, too, can hold back its tendency to oxidize. But at high yields of 100hl/ha or more it is the sort of wine that gave the Midi a poor reputation.

In South Africa, where in 1999 it took up 1.3 per cent of the vineyard, it is regarded as a low-alcohol variety, presumably because yields are high. It is usually blended. The Blanquette of Australia's Hunter Valley is Clairette, and should not be confused with the Blanquette variety found in the Languedoc region of Limoux, which is Mauzac. Clairette is also a synonym of several other grapes in the south of France, including Ugni Blanc (known as Clairette Ronde in the Languedoc) and Bourboulenc. There is also a Clairette Gris, which has a pink tinge to its skins. Best producers: (France) Achard-Vincent, Die co-op, Faure, de Magord.

## CLEVNER

Clevner (or Klevner) is the Alsace name for members of the Pinot family, usually Pinot Blanc. In Switzerland's Zurich region Klevner is the local name for Pinot Noir. See pages 174–185.

## COLOMBARD

The French variety Colombard seems to have been the result of a crossing of Gouais Blanc and Chenin Blanc, and to have originated in the Charente region north of Bordeaux, where it was long made into white wine, and later distilled into Cognac. The slowdown of interest in brandy worldwide has meant that many of the grapes are now only wanted for making into still wine – which Colombard is in fact much better suited for than brandy since it is higher in alcohol and lower in acidity than the usual brandy grape, Ugni Blanc (the best brandies are from neutral, low alcohol, high acid wines).

Colombard is particularly effective in contributing peach and nectarine fruit and citrus lemon perfume to Vin de Pays des Côtes de Gascogne (from the Armagnac region) and, to a lesser extent, Vin de Pays Charentais (from Cognac). It is also planted in other parts of south-west France, and pops up in many white blends – in lesser Bordeaux regions such as Blaye, Colombard actively improves the perfume and acidity of whites.

Colombard is a useful, warm-climate grape, too, but few producers have maximized its potential. Even so, it's on the rise in South Africa – from 6.7 per cent of the vineyard area in 1985 to 11.2 per cent in 1999. It maintains its acidity well and gives good crops of fresh, grapefruit and peach-flavoured whites for early consumption. One or two good examples have also appeared in Australia.

California's Colombard, known as French Colombard, used to be prized for its good acidity in hot conditions, but it is in decline there. Until 1991 it was the state's most widely planted white vine (in 1999 this was Chardonnay with 253,442 acres (102,568ha) as opposed to Colombard's 110,165 acres (44,584ha). Most but not all plantings are in the Central Valley. Yields are very high – up to 12 tons per acre (210hl/ha) – and the wine is usually blended into everyday jug wines. Best producers: (France) Jean Aubineau, Meste-Duran, du Tariquet, Producteurs Plaimont; (South Africa) Graham Beck, Longridge, Robertson Winery, Swartland Wine Cellar.

## COMPLEXA

A grape found on the Portuguese island of Madeira, where it was planted in the 1960s. Used for table wine, it gives wine that is darker in colour than the more widely planted Tinta Negra Mole, and less astringent, but it is very susceptible to rot.

## CONCORD

This *Vitis labrusca* grape was the one that coaxed many North Americans into drinking wine, until the California wine revolution introduced them to Chardonnay in the 1980s. It is grown widely in the north-eastern states, particularly New York State, in Canada and to some extent in Brazil. It is better than most *Vitis vinifera* varieties at withstanding harsh winters (conditions in Brazil are unfavourable in the opposite direction), but has the strangely aromatic flavour typical of *labrusca* vines. This flavour is usually referred to as 'foxy', but is in fact closer to the smell of mayblossom and nail varnish, and is a long way from any wine aroma familiar to drinkers of *vinifera* wines.

The vine got its name from Concord, Massachusetts, where one Ephraim W Bull planted the seeds of a wild vine in 1843. It is used for grape juice and jelly as well as for wine, which is often very sweet.

## CORTESE

In spite of the high prices charged for Cortese's most famous wine, Gavi, this Italian grape is seldom more than pleasant. It never reaches top quality, but it does have the virtue of retaining its acidity even in hot summers, which makes it a good bet in its homeland, north-west Italy. Plantings are concentrated in Piedmont, and spill over the border into Lombardy. Gavi from the town of Gavi, formerly called Gavi di Gavi and now Gavi del commune di Gavi is sometimes, though not necessarily, better than plain Gavi. In other parts of Piedmont and Lombardy it produces DOC wines including Cortese del Alto Monferrato and Colli Tortonesi; it is also part of the blend in Bianco di Custoza.

At its best it has good body and a nose of limes and greengages, but yields must be kept down if the wine is to have sufficient body to balance the acidity. In cool summers the acidity dominates and the wine can be unattractively lean. Some producers use barrique fermentation or put a proportion of the wine through the malolactic fermentation in an attempt to rectify this. Best producers: (Italy) Banfi, Gian Piero Broglia, Chiarlo, La Giustiniana, Franco Martinetti, La Scolca, Castello di Tassarola, Villa Sparina.

## CORVINA

Corvina is the mainstay of Valpolicella and Bardolino, both of them light red wines from north-eastern Italy. It will therefore come as no surprise to learn that Corvina seldom has much colour or tannin, but what it does bring to the party are aroma and acidity. When it is overcropped, as it all too often is, its wines are insubstantial and poor quality, as any drinker of inexpensive Valpolicella will testify. But on good hillside sites, and on the right soil – volcanic *toar*, which gives the most perfumed wines, or chalk, or alluvial – it can produce wines of considerable interest and floral, cherryish perfume. Controlling yields can be a problem, because the first few buds on the cane don't fruit; it therefore needs a long cane, and pergola or *spalliera* training, to produce a crop at all.

Corvina really comes into its own as a grape for drying to make *recioto* and *amarone* wines. It has small berries with thick skins, the latter helping to protect the grapes from rot while they shrivel. Best producers: (Italy) Accordini, Allegrini, Bertani, Boscaini, Brunelli, Tommaso Bussola, Dal Forno, Masi, Quintarelli, Le Ragose, Le Salette, Speri, Tedeschi, Tommasi, Zenato.

## CORVINONE

It is not clear whether Corvinone is a sub-variety of Corvina (see above) or a different variety altogether. Certainly the two varieties are thoroughly mixed up in the Valpolicella vineyards in north-eastern Italy, with many growers using the name of Corvina for both. But there are differences: Corvinone has larger berries (as the name suggests), more colour and tannin and reaches higher sugar levels, and it does not have Corvina's problem of infertile buds at the base of the cane.

## COT

Cot is the proper French name for Malbec, see pages 118–119.

## COUNOISE

A southern French grape that plays a small but valued part in many reds of the southern Rhône, Provence and Languedoc. Its close resemblance to Aubun makes it difficult to identify in the vineyard, though the wine is of better quality. It has a peppery, spicy character, damson fruit and good acidity, though not great colour or tannin, and it is tricky to grow, being low yielding and late ripening. But it is disease-resistant and adds body and fruit to blends. Best producers: (France) de Beaucastel, du Mas Blanc, Romanin.

## CRIOLLA GRANDE

Low-quality, pink-skinned grape grown in Argentina for the most basic white wine. It was one of the first vines to be planted, probably from seed, in the Americas by the first European settlers and it still covers huge areas, especially in Mendoza. There is also a lighter-skinned vine called Criolla Chica in Argentina; this is the same as the País of Chile, and the Mission of California.

## CROATINA

The attractive, soft, plummy flavour of Croatina is the simplest thing about it. Sorting out what Croatina actually is is more complex. It grows in north-western Italy, and plays a part in Gattinara and Ghemme, alongside Nebbiolo (alias Spanna) and Bonarda. That's the easy bit. In Oltrepò Pavese (Lombardy) and Colli Piacentini (Emilia), however, it takes the name of Bonarda (see page 43). The wine is frequently juicy, perfumed and delightful.

## CROUCHEN

A neutral French grape now abandoned by France, but still constituting 3 per cent of South Africa's vineyards. It is concentrated in the Paarl and Stellenbosch regions and at its best the wine is fairly steely. It is known there as Cape Riesling, South African Riesling or Paarl Riesling. (Proper Riesling is called Weisser or Rhine Riesling in South Africa.) Crouchen has almost vanished from Australia, where it was called Clare Riesling until 1976 when its identity was established by ampelographer Paul Truel.

## CYGNE BLANC

Cygne Blanc is a white seedling of Cabernet Sauvignon, discovered by grower and winemaker Dorham Mann in his garden in Western Australia's Swan Valley region – hence, of course, the name, *cygne* being the French for 'swan'. Mann sought A$23m to develop the vine commercially and we wait with bated breath to see what happens.

## DELAWARE

American hybrid named after the Ohio town where it was first propagated in 1849. Its flavour is less foxy than that of Concord, and it ripens early, which makes it useful in New York State and Japan. In both these places it is widely planted.

## DIMIAT

Bulgarian grape that covers large areas in the south and east of that country. Its wine is aromatic and simple, and usually off-dry to sweet. It is said to be named after a town in the Nile Delta, from where it was taken to Thrace at the time of the Crusades.

# Dolcetto

'Little sweet one' is what Dolcetto means in Italian, but, as I try to find a Dolcetto which fits its description, I often feel like a mother shaking her head in long disapproval at a naughty son who is anything but a 'little sweet one'. Dolcetto wine *can* have all the sweetness and delightful winsomeness of a mischievous favourite son, but such examples are anything but common, and although it should produce the bright refreshing everyday reds of Piedmont, tannin, acidity and coarseness get in the way surprisingly often. Ideally Dolcetto should have moderate acidity, unintrusive tannin (at any rate, compared to Nebbiolo, Piedmont's most famous grape), a distinctive suggestion of orchard blossom perfume, and an appetizing bitter-sweet twist at the end. And it should be drunk at a year or two, when it is hopefully still brimful of fruit. There are producers who make a more tannic, richer, oakier style – but in Piedmont at the moment there are producers who try to make grand statements out of every grape they can find.

Dolcetto isn't really suited to the prestige treatment – it just isn't that sort of grape, and traditionally it has known its place. Because it is not as prestigious as Barbera or Nebbiolo – and also because it ripens a couple of weeks before Barbera, and up to four weeks before Nebbiolo – it gets planted in the cooler sites not suitable for the other two – OK in a good year, pretty iffy in a poor one. It is regarded as being easy to grow in Piedmont and consequently few growers lavish much attention on it. Dolcetto is traditionally best and most characterful in the Ovada and Alba zones and Alba is now the chief production area.

Decent Dolcetto is quite common. Exciting Dolcetto depends on talented and determined winemaking. When Ovada's greatest Dolcetto maker died, no-one else was able to pick up the baton. There's a little Dolcetto elsewhere in Italy – Liguria, for instance, calls it Ormeasco and produces some interesting examples. Savoie calls it Douce Noire, and it produces deep, dark, chunky, chocolaty stuff in California under the soubriquet Charbono.

Otherwise only Australia seems to have any – and we're talking just a few vines – but since the oldest ones go right back to the 1860s we should take them seriously, because they are probably the oldest Dolcetto plantings in existence in the world.

**The taste of Dolcetto**

Cherry flavours are typical of Dolcetto: ripe black cherries on the nose and palate, and bitter cherries on the finish for that characteristic Italian twist. But there can be flavours of prunes as well, and licorice. If you're lucky you may also find a wine with intriguing perfume, but, in spite of its name, Dolcetto wines are dry. The grapes are not even notably high in sugar.

*Aldo Vajra's single-vineyard Dolcetto d'Alba Coste e Fossati is as concentrated as Dolcetto gets – this is a Dolcetto to be taken seriously. It's fascinating and wild when young, but benefits enormously from three to four years' aging in bottle.*

Marcarini

*Marcarini's punchy, powerful Dolcetto d'Alba comes from 100-year-old vines planted, unusually, on their own roots. The wine is unoaked – and is all the better for it.*

Quinto Chionetto

*An outstanding example of Dolcetto from the Dogliano zone of Piedmont. This single-vineyard wine comes from low-yielding, old vines.*

***Left:*** *Harvesting Dolcetto in the vineyard of Aldo Conterno at Bussia near Monforte d'Alba in Piedmont. Dolcetto is generally regarded as the third-best red variety in Piedmont, after Nebbiolo and Barbera. And as it is earlier ripening than either, it's often used on sites too cool for the top two.*
***Above:*** *Dolcetto grapes are tremendously dark in colour, and need only a short maceration on the skins to produce equally dark wine. This short maceration is the reason why the wines are light in tannin: the grapes have as much tannin as any other, should a winemaker wish to extract it. Dolcetto is now being seriously studied for planting in Australia, and it will be fascinating to see what the Australians do with it if they go for it in a big way.*

## CONSUMER INFORMATION

**Synonyms & local names**
Known as Ormeasco in the Riviera Ligure di Ponente zone in western Liguria. The French region of Savoie calls it Douce Noire and California Charbono.

**Best producers**
**ITALY/Piedmont/Dolcetto d'Acqui** Viticoltori dell'Acquese, Villa Sparina; **Dolcetto d'Alba** Alario, Altare, Ascheri, Azelia, Enzo Boglietti, Bongiovanni, Brovia, Cà' Viola, Cigliuti, Domenico Clerico, Aldo Conterno, Giacomo Conterno, Conterno-Fantino, Corino, Gastaldi, Ettore Germano, Bruno Giacosa, Elio Grasso, Marcarini, Marchesi di Gresy, Bartolo Mascarello, Giuseppe Mascarello, Moccagatta, Fiorenzo Nada, Oddero, Armando Parusso, Pelissero, Ferdinando Principiano, Prunotto, Renato Ratti, Albino Rocca, Bruno Rocca, Luciano Sandrone, Scavino, La Spinetta, G D Vajra, Eraldo Viberti, Vietti, Vigna Rionda, Gianni Voerzio, Roberto Voerzio; **Dolcetto d'Asti** Brema; **Dolcetto di Diano d'Alba** Alario, Bricco Maiolica, Fontanafredda; **Dolcetto di Dogliani** M & E Abbona, Francesco Boschis, Quinto Chionetti, La Collina, Del Tufo, Devalle, Luigi Einaudi, Gillardi, Marenco, Pecchenino, Carlo Romana, San Fereolo, San Romano, Schellino, Giovanni Uria; **Dolcetto di Ovada** La Guardia, Giuseppe Ratto/ Cascina Scarsi, Annalysa Rossi Contini, Terre da Vino; **Liguria** Lupi, Lorenzo Ramò. **USA/California** Duxoup, Kent Rasmussen.

### RECOMMENDED WINES TO TRY

**Ten Dolcetto d'Alba/Dolcetto di Diano d'Alba wines**
Alario *Dolcetto di Diano d'Alba Costa Fiore*
Brovia *Dolcetto d'Alba Solatio Brovia*
Ca' Viola *Dolcetto d'Alba Barturot*
Conterno-Fantino *Dolcetto d'Alba Bricco Bastia*
Marcarini *Dolcetto d'Alba Boschi di Berri*
Armando Parusso *Dolcetto d'Alba*
Pelissero *Dolcetto d'Alba Augenta*
Albino Rocca *Dolcetto d'Alba Vignalunga*
Luciano Sandrone *Dolcetto d'Alba*
G D Vajra *Dolcetto d'Alba Coste e Fossati*

**Ten Dolcetto di Dogliani wines**
Marziano & Enrico Abbona *Dolcetto di Dogliani Papa Celso*
Francesco Boschis *Dolcetto di Dogliani Vigna dei Prey*
Quinto Chionetti *Dolcetto di Dogliani Briccolero*
Antonio Del Tufo *Dolcetto di Dogliani Vigna Spina*
Luigi Einaudi *Dolcetto di Dogliani Vigna Tecc*
Gillardi *Dolcetto di Dogliani Vigna Maestra*
Pecchenino *Dolcetto di Dogliani Sirì d'Jermu*
Carlo Romana *Dolcetto di Dogliani Bric dij Nor*
San Fereolo *Dolcetto di Dogliani San Fereolo*
San Romano *Dolcetto di Dogliani Vigna del Pilone*

**Five other Dolcetto wines**
Duxoup *Napa Valley Charbono (California)*
Lupi *Riviera Ligure di Ponente Ormeasco Superiore Le Braje (Italy)*
Lorenzo Ramò *Riviera Ligure di Ponente Ormeasco (Italy)*
Kent Rasmussen *Napa Valley Dolcetto (California)*
Villa Sparina *Dolcetto d'Acqui Bric Maiola (Italy)*

### DOÑA BLANCA

Iberian grape grown (as Doña Branca) in the north of Portugal, and in Galicia across the border in Spain. It forms part of the white port blend but is also grown for unfortified wines which are strongly perfumed.

### DORNFELDER

Some of Germany's most attractive red wines are made wholly or in part from Dornfelder. True, they may not be great wines in the world-class league, but they are more often than not far preferable to over-ambitious and over-oaked attempts at Pinot Noir. There's a certain honesty about Dornfelder: it doesn't pretend to be more than a well-coloured, juicily-fruited grape for short- to medium-term drinking, and it fulfils that role very well. Plantings in Germany are concentrated in the Pfalz, Rheinhessen and Württemberg, where even at high yields of 120hl/ha it gives decent colour. Producers who want to age their Dornfelder in oak and make a wine that will age for a few years will opt for lower yields than this.

Dornfelder is the 1955 offspring of Helfensteiner and Heroldrebe, which are themselves crossings bred in Germany in the 20th century. Dornfelder's pedigree is exceedingly complicated, and includes all of Germany's major red vines; it is also the parent of several new crossings, bred at Württemberg and introduced in 1999: Acolon is Lemberger x Dornfelder; Cabernet Dorsa is Dornfelder x Cabernet Sauvignon, and Cabernet Dorio is another Dornfelder-Cabernet Sauvignon crossing.

Dornfelder is also found in England, where its wines are often blended with Pinot Noir and, as well as decent rosé, it gives some of England's few full-flavoured reds. Best producers: (England) Chapel Down, Denbies, Valley Vineyards; (Germany) Graf Adelmann, Drautz-Able, Lingenfelder.

### DURAS

Deep-coloured grape found in south-west France, where it is blended with such grapes as Négrette and Fer. The wine is peppery and structured. Plantings are slowly increasing.

### DURIF

A Dr Durif propagated this vine in the 1880s, and it was originally grown in the south of France for its resistance to downy mildew – though not for its quality in any other respect. It produces coarse, rustic red wine and has virtually disappeared from French vineyards.

In recent years it has been most famous as the possible alter ego of California's Petite Sirah, though DNA fingerprinting has shown that the two varieties are different, with Durif being a crossing of Peloursin and Syrah. The confusion in California seems to date from the 1920s, when officialdom lumped together Durif, Petite Sirah, Syrah and several other vines under the name of Petite Sirah. Much of the fruit called 'Petite Sirah' in California is in fact Durif, and suffers from rot here far less than it does in its native France. See Petite Sirah page 168 for more details of this vine in California.

In Australia Durif is grown under its own name, and produces dry, solid, four-square wines in warm climates that supposedly age for ever – though they're so impenetrable to start with, what they evolve into I've never been able to wait long enough to see. Best producers: (Australia) Campbells, Morris.

### DUTCHESS

An American hybrid found in New York State and in Pennsylvania. It makes rather 'foxy'-tasting wine, and only a few hundred acres are grown.

### EHRENFELSER

German crossing bred in 1929 from Riesling and Silvaner. It was intended, like so many such crossings, to have all the advantages of Riesling (elegance, finesse, complexity, longevity) with a few more thrown in for luck, mainly earlier ripening and higher yields. The wine is actually fairly good, though acidity is low and quality is not nearly as high as that of Riesling. It made some impact in the Pfalz and Rheinhessen regions, but its success was never notable.

### ELBLING

Elbling dominated the vineyards of medieval Germany and no doubt produced wines that were every bit as painfully acidic as they are today. Nowadays it is found mostly in the uppermost reaches of the Mosel Valley and over the border in Luxembourg, where both a red and a white version are known. Its main use is for sparkling wine. It yields generously but reaches only low sugar levels.

### ENCRUZADO

The principal white grape of the Dão region, in Portugal. Growers are experimenting with barrel fermentation and lees stirring to bring out its character and a little oak for aging seems to suit it rather well. At its best the wine is quite elegant, with some leafy aroma, and

*Dornfelder does well in cooler climates and many of England's better red wines owe a lot to this variety, seen here growing at Denbies Vineyards in Surrey.*

excellent balance. Best producers: (Portugal) José Maria da Fonseca, Quinta dos Roques, Quinta de Saes, Sogrape.

## ERBALUCE

The earliest written record of Erbaluce in Piedmont is in 1606, and the vine seems to have originated in the alpine foothills here. Its piercing acidity makes it ideal for sweet wines, and Caluso Passito, made from dried grapes, is its finest incarnation. There is also some dry wine called Erbaluce di Caluso, but this has to be very ripe to combat the acidity. It's a good, often very interesting grape, with attractive appley fruit. Best producers: (Italy) Antoniolo, Luigi Ferrando, Orsolani.

## ERMITAGE

Northern Rhône synonym for Marsanne (see pages 124–125).

## ESGANA CÃO

The Sercial of Madeira is grown in many parts of Portugal, and on the mainland is generally known as Esgana Cão. Its unfortunate effect on canines – the name means 'Dog strangler' – is thought to be the result of its high acidity. It is found in Vinho Verde, Bucelas, where it is usually blended with Arinto, and the Douro Valley. See Sercial page 240.

## ESPADEIRO

A source of light red Vinho Verde in northern Portugal. The Espadeiro grown further south around Lisbon is, in fact, Tinta Amarela.

## FABER

Also known as Faberrebe. A generally uninspiring crossing, either of Weissburgunder (Pinot Blanc) and Müller-Thurgau or of Silvaner and Pinot Blanc, bred in the 1920s and favoured in Germany's Rheinhessen region for its high sugar levels, high acidity and ability to ripen in cooler sites than those needed by Riesling. It nevertheless lacks the cardinal virtues of flavour and character, and is now on the decline.

## FALANGHINA

High quality but little planted Italian variety found in Campania. It makes Capri's DOC Bianco, can make good sweet *passito* wines, and may have been the grape responsible for Falernian, one of the most prestigious wines of the ancient world. It seems to be attracting more notice lately. Best producers: (Italy) Di Majo Norante, Mastroberardino, Terredora di Paolo, Villa Matilde.

## FAVORITA

Favorita, with its large berries, can double as a table grape, though now its popularity as a wine grape seems to be gently increasing again, the table might have to do without. Its home is in Piedmont, where it grows in the Roero and Langhe zones.

Its wine is well-structured and with good acidity, but without much aroma except on the rare occasions it evokes a fleeting memory of pears. At its weightiest and ripest it resembles good Vermentino, a variety to which it is sometimes said to be related, even to the point of being the same vine. However, the Italian Ministry of Agriculture in 1964 determined that there were substantial differences between the two in leaves, buds and clusters, and that therefore they were entirely separate varieties. It is late ripening and is sometimes blended with Nebbiolo to soften the latter. Best producer: (Italy) Gianni Gagliardo.

## FENDANT

Chasselas is known by this name in Switzerland's Valais region. See page 59.

## FER

Fer, or Fer Servadou, lends its perfumed, redcurrant fruit to a variety of blends in south-west France. It is a minority grape almost everywhere, though plays a more substantial part in the Marcillac, Entraygues and Estaing vineyards. It is also found in Madiran, though only to a small extent. It is actually quite a good grape, with concentration and character, and for once it's not on the decrease. In fact its suppleness means that it is useful in softening Tannat, and making the latter's wines approachable earlier.

There is a vine called Fer grown in Argentina, but it is believed to be a clone of Malbec. Best producers: (France) du Cros, Producteurs Plaimont, Vallon co-op.

## FERNÃO PIRES

A fairly aromatic and very versatile variety found all over the country: in the Bairrada region it is known as Maria Gomes. It is probably Portugal's most planted white grape, and can make anything from sparkling wines to still dry ones to botrytized sweet ones, and can be successfully oak aged provided the oak isn't overdone. The wine is best drunk young and some examples tire within the year. Best producers: (Portugal) Quinta da Boavista, Quinta do Carmo, Quinta do Casal Branco, Quinta das Setencostas.

## FETEASCA

The soft, vaguely Muscaty wine produced from this old eastern European grape is generally of reasonable, though not high, quality. It can be low in acidity in warm climates and ultimately lacks much character, though its gently peachy wine is agreeable enough.

Romania boasts two Feteascas: Alba, and Regala, which is a crossing of Feteasca Alba and Grasa, and dates from the 1920s. Feteasca Regala is the later ripening of the two, and by far the most planted. Its wine has more finesse and makes good rich late-harvest versions.

Romania has large areas planted with both Feteascas, and there is also some in Hungary, Bulgaria, the Ukraine and Moldova. Feteasca Alba is also known as Leányka, and Feteasca Regala is also known as Királyleányka. There is a dark-skinned version called Feteasca Neagra, which is less widely grown. Best producer: (Romania) Cotnari Cellars.

## FIANO

Extremely interesting southern Italian grape responsible for the aromatic Fiano di Avellino of Campania. At their best the wines are weighty and honeyed with notes of flowers and spice, and have the potential to improve in bottle. Best producers: (Italy) Colli di Lapio, Di Majo Norante, Feudi di San Gregorio, Mastroberardino, Paternoster, Giovanni Struzziero, Terredora di Paolo, Vadiaperti.

FEUDI DI SAN GREGORIO
*This is Feudi di San Gregorio's straightforward Fiano di Avellino, and it has concentrated flavours of peaches and nuts. The estate also makes a partly botrytized, late-picked version.*

## FOLLE BLANCHE

Folle Blanche is rapidly becoming superfluous to requirements in western France, where once it flourished. In the Gers it is known as Piquepoul or Picpoul, but it is unrelated to true Picpoul (see page 169). It was widely grown for distillation into Cognac and Armagnac until the onset of phylloxera, but is susceptible to rot and has since been largely supplanted in these regions by other varieties, notably Ugni Blanc (although brandy from Folle Blanche is generally better than that from Ugni Blanc). As Gros Plant it is grown for the VDQS wine of the same name near Nantes in the western Loire, but here too demand is falling, and growers are replacing it with Chardonnay for Vin de Pays du Jardin de la France. It is late ripening, and must be picked two weeks after Melon de Bourgogne, alias Muscadet, the main grape of the area; its acidity is very high, and its flavour neutral. All of which implies that its wine is pretty horrible. Indeed, it is – until you plonk it down next to a plate piled high with *fruits de mer* – and then, just sometimes, it's the perfect wine.

Folle Noire is not related: this is a French synonym for several varieties, Négrette and Jurançon among them.

*Bottles of Tokaji in the cellars of Oremus at Tolcsva, Hungary. The blend is Furmint, Hárslevelü and sometimes a little Muscat.*

## FRANCONIA

The Austrian grape Blaufränkisch (see page 42) sometimes takes this name in Friuli, in north-eastern Italy.

## FRAPPATO NERO

Soft, low-tannin Sicilian grape that makes wines of considerable charm and some aroma. It is usually blended with other red grapes, particularly Calabrese, Nerello and Nocera. It could have potential as a Merlot lookalike.

## FREISA

A love-it-or-hate-it Piedmontese grape, high in strawberry and raspberry aromas and acidity and sometimes quite high in tannins. The problems arise with the bitterness detectable on the finish, and the residual sugar found in many examples which can be too much for some people though others find it quite irresistible. I err on the side of the believers.

It is thought to have originated in the hills between Asti and Turin, and most authorities differentiate between Freisa Grossa, which has large berries and large clusters, and Freisa Piccola, which has small ones. Whether they are, in fact, different sub-varieties or not is unclear. They certainly seem to produce different styles, with Freisa Piccola giving more character and perfume. Freisa di Chieri is a small-berried, deeper-coloured and more tannic version and has its own DOC. As so often happens with Italian vines, the closer one looks the more the picture is obscured.

Styles of wine are equally varied: Freisa can be *frizzante* (both sweet or dry, like super-Lambrusco), or dry, still and more serious, but not necessarily better. Best producers: (Italy) Caudrina, Podere Colla, Luigi Coppo, Piero Gatti, Giuseppe Mascarello, Cantina del Pino, Giuseppe Rinaldi, Scarpa, Vajra, Rino Varaldo, Gianni Voerzio.

## FRONTIGNAC

Australian synonym for Muscat Blanc à Petits Grains (see pages 144–153).

## FURMINT

Very high quality grape that, having survived the poor handling it suffered in Hungary's Tokaj region under Communism, is now at last beginning to come into its own.

Whether it originated in Hungary or not is unclear. It was certainly much grown for sweet Ausbruch wines in Austria's Burgenland region in the past, and is enjoying a small revival there now. Both green and yellow versions exist there, with the yellow being more highly prized. In Burgenland it is called Zapfner and further south, in Steiermark, it is called Mosler. It is said to be the same grape as Slovenia's Sipon, Romania's Grasa (though see page 112) and Croatia's Posip. Tokaj, in Hungary, though, is now its homeland, and it forms the major part of plantings there.

Its advantages are its complexity of flavour, its finesse, its longevity and its high acidity. Young dry Furmint has flavours of steely smoke, lime peel and pears. Sweeter wines, affected to a greater or lesser degree by botrytis, taste of apricots and marzipan, barley sugar and blood orange and become nutty, smoky and spicy, with flavours of tea, chocolate and tobacco and sometimes with a distinct note of cinnamon, as they age.

The problem with dry Furmint is in expressing these flavours. The Disznókö estate, so far Tokaj's most successful exponent of dry Furmint, has found that it is necessary to pick non-botrytized grapes for dry wine after picking botrytized ones for sweet wine – the opposite of what one would normally expect. If done the other way round the acidity in the dry wines is painfully high and flavour is lacking.

The other problem is that the new(ish) joint venture companies, who entered the region after the end of Communism, have been mostly interested in creaming off the 10 or 15 per cent of the crop that is affected by botrytis. The aszú wines for which these are used are of extremely high quality, and in terms of longevity seem to be immortal. There has been little interest in making good dry Furmint, though that seems to be changing now that Crown Estates, which previously had the monopoly of Tokaji production, is investing in outside enologists. With its 2800 contracted growers, it is the only company with a vested interest in making good dry Furmint, so one may hope for great improvements there. Best producers: (Austria) Wenzel; (Hungary) Château Megyer, Château Pajzos, Disznókö, Oremus, Royal Tokaji Wine Co.

## GAGLIOPPO

An ancient variety which is a source of sturdy, red wine in Calabria, Abruzzo, the Marche and Umbria. Its most famous incarnation is as Cirò, on Italy's east coast – deep-coloured, alcoholic, weighty and often very good, especially if the producer has had only a light touch with the oak. Examples with no oak at all can be even better. Gaglioppo is currently attracting interest from winemakers, so we can expect to see more from this grape. Best producers: (Italy) Caparra & Siciliana, Librandi, Odoardi, San Francesco, Statti.

## GAMAY

Currently out of favour with wine drinkers outside its homes of France and Switzerland, Gamay Noir à Jus Blanc (its full name) is nevertheless a grape of many attractions for those who value lightness and aroma in red wine. Gamay is never blockbusting, but it can have considerably more character than the underweight Merlots or clumsy Pinot Noirs that have replaced it for many English-speaking drinkers. Its aroma is of pear drops, bananas, raspberries, black pepper and cherries; it has very low tannin and seldom high alcohol, unless a producer has been heavy-handed with the chaptalizing sugar at fermentation time. Its acidity is good, making it one of the most refreshing of reds.

Its home is in Beaujolais, north of Lyons, where a distinction is drawn between Beaujolais Nouveau, which is released on the third Thursday of each November, and the other Beaujolais wines. At the height of its fashionability Nouveau accounted for over half the whole crop, though thank goodness it doesn't any more. Other distinctions are made between plain Beaujolais, Beaujolais-Villages and the Crus (Brouilly, Chénas, Chiroubles, Côte de Brouilly, Fleurie, Juliénas, Morgon, Moulin-à-Vent, Regnié and St-Amour): plain Beaujolais comes from the flatter southern part of the region, where the soil is sedimentary clay and limestone, while Villages wines (39 villages may call their wine Beaujolais-Villages) and the Crus come from the granite hills of the North. Almost all the wines are made by a modified form of carbonic maceration, and the vines are trained in gobelet form, which tends to restrain their natural vigour.

All Beaujolais, generally speaking, is meant to be drunk young, and that applies equally to Gamay from anywhere else. A few Cru wines will improve in bottle for up to ten years: the longest-lasting Beaujolais are usually Moulin-à-Vent, Morgon, Chénas and Juliénas. And generally the balance of aroma and weight in the wine depends on the vintage: the most aromatic years are those when the grapes only just manage to ripen. Such years may be particularly good for Nouveau, though poor for the Crus.

North of Beaujolais in the Mâconnais and Côte Chalonnaise, the vine has been losing ground to Chardonnay. In the Loire Valley, in Touraine and in regions to the west of it, it flourishes on flinty silex soil, producing light, peppery wines with good aroma. In Switzerland it is often blended with Pinot Noir to the advantage of neither: the resulting blend is called Dôle and often tastes rather unclear, thick and dull.

Gamay was introduced to Italy in 1825, but there is only a little there now. It can be found, however, throughout eastern Europe, especially in the countries of the former Yugoslavia. It is often confused with Blaufränkisch (see page 42). Its early-budding, early-ripening nature makes it suitable for cool climates, though spring frosts can be a problem. There are occasional plantings in Canada and New Zealand, both countries where it could be interesting.

California boasts two grapes, one called Gamay Beaujolais and other called Napa Gamay. Gamay Beaujolais is not Gamay at all, but a poor clone of Pinot Noir. Plantings are gradually being replaced by Napa Gamay or better Pinot Noir, and the name Gamay Beaujolais will no longer exist after 2007.

Napa Gamay is more complicated: long thought to be true Gamay, it is now known to be Valdiguié, a French grape so poor it has been pretty well kicked out of its homeland. Napa Gamay, too, is on the decline in California, but the occasional juicy, herby example makes you wonder first whether Gamay doesn't have a future there after all, and second, what the grape variety really is in California.

Touraine also has some small plantings of Teinturier Gamays, with deep red flesh and juice. These include Gamay de Chaudenay, Gamay de Bouze, Gamay de Castille, Gamay Mourot and Gamay Fréaux. Their wine is robust, solid and unaromatic – quite unlike Gamay Noir à Jus Blanc. Best producers: (France) Aucoeur, Aujoux, Berrod, Cellier des Samsons, Charvet, Duboeuf, Henry Fessy, Sylvain Fessy, Fuissé, Thivin, Thorin, Pelletier; (Switzerland) Caves Imesch, Caves Orsat.

## GAMZA

Hungary's Kadarka grape (see page 115) is known as Gamza in Bulgaria.

## GARGANEGA

This late-ripening, highly vigorous vine is the main grape behind Soave, Gambellara and other Veneto whites. It spills over into Friuli and Umbria as well, but Soave is the wine most associated with it, for good or bad. Good Soave, and good Garganega, is exceedingly good. It has both delicacy and structure, finesse and just enough weight, and a flavour reminiscent of almonds, greengage plums and citrus fruit. When made as a sweet Recioto, from raisined grapes, the wine is intensely sweet with good though not piercing acidity; Recioto wines will improve in bottle for a decade or more, and even good single-vineyard Soaves from top producers in the Classico zone can sometimes improve for nearly as long.

Garganega's problem is that its vigour has encouraged too many producers to plant it on ultra-fertile soils in the flatlands outside the Classico zone, and allow it to yield grossly. The wine then is at best thin, neutral and dull – a description which applies to too much Soave.

It is said to be related to Sicily's Grecanico, a vine brought by early Greek settlers. The author Nicolas Belfrage points to similarities in the cluster, berry and leaf shapes between the two vines. Best producers: (Italy) Anselmi, Ca' Rugate, Coffele, Guerrieri-Rizzardi, Inama, Masi, Pieropan, Pra, Suavia, Tamellini, Tedeschi.

## GARNACHA BLANCA

The Spanish name for Grenache Blanc (see page 113) is the oldest and therefore the correct one, but there is approximately as much in southern France as there is in Spain. It is found in Spain's North-East, where it is the main grape in the Alella DO. It is permitted in Priorat, Tarragona and Rioja, but only tiny plantings exist. Best producers: (Spain) Celler de Capçanes, Masía Barríl, Bàrbara Fores, de Muller, Scala Dei, Costers del Siurana; (France) de l'Amarine, de Casenove.

## GARNACHA TINTA

See pages 92–101.

## GARNACHA TINTORERA

A Teinturier grape that isn't actually Garnacha at all: instead it is another name for Alicante Bouschet (see page 35). It is found in many parts of Spain, and plays a role in many blends.

## GELBER MUSKATELLER

The German and Austrian synonym for Muscat Blanc à Petits Grains (see pages 144–153).

# Garnacha Tinta/Grenache Noir

A LOT OF WINE EXPERTS TURN THEIR NOSES UP at Grenache Noir, or Garnacha Tinta as it's known in Spain. They dismiss it with a faintly uneasy wave of the hand as a coarse, classless interloper into their rarefied world of cool, self-contained classic grapes. Well, they dismiss Grenache at their peril.

But I'm not annoyed at them, I'm sorry for them, because good Grenache is one of the great wine experiences. It has a wonderful raw-boned power that sweeps you along in its intoxicating wake. It exudes a blithe bonhomie and a taste – all ruddy cheeks and flashing eyes, and fistfuls of strawberry fruit – that seduces you yet makes you think that surely it's all harmless fun. But it isn't. And as your head spins from one glass too many – and it could be just your second – you realize woefully that you've been had again and when will you learn, but you hurl yourself anyway into the fandango of delight that is Grenache.

*High on the skyline loom the gaunt ruins of the castle built by the popes during their sojourn at Avignon in the southern Rhône, instead of Rome. This was their new castle, their* château neuf *– and the vineyards spread around the castle walls are those of Châteauneuf-du-Pape. Gnarled old Grenache vines grow in a soil covered in large white* pebbles *or* galets roulés *that retain the heat of the southern sun long into the night. The pebbles make good homes for lizards, too.*

Grenache is for me the wild, wild woman of wine, the sex on wheels and devil take the hindmost, the don't say I didn't warn you.

Grenache is the world's most widely planted red variety but most of the vines are in one country – Spain – and its colonization of the rest of the world has been decidedly patchy. The only French classic wine in which it is included as part of the blend is Châteauneuf-du-Pape in the southern Rhône; elsewhere Grenache has always been most famous as a reckless provider of alcohol in a blend.

So long as you had some good hot weather you could plant Grenache, and every time the sun winked – hey presto, another degree of alcohol. Grenache can easily ripen to 16 per cent alcohol all by itself – which meant that blenders loved it, and fortified winemakers liked it – but they were all a bit bashful about admitting they used it. 'Junk grape' some winemakers called it. But that didn't stop big plantings all round the Mediterranean basin, in Australia, in California and further south, yet nobody ever really admired it.

The wandering New World winemakers known as 'flying winemakers' changed all that in the early 1990s. They turned up in forgotten Spanish areas like Calatayud and Cariñena and saw that there were vast plains full of super-ripe Garnacha that nobody seemed to want, but which they knew they could turn into juicy grog full of alcohol and fun if only they applied a bit of New World know-how. Which they successfully did.

At the other end of the price scale, one of the slumbering legends of Spanish wine decided it was time to finish hibernating and start making a bit of a noise. Priorat, a dense, brooding red of enormous alcohol (as high as 18 per cent) made from tiny yields of primarily Garnacha grapes (as low as 5 hectolitres per hectare) had been around for 800 years or so, but it took the rise of Catalan self-awareness and a bunch of ambitious young growers to revive its reputation during the 1990s.

And, as so often, there were the Aussies. Once Shiraz got famous, they looked around and realized they had piles of Shiraz's Rhône stablemate Grenache in their vineyards – and what's more these were often enviably old vines, giving concentrated wines of great depth. Junk grape no more. The Aussies gave it the sexy, lush, fruit-first, high-alcohol treatment – and one more irresistible, irrepressible party animal was born.

**Garnacha Tinta/Grenache Noir: from Grape to Glass**

*Geography and History page 94; Viticulture and Vinification page 96; Garnacha Tinta around the World page 98; Enjoying Garnacha Tinta page 100*

# Geography and History

GARNACHA TINTA IS A MEDITERRANEAN GRAPE par excellence. It clings to those warm lands as tenaciously as a tourist from northern Europe on holiday and like that tourist, it only has one thing on its mind: pleasure, as undemanding yet warming as a few hours baking on the beach. What, sunburn too? No – but a hangover? Yes! Wherever Grenache grows, high alcohol is the objective, and those innocuous little rosés it makes in Spain, France and Italy are all far more potent than they seem – as anyone who's slumbered through the afternoon after a seemingly harmless few glasses of Provence rosé or Navarra rosado can tell you.

Like a package tourist, it has not travelled far from the sun, although it changes its name regularly: Grenache Noir in France, Cannonau in Sardinia. Even so, it has ventured

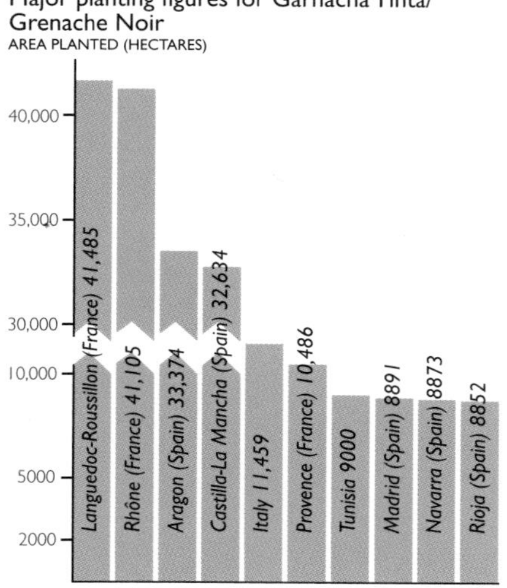


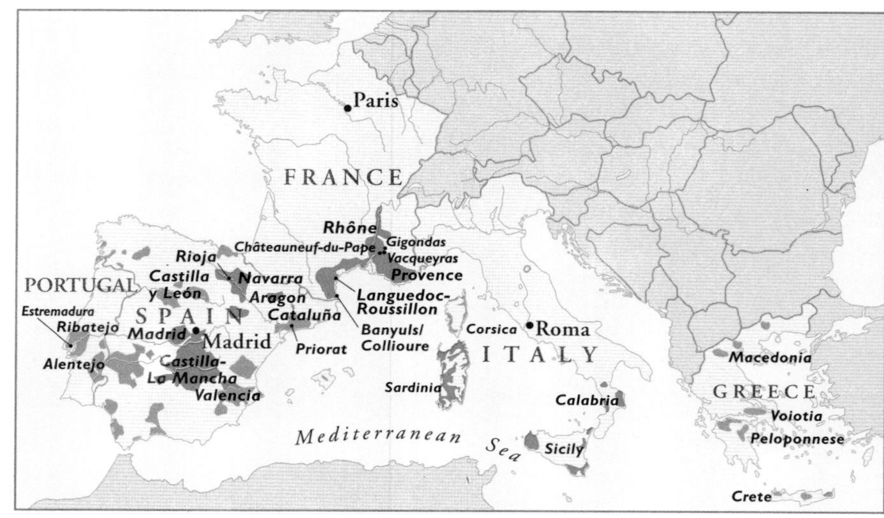

right the way round the Mediterranean coast via Sardinia to Greece, Israel, Cyprus and North Africa in search of sun-drenched fields. It suits equally sunny spots in the New World, and could make very good wine in California and South Africa, if it was given the chance.

Strength is Garnacha's strong suit, more than subtlety. But the heady, upfront style of most Garnacha makes aging unnecessary – a good thing, because most Garnacha doesn't improve with maturity. So it is often blended with other grapes: with Tempranillo in Rioja, with Mourvèdre, Syrah and others in the southern Rhône, with Cinsaut in Tavel. But because of its easy lushness at an early age (now now, no more tourist jokes) it is rapidly becoming extremely fashionable. Plantings in France are rising: it is the fourth most popular vine at French nurseries, with 23,399,256 cuttings being sold in 1998, according to the ampelographer Galet.

In Australia it is the comeback kid: along with Shiraz and Mourvèdre it had dominated plantings when fortifieds ruled the market, but fallen right out of favour before steaming back with a shout of raucous joy in the late 1990s.

### Historical background

Garnacha Tinta may be better known to English- and French-speaking wine lovers as Grenache Noir, but we give it its Spanish name here because all the evidence suggests that it is a Spanish grape that moved across the border to France and beyond, rather than vice versa.

For once there seem to be no legends linking the vine with the Romans. Instead the Spanish are given full credit for cultivating the original Garnacha from scratch, probably somewhere on the east coast, in Aragon or Cataluña. One of its synonyms, indeed, is Tinto Aragonés – which should not be confused with Aragonez, a Portuguese name for Tempranillo. Aragon can, it seems, lay claim to having been the birthplace of both Spain's major red grape varieties.

From there it was but a short hop to France – and even shorter when one considers that until France annexed it in 1659, Roussillon was part of Spain and Garnacha was probably already established there in the Middle Ages, before changing its name to Grenache and marching off through the Languedoc to the Rhône Valley. It was one of the first varieties to be planted in Australia at the end of the 18th century.

Garnacha Blanca/Grenache Blanc is the white-berried form of the vine, discussed separately on page 91. There are also grey and pink versions (Grenache Gris and Grenache Rose), which cover some 2926ha in France, and which are blended in the southern Rhône and the Midi with Grenache Noir to make fortified wines or *vins doux naturels* (see page 97). The downy-leaved red grape Garnacha Peluda, may or may not be related.

Garnacha Tintorera, however, is not Garnacha at all. Instead, it is a synonym for Alicante, a Teinturier grape (one with red, rather than colourless pulp) which is found all over Spain and goes into many blends. Some DO regulations in Spain list the two names separately, others give only one name.

*New Garnacha plantings at Clos l'Ermita in Gratallops, Priorat. This dramatically high, rugged region of Cataluña was once an inland sea. Now the vines on these mountains are planted up to 700m above sea level, and land prices are soaring. These are new vines, but the very best Garnacha grapes, from the oldest, centenarian vines, are in huge demand for wines that enjoy cult status with drinkers around the world.*

*Autumn at Gratallops. The crumbly schist soil of this area retains water remarkably well – an advantage in a climate where there is very little rain.*

*René Barbier, one of the Gratallops pioneers, at Clos Mogador with a piece of the schist that is the secret of the best wines of Priorat.*

# Viticulture and Vinification

GARNACHA IS THE MOST CLUBBABLE of grapes, fitting comfortably into many different blends. Indeed it is usually happier in a blend than solo – which is not preventing more and more producers, in search of a new grape name to put on the label, from making it as a varietal.

In some ways it is a useful grape in the vineyard, particularly in dry climates, since it laughs at drought. But it is an irregular cropper, and in irrigated vineyards where drought is not part of the gameplan, Garnacha may respond by producing high yields of pale, dilute, forgettable wine – though the alcohol level may still be high.

To get serious wine from Garnacha, you must treat it seriously, and low-yielding, old bush vines on poor soil have produced all the best results so far.

## Climate

Having said that Garnacha shrugs off drought, one should add that it needs a drought-resistant rootstock to do so. Otherwise its natural resistance to dry weather is somewhat undermined.

It is a late ripener and loves warmth. Even fierce, dry winds like the Mistral of southern France do not seem to worry it unduly. Some Australian producers like to plant it on hilltops, where it gets whatever harsh weather is going. It seems to produce far better wine when it is stressed, perhaps because it is naturally very vigorous.

But vigour does not necessarily go hand in hand with resistance to all disease. Garnacha suffers from *coulure*, or floral abortion, which can cut yields to unpredictable levels; and is susceptible to downy mildew, and to bunch rot, because of its tight clusters. Marginal climates increase the risk of *coulure* at one end of the growing season and of bunch rot at the other end, if the grape's late ripening pushes it into autumn rains.

## Soil

Garnacha is irrevocably associated with hot, dry soils, preferably poor and well-drained. Apart from those basic requirements it is not over-fussy, though the best French Grenache often comes from schist or granite soil, or relatively high-altitude sites.

Châteauneuf-du-Pape has its famous, heat-retentive stones, its *galets roulés*, on the higher plateaux; opinions vary as to whether the *galets roulés* soils give the best wines. Schist is also important in Rioja, and especially in Priorat, where Garnacha produces some of its very best wines. Vineyards here are planted up to 700m above sea level.

## Cultivation and yields

Gobelet or bush training, with four or more arms, and spur pruning, with two spurs per arm, seem well suited to this vigorous, upright vine, though guyot or royat training on wires is increasingly popular. Garnacha needs to be pruned hard and debudded if yields are to be kept within bounds: yields of under 35hl/ha will give very different wine even from yields of 50hl/ha (the base yield for AC Côtes du Rhône).

But training on wires and using irrigation sensibly can give higher quality at higher yields than bush training with no irrigation. In Priorat yields from very old vines may be as low as 5–6hl/ha while at Château Rayas in Châteauneuf-du-Pape yields average 15–20hl/ha. Charles Melton in the Barossa Valley recommends no more than 1.5 tons/acre (27hl/ha). By contrast, Garnacha/Grenache vineyards planted for high volume and low or lowish quality, like those in California's Central Valley, are likely to give much more than 50hl/ha; it is on such wines that the grape's reputation for uninteresting quality rests. Low yields give structure; at high yields the grape's tendency to low acidity is exacerbated, and colour and flavour disappear.

Anyone tasting a Grenache-based Priorat, or the Grenache-based Châteauneuf-du-Pape Château Rayas, would find it hard to believe that such powerful mouthfuls have anything to do with the lighter, lifeless jug wines of much of southern France, Spain and California.

*The* galets roulés *stones of Châteauneuf-du-Pape, shown here with bush-trained vines, absorb heat during the day and give it out during the night. Not surprisingly, the grapes reach super-ripeness and give wines with high levels of alcohol. But the subsoil, of red clay and ferruginous sands, is probably more vital to quality.*

Irrigation must therefore be treated with caution if quality is the aim. In Châteauneuf irrigation is permitted, though officially limited; in practice, things being what they are, overwatering is not uncommon.

Garnacha/Grenache reaches high levels of ripeness easily in warm climates. Picking late, at about 15 per cent potential alcohol, seems to give the best balance and flavour; picking earlier for (supposedly) better acidity and elegance doesn't seem to work: you simply get green flavours and poor colour. The problem is, therefore, to get physiological ripeness without losing too much acidity: growing at higher altitudes can help, since cooler nights will help the retention of acidity. The other solution is to blend with some other grape for which acidity is less of an issue. It is not unknown for the alcohol content of Garnacha table wines to rise to 18 per cent without the aid of fortification, though a slightly more drinker-friendly 14.5 per cent is increasingly favoured.

*Carlos Pastrana at his winery, Costers del Siurana, in Priorat, Cataluña. Pastrana was one of the driving forces of the Gratallops projects in the Priorat region, and brought together the five pioneering companies. The others are Alvaro Palacios (Clos l'Ermita), Mas Martinet, René Barbier (Clos Mogador) and Clos & Terrasses.*

### Clones

These vary widely in quality. Some clones are highly productive, others less so; some have better colour than others, or produce more or less irregularly. Galet (1998) lists clone 362 as being particularly good for *vin doux naturel* because of the high degree of ripeness it attains.

### At the winery

Garnacha/Grenache must be handled gently: it oxidizes with extreme ease, and loses colour if care is not taken. The wine can have a tendency to green, herbaceous flavours, which can be made worse by the inclusion of too many stems in the fermentation vat. Over-harsh pressing or over-hot fermentation, both designed to extract more tannin than is natural to the wine, tend to give astringency. A long, slow fermentation, followed by a long maceration to extract tannin, is best, followed by as little racking as possible, to prevent oxidation.

In the southern Rhône and in Spain, old oak barrels are usual for aging. New oak, inevitably, is creeping in, and can help to prevent oxidation and fix the colour. Some think the flavour of new oak an aberration in Garnacha/Grenache; others welcome it. It is, perhaps, a matter of taste, but I don't think more than a small percentage of new oak adds anything to a good Grenache's unmistakably fruity style. If anything it masks the fruit and strips it of individuality.

### FORTIFIED WINES

Cast your mind back, if you will, to 1299. In that year the king of Mallorca granted a patent to a Catalan alchemist called Arnaldus de Villanova; the patent was for the process, which he had developed but not invented, of stopping the fermentation of grape juice by the addition of grape spirit. (One can imagine the conversation: 'Well, that's great, Arnaldus, but, like, what's it for?')

The king of Mallorca's realm included Roussillon, and Roussillon went on to become a centre for the production of what is now called *vin doux naturel* (VDN). It still is; and there is a long tradition both in Roussillon and in neighbouring Cataluña (the two regions were both under Spanish rule until 1659) of using Garnacha/Grenache for these sweet concoctions of partly fermented grape juice and spirit. These wines may be made *rancio* by leaving them outside in glass demi-johns (or *bonbonnes*) or wooden barrels exposed to the air and to hot daytime temperatures for several years, until they acquire a maderized, sour *rancio* tang of nuts and raisins and cheese.

In Roussillon, Grenache-based VDNs are made in Maury and Banyuls. Banyuls is generally higher in quality than Maury, and, in addition, there is a rarely seen Banyuls Grand Cru appellation, for which 75 per cent of the blend must be Grenache Noir, compared to 50 per cent for normal Banyuls. Styles in Banyuls vary enormously: some wines are fruity, some dark and concentrated, some *rancio*. Some may be 20 or 30 years old.

In the southern Rhône, the village of Rasteau specializes in a *vin doux naturel* made from Grenache. Cataluña, in particular Tarragona, also makes sweet fortified Garnacha, allowing the juice to ferment for three days, then adding grape spirit to bring the strength up to 15–16 per cent alcohol. In Sardinia, too, where Garnacha is known as Cannonau, some fine fortifieds are made. This practice spread to the New World, in particular to Australia, where Grenache, sometimes blended with Shiraz or Mourvèdre, provided the backbone for the country's 'port' industry and produced many super examples.

Garnacha may be blended for these purposes with other grapes, often Cariñena and Tempranillo in Spain, and Syrah, Cinsaut, Carignan, Grenache Gris or Grenache Blanc in France.

# Garnacha Tinta/Grenache Noir around the World

GARNACHA CAN BE GREAT, simple and charming, or dull and uninteresting. Only a very few unblended ones are great – but they should be enough to inspire the others. But are varietal examples necessarily in Garnacha's best interests? Not unless they're cleverly grown and made. The last thing the world needs is more overcropped, anonymous red.

### Priorat

This ancient vineyard region in Cataluña has suddenly, and dramatically, leapt to the forefront of Spanish red wines (after languishing in obscurity for centuries), making dark, heady wines to be sold at crazy prices. Priorat winemaking can be traditional or it can be new wave, but either way yields (for the best wines) are extremely low and the wines very concentrated. But beware: quality does vary.

The old way here, and it still just about exists, is to make powerfully alcoholic wines that are almost black in colour and might finally soften in time for your grandchildren to enjoy them. The new way is to make wines which have huge blackberry fruit in youth, with perhaps a bit of Cabernet or Merlot added for aroma; these wines are drinkable pretty much on release.

Of the nine villages in the DO, Gratallops is in the lead as far as new-wave wines are concerned. The pioneers of the new style arrived here in 1986, planted Syrah, Cabernet Sauvignon and Merlot alongside the existing Garnacha (which covers about 40 per cent of the Priorat vineyard) and Cariñena vines, installed drip irrigation and modern winemaking equipment, and started to make wines with more fruit and more new oak than was traditional.

With time all Priorat seems to develop a flavour of tarry figs; whether you think it should be drunk young or left to age is very much a matter of taste.

Vines are planted at between 100 and 700m up, on terraces if the angle of the slope defies gravity too much. In Gratallops the soil is schist, but the typical Priorat soil is Llicorella, which sounds like a drink aimed at the youth market, but is in fact a rock composed of stripes of slate and quartzite that glitters black and gold in the sun.

### Rioja

Curiously, Garnacha is less well regarded in Rioja than in many other parts of Spain, and it is the only red variety in decline in the region. Most of Rioja's 9000 ha of Garnacha are in Rioja Baja, where the greater heat and lower rainfall suits it. Soils are mostly sandy here, and most bodegas like to add a proportion of Garnacha – 15 to 20 per cent is common – to their Tempranillo, Mazuelo and Graciano for the body and alcohol it brings to the blend. However, it tends to oxidize before the other grape varieties, and yields must be kept low if it is to age well.

### Navarra

Rosado used to be what Spaniards drank when they wanted something that wasn't red and/or oaky; real men didn't drink white wine, apparently. And Garnacha makes excellent pink wine: juicy, soft and only good young. Navarra makes a speciality of Garnacha rosado, particularly from the sandy-soiled, dry south of the region. Some 54 per cent of Navarra's vineyards are planted with Garnacha, though the proportion is falling as more red and less rosado is produced, and the growers are encouraged to replant with Tempranillo. The official aim is to have 35 per cent Garnacha and 31 per cent Tempranillo: the figure for Tempranillo currently stands at 23 per cent.

### Rest of Spain

Garnacha's stronghold in Spain is in the north and east of the country. In Calatayud it covers about 65 per cent of the vineyard, much of which makes rosado; in Campo de Borja it accounts for 75 per cent of the vineyard. It is more than half the vineyard in Cariñena and takes up a good chunk of most other DOs in this part of Spain.

Versatility is its key: it can turn out rosado and attractive *joven* or young reds if you ask it to, but if you put it into a barrel and blend in some backbone with some other grape, then you have a sturdy, oaky red. Quality is usually pleasant rather than anything more, but these regions are not aiming for world domination of the blue chip investment wine market. They're just out to provide a decent glass of grog at a fair price – and that is a noble calling.

### Southern Rhône and the Midi

The secret of good Châteauneuf-du-Pape lies partly in Grenache, which forms the bulk of the blend, but partly also in Mourvèdre, which adds tannin and earthy, savoury

MAS MARTINET

*Cabernet Sauvignon, Merlot and Syrah are blended with Garnacha in this top-notch wine from Priorat. It needs at least five years' aging.*

JULIAN CHIVITE

*Navarra makes a speciality of Garnacha-based rosados. This example, from Julian Chivite, is strawberryish and fresh.*

DOMAINE DU PÉGAU

*Paul Feraud and his winemaker daughter Laurence produce spicy, earthy, chocolaty and cherry-fruited Châteauneufs from a Grenache-based blend.*

flavours to the wine, and Syrah, which brings structure and fantastic perfume. It is often said, particularly by growers of Châteauneuf and Gigondas, that the preponderance of Grenache in their vineyards is the legacy of domination by the Burgundian merchant houses, with their ceaseless demand (in the past, of course) for wine of high alcohol to beef up their pallid brews. AC yields here, at 35hl/ha, are low, and minimum alcohol, at 12.5 per cent with no chaptalization, is high. With Grenache, achieving the second is easy if the first is adhered to. But quality is mixed, and overproduction not uncommon. The most usual fault is high alcohol without the backbone to support it.

There is also some carbonic maceration done alongside more traditional vinification, which produces wines lighter and fruitier than the rich, spicy Châteauneuf of most drinkers' imagination; as a part of the blend, these fruitier wines can be very attractive.

The soil of Châteauneuf is famously stony in parts: the big round *galets roulés* or pudding stones do not cover the entire vineyard area, and they are not essential to good Châteauneuf. Certainly, they retain heat during the day and give it out at night, but lack of heat is not really an issue in this part of the southern Rhône.

Grenache is also the foundation of Gigondas, where it can form up to 80 per cent of the vineyard, and Vacqueyras. Officially, the maximum yield in both is 35hl/ha, and while wines with a fair amount of Syrah and Mourvèdre in the blend have better structure and more substance, if Grenache gets up a head of steam out in those torrid, rocky vineyards, it can make as burly, as chewy, as intoxicating a red as any in the Rhône Valley. Grenache is traditionally aged in large, old wooden barrels; where smaller ones are creeping in they are often kept for the other grapes.

*Wine from these old vines in Australia's Clare Valley will go into Tim Adam's The Fergus Grenache. Adams is dedicated to the Clare Valley, choosing to buy in grapes rather than own vineyards so that growers as well as winemakers in the region can make a decent living. Of course this encourages them to keep their old vines – and Adams gratefully hoovers up any available fruit.*

In Lirac Grenache must take up at least 40 per cent of the vineyard, and makes reds and rosés; in Tavel it makes only rosés. The rosés from both are refreshing if you get them young; as is the case elsewhere, their structure depends largely on the proportion of other grapes in the blend.

Grenache is also the staple grape of Côtes du Rhône (apart from those sections of the appellation in the Syrah-only North) and Côtes du Rhône-Villages, and plays varying parts in the wines of Provence, Languedoc, Minervois, Corbières, Fitou and Roussillon.

### Australia

Time was if you saw a vine in Australia, particularly in South Australia, it was quite likely to be Grenache. But that was before the grape fell victim to the Cabernet–Chardonnay fashion. The vine-pull scheme of the 1970s ensured that many old bush-trained, low-yielding vineyards of old Grenache were uprooted along with Shiraz vines of equal quality; but luckily some survived. They are now much in demand as Australia rediscovers the virtues of Grenache. Plantings are rising again, and while blends of Grenache, Shiraz and Mourvèdre are becoming classic, there are also an increasing number of varietals. Some of these are excellent, though some lack backbone – the usual Grenache problem. Barossa Valley versions are intense, even jammy; McLaren Vale Grenache is spicy and lusciously rich.

### USA

Grenache has yet to become fashionable here, perhaps because plantings are concentrated in California's Central Valley, and are intended for nothing more than inexpensive jug wines. Individual Grenaches and Rhône blends, from producers like Alban and Bonny Doon, show what can be done if the grape is planted in the right place.

### Rest of the world

In Sardinia, as Cannonau, it produces some earthy reds, often of very interesting quality; there is also a sweet, porty, high alcohol version under the name of Anghelu Ruju from Sella & Mosca. There is a little in both Greece and South Africa, and it may yet follow Syrah into Chile as producers there look for new grapes to plant. In North Africa, in common with every other planted variety, it awaits a revival of interest in winemaking.

**Celliers des Templiers**

*A wine somewhere between port and oloroso sherry in style, Celliers des Templiers' Cuvée Président Henry Vidal nevertheless has a cleansing, dry finish.*

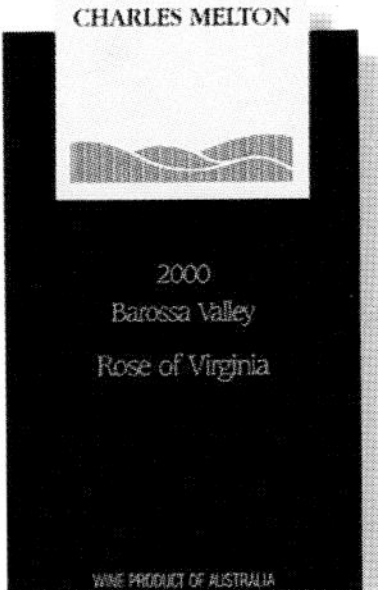


**Charles Melton**

*A deep, dark rosé with a touch of tannin which should be drunk young. Charles Melton is one of the leaders of the Barossa's Grenache revival.*

**Rosemount**

*'GSM' is Rosemount's shorthand for a classic Aussie Châteauneuf-inspired blend from Grenache, Shiraz and Mourvèdre.*

# Enjoying Garnacha Tinta/Grenache Noir

IT IS TOO EASY TO ANSWER the question of how Garnacha/Grenache ages with the simple answer that it doesn't. But some Garnacha/Grenache ages brilliantly, though it has to be admitted that the great majority is best drunk young. Most Garnacha/Grenache wines oxidize fast, and so ought to be drunk in the first ruddy bloom of youth.

The exceptions are the best wines of Châteauneuf-du-Pape and of Priorat, and good examples of these can both happily age for a decade or more. Of course, quite a few Riojas age well, but these are usually mostly made from Tempranillo: in general Garnacha/Grenache ages much better if it has another grape to wrap its fat juiciness around.

A good Châteauneuf, made from low-yielding vines so that the Grenache has structure and concentration, can last from five or six years up to about 20: nearly as long as a good Côte-Rôtie. A low-yield Châteauneuf superstar like Château Rayas can last a good deal longer than that, but commercial blends of Châteauneuf and Gigondas won't improve much beyond five or six years.

Many modern Priorats, despite their powerful flavours, are made to be drinkable early: they don't, on the whole, have to be tucked away for very long before you can open them. Nor will they last all that long: ten years or so is usually the limit. Traditional massive, black Priorats will last much longer, even up to 25 years.

The *vins doux naturels* of Roussillon, since they are effectively mummified by the oxidation process they undergo in barrels and demi-johns, will happily last 20 or 30 years, and don't seem to change very much over the decades.

Australian 'ports', where there is also some Shiraz blended in with the Grenache, can age for decades, but they are drinkable sooner than their Portuguese counterparts, usually being a little sweeter and less fierce in their youth, and achieve a tawny maturity much sooner.

**The taste of Garnacha**

Good Garnacha/Grenache has wild, unexpected flavours: roasted nuts, leather, blackcurrants, honey, gingerbread, black cherries, pepper, coffee, spices, even tar and black olives. As yields are progressively increased these fade into a gentle, soft leathery earthiness; the shock is that they fade rather early. That leatheriness is a telltale sign that the fruit won't be there for ever. Young, vigorous Grenache tastes of strawberries and raspberries, as well, with often a certain dustiness to the fruit that speaks of those arid vineyards in Spain and southern France. Too fanciful a connection? Probably. Perhaps not.

And Priorat: I mentioned earlier that it seems to attain a tarry, figgy character with age. The blackberry fruit fades, and you're left with something not unlike a Spanish version of Italian amarone. It must be something in the terroir as well as the grape, because you find it in Priorat of both old and new styles.

*Rancio* Garnacha/Grenache is leathery, certainly, but also nutty like tawny port and sour like cheese. Young rosado/rosé wines, by contrast, are all strawberries and cream.

*There is no legal definition of how old vines must be to be described as 'old', but at the Clarendon Hills winery in McLaren Vale, South Australia, the Grenache from the Blewitt Springs vineyard is from vines that are about 90 years old.*

*At Château Rayas in the heart of the Châteauneuf-du-Pape appellation the vines are much younger: the average age is around 35 years and the wine is perhaps the finest red Châteauneuf of all.*

**Matching Garnacha/Grenache and food**

This grape comes in so many styles that it's quite difficult to think of any meat dish it clashes with. Soft, low tannin versions can be good with spicy Indian food, where tannin doesn't work. Barbecued food also matches these well, as do stuffed peppers and aubergines. Bigger, more tannic versions will go with roast beef and lamb, pheasant and duck and with the most flavoursome casserole in your repertoire.

Chilled young Garnacha/Grenache can be a good summer red, while pink versions are great with vegetarian dishes and strongly flavoured fish such as grilled sardines.

## CONSUMER INFORMATION

**Synonyms & local names**
Also known in Spain as Aragón, Aragonés, Garnacha Tinta, Garnacho Tinto, Garnatxa, Lladoner and Tinto Aragonés; as Alicante in France, as well as Grenache Noir; as Granaccia or Granacha in Italy and Cannonau in Sardinia.

**Best producers**
**FRANCE/Rhône/Châteauneuf-du-Pape** Beaurenard, Henri Bonneau (Réserve des Celestins), Bosquet des Papes, Chapoutier (Barbe-Rac), Clos du Caillou, les Cailloux (Centenaire), Clos des Papes, Clos du Mont Olivet, Jean-Luc Colombo, Font de Michelle, Fortia, Galet des Papes, de la Gardine, Guigal, Grand Tinel, la Janasse, Marcoux, Mont-Redon, Montpertuis, de la Mordorée, de Nalys, la Nerthe, du Pégaü, Père Caboche, Roger Perrin, Rayas, la Roquette, Roger Sabon, Tardieu-Laurent, Pierre Usseglio, J-P Usseglio, Raymond Usseglio, Vieux Donjon, Vieux Télégraphe, de Villeneuve; **Côtes du Rhône-Villages** Achiary, D & D Alary, Brusset, Gramenon, du Grand Moulas, de l'Oratoire St-Martin, Rabasse-Charavin, de la Réméjeanne, Richaud, Ste-Anne, la Soumade, du Trapadis; **Gigondas and Vacqueyras** des Amouriers, la Bouissière, Brusset, de Cassan, de Cayron, Clos des Cazaux, Clos du Joncuas, de la Charbonnière, Couroulu, Cros de la Mûre, des Espiers, Delas, Font-Sane, la Fourmone, les Goubert, Gour de Chaule, Jaboulet, de Montmirail, Moulin de la Gardette, les Palleroudias, les Pallières, Raspail-Ay, Redortier, St-Cosme, St-Gayan, Santa Duc, Tardieu-Laurent, de la Tourade, des Tours, du Trignon; **Southern French vins doux naturels** Casa Blanca, Cazes, Celliers des Templiers, des Chênes, Clos des Paulilles, l'Étoile, de Jau, Mas Amiel, du Mas Blanc, les Vignerons de Maury, de la Rectorie, Sarda-Malet, la Soumade, la Tour Vieille, Vial-Magnères.
**SPAIN/Cataluña/Priorat** Clos Erasmus, Clos Mogador, Costers del Siurana, Fuentes, Mas Martinet, Alvaro Palacios, Rotllan Torra, Scala Dei; **other Spanish producers** Borsao, Celler de Capçanes, Martinez Bujanda.
**ITALY/Sardinia/Cannonau di Sardegna** Argiolas, Sella & Mosca.
**USA/California** Alban, Bonny Doon, Jade Mountain, Sine Qua Non; **Washington State** McCrea Cellars.
**AUSTRALIA** Tim Adams, Cimicky, Clarendon Hills, Coriole, D'Arenberg, Hamilton, Hardys, Charles Melton, Mitchelton, Penfolds, Rockford, Rosemount, Seppelt, Tatachilla, Veritas, Yalumba.

## RECOMMENDED WINES TO TRY

**Châteauneuf-du Pape wines**
See Best Producers left.

**Ten other southern French reds**
Domaine de Cayron *Gigondas*
Domaine le Clos des Cazaux *Vacqueyras Cuvée des Templiers*
Domaine les Goubert *Gigondas*
Domaine Gramenon *Côtes du Rhône Ceps Centenaires*
Domaine Lafon-Roc-Épine *Lirac*
Domaine de l'Oratoire St-Martin *Côtes du Rhône-Villages Cuvée Prestige*
Domaine de la Rectorie *Collioure la Coume Pascal*
Domine St-Gayan *Gigondas*
Domaine Santa Duc *Gigondas*
Domaine la Tour Vieille *Collioure*

**Ten other European reds**
Argiolas *Turriga (Italy)*
Bodegas Borsao *Campo de Borja (Spain)*
Celler de Capçanes *Tarragona Costers del Gravet (Spain)*
Clos Mogador *Priorat (Spain)*
Costers del Siurana *Priorat Clos de l'Obac (Spain)*
Martínez Bujanda *Rioja Garnacha Reserva (Spain)*
Mas Martinet *Priorat Clos Martinet (Spain)*
Alvaro Palacios *Priorat l'Ermita (Spain)*
Rotllan Torra *Priorat Amadis (Spain)*
Sella & Mosca *Anghelu Ruju (Italy)*

**Ten New World Grenache-based reds**
Tim Adams *The Fergus (Australia)*
Charles Cimicky *Grenache (Australia)*
Clarendon Hills *Old Vine Grenache Blewitt Springs Vineyard (Australia)*
D'Arenberg *The Custodian (Australia)*
Hardys *Tintara Grenache (Australia)*
McCrea Cellars *Tierra del Sol (Washington State)*
Charles Melton *Grenache (Australia)*
Rockford *Dry Country Grenache (Australia)*
Sine Qua Non *Red Handed (California)*
Tatachilla *Keystone Grenache/Shiraz (Australia)*

*The colour looks wonderfully dark – but get Grenache into the winery and that colour can disappear with disconcerting ease. It's not the simplest of grapes to handle, and needs to be treated seriously if it is to produce serious wine.*

**Maturity charts**
Grenache is increasingly beloved of wine producers who want to make rich red wines for early drinking but will keep too.

**1999** Priorat (new style)

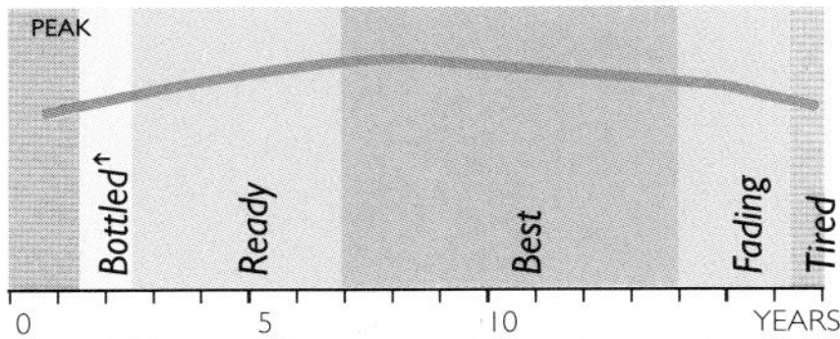


*Incredible when first released, the wine has huge blackberry fruit which goes tarry two to three years later, and it develops an almost amarone style.*

**1998** Châteauneuf-du-Pape

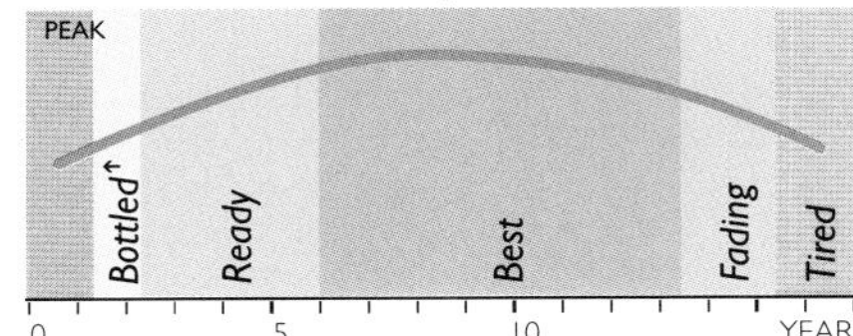


*1998 was a tremendous vintage in Châteauneuf, with wines of power and ripeness. The lightest are already drinking; most will be drinkable at five years.*

Vin doux naturel Banyuls Grand Cru

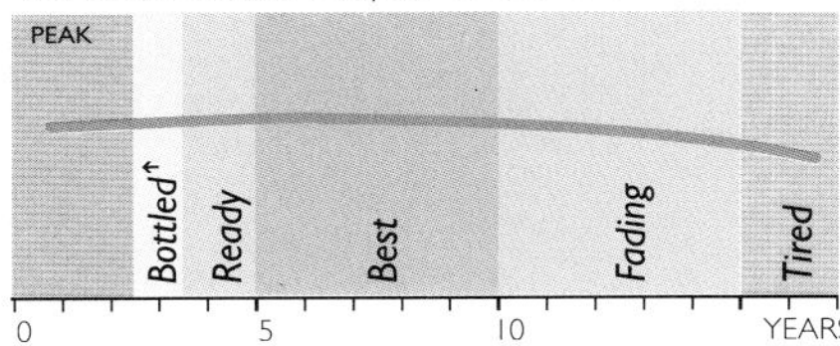


*Vin doux naturel wines are bottled at various ages when they are ready to drink: they require no further aging and will not particularly benefit from it.*

Termeno
E·R

# Gewürztraminer

It must be tough being a Gewürztraminer sometimes. All your life, all you've ever wanted to do is to please people. And not just the odd person. You want to please everybody, and you're willing to use every weapon in your armoury to win them over. So you doll yourself up to the nines. You preen in front of the mirror and spread out your vast panoply of make-up, especially mascara and rouge. And you ladle them all on to emphasize every single feature of your face. Might some features perhaps have been better left unadorned, possibly even muted a bit? Do you think your cheeks should be quite so pink, your lips quite so pouting and bright? Well, well, maybe not, but once you get started you sort of can't leave anything alone.

And scent. Aah, scent. What should it be? A dab of something mineral and restrained behind the ears? But who would smell that? Something high-toned and floral with citrus notes giving way to autumnal orchard ripeness? Mmm, yes, yes, but you want more. You want the sultry tones of passion and seduction giving way to the earthy notes of exhaustion and sleepy satisfaction. You want to guarantee that nostrils will quiver from the first moment you sweep through the door, you want clouds of Yves Saint Laurent's Opium, Calvin Klein's Obsession and Giorgio of Beverley Hills to billow out before you, announcing the arrival of the one grape no-one can resist. Yourself!

*The Gothic-style, carved wooden spice cabinet, stork, and buttery, knot-shaped sweet pretzels suggest Alsace where the grape achieves its highest fame. The pomander, cloves, nutmeg and spice mortar are references to the word 'Gewürz' which means spice in German – but Gewürztraminer is a complex story. The wines are far more than just being fat and spicy, famous only for their heady perfume.*

Ah, if only. Poor old Gewürz. Not all wine lovers can take it. For some, Gewürztraminer is a parody of perfume and powder that sashays around them. For them anything as scented, as seductive, as voluptuous as Gewürztraminer offers far too thrilling an experience to possibly be any good.

It is strange, isn't it? You get a grape that combines a most irresistible scent of lychees and tea rose petals with the lushness of tropical fruit, the bite of black pepper, and the intimate dressing room aroma of Nivea Crème, and instead of getting excited by the sheer sensuousness of the experience, they call it overblown, blowzy, boorish, clumsy and so on.

It's true that Gewürztraminer does have its problems for the winemaker. Do you go all out for perfume and lushness like Alsace? Or will that put people off? If you make the grape exercise a little restraint, will it then come over as half-hearted? And is there any point in being half-hearted? So far the world has yet to find a more convincing model of Gewürztraminer than the Alsace one. Most others, even if they're good, are simply too polite, as if they started off wanting to be Italian Pinot Grigio. There's no point in growing Gewürztraminer in that case.

It's true that bad Gewürztraminer can be fat and blowzy. It can have a strange, disappointingly hard edge to it, and the fruit can resemble leftover marmalade on the edge of the breakfast plate. But bad Chardonnay is tasteless, lifeless stuff. Unripe Cabernet is stalky and tart. Poorly made Riesling is sulphurous and stale. Yet we don't judge these varieties on their bad examples but on their good ones.

So let's do the same with Gewürztraminer. She's not an easy-going grape; overcropping dilutes the lovely perfumed richness of the wine. But all over the world – quite often in France's Alsace, sometimes in Germany, Italy and Austria, occasionally in the Americas and Australia, and definitely in New Zealand – Gewürz makes some heavenly scented wines we should be grateful for. Now go and buy a bottle and find out for yourself.

**Gewürztraminer: from Grape to Glass**

*Geography and History page 104; Viticulture and Vinification page 106; Gewürztraminer around the World page 108; Enjoying Gewürztraminer page 110*

# Geography and History

IF YOU WANT YOUR WINE ALL POLITE and well-mannered, you shouldn't really be drinking Gewürztraminer at all. The magic of Gewürztraminer only really shines out when its sumptuous, exotic perfumes make your head spin and your thoughts go giddy with desire.

But for those of you who find such self-indulgence a bit embarrassing, there are some milder forms of Gewürztraminer available from all across the globe, as the map shows. But despite being spread widely, it is also spread thinly. In no region is Gewürz dominant, not even in Alsace, where it has a mere 17.8 per cent of vineyard plantings; yet this small strip of hillside in north-east France produces nearly all the great examples of the grape – luscious, weighty, laden with scent.

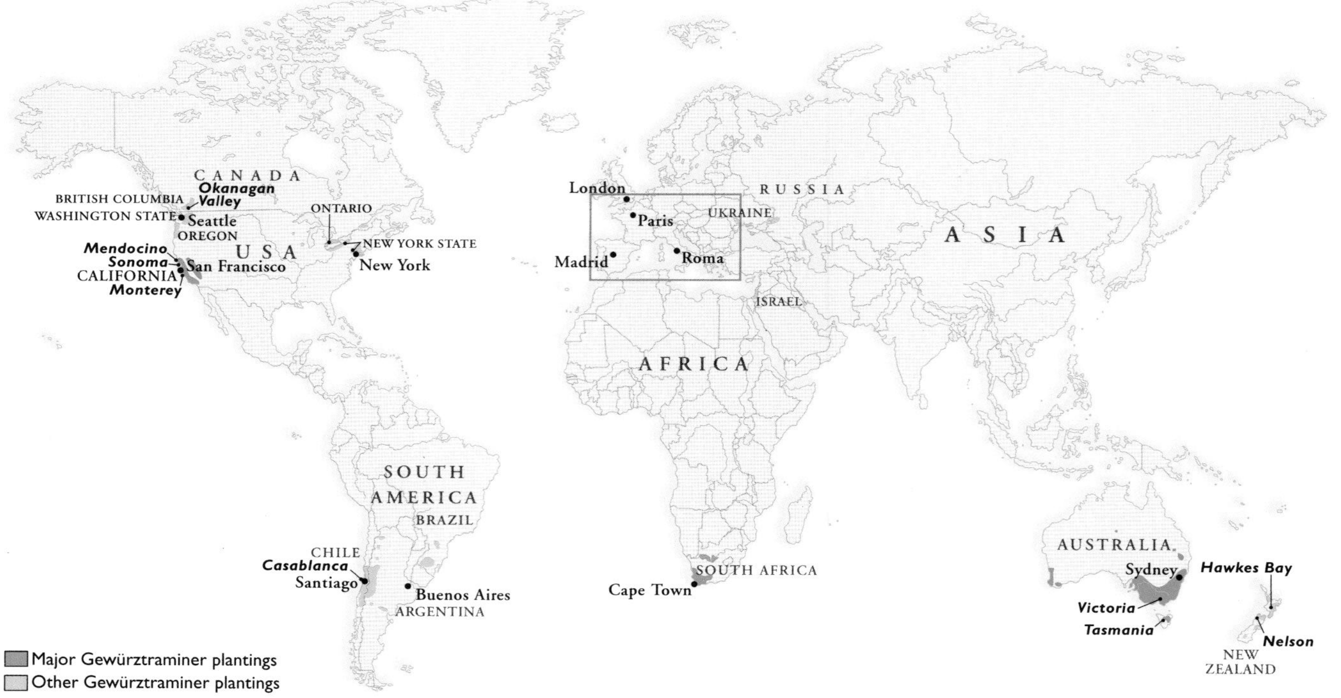

Major Gewürztraminer plantings
Other Gewürztraminer plantings

Major planting figures for Gewürztraminer
AREA PLANTED (HECTARES)

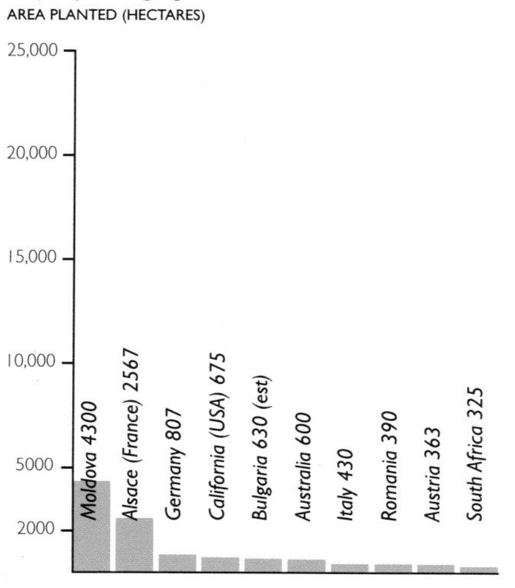


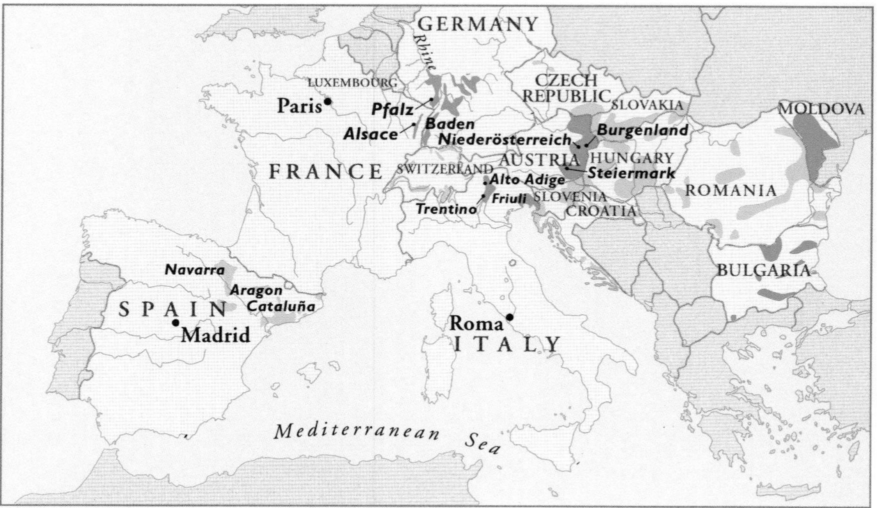

That very weight, that uncompromising, flavour-driven personality, is both Gewürztraminer's calling card and its big disadvantage. Tone it down and you get a wine hardly worthy of the name of Gewürztraminer. Play it up and you get a wine that outside its devoted fan club is difficult to sell. It's easier to sell high-priced Chardonnay than Gewürztraminer of equal quality, even at a much lower price.

So world plantings are not rising. It pops up all over eastern Europe, but results are mixed, to put it mildly. In the vine's birthplace, the Alto Adige in north-eastern Italy, its wines are lighter and more acidic, and much less exuberant and flavour-laden than in Alsace. New Zealand makes some very good versions, and the first few efforts from Chile are promising.

But nobody makes good Gewürztraminer by accident. Those who choose to grow it do so because they love its richness and are prepared to accept low yields and relatively low returns. It's unlikely ever to cover the map: it is not, and never can be, an all-purpose wine. But with such an in-your-face personality, how could it ever be?

### Historical background

There are two questions with Gewürztraminer. One is whether or not it is the same grape (a mutation, perhaps) as Savagnin Rose; the other is where Gewürztraminer ends and Traminer begins, since both names are used around the world. Galet (1998) says that Traminer is the same grape as Savagnin Rose; the Office International de la Vigne et du Vin believes them to be extremely similar but separate. Morphologically they are almost identical; but the flavour is different.

What is generally accepted is that Traminer originated in the Alto Adige village of Tramin, or Termeno if you prefer the Italian name. The earliest record of it there dates from 1000, and it was planted widely there until the 16th century, when growers decided that they preferred the more generous yield of the red Vernatsch, or Schiava.

Its route to Alsace, where it has been grown since the Middle Ages, is believed to have been via Germany's Pfalz region. Its 'gewürz', or 'perfumed' prefix, was not adopted in Alsace until after 1870. Galet mentions two forms existing side by side in the Pfalz: Savagnin Rose *non musqué*, or Heiligensteiner Klevener, and the *musqué* form, called Gewürztraminer. The two look very similar, the only apparent difference being that Savagnin Rose grapes have translucent skin just before *véraison*, while Gewürztraminer grapes are opaque.

To what extent the names have always been used for the different forms is hard to say. In Germany the wine may be named according to its nose, with spicier ones being dubbed Gewürztraminer, less spicy ones plain Traminer. In Alsace, too, until the 1970s, the two names were in use, with Gewürztraminer being used for the 'better', more aromatic wines. In 1973 the name Traminer was discontinued in Alsace, except for the area of Heiligenstein which uses the pink-berried but less aromatic Savagnin Rose to make a wine called Klevner de Heiligenstein.

*The Grand Cru Zinnkoepflé high above the village of Soultzmatt in the Haut-Rhin, Alsace. Gewürztraminer occupies less than one-fifth of the total Alsace vineyard area, and deep, rich, marly soils with some chalk, like the soil of Zinnkoepflé, suit it well. It is particularly happy if planted towards the lower part of a slope, where the soil is richest and deepest, but it also needs good drainage, so it won't thrive on the valley floor.*

*Ripe Gewürztraminer in the Grand Cru Kitterlé at Guebwiller in Alsace. It flowers early and ripens in mid-season, before Riesling; late-picked wines have concentrated sweetness and sensational perfume.*

*The fortified village of Bergheim, with its brightly painted, medieval, timber-framed houses, is famous for its magnificent Gewürztraminer, especially from its two Grands Crus, Altenberg and Kanzlerberg.*

# Viticulture and Vinification

GEWÜRZTRAMINER IS NOT THE EASIEST of grapes to grow or handle. Its perfume is its raison d'être – all roses, face cream and lychees – but it must have some structure to back it up. Without structure Gewürztraminer speedily tumbles into cloying blowziness. And low acidity is its besetting fault.

It needs a long ripening season, and loves sunny and dry weather, but too much heat exacerbates the low acidity problem. All that perfume does not suit the flavour of new oak, but it can make glorious sweet wines, even though the action of *Botrytis cinerea* tends to destroy its varietal perfume.

## Climate

If Alsace is the template for Gewürztraminer, then it needs mild springs, sunny summers and low rainfall. The last is the work of the Vosges mountains, which protect the vineyards from westerly rain-bearing winds: Colmar rates as one of the driest spots in all France. The sunshine continues well into the autumn, and there can be snow on the highest peaks of the Vosges while *vendange tardive* (late-harvest) grapes are picked lower down. Gewürztraminer buds early, which makes it susceptible to late frosts, and ripens in mid-season; too much heat will bring it on too quickly, resulting in a loss of acidity and aroma. But that's the beauty of Alsace: for all its sunshine, it still has a relatively cool climate – and loads of sunshine and a cool climate equals loads of sugar in the grapes but appetizing acidity and aroma.

Much of New Zealand also has a cool climate with loads of sun – although rainfall is more erratic here and there can be unwanted autumn deluges. During its first re-birth as a wine nation in the 1960s, visiting German consultants recommended Germanic grapes as the route to success in New Zealand and that is why a fair amount of Gewürztraminer was planted. Indeed, New Zealand's first world-class wines may have been the early Matawhero Gewürztraminers from the warm, sunny though rainy, Gisborne region. With the current advance of Chardonnay and Sauvignon Blanc there's not a mass of Gewürz left – the planted area decreased by more than 50 per cent during the 1990s – but Gisborne can still excel.

Germany's more northerly vineyards can be too cool; Gewürztraminer is a more natural choice for the warmer lands of the Pfalz and Baden. On the other hand the high mountain vineyards of Italy's Alto Adige – again loads of sun but cool – are ideal; however, few growers limit the yields enough to make more than mildly aromatic examples. Most New World countries have a little Gewürztraminer, but the conditions are almost always too hot to avoid flabbiness. In Australia, Riesling is blended in to sharpen up the wine. Chile has used Sauvignon Blanc as an effective blender – again, to avoid acidity.

## Soil

Gewürztraminer is fairly unfussy about soil, though it may be that the richness of Alsace versions comes from the soil above all else. There is certainly a spiciness that pervades all Alsace's white varieties. Alsace producer Olivier Zind-Humbrecht MW, who has made a speciality of terroir, states that it is the high percentage of calcareous or limestone soil in Alsace that gives the region's wines their spicy character. Gewürztraminer also seems to like a proportion of clay, and is best suited to rich, deep but well-drained soils, with plenty of minerals. For every Alsace grower who stresses the importance of limeestone there is another who stresses the clay: Albert Mann, for example, believes that the reason that acidity in Alsace Gewurztraminer can be low is because so many of the vines are planted on clayey soils.

Some of the best Grand Cru vineyards for Gewürztraminer are Altenberg and Kanzlerberg (in Bergheim), Eichberg (Eguisheim), Hengst (Wintzenheim), Kessler

*Gray Monk vineyard and winery above Lake Okanagan in British Columbia, Canada. Gray Monk has just over half an acre/0.2ha devoted to Gewürztraminer. The harshness of the Canadian winter is mitigated by the proximity of the lake and the angle of the hillside, which helps cold air drain from the vineyards.*

and Kitterlé (Guebwiller), Kirchberg (Barr), Mambourg (Sigolsheim), Schlossberg (Kientzheim), and Zinnkoepflé (Soultzmatt and Westhalten). Affenberg, Bollenberg, Kaefferkopf and Zahnacker are other good vineyards. Zind-Humbrecht adds that the soil type can affect the skin colour of Gewürztraminer as well as the taste: in his experience limestone soils give more orange skins with visible darker orange lines; acidic or gravelly soils give more purple colours.

### Cultivation

The vine is naturally vigorous, and problems of balance and structure become far worse if it is allowed to yield generously. Serious Gewürztraminer producers are all agreed – there is a direct correlation between low yield and intense perfume. Many would put the crisis point for loss of quality at no higher than 40hl/ha. Alsace Grand Cru yields, now reduced to 55 hl/ha from 60hl/ha, are still too high for good quality, and top growers take much less. Also, 55hl/ha is only the base yield (*rendement de base*) for Alsace Grand Cru: there is now a 'buffer' yield set at 66hl/ha, which has been introduced as a sop to the producers because an even bigger dodge has been removed. This was the system by which the growers could produce higher yields of one grape variety to compensate for low yields in another, providing the total did not exceed 60hl/ha.

*Freshly ploughed soil ready for replanting in the Grand Cru Altenberg de Bergheim in Alsace. The mix of chalk and clay here is ideal for Gewürztraminer.*

Many of the rootstocks in Alsace encourage vigour, while in Germany high-yielding clones usually compromise quality. Excess vigour increases the risk of stem rot, to which the vine is very susceptible.

In New Zealand high-quality Gewürztraminer cannot be cropped over 2–3 tons per acre. One grower points out that if you are a commercial grower, you are probably going to be able to sell your Gewürztraminer grapes for NZ$1000–1200 per ton. It is obviously far more profitable to get paid $2000 per ton for similar yields of the Mendoza clone of Chardonnay – or even $1400 per ton for Sauvignon Blanc that you have cropped at 6 tons plus per acre. The tendency is therefore to try and crop Gewürztraminer at higher yields, to get a higher return. But then quality suffers, and you get into a vicious circle of overcropped, faintly perfumed wine in a market that does not know much about Gewürztraminer anyway. This, basically, is why plantings are falling in New Zealand.

### Clones

There is not a great deal of clonal variation reported, which is perhaps surprising in view of the difficulties of deciding between what is Savagnin Rose, Gewürztraminer, and Traminer. The colour of the grapes is not necessarily a result of clonal variation, although Alsace producer Olivier Zind-Humbrecht points out that modern, high-yielding clones tend to have darker, more purple colours, and softer, thinner skins. But he attributes the yellow or grey colour of some Gewürztraminer to different growing conditions and insufficient time to mature the colour of the skins. Potential alcohol levels can be over 13 or 14 per cent, but if a hot climate has encouraged too rapid a development of sugar, the grapes will not be physiologically ripe at that level and the seductive hothouse scent simply won't have developed in the grapes.

However, Gewürztraminer ripens unevenly. It is not unusual to find berries with 15 per cent potential alcohol on the same cluster as green berries. This makes vinification remarkably difficult, even if the average ripeness looks satisfactory.

### VINIFICATION

Olivier Zind-Humbrecht of Alsace describes his winemaking thus:

'All the quality of Gewürztraminer is in the skin, so it pays to press the grapes slowly and gently, in order to extract the maximum of flavour. But lack of flavour is not the main worry, so one very rarely finds skin contact in Alsace. Gewürztraminer tends to ferment quickly to high alcohol levels, so it is important to keep a clear must, ideally from whole cluster pressing, to avoid the extraction of tannin from the stems, and too much sediment. I am not a big fan of cold fermentations, and I believe that any wine must reach a certain temperature in order to express complexity and not just banana or strawberry character. The richer and more complex a wine is, the more it can stand higher temperatures of up to 23° or 24°C. Poor grapes must ferment cooler in order to preserve the little flavour they have. But temperature control is very important, even if it is used for only a day or two. Gewürztraminer can ferment very quickly and reach temperatures of over 25°C that could damage the wine.

'Instead of lees stirring, I prefer to have fermentations that last three months or more, so that the lees are naturally in suspension in the wine, and stirring is not necessary. I prefer to have yeast alive in suspension in the wine, rather than dead yeasts that have to be stirred regularly.

'The malolactic fermentation is usually prevented in Gewürztraminer, but I let the wine do what it wants to do. If the grapes are very ripe and concentrated, and if the malo happens only late and slowly, perhaps after Christmas, it often makes better wines. I have often had trophy-winning wines that had been through the malo.'

Many New World fermentations are at far cooler temperatures than those in Alsace. But not all: in New Zealand, Dr Neil McCallum of Dry River likes to ferment his Gewürztraminer at 16–17°C, and lets it rise to 20°C as the fermentation progresses. Says McCallum, 'the Cool School [of fermentation temperatures] believe they retain more aromatics at low temperatures, and this therefore is in the wine's interest. On the other hand most Europeans feel that fermentations at higher temperatures yield a wine with more body, while not sacrificing undue varietal flavours. Studies of about five years ago seem to indicate that the major amount of volatile losses are in the first two to three days of the fermentation and it therefore seems possible to embrace both schools to some extent. Natural ferments in Alsace would mostly start at low temperatures.'

# Gewürztraminer around the World

GEWÜRZTRAMINER'S FAME IS FIRMLY BASED on its performance in Alsace, the thin sliver of vineyard land in north-east France bordered by Germany and the river Rhine. Here, Gewürztraminer can – when the most talented growers and producers put their minds to it – produce one of the most astonishing scented wine styles in the world. Such intensity is rarely repeated elsewhere, but the vine does turn up all over the world with varying degrees of success.

## Alsace

The usual planting density here is 4400 to 4800 vines per hectare; the legal minimum is 4500. (The yields per vine in Alsace are high when compared with the 10,000 vines per hectare of the Côte d'Or.) Under new legislation machine-harvesting will no longer be allowed for Grand Cru wines. But over-production is a problem in Alsace, even for the relatively low-yielding Gewürztraminer, and so measures have been introduced to try and curb this. Instead of leaving the over-production on the vine, all vineyards must now be fully picked – the idea being that the thought of paying for all that unnecessary picking might encourage such growers to control their yields better. As always, though, bureaucracy is less inventive than quantity-minded growers, who have sometimes simply uprooted one row in three of their (non-Grand Cru) vineyards in order to allow space for mechanical harvesters and have higher yields per vine.

Acidification is not permitted in Alsace. Small amounts of sorbic acid, ascorbic acid or citric acid are permitted but these are used as anti-oxidants, not as acidifying agents.

As well as intensely aromatic dry wines, Gewürztraminer makes some of Alsace's finest sweet wines, but it is not in fact very susceptible to *Botrytis cinerea*. It is easy to ripen Gewürztraminer to *vendange tardive* richness – the grape can get to 16 per cent potential alcohol without a scrap of noble rot – but it has rather thick skins, which tend to protect it against botrytis.

## Germany

From being a speciality of the Pfalz in the 19th century, Gewürztraminer had declined to just 807ha in the whole of Germany by 1998, of which 349ha were in the Pfalz. Along with other Pfalz styles, the wines are no longer the oily, somewhat blowsy creatures they once were: structure and acidity are generally much better. Flavours tend to be fruitier and more flowery than those of Alsace, seldom reaching such heady spiciness. In Baden, which is effectively a continuation of Alsace across the Rhine (or vice versa, depending on your outlook) there is more Traminer planted in the village of Durbach than in any other in the whole of Germany; it is known locally as Clevner.

## Austria

The grape is more likely to be called Traminer here than Gewürztraminer, though its flavour can be as redolent of roses and spice as that of Alsace. Sunny Steiermark produces both lean, bone dry and barely aromatic versions and rather riper ones. Traminer makes powerful sweet wines in Burgenland and both dry and sweet ones in Vienna and Niederösterreich. For most of its growers, however, it is a minority grape.

## Italy

It is nice to report that Traminer Aromatico, or Gewürztraminer, is doing nicely in its birthplace of Alto Adige, although precious little is found outside the immediate area of Tramin itself. Quality has improved, perhaps because yields are less, and perhaps because fewer vines are trained on pergolas and more on wires, Guyot-fashion. In spite of its Aromatico suffix, few wines have traditionally been very scented here, though they can be attractively elegant to compensate. The vine comes in red, white and pink versions and local wisdom is that 'roter' or red Traminer is generally less aromatic but plumper.

## Rest of Europe

Traminer or Gewürztraminer adopts various names throughout eastern Europe, including Tramini in Hungary and Traminac in Slovenia. In the old 'Yugoslavian' days, Slovenian Traminer was fat and scented and things should only improve as Slovenia modernizes its wine industry. It is widely grown by Lake Balaton in Hungary, where soils are rich and volcanic; and, as in Slovenia, it can give full, fiery, lychee- and mango-flavoured dry whites of considerable character. It is also grown in Romania, Russia, the Ukraine, Moldova and Slovakia. It would probably make some very attractive wines here – light and elegant in Slovakia, more substantial and scented in Romania and Moldova. But the

DOMAINE WEINBACH

*This domaine makes exemplary Gewurztraminer that is always among the best in Alsace. Each cask of wine is bottled and sold separately.*

DOMAINE MARCEL DEISS

*The Deiss style of Gewurztraminer is less flamboyant than Domaine Weinbach's, but there is the same emphasis on the importance of terroir.*

ANDREAS LAIBLE

*Probably the best producer in Durbach, Andreas Laible makes Traminer in a weighty Baden style but with more elegance than is customary.*

*The vineyards around the village of Tramin or Termeno in Italy's Alto Adige, where the grape is thought to have originated. Termeno Aromatico is still one of the grape's synonyms, but it has been edged out of most vineyards in its birthplace by other, higher-yielding varieties. What is left makes light but fragrant wine.*

serious socio-economic problems of these newly emergent democracies mean that the grape is regularly overcropped, leading to dull, dilute flavours, and is poorly vinified, leaving unattractive grubby odours where there should be seductive scent. But in countries where expertise and finance are in very short supply, all efforts are likely to be concentrated on easy-to-sell Chardonnay, Sauvignon, Cabernet and Merlot rather than unfashionable old Gewürz.

There are small amounts grown in Switzerland, Luxembourg and Spain.

### New Zealand

New Zealand's cool but sunny climate seems to offer a perfect fit to Gewürztraminer. The drawback, however, is that the wine is difficult to sell, and so plantings fell from 182ha in 1991 to 103ha in 1999. However, figures are projected to creep up again, with an expected 142ha by 2003.

Style-wise, it is the direct opposite of New Zealand's star grape, Sauvignon Blanc; but whereas consumers seem happy to accept the high acidity of Sauvignon as intrinsic to the grape (albeit with a few grams of residual sugar to balance), they seem unwilling to embrace Gewürztraminer's low acidity with the same enthusiasm. I suspect the main problem is that people relate its soft, perfumed style to the cheap, off-dry Müller-Thurgaus that were the mainstay of New Zealand's early wine revival and which are now regarded with disdain. A pity. Acidification is quite common, and some producers put the wines through the malolactic fermentation, either wholly or partially, to produce wines of a different style. Most are dry or off-dry, and there are some late harvest wines, sometimes fermented in new oak.

### USA

Plantings here are falling and, as in New Zealand, the fundamental problem seems to be that while it is a relatively expensive grape to grow, with the usual problems of uneven ripening and low yields, you simply can't get as much money for it as for the more popular, easier to grow varieties like Chardonnay. In the mid-1980s there were nearly 4000 acres/1620ha of Gewürztraminer, but by the late 1990s the figure was around 1700 acres/700ha. In California it suits the cooler areas of Monterey, Sonoma, Mendocino and Russian River Valley well, and styles are usually dry or off-dry, though Anderson Valley seems able to produce some promising dessert versions. It is found also in Oregon and Washington State, though many wines can be a little too dilute and a little too sweet.

### Rest of the world

In Australia the grape may be called either Traminer or Gewürztraminer; as Traminer, particularly from warmer, highly productive regions, it is often blended with Riesling to improve its acidity. As Gewürztraminer, from cooler areas of Tasmania and Victoria, it can be thrillingly scented. Varietal wines range from dry through off-dry to late harvest.

Canada has the potential to make good Gewürztraminer, and so does Chile: there are one or two distinctly promising Chilean examples and some enterprising blending with Sauvignon Blanc to improve acidity. There are some rare sweet versions from South Africa.

SEIFRIED ESTATE

*One of the New Zealand estates still persevering with Gewürztraminer, Seifried makes wines of elegance, balance and aroma.*

MISSION HILL

*This Icewine was picked on 2 January 1998, when temperatures in British Columbia's Okanagan Valley fell to –14°C, freezing grapes on the vine.*

VIÑA CASABLANCA

*Cool Casablanca seems to be Chile's best region so far for crisp, clean Gewürz like this, though the grape has only 1% of plantings in the valley.*

# Enjoying Gewürztraminer

GEWÜRZTRAMINER SEEMS TO BE ABLE to offer exceptions to the rule that white wines need high acidity in order to age well. Admittedly, in those Alsace vintages where the sun shone too warmly and the wines verge on flabbiness, long aging is not recommended because the wine can quickly develop a bitter edge and a taste of breakfast marmalade. It does need relatively high acidity to make it past its third or fourth birthday.

In years when the balance is good, and providing that yields are kept down, a Grand Cru Gewürztraminer can happily age for ten years. At the end of that period it has lost its exuberant perfume and become winier and more subtle – less like Gewürztraminer, perhaps more like Pinot Gris. But it gains enough in honey and complexity to make the loss of roses and lychees worthwhile.

Few New World wines are made to be aged, though the best will improve for a few years. Late-harvest wines can age, though are so delicious young that it is tempting to drink them then and there. The most concentrated Vendange Tardive (late-harvest) and Sélection de Grains Nobles (super-ripe, sweet, usually botrytized) wines from Alsace will live for 20 years or more – but again, they need respectable, even if not high, levels of acidity to begin with.

### The taste of Gewürztraminer

Ripe, concentrated Gewürztraminer tastes like no other grape, though underripe or overcropped versions can resemble second-rate Muscat, and less aromatic versions, if underripe, can be no better than third-rate Riesling. But we are only interested in good Gewürztraminer here: low-cropped, cleanly made and with enough structure and acidity to give form and balance to all that headiness of perfume.

So what does this fabled perfume consist of? Lovely freshly plucked tea roses, the petals rubbed lasciviously between thumb and forefinger before you gorge yourself on their heady scent. And lychees too, fresh or tinned, it doesn't matter, their flesh almost slithery as your tongue and teeth worry it free of the stone and your palate is amazed by its scented fruit. Face cream, too – particularly Nivea Crème – and a whole range of other scents – cinnamon or lilac, orange blossom and citrus peel, tea and bergamot and honeysuckle.

And, intriguingly, really good Alsace examples also have a fierceness like really fresh ground black peppercorns. In simple wines these perfumes are muslin-light; in weightier ones they can become oily. If acidity is lacking they become flabby and slightly greasy, and one is reminded of butter melting in the sun and marmalade left uneaten at the side of the breakfast plate; unbalanced Gewürztraminer can be every bit as unappealing and unrefreshing as that.

Late-harvest Gewürztraminers have a combination of sweetness and heavy perfume that can make them tricky to match with food, though bottle age helps here, by toning down the perfume to more manageable levels. Botrytis, where it is part of the picture, adds complexity and richness at the same time as it reduces the roses and lychees element.

*In Alsace, Olivier Zind-Humbrecht MW makes some of the weightiest wines in the region, all big and brilliantly complex. Committed Traminerphile Dr Alan Limmer of Stonecroft in New Zealand's Hawkes Bay region makes powerful, opulent examples. 'It's not an easy grape to get right,' he says, 'because the inherent balance is so different to most wines. The simple answer is to make it like all your other wines – but then it's not Gewürz.'*

### Matching Gewürztraminer and food

Traditional Alsace food and wine partnerships include rich, smooth duck or chicken liver pâté (with either dry or sweet examples), onion tart, smoked fish, roast goose, and pungent washed rind cheeses such as Munster.

Young Gewürztraminer is an absolute delight to drink by itself, but it is also excellent with Indian and Chinese dishes, or fusion food redolent of ginger or other spices. South-East Asian food, with its use of lemon grass, coriander and coconut, is often beautifully matched with Gewürztraminer. When the spicing is more subtle, mature Gewürz is an excellent partner for chicken.

## CONSUMER INFORMATION

### Synonyms & local names

Also known as Traminer; as Rotclevner, Traminer Musqué, Traminer Parfumé and Traminer Aromatique in Alsace; as Roter Traminer, Clevner or Klavner in Germany and Austria; as Traminer or Termeno Aromatico, Traminer Rosé or Traminer Rosso in Italy; in Switzerland as Heida, Heiden or Païen; in Eastern Europe as Tramini (Hungary); Traminac (Slovenia); Drumin, Pinat Cervena, Princ or Liwora in the Czech and Slovak Republics; Rusa in Romania; Mala Dinka in Bulgaria; and Traminer in Russia, Moldova and Ukraine.

### Best producers

**ALSACE** Adam, Albrecht, Allimant-Laugner, Barmès-Buecher, Bechtold, Bernhard-Reibel, Beyer, P Blanck, Boesch, Bott-Geyl, Boxler, Burn, Deiss, J-P Dirler, Dopff & Irion, Dopff au Moulin, Eguisheim co-op, Pierre Frick, Albert Hertz, Hugel, Hunawihr co-op, Josmeyer, Kientzler, Kreydenweiss, Kuentz-Bas, Seppi Landmann, J-L Mader, Albert Mann, des Marronniers, F Meyer, Meyer-Fonné, Mittnacht-Klack, René Muré, Ostertag, Pfaffenheim co-op, Rieflé, Rolly Gassmann, Eric Rominger, Schaetzel, A Scherer, Charles Schleret, Schlumberger, Schoffit, Gérard Schueller, Jean Sipp, Louis Sipp, Bruno Sorg, Pierre Sparr, J-M Spielmann, Tempé, Trimbach, Turckheim co-op, Weinbach, Zind-Humbrecht.
**GERMANY** Fitz-Ritter, A Laible, U Lützkendorf, Heinrich Männle, Wolff-Metternich, Klaus Zimmerling.
**AUSTRIA** Fritz Salomon.
**ITALY/Alto Adige** Abbazia di Novacella, Caldaro co-op, Colterenzio co-op, Cornaiano co-op, Franz Haas, Hofstätter, Prima & Nuova/Erste & Neue, San Michele Appiano co-op, Termeno co-op, Elena Walch; **Trentino** Cesconi, Pojer & Sandri.
**SPAIN** Enate, Raimat, Torres, Viñas del Vero.
**USA/California** Adler Fels, Bouchaine, Thomas Fogarty, Lazy Creek, Navarro; **Oregon** Amity, Bridgeview, Eola Hills, Foris, Henry Estate, Tyee.
**CANADA** Gray Monk, Konzelmann, Mission Hill, Sumac Ridge.
**AUSTRALIA** Delatite, Henschke, Knappstein, Moorilla Estate, Piper's Brook, Seppelt, Skilloglee, Audrey Wilkinson.
**NEW ZEALAND** Corbans, Dry River, Hunters, Lawson's Dry Hills, Revington, Rippon Vineyard, Seifried, Stonecroft.
**CHILE** Concha y Toro, Viña Casablanca.
**SOUTH AFRICA** Delheim, Neethlingshof, Simonsig Estate.

### RECOMMENDED WINES TO TRY

#### Ten Alsace dry or off-dry wines

Léon Beyer *Cuvée des Comtes d'Eguisheim*
Paul Blanck *Furstentum Vieilles Vignes*
Ernest Burn *Cuvée de la Chapelle*
Marc Kreydenweiss *Kritt*
Meyer-Fonné *Wineck-Schlossberg*
Mittnacht-Klack *Schoenenbourg*
Schaetzel *Kaefferkopf Cuvée Catherine*
Domaines Schlumberger *Cuvée Christine*
Trimbach *Seigneurs de Ribeaupierre*
Turckheim co-op *Brand*

#### Ten Alsace Sélection de Grains Nobles

Deiss *Altenberg*
J-P Dirler *Spiegel*
Hugel
Kuentz-Bas *Pfersigberg Cuvée Jeremy*
Albert Mann *Furstentum*
René Muré *Vorbourg Clos St-Landelin*
Schoffit *Clos St-Théobald Rangen*
Bruno Sorg
Weinbach *Cuvée d'Or Quintessence*
Zind-Humbrecht *Rangen*

#### Five Italian wines

Cesconi *Trentino Traminer Aromatico*
Hofstätter *Alto Adige Gewürztraminer Kolbenhof*
Caldaro co-op *Alto Adige Gewürztraminer Campaner*
San Michele Appiano co-op *Alto Adige Gewürztraminer Sanct Valentin*
Elena Walch *Alto Adige Gewürztraminer Kastelaz*

#### Ten New World wines

Amity Vineyards *Gewürztraminer (Oregon)*
Delatite *Dead Man's Hill Gewürztraminer (Australia)*
Foris *Rogue Valley Gewürztraminer (Oregon)*
Knappstein *Dry Style Gewürztraminer (Australia)*
Henschke *Joseph Hill Gewürztraminer (Australia)*
Moorilla Estate *Gewürztraminer (Australia)*
Rippon Vineyard *Gewürztraminer (New Zealand)*
Stonecroft *Gewürztraminer (New Zealand)*
Eola Hills *Gewürztraminer Vin d'Epice (Oregon)*
Mission Hill *Grande Reserve Gewürztraminer Icewine (Canada)*

*Gewürztraminer berries may be more orange, more purple or more red than these, depending on the growing conditions. The wine generally has a deep golden colour, though Traminers in the northern Italian mould may be paler.*

### Maturity charts

Most simple Gewürztraminer follows the New Zealand pattern: drink the wines early. Only top Alsace examples age successfully in bottle.

**1998** Alsace Gewurztraminer Grand Cru (dry)

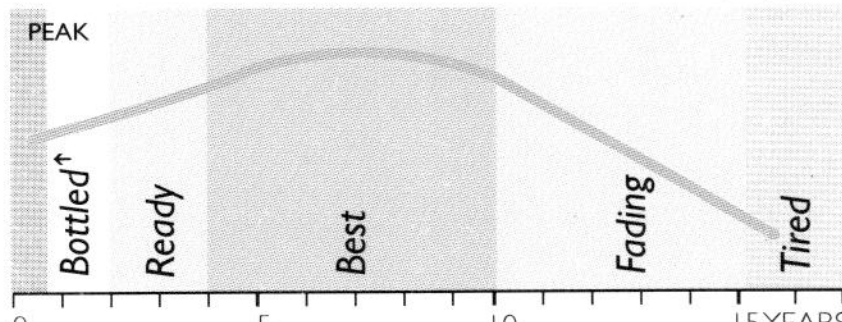


*A difficult year in which some producers excelled at Gewurztraminer and some at Riesling. Good quality depended on low yields and not picking too early.*

**1998** Alsace Gewurztraminer Vendange Tardive

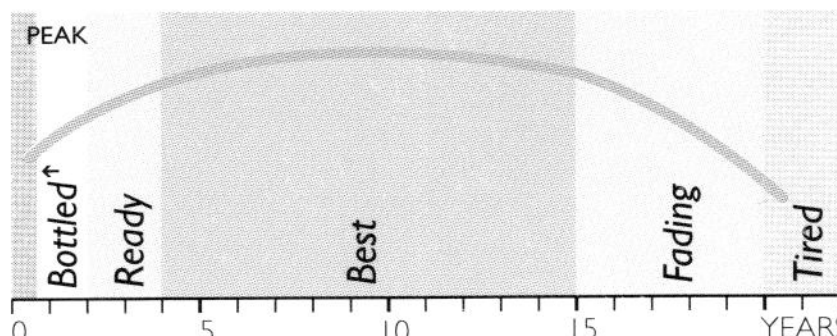


*Even* Vendange tardive *wines may have some noble rot in this botrytis-rich year. A terrific vintage, for these wines, with richness and balancing acidity.*

**2000** New Zealand North Island Gewürztraminer

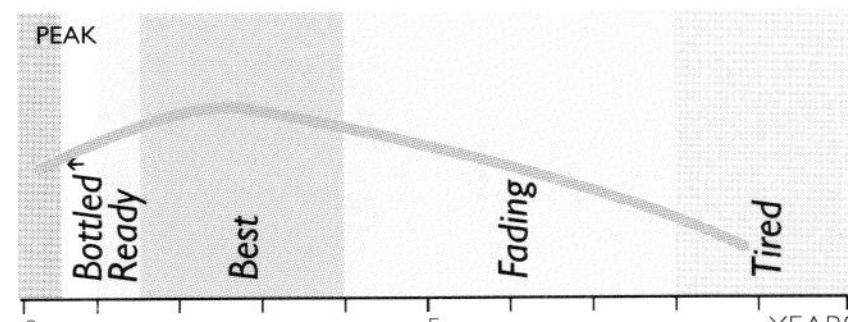


*2000 produced pretty and aromatic Gewürztraminer Like most examples of the grape from New Zealand, it is best without further bottle age.*

## GODELLO

This rich, aromatic, high-quality Spanish grape is almost certainly the same as Portugal's Verdelho (see page 272). As Godello it is the best white grape in Valdeorras, though still only manages to account for 3 per cent of the vineyard: Palomino and Valenciana, or Doña Blanca, far outdo it in terms of area planted. It also appears in other DOs of the far North-West, including Ribeira Sacra, Ribeiro, Monterrei and Bierzo.

It must surely be in the running for the title of Spain's most interesting white grape – a title for which, admittedly, there are but few contenders. It shares the softly aromatic apricot character of Albariño but has a silkier texture, rather like good Viognier. It seems to be native to the gorges of the Sil Valley, but nearly became extinct in the 1970s, and only determined efforts by the Consejo Regulador of Valdeorras saved it. It can be made as a varietal, but also blends well with other grapes, including Albariño, Treixadura and Doña Blanca. Best producers: (Spain) Godeval, La Tapada, Jesús Nazareno co-op, Santa Maria dos Remedios co-op, Señorío.

LA TAPADA
*Excellent Godello from La Tapada in Galicia, north-west Spain – all smoky, mineral and floral fruit with nicely balanced oak – shows exactly why the grape was worth saving from extinction.*

## GOLDMUSKATELLER

A strain of Muscat Blanc à Petits Grains (see pages 144–153) grown in Italy's Alto Adige is known as Goldmuskateller. Best producers: (Italy) Thurnhof, Tiefenbrunner.

## GOUAIS BLANC

An obscure French grape now shown, via the wonders of DNA fingerprinting, to be the ancestor, together with Pinot Noir, of no fewer than 16 modern grape varieties (including Aligoté, Aubun Vert, Auxerrois, Bachet Noir, Beaunoir, Chardonnay (yes, that Chardonnay), Franc Noir de la Haute-Saône, Gamay Blanc Gloriod, Gamay Noir, Knipperlé, Melon, Peurion, Roublot and Sacy. Gouais Blanc itself no longer exists in the regions of central and north-eastern France where it once grew – indeed, many of these wine regions have disappeared. The wine was never admired, and as early as the 16th century the authorities were ordering growers to uproot it. Today some exists in Rutherglen in north-east Victoria, Australia, in a vineyard owned by Chambers, and is blended with Riesling.

## GOUVEIO

The local name in Portugal's Douro Valley for Verdelho (see page 272). Best producers: (Portugal) Churchill, Martinez, Niepoort, Sogrape.

## GRACIANO

A spicy, aromatic, intensely flavoured grape that is far and away the most interesting red vine in Rioja, but which is unfortunately far more capricious to grow than the region's other vines. There are some 400ha planted in Rioja (out of a total vineyard of around 50,000ha), but that's double the area there was five years ago. According to one Spanish authority, however, 'a lot only exists on back labels'.

A couple of bodegas make a licorice-scented varietal Graciano, and just 15 per cent in a blend will add perfume. However, it is less tannic than Tempranillo and tends to oxidize easily: it is not a grape designed for long ageing. It is late budding, susceptible to downy mildew and low yielding – the last two being characteristics not likely to endear the vine to the average grower. It prefers cool, clay-chalk soils and mild, damp climates; it dislikes drought and strong summer heat.

As Morrastel it is sometimes found in the south of France, but in Spain the name of Morrastel is a synonym for the very different Monastrell, or Mourvèdre (see pages 140–141). What is called Morrastel in North Africa can be either grape. Portugal's Tinta Miúda, which is found in the Ribatejo and in Estremadura, is thought to be the same grape as Graciano. Best producers: (Spain) Artadi, Campillo, Contino, CVNE, Viña Ijalba, La Rioja Alta.

## GRANDE VIDURE

The old Bordeaux name for Carmenère (see pages 60-61), a grape almost extinct there but flourishing in Chile. Some wineries in Chile opted to call their Carmenère by this name but since 2001 its use has been banned under EU legislation.

## GRASA

Some authorities maintain that this Romanian vine is the same as Furmint (see page 90), though Grasa, unlike Furmint, is somewhat low in acidity. The leaves, however, are practically identical. It is concentrated in the Cotnari region, where it makes rich, sweet, nobly rotten wines, often given a bit of extra zest by an admixture of Tamaîoasa Românească. Best producer: (Romania) Cotnari Cellars.

## GRASEVINA

Welschriesling (see page 284) may be called by this name in Croatia.

## GRAUBURGUNDER

German synonym for Ruländer (see page 206), or Pinot Gris (see pages 172–173). Best producers: (Germany) Bercher, Schlossgut Diel, Drautz-Able, Dr Heger, Müller-Catoir, Salwey.

## GRECANICO DORATO

Plantings of this Sicilian vine are on the increase, and the wine has very good flavour, with some grassy pungency. Best producer: (Italy) Planeta.

## GRECHETTO

An interesting central Italian grape that adds structure, richness and a pleasing nuttiness and leafiness to many Umbrian whites. Good Orvieto owes a lot to Grechetto; so do the Antinori's Torgiano and Cervaro whites. It is generally blended: with Chardonnay in Cervaro, with Trebbiano, Malvasia and Verdello in others. It is low yielding but sufficiently disease-resistant to compensate. It makes good Vin Santo. Best producers: (Italy) Antonelli, Avignonesi, Barberani-Vallesanta, Arnaldo Caprai, Falesco, Palazzone, Castello della Sala.

## GRECO BIANCO

Greco Bianco has been cultivated in the south of Italy for some 2500 years, at which time it was imported from Greece, as its name suggests. Whether it is the same as any vine currently found in Greece is unknown; it is also not clear whether Greco Bianco is one variety or more than one. In any case, it is often rather good, in a neutral sort of way. The flavour is slightly peachy and leafy, fresh and brisk. Greco di Tufo is its best known form; semi dried grapes can also be used to make sweet Greco di Bianco which comes, confusingly, from a Calabrian town called Bianco. Best producers: (Italy) Botromagno, Feudi di San Gregorio, Mastroberardino, Giovanni Struzziero.

*Greco Bianco growing near the town of Bianco in Calabria, Italy for – yes – Greco di Bianco.*

## GRECO NERO

A dark-skinned version of Greco Bianco, also found in southern Italy. Best producer: (Italy) Odoardi.

## GRENACHE BLANC

Intrinsically a rather dull grape which, like its black counterpart, Grenache Noir, oxidizes easily and is rather low in acidity. However, it contributes a certain softness and fleshiness to a blend and is widely planted across the south of France. It features heavily in much white Châteauneuf-du-Pape and Côtes du Rhône and to a limited extent (10 per cent, to be precise) in some Côtes du Rhône-Villages. It also appears in some Rivesaltes Vin Doux Naturel from Roussillon.

It is also now being used as a varietal, and modern winemaking wizardry can work wonders in turning it into very attractive, though rarely thrilling wine. Low temperature fermentation is essential, and can produce a fresh, dill-scented wine to drink very young. Alternatively the whole gamut of maceration on the skins, and new oak aging and/or fermentation, lees aging and perhaps a dash of something more aromatic like Muscat, can give it more memorable quantities of flavour.

As Garnacha Blanca (see page 91) it is found in north and north-east Spain, where, like Garnacha Tinta, it originated. Best producers: (France) de Beaucastel, de Caraguilhes, Clavel, Font de Michelle, la Gardine, de Mont-Redon, la Nerthe, Rayas.

## GRENACHE NOIR

See Garnacha pages 92–101.

## GRIGNOLINO

A pale but not hugely interesting red grape that originated in Piedmont in north-west Italy and has never strayed; and the wine itself is made and mainly drunk locally.

The acidity and tannin are high, and the flavour varies between fresh leafiness and herbal rasp, and tired vegetable stock, according, apparently, to the luck of the draw. There is great clonal variation, which may account for some of the differences in style. The vine also tends to reflect its soil type quite strongly. At its best it is quite attractive though curious – not exactly a wine you'd want every day. It is best drunk young. There are three DOCs: d'Asti, del Monferrato Casalese and Piemonte. For some reason the Heitz vineyard in California's Napa Valley grows a bit – don't ask me why. Best producers: (Italy) G Accornero & Figli, Braida, Bricco Mondalino, Dante Rivetti, Cantine Sant' Agata, Vinchio e Veglio Serra co-op.

## GRILLO

One of the better grapes for Marsala, Sicily's fortified wine, Grillo reaches high levels of potential alcohol – always a plus point for makers of fortified wine, because it saves on (expensive) brandy. But plantings are nevertheless on the decline because of the shrinking market for good Marsala. It's a potentially interesting grape, though, with lemony fruit, robust weight and structure. Best producers: (Italy) De Bartoli, Florio, Pellegrino.

## GROLLEAU

The most interesting thing about Grolleau or Groslot is that, according to Galet, its name is derived from the old French world *grolle* or *grole*, meaning 'crow'; the comparison is with the deep black of Grolleau's berries. It is curious but appropriate that while the more aristocratic Merlot grape is said to be named after the melodious blackbird, or *merle*; poor old plebeian Grolleau gets the crow. Grolleau, perhaps depressed by all this, is on the decline in its Loire homeland. It is only allowed into rosé wines, not red, and plays a part in Anjou, Saumur and Touraine Rosé. Its wines are light, low in alcohol, a bit earthy and high in acidity. Grolleau Gris is a pink-berried version. Best producers: (France) Bouvet-Ladubay, Pierre-Bise, Richou.

## GROS MANSENG

An intensely flavoured grape of south-west France, now beginning to be revived as winemakers learn how to get the best from it. It is found in Béarn, Jurançon and Gascony, where it is considered a *cépage améliorateur* for its high acidity and floral, spicy, apricot-quince fruit. It is related to Petit Manseng (see page 167), and there are forms of the vine that appear to be between the two varieties.

It is quite a good yielder, and gives up to 70 or 80 hl/ha without any adverse affect on quality. What problems it does have are in the winery. In colour it is *gris* rather than white, with deep, golden-coloured grapes, and the skins have high levels of tannins, colour and polyphenols. If the grapes are treated at all roughly in the winery the results will be coarse: it must be pressed extremely delicately, and carefully settled so that only clean juice goes to the fermentation vats.

It can produce interesting, though different results at a wide spectrum of ripeness. If picked at 11.5–12 per cent potential alcohol the wine will be fresh, flowery and pretty. The grapes are also easier to work with at this level. But to get the full potential out of the grape it is necessary to pick at 12.5–13.5 per cent potential alcohol. Then the wines are much more powerful, but still with high acidity, and still dry.

It can also make very good sweet wines, and recent experiments with new oak have shown wines full of personality not unlike oaked Sémillon, much valued locally for drinking with foie gras. These wines may be from late-harvest or noble-rotted grapes.

It reflects its terroir very strongly, giving fresh wines on chalky clay, richer, fatter wines on sandy clay, and light, delicate wines on sand. Best producers: (France) d'Aydie, Brana, Cauhapé, Clos Lapeyre, Clos Uroulat, de Souch.

## GRÜNER VELTLINER

A versatile Austrian grape that is responding interestingly to the huge improvement in winemaking that has taken place there recently. It covers over a third of the Austrian vineyard, so quality is unlikely to be consistent. At its simplest it forms Austria's everyday jug wine, is fresh and peppery, and best drunk within months of the vintage. The best growers in the Wachau and Kamptal regions of Niederösterreich produce much more structured, ripe wines, however, which bring out the lentil, celery and white pepper aroma of the grape, with sometimes a touch of citrus and honeyed weight. It is never very aromatic, but can age well in bottle, becoming more honeyed after a few years and yet staying peppery. It is site sensitive and favours loess soil but is planted pretty well anywhere. Yields are high, up to 100hl/ha for everyday wine.

Grüner Veltliner is also found in the Czech Republic and Hungary. Its names here are Veltlin Zelene or Veltlinkske Zelené, and Zöldveltelini, respectively.

Austria also has three other Veltliner vines: Roter, Brauner and Frühroter. All are dark-skinned and while the first two are planted only in very small quantities, Frühroter Veltliner (red-skinned rather than black) is slightly more widely planted, mostly in the Weinviertel region. It has lower acidity than Grüner Veltliner. Best producers: (Austria) Bründlmayer, Fred Loimer, Franz Hirtzberger, Josef Jamek, Emmerich Knoll, Lenz Moser, Familie Nigl, Nikolaihof, Willi Opitz, Franz Xaver Pichler, Prager, Dr Unger, Freie Weingärtner Wachau.

**BRÜNDLMAYER**

*This single-vineyard Grüner Veltliner, with flavours of yellow plum, white pepper and white chocolate fruit, shows how impressive and complex top examples can be.*

## GUTEDEL

Gutedel, or Chasselas, as it is better known, is far less famous than its offspring: by virtue of an unscripted encounter with Riesling it is a parent of Müller-Thurgau. 'Unscripted' because until recently the parents of Müller-Thurgau were believed to be Riesling and Silvaner, which is presumably what Dr Müller of Thurgau himself thought. DNA fingerprinting has demonstrated the truth: clearly there is no limit to what vines will get up to, left to themselves. Gutedel is less important in Germany than Chasselas is in Switzerland (see page 59). It is found almost entirely in the Markgräferland region of Baden where it makes typically neutral, slightly earthy wine. There is also a dark-berried version, Roter Gutedel. Best producers: (Germany) Blankenhorn, Dörflinger, Pfaffenweiler co-op, Schneider.

## HANEPOOT

Muscat of Alexandria is known by this name in South Africa. See pages 144–153.

## HARRIAGUE

Tannat (see page 241) may be called by this name in Uruguay, though the ampelographer Galet suggests that the Harriague of Uruguay might in fact be Graciano (see page 112).

## HÁRSLEVELÜ

Tokaj's second main grape variety is found in other parts of Hungary as well, but so far few examples have the delicate pollen scent and elderflower fruit which the grape can boast when made as a dry wine. In Tokaj its sweet wines are typically smoky and spicy, robust and intense, and fat sometimes to the point of being slightly oily. Hárslevelü is also found in Slovakia and in South Africa. Best producers: (Hungary) Château Megyer, Château Pajzos, Disznókö, Oremus, Royal Tokaji Wine Co.

## HEIDA/HEIDEN

Traminer (see page 270) takes this name in Switzerland.

## HONDARRABI BELTZA

The dark-skinned grape of Spain's Basque country, where it features in the Chacolí de Guetaria and Chacolí de Vizcaya DOs (or Getariako Txakolina and Bizkaiko Txacolina if Basque is your preferred tongue). It covers about 11 per cent of the vineyard in the former and nearly 20 per cent in the latter: both regions are dominated by white wines rather than red or rosado. Hondarrabi Beltza's wines are fresh, acidic and best drunk young. The vine is believed to have originated in the Pamplona Valley, and according to the ampelographer Galet could be related to Cabernet Franc. Best producer: (Spain) Txomin Etxaniz.

## HONDARRABI ZURI

White Hondarrabi dominates the vineyards of Chacolí de Guetaria /Getariako Txakolina and Chacolí de Vizcaya/Bizkaiako Txacolina, covering some 89 per cent of the planted area in the former, and about 80 per cent in the latter. The former is slightly fuller and richer than the latter, but being light, lemony and acidic both are ideal for the local seafood. They should always be drunk young.

The amount of land under vine in the Basque country is far less than it was before phylloxera, and while the wines are prized locally they are seldom seen outside the region. White Hondarrabi seems to have originated here and never travelled anywhere else. The vines are trained on a modified version of pergolas; yields average less than the legal maximum of 93.6hl/ha, and generally work out at around 60hl/ha. The wines generally contain an admixture of Folle Blanche and perhaps a little Chardonnay, Sauvignon Blanc or Riesling. Best producers: (Spain) Eizaguirre, Txomin Etxaniz, Virgen de l'Orea.

## HUMAGNE BLANCHE

One of the less aromatic Swiss varieties. The wine does, however, have some attractive plumpness and freshness. It seems to have been cultivated in the Valais region since at least the 12th century.

## HUMAGNE ROUGE

Rustic Swiss grape grown in the Valais region, but only to a small extent. The wines have relatively high tannin. The vine doesn't seem to be related to Humagne Blanche.

## HUXELREBE

Perfumed German crossing, propagated in 1927, that produces grapes with high sugar levels and plenty of flavour, but not much elegance. It is found in declining though still substantial quantities in the Pfalz and Rheinhessen. It is capable of producing enormous crops, and if it does the flavour will be accordingly diluted. It is also grown in England where it can produce delightful grape juice and elderflower-scented, relatively dry wines in these cooler conditions. Best producers: (Germany) Kurt Darting, Geil, Koehler-Ruprecht, Johann Ruck, Schales; (England) Biddenden, Carr Taylor, Staple St James, Three Choirs.

## INZOLIA

A good-quality, low-yielding grape found in Sicily, where it is part of the blend, along with Catarratto and perhaps Trebbiano, in many of the island's white wines. Its wine is fresh and rather racy at its best. It is part of the Marsala blend (at least, the genuine Marsala blend as understood by leading producer De Bartoli), along with Grillo, and the grapes may be partially dried and made into sweet wine. It is also planted in Tuscany, and is sometimes known as Ansonica or Anzonica. Best producers: (Italy) Colosi, De Bartoli, Florio, Pellegrino, Duca di Salaparuta, Tasca d'Almerita.

## IRSAY OLIVER

There's nothing very complex about this eastern European crossing of Pozsony with Pearl of Csaba – and if its parents sound unfamiliar it is because it was bred as a table grape. It has a pleasant, rather Muscatty perfume, and lowish acidity. It is found in Slovakia and Hungary, where it is called Irsai Olivér. If you like grapy dry whites, these are never expensive. Best producers: (Hungary) Balatonboglar Winery; (Slovakia) Nitra Winery.

## JACQUÈRE

Jacquère is found in the French region of Savoie, where it produces light, mountain-fresh white wines for drinking young. Best producers: (France) Blard et Fils, Pierre Boniface, Charles Gonnet, Philippe Monin, Jean-Claude Perret.

## JAEN

Early ripening, easy-to-grow variety that is the second most planted grape, after Baga, in Portugal's Dão region. Its colour is good but its acidity and tannin are low, so it is usually blended with other varieties, often the more aristocratic Touriga Nacional, or Alfrocheiro Preto. It is believed to be the same vine as Spain's Mencía, and therefore of Spanish origin. Best producers: (Portugal) Quinta das Maias, Quinta dos Roques, Sogrape.

## JOÃO DE SANTARÉM

Another name for Castelão (see page 58) or Periquita (see page 167). Best producers: (Portugal) Horta de Nazaré, Quinta de Lagoalva de Cima, Herederos de Dom Luís de Margaride.

## JOHANNISBERG

Swiss name for Silvaner (see pages 242–243) from the Valais region, where it is sometimes planted in the best sites. The wines can be rather plump and tasty. Best producers: (Switzerland) Caves Imesch, du Mont d'Or.

## JOHANNISBERG RIESLING

The name, or one of the names, given to Riesling in California. Best producers: (USA) Callaghan Vineyards, Chateau St Jean, Gainey. See pages 190–201.

## KADARKA

This Hungarian vine is on the decline partly because it is late ripening and prone to grey rot, and partly because the export market has turned so decisively against what was its main wine, Bull's Blood from Eger. It can make rather good weighty, tannic red but it needs careful cultivation for this, and its natural vigour needs to be kept in check. If yields are too high, which they almost always are, and if it is picked before it is fully ripe, which is almost always is, it makes dull, dilute wine. It is still grown in Hungary, especially on the Great Plain and in Szekszárd, but plantings are smaller than they were.

It is found in small quantities in Austria's Neusiedlersee, in Romania (as Cadarca) and in Vojvodina. It plays a bigger part in Bulgaria, where it is called Gamza and is found in the north. There is also a white grape called Izsáki, or White Kadarka, which is grown on Hungary's Great Plain too. Whether it is related is unclear. Best producers: (Hungary) Hungarovin; (Bulgaria) Domaine Boyar.

## KÉKFRANKOS

The Hungarian name for Austria's Blaufränkisch (see page 42). It produces reasonable quality in Hungary. Best producers: (Hungary) Hungarovin, Akos Kamocsay.

## KÉKNYELU

Potentially aromatic and high quality Hungarian grape, now hardly grown.

## KÉKOPORTO

Blauer Portugieser (see page 42) takes this name in Hungary. It is also sometimes known as Oporto. Best producer: (Hungary) Hungarovin.

## KERNER

A reasonably good quality vine bred in Germany in 1969. It is one of the better modern crossings (Trollinger x Riesling), which perhaps is not saying a great deal, and German growers have loved it from the start for its combination of high sugar levels and high yields. From the consumer's point of view, the flavour is not bad, being less vulgarly perfumed than most crossings and rather nearer to Riesling in its balance and character. But it does not hold a candle to good Riesling. It is concentrated in the Pfalz and Rheinhessen, where it is one of the better ingredients in Liebfraumilch, but at its best is worth bottling as a varietal. There are also plantings in Württemberg and Mosel-Saar-Ruwer. Best producers: (Germany) Jürgen Ellwanger, W G Flein, Geil, Karl Haidle, Jan Ulrich.

## KLEVNER

A name applied in Alsace to several grape varieties, among them the Pinot family and Chardonnay; the same usage is common in German-speaking regions. It is not the same as Alsace's Klevener de Heiligenstein, which is Savignin Rose (see Gewürztraminer pages 102–111). Best producers: (France) Marc Kreydenweiss; (Germany) Andreas Laible.

## KOTSIFALI

Cretan variety giving soft, spicy, broad wines and best blended with something with a bit more backbone. Best producers: (Greece) Archanes co-op, Peza co-op, Miliarakis Bros.

## LAGREIN

Interesting, mouthfilling Italian variety found in Trentino-Alto Adige. It has a flavour of sour plums with a touch of grass and bitter cherries, some black chocolate richness and a deep, dark colour, but it is not very tannic, and so it's criticized for not being ageworthy.

Well, first, how long do you want it to age? Its low tannin gives it that rare combination of good chewy dark fruit and easy drinkability. And second, I've seen deep, satisfying chocolate and plum 10-year-old examples that had lost nothing and gained quite a bit by the sojourn in bottle. Some producers give the wine more substance by aging it in barrique, but it doesn't really need it. Without some added oak the style of Lagrein is very much in the mould of other north-eastern Italian reds, such as Refosco dal Peduncolo Rosso, Marzemino and Teroldego. If you like them, you'll have a soft spot for Lagrein, too.

Lagrein Dunkel or Scuro is the name of the red wine; Kretzer or Rosato is pink, though usually pretty dark pink. Lagrein di Gries or Grieser comes from just outside Bolzano in Alto Adige. One Italian consultant winemaker sees Lagrein as having potential in New Zealand. Best producers: (Italy) Viticoltori Alto Adige co-op, Barone de Cles, Colterenzio co-op, Muri Gries, Gries co-op, Josephus Mayr, Niedermayr, Hans Rottensteiner, Santa Maddalena co-op, Thurnhof, Tiefenbrunner, Peter Zemmer.

## LAIRÉN

Airén may be encountered under this name in southern Spain (see page 34).

## LAMBRUSCO

The Italian grape and the wine have the same name, and in addition the grape has numerous subvarieties, some of which may have their names specified on the label for DOC wines. But as well as differences between different Lambrusco vines, there are also differences between different Lambrusco wines. The main division is between 'proper' Lambrusco – dry, frothy, strawberry-fruited and with a bitter twist on the finish, low in tannin and ideal for drinking with the fatty, pork-based diet of the Modena region in Emilia-Romagna – and industrial Lambrusco, sold in screwcapped bottles, often sweetened, and with only a distant memory of the refreshing character that makes real Lambrusco so desirable.

Proper Lambrusco comes in cork-stoppered bottles, and is DOC. It is also more expensive and more difficult to find outside the region, and especially outside Italy: sweetened Lambrusco was, from the late 1970s, exported to the USA in huge quantities. Pink, white and low alcohol versions were also invented. There is no white Lambrusco grape: Lambrusco Bianco is made by vinifying the grapes without the skins.

On to the vines. It (or they) seems to be native to the region, and has been there long enough to develop a lot of clonal variation. The best sub-varieties (and there are many) are: Grasparossa, which is found around the village of Castelvetro, south of Modena; Salamino (its clusters are supposed to resemble the shape of salami sausages), which is found at Santa Croce, east of Sorbara; and the most highly prized, Sorbara, from the village of that name.

There are four DOCs which, needless to say, do not coincide precisely with the sub-varieties. DOC Lambrusco di Sorbara must be made from Sorbara and Salamino grapes; DOC Lambrusco Grasparossa di Castelvetro must be 85 per cent Grasparossa; DOC Lambrusco Salamino di Santa Croce must be 90 per cent Salamino; DOC Lambrusco Reggiano is made from a blend of Salamino with the lesser Marani, plus the even lesser Maestri and Montericco. DOC Lambrusco Reggiano is often made *amabile*, or lightly sweet, this sweetness coming from up to 15 per cent of a grape called Ancellotta. There is a lot of Ancellotta in the region; more, in fact, than of any individual strain of Lambrusco. Lambrusco Reggiano, the most produced, is seldom the most interesting of the DOCs, though even so it can be good. Best producers: (Italy) Casimiro Barbieri, Francesco Bellei, Casali, Cavicchioli, Moro Rinaldo Rinaldini, Vittorio Graziano.

*Lambrusco vines are trained high, on pergolas, for the eponymous wine. In times past they used to be trained up tree trunks: even today, poplars with vines climbing up them are a familiar sight in Italy.*

## LASKI RIZLING

The name under which Welschriesling (see page 284) is most often encountered in the countries of the former Yugoslavia. The branded version exported for many years by the state-run winemaking operation gave the grape a bad name; undeservedly so, since the style of that wine owed more to poor equipment, bad handling and the desire to sell at the lowest possible price than it did to the grape's potential. Best producers: (Slovenia) Slovenjvino, Stanko Curin.

## LEÁNYKA

Hungary's name for the Fetească Albă of Romania (see page 89). Best producer: (Hungary) Hungarovin.

## LEMBERGER

Blaufränkisch (see page 42) is known by this name in Germany and in Washington State. Limberger is an alternative German name but less common. In Germany it is found in Württemberg, where it produces pale-coloured wines that are light but have decent acidity. It is often blended with Trollinger (see page 270). Best producers: (Germany) Graf Adelmann, Drautz-Able, Fürst zu Hohenlohe-Öhringen; (USA) Covey Run, Hogue Cellars, Hoodsport.

## LEN DE L'EL

The name of this grape, native to south-west France, means 'far from sight' – it is a corruption of 'Loin de l'oeil'. A curious name for a vine, you might think: the clusters have long stalks, and so are a long way (well, relatively) from the eye, or bud, from which they sprang.

It's true that plantings have declined, but there is a minimum (15 per cent) of Len de l'El required in the blend for white Gaillac (although Sauvignon Blanc is allowed as an alternative) so it is to be hoped that this characterful grape will not disappear completely from view. It is low in acidity and the grapes rot easily, but the wine is powerful, weighty and good. Best producers: (France) de Causses-Marines, de Labarthe, Labastide-de-Lévis co-op.

## LIMBERGER

A less common German name for Lemberger (see above) and the Blaufränkisch of Austria (see page 42).

## LIMNIO

One of Greece's most important red vines, this is still found on its original home of

Lemnos, though for some reason is not used for appellation wine there. It is also found in the north-east of the country, where it contributes colour, weight, acidity and a flavour of bay leaves to a blend. Best producer: (Greece) Domaine Carras.

### LISTÁN

Another name for Palomino, one of the world's most boring grapes – but one that paradoxically can produce extraordinarily complex flavours when turned into sherry. It is found also in France's Languedoc and, further west, in the Armagnac vineyards of Gascony, but it is being replaced here by better varieties. See Palomino Fino page 165.

### LISTÁN NEGRO

The black version of Listán, found on the Canary Islands, where it is blended with such others as Negramoll (the Tinta Negra Mole of Madeira), Prieto, Tintilla, Malvasia Rosada and others. In Tacoronte-Acentejo on Tenerife it may make up most of the blend, or be made as a varietal.

It is the main red vine on Lanzarote, where the vines are planted in hollows in the black volcanic soil, and sheltered from the strong humid winds by low stone walls. Its wines are mostly soft and appealing, without tremendous weight, and may be made more substantial by oak aging, or softer and lighter by carbonic maceration. There is also a small amount of sweet red produced on Tenerife from Listán Negro. Best producers: (Spain) Insulares Tenerife, La Isleta, Monje, La Palmera.

### LLADONER PELUT

Another name for Spain's Garnacha Peluda, or hairy Garnacha. A southern French name is Grenache Poilu or Velu. The only difference is that the underside of the leaves have more down on them than normal Garnacha/Grenache. It is specified separately in many ACs in Languedoc Roussillon, and the vine is less prone to rot, but the flavour of the wine is very similar to that of Garnacha/Grenache (see pages 92–101).

### LOUREIRO

Aromatic grape used for Vinho Verde in the north of Portugal and, as Loureira, for the very similar wines of Rías Baixas, over the border in north-west Spain.

Its aroma is reminiscent of bay leaves, with high acidity and low alcohol. It seems especially suited to the cooler parts of the region around Braga and near the coast; the flavour can sometimes remind one of young Riesling. Yields are high and the wine is either used as a varietal or blended with Trajadura and Paderña in Vinho Verde, or, in Rías Baixas, made as a varietal or blended with Albariño and/or Treixadura. Best producers: (Portugal) Quinta da Aveleda, Quinta da Franqueira, Sogrape, Solar das Bouças, Quinta do Tamariz.

*Black grapes, black soil. These vines are Listán Negro, planted on the island of Lanzarote, where stone walls are necessary to protect the vines from the strong humid winds.*

### MACABEO

A non-aromatic variety found across northern Spain and, as Maccabéo or Maccabeu, in France's Languedoc-Roussillon. It brings neutrality, a resistance to oxidation and sometimes good palate weight to many white blends, but not flavour or character.

As Viura (see page 284) it is the backbone of white Rioja, along with lesser quantities of Malvasia and Garnacha Blanca, but white wine has always been something of an afterthought in Rioja, and it was used in the past as a softener for blending in with the red.

Macabeo is used for Cava both in Cataluña and elsewhere in northern Spain: Catalan Cava contains an admixture of Parellada and Xarel-lo, though the small amount made in other regions tends to be pure Macabeo.

In Languedoc-Roussillon Macabeo is used for Vins Doux Naturel, and is blended into many white wines in Minervois and Corbières. Its affinity for hot, dry climates – it is prone to rot in wet ones – means that it has also been grown in North Africa. Best producers: (Spain) Agramont, Cosecheros Alaveses, Castellblanch, Cavas Hill, Codorníu, Franco Españolas, Freixenet, Marqués de Monistrol, Masía Barríl, Pirineos; (France) Casenove, Vignerons Catalans, Les Vignerons du Val d'Orbieu.

### MADELEINE ANGEVINE

Dual-purpose table and wine grape vine grown in England for quite aromatic, grassy wine. It is also found in Kyrgyzstan. It is a name that embraces several different vines. Best producers: (England) Three Choirs, Valley Vineyards.

### MALAGOUSIA/MALAGOUSSIA

Greek vine that has recently been pulled back from the verge of extinction. Its acidity is low, but its flavour is ripe and peachy. It is possibly related to Malvasia.

### MALBEC

See pages 118–119.

### MALBECK

An occasional Argentinian spelling of Malbec (see pages 118–119).

### MALMSEY

More a style of wine than a grape, Malmsey is a corruption of Malvasia (see pages 120–121), and was in the past used generically for the sweet wines of Greece and the eastern Mediterranean. It is now used specifically for the sweetest style of Madeira.

However, true Malmsey or Malvasia vines have almost disappeared from the island. Only 13–14 per cent of the vines on Madeira are noble varieties – that is, Sercial, Verdelho, Bual or Malmsey – with the rest being Tinta Negra Mole or hybrids. True Malmsey Madeira is thus a fairly rare wine, though any bottle that calls itself Malmsey on the label should be made from Malvasia grapes. When young the wine has an attractive floral orange blossom smell; after the characteristic Madeira aging process it becomes pungent, intense and caramelly, rich and smoky, but never loses its acid tang.

The usual Malvasia grown on Madeira is Malvasia de S. Jorge. It is subject to irregular fruit set, is early ripening and picked at between 11.5 and 15 per cent potential alcohol. Best producers: (Portugal) Barros e Sousa, Blandy, Cossart Gordon, Henriques & Henriques, Leacock, Rutherford & Miles.

# MALBEC

WELL, THIS GRAPE IS A NATIVE of south-west France. But try telling that to a modern wine drinker. It's far more likely that today's enthusiast will say 'Argentina'. And they'd be right. But five years ago when asked to respond to the question – 'What is Malbec?', you'd have been met with a blank stare and a distinct lack of interest. Ah, the wonderful modern world of wine, where everything can change in a vintage. Malbec is now a seriously thrilling rising star in the red wine firmament. And that's entirely thanks to Argentina.

In its birthplace of Bordeaux and its traditional French home base of Cahors it was going nowhere – fast. It's a soft, juicy grape that gives lovely dark, damsony, perfumed, purple wine in a dry warm climate – but it never recovered from the severe frost of 1956. So its ability to soften the Bordeaux blend has been supplanted by the Merlot – which is less susceptible to *coulure* and rot and gives a more regular crop. To be honest, it was never grown in very large quantities, and few estates had more than 10 or 15 per cent.

It is a blending component in many south-west reds, but only leads the field in Cahors, where it must comprise at least 70 per cent of the blend. In a warm dry year (not that common) it can give deep dark wines tasting of damson skins and tobacco leaf. It is grown to a limited extent in the Loire, where it is called Cot, but it rarely ripens fully there.

It was first propagated in Argentina in 1852, but the cuttings came from Bordeaux, not from Cahors, and the Malbec in Argentina now seems to be rather different from that in Cahors. Massal selection in both places has taken the vine in different directions. The berries tend to be smaller in Argentina and the tannins ripe, and the wine is capable of successful aging rather more regularly. In warmer parts of Argentina the acidity can plummet, and the fruit can become somewhat flabby and unfocussed.

A vine pull scheme in Argentina, in the ten years to 1993, uprooted a lot of Malbec. At the time Argentina was in political turmoil and high-yielding vines to provide oceans of cheap wine seemed the way forward. Oh dear. Argentina lost the majority of its Malbec, most of it over 50 years old. Just in time, a few wise wine men cried 'Stop! That's our heritage'. Malbec is likely, because of Argentina's success, to be planted in other warm, dry areas. Chile already has some, and there are some historic plantings in South Australia.

**The taste of Malbec**

At its best, carefully grown and skilfully vinified in Argentina, Malbec has a dark purple colour, a thrilling damson and violet aroma, a lush fat rich fruit flavour and a positively soothing ripe tannic structure. It can take new oak aging, but it's a pity to smother its natural delicious ripeness with wood. In Cahors the flavour is more likely to be raisins, damson skins and tobacco. In both Chile and Australia you occasionally get the violet perfume and you usually get the soft ripe lush texture.

*The Argentinian company Fabre Montmayou is part of the French-owned Domaine Vistalba group, and the wines show a strong Bordelais influence but with extra richness. Luján de Cujo, in the province of Mendoza, where the vineyards for this wine are located, is one of Argentina's newest and highest regions. The big difference between day and night temperatures here means that the grapes can ripen slowly, and hang longer on the vines – several weeks longer compared to the more traditional, lower-altitude vineyards. That means better balance and more concentration in the wine.*

CHÂTEAU DU CÈDRE

*Malbec, alias Auxerrois, is the mainstay of Cahors in south-west France: this cuvée, called Le Cèdre, is 100% Auxerrois from 30-year-old vines. Aged in new oak barrels for 20 months, it will last for at least 10 years.*

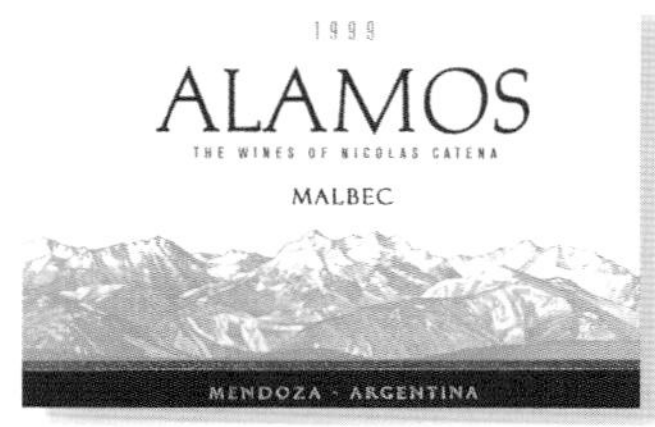


CATENA

*Dr Nicholas Catena was one of the first winemakers in Argentina to take note of what foreign consumers want to drink: this wine is determinedly international in style.*

***Left:*** *Malbec vineyards in the up-and-coming region of Luyán de Cujo in the province of Mendoza, Argentina with the Andes in the background. The cold nights found at high altitudes here extend the growing season and give crucial extra ripeness while still conserving acidity. And while Malbec rots in wet weather, there is precious little rain this side of the Andes, so rot is not a problem.*
***Above:*** *Malbec may be blended with Cabernet and Merlot in small amounts in New Zealand and Australia, for extra aroma and fatness of texture.*

## CONSUMER INFORMATION

### Synonyms & local names

Known in the Loire Valley and the South-West of France as Cot or Côt – in Cahors it is called Auxerrois but this is not the same as the white Auxerrois of Alsace. On Bordeaux's Right Bank around Libourne it is called Pressac. In Argentina where it is called Malbec the spelling Malbeck is also found.

### Best producers

**FRANCE** des Aires Hautes, du Cayrou, du Cèdre, Clos Carreyrès, Clos la Coutale, Clos de Gamot, Clos Rességuier, Clos Triguedina, Eugénie, de Gaudou, Gautoul, de Haute-Serre, Les Ifs, Lagrezette, Lamartine, du Peyrié, Pineraie, Primo-Palatum, les Rigalets, des Savarines.
**ARGENTINA** Alamos Ridge, Alta Vista, Altos de Medrano, Altos de Temporada, Anubis, Bodegas Balbi, Luigi Bosca, Humberto Canale, Catena, Cavas de Weinert, Etchart, Fabre Montmayou, Finca El Retiro, La Agricola, La Rural, Medalla, Nieto Senetiner, Norton, Peñaflor, Salentein, Terrazas de Los Andes, Michel Torino, Trapiche, Weinert.
**CHILE** Montes, MontGras, Morandé, Valdivieso, Viña Casablanca.
**AUSTRALIA** Jim Barry, Henschke, Leasingham, Taltarni, Wendouree.
**NEW ZEALAND** Esk Valley, Kumeu River.

## RECOMMENDED WINES TO TRY

### Ten top Argentinian wines

Alta Vista *Alto*
Altos de Medrano *Vigna Hormigas*
Luigi Bosca *Malbec*
Humberto Canale *Rio Negro Black River Gran Reserva Malbec*
Catena *Alta Angelica Vineyard Malbec*
Fabre Montmayou *Malbec*
Nieto Senetiner *Cadus Malbec*
Norton *Reserva Malbec*
Salentein *Primus Malbec*
Terrazas de Los Andes *Mendoza Gran Malbec*

### Five good-value Argentinian wines

Anubis *Mendoza Malbec*
Catena *Alamos Malbec*
Finca El Retiro *Malbec*
La Agricola *Santa Julia Malbec Reserva*
Peñaflor *Bright Brothers Malbec*

### Ten Cahors wines

Ch. du Cayrou *Cahors*
Ch. du Cèdre *Cahors le Cèdre*
Clos la Coutale *Cahors*
Clos de Gamot *Cahors Cuvée Vignes Centénaires*
Clos Triguedina *Cahors Prince Probus*
Ch. Gautoul *Cahors Cuvée d'Exception*
Ch. de Haute-Serre *Cahors*
Ch. Lagrezette *Cahors*
Ch. Lamartine *Cahors Cuvée Particulière*
Ch. Pineraie *Cahors*

### Twelve other wines containing Malbec

Domaine des Aires Hautes *Vin de Pays d'Oc Malbec (France)*
Jim Barry *McCrae Wood Cabernet Sauvignon Malbec (Australia)*
Esk Valley *Hawkes Bay Reserve The Terraces (New Zealand)*
Henschke *Keyneton Estate (Australia)*
Kumeu River *Melba (New Zealand)*
Montes *Reserve Malbec (Chile)*
MontGras *Reserva Malbec (Chile)*
Morandé *Chilean Limited Edition Malbec (Chile)*
Taltarni *Malbec (Australia)*
Valdivieso *Reserve Single Vineyard Malbec*
Viña Casablanca *San Fernando Miraflores Estate Malbec*
Wendouree *Cabernet/Malbec (Australia)*

# Malvasia

I THINK THE MOST GORGEOUS MALVASIA I ever had was pink and frothy, heady with the scent of rose petals and hothouse grapes, and it was from northern Italy. We were all entranced by it, but nobody could really say why it was like that – or even if it had ever been like that before. But that's the thing about Malvasia – it's not so much a single grape as a whole family. France uses the name of Malvoisie as a synonym for a whole range of grapes, none of which are Malvasia Bianca. But most proper Malvasias are subvarieties of Malvasia Bianca even though the grapes can be white, pink or red.

Italy and Iberia are its main homes. As a dry white it is blended with Trebbiano all over central Italy, generally to Trebbiano's benefit because while Trebbiano is often marked by a singular lack of flavour or texture, Malvasia is generally quite fat and soft and should add a bit of peachy plumpness when young and nutty, depth when mature. But not too mature: one or two years old is mature in these sorts of wines.

In Lazio's Frascati, where Malvasia certainly improves the blend, the local version is Malvasia di Candia, though the Malvasia Puntinata or Malvasia di Lazio is regarded more highly and makes wines with less of a tendency to flabbiness. Emilia's Malvasia is also called Malvasia di Candia. There is also a Malvasia di Toscana, which is usually only fit for uprooting. In Friuli there is Malvasia Istriana, which is well structured and flavoursome, and is used for lightly sparkling wines with a touch of sweetness and delightful grapy scent, as well as still dry ones.

*Passito* or fortified sweet white Malvasia – the latter have more aroma – is a speciality of southern Italy, though many wines are almost extinct. That's a pity, because some of the few survivors are really good. Malvasia delle Lipari is the best known; others are Malvasias di Bosa, di Grottaferrata, di Planargia and di Cagliari. Black Malvasia Nera is found in Puglia and Tuscany, and adds a very interesting dark, rich quality to Tuscan Sangiovese reds as well as a chocolaty, sometimes grapy warmth to Puglian blends. There is also some in Piedmont.

Malvasia is found in Rioja and Navarra in Spain and old-fashioned, barrel-aged white Riojas benefit from the Malvasia fatness fleshing out the skinny Viura. You find it in different forms in Portugal, especially the Douro (Malvasia Fina and the inferior Malvasia Rei, which could be the same as Palomino), Beiras, Estremadura (where Malvasia Fina is called by its alias of Vital) and Madeira, where it produces the rich, smoky, acidic Malmsey (see page 117). There is some in California, and a little in Greece and the countries of the former Yugoslavia, and knowing Malvasia, it'll keep turning up in a good few more places yet.

**The taste of Malvasia**

With so many subvarieties it is difficult to generalize about Malvasia's flavour, but where lesser versions can be flabby and low in acidity, tending to oxidize easily, better Malvasias are characterful and aromatic, though not as exotically so as Muscat, with flavours of peaches, apricots and white currants. Sweet ones are apricotty and rich, red ones dark and chocolaty. In Madeira the taste is intense, smoky and treacly but cut with acidity.

*Malvasia delle Lipari comes from a small group of volcanic islands off the coast of Sicily. Until phylloxera destroyed the vineyards and the patterns of viticulture it was made in huge quantities. This current example from the producer Colosi makes me regret that more is not made nowadays: it's sweet, fresh and wonderfully aromatic.*

SCHIOPETTO

*This producer in Friuli-Venezia Giulia is known for wines of great purity and concentration of flavour. They are invariably slow developing and long-lived.*

HENRIQUES & HENRIQUES

*Madeira should really be at least 10 years old to show its distinctive character and 15 years, like this intensely nutty and treacly Malmsey, is even better.*

*Left:* *Barrels of Malmsey at the Madeira Wine Company. Because Madeira gains its character from the aging process more than from the grape or from the winemaking, Malmsey Madeiras do not have the peachy, apricot aromas of most Malvasias. Malmsey is the sweetest style of Madeira but even so never loses its acid bite and smoky tang. In Madeira several different strains, or even different grapes, go by the name of Malvasia. Plantings of all Malvasias put together on the island are very small.*
***Above:*** *Malvasia is one of the hardest of all vines to pin down, simply because it isn't just a single vine, and not all vines called Malvasia are necessarily related.*

## CONSUMER INFORMATION

### Synonyms & local names

Widely planted in Italy and Iberia, there are many different subvarieties of Malvasia Bianca, usually called Malvasia this or that. Not to be confused with the name Malvoisie which is rarely used as a synonym for Malvasia. Malmsey is a style of wine using Malvasia grapes on the island of Madeira.

### Best producers

**ITALY/Friuli-Venezia Giulia** Borgo del Tiglio, Paolo Caccese, Ca'Ronesca, Cormons co-op, Sergio & Mauro Drius, Edi Kante, Lorenzon/I Feudi di Romans, Eddi Luisa, Alessandro Princic, Dario Raccaro, Schiopetto, Villanova; **Emilia-Romagna** Forte Rigoni, Luretta, Gaetano Lusenti, La Stoppa, La Tosa, Vigneto delle Terre Rosse; **Tuscany** (Vin Santo) Capezzana, Isole e Olena, Pieve Santa Restituta, Le Pupille Rocca di Montegrossi, San Giusto a Rentennano, Selvapiana; **Lazio** Castel del Paolis; **Sicily** Colosi, Carlo Hauner.
**SPAIN** Carballo, El Grifo, Abel Mendoza.
**PORTUGAL/Madeira** Barbeito, Barros e Souza, H M Borges, Henriques & Henriques, Vinhos Justino Henriques, Madeira Wine Company (Blandy, Cossart Gordon, Leacock, Rutherford & Miles), Pereira d'Oliveira.
**USA/California** Bonny Doon, Robert Mondavi, Sterling Vineyards.

### RECOMMENDED WINES TO TRY

### Ten dry Italian wines

Borgo del Tiglio *Collio Malvasia*
Castel del Paolis *Frascati Superiore Vigna Adriana*
Sergio & Mauro Drius *Friuli Isonzo Malvasia*
Forte Rigoni *Colli di Parma Malvasia*
Edi Kante *Carso Malvasia*
Luretta *Colli Piacentini Malvasia Boccadirosa*
Dario Raccaro *Collio Malvasia*
Schiopetto *Collio Malvasia*
La Tosa *Colli Piacentini Malvasia Sorriso di Cielo*
Vigneto delle Terre Rosse *Colli Bolognesi Malvasia*

### Five sweet Malvasia-based Italian wines

Colosi *Malvasia delle Lipari Passito di Salina*
Carlo Hauner *Malvasia delle Lipari Passita*
Isole e Olena *Vin Santo*
La Stoppa *Colli Piacentini Malvasia Passita Vigna del Volta*
San Giusto a Rentennano *Vin Santo*

### Five other Malvasia wines

Bonny Doon *Ca' del Solo Malvasia Bianco (California)*
Carballo *La Palma Malvasía Dulce (sweet) (Spain)*
El Grifo *Lanzarote Malvasía Dulce (sweet) (Spain)*
Abel Mendoza *Rioja (Blanco) Fermentado en Barica (Spain)*
Sterling Vineyards *Carneros Malvasia (California)*

### Five Madeira wines

Blandy *10-year-old Malmsey*
Blandy *15-year-old Malmsey*
Cossart Gordon *1920 Malmsey*
Henriques & Henriques *15-year-old Malmsey*
Henriques & Henriques *Vintage Reserva Malvasia*

## MALVASIA NERA

A black form or forms of Malvasia (see pages 120–121). In Italy Malvasia Nera is found in the province of Bolzano in Alto Adige, and in Puglia, Sardinia, Basilicata, Calabria, Tuscany and Piedmont. It may be blended with other local red grapes: Negroamaro in Puglia, and Sangiovese in Tuscany. It often has a black plum richness and hint of floral scent that I rather like. In Piedmont it has two DOCs, Malvasia di Castelnuovo Don Bosco and Malvasia di Casorzo. Both wines can be sweet, sparkling and gently aromatic. Best producers: (Italy) Francesco Candido, Leone de Castris, Roda del Golfo, Cosimo Taurino.

## MALVOISIE

This ought to be a French synonym for Malvasia, but it isn't. Instead, it is a name applied locally to several different varieties in France, among them Pinot Gris (Malvoisie is also the name by which Pinot Gris is known in Switzerland's Valais), Maccabéo, Bourboulenc, Clairette, Torbato and Vermentino (also known by the name of Malvoisie in Spain and Portugal). Best producers: (France) Vignerons Catalans, Jacques Guindon.

## MAMMOLO

'Mammole' is the Italian word for violets, and Mammolo is said to smell of them. It is planted in such small quantities now that few Tuscan reds owe much to it. It is permitted in Chianti but has largely been bypassed in the rush to improve Sangiovese. There is also a little planted in the Vino Noble di Montepulciano region. Best producers: (Italy) Antinori, Boscarelli, Contucci, Dei, Poliziano, Castello di Volpaia.

## MANDELARI

Widely planted Greek vine that makes dark, powerful, tannic wine, and is found on various islands, including Crete. Best producers: (Greece) CAIR, Archanes co-op, Paros co-op.

## MARÉCHAL FOCH

Early ripening and with good resistance to winter cold, this French hybrid is popular in Canada and New York State, where it may be made as a varietal. It makes attractively soft, sometimes jammy, sometimes smoky red wines, and may be aged in oak for more substance. Carbonic maceration will, conversely, make lighter wines. It is named after the First World War French general. Best producers: (Canada) Stoney Ridge, Quails Gate; (USA) Wollersheim.

## MARGARITA

A *Vinifera labrusca* crossing bred in, of all places, Azerbaijan. The grapes are large and spherical and so bear no resemblance whatever to their namesake, this book's svelte co-author.

## MARIA GOMES

The name given to Fernão Pires (see page 89) in the Bairrada region of northern Portugal. Best producers: (Portugal) Caves Aliança, Luís Pato, Messias, Quinta de Pedralvites.

## MARSANNE

See pages 124–125.

## MARZEMINO

Don Giovanni operatically enjoys Marzemino just before being swallowed up in Hell, and while Marzemino may not be everyone's idea of an ideal last wish, it is very attractive in less desperate circumstances.

It is found in Trentino and to a lesser extent in Veneto and Lombardy, and has typically northern Italian grassy acidity and cherryish fruit, combined with good colour and ripeness. In Trentino it is usually made as a varietal wine for drinking young; in Lombardy it may also be blended with Sangiovese, Barbera, Groppello and Merlot, though probably not all at once. Sweet *passito* versions also exist, perhaps blended with other grapes. Best producers: (Italy) Battistotti, La Cadalora, Cavit, Concilio Vini, De Tarczal, Isera co-op, Letrari, Mario Pasolini, Simoncelli, Vallarom, Vallis Agri.

## MATARO

An Australian name for Mourvèdre (see pages 140–141), now gradually slipping out of use as the grape becomes increasingly fashionable under its French name.

## MAUZAC

A grape of south-west France with a rustic, green apple skins flavour that is not always delicious, Mauzac is found principally in the Gaillac and Limoux regions. It probably takes its name from either the town of Mauzac in Haute-Garonne, or from that of Meauzac in Tarn-et-Garonne. It is usually blended with other grapes – Len de l'El in Gaillac, Chardonnay and Chenin Blanc in Limoux. In Gaillac it appears in almost every guise, from dry to sweet, still to sparkling. In Limoux it has been partly supplanted by Chardonnay.

In Limoux, too, it is made into sparkling Blanquette de Limoux and Crémant de Limoux, though it is more normal for standard sparkling wine techniques to be used than for the old *méthode ancestrale* to be employed. This method basically relied on the fermentation dying down in the autumn – Mauzac ripens late and so the fermentation

*Malvasia Nera vines at Squinzano in Puglia, southern Italy. Black Malvasia vines are nearly as diverse as white, and different strains are found in different places. In Puglia much of it is blended with Negroamaro.*

was not finished before the winter – and then starting again in the spring by which time the only partially fermented wine would have been bottled. Now the grapes are more likely to be picked earlier for acidity and the Champagne method is used to make a good sparkling wine. Mauzac can produce berries that vary in colour from green through russet and pink to black; Mauzac Noir, however, is a different vine altogether. Best producers: (France) de l'Aigle, de Gineste, Maison Guinot, de Labarthe, Robert Plageoles.

## MAVRO

The main red grape of Cyprus is of pretty low quality, and there may be limits to what the most inventive modern winemaking can do with it. Since the name merely means 'black' it may however cover several different grapes; little research has yet been done on the varieties found in Cypriot vineyards.

## MAVRODAPHNE

A Greek vine which makes sturdy, rich red wines, in particular the porty Mavrodaphne of Pátras. It is nearly always made sweet, and treating it this way seems to bring out all its aroma; dry versions are usually blended. Best producer: (Greece) Achaia-Clauss.

## MAVRUD

Mainly found in Bulgaria, Mavrud makes weighty, solid red wines of character but no particular elegance. It is high in tannins and colour and takes well to oak aging. Best producer: (Bulgaria) Domaine Boyar.

## MAZUELO/A

The Rioja name for Carignan. It brings acidity, colour and tannin to the blend, but is understandably less prized than Tempranillo. Best producers: (Spain) Amézola de la Mora, Berberana, Martinez Bujanda, CVNE, Muga, La Rioja Alta, Marqués de Riscal.

## MELNIK

Bulgarian vine capable of producing wines of quite high quality. It is found particularly around the town of the same name, near to the Greek border. Its wines have very good colour, tannin and balance, and a certain rich warmth, particularly if given oak aging, that marks them out from the general run of Bulgarian reds. They can be smoky, coffeeish and toffeeish and very appealing. They can also improve in bottle. Quality, however, is not uniform, to say the least. Best producer: (Bulgaria) Domaine Boyar.

*Picking Muscadet grapes at Clisson in the Muscadet de-Sèvre-et-Maine AC. Muscadet wines have been out of fashion for a decade but the best now deserve to be looked at afresh.*

## MELON DE BOURGOGNE

This is the one and only grape behind Muscadet, and so closely is it identified with the wine that the grape's other name is Muscadet. But as one might gather from this name, it originally came from Burgundy and travelled west to reach the western end of the Loire Valley around the city of Nantes. It had been much planted in Burgundy, until its destruction was ordered in the early 18th century whereupon, according to the ampelographer Galet, the growers, in order to save such a useful vine, engineered confusion with Chardonnay and managed to get some Chardonnay torn up in its place. Also according to Galet, there is still some, albeit only a little, planted on the Côte d'Or.

The Burgundy growers favoured Melon so much because of its resistance to cold – although it buds early, so is apt to be hit by late frosts – and its generous and regular yields. Its secondary buds are fertile, so even after frost it may be able to produce some sort of a crop. It is, however, subject to some of the grower's worst nightmares, in the form of downy and powdery mildew and grey rot, and may have to be picked before complete maturity to avoid the worst effects of these.

In the Pays Nantais, where it was introduced after the freezing winter of 1709 which pretty much destroyed whatever grapes were there, it has gained itself a poor reputation for high acidity and low flavour, and the best that is usually said of it is that it goes well with the local *fruits de mer*. But in fact winemaking standards have been rising in the Pays Nantais. Skin contact, lees stirring and barrel fermentation for the best wines (not usually new barrels, which can unbalance the flavours) all help to give greater weight and richness. There are even some *vins de garde* being made, from vines cropping at about 45hl/ha instead of the more usual 55-60hl/ha, and such wines can improve in bottle for a decade, becoming honeyed with flavours of quince and ripe greengage.

Such wines are, of course, not the norm. Muscadet is still at heart a simple wine for drinking young, and with 80 per cent of the market controlled by the *négoçiants*, few supermarket wines have much aspiration to quality. Growers' wines are the best bet if you want something better than basic.

*Sur lie* wines, bottled off their lees, have the greatest depth, and the schist soils of the Sèvre-et-Maine region give the best quality. The Coteaux de la Loire appellation has good terroir but is small, and the sandy soil of the Côtes de Grand Lieu appellation produces light wines for very early drinking.

In California and Australia there has been considerable confusion between Melon and Pinot Blanc, with even cuttings from the University of California at Davis being wrongly identified as Pinot Blanc. Plantings are more or less sorted out now, however – presumably to relief all round, since Pinot Blanc is certainly the better grape. Best producers: (France) Chéreau-Carré, Luc Choblet, de l'Ecu, Pierre Luneau, Louis Métaireau.

## MENCÍA

This grape of north-western Spain gives light, fresh, acidic reds with a raspberry and blackcurrant leaf flavour not unlike a slightly raw Cabernet Franc, and good tannin – altogether as you might expect of a red vine that grows alongside such white ones as Albariño (in Rías Baixas) and Godello, Treixadura and Loureira (in Ribeira Sacra).

It is also found in Valdeorra, Ribeiro and, above all, in Bierzo, where it is the main red grape and where it may be blended with Garnacha Tinta. Mencía is a grape with some potential, though it is usually best drunk young. It may be the same vine as Portugal's Jaen (see page 115). Best producers: (Spain) Jesús Nazareno co-op, Moure, Priorato de Pantón, Vire dos Remedios.

# MARSANNE

I USED TO THINK THAT MARSANNE and Roussanne were the Siamese twins of white Rhône grapes; one was never mentioned without the other, but one was also clearly more favoured. Roussanne was always described as far more frangrant and refined, and the clumsier Marsanne was chided for marching in and taking over most of Roussanne's best sites. Well, it's true Marsanne has largely supplanted Roussanne in the northern Rhône – it's a far more reliable cropper which inevitably influences growers' decisions – but it's not a bad grape at all, and, now that I've had time to compare single varietal versions from France and elsewhere, I'm not really that sure it's even inferior. Anyway, it dominates the white blends in Hermitage, Crozes-Hermitage, St-Joseph and St-Péray but is not permitted in Châteauneuf-du-Pape, whereas Roussanne is.

If Marsanne is to have character it must be grown in the right place, and vinified with care. In too hot a climate it becomes flabby; in too cool a spot it has an undeveloped simple, bland flavour. And it's another of those grapes you can't overcrop – it becomes neutral and characterless. As for new oak, more than just a splash squashes the grape's personality.

Most Marsanne is made to be drunk young, within just a few years of the harvest. Grapes for this style are often picked before full ripeness to retain acidity, but you risk having a wine without any real taste. That's why in the Languedoc Marsanne is often blended with the more aromatic Viognier. If you do let the grapes ripen fully on the vine, as they often do in Hermitage and sometimes in Australia, you get high alcohol wines that can age a good while, and though they often go through a dumb phase after a few years when they taste flat and unforthcoming, after about ten years they emerge darker in colour and more complex in flavour, and with an oily, weighty, honeyed character accompanied by nuts and quince fruit. Hermitage and the better wines of the Rhône are the best candidates for this sort of aging; there are also a few suitable examples in Australia, where it is planted in small quantities in Victoria for high quality wine.

There is some Marsanne in Switzerland, where it produces good wine called Ermitage in the Valais region, and there's some in France's Savoie, where it is known as Grosse Roussette. California has a bit of Marsanne, but so far the wine tends to be dull and gluey in flavour.

**The taste of Marsanne**

When young it has a minerally edge, often with a citrus, peachy flavour. With age this matures into a rich palate of honeysuckle and jasmine, acacia honey and perhaps a touch of apricot or quince; it is aromatic, quite oily, nutty, and surprisingly heavyweight. Accordingly it can partner rich food very well when it's mature. As a young wine it's equally good with or without food.

*The Chateau Tahbilk winery in central Victoria is one of the most beautiful in Australia, and has Australian National Trust classification. Tahbilk's association with Marsanne can be traced as far back as the 1860s and although none of these plantings have survived vines dating from 1927 are still in use and reputed to be the oldest Marsanne vines in the world. Chateau Tahbilk Marsanne was served to the young Queen Elizabeth II when she visited the winery in 1953. The wine enjoys a distinct honeysuckle aroma and flavour and is very good young as well as improving with several years' aging in bottle.*

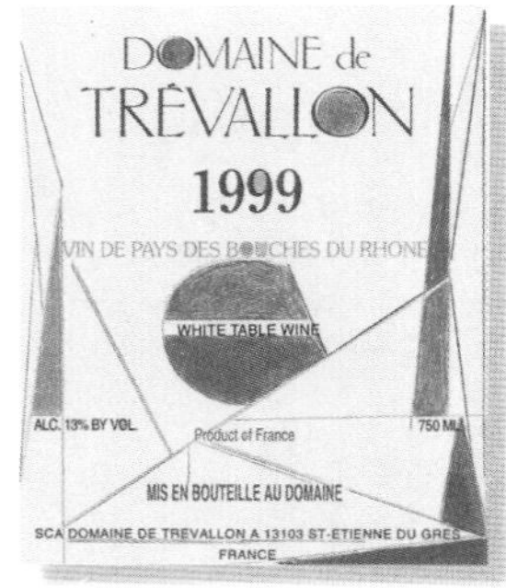


DOMAINE DE TRÉVALLON

*This intense wine from Provence is a blend of 45% each Marsanne and Roussanne and 10% Chardonnay; since it fits into no AC rules, it is classified as Vin de Pays.*

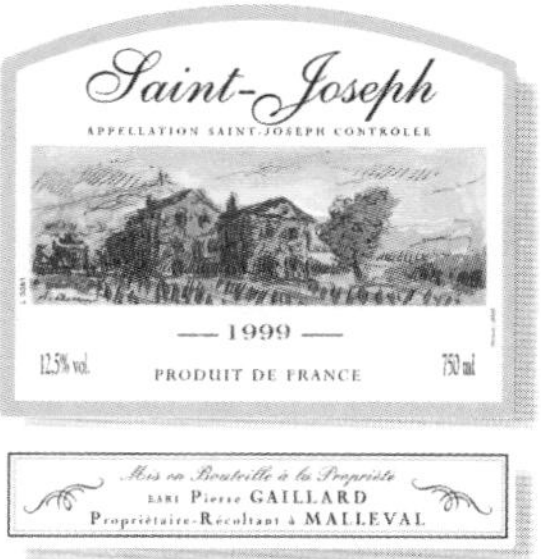


PIERRE GAILLARD

*This 100% Marsanne St-Joseph Blanc, vinified in barriques, is made by a producer who used to run other people's vineyards before setting up on his own.*

***Left:** Picking Marsanne grapes in Chapoutier's Chante-Alouette vineyard at the top of the Hermitage hill in the northern Rhône. Chapoutier uses Marsanne to make several single-vineyard Hermitage wines as well as a Hermitage Vin de Paille. For the vin de paille the grapes are picked extra ripe and then left to dry indoors for two months before being fermented into an intensely sweet dessert wine.* ***Above:** Marsanne can be a bit of a love-it-or-hate-it grape; but much depends on how well it is handled. Bland, gluey Marsanne is no-one's idea of a good time; but get the balance right and you'll be rewarded with one of the most unusual whites of all, with weight and complexity and the ability to transform from a bright honeysuckle, fresh youth to a deep, waxy, quincy maturity.*

## CONSUMER INFORMATION

**Synonyms & local names**
Known in Savoie as Grosse Roussette and in Switzerland as Ermitage or Ermitage Blanc.

**Best producers**
**FRANCE/Rhône Valley** de Beaucastel, Belle, Chapoutier, Chave, Coursodon, Yves Cuilleron, Delas, des Entrefaux, Florentin, Gaillard, Gripa, Grippat, Guigal, Jaboulet, Jean Lionnet, de l'Oratoire St-Martin, Perret, Pradelle, des Remizières, Marcel Richaud, Sorrel, Tain l'Hermitage co-op, Trollat, François Villard; **Provence** Clos Ste-Magdelaine, de la Ferme Blanche, de Pibarnon, de Trévallon; **Languedoc-Roussillon** F Alquier, des Estanilles, Fabas Augustin, de Jau, de Lascaux.
**SWITZERLAND** Gilliard, Caves Imesch.
**ITALY** Bertelli, Casòn Hirschprunn.
**SPAIN** Celler Mas Gil.
**AUSTRALIA** All Saints, Chateau Tahbilk, Cranswick Estate, Marribrook, Mitchelton.
**USA/Virginia** Horton Vineyards.

## RECOMMENDED WINES TO TRY

**Ten northern Rhône white wines**
Chapoutier *Ermitage Cuvée de l'Orée*
Chave *Hermitage Vin de Paille*
Pierre Coursodon *St-Joseph Blanc le Paradis St-Pierre*
Yves Cuilleron *St-Joseph Blanc le Lombard*
Delas *Hermitage Blanc Marquise de la Tourette*
Pierre Gaillard *St-Joseph Blanc*
Grippat *Hermitage Blanc*
Guigal *Hermitage Blanc*
Jaboulet *Crozes-Hermitage Blanc la Mule Blanche*
Sorrel *Hermitage Blanc les Rocoules*

**Ten other French wines containing Marsanne**
Frédéric Alquier *Vin de Pays d'Oc Roussanne/Marsanne*
Ch. de Beaucastel *Côtes du Rhône Blanc Coudoulet de Beaucastel*
Clos Ste-Magdelaine *Cassis*
Ch. des Estanilles *Coteaux du Languedoc Blanc*
Ch. Fabas Augustin *Minervois (Blanc)*
Ch. de Jau *Côtes du Roussillon Blanc de Blancs*
Ch. de Lascaux *Coteaux du Languedoc Pierres d'Argent*
Domaine de l'Oratoire St-Martin *Côtes du Rhône-Villages Blanc Haut-Coustias*
Ch. de Pibarnon *Bandol (Blanc)*
Domaine de Trévallon *Vin de Pays des Bouches-du-Rhône (Blanc)*

**Five other European wines including Marsanne**
Bertelli *St-Marsan Bianco (Italy)*
Casòn Hirschprunn *Contest (Italy)*
Robert Gilliard *Ermitage Réserve Choucas (Switzerland)*
Caves Imesch *Ermitage du Valais (Switzerland)*
Celler Mas Gil *Clos d'Agon Blanco (Spain)*

**Five Australian Marsanne wines**
All Saints *Victoria Marsanne*
Chateau Tahbilk *Goulburn Valley Marsanne*
Cranswick Estate *Nine Pines Marsanne*
Marribrook *Great Southern Marsanne*
Mitchelton *Goulburn Valley Reserve Marsanne*

# Merlot

It's not often that a grape variety has a TV show to thank for propelling it from the ranks of perennial bridesmaid to superstar, but that's the case with Merlot. In 1991 the prime time USA television show '60 Minutes' covered a phenomenon they titled 'The French Paradox'. This marvellous paradox is based on the fact that the French eat large amounts of fat and yet have far less heart disease than nations who eat less.

Ever since Prohibition ended in 1933 American wine producers have been trying to turn the USA into a nation of wine drinkers – with limited success. But if the man on the telly tells you you must knock back a couple of glasses of red a day – suddenly millions of Americans who have never drunk wine before are queuing up for their daily dose. American consumption of red wine quadrupled within the year. But these were novice wine drinkers. They didn't necessarily like the flavour of red wine very much – and what they needed was something that was soft, easy and mild, and yet discernibly, undeniably red – and one grape fitted the bill perfectly – Merlot. Suddenly Merlot's perceived weaknesses were its unique selling point.

*Merlot's name is said to be derived from* merle, *French for blackbird, which apparently loves its sweet, early-ripening fruit. Being planted like fury around the world, Merlot has done particularly well in the Napa Valley where the California poppy, as orange as a blackbird's bill, grows wild.*

Merlot has traditionally been Bordeaux's 'other' red grape. Over the generations connoisseurs and writers have favoured wines made with Cabernet Sauvignon – the deep, dark, haughty aristocrat of red grapes, and the main variety in Bordeaux's Pessac-Léognan/Graves region and in the Médoc region just to the north with its famous villages of Pauillac, St-Julien and Margaux. Merlot was necessary to soften the toughness of the Cabernet, but it was always talked of as an inferior variety, with no great personality of its own. Where it did come into its own and show a full, ripe, luscious personality that was quite surprising in its come-hither juiciness was in St-Émilion and Pomerol. These regions were less well known, and it wasn't until the leading American critics discovered them during the 1980s and realized that their ability to mix the fabled reputation of Bordeaux with a wine you could glug back at only a couple of years old would play brilliantly back home in the States that Merlot finally stepped out from Cabernet's shadow.

And it hasn't stepped back into the shade. Merlot was already on the move in California before the '60 Minutes' show, but compare the state's 1985 plantings of 2000 acres (800ha) with its 1996 plantings of 33,000 acres (13,400ha) and the latest figures showing almost 48,000 acres (19,400ha) and you can see that it's struck a chord – soft, mellow Merlot suddenly showing that it better reflects the general population's wishes in a red wine than technically more interesting but tougher Cabernet Sauvignon. The marketing men in the USA pride themselves on knowing what the general public wants – or perhaps deciding for them what the general public wants. Merlot was 'it'. And they weren't wrong.

Boosted by American enthusiasm, Merlot plantings in France's Languedoc doubled in the decade to 1998. Merlot is booming in Italy and the countries of Eastern Europe. Everywhere it's the same story. And, of course, it's become the star grape of Chile – though the vineyards there are so delightfully confused that they're not quite sure what is Merlot and what is Carmenère. Even Australia, the home of the crowd-pleasing red, has fallen for it. That should keep them busy for a while because they already own the bragging rights to an even bigger crowd pleaser than Merlot – Shiraz.

**Merlot: from Grape to Glass**

*Geography and History page 128; Viticulture and Vinification page 130; Merlot around the World page 132; Enjoying Merlot page 136*

# Geography and History

At first glance the map showing where Merlot is found in the world might look remarkably similar to that of Cabernet Sauvignon. And why not? After all, in Bordeaux they collaborate to excellent effect, and many a Cabernet Sauvignon grower in South Africa, California, Australia and Eastern Europe has found that the traditional Bordeaux blend seems to work pretty well.

And yet there are differences. Merlot is saved from a role of eternal bridesmaid to Cabernet Sauvignon by two factors: its liking for cooler climates than Cabernet Sauvignon, and its ability to produce soft, rich-textured, low acid, low tannin red. The latter appeals to many red wine drinkers who don't care for the rasp of tannin and acidity, but love the appetizing black cherry and chocolate and fruitcake flavours

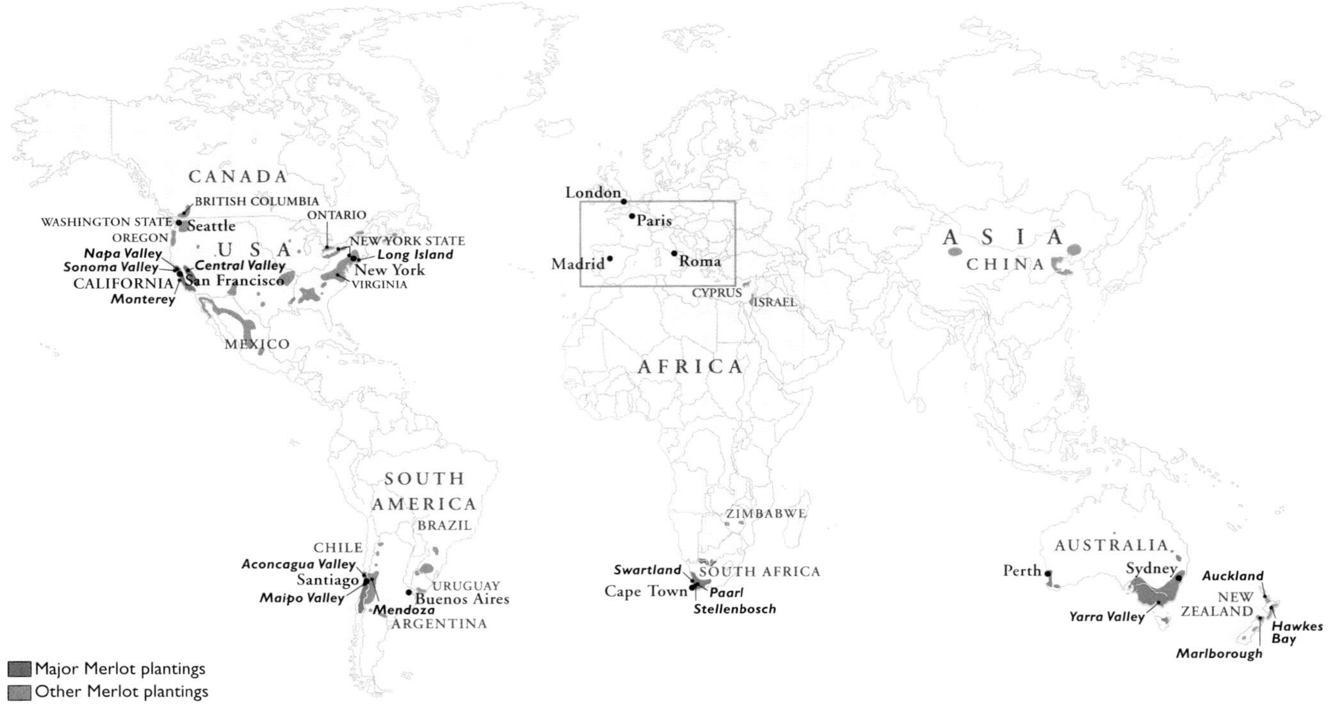

Major Merlot plantings
Other Merlot plantings

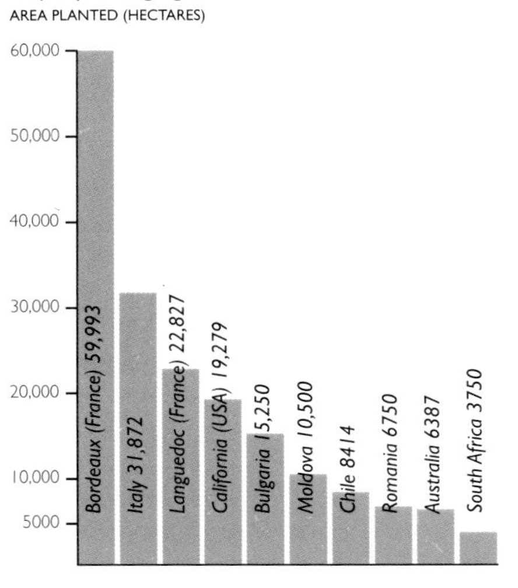


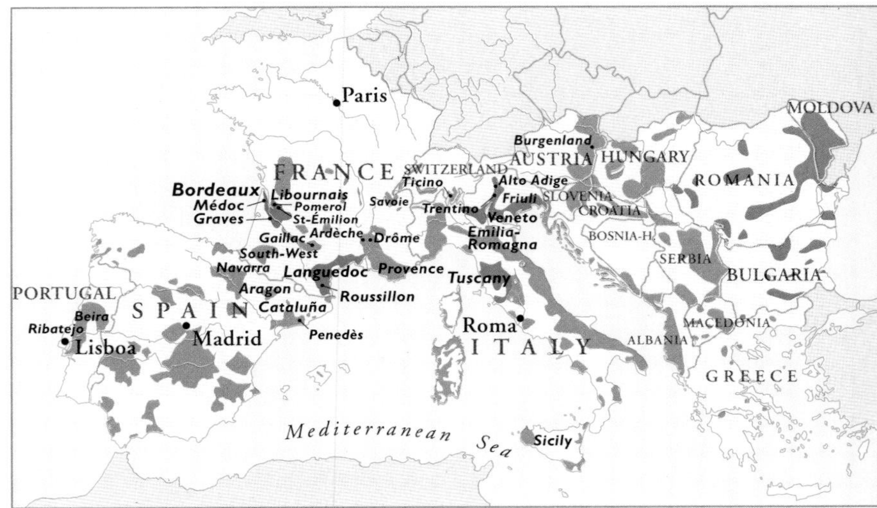

with which Merlot is packed. The former appeals to winemakers in climates that are just too cool for Cabernet Sauvignon. In these sites, where Cabernet Sauvignon will turn out too green and raw for fun, Merlot can ripen to soft, silky sweetness.

So while the maps of the two varieties might look much the same, they disguise differences of emphasis. Above all, Merlot inhabits the cool fringes where growers look at Cabernet Sauvignon every year and wonder whether to persevere with it. Here Merlot excels.

Merlot likes warmer conditions too, but doesn't prosper with too much heat because it actually ripens too fast and never develops its lovely plummy personality. And it needs a firm hand in the vineyard, since it will overcrop like mad if you let it. If you've ever stared in disappointment at a glass of pale, flavourless north Italian Merlot, you're looking at the results of a massively overcropped vineyard. It is far from being as forgiving of unskilful viticulture as Cabernet Sauvignon. Merlot may have wide appeal, but it will never be an all-purpose grape for lackadaisical producers.

## Historical background

Merlot is a mystery grape. I mean, everyone reckons that it is native to Bordeaux, but until the 19th century it gets scant attention from commentators on the region. The earliest mention seems to be from 1784, when a local official called Faurveau named it as one of the best vines in the Libournais region on the Right Bank of the Dordogne; by the mid-19th century it was being planted in the Médoc on the Left Bank of the Gironde for the sake of its ability to blend well with Cabernet Sauvignon and Malbec.

Where did it spring from? DNA fingerprinting studies at the University of California at Davis (UCD) were not complete at the time of writing, though Dr Carole Meredith of UCD conjectures that it could be an offspring of Cabernet Franc. Interestingly, she surmises the same of Carmenère, which could thus be either a full or half-sibling of Merlot, depending on what the other parent of each is. These are the two vines with which Merlot is most easily confused, and in countries like Chile the confusion in the vineyard can continue to this very day.

Merlot seems to have been noted in Italy at much the same time as it was moving into the Médoc. It is first recorded in the Veneto (as Bordò) in 1855, and in the Swiss canton of Ticino between 1905 and 1910. However, it was brought to Switzerland from Bordeaux, not from Italy.

Its Bordelais name (or one of them; it has had many synonyms) is said to be derived from the blackbird, or *merle*, which apparently loves its sweet, early-ripening fruit. (Merlau is another synonym for the grape.)

*The limestone côte on the edge of the town of St-Émilion is one of the best spots for Merlot in the whole of Bordeaux. Merlot is often thought of as a soft, mild wine, but from these steeply sloping vineyards the wines can rival those of the Haut-Médoc on the other side of the Gironde for complexity and longevity and yet they do retain a delightful suppleness from youth through to maturity.*

*Sorting Merlot at Château Lafite-Rothschild in Pauillac. Here on Bordeaux's Left Bank Merlot is an essential softener for Cabernet; ripening earlier, it is also a useful insurance against poor weather.*

*The Clos de l'Oratoire barrel* chai *in St-Émilion. Merlot comprises about 80 per cent of the blend here, the balance being Cabernet Franc, and the result is a juicy, fleshy, robust wine.*

# Viticulture and Vinification

IT IS UNUSUAL FOR A GRAPE to become so fashionable without there being any real consensus about how it should be grown and made. If there is a benchmark style it is presumably that of Pomerol and St-Émilion, but in much of Bordeaux Merlot has long been as much of an insurance policy as a grape appreciated for its own distinct virtues. Its failings as a varietal wine include poor flavour profile: it can be a doughnut wine, with a hole in the middle of the palate.

Clonal variation is important, and there is undoubtedly more work to be done here. It is possible that only 60 or 70 per cent of Merlot's potential has so far been realized, as anyone who has suffered a lean, mean Merlot vin de pays will testify.

## Climate

The climate in the Libournais – the Right Bank of the Dordogne in Bordeaux, where Merlot is concentrated – is actually warmer than that of the Médoc across the Gironde on the Left Bank. The Libournais has less rain and warmer days, but more chance of frost and colder nights. One might think that such conditions would favour Cabernet Sauvignon, with its greater resistance to cold, and greater need of warmth for ripening, and no doubt the French wine authorities reasoned the same way when they insisted on Cabernet Sauvignon being planted on the lower slopes here in the 1970s. But warmer weather can't compensate for colder soil – at least, not in Bordeaux – and the soils in the Libournais are mostly cold, damp clay. Merlot's early budding, early ripening habit means that it can cope with cold soils and still ripen, though early budding increases the risk of frost damage. It is also sadly subject to rot because of its thin skin: the vintages of 1963, 1965 and 1968 were even more rot-ridden in St-Émilion and Pomerol than elsewhere in Bordeaux, and if autumn rains threaten, growers fearfully eye their Merlot crop since the grapes could begin to split and rot within a couple of days. Its thin skin also makes it dangerously susceptible to rot in humid climates.

But at least it buds and ripens up to two weeks earlier than Cabernet Sauvignon, which makes it a better bet in most parts of New Zealand; the same is true of Austria, northern Italy, Switzerland, New York State and Canada.

Though its reputation is for soft, plummy flavours, it can all too easily exhibit a herbaceous green streak. A touch of herbaceous greenness was long thought of as part of the St-Émilion style, until the present vogue for lush, ultra-ripe wines took over; Washington State, New Zealand, Switzerland and the Veneto in northern Italy have all had problems with greenness, which can often be overcome by better canopy management. Even Australian examples are far from being perfectly ripe all the time. Much Californian Merlot avoids any risk of underripeness by being planted in the hot Central Valley – but the downside is that the wines frequently taste baked and lacking in freshness.

## Soil

According to Galet (1998), Merlot likes cool soils that retain their humidity in summer; in dry soils, says Galet, the grapes don't fill out properly. Jean-Claude Berrouet, technical director of Château Pétrus in Pomerol, believes that, on the contrary, water stress is more important than soil temperature for Merlot, and that it performs far better on well-drained soil than at the foot of a slope. Well, they're both experts. Take your pick. Château Pétrus itself is located on a tiny blister of clay that bulges through the topsoil of ferruginous sand and gravel, and that imparts more than usual structure to the wine; it is clay and iron that give the wines of

*Christian Moueix (above, standing next to the hopper) at his property, Château Pétrus in Pomerol. Pétrus is one of the world's most famous red wine properties, renowned for an irresistible, lush quality, yet Christian Moueix is at pains to create a wine with as much restraint as its brilliant super-ripe grapes will allow, rather than something in the highly extracted, extra-rich modern style currently in vogue. So, however sumptuous a Pétrus is, you never tire of the taste.*

Pomerol the backbone that most Merlots lack – helped, usually, by a little Cabernet Franc for perfume and acidity. The percentage of clay in some Pomerol vineyards is as high as 60 per cent.

In the sand, clay–limestone (*argilo-calcaire*) and gravel over limestone soils of St-Émilion, too, the traditional blend is with Cabernet Franc. Styles vary with the terroir: sandier soils give looser-knit wines; clay gives more substance, limestone more elegance and perfume.

### Clones – and confusion

In much of the New World, clone and climate currently make more difference to style than soil type. Stephan von Neipperg of Château Canon-la-Gaffelière in St-Émilion believes there are only two or three really good clones of Merlot, though many more than that are available. Some give structure; others have larger berries, larger yields and give soft, early-drinking wine. The right choice of rootstock can also help to control the vigour of this naturally vigorous vine and help keep its cropping under control.

Some plantings in Australia and Austria are strongly suspected of being Cabernet Franc rather than Merlot, and others in California have been proved to be so: in the Napa Valley, Duckhorn's best block turned out to be Cabernet Franc; and Christian Moueix of Bordeaux confirms that as long ago as the early 1980s he'd thought that some of the Merlot in California was Cabernet Franc. One grower even reckons he can't tell the difference between the two varieties until their third year.

In Chile the confusion has been with Carmenère (see pages 60-61 and 134).

### Yields

Overcropped Merlot produces a thin, stringy, weedy wine, light in colour and vegetal in flavour. Some *vins de garage* in St-Émilion may have yields of as low as 26hl/ha; normal Bordeaux yields reach 60hl/ha, and are generally higher than those of Cabernet Sauvignon unless reduced by the vine's tendency to poor fruit set. In the Midi yields may rise to 80hl/ha; over 100hl/ha quality falls rapidly. In New Zealand yields are between 8–10 tonnes/ha; in California's Napa Valley, between 4–4.5 tons/acre; in Argentina, where yields are generally high, 12 tonnes/ha is considered low, and 18 tonnes/ha is not unusual.

*Michel Rolland, Merlot guru and consultant to (it sometimes seems) most of the world, extols ripe fruit and great lushness. His best wines are phenomenally rich and mouthfilling and even when he is dealing with less than perfect vineyards around the world he creates a delicious ripeness of style that is currently very much in fashion.*

Michel Rolland, worldwide consultant winemaker, Merlot specialist, and advocate of ultra-ripe, lush wines, believes that yields of the grape throughout the world are too high. 'It doesn't like too much sun, and it is difficult to make very expressive, complex Merlot if the climate is too hot. But if you reduce the yield you can improve quality.'

### When to pick

This is one of the main focuses of debate between winemakers around the world. A characteristic of Merlot is that once ripe, it must be picked quickly. It will rush from ripeness to overripeness in a matter of days; in this it is less accommodating than Cabernet Sauvignon.

There are two schools of thought. Rolland favours late picking, and welcomes fruit with an element of overripeness. 'The picking date is vital. Get that wrong and everything else will be wrong; you will lose 80 per cent of the potential. Many properties in Bordeaux pick between a few days and a week too early. But the crop level is the main factor. If the crop is too big, you can wait until Christmas, and it won't ripen. The ideal crop is 40–50hl/ha. If you delay picking for eight to ten days more, you will get good results at that level. And if you remove some of the leaves, there is less risk of rot, so you can wait.'

The other side of the Merlot coin, that which advocates traditional elegance, finesse and potential for longevity above richness and lushness, is represented by Christian Moueix and Jean-Claude Berrouet, the team responsible for, among others, Château Pétrus. They point out that overripeness means loss of acidity. 'We want light, elegant wines for drinking, not body-building wines for winning tastings.'

### WINEMAKING: THE GREAT DEBATE

At the higher quality levels, Merlot is a wine with an identity crisis. It needs elegance, complexity, fine tannins and silky texture – but can winemakers best achieve that ideal by treating the grape like Pinot Noir, or by treating it like Cabernet Sauvignon?

The issues revolve around extraction and oak. Should Merlot be heavily extracted and tannic, and aged in a high proportion of new oak? Or should it be lighter, more delicate? Fashion dictates that it should be the former: light reds are currently out of vogue. But not all Merlot is naturally as structured as the example from Château Pétrus, or as concentrated as Valandraud. And if it's not, it cannot be made so by long aging in new oak.

Consultant winemaker Michel Rolland recommends extracting the phenols (colour and flavour compounds) at the beginning of the fermentation process with a shorter, faster maceration with frequent pumping over. At the time of writing researchers in Bordeaux are trying to build up a better picture of Merlot by analysing the structure of its tannins and phenolic compounds. By compiling a dossier of the aromas and flavours most frequently found in the wine they hope to arrive at a clearer understanding of its potential.

I have to say, some of the most beautiful St-Émilions I've enjoyed have been made in the lighter style, because Merlot gains a gorgeous soft butter and toffee smoothness with age if you don't try to force more weight and colour out of the grape than it naturally possesses. However, if you're after something altogether more powerful and lush, Michel Rolland clearly has the answer, since he consults to many of the top rated, most fashionable projects in this and other regions around the world.

# Merlot around the World

HOW MANY GREAT VARIETAL MERLOT WINES are there in the world? A lot depends on what you mean by 'great'. If you mean dark, muscle-bound bruisers built for the long haul, or slow-developing intellectual beauties that take a generation to reveal their treasures, then the answer is – there aren't many. But if greatness can be measured in a hedonistic way, if succulence, sensuous fruit and a perfume that fills your head with sweet ripeness could be called great, there'd be quite a few. High alcohol, dark colour, low tannins and acidity and a cauldron-full of nice, ripe fruit is what Merlot is best at. Pleasure, in other words. But are they great? Look. I'm too thirsty to argue. Hand me the bottle. I'll drink, you fret.

## Bordeaux

Merlot has quietly been staging a takeover in Bordeaux, spurred on by the worldwide demand for softer, earlier-drinking wine. It now constitutes some 56 per cent of plantings overall – but that figure disguises its minority role in the better parts of the Médoc and Graves, where it makes up an average of 25 per cent of the blend, and its dominance in St-Émilion and Pomerol. In the outlying areas of Bordeaux – the areas collectively known as the Côtes – it is very often the majority grape.

In St-Émilion it accounts for over 60 per cent of plantings; in Pomerol about 80 per cent. In both these appellations it is commonly blended with some Cabernet Franc; there is a little Cabernet Sauvignon in St-Émilion, and some, though less, in Pomerol.

The standard Bordeaux blend of Merlot, Cabernet Sauvignon and Cabernet Franc, with or without the addition of Malbec or Petit Verdot, has a lot to recommend it. Merlot adds softness and fleshiness, Cabernet Franc adds perfume and Cabernet Sauvignon adds the all-important structure. Merlot on its own is generally less long-lived, although St-Émilion and Pomerol can be tight and closed when young, and like Médoc and Graves reds, can need some years to open.

It is the *crasse de fer* – the iron-rich clay subsoil – that imparts structure to Pomerol's Merlot; the very structure that can be so wanting elsewhere. Château Pétrus is made from about 95 per cent Merlot in most years; Le Pin and Gazin are 90 per cent Merlot. At some Pomerol châteaux, for example Lafleur, the amount of Merlot in the vineyards goes down to 50 per cent.

In St-Émilion the rapid growth in the number of *vins de garage* – wines made on so small a scale that one could envisage making them in one's garage – has been fuelled by Merlot. Yields of 30 hl/ha or less are the rule here, from vineyards of around two hectares in size (Valandraud is 2.5ha; Clos Nardian is 1.52ha; La Mondotte is positively large at 4.3ha). Picking may be by the individual berry; viticulture may be organic or biodynamic; oak will almost certainly be 100 per cent new, and may even be 150 per cent new, this remarkable figure being achieved by putting the wine into 100 per cent new oak to begin with, and then racking it into more new oak.

These wines have enormous and seductive richness, but sceptical questions are sometimes raised about the quality of their terroir, and their longevity. Neither question seems to count with the buyers who queue to empty their wallets.

*Vins de garage* could be made anywhere: they're the equivalent of a bigger estate creaming off the very best vat, or couple of vats; something that bigger estates are usually loath to do because of its detrimental effect on the rest of the blend. However, the first *vin de garage* is now being made in the Médoc, by proprietors who bought land for exactly that purpose, and the principle is becoming accepted by winemakers and growers (not necessarily of Merlot) across the globe who want to make a fast impact.

## Other French Merlots

Merlot is France's third most planted black grape, after Carignan and Grenache, but is in more demand than either at vine nurseries: 43.6 million Merlot vines were planted in France in 1998. It is a recommended variety for appellation contrôlée wines and vins de pays in Provence, the Languedoc, the South-West, the Ardèche, the Charente, the Corrèze, the Drôme, Isère, the Loire, Savoie and Vienne, though in the South it is a fairly recent arrival, having only been recommended there since 1966. In the South-West it is a traditional part of many appellation contrôlée blends. In the Languedoc most Merlot goes into vins de pays, both varietals and blends. Most are soft, fruity wines for early drinking.

CHÂTEAU PALMER

*Palmer often has 40 per cent Merlot in the blend, high for a top Haut-Médoc wine, and the result is classic Margaux perfume married with seductive softness of texture.*

CHÂTEAU PÉTRUS

*If anyone doubts that Merlot can age, show them a mature Pétrus: it's usually at least 95 per cent Merlot. But it does come from a unique, Merlot-friendly terroir.*

DOMAINE DE LA BAUME

*This is Merlot for those of us who can't afford Pétrus: south of France fruit plus Australian winemaking equals a soft, plummy glassful.*

*Merlot vineyard at Tenuta dell' Ornellaia owned by Marchese Lodovico Antinori at Bolgheri, so near the sea that locals in this part of Tuscany used to reckon that the wines from round here tasted of salt. That was before the super-Tuscan brigade got cracking. This Merlot, called Masseto, is one of the world's best.*

## Italy

Merlot fever has hit Italy in a big way: in the mid-1990s it was Italy's fifth most planted grape, but current plantings seem set to push it further up the ranking. In the north-east it has traditionally been regarded as a high-volume producer, and clones were chosen primarily for their yield. It wasn't given the best sites, and the wine was easy-drinking stuff, not intended to be taken seriously. This is true of the Veneto, Trentino and Alto Adige, though Merlot from Alto Adige is benefiting from better clones and more attention to site. The best wines come from Friuli, Tuscany and Umbria, and they are very good indeed. Some is varietal; a lot is blended with Sangiovese, and is valued for adding richness without dominating the blend in the way that Cabernet Sauvignon does. In Tuscany it is increasingly being planted in the warmer Maremma, and alcohol levels can reach 14 per cent, though acidity doesn't seem to suffer; perhaps it is added. In more northerly parts of Tuscany, Merlot's low acidity is often seen as a useful softener for the more acidic local grapes.

## Rest of Europe

In Switzerland's canton of Ticino, Merlot accounts for 85 per cent of production. It needs warm sites, and vineyards no higher than 450m, to produce wines of any substance; the best are well structured, and have some oak aging. Some is made 'white' – pale pink, in other words.

Merlot in Europe is concentrated in the cooler areas: it has made little impact in Iberia, though there are moves to get it on to the list of approved grapes in Rioja. (There are moves to get practically everything on to the list of approved grapes in Rioja.) There is some in Slovenia and the former Yugoslavia, some in Austria (where it is being planted in Burgenland at the expense of lesser white varieties like Welschriesling) and more in Romania, Bulgaria, Hungary and Moldova; much is blended with Cabernet Sauvignon.

## California

Experience is telling here. More and more Merlot is being planted, and once the current plantings come on stream, California will be the world's fourth largest producer of Merlot, after Bordeaux, Italy, the Languedoc and Romania. Post-phylloxera plantings are on more suitable rootstocks than before, and much has been learnt about trellising: Merlot grows particularly long shoots, which need to be tied, often to an extra moveable wire. Site selection is also better. Cooler spots that wouldn't suit Cabernet Sauvignon are being chosen – Carneros, for example, gives good bright flavours to Merlot, though Cabernet doesn't ripen well there. Parts of the Napa, like Oakville and Stags Leap, give richer wines, and hillside sites like Howell Mountain, where Cabernet is structured and long-lived, impart similar qualities to Merlot.

These, and other parts of California like Mendocino, Alexander Valley, Dry Creek

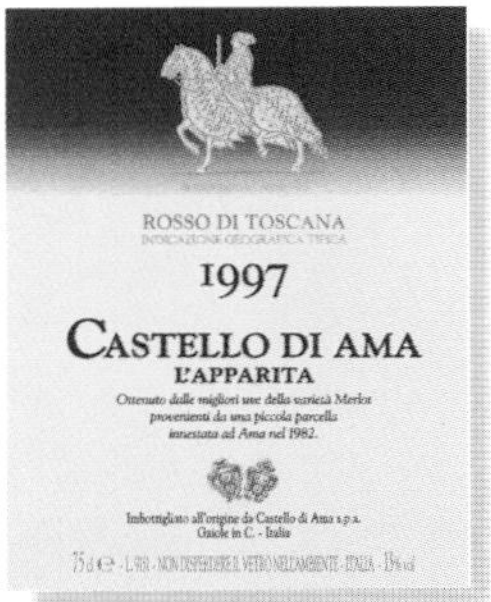


CASTELLO DI AMA

*Castello di Ama already makes one of the most sensuous and approachable of Chianti Classicos. This Merlot is in the same delicious vein.*

MATANZAS CREEK WINERY

*This wine is about as expensive as California Merlot gets – and that's pretty expensive. It'll age for some years, but as successfully as Pétrus? Only time will tell.*

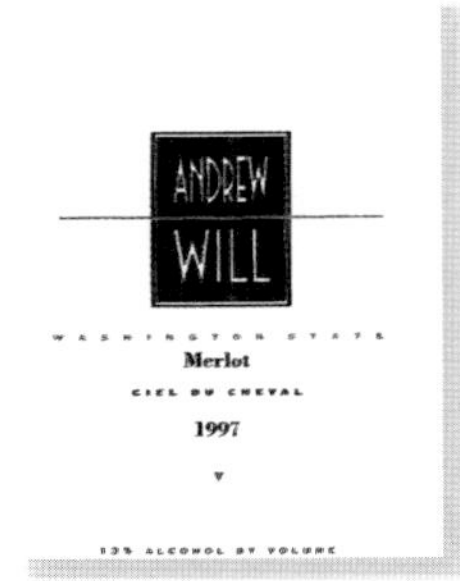


ANDREW WILL

*Many experts predict great things for Washington State Merlot. Certainly this example has sumptuous fruit and perfume as well as quite a beefy structure.*

Valley, Sonoma Valley and the warmer areas of Monterey, are where the serious Merlots are coming from. Some are huge wines, so massive that they make Pomerols look feeble. Extreme ripeness – the US market recoils at any hint of herbaceousness – is obtained by leaving the grapes on the vines for as long as possible, and alcohol levels are high. But those who want light Merlot (or perhaps Merlot lite) are also being catered for. A lot is being planted in the hot Central Valley for high-volume, low-intensity wine. And by low, I mean low.

### Washington and rest of the USA

Merlot is the most popular red grape in Washington State, with 5600 acres/2266ha in 1999; it comes a close second to Chardonnay in terms of plantings. It gives richer flavours than Cabernet Sauvignon, but has nevertheless suffered from the sort of greenness of taste that the US market so dislikes. However, after ten years of experience of growing Merlot the wines are riper and plummier than they were. Site selection, control of irrigation and later picking have all helped. Cold winters remain a problem for the variety. Classic Bordeaux blends are also successful, and at the moment seem the best way of using Cabernet here.

Oregon seems generally too cool, and fruit set is poor. New York State is more promising: Long Island's long growing season means that Merlot ripens reliably here.

### Chile and Argentina

The term 'Chilean Merlot' is somewhere between a euphemism and a nickname, and will remain so until all Chile's 'Merlot' vineyards are sorted out. The reason is that nearly all Chilean Merlot is a field blend of Merlot and Carmenère: the original cuttings were taken to Chile from Bordeaux before phylloxera and propagated as Merlot ever since. Nobody had thought to question the vines' identity until it was realized in the early 1990s that a lot of Chilean Sauvignon Blanc was in fact the less aromatic Sauvignonasse, or Sauvignon Vert. French ampelographers were then invited to look over the vineyards, and came up with the unsettling discovery that a great deal of the 'Merlot' was not Merlot at all.

The two grapes are being separated in new plantings, but it will be years – decades, even – before the existing mixed plantings are superseded. There are already a small number of genuinely varietal Merlots on the market, but they will remain a minority for a long time yet.

Nobody knows how much of the old plantings are actually Merlot. The official figure is that in 1999 there were 8414ha of Merlot and 1167 ha of Carmenère, but no Chilean producer seems to take this seriously. Most estimates put the amount of Carmenère at between 60 and 90 per cent. One company's 1999 Reserve Merlot is about 10 per cent Merlot, the rest being Carmenère, Alicante Bouschet, Cabernet Sauvignon and Zinfandel.

The problem is not that Carmenère is an inferior grape – it may in fact turn out to have greater potential for quality than Merlot – but that they ripen at different

#### Merlot in Chile: a tale of two vines

*'We were the origin of the discovery that a lot of Chilean Merlot is actually Carmenère', say Paul Pontallier and Bruno Prats. These leading Bordeaux winemakers (Pontallier makes Château Margaux, and Prats used to own Cos d'Estournel) joined forces in the late 1980s to make wine near Santiago in the Maipo Valley – the estate, now called Domaine Paul Bruno, is located at 700–800m.*

*They bought two batches of Merlot vines to plant, and soon realized that if one was Merlot, the other had to be something else. An ampelographer friend from France eventually identified the mystery vine as Carmenère. 'The latter blends well into Chilean reds', says Pontallier, because 'Merlot doesn't behave in Chile as it does in Bordeaux: it is harsher than in Bordeaux, whereas Cabernet Sauvignon is softer here. So the classic Bordeaux blend may not be the best thing for Chile.'*

VALDIVIESO

*Valdivieso's Merlot, like most others in Chile, is a blend of Merlot and Carmenère. The Carmenère adds real savoury personality to the milder-mannered Merlot.*

times. Merlot ripens some three weeks before Carmenère, so unless you have each vine marked in the vineyard and can pick the two varieties separately, you will either end up with ripe Merlot and underripe Carmenère, or overripe Merlot and ripe Carmenère. Neither makes for good balance, and unripe Carmenère has a strong flavour of green peppers, which sits awkwardly with ripe Merlot's chocolaty fruit.

Site selection is only just beginning in Chile, and no region has revealed itself as being better suited than any other to Merlot – though cool regions like Casablanca have largely ruled themselves out. Maipo is still the principal red wine area, even though Aconcagua, Colchagua and Curicó, producing wines with sweeter tannins and richer colours, are attracting more attention and most of the new plantings.

Merlot in Argentina is less planted than Cabernet Sauvignon or Malbec, and can be trickier to grow than either. High yields may well be part of the problem. The best, most concentrated wines seem to come from plantings in the Mendoza hills at altitudes of around 1100m.

*Hawkes Bay is one of New Zealand's best regions for Merlot., the country's most successful 'Bordeaux' red variety. It ripens earlier and more completely than the other varieties, and provides crucial richness and softness to blends.*

### Australia

Merlot comprises around 3 per cent of production in Australia, but recent years have seen a 100 per cent increase in plantings: Merlot mania has hit here, as well. At the moment a lot is still blended with Cabernet Sauvignon, but the number of varietal wines will increase as growers learn to understand it better. Paul Lapsley of Coldstream Hills in the Yarra Valley, believes that the clones available in Australia are particularly poor. But new clones, due for release over the next few years, should help to improve flavours and the bane of the Merlot grower's life, poor fruit set.

### New Zealand

New Zealand growers like Merlot because, unlike Cabernet Sauvignon, it's a reliable ripener in cool climates. Plantings have shot up, and Merlot is now the country's third most planted grape; it is at its best in the North Island, with the South Island being a little chilly even for Merlot. Over half is in Hawkes Bay, with most of the rest in Auckland and Marlborough: indeed, it could be the key to red wine production in cool-climate Marlborough, where Cabernet Sauvignon is a poor ripener. Blends are usually with the other Bordeaux varieties, with some producers leaving Cabernet Sauvignon out of the list in favour of Cabernet Franc and Malbec – again because of Cabernet Sauvignon's often green flavours.

### South Africa

There were 2134ha of Merlot in South Africa in 1997, the most recent year for which figures are available. That accounted for just 2.2 per cent of the total area under vine; Cabernet Sauvignon, by contrast, had 5.1 per cent. Since there was no Merlot planted at all until the 1980s, that is quite a rapid increase – and expect plantings to continue to rise, because Merlot is suddenly seriously trendy in the Cape. But, as in other warm regions of the world, the quick-ripening Merlot is not an unqualified success, and grapes from many vineyards have soared to about 14 per cent alcohol without ripening their flavours. Even so, the skilful producers are now working out where to grow it – often in the cooler regions – and how. Two basic styles are emerging: the soft, juicy, easy-drinking version, and something altogether more impressive, powerful and potentially long-lived. Most is in Stellenbosch, with quite a lot also in Paarl. New virus-free clones have helped its advance.

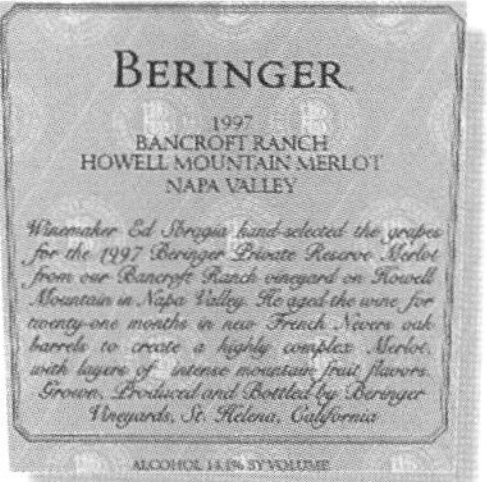


**BERINGER**

*The Howell Mountain AVA in the Napa Valley produces Merlots (and Cabernets) that are rich and perfumed. Beringer's Private Reserve Bancroft Ranch has wonderful fruit as well as structure.*

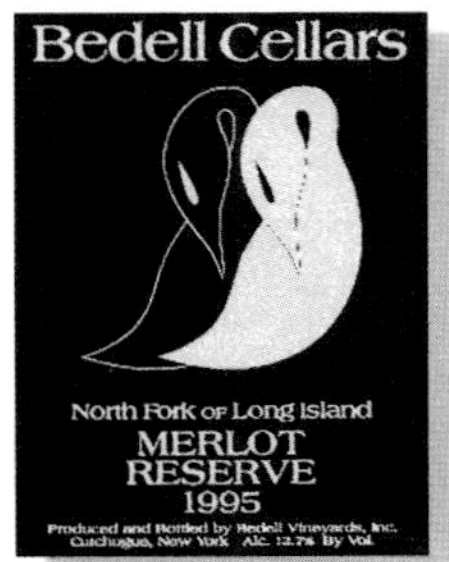


**BEDELL CELLARS**

*Bedell Cellars' winery is in a converted potato barn. No jokes about chips with everything, please: this Reserve is one of Long Island's top reds.*

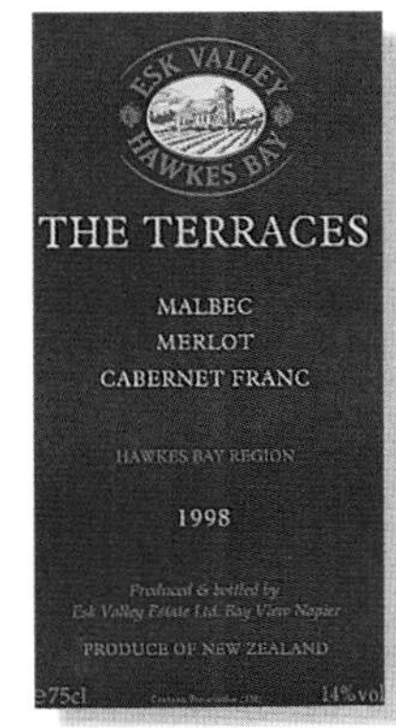


**ESK VALLEY**

*The Terraces is a tip-top blend of Malbec, Merlot and Cabernet Franc from one of the warmest vineyard sites in New Zealand's Hawkes Bay.*

**SPICE ROUTE**

*A dense, powerful Merlot from unirrigated bush vines in Swartland. Spice Route is a young company owned by Charles Back of Fairview, one of South Africa's most socially progressive producers..*

# Enjoying Merlot

A GREAT DEAL OF THE TIME, the question of how well Merlot ages is irrelevant. Most is made to be drunk young, even immediately: modern Merlot is the epitome of the red wine that gets all the bottle age it needs on the way back from the wine shop.

That is true both of simple, juicy Merlots and of big, concentrated ones. Merlot's low tannins and low acidity make it a less than obvious wine for the cellar; the majority of Merlots not only don't need bottle age, but often positively don't want it. Age these wines and your patience will be rewarded by fading fruit and disintegrating structure.

Yet Merlot can age perfectly well as part of the Bordeaux blend – and not just because it has Cabernet Sauvignon to prop it up. St-Émilions and Pomerols, usually with little or no Cabernet Sauvignon in the blend, will easily last a decade, and the very best wines will see in their 30th birthday. Such wines need time to come round in the first place: from two or three years to longer, depending on the property and the vintage. Top Pomerols and St-Émilions are, however, exceptions even in Bordeaux: normally if a Bordeaux proprietor wants to make earlier-maturing wine, he plants more Merlot.

The structure of Merlot from Bordeaux is also found in the best examples from California, Italy and elsewhere. There is no question about these balanced wines having the grip and backbone to last at least ten years and perhaps more; the question is how many get the chance.

### The taste of Merlot

This is where one has to raise the biggest question mark of all. Merlot tastes of all sorts of things: it can be anything you want, from light and juicy through Pinot Noir-silky to Cabernet-oaky and extracted. What should it taste like? That's the very question that winemakers across the world are asking themselves. The lack of benchmarks means that there is no easy answer.

At its best it is succulent and silky, with velvety tannins: 'smooth' is the word that has cropped up most often in US market research. Fruit flavours can be of strawberry, raspberry, black cherry, blackcurrants, plums, damsons, figs or prunes. Fruitcake flavours are often found in Bordeaux. There can be spice, too: cinnamon and cloves and a touch of sandalwood, and truffles, tobacco, licorice and toasted nuts. Warm climates can give stewed fruit flavours; cool ones, or over-cropped vines, can give minty, lean, herbaceous notes. It can be gamy, chocolaty or coffeeish; yet for every Merlot that exhibits such appetizing complexity there are 20 examples that aim for no more than sweet juicy fruit and that seductively velvety texture.

To many a Californian winemaker, Merlot is about texture more than flavour. Texture is what they believe consumers notice most; and if Merlot mania is a fair guide, perhaps they're right. But what they are really saying is 'lack of texture': blandness, total smoothness, with a flavour as mild as can be, sometimes helped along with a little sugar. This is lowest common denominator 'wine without tears', and the phenomenon has been massively boosted by the claims of red wine being good for your heart. These claims meant that millions of people who didnt' like red wine, and never drank it, would drink it for their health – so long as it didn't taste like red wine. Many of the budget-priced commercial American Merlots are merely trying to satisfy that market. Good luck to them. May it keep their customers healthy. I'll stay healthy by drinking equal amounts of something slightly more interesting.

*Right: the new Bordeaux next to the new Chile. Château Angélus in St-Émilion, where around 50 per cent of the blend is Merlot, was promoted to Premier Grand Cru Classé in 1996, after a decade or so of hard work by its owners to improve quality. Casa Lapostolle's objective in its Cuvée Alexandre Merlot is to match Angélus drop for drop in terms of lushness, spice and exuberance. Consultant winemaker Michel Rolland provides a link: he is responsible for the winemaking at Casa Lapostolle, and has been advising at Angélus for the past two decades.*

### Matching Merlot and food

Merlot makes soft, rounded, fruity wines that are easy to enjoy without food, but it is also a good choice with many kinds of food. Herby terrines and pâtés, pheasant, pigeon, duck, goose and spicier game dishes all blend well with Merlot. It can also partner subtly spiced curries and tandoori dishes. Meaty casseroles made with wine are excellent with top Pomerol châteaux; and the soft fruitiness of the wines is perfect for savoury foods with a hint of sweetness such as ham, savoury pancakes and gratins.

## CONSUMER INFORMATION

**Synonyms & local names**
Also known as Merlot Noir, and in Hungary as Médoc Noir. Not to be confused with the unrelated Merlot Blanc, also found in Bordeaux but now almost disappeared.

**Best producers**
**FRANCE/Bordeaux/St-Émilion** Angélus, Ausone, Beau-Séjour Becot, Magdelaine, le Tertre-Roteboeuf, Troplong-Mondot; **Bordeaux/Pomerol** le Bon Pasteur, l'Église-Clinet, l'Evangile, la Fleur-Pétrus, Lafleur, Pétrus, le Pin, Trotanoy, Vieux-Château-Certan; **Bergerac** de la Colline, Tours des Verdots; **Languedoc** de la Baume, Cazes, de Limbardié, Skalli/Fortant de France.
**ITALY/Veneto** Maculan; **Friuli** Livio Felluga, Russiz Superiore; **Tuscany** Castello di Ama, Avignonesi, Castelgiocondo, Le Macchiole, Ornellaia, Petrolo, Tua Rita; **Lazio** Falesco; **Sicily** Planeta.
**SPAIN/Penedès** Can Ràfols dels Caus, Torres (Atrium); **Somontano** Viñas del Vero.
**SWITZERLAND** Christian Zündel.
**USA/California** Arrowood, Beringer, Cafaro, Chateau St Jean, Clos du Bois, Cosentino, Duckhorn, Ferrari-Carano, Forman, Harrison, Havens, Jade Mountain, Lewis, Matanzas Creek, Newton, Pahlmeyer, Paloma, Ridge, St Francis, Selene, Shafer, Swanson, Truchard; **Washington State** Andrew Will, Chateau Ste Michelle, Columbia Crest, Columbia Winery, Foris, L'Ecole No. 41, Leonetti; **New York State** Bedell, Hargrave, Palmer.
**AUSTRALIA** Brookland Valley, Charles Cimicky, Clarendon Hills, James Irvine, Katnook Estate, Maxwell, Pepper Tree, Petaluma, Tatachilla, Yarra Yering.
**NEW ZEALAND** Brookfields, Corbans (Reserve), Kim Crawford, Esk Valley, Goldwater, Morton Estate, C J Pask, Sileni, Villa Maria.
**SOUTH AFRICA** Meinert, Morgenhof, Radford Dale, Saxenburg (Private Collection), Spice Route, Thelema, Veenwouden, Vergelegen.
**CHILE** Bisquertt (Casa La Joya), Caliterra (Arboleda), Carmen (Reserve), Casa Lapostolle (Cuvée Alexandre, Clos Apalta), Casa Silva, Cono Sur (20 Barrel), Errázuriz, La Rosa (La Palmería), San Pedro, Santa Rita (Unfiltered), Valdivieso, Viña Casablanca (Santa Isabel), Viña Gracia, Viu Manent.

## RECOMMENDED WINES TO TRY

**Ten more classic Bordeaux red wines (as well as the Best producers left)**
Ch. Canon-de-Brem *Canon-Fronsac*
Ch. Canon-la-Gaffelière *St-Émilion*
Clos Fourtet *St-Émilion*
Ch. la Conseillante *Pomerol*
Ch. Gazin *Pomerol*
Ch. Grand-Mayne *St-Émilion*
Ch. Fontenil *Fronsac*
Ch. Latour-à-Pomerol *Pomerol*
Ch. Pavie-Macquin *St-Émilion*
Ch. Roc de Cambes *Côtes de Bourg*

**Five good-value Bordeaux reds**
Ch. Annereaux *Lalande-de-Pomerol*
Ch. Brulesécaille *Côtes de Bourg*
Ch. Carsin *Premières Côtes de Bordeaux Cuvée Noir*
Ch. la Prade *Côtes de Francs*
Ch. Segonzac *Premières Côtes de Blaye*

**Five top Italian Merlots**
Castello di Ama *Vigna l'Apparita*
Falesco *Montiano*
Livio Felluga *Colli Orientali del Friuli Rosazzo Merlot Riserva Sossó*
Tenuta dell'Ornellaia *Masseto*
Russiz Superiore *Collio Rosso Riserva degli Orzoni*

**Ten New World lookalikes**
Andrew Will *Ciel du Cheval Merlot (Washington)*
Beringer *Bancroft Vineyard Merlot (California)*
Casa Lapostolle *Cuvée Alexandre Merlot (Chile)*
Charles Cimicky *Merlot*
Goldwater Estate *Merlot (New Zealand)*
Leonetti *Merlot (Washington)*
Matanzas Creek *Journey Merlot (California)*
Newton *Merlot (California)*
Spice Route *Flagship Merlot (South Africa)*
Veenwouden *Merlot (South Africa)*

**Ten good-value Merlots**
Brookland Valley *Merlot (Australia)*
Clos du Bois *Merlot (California)*
Columbia Crest *Merlot (Washington)*
Kim Crawford *Te Awanga Merlot (New Zealand)*
Errázuriz *Merlot (Chile)*
La Palmería *Gran Reserva Merlot (Chile)*
San Pedro *35 South Merlot (Chile)*
Santa Rita *Reserva Merlot (Chile)*
Viñas del Vero *Merlot (Spain)*
Viu Manent *Reserva Merlot (Chile)*

*Merlot magic: put it on the shop shelf and watch it disappear. But Merlot's commercial success goes hand in hand with an identity crisis. Merlot winemakers around the world can never agree about what the wine should taste like.*

**Maturity charts**
Merlot, even premium Merlot from Napa or Sonoma, does not fit into a single pattern. Not all will age well in bottle, or be meant to.

**1998** St-Émilion Premier Grand Cru Classé and top Pomerol

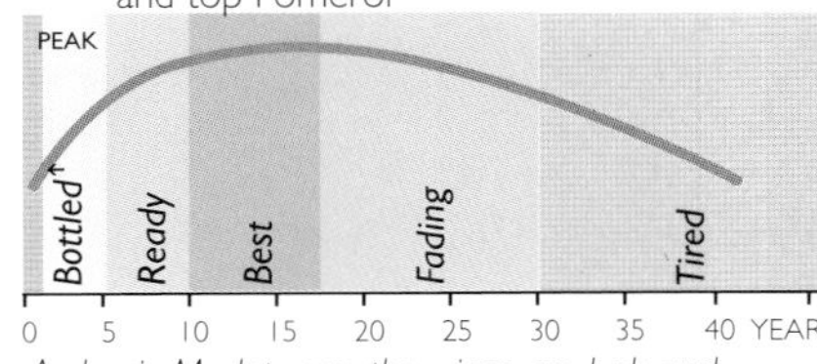


*A classic Merlot year: the wines are lush and drinkable early, but their excellent structure ensures a long peak of pleasure and a gentle decline.*

**1999** Top Napa/Sonoma Merlot (Reserve)

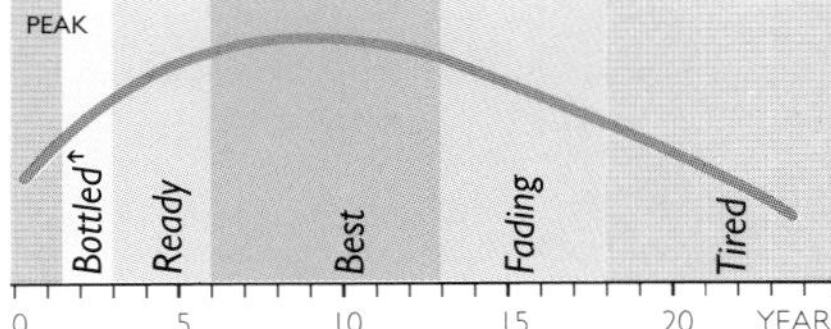


*A late vintage, in which the longer hang time produced grapes of excellent colour and concentration. The wines are rich, dark and supple.*

**2000** Chile Merlot (Reserve)

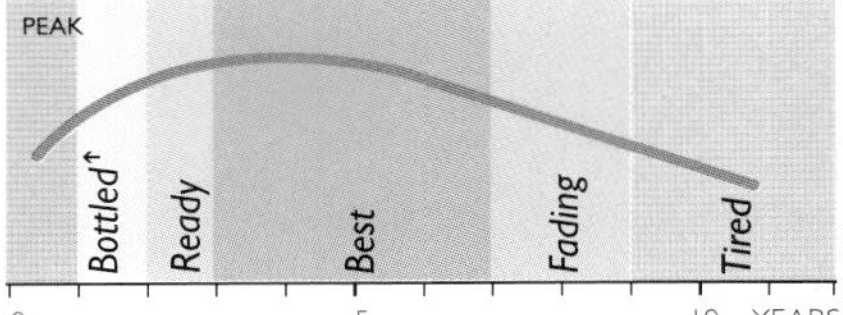


*A year of intense flavours and good concentration. Acidity is also good, and the wines have excellent balance and varietal flavours.*

## MEUNIER

Meunier's most common synonym, Pinot Meunier, bears witness to its probable origin as a mutation of Pinot Noir. True, the leaf now looks very different and far more indented. The leaves are also downy on the underside, giving them the floury appearance from which the vine takes its name: 'Meunier' means 'miller', and many of the the vine's other synonyms – Farineux or Noirin Enfariné, or Müllerrebe (see page 142) or Müller-Traube in Germany, right down to Dusty Miller in England – derive from the same characteristic. But it is possible to find canes bearing completely hairless leaves; this is taken by ampelographers as evidence of an origin in Pinot Noir.

It is best known as a blending partner for Pinot Noir and Chardonnay in Champagne, and it is popular with growers there because it buds late and ripens early. Both of these are useful attributes in chilly Champagne, and it is found in the cooler, more frost-prone parts, especially in the Aube, where neither Chardonnay nor Pinot Noir would be a safe bet. It is useful in the blend because it matures much faster than the other two wines and provides softness, fatness and appealingly round fruit at an early age: ideal for wines intended to be sold and drunk young.

Generally it is not thought to age well, and is not regarded as having as much finesse or quality as the other two grapes. As a result most producers are somewhat shy of talking about the amount of Meunier they include in their blends. The exception is Krug, which uses a fair proportion (though still much less than either of the other grapes) in its very long-lived Champagne.

In the 19th century it was the great standby of vineyards all over the north, from the Paris basin as far as Lorraine – regions where no vineyards exist now. It is still occasionally found in the Loire and makes a pleasant smoky pale pink Vin Gris near Orléans, but there is much more in Germany, where it grows in Württemberg, Baden, the Pfalz and Franken under the names of Müllerrebe, Müller-Traube or Schwarzriesling. In Württemberg it is used for the local pink speciality, Schillerwein; it also makes white sparkling and still red in Germany. Colour is fairly light – lighter than that of Pinot Noir – and the wine is often slightly higher in acidity and smoky in taste. In Germany there is a local variant, called Samtrot, found in Württemberg.

There is some in Austria and in German-speaking cantons of Switzerland.

It has been cultivated on a small scale for many years in Australia. Some slightly jammy reds are made and it is also used in Champagne-style fizz. New Zealand also has small plantings. Best producers: (Australia) Best's, Seppelt, Taltarni; (Canada) Tinhorn Creek; (France) Billecart-Salmon, Blin, Charles Heidsieck, Alfred Gratien, Krug, Laurent Perrier; (USA) S Anderson, Roederer Estate, Schramsberg.

## MISKET

An Bulgarian grape that in spite of its name and its grapy aroma has nothing to do with Muscat (see pages 144–153). Instead it is the result of crossing Dimiat with Riesling; there is also a red version which is vinified without the skins for white wine. Best producer: (Bulgaria) Domaine Boyar.

## MISSION

If you were an early Spanish settler in the Americas, Mission was your everyday wine. It may seem odd that the Franciscan missionaries who brought it, presumably from Spain, didn't take the trouble to bring something better. But the vine found in the Americas (it is the same as the Criolla Chica of Argentina; some authorities maintain that the País of Chile and the Negra Corriente of Peru are also the same vine; others believe them to be different) was probably grown from seed rather than from cuttings, and vine seedlings notoriously do not turn out like their parents.

In any case, it has the advantage of producing good crops in drought conditions, and because of the wine's high resveratrol content it is said not to turn easily to vinegar. Both of these attributes could account for its survival in large quantities to the present day. In Chile, País is mostly grown in the south of the country, in Maule and Bío-Bío, where it may also be called Negra Peruana. The vines are bush trained and head pruned, and can yield 3000 kilos per hectare of grapes on the unirrigated hills, and ten times that amount on the irrigated plains. Attempts by one co-operative winery to produce serious organic País show that, while it can have some character, and soft plums and sweet chocolate fruit, it is never better than rustic and has no pretensions to finesse.

In Argentina the lighter-skinned Criolla Chica is less commonly planted than Criolla Grande, and is of marginally higher quality.

Mission can still be found in California, but it is in decline. It would be nice to think that some will always be made, however: it is part of California's history. Best producers: (USA) Sobon, Story.

## MOLETTE

A neutral grape grown in the Savoie region of France, albeit in declining quantities, for sparkling wine. Roussette (see page 206) is often added to the blend to give more character. Best producer: (France) Varichon et Clerc.

## MOLINARA

A minority grape in the Valpolicella blend, Molinara is easy to drink and light in body but pale in colour and contributes little except acidity. Quality-conscious producers use it less and less. Best producer: (Italy) Quintarelli.

## MONASTRELL

The Spanish name for Mourvèdre (see pages 140–141) – although, since the variety originated in Spain, it would be more accurate to say that Mourvèdre is the French name for Monastrell. Be that as it may, the grape is best known by its French name.

Quality in Spain has risen, having been pretty dismal for many years. The greatest efforts have been in Jumilla, and were spurred by the arrival of phylloxera in 1988-89, necessitating massive replanting. The use of new and better virus-free clones and some serious investigation into how best to handle Monastrell – when to pick it, how long to macerate it, how hot or cool to ferment it and when to bottle it – have paid off. The grape may be made as a varietal (perhaps with carbonic maceration for early drinking) or it may be blended with Tempranillo, Cabernet Sauvignon and Merlot. The wines of Yecla, where Monastrell is also the majority grape, are not yet up to those of Jumilla, but improvements continue to be made. Garnacha and Merlot are commonly blended in here. Best producers: (Spain) Agapito Rico, Castaño, Salvador Poveda, Primitivo Quilés, Señorio del Condestable.

## MONDEUSE NOIRE

It is sometimes said that this French grape from Savoie is the same as the larger berried form of Syrah, Gros Syrah. What is certain, however, is that Mondeuse Blanche, which is not a mutation of Mondeuse Noire but is also found in Savoie and Bugey, is perhaps one of the parents of Syrah, the other possibly being Dureza.

But back to Mondeuse Noire. There is another theory, also disputed, that it could be the same as Friuli's Refosco. It certainly can't be both Syrah and Refosco and most likely it is neither, though its wine is not unlike that of Refosco, with the same Italianate intensity of dark-scented plum fruit and a bitter cherry twist to the finish. It has good deep colour and firm structure, but for all its virtues plantings are falling, which is a pity. Best producers: (France) des Rocailles.

## MONICA

Monica is found in Sardinia, where it makes easy-to-drink everyday red, often with low tannin and low acidity. It is sometimes said to be the same as Mission (see left) and País and to have been taken to the Americas from Spain. Well, it certainly doesn't taste like Mission to me, because it produces nice tart, cranberryish red wine to drink young. It is no longer found in Spain, but Spain was at one time the overlord of Sardinia, so that's presumably how it got there. Best producers: (Italy) Argiolas, Dolianova co-op, Santadi co-op, Trexenta co-op.

## MONTEPULCIANO

This should not be confused with Vino Nobile di Montepulciano, which is an altogether finer wine made from Sangiovese from an area around the town of Montepulciano in Tuscany. Montepulciano the grape has nothing to do with this wine; instead it is planted in central Italy (including Umbria and southern Tuscany) and the South. Its best known form is as Montepulciano d'Abruzzo, round, plummy and weighty red with ripe tannins, good acidity and a low price tag. Its most useful contribution is probably as a dominant partner in Rosso Conero in the Marche.

It is also found blended into many reds in much of the southern half of Italy. It is late ripening and perhaps would not be successful north of Tuscany and Umbria. Despite being hailed as a splendid performer I have to admit I've had few exciting examples. Best producers: (Italy) Boccadigabbia, Cornacchia, Garofoli, Illuminati, Masciarelli, Moroder, Nicodemi, Saladini Pilastri, Le Terrazze, Umani Ronchi.

## MORELLINO

A synonym for Sangiovese in the Scansano region just inland of the Maremma in southern Tuscany. Morellino di Scansano is traditionally an appealingly fruity, cherryish, fairly light red with soft tannins; not as elegant or refined as the wines of the best Chianti estates, but with plenty of charm. However, some producers are now making more gutsy examples with sweet sour fruit and a herbal rasp.

Scansano has the advantage of an almost perfect mesoclimate for Sangiovese: it ripens easily here, and gives good concentration and structure providing yields are kept well under the legal maximum of 84hl/ha. True, the summers can be very hot and dry, but irrigation is a possibility for the future, which should solve any ripening problems induced by drought. Up to 15 per cent of other varieties are permitted under the DOC rules: Alicante and Spagna are the vines most commonly grown in the area, apart from Sangiovese; Merlot and Cabernet are also used for more international flavours. Best producers: (Italy) Erik Banti, Carletti, Cecchi, Mazzei, Montellassi, Moris Farms, Le Pupille. See also Sangiovese pages 208–217.

## MORILLON

Chardonnay goes by this name in Austria's Steiermark region; Morillon used also to be a synonym for Chardonnay in much of France. Best producers: (Austria) Frühwirth, Neumeister, Polz, Erwin Sabathi, Manfred Tement.

## MORIO-MUSKAT

This over-scented German crossing is now mercifully on the decline. It used to be a standby of the Liebfraumilch industry, and was much grown in the Rheinhessen and Pfalz, where most Liebfraumilch originates. But demand for Liebfraumilch is declining and much Morio-Muskat has been uprooted. It has a coarse aroma and flavour, as grapy as Muscat but without any of the latter's elegance or lightness, and in fact is not related. It is the result of crossing (in 1928) Silvaner with Weissburgunder, which just goes to show you can't be too careful: one could never have predicted such an offspring from such inoffensive parents.

It is at its most aggressive, mixing musky perfume with grapefruit acidity, when picked slightly unripe, which it usually is because it needs good sites to ripen properly, and good sites are quite rightly reserved for the best grapes. It is subject to downy mildew, oidium and rot. Very small amounts are planted in South Africa and Canada. Don't ask me why.

## MORISTEL

Light-coloured red grape found in Somontano, Spain. It may or may not be a synonym for Morrastel (see below). Best producers: (Spain) Borruel, Monclús, Pirineos.

## MORRASTEL

Either the French name for Graciano, or a Spanish name for Mourvèdre (see pages 140–141), depending on where you are.

## MOSCADELLO

A local variant of Moscato, or Muscat (see pages 144–153) found in Tuscany, where it has a long history. It makes fairly sweet grapy fizz, somewhat weightier than that of Piedmont. There is also some still, sweeter, late-harvest Moscadello. The Muscat used is a version of the superior Muscat Blanc à Petits Grains. Best producers: (Italy) Altesino, Col d'Orcia, Il Poggione.

## MOSCATEL

Iberian name for Muscat. Wines labelled simply as Moscatel are basic, sweet and dark and usually made from Muscat of Alexandria, or Moscatel de Alejandría; another name for this grape is Moscatel Romano. Muscat Blanc à Petits Grains is Moscatel de Grano Menudo. Moscatel de Malaga is believed to be a variety local to southern Spain. Best producers: (Spain) Gandía, López Hermanos, Ochoa.

## MOSCATO DI ALEXANDRIA

Muscat of Alexandria in Italian. See Muscat (pages 144–153) for more details.

## MOSCATO BIANCO

Italian name for Muscat Blanc à Petits Grains (see pages 144–153). It is found pretty well all over Italy, and most regions have a tradition of lightly sweet, aromatic wines that are low in alcohol. The South and the islands produce sweeter, richer wines, and in the North, Trentino-Alto Adige has good crisp, dry versions.

## MOSCATO GIALLO

A high-quality, golden-berried strain of Muscat found in north-eastern Italy. See Muscat (pages 144–153) for more details.

## MOSCATO ROSA

A high-quality pink-berried strain of Muscat found in north-eastern Italy. See Rosenmuskateller (page 203) for more details.

## MOSCHOFILERO

An aromatic Greek grape of rather good quality, used either as a varietal or as part of a blend. Its aroma is somewhere between that of Gewürztraminer and that of Muscat, but with better acidity than either. Its grapes are pink, and it is used either for white wine or for pink. Best producer: (Greece) Cambas.

## MOURISCO

One of the lesser Portuguese grapes used in the Port blend and best suited to warm spots. The wine is light in colour and quite high in acidity; it is not one of the five favoured varieties planted in the newer vineyards of the Douro Valley.

# Mourvèdre

The French call it Mourvèdre, the Australians and Californians call it Mataro, but the Spanish call it Monastrell. Which is odd, because the grape originated in Spain either near Murviedro in Valencia or near Mataró in Cataluña. Perhaps local pride meant that both areas claimed the grape so fiercely that Monastrell was chosen so as not to offend anyone. Well, no-one except the Spanish uses the Monastrell name and frankly, Mourvèdre is much more honoured away from its birthplace. In Spain it's the key to many reds from south of Madrid, giving high alcohol and plentiful tannins but so far not bringing much distinction in its wake. The growers here are being persuaded to time their picking better and recently there has definitely been an improvement in fruit flavour, injecting much needed life into what has been a forgotten part of vinous Spain.

Mourvèdre is, however, best known by its French name. It is very much a grape of the South – it won't ripen north of Châteauneuf-du-Pape, and doesn't even ripen properly there in cool years. But in Bandol on the Mediterranean coast, which is some 5°C warmer, it produces big, hefty wines which are nevertheless not without finesse.

It is a thoroughly finicky grape to get right: it needs the warmest, south-facing sites, but cool, shallow, clay soil to tame its vigour. Growers must get a minimum of 13 per cent alcohol because it simply has no flavour at a lower strength; they have to pick in a small window when the grape finally has body and fruit but before it becomes too pruney.

In the winery and in bottle it also has its idiosyncrasies. It is very reductive, by which I mean it smells distinctly farmyardy more often than not. This means that it's a surprisingly good partner for the lush fat Grenache, which is prone to oxidation and which benefits from the rather meaty aromas of Mourvèdre. What fruit it has is blackberryish and there's generally quite a rasp of herbs. This means it's easy to think a young Mourvèdre is faulty, but after an even more farmyardy middle age, it usually emerges into a rich, leathery maturity after five years or more.

It's only quite recently that people have started to take Mataro/Mourvèdre seriously in California and Australia. There are lots of old bush vines there giving dark, herby, rich but rustic reds that are a bit of a handful to start with, but age well and add grit and muscle to many red blends.

### The taste of Mourvèdre

Young Mourvèdre, picked at low yields, has a fairly wild mix of rasping hillside herbs, more than a hint of the farmyard, and, if you're lucky, flavours of blackberries and bilberries. It is solid in style, with high alcohol and tannin. Most is blended, often with other southern French grapes like Grenache and Syrah. Blended or not, Mourvèdre will always add a farmyardy, herby roughness for a few years before developing flavours of leather, gingerbread and game.

*Californian examples of this grape tend to be less broodingly tannic than examples from Bandol in southern France, but Jade Mountain's Napa County version is nevertheless dense and robust to balance its rich fruit. As well as this varietal Mourvèdre, Jade Mountain blends the grape with Cabernet Sauvignon and Syrah in other wines.*

**Domaine Tempier**

*One of the two top Bandol estates (the other is Pibarnon), Tempier uses a high proportion of Mourvèdre in all its reds; this one is 100% Mourvèdre.*

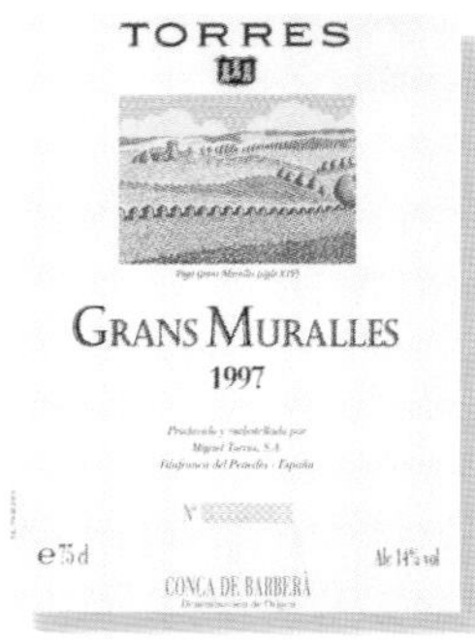


**Torres**

*Monastrell is the largest single grape variety in this big, rich blend from Torres. The other grapes are Garnacha Tinta, and two old Catalan varieties, Garró and Samsó.*

*Left: Mataro/Mourvèdre growing in the Barossa Valley in South Australia. The name Mourvèdre is supplanting that of Mataro in Australia as demand for old vine fruit soars. Aficionados reckon that if you call it Mataro, you don't take it seriously. And wines from old bush vines like these are very serious indeed: they're not as immediately likeable as Shiraz or Grenache but can add a fascinating extra something to red blends.* ***Above:*** *Mourvèdre is not a variety that is terribly easy to grow successfully: it is said to like its face in the hot sun and its feet in the water, which is a tall order even for much of the Bandol vineyards perched on the cliffs high above the Mediterranean. It's an even taller order for most of Australia.*

## CONSUMER INFORMATION

**Synonyms & local names**
Called Monastrell, Morrastel or Morastell in Spain. Mataro is a synonym used in the New World (Esparte is also used in Australia). One of its more picturesque French names is Estrangle-Chien or 'Dog strangler'.

**Best producers**
**FRANCE/Southern Rhône** de Beaucastel, de l'Oratoire de St-Martin, Richaud, Dom. Ste-Anne; **Provence** la Bastide Blanche, de la Courtade, d'Esclans, de Frégate, Galantin, Jean-Pierre Gaussen, Gavoty, de l'Hermitage, Mas Redorne, Moulin des Costes, de la Noblesse, de Pibarnon, Ch. de Pradeaux, Ray-Jane, Roche Redonne, Romassan, Ch. de la Rouvière, Ch. Ste-Anne, Salettes, Ch. Simone, la Suffrène, Tempier, de Terrebrune, de la Tour de Bon, Vannières; **Languedoc-Roussillon** Jean-Michel Alquier, Canet-Valette, de la Grange des Pères, de l'Hortus, du Mas Blanc.
**SPAIN** Agapito Rico, Bodega Balcona, Castaño, Julia Roch Melgares, Torres; **Alicante Fondillón** Bodegas Alfonso, Bocopa, Bodegas Brotons, Salvador Poveda, Primitivo Quiles.
**USA/California** Bonny Doon, Cline Cellars, Edmunds St John, Jade Mountain, Qupé, Ridge, Sine Qua Non, Sean Thackrey, Zaca Mesa; **Virginia** Horton.
**AUSTRALIA** Cascabel, D'Arenberg, Hewitson, Charles Melton, Penfolds, Pikes, Rosemount, Seppelt, Torbreck, Veritas, Wendouree, Yalumba.
**SOUTH AFRICA** Fairview.

## RECOMMENDED WINES TO TRY

**Ten southern French wines**
Domaine de la Bastide Blanche *Bandol Cuvée Fontanieu*
Ch. de Beaucastel *Châteauneuf-du-Pape*
Domaine de l'Hermitage *Bandol*
Ch. de Pibarnon *Bandol*
Ch. de Pradeaux *Bandol*
Domaine Ray Jane *Bandol*
Ch. de la Rouvière *Bandol*
Domaine Tempier *Bandol Migoua*
Domaine de Terrebrune *Bandol*
Ch. Vannières *Bandol*

**Five New World Mourvèdre wines**
Bonny Doon *Old Telegram (California)*
D'Arenberg *The Twenty Eight Road Mourvèdre (Australia)*
Hewitson *Barossa Valley Old Garden Mourvèdre (Australia)*
Jade Mountain *Mourvèdre (California)*
Ridge Vineyards *Evangelo Mataro (California)*

**Ten New World wines containing a significant percentage of Mourvèdre**
Bonny Doon *Le Cigare Volant (California)*
Cline Cellars *Oakley Cuvée (California)*
Edmunds St John *Les Côtes Sauvage (California)*
Jade Mountain *La Provençale (California)*
Charles Melton *Nine Popes (Australia)*
Qupé *Los Olivos Cuvée (California)*
Rosemount *GSM (Australia)*
Sine Qua Non *Red Handed (California)*
Yalumba *Antipodean (Australia)*
Zaca Mesa *Cuvée Z (California)*

## MÜLLER-THURGAU

It is a curious fact that all the world's greatest grape varieties have happened by accident. Man may be a dab hand at the selective breeding of pigs or cattle, but when it comes to grape vines nature has always (so far, anyway) done better on her own.

Müller-Thurgau is a case in point. It was bred in 1882 by Dr Hermann Müller-Thurgau, working at the Geisenheim Institute in Germany. His aim was, as usual in Germany, to produce a crossing that would combine all the advantages of Riesling – high quality, elegance, complexity – with the earlier ripening of Silvaner. Müller-Thurgau proved so popular that by the early 1970s it was the most planted grape in Germany – though it is hard to believe that anyone thought the vine had one iota of the quality of Riesling. It is now in decline, but has a long way to go yet.

Ironically, it has now been shown not even to be the Riesling x Silvaner cross that it was long believed to be. DNA fingerprinting has revealed its parents to be Riesling and Chasselas de Courtillier. It lacks both the intensity of the one and the delicacy of the other, and is merely dull, aromatic in a rather unspecific way, and with little structure. But it is easy to grow and yields extremely generously, easily giving 200hl/ha, and that, in an era of Liebfraumilch and associated wines – Piesporter, Niersteiner and Bernkasteler – was a great attraction for the growers of most German regions. It even made inroads to the tune of 20 per cent in the Mosel-Saar-Ruwer.

It is true that if yields are restricted and the vineyards are well managed then Müller-Thurgau is capable of producing respectable wine, attractively scented and with a pleasant acidity. But the problem is that to produce a top Müller-Thurgau you need to plant it in a top Riesling site, which is a terrible waste of a top Riesling site. In a third-rate site you'll get a third-rate Müller-Thurgau – but then Riesling from such a site wouldn't be worth drinking at all. It is also a grape that seems to be at its best in categories not higher than Spätlese since it seems to lose what definition it has as it gets riper. For most cheap German wines it is blended with something more assertively aromatic, like Morio-Muskat.

Its disadvantages (apart from the above) in Germany are that it is less resistant to winter cold than Riesling, and susceptible to rot, downy mildew and a painful sounding fungal affliction called Rotbrenner.

A great deal of Müller-Thurgau was planted, on German advice, in New Zealand in the 1950s and 1960s, when that country was just beginning to think seriously about modernizing its wine industry. It is gradually being replaced – plantings have declined from 1327ha in 1991 to a projected 353ha in 2001 – but in fact New Zealand examples can have reasonable balance, are less overcropped than German ones and can be quite pleasant.

The same can be said of Müller-Thurgau from northern Italy, where the high-altitude vineyards of the Alto Adige region keep good acidity in the wine.

It is found across Europe from Austria eastwards, but seldom produces anything of note. England and Luxembourg (where it is called Rivaner), on the other hand, have cool climates, and produce some quite nice snappy wines. Oregon and Washington State have produced some decent ones, but are unlikely to continue doing so for much longer. Best producers: (England) Chapel Down, Valley Vineyards; (Germany) Karl-Heinz Johner, Juliusspital, Dr Loosen, Markgraf von Baden (Salem/Durbach); (Italy) Nilo Bolognani, Caldaro co-op, Casarta Monfort, Cortaccia co-op, Graziano Fontana, Pojer e Sandri, Pravis, Enrico Spagnoli, Tiefenbrunner, A & R Zeni; (New Zealand) Giesen, Nobilo; (USA) Tualatin.

## MÜLLEREBE

A German name for Meunier (see page 138). The name – 'Miller's grape' – means the same as the French 'Meunier'; both names derive from the floury appearance of this downy-leaved vine. Best producers: (Germany) Drautz-Able, Fürst zu Hohenlohe-Öhringen, von Neipperg.

## MUSCADELLE

Muscadelle is unrelated to Muscat, although its flowery, grapey aroma might suggest some connection. It is found in Sauternes as a minor part of the blend: 2–3 per cent Muscadelle added to the Sémillon and Sauvignon Blanc can give this sweet wine an attractive lift and some extra finesse. In Monbazillac it is the second most planted grape after Sémillon, but is losing ground to the latter.

In dry white Bordeaux from Entre-Deux-Mers, Premières Côtes de Bordeaux and other areas it can form up to 40 per cent of the blend but seldom does, since it is an irregular yielder

*Müller-Thurgau grapes being harvested at Lamberhurst Vineyard Park in Kent. The cool climate of England suits Müller-Thurgau, and it can produce attractive wine when well cultivated. But it never has been, and never will be, one of the world's great grapes. In most countries it is in decline.*

and rots easily. In the gravelly south of Entre-Deux-Mers it can be picked earlier and is often quite good, but these regions are mad for red wine these days and are busy uprooting their white vines of all types.

Where Muscadelle reaches its apogee, however, is in north-eastern Victoria, where it is known as Tokay and makes dark, sticky fortifieds that are the counterpart to those produced there from the Muscat grape. Both wines these days come under the umbrella name of Rutherglen Muscat. The quality categories range from basic Rutherglen Muscat, to Classic, then Grand, then Rare. With age, Muscat and Tokay become more like each other, though Tokay always has a more malty, butterscotch flavour which contrasts with the figginess of Muscat, and Tokay is the less fleshy of the two wines. See Muscat, pages 144–153, for more details of Rutherglen Muscat. Best producers: (Australia) All Saints, Baileys, Campbells, Chambers, Morris.

*The colour of these old casks of Liqueur Muscats and Tokays in the corrugated iron winery of Morris Wines, Rutherglen, Victoria, Australia, echoes the colour of the wine. It can stay in cask for decades, becoming ever blacker, more concentrated and stickily sweet.*

## MUSCADET

This grape is just as well known by this name as it is by the name of Melon de Bourgogne (see page 123). It is the grape that makes Muscadet wine, and although it originally arrived in the Pays Nantais having travelled from Burgundy, it is now associated almost entirely with the flat lands at the mouth of the Loire around the city of Nantes. Best producers: (France) Chéreau-Carré, Luc Choblet, de l'Ecu, Pierre Luneau, Louis Métaireau.

## MUSCARDIN

This grape comprises a small part of the blend for red Châteauneuf-du-Pape in the southern Rhône Valley. It appears to have no synonyms, which is extremely rare for a vine, and nothing is known of its origins. The wine is light in colour, but with an appealingly floral aroma.

## MUSCAT

See pages 144–153.

## MUSCAT OF ALEXANDRIA

Muscat of Alexandria has less finesse than Muscat Blanc à Petits Grains, but is widely grown throughout the world. See Muscat pages 144–153.

## MUSCAT BLANC À PETITS GRAINS

The strain of Muscat that almost invariably produces the best, most elegant, most finely aromatic wines. See Muscat pages 144–153.

## MUSCAT CANELLI

Californian synonym for Muscat Blanc à Petits Grains derived from the name Moscato di Canelli or Moscato Bianco. See Muscat pages 144–153.

## MUSCAT DE FRONTIGNAN

French synonym for Muscat Blanc à Petits Grains when used for Muscat de Frontignan, the Languedoc's main fortified wine. See Muscat pages 144–153.

## MUSCAT GORDO BLANCO

Australian name for Muscat of Alexandria. See Muscat pages 144–153.

## MUSCAT OF HAMBURG

A dual-purpose wine and table grape, but more often grown for the latter. As a wine grape it is found mostly in eastern Europe. See Muscat pages 144–153.

## MUSCAT OTTONEL

An addition to the Muscat family bred in 1852 in the Loire Valley, and with less intensity than either Muscat Blanc à Petits Grains or Muscat of Alexandria. It is the main Muscat variety grown in Alsace. See Muscat pages 144–153.

## MUSKADEL

South African name for Muscat. It generally refers to Muscat Blanc à Petits Grains. See Muscat pages 144–153.

## MUSKAT-SILVANER/SYLVANER

Sauvignon Blanc (see pages 218–227) is often called by this name in German-speaking countries. Best producers: (Austria) Alois & Ulrike Gross, Lackner-Tinnacher, Erich & Walter Polz, Otto Riegelnegg, Sattlerhof.

## MUSKATELLER

German name for Muscat. It generally refers to Muscat Blanc à Petits Grains: Gelber Muskateller, the gold-skinned version found in Austria's Steiermark region and in Slovenia, and the red-skinned Roter Muskateller are both forms of this strain of Muscat. Weisser Muskateller and Grüner Muskateller are other possible German language synonyms. See Muscat pages 144–153.

## MUSKOTÁLY

Muscat takes this name in Hungary: most of the local Muscat there is Muscat Ottonel, but there is some Muscat Blanc à Petits Grains, which goes by the name of Muscat Lunel. See Muscat pages 144–153.

E·R

# MUSCAT

WE HAD THIS OLD GREENHOUSE at home when I was a kid. Out past the potting shed and the wood pile, just before you got to the potato plot. For most of the year I didn't bother with it much; it was grimy, unkempt, with a great sprawl of branches and shoots pressing against the windows and cutting out most of the light. I'd venture in once or twice in the summer, when the place was like a forest and a canopy of broad green leaves added to the gloom. But come the autumn, just before I had to go back to school, that's when I'd creep inside and breathe in an air so muggy and hot and thick with scent I could hardly open my throat wide enough to draw it in. The perfume was so heady, so exotic you'd think it should have been coloured purple and gold and been as impenetrable as clouds. But it wasn't. There was just enough light for me to look above my head and see vast clusters of purple black grapes, so ripe and fat that they threatened to tumble and crush me if I so much as touched them.

*The wine-dark sea around the island of Samos, a home of the Muscat grape, lies in the distance. Was this the wine drunk by Dionysus, the Greek god of wine? The ancient Greeks clearly loved wine and all its paraphernalia. Here is a mixing jar, a wine jug and a beautiful* kylix *or two-handled wine cup. And, goodness me, some hooligan has been scrawling his name on the side of the cup. Disgraceful.*

These were our Black Muscats or Muscats of Hamburg. Just once a year, I'd gorge myself on their succulent wildly aromatic flesh before stumbling back to the house with no plausible excuse as to why I was too full to eat tea. And the next day the man from the Savoy Hotel would come and measure the grapes. All the complete bunches of all the biggest grapes, he'd buy. It was usually worth making myself pretty scarce when my mother discovered how many he *wasn't* going to buy. No money for the cinema *that* week.

But the perfume in that hothouse and the flavour of those succulent grapes has never left me. The scent of Muscat haunts me in an entirely pleasurable way, and I can't smell a top-notch Muscat wine without being wreathed in the happy memories of childhood. And it makes me wonder. If Muscat was the original wine grape – and there's a fair amount of opinion suggesting that it was – what on earth made producers move on to other lesser, leaner varieties when they had this irresistible headspinning grape? As it was, Muscat led the wine grape's conquest of the Mediterranean – spreading from the East through Greece and Italy, then on to southern France and down the Spanish coast to Málaga, within sight of the shores of Africa. And for hundreds of years it was highly prized for its fabulous perfume and for the fact that, of all the wine varieties, it was just about the only one that made wine that tasted of grapes.

And yet today it's treated with casual disregard by the majority of wine drinkers. This marvellously adaptable grape with the unmistakeable scent of the vine in full bloom is even sneered at. What does a grape have to do to earn a little respect? Producing utterly delicious wines for anyone with half a palate and a heart that beats obviously isn't enough. Having a gorgeous hothouse scent that makes you swoon with delight doesn't seem to do the trick. Being so versatile that it can make outstanding wine sweet, medium or dry, sparkling or fortified – is that enough? No, it doesn't seem to be.

I'm afraid Muscat is the victim of a current white wine obsession driven by the distinctly unexotic charms of Chardonnay, the tangy attack of Sauvignon and the all-pervasive influence of vanilla-scented oak. But when these fashions have all faded away, Muscat, the original wine grape, will still be there to remind us why this whole thing got started in the first place.

**Muscat: from Grape to Glass**

*Geography and History page 146; Viticulture and Vinification page 148; Muscat around the World page 150; Enjoying Muscat 152*

# Geography and History

IF THERE IS A MORE COMPLICATED family of grapes than the Muscats, it has yet to be discovered. Not only are there three quite distinct kinds of Muscat – Muscat Blanc à Petits Grains, Muscat of Alexandria and Muscat Ottonel – but their capacity for mutation can make the Pinot family look positively regimented. Their berries can be golden, pink or black, and Muscat Blanc à Petits Grains is notorious for its ability to produce one colour one year and another the next. If they were human the Muscat family would be regarded as hopelessly feckless. Which just goes to show that orderliness is not everything.

Muscats of different sorts are planted all around the Mediterranean and across Central and Eastern Europe, as well as in Australia, New Zealand, parts of North and South

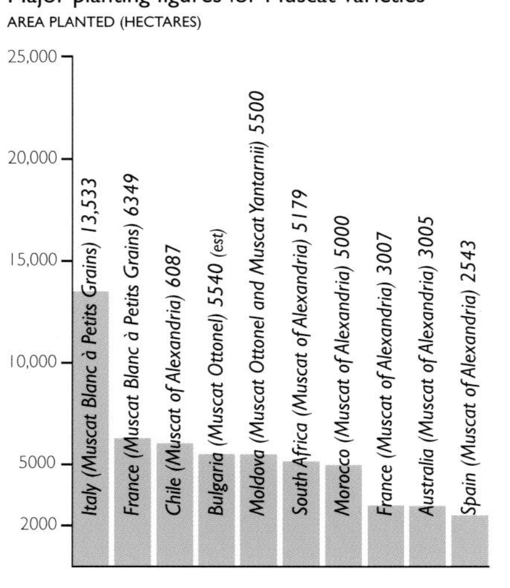

America, and South Africa. They make light dry wines, light, sweet sparkling ones, late harvest and sweet fortified wines. In theory Muscat could make nobly rotten sweet wines, though in practice it generally doesn't because the *Botrytis cinerea* fungus would destroy its extraordinary aroma.

But don't think that this blanketing of the vinous globe with Muscat means that it produces some of the world's most popular wines. It doesn't. Some of the world's most delicious wines – yes. Wines that express the pure perfumed essence of the grape in a way no other variety can approach – yes. But these wildly aromatic styles – dry, medium and sweet – are not in fashion right now. And this is where Muscat is lucky, because its superb, heady, grapy fragrance makes it an excellent table grape – one of the very few wine grapes that makes good eating. It also makes pretty handy raisins. Indeed Muscat of Alexandria and Muscat Ottonel are widely grown for table grapes or raisins, and these plantings can distort the picture if one is only looking at wine grapes. Much of the Muscat of Alexandria in Chile and Peru, in addition, is grown for distillation into pisco brandy.

## Historical background

Muscat Blanc à Petits Grains may well be the oldest grape variety we have. The ampelographer Galet (1998) identifies it with the *Anathelicon moschaton* vine of the Greeks, and the *Apianae* vine of the Romans because of the fondness of insects, particularly bees (*apis* in Latin) and wasps, for the grapes. These names might have been given with rather grim satisfaction: bees and wasps will feverishly devour the flesh of the grapes, leaving nothing but skin and pips.

This Muscat, the oldest member of the family, was probably brought to the province of Narbonne by the Romans, though the Greeks seem to have known it earlier and might have brought it to Marseille. Frontignan was exporting its Muscat wines by the time of Charlemagne. As Muskateller it is recorded in Germany in the 12th century, and it had certainly reached Alsace by the 16th century.

Muscat of Alexandria was also spread around the Mediterranean by the Romans, having presumably originated, as its name suggests, in Egypt. The Australian synonym Lexia is not that far from Alexandria.

Muscat Ottonel is a more recent arrival, having been bred in the Loire in 1852. Its parents were Chasselas and the undistinguished Muscat de Saumur.

The most recent addition to the family is a cross between Muscat Ottonel and Pinot Gris, bred (for greater resistance to *coulure*) in Alsace and so known, for the time being at least, as Muscat de Colmar. The grape is still at an experimental stage, and the first vines are not likely to be available to growers until 2005 or 2006. One just hopes fervently it was bred with beauty and fragrance in mind, not just high yield and disease resistance.

The red Italian grape, Aleatico (see page 35), grown on Corsica and in Chile as well as, curiously, in Kazakhstan and Uzbekistan, may be a dark-berried variant of Muscat Blanc à Petits Grains.

*Muscat vines on the slopes of the Dentelles de Montmirail at Lafare, Beaumes-de-Venise in the southern Rhône. The appellation's name is a combination of the old Provençal word for grotto – 'beaumes' – and a corruption of 'venaissin', an Old French word denoting church property. In the Middle Ages the Avignon Popes claimed all this area as their property.*

*A sign for the local Muscat at Lunel in France's Languedoc. According to the local growers, Napoleon drank Muscat de Lunel in exile on St Helena. Well, it's a nice story, but it didn't seem to do him much good.*

*Harvesting Muscat at Beaumes-de-Venise. The most aromatic wines come from the highest altitudes; the wines from Beaumes-de-Venise are in any case lighter than other fortified Muscats.*

# Viticulture and Vinification

ONE MIGHT THINK, from the spread of Muscat across the globe, that it is easy to please in terms of soil and climate. In some ways this is true, since the same type of Muscat will make different styles of wine, from light and dry to rich and fortified, according to where it is grown. But it is far from being an easy grape to grow. It tends to crop erratically, and acidity is low. In the winery new oak is anathema, and blending with other grapes unusual, though a little Muscat may well be blended in with other varieties – red or white – when a bit of fruit and fragrance is needed.

Muscat is the opposite of Chardonnay: flamboyantly itself where Chardonnay is all things to all winemakers; yet temperamental and sickly where Chardonnay is robust. And it's far from being a name on every wine drinker's tongue.

## Climate

The only generalization one can make is that all Muscats like a fair amount of warmth. Mediterranean climates, especially those of the south of France, Spain, southern Italy and Greece, are its natural habitats, and its most famous New World incarnations, Australia's fortified Muscats, come from a warm, though not scorching hot, region.

The varying susceptibilities of the different Muscats limit their suitability to different regions: Muscat Blanc à Petits Grains buds early and ripens late, and so needs a long growing season. It is a martyr to leaf roll, oidium, downy mildew and grey rot, and dislikes too much humidity; Muscat of Alexandria needs warm weather during flowering to avoid *coulure* and *millerandage*, and is also susceptible to mildew, though it endures drought well so is happy in hot, dry regions. Muscat Ottonel is the earliest ripening of the three, and so is often planted in relatively cool Alsace (cool relative to the south of France, that is), though it nevertheless suffers from *coulure* here. In Alsace all Muscat needs sites with the best exposure to the sun.

## Soil

It is not hard to guess that this multi-faceted grape is planted on a huge variety of soils throughout the world. In Alsace it favours loamy, sandy or calcareous soils, and Muscat Ottonel likes soils that are deep and rather damp; in Mireval and Frontignan the soils are red clay and limestone. In Piedmont it grows on chalk and limestone; in Australia's Rutherglen and Glenrowan soils vary from alluvial loam over gravel to deep, friable granite-derived soils. Soil type probably has little influence on the fortified Muscats of Rutherglen, however; much of the character of this wine comes from the aging process, and the availability of well aged stocks of rich, sticky wine to blend in with younger stuff is more important.

## Cultivation and yields

The maximum legal limit for French *vins doux naturels* is 28–30hl/ha, though on deep fertile soils even the medium-vigour Muscat Blanc à Petits Grains can produce double this. However, if it does its sugar level falls and its aroma disappears, and both aroma and sugar are essential to these wines. In Rutherglen, in north-east Victoria, some growers take around 2.5 tonnes per hectare, while others irrigate and get much higher yields. It's not difficult to tell which producer does which when you taste a sumptuous mouthful of grapy gorgeousness next to a simple, sweet quaffer, both from the same region, yet poles apart in pleasure.

Muscat of Alexandria is more vigorous, more robust and higher yielding, which is why some growers prefer it to the superior-tasting Muscat Blanc à Petits Grains; Muscat Ottonel is low vigour.

## At the winery

Muscat may be made into a dry wine, as in Alsace; a light, sparkling one, as in Asti; or a sweet, fortified one, as in the *vin doux naturel* of Beaumes-de-Venise or the Liqueur Muscat of Rutherglen. It may also be made into a late-harvest sweet wine, or made sweet from *passito* (dried and shrivelled) berries. You very occasionally see a noble rotted (Sélection de Grains Nobles) sweet Muscat from Alsace,

*Asti and Moscato d'Asti have never pretended that they're made by anything other than industrial methods. And why should they? They wouldn't be nearly so refreshing if these large pressurized steel tanks, known as autoclavi, didn't keep the partially fermented juice fresh until it's ready to finish, bottle and ship.*

*Brown Muscat grapes starting to raisin at Rutherglen in North-East Victoria, Australia. Nobody really knows why Muscat was first planted in this flat, unpromising region, but probably the gold rush was the reason: goldminers and prospectors tend to be thirsty, and where they go wine tends to follow.*

and it's always nice to be able to find an exception to prove the botrytis-and-Muscat-don't-mix rule.

Neither the sparkling and semi-sparkling Moscato wines of Piedmont are made by the traditional method (which, thanks to the hyperactive protectionism of the Champenois we are no longer able to call the Champagne method, though that, please note, is what we mean). Instead the must, fermented to up to 2–3 per cent alcohol, is kept in pressurized steel tanks called autoclaves at about 0°C until it is needed; the fermentation is then completed. The grapes are picked at a potential alcohol level of around 11 per cent, and over half the sugar is fermented; the rest remains in the wine as sweetness. Asti is fully sparkling, Moscato d'Asti semi-sparkling, and the minimum alcohol content of Asti is 7 per cent. Lightness is everything with these wines: light sweetness, elegant fruit. They should be full-flavoured, yet always delicate.

The southern French fortified wines or *vins doux naturel* (VDN) are made by *mutage*, which involves adding spirit halfway through fermentation to knock out the yeasts – the same technique that is used to make port. For VDN the fermentation is allowed to continue until the alcohol has reached about 6 per cent, and the addition of alcohol brings the total to 15 per cent or slightly over. At around 95 per cent, the alcohol used for VDN is stronger than that used in the Douro Valley for port, which is usually 77 per cent.

Some skin contact may be used, particularly in Rivesaltes, where the vineyards are dominated by Muscat of Alexandria. This has the effect of increasing the perfume, but must be used sparingly if heaviness, even bitterness, is not to result.

Rutherglen's methods of producing its fortified Muscats are difficult to codify. Some producers pick their grapes at potential alcohol levels of 16 per cent, others prefer to pick at 20 per cent. In some years 30 per cent is possible. Some like to pick raisined grapes, some don't.

A typical method of winemaking is to ferment the must for a mere 24 hours before fortifying; hogsheads or larger casks are used for aging. Some producers operate a solera system, or a version thereof, generally employing hogsheads or large casks for aging, though as the wine ages and loses volume to evaporation, barrels are likely to get smaller and smaller; others blend selected wines before bottling. Some add alcohol to the unfermented must, so that no fermentation at all takes place. Whatever the method, the final figures are around 17–18 per cent alcohol and 9–14° Baumé residual sugar. Aging, usually in pretty torrid conditions under the tin roofs of the wineries, may be for years, decades or even longer: some wineries boast a barrel or two of dark, treacly wine from the 19th century, used in tiny quantities to give depth and viscosity to younger wines.

## WHICH MUSCAT?

Of the three main Muscat grapes used for winemaking, Muscat Blanc à Petits Grains is the star. It produces the most refined wines, with the greatest elegance, and with aromas that are both the most intense and the most delicate and complex. Its name, however, is deceptive: its berries are certainly small, but to describe them as white flies in the face of all the evidence. Golden yellow is their most common colour, although they can be all shades of pink and can even have dark skins, though seldom dark enough to make red wine.

Australians, typically, do not beat about the bush: the Muscat Blanc à Petits Grains that grows in Rutherglen is known simply as Brown Muscat. The astonishingly rose-scented deep pink or red Rosenmuskateller of Italy's Alto Adige and Trentino is also a variety of Muscat Blanc à Petits Grains. Galet lists a Muscat de Rivesaltes which has downy leaves; this is a mutation of Muscat Blanc à Petits Grains.

Muscat of Alexandria is less refined. Its wines are sweet and intense, but they lack the complexity of Muscat Blanc à Petits Grains, and they tend to be clumsier, lacking the balance and focused perfumed beauty of the best Muscat. Muscat of Alexandria's more assertive and less subtle aroma is accounted for by a higher proportion of geranium-scented geraniol, and a lower proportion of fragrant nerol. It is much used for grape concentrate, raisins and brandy, as well as for table grapes.

Muscat Ottonel has less powerful aromas than the previous two Muscats. It is also paler in colour.

Muscat of Hamburg is always black, and is best as a table grape. Its quality is lower than the previous three, though it is planted to some extent in Eastern Europe for light red wine, and is found elsewhere as well.

The names of all these Muscats seem to have mutated at an even faster rate than the vines. We list the main synonyms of each variety on page 153; but it is worth noting that while anything called Moscato, Moscatel, Moscadello, Muskateller or Muscadel is Muscat, any vine called Muscadelle, Muscadet or Muscardin is not. Muscat Bailey A is a hybrid grown in the Far East. Morio Muskat is a flashily-scented German variety unrelated to Muscat. Muskat-Sylvaner is an Austrian synonym for Sauvignon Blanc.

# Muscat around the World

THERE ARE VAST SWATHES OF MUSCAT OF ALEXANDRIA growing in the warm vineyards of the world. But remember, a considerable percentage is not destined for winemaking but for eating. There's less Muscat Blanc à Petits Grains planted, but most of it is destined for winemaking and it makes some of the most exotically scented wines in the world.

## France

Both Muscats are used in southern France's cluster of *vins doux naturels*. Several of the ACs specify that Muscat Blanc à Petits Grains must be used: it should be the only variety in the delicious, scented Muscat de Beaumes-de-Venise (whose residual sugar level is 110g/litre instead of 125g/litre in the other ACs). Mireval, St-Jean-de-Minervois, Lunel and Frontignan should also be Muscat Blanc à Petits Grains but aren't always. Muscat de Rivesaltes is dominated by Muscat of Alexandria grapes, although Muscat Blanc à Petits Grains is on the increase here thanks to new clones which yield more reliably. Rivesaltes also produces VDNs called simply Rivesaltes, from blends of grapes: both Muscats may be mixed with Grenache, Maccabéo and Malvoisie du Roussillon (alias Torbato), and these wines turn out every imaginable colour, and in styles that range from young and fresh to *rancio*. Naturally, they do not have the pure Muscat aroma of Muscat de Rivesaltes.

There is also some dry, unfortified Muscat now being produced in the South, usually with more body and substance but less ethereal perfume and crunchy fruit than the dry versions from Alsace.

Muscat Blanc à Petits Grains also dominates the blend in light, sweet and sparkling Clairette de Die Tradition, often to the virtual exclusion of Clairette and to the great benefit of the wine.

In Alsace, Muscat takes up only about 3 per cent of the vineyard, and while Muscat Ottonel is the preferred variety, it never produces such delightful wine as Muscat Blanc à Petits Grains. Growers are no doubt waiting to see how the new Muscat de Colmar (see page 147) will turn out. So am I. But if it isn't wildly, wickedly scented, I don't want to know, however resistant to disease it is.

## Italy

Moscato (mainly the superior Muscat Blanc à Petits Grains) is grown all over the peninsula. In Piedmont, where it is grown for light, sweet, sparkling Asti and Moscato d'Asti, Muscat Blanc à Petits Grains is planted up to 550m, with the higher, steeper vineyards giving more acidity and the lower vineyards more body. But even at relatively low potential alcohol levels (about 11 degrees for Moscato d'Asti) there is no shortage of aroma. Moscato Rosa, or Rosenmuskateller, is a variation of the same Muscat, used for intensely rose-scented (and even rose-coloured, if the roses in question are red) still wines in Trentino and Alto Adige; Moscato Giallo or Goldmuskateller is another variant, and makes orange-blossom scented dry or sweet wines in Trentino and Alto Adige. Sweet red Aleatico may be a mutation of Muscat Blanc à Petits Grains.

On the island of Pantelleria, between Sicily and Tunisia, Muscat of Alexandria, often under the name of Zibibbo, is grown for amber-coloured Moscato di Pantelleria, made from *passito* grapes. Sardinia, Basilicata and Puglia also have traditions of sweet Moscato, though they're hardly flourishing.

## Spain

Simple, sweet, grapy Moscatel is grown all over Spain, and usually the grape in question is Muscat of Alexandria (Moscatel de Alejandría), though there is some Muscat Blanc à Petits Grains (Moscatel de Grano Menudo) in the North. Moscatel de Málaga is a speciality of the South, and is probably related to Muscat of Alexandria. Only about half Spain's Moscatel makes wine; the rest is used for raisins or table grapes.

## Portugal

Moscatel de Setúbal is Portugal's best known Muscat wine, made from Muscat of Alexandria, but since the 1980s crisp, dry Muscats have been made in the same area.

## Austria

Gelber Muskateller or Muscat Lunel (both synonyms for Muscat Blanc à Petits Grains) is the grape behind Austria's best Muscats, though Muscat Ottonel takes up more of the vineyard. Muscat Ottonel is however currently being uprooted to make way for red vines. Muscat Ottonel makes good late-harvest wines in the Neusiedlersee, while Muscat Blanc à Petits Grains appears as dry wine in the Wachau and Südsteiermark. There is also some in Burgenland.

DOMAINE DE DURBAN

*One of the principal producers of Muscat de Beaumes-de-Venise, Domaine de Durban's wine has a style that is delicate but beautifully scented.*

JEAN-PIERRE DIRLER

*Muscat is one of the four noble varieties that can be planted in Alsace's Grand Cru vineyards. The sandy soil here gives Dirler's Muscat lightness and perfume.*

LA SPINETTA

*Bricco Quaglia is the Rivetti family's top Moscato from its La Spinetta estate near Asti. 'Bricco' is a Piedmontese dialect word meaning 'hilltop'.*

### Greece

Muscats Blanc à Petits Grains and Alexandria are grown here, though only the former is used for the high-quality sweet wines of Samos, Pátrai and Kefallinía; these may be unfortified, fortified in the manner of *vin doux naturel*, or made from grapes dried in the sun. Muscat of Alexandria rules in Cyprus. Greece has obvious potential, given its current talent for making crisp, dry, characterful whites, for aromatically dry Muscat.

### Rest of Europe

Muscats are grown in Bulgaria, Romania, Slovenia, Moldova, Russia, Uzbekistan, Kazakhstan, Tajikistan and Turkmenistan, and in Ukraine, where the Massandra winery continues its tradition of making both white and pink Muscat. The former is aged for six years before release, the latter for 20. Hungary has some Muscat in Tokaj, which is picked late (and, yes, botrytized) for sweet Tokaji, but which also has potential as a dry wine.

### Australia

Of the two great fortified wines made in Rutherglen and Glenrowan, in north-eastern Victoria, one is made from Muscat grapes. This is, not surprisingly, known as Rutherglen Muscat, or Liqueur Muscat; the other wine, made from Tokay, alias Muscadelle, is usually known as Liqueur Tokay.

Both wines darken with age until they are thick, black and viscous, but both start out much paler. Young Muscat tastes grapy/raisiny and rich, often with a tea-rose scent; with increasing maturation in barrel it becomes more concentrated and complex, and takes on flavours of figs and blackberries, coffee and chocolate, and the acidity at length becomes more apparent, though Muscat always tastes sweeter and fleshier than

*Terraced vineyards on the Greek island of Samos. Muscat is grown at altitudes of up to 800m here, and grapes from the highest, coolest spots may be picked as much as two months after those grown lower down.*

Tokaji. A new four-tier quality system has recently been introduced, with Rutherglen Muscat being the basic one. Above it, in ascending order, are Classic, then Grand, then Rare. There are no specific criteria for any of these categories; peer pressure, in this small community of growers, is supposed to ensure high quality at every level.

Not all Australian Muscat is used for such styles. Both Muscat of Alexandria (Muscat Gordo Blanco, or Lexia) and Muscat Blanc à Petits Grains (White Frontignac or Brown Muscat) are used to make light, fruity bulk wine sold in wine casks. White Frontignac covers about 270ha in all, or 0.2 per cent of the total vineyard; Muscat Gordo Blanco has 3005ha, or 2.4 per cent.

### South Africa

Hanepoot, alias Muscat of Alexandria, is concentrated in Worcester and Olifants River, but is found in most of the country's wine regions. It accounts for less than 5 per cent of total plantings, and is used for everything a grape can be used for: dry, sweet and fortified wine, table grapes, raisins and grape juice. Quality (of the wine, at least) is generally dull, excepting a few top fortifieds. Muscadel (Muscat Blanc à Petits Grains) is the grape behind Jerepigo, 17 per cent alcohol *mistelle* made by adding grape spirit to unfermented grape juice.

### USA

Plantings of Muscat of Alexandria outnumber those of Muscat Blanc (Muscat Canelli, or Muscat Blanc à Petits Grains), but are falling fast. It grows mostly in California's Central Valley, where it was planted for raisins and also makes bulk wine for blending. Muscat Canelli is planted both in the Central Valley and in the coastal regions, and is generally made light, medium-sweet and slightly sparkling. There are also some fortified versions.

California also has small amounts of orange-flower-scented Orange Muscat, which appears to be unrelated to Muscat Blanc à Petits Grains, and some rose-scented Black Muscat, which is presumably Muscat Hamburg. Both can make exciting sweet fortified wines in the right hands.

### Rest of the world

New Zealand's small plantings of the table grape Muscat Dr Hogg are used for sweetish sparkling wines, medium-sweet still wines, or for blending with Müller-Thurgau in bulk wines. Tunisia makes some dry rosé from Muscat of Alexandria. South America grows large quantities of Moscatel of different sorts for pisco brandy, table grapes and bulk wine.

**MARCO DE BARTOLI**

*Moscato grapes are allowed to dry naturally in the autumn sunshine for Marco De Bartoli's passito wine. Zibibbo is the local name for Muscat of Alexandria.*

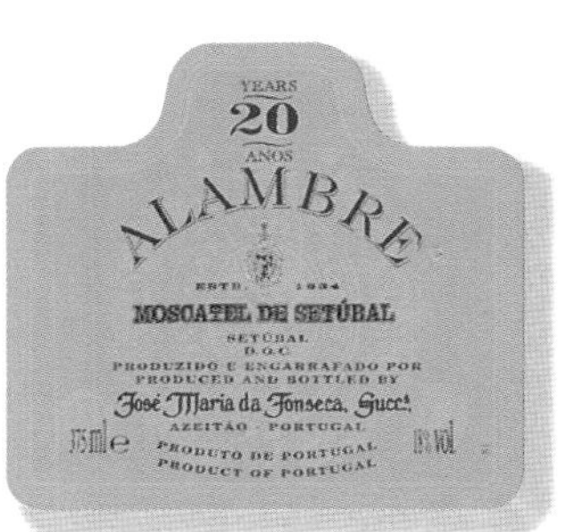


**JOSÉ MARIA DA FONSECA**

*This wine is a blend of Moscatel with Bual and Malvasia; other Moscatels de Setúbal from the same house may be different blends, or pure Muscat.*

**MORRIS WINES**

*Mick Morris makes some of Rutherglen's most intense fortified Muscats in his tin-shed winery – which also houses some barrels of pre-phylloxera wine.*

# Enjoying Muscat

ENJOYING MUSCAT IS SO EASY, I feel a bit silly even considering writing down any lists of dos and don'ts. But it is that very fresh, fragrant, blossomy beauty which is at both the heart of the pleasure and the core of the decline. It rarely lasts. It fills the glass with the heady scents of summer ripeness, the sweet perfumes of a Mediterranean bower laden with fruit – and then, just as summer always crumbles into autumn, the scent becomes dulled, the youthful beauty becomes coarsened and fat. Summer is gone. And so is the succulent, sybaritic fleeting beauty of Muscat. It would be hard to bear were there not another harvest every year: another chance to revel in young Muscat yet again.

Well, this is the case with most Muscat. Most dry examples have a beautiful but fleeting perfection – though some Alsace ones can age to a fascinating scented decadence. Even the sweet *vins doux naturels* of southern France gain nothing from aging, though they hardly fall flat on their faces.

Lack of acidity is the problem. Perfume almost always needs balancing acidity in wine to develop past its first bright-eyed flush of youth, and Muscat simply doesn't have the requisite acidity. If the aroma tires, Muscat wines quickly fade to an oily sullenness. Yet there are a few exceptions – fortified wines made from small crops of grapes, their flavours intensified by overripeness and shrivelling on the vine, their dark sensuous fruit arrested and preserved by the fortification process, and their intoxicating scent married with the mystical tastes of age and decay by long barrel-aging and blending. The greatest of these are the Muscats of Glenrowan and Rutherglen in Australia, though South Africa can occasionally come up with the goods too.

### The taste of Muscat

Light young Muscat tastes of rose petals spiked with orange blossom, sometimes elderflower, and perhaps a touch of orange zest; it is this slight citrus tang that can help to balance the wine's low acidity. Sometimes the roses dominate; sometimes there is a peachy richness; sometimes the orange comes to the fore. With so much variation even within Muscat Blanc à Petits Grains vines, it is hardly surprising that the wines can taste different.

Oh! And grapes. There is always a crisp, crunchy grapiness about good examples of Muscat Blanc à Petits Grains, and it is this crispness, as well as greater finesse and subtlety, that makes Muscat Blanc à Petits Grains wines such sublime drinks. Muscat of Alexandria is simpler, broader, less perfumed, less beautiful; Muscat Ottonel less powerfully scented, sometimes verging on the vegetal.

Dark, aged Muscats that have been sitting in old oak barrels for years acquire an oxidative bouquet of figs, blackberry and coffee, prunes, treacle, nuts and chocolate, immensely complex and pungent. These wines can be so thick and viscous that they coat the glass with a thick film of translucent brown; the smell will remain in the glass long after the wine has been drunk.

And, of course, most Muscat is sweet. Sweetness is certainly what such a perfume leads you to expect – and it is generally what you get – a sweetness so fruity that such wines are the best all-purpose wines to go with dessert. Even so, with Alsace in the lead, the number of wines of crystalline fragrant dryness is increasing.

*Campbells Merchant Prince Muscat is dark, rich and very sweet indeed, although it is by no means the most heavyweight Muscat to come from Rutherglen in north-east Victoria; by contrast, René Muré's Muscat Vendange Tardive Clos St-Landelin is a rare medium-sweet example of the grape from Alsace;.*

### Matching Muscat and food

These fragrant, grapy wines range from delicate to downright syrupy. The drier ones are more difficult to pair with food, but are a good choice with delicately spiced Thai or Indian food. The sweeties come into their own with most desserts (try chilled Liqueur Muscat with really good vanilla ice-cream). Chilled Moscato d'Asti, delicious by itself, goes well with rich Christmas pudding or mince pies.

## CONSUMER INFORMATION

### Synonyms & local names

Muscat Blanc à Petits Grains, the most important variety, is also called Muscat de Frontignan, Muscat Lunel, Muscat Blanc, Muscat d'Alsace, Muscat Canelli, Muskateller in Germany, Brown Muscat or Frontignac in Australia and Muskadel or Muscadel in South Africa; the Italians call it Moscato or Moscato Bianco, Moscato Canelli or Moscato d'Asti. Moscato Giallo (Goldmuskateller) and Moscato Rosa (Rosenmuskateller) are mutations with darker berries from Alto Adige and Moscadello is a Tuscan variant. Spanish synonyms are Moscatel de Grano Menudo or Moscatel de Frontignan. Tamîioasa is the Romanian name. The less thrilling Muscat of Alexandria or Muscat Romain is called Moscatel in Spain (local names include Moscatel de Alejandría, Moscatel de España, Moscatel Gordo Blanco and Moscatel de Málaga), Moscatel de Setúbal in Portugal and Moscato di Alexandria in Italy (Zibibbo in Sicily). It is called Muscat Gordo Blanco in Australia and Hanepoot in South Africa. Muscat Ottonel is Muskat-Ottonel in Germany and Muskotály in Hungary. None of these should be confused with Muscadelle or Muscadet.

### Best producers

**FRANCE/Alsace** Adam, Albrecht, Becker, Bott-Geyl, Boxler, Burn, Théo Cattin, Deiss, Dirler, Jung, Kientzler, Kuentz-Bas, Mann, René Muré, Ostertag, Pfaffenheim co-op, Rolly Gassmann, Schaetzel, Schleret, Schoffit, Sorg, Trimbach, Weinbach, Zind-Humbrecht;
**Rhône/ Southern France** Achard-Vincent, Aréna, des Bernardins, Cazes, Clairette de Die co-op, Clos Nicrosi, Durban, Gentile, Jaboulet, de Jau, Laporte, Leccia, de la Peyrade, St-Jean-de-Minervois co-op, Sarda-Malet, Vidal-Fleury.
**GERMANY** Bercher, Bürklin-Wolf, Müller-Catoir, Rebholz.
**AUSTRIA** Gross, Alois Kracher, Lackner-Tinnacher, Opitz, F X Pichler, E & M Tement.
**ITALY** Bera, Braida, Castel de Paolis, Cascina Fonda, Caudrina, Col d'Orcia, Colterenzio co-op, G Contratto, La Crotta di Vegneron, De Bartoli, Forteto della Luja, Gancia, Franz Haas, Lageder, Icardi, Ivaldi, Conti Martini, Murana, Perrone, La Spinetta, Saracco, Schloss Salegg, Tiefenbrunner, Vignalta, Voyat.
**SPAIN** Bocopa co-op, Camilo Castilla, Gutiérrez de la Vega, Lustau, Enrique Mendoza, Ochoa, Miguel Oliver.
**PORTUGAL/Moscatel de Setúbal** José Maria da Fonseca, J P Vinhos.
**GREECE** Samos co-op.
**USA/California** Bonny Doon, Robert Pecota, Quady, Philip Togni; **Washington State** Andrew Will, Kiona.
**AUSTRALIA** All Saints, Baileys, Brown Brothers, Campbells, Chambers, McWilliams, Morris, Seppelt, Stanton & Killeen, Wendouree.
**SOUTH AFRICA** Delheim, Klein Constantia, Nuy Wine Cellar.

### RECOMMENDED WINES TO TRY

**Five Alsace Muscat wines**
Albert Boxler *Alsace Muscat Brand*
Ernest Burn *Alsace Muscat Goldert Clos St-Imer Vendange Tardive*
J-P Dirler *Alsace Muscat Spiegel*
René Muré *Alsace Muscat Vorbourg Clos St-Landelin Sélection des Grains Nobles*
Schoffit *Alsace Muscat Vendange Tardive*

**Five other European Muscat wines**
Bercher *Burkheimer Feuerberg Muskateller Spätlese Trocken (Germany)*
Alois & Ulrike Gross *Steiermark Muskateller Kittenberg (Austria)*
José Maria da Fonseca *Moscatel de Setúbal 20-year-old (Portugal)*
Alois Kracher *Muskat Ottonel TBA (Austria)*
Salvatore Murana *Moscato Passito di Pantelleria Martingana (Italy)*

**Five French Vins Doux Naturels**
Domaine des Bernardins *Muscat de Beaumes-de-Venise*
Durban *Muscat de Beaumes-de-Venise*
Jaboulet *Muscat de Beaumes-de-Venise*
Sarda-Malet *Muscat de Rivesaltes*
St-Jean de Minervois co-op *Muscat de St-Jean-de-Minervois*

**Five Australian fortified Muscat wines**
All Saints *Rutherglen Museum Release Muscat*
Campbells *Rutherglen Merchant Prince Muscat*
Chambers *Rutherglen Special Liqueur Muscat*
Morris *Rutherglen Old Premium Muscat*
Seppelt *Rutherglen Show Muscat DP63*

**Five Italian sparkling wines**
Braida *Moscato d'Asti Vigna Senza Nome*
Caudrina *Moscato d'Asti La Caudrina*
Forteto della Luja *Loazzolo Vendemmia Tardiva*
G Contratto *Asti Spumante De Miranda*
La Spinetta *Moscato d'Asti Bricco Quaglia*

*Muscat is the most protean of vines, and exists in many local variants around the world. The 'standard' colour of the berries may be white – though golden would be a better description – but many Muscats have pink, red or even brown berries.*

### Maturity charts

This is not a grape to age in bottle. Fresh, aromatic young Muscats will only lose their aroma and gain nothing in its place.

**1998** Alsace Muscat (dry)

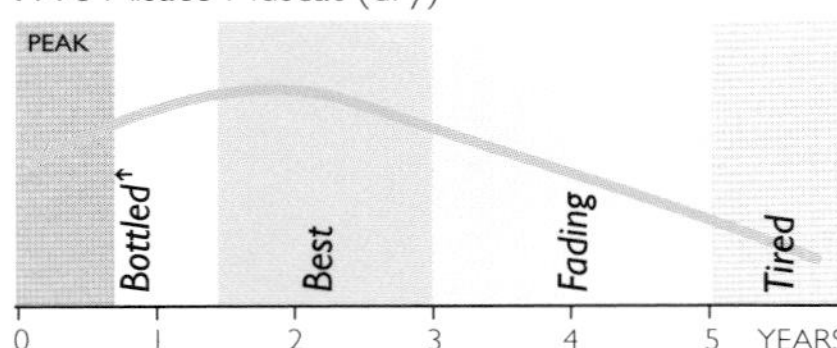


*1998 produced silky, ripe wines that, like all dry Alsace Muscat, will not improve in bottle. All should be drunk within two or three years.*

**1999** Southern France *vin doux naturel* Muscat

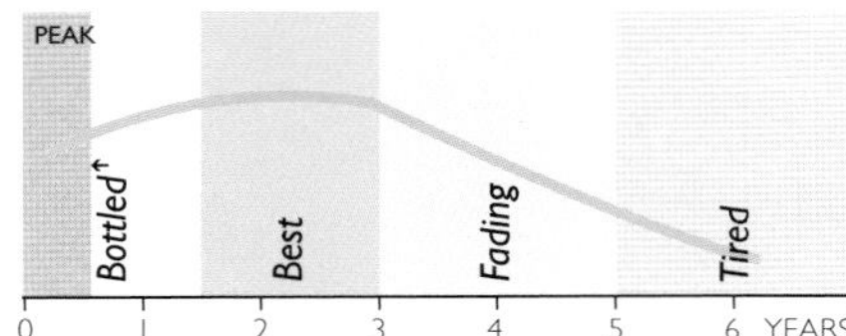


*Another Muscat wine for which bottle age is not an advantage. Again, these wines lose aroma and freshness if kept.*

North-East Victoria Muscat (Rutherglen)

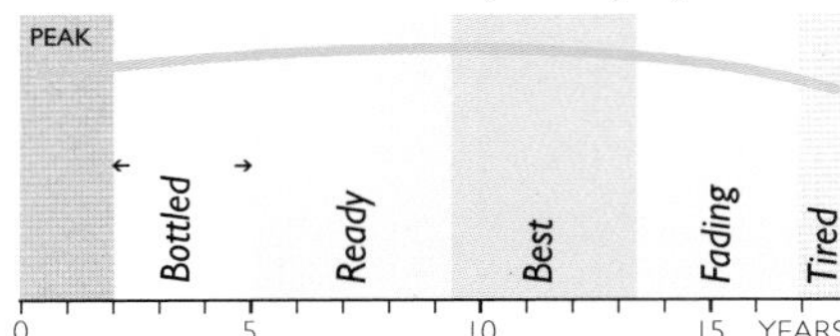


*These wines hardly change in bottle. Dark, sticky and super-raisiny, they will stay much as they are for a couple of decades or more.*

# NEBBIOLO

I DON'T KNOW. I TRIED HARD ENOUGH. But for years I just couldn't get to grips with Nebbiolo. Oh, I used to leap at the chance to taste it – ideally in its Barolo or Barbaresco manifestations. I've hummed and hawed, nodded my head as I listened to my elders and betters, all the while trying to stifle a grimace on my face as my gums withered at the brutal assault of tannin and acidity from whatever wine they'd brought. Phew! Expectoration was never such sweet pleasure. And slowly the life would creep back into my battered palate. What were those people going on about, with all their reverential verbiage about tar and roses and blackberry and sloes. And how it was time for lunch.

*Dusty purple grapes and bright red leaves seen against a background of snow and fog – this is the Piedmontese town of Alba, famous for red wine and white truffles, in autumn. Against a backdrop of snow-capped Alps, the town's medieval towers emerge from the late autumn fogs or* nebbie *that have given the Nebbiolo grape its name. In late autumn truffle hunters and their specially trained dogs set off into the oak forests bordering the vineyards in search of the revered delicacy.*

Well, in my callow, know-all state of mind, I rarely stayed for the food. I should have done. For two reasons. First, many of the old-timers would never dream of bringing out their true treasures – their 20- and 30-year-old Riservas that had lain undisturbed at the back of the cellars since they were bottled – without platefuls of Piedmontese fare to accompany them. Barolo and Barbaresco, more than almost any other great wines, demand the convivial company of hearty food – *bollito misto*, pheasant, pigeon or hare, great chunks of local beef smouldered in lakes of rich red wine sauce – and did I mention white truffles?

And of course – age. That unfashionable, curmudgeonly state of mind that is known as age. Did I say 20 or 30 years? Certainly – and more when the vintage was perfect and the producer devoted to his craft. Then, at last the crusty, argumentative, abrasive nature of Nebbiolo slips quietly from the wine to reveal something old but magnificent, prepared to give of itself all if the food is good and the company warm – and suddenly you sit bolt upright with the glass halfway to your lips – and there it is – the haunting, beautiful aroma of sloes, and blackberry, and roses, and tar.

So the old timers were right. And thankfully, I did occasionally get lucky over time. The first Nebbiolos that I got excited about, strangely, were rare Gattinaras from Dessilani and Vallana – well, I presume they were Nebbiolos, but in those days when you asked what grapes went into a bottle of Gattinara you were met with much winking and fingers to the nose. Anyway, the wines were lovely, plummy, perfumed and ripe – however they got there.

And then I would stumble over a mature Aldo Conterno Barolo, or a well-aged Bruno Giacosa Barbaresco, and later on I'd excitedly sample a Mascarello, an Altare, a Voerzio or a Vajra, and so the legend grew. I still tasted numbers of unfriendly Nebbiolos – and from the large merchant houses and co-operatives (excepting the one at Barbaresco) many supremely unattractive wines – but I knew great Nebbiolo was possible, did exist, and could be found again.

Now things are different. Piedmont has been through years of ferment as it tried to determine how best to coax out Nebbiolo's undoubted beauty. Traditionalists and modernists have raged at each other as they have tried to unlock Nebbiolo's splendour without resorting to the all-too-common international methods of over-extraction, overripeness and over-oaking. Ancient faults of dirtiness and volatility have been replaced by the modern faults of excess of zeal and excess of pride.

And out of all the tumult will come more good, and some great wine – but not that much great wine. Nebbiolo is not an easy grape to handle nor will it ever be and producers with the sensitivity, the subtlety and the skill to tease the grape into greatness are few and far between.

**Nebbiolo: from Grape to Glass**

*Geography and History page 156; Viticulture and Vinification page 158; Nebbiolo around the World page 160; Enjoying Nebbiolo page 162*

# Geography and History

BRITISH WINE MERCHANTS, IT SEEMS, 'discovered' the Nebbiolo wines of Piedmont in north-west Italy early in the 18th century, when Britain and France were at war and French wine was, to some extent, out of bounds to British drinkers. But the local duties were prohibitive, transport all but impossible, and the project of exporting Nebbiolo to Britain was abandoned. But if the opposite had happened, if there had been an accessible port and the authorities had been amenable, would Nebbiolo now be established in a much wider area than its Piedmontese stronghold? There's every reason to believe so. The great wines of Burgundy, Bordeaux, sherry and port became famous through being disseminated along the world's trade routes. France sold wine to all of northern Europe; Piedmont to almost no-one. It's no

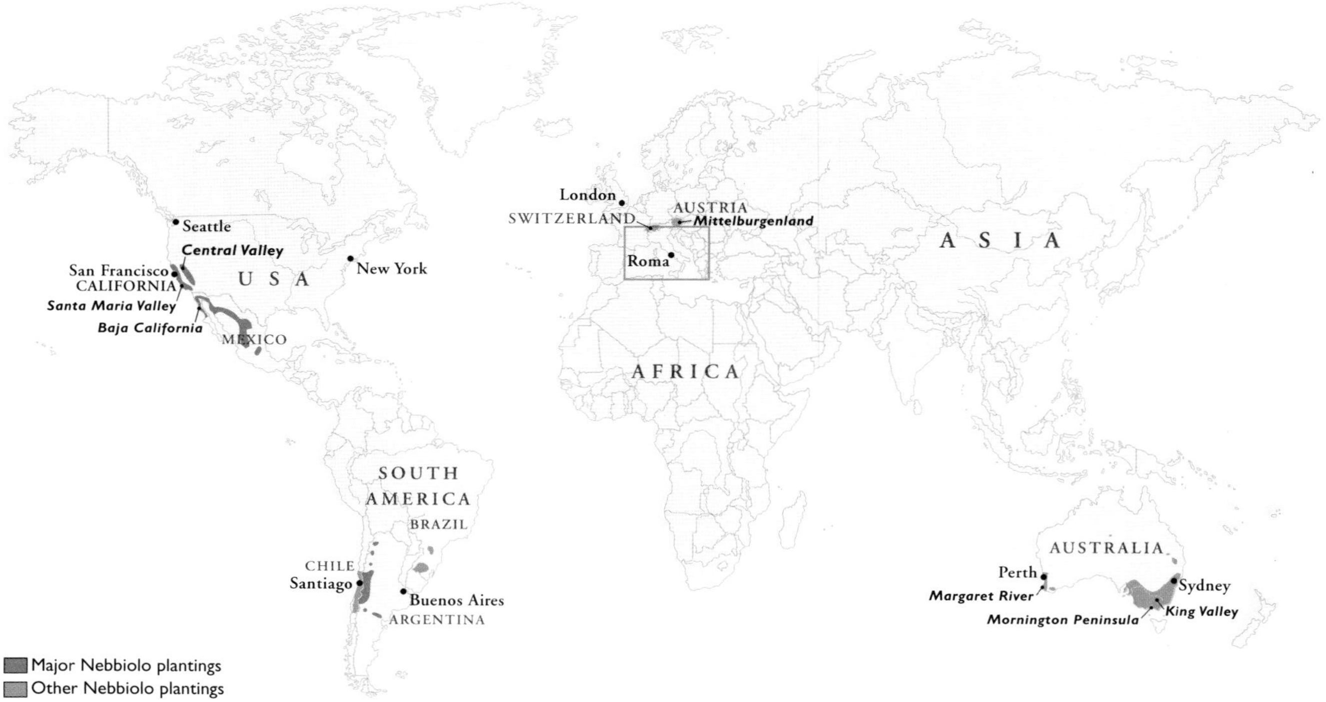

London
SWITZERLAND
AUSTRIA
Mittelburgenland
Roma
ASIA
Seattle
Central Valley
San Francisco
CALIFORNIA
USA
New York
Santa Maria Valley
Baja California
MEXICO
AFRICA
SOUTH AMERICA
BRAZIL
CHILE
Santiago
Buenos Aires
ARGENTINA
AUSTRALIA
Perth
Margaret River
Sydney
King Valley
Mornington Peninsula
Major Nebbiolo plantings
Other Nebbiolo plantings

Major planting figures for Nebbiolo
AREA PLANTED (HECTARES)

6000
5000
4000
3000
2000
1000
500
Italy (total) 5247
Piedmont (Italy) 2741
Lombardy (Italy) 1250
Argentina 189
Mexico 150
California (USA) 89
Valle d'Aosta (Italy) 70

Valle d'Aosta
Valtellina
Carema
Lombardy
Piedmont
Milano
Torino
Roero
Barbaresco
Barolo
Adriatic Sea
Roma
ITALY
Mediterranean Sea

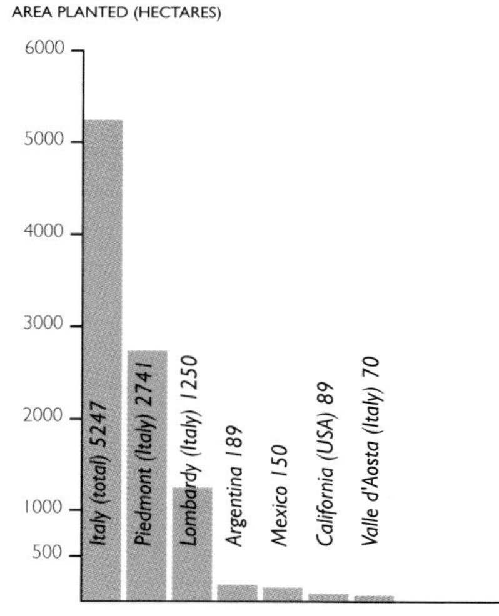

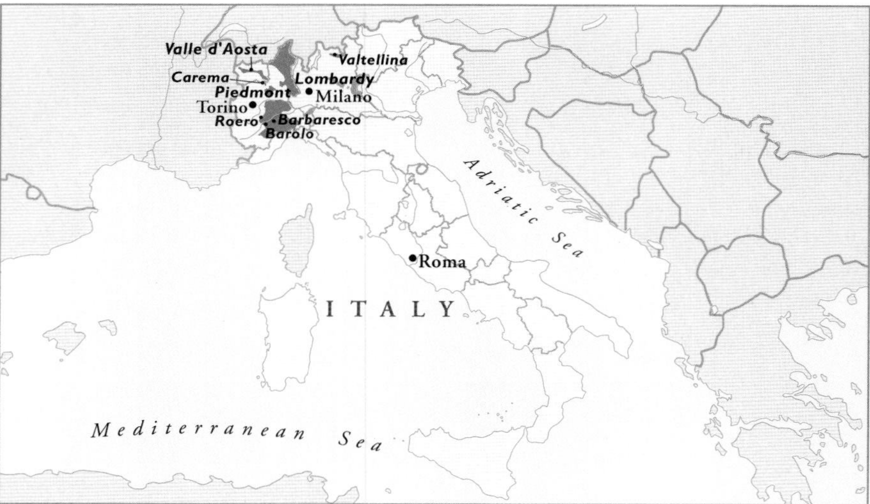

surprise that France became the undisputed world leader in wines while Piedmont became a connoisseur's curiosity.

Nebbiolo is grown in North and South America and in Australia as well. But don't let the map imply there are massive plantations. There aren't. And very few of the smattering of examples have so far shown much class. I can't think of any examples that have any of Barolo's fascination and haunting flavours. Of the great Italian red grapes currently being studied by winemakers everywhere in their search for new flavours, Nebbiolo is proving the most awkward.

This is not to say that great Nebbiolo will never be made outside Piedmont. Undoubtedly it will; but it will take time to learn which are the best sites and which the best winemaking techniques to apply to which clones, since these matters are still heatedly argued over in Piedmont itself. It is a vine that bears more than a little resemblance to Burgundy's Pinot Noir in its refusal to be treated casually in vineyard or winery; it took New World winemakers a generation to crack Pinot's code. I've a feeling Nebbiolo will prove even more difficult.

### Historical background

Genetically unstable vine varieties that mutate readily are often old varieties; and Nebbiolo can demonstrate considerable clonal variation. It has been with us for a long time.

What it cannot demonstrate, of course, is precisely when it came into being. The name is usually said to derive from the word for fog – *nebbia* – the connection being that fog is common in September and October in Piedmont's hilly districts, when the late-ripening Nebbiolo is at last maturing. It seems to have originated in the Novara hills, between Turin and Milan in northern Piedmont, and the first written record of it by name was in 1268, when 'Nibiol' is mentioned as growing at Rivoli, which is now a suburb of Turin. In 1303 Canale d'Alba names it as 'nebiolo', and in Petrus de Crescentiis' 1304 *Liber ruralium commodorum* there is a mention of 'Nubiola'. In the 15th century anyone in the commune of La Morra guilty of cutting down a Nebbiolo vine was liable to be fined, have a hand amputated, or be hanged.

In the 19th century it was far more widely planted in Piedmont than it is today – and not just because it became possible to get rid of it without being hanged or having your hand chopped off. (Appellation laws in Europe have become more enlightened over the years, believe it or not.) When phylloxera wiped out the vineyards, the growers took the opportunity to replant with Barbera, which yields more generously and more reliably. Nebbiolo now covers only about 6 per cent of the vineyard in Piedmont, yet is far and away the most prestigious vine of the region, being the sole variety in the majestic DOCG wines of Barolo and Barbaresco, from the Langhe hills around the town of Alba. It almost always hogs the warmest sites – and it needs them: its colleagues Barbera and Dolcetto can be picked and fermented before Nebbiolo is ready to harvest.

*The commune of Castiglione Falletto with, sloping down from the village, part of the Rocche subzone. Top Barolo these days comes from single vineyards, or crus, within the subzones, though a great deal of local politicking remains to be done before everyone agrees on an official classification.*

*Aldo Conterno, here in La Cicala vineyard at Bussia, Monforte d'Alba, is one of the great Barolo names. In some years he bottles Cicala separately; in others it is blended with another Bussia cru, Colonello.*

*Roberto Voerzio is a more recent arrival in the Barolo firmament, known for his fastidious attention to detail in the vineyards. In the winery he is a leading modernist, producing supple, rich wines.*

# Viticulture and Vinification

NEBBIOLO IS IN SOME WAYS a grape at odds with its reputation. For many Barolo is the epitome of the black, tannic wine that needs years in bottle before it is tamed into relative politeness. Modern Barolo, and modern Nebbiolo, is a very different creature. It is softer and more forward but still keeps its tannic rasp. It can exhibit a divine rose petal scent, black chocolate and prunes richness, tarry intensity – all lovely, but all in danger of becoming just one more braggart on the new-oaked circuit of the international superstars. I have to say that almost without exception I prefer the new styles. And yet, shining among a sludge of bitter, fruitless old-style Barolos and Barbarescos, I have tasted a few ancient jewels far more thrilling, more individual, more indelible in taste, than anything the modernists have yet produced.

## Climate

Nebbiolo is extremely fussy about its climate, and getting the climate wrong is probably the biggest single reason why so few New World Nebbiolos have been successful. In Piedmont it needs the best exposed, south- and south-west-facing sites, at altitudes of between 150 and 300m. Higher than that is too cool. Nebbiolo suffers horribly from *colatura* (which sounds like a plague of sopranos, but is merely poor fruit set at flowering) and so must be sheltered from cold and wind, especially since it is the earliest of all Piedmontese varieties to bud. Wet weather at flowering doesn't help, either.

Nor does wet weather in September and October help. These are relatively rainy months in the Langhe and Roero hills of southern Piedmont, and it is notable that the best Nebbiolo vintages are those in which September is fairly dry. And this makes perfect sense: while weather conditions in spring and summer affect the quantity of a vintage, it is weather after *véraison* that affects quality.

The Piedmont climate is markedly continental, with hot summers and cold winters, and its shortish growing season means that Nebbiolo (early budding, late ripening) is at the limit of its cultivation here. Experience in Australia, however, suggests that planting in much warmer climates doesn't work. Research there points out that Piedmont has a curiously contradictory climate: continental, but with relatively little variation from year to year. Australian growers trying to match Piedmont's growing conditions have a difficult task on their hands.

## Soil

Nebbiolo seems to be able to cope with a wide variety of soils in north-west Italy, though it produces its greatest quality only on the clay and limestone of the Albese, the area around Alba, on the south bank of the Tanaro river. The soil type, as well as the climate, is reflected vividly in the wine: soils in Roero, on the other side of the Tanaro from the Langhe, are sandy, with a neutral or slightly acid pH, and produce earlier-maturing wine. The hills between Novara and Vercelli, where lighter Gattinara and Ghemme come from, are porphyry and acidic; Valtellina's soils are schist, and those of the lower Aosta Valley are granitic. Good drainage is important, and the presence of ores such as potassium and magnesium can affect the style of the wine.

## Cultivation and yields

Around 40 different clones of Nebbiolo have been identified, but the three that are most widely grown are Lampia, Michet and Rosé. Lampia has large three-lobed bunches with good colour and yields well, sometimes too well. Michet has smaller, more compact bunches and gives lower yields; it is a genetic mutation of Lampia, the mutation having been caused by a virus. Rosé is paler in colour, as its name suggests, and is unpopular for this reason. Nebbiolo is often problematically light in colour even without this clone. New clones based on these three have been developed, and almost all growers prefer to grow a mix of clones rather than confine themselves to one or two. Different regions have their specialities, too: in Aosta and the far north of Piedmont the local clone is called Picoutener,

*Angelo Gaja was one of the first in Piedmont to turn to international winemaking practices, determined to make the world take notice of his Barbaresco. He was also one of the first to charge sky-high prices – and achieve them. And he's a fanatic for detail: the extra-long corks he specifies proved such a problem for restaurant wine waiters that he had to become the Italian agent for Screwpull.*

or 'tender stem', and in Valtellina, in Lombardy, it is the Chiavennasca.

Unless weakened by viruses Nebbiolo is vigorous, but since the basal buds are not very fertile it is usual to leave a long cane with ten to twelve buds on a guyot training system. But the better growers of the Langhe maintain that about 40–60 quintals (4–6 tonnes) per hectare is the maximum for top quality, even though the DOCG regulations allow 80 quintals. So they might cut the cane back to seven to ten buds. A smaller crop also has the advantage, in this marginal climate, of ripening earlier.

## At the winery

In terms of vineyards and clones, the Nebbiolo of Barolo is probably further ahead than the Sangiovese of Tuscany – but then it had less far to come. The quality of the grapes is generally very good. All the controversy hinges, as it has done for the last couple of decades, on the winemaking. In Barolo (and, indeed, in Barbaresco) you are a traditionalist or a modernist. Time was when the divisions between the two were as sharp, and as sharply felt, as between Montagu and Capulet. One winemaking family even built a wall physically to divide the traditionalist sibling from the modernist.

Now modernism has affected (infected, some might say) most of the traditionalist producers to some extent. Most Barolo is, accordingly, more forward and fruity than it was. To see how and why, let us first examine Barolo winemaking as it was before anything changed.

Because Nebbiolo is late ripening, by the time it is picked the weather, and the cellars, are getting cold. Starting the fermentation was not easy in the past: it could take eight or fifteen days before there was any sign of action in the must. Delayed fermentations like this, in old wood or cement where hygiene is not absolute (which is not the case with today's traditionalist producers), can lead to bacterial infection. That was the cause of the off-flavours that necessitated opening the bottles 24 hours in advance.

Once the fermentation began the temperature of the must could rise extremely high, to 35° or 38°C; this tended to reduce aroma and fruit flavours. But the weather was getting colder all the time: the fermentation would slow, and drag out for two or three months. The amount of tannin extracted was enormous, and had to be softened by leaving the wines in big old casks for five years or more, which in turn could cause problems of volatile acidity.

*Traditional large* botti *and modern small barriques in the La Morra cellars of Renato Ratti. Ratti is no barrique fanatic, but was one of the first in Barolo to tackle the problems of oxidation in wine.*

Today's traditionalism is based on long macerations of 20 to 30 days as well as long aging. Technical improvements such as getting the fermentation off to a swift start, and hygiene, can be taken for granted. Modernism, meanwhile, centres on the use of barriques, on shorter fermentation and maceration times and on earlier bottling. The idea is that with temperature control you can keep the temperature down to 28° or 30°C to preserve fruit and aroma. Rotovinifiers are currently fashionable in Barolo: 1996 was first the vintage in which they were generally used. They give better extraction, so that a short maceration of seven to ten days may be all that is necessary. Heating the cellars towards the end of the alcoholic fermentation will encourage the malolactic one, which in turn means less aggressive acidity.

The point of small oak barriques, for fermentation, aging or both stages, is not to add tannin but to soften the wine for two years or so with their gentle oxidative effect. But they do, of course, give a certain sameness to the wines. Barolo is cleaner than it was, easier to understand, fruitier, more supple and less unpredictable to buy and drink. But new oak can cover the wonderful rose scents of Nebbiolo with its vanilla richness. And tamer wines, by definition, are less exciting wines. Modernism is both a gain and a loss.

### BLENDING NEBBIOLO

In Piedmont most Nebbiolo is made as a varietal wine. Barolo is 100 per cent Nebbiolo; so is Barbaresco. Across the Tanaro river from Barolo and Barbaresco comes Roero which is supposed to contain between 2 and 5 per cent Arneis, but often doesn't these days. In Gattinara, Ghemme, and the less familiar wines of northern Piedmont, blending is also officially required, the other grapes being Vespolina, Croatina and Bonarda, but not all producers opt for obeying the law on this point. In Lombardy's Valtellina, Nebbiolo (called Chiavennasca here) may be blended with Rossola, Pignola, Prugnolo, Pinot Nero and Merlot for basic Valtellina, and with other red grapes up to 5 per cent for Valtellina Superiore. That's the dull bit.

The interesting bit, for those curious about how things are really done, as opposed to how they're supposed to be done, is whether or not other grapes are blended into Barolo and Barbaresco. Officially, of course, this is completely forbidden, and if you want to add a dash of Barbera or whatever, you must demote the wine to Vino da Tavola (and probably charge an astonishingly high price for it). But some tasters have noted smells of Cabernet Sauvignon or Syrah in Barolo; others have pointed to unusually deep colours in wines from this rather poorly coloured grape. Some go so far as to point out that Barbera was always the traditional blending grape in Barolo, because of its greater colour and lower acidity. Purely hypothetical, of course. No-one ever admitted to doing it, though an amused smile could be seen playing on the lips of some successful producers.

In 1998 there was a proposal to make it legal to add 10–15 per cent of other grapes to Barbaresco – the grapes could have been Cabernet Sauvignon, Merlot or anything else. They would have pepped up the wine in a bad vintage, and made it rounder and richer; more saleable, in other words. The growers of Barolo were invited to join in the party, but they declined, and then the idea got a bad press in the Italian media. It was then discarded in the interests (as the official line went), of maintaining the individuality of Barbaresco in the face of increasing international standardization. Well, good for them for making a stand. Especially if, as some commentators believe, they're doing it on the quiet anyway.

# Nebbiolo around the World

ATTEMPTS AT MAKING NEBBIOLO WORLDWIDE seem so far to be having only limited success, and while it's true that even in its native Piedmont quality can vary hugely, it is Barolo that is the ultimate target of every ambitious producer in California or Australia. At the moment I'd be thrilled if they could make anything as good as simple Langhe Nebbiolo.

## Italy

Nebbiolo is extremely sensitive to differences in soil and mesoclimate, yet pinpointing the differences between Barolo and Barbaresco is not easy. Barbaresco is said to be the lighter, softer wine, though differences between communes in Barolo can be as great as differences between Barolo and Barbaresco. The regulations are a little less stringent in Barbaresco: slightly shorter minimum aging (21 months, including nine months in oak for *normale*, compared to Barolo's three years total, and one year in wood; Barbaresco Riserva must have 45 months' total aging, compared to Barolo Riserva's four years and nine months) and a slightly lower minimum alcohol level (12.5 per cent to Barolo's 13 per cent).

Within Barolo – which is a tiny zone, though three times the size of Barbaresco – Serralunga is the latest-ripening commune. Picking begins here some 15 days after it starts in La Morra, even though the latter is higher; the closer proximity of the river in La Morra makes for a slightly warmer mesoclimate. The wines of Serralunga are among the weightiest in the Barolo region; those of La Morra are at the other extreme: silky, and appealing when young. The other communes in the Barolo zone are Castiglione Falletto, Grinzane Cavour, Monforte, Novello, Verduno, and of course Barolo itself. Castiglione Falletto makes concentrated, powerful wines of great finesse; Monforte wines are firm and long-lived; Barolo wines combine structure with supple fruit. Well, that's the theory anyway.

The main difference in soil is between the sand and limestone in the east, in the communes of Castiglione Falletto, Monforte and Serralunga, which give burly, aggressive wine, and the chalky marl in the west, in La Morra and Barolo, where the wines are less robust and more perfumed. But clay is found everywhere, and it is the clay of Barolo (and Barbaresco, too) which gives the wines their remarkable tannin. Elsewhere Nebbiolo is seldom quite as tannic.

The best (and certainly the most expensive) wines are from single vineyards or crus. There is no established cru system, though this has been mooted for a while, and vineyard names are used on labels. So far only individual subzones in the communes are legally recognized.

In theory Nebbiolo is planted on the best sites here. But by 2000 optimism was running so high in the region that other, less suitable sites were also being planted – the lower slopes, which are usually given over to Barbera, or vineyards at too high an altitude, or with poor exposure. The same thing has happened before, in times of boom; such vineyards were abandoned later, and will no doubt be abandoned again.

Roero has traditionally made lighter, less tannic, more supple wines than those from Barolo and Barbaresco south of the Tanaro river. Occasional examples are bigger and more complex, however. Nebbiolo d'Alba, from the same area, is simpler and earlier-maturing; Langhe Nebbiolo is a catch-all DOC that covers anything that doesn't fit into the other DOCs.

In the Novara and Vercelli hills in norther Piedmont, Nebbiolo is known as Spanna, and in Carema it goes by the name of Picoutener. Gattinara and Ghemme, from the former, tend to be somewhat earthy; Carema can be more perfumed, but also less ripe. In Valtellina Nebbiolo becomes Chiavennasca, and all too often becomes a touch stringy. Greater ripeness would help, and perhaps lower yields. Frankly, I think a bit of blending with softer, earlier-ripening varieties would also help, but there you go. There is also an Amarone-type red called Sfursat or Sforzato which, when it's good, is very good. But the wines from here are seldom exported further than Switzerland.

Some is grown in Lombardy for Franciacorta Rosso, for which it is blended with Cabernet Franc, Barbera and Merlot.

## Australia

Nebbiolo should have great potential here, but finding the right sites for this very site-specific vine has been tricky. Clones have also been a problem, usually giving too much vigour and not enough colour. Victoria's King Valley, where the climate is marginal for Nebbiolo, is producing some interesting examples, and

GIACOMO CONTERNO

*Unlike his brother Aldo (see page 157), Giovanni Conterno is a traditional Barolo producer and this wine will last for decades.*

ELIO ALTARE

*A modernist par excellence, Altare seeks balance above all else in his wines that are supple in youth yet will age. This cru spends time in both barrique and botte.*

BRUNO GIACOSA

*Giacosa is a semi-traditionalist: his macerations are long, but shorter than they used to be, and he ages his wines in French, not Slavonian, botti.*

*Jim Clendenen's Bien Nacido vineyard in California's Santa Barbara County. Barbera is planted in the foreground: the block high up the hillside on the right of the picture is Nebbiolo. Clenenden sells wine from Italian grape varieties under his Il Podere dell'Olivos label; his wines from French varieties like Chardonnay and Pinot Noir go under the Au Bon Climat label.*

according to research in Australia by Alex McKay, Garry Crittenden, Peter Dry and Jim Hardie, Margaret River (Western Australia) and Mornington Peninsula (Victoria) can match the Langhe in terms of sunshine hours, relative humidity and rainfall during the growing season. If heat summation is to be matched, however, then Clare Valley, Mudgee and Bendigo might be appropriate, but with Nebbiolo I only believe it when I see it.

### USA

Given the Italian influence in establishing America's vineyards, there was a reasonable amount of Nebbiolo in the old days. But it never had much success. I've done a tasting of just about every varietal Nebbiolo in California, mostly produced by a variety of old codgers I'd never heard of, and I might as well have stepped back 50 years, the wines were so lumpen and stewed. A bit like traditional Barolo, perhaps, but not at all like modern Barolo. It's strange: you'd expect California to be able to apply first-rate winemaking principles to the grape, and you'd have thought they could identify the right location and viticultural techniques to get concentration, colour and ripe tannins in the finished wine. But no – in this Cabernet and Merlot obsessed world little thought has been given to the notoriously moody Nebbiolo. Only about 130 acres/53ha of the original plantings remain, and most are scattered forlornly about the bulk-producing Central Valley. The best wine so far is from Jim Clendenen at Au Bon Climat in the Santa Maria Valley; he has planted lower-yielding clones and hopes in the next few years to move from making good wine to making great wine. Hope for the future may rest with the Consorzio Cal-Italia, a group of around 50 California wineries producing wines from classic Italian grape varieties. This consorzio is a logical extension of the Cal-Ital phenomenon, the fusion of Italian and West Coast cooking styles.

Nebbiolo has been equally problematical in Washington State, where the climate ought at least to be warm enough for it. At the moment Nebbiolo seems to be the least promising of all the Italian varieties currently being tried in the USA.

### Rest of the world

Experiments with Nebbiolo in Chile have produced wines with poor colour and high acidity, perhaps partly because the clones were less than ideal. Argentina has a little planted, but yields are high. Mexico has made some rather attractive examples in Baja California, up towards the US border. There is a little planted experimentally in Austria's Mittelburgenland region.

NEBBIOLO D'ALBA
Denominazione di Origine Controllata
1998
LA VAL DEI PRETI
Imbottigliato all'origine dal viticultore
Estate Bottled by Matteo Correggia
Canale - Italia
Alc.14% by vol. Net. cont. 750 ML ℮

**Matteo Correggia**

*The Roero hills do not have the same reputation as Barolo or Barbaresco across the river Tanaro but this wine has depth and complexity to rival all but the very best.*

**Nino Negri**

*This 100% Nebbiolo, aged in barrique, is one of the stars – in fact it's probably the greatest star – of the Valtellina region in northern Lombardy.*

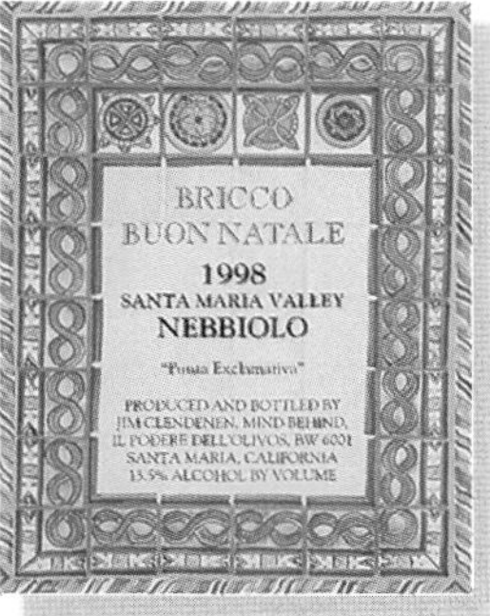


**Il Podere dell'Olivos**

*Cal-Ital style from California's Santa Maria Valley. Winemaker Jim Clendenen is a fanatic for Italian grapes, and Nebbiolo was the one with which he started.*

# Enjoying Nebbiolo

THE TRADITIONAL VIEW IS that not only does Nebbiolo age well, but that it positively must be aged for decades if it is to be even drinkable. Well, there's no doubt Nebbiolo can age well. If it had the ripeness in the first place and wasn't wrecked by incompetent winemaking – this still happens, but far less than before – Nebbiolo should last a decade at the least. But tough, dry Nebbiolo will usually remain just that – tough and dry, but older. However, you don't have to age modern Nebbiolo. The grape actually has a rather beautiful perfume and beguiling black cherry, sloes and damson fruit when carefully made. Add a splash of new oak to this and you've got a very attractive though sturdy drink that is challenging but enjoyable at only three or four years old.

The lighter sort of Nebbiolo – from Carema or Langhe – can certainly be drunk within a few years of the vintage and will last from five to eight years or so; perhaps longer in the case of the most concentrated Caremas. Gattinara is slower maturing and longer lived: it often needs six to eight years before it reaches its best, and will live in bottle for another ten or more. Roero is nearly always much lighter, and should be drunk within a few years, but you'll sometimes find weightier examples from adventurous producers.

Barbaresco is supposed to be lighter than Barolo, but there are as many differences between wines and communes in Barolo as there are between Barolo and Barbaresco themselves. Lighter wines from both regions may be at their best before their tenth birthday; the top, most concentrated cru wines should probably not be opened until they are about eight years old, and will easily live for 20 or 30 years.

Experience in Australia and California is too short for any rules on aging to be formulated. The wines made so far have aged relatively rapidly compared to the wines of Piedmont.

### The taste of Nebbiolo

The classic description of the scent of Nebbiolo is tar and roses, immortalized in Michael Garner's and Paul Merritt's 1990 Barolo study of the same name. It sounds an improbable combination, but then Nebbiolo is an improbable grape, combining as it does high levels of tannin with high acidity and (in its best examples) some of the most complex, exotic scents to be found on any red wine.

To tar and roses one might add cherries, damsons and mulberries, leather, herbs both fresh and dried, spice, licorice and dried fruit. The tannins mature to a powdery softness; the acidity should be ripe, and there should be sufficient weight of fruit and alcohol to balance the tannin and acidity while the wine ages. There is often a certain austerity about Barolo and Barbaresco, though this is less common in wines from modernist producers where the fruit is fleshier and rounded out by aging or fermentation in new oak barriques. Even in the most traditionalist examples, mature Barolo should be supple.

It is a pity if Barolo or Barbaresco smells strongly of new oak. Any wine from anywhere in the world can do this: a whiff of new oak is one of the easiest of all aromas to achieve in the winery. But only Nebbiolo has this extraordinary, haunting aroma. To mask it with add-on scents is to lose the point of the wine completely.

*Michele Chiarlo uses no barriques for his regular Barolo bottling, though cru bottlings may get the barrique treatment, and other wines may be blends of Nebbiolo with Barbera or Cabernet. Bricco Maiolica's Nebbiolo d'Alba Il Cumot is also modern in its suppleness and concentration. Different producers of Nebbiolo, in other words, approach modernism in different ways.*

### Matching Nebbiolo and food

Fruity, fragrant, early-drinking styles of Nebbiolo wine are best with local salami, pâtés, *bresaola* (wind-dried beef sliced wafer-thin) and lighter meat dishes. The best Barolos and Barbarescos need big, substantial food to stand up to them; *bollito misto* (a hotchpotch of boiled meats served with a green garlicky sauce), jugged hare, spiced beef casseroles and *brasato al Barolo* (a large piece of beef marinated then braised slowly in Barolo) are just the job in Piedmont, or anywhere else for that matter.

## CONSUMER INFORMATION

**Synonyms & local names**
Also known in Piedmont as Spanna (especially in the north in the provinces of Novara and Vercelli) and Picoutener (especially in Carema and the Aosta Valley); and as Chiavennasca in Lombardy.

**Best producers**
**ITALY/Piedmont/Barolo** Claudio Alario, Gianfranco Alessandria, Altare, Azelia, Enzo Boglietti, Brovia, Ceretto, Michele Chiarlo, Domenico Clerico, Elvio Cogno, Aldo Conterno, Giacomo Conterno, Paolo Conterno, Conterno-Fantino, Luigi Einaudi, Ettore Germano, Attilio Ghisolfi, Bruno Giacosa, Elio Grasso, Silvio Grasso, Bartolo Mascarello, Giuseppe Mascarello, Monfalletto-Cordero di Montezemolo, Andrea Oberto, Armando Parusso, Pio Cesare, Enrico Pira, Luigi Pira, Ferdinando Principiano, Prunotto, Renato Ratti, Rocche Costamagna, Rocche dei Manzoni, Sandrone, Paolo Scavino, Mauro Sebaste, Vajra, Mauro Veglio, Giovanni Viberti, Vietti, Vigna Rionda, Gianni Voerzio, Roberto Voerzio; **Barbaresco** Produttori del Barbaresco, Ceretto, Cigliuti, Stefano Farina, Fontanabianca, Gaja, Bruno Giacosa, Marchesi di Gresy, Moccagatta, Fiorenzo Nada, Castello di Neive, Paitin, Pelissero, Pio Cesare, Prunotto, Albino Rocca, Bruno Rocca, Sottimano, La Spinetta, Vietti; **Roero** Matteo Correggia; **Lombardy/ Valtellina** La Castellina, Fay, Nino Negri, Rainoldi, Conte Sertoli Salis, Triacca, Enologica Valtellinese.
**USA/California** Il Podere dell'Olivos, Renwood, Viansa; **Washington State** Cavatappi.
**MEXICO** L A Cetto.

## RECOMMENDED WINES TO TRY

**Twenty classic Barolo wines**
Altare *Barolo Vigneto Arborina*
Azelia *Barolo San Rocco*
Brovia *Barolo Villero*
Ceretto *Barolo Bricco Rocche Prapò*
Domenico Clerico *Barolo Ciabot Mentin Ginestra*
Aldo Conterno *Barolo Gran Bussia*
Giacomo Conterno *Barolo Riserva Monfortino*
Conterno-Fantino *Barolo Sorì Ginestra*
Luigi Einaudi *Barolo Nei Cannubi*
Bruno Giacoso *Barolo Villero di Castiglione Falletto*
Elio Grasso *Barolo Ginestra Vigna Casa Maté*
Giuseppe Mascarello *Barolo Monprivato*
Monfalletto-Cordero di Montezemolo *Barolo Enrico VI*
Armando Parusso *Barolo Vigna Rocche*
Luciano Sandrone *Barolo Cannubis Boschis*
Paolo Scavino *Barolo Bric del Fiasc*
Vietti *Barolo Riserva Villero*
Vigna Rionda *Barolo Riserva Vigna Rionda*
Gianni Voerzio *Barolo La Serra*
Roberto Voerzio *Barolo Cerequio*

**Ten classic Barbarescos (or equivalent)**
Ceretto *Barbaresco Bricco Asili*
Cigliuti *Barbaresco Serraboella*
Gaja *San Lorenzo* and *Sorì Tildìn*
Bruno Giacosa *Barbaresco Santo Stefano*
Marchesi di Gresy *Barbaresco Martinenga Gaiun*
Pio Cesare *Barbaresco Il Bricco*
Prunotto *Barbaresco Montestefano*
Albino Rocca *Barbaresco Vigneto Brich Ronchi*
La Spinetta *Barbaresco Vigneto Gallina*

**Five other classic Italian Nebbiolo-based wines**
Conterno-Fantino *Langhe Monprà*
Matteo Correggia *Nebbiolo d'Alba la Valle dei Preti*
Luigi Ferrando *Carema Black Label*
Gaja *Langhe Sperss*
Nino Negri *Valtellina Sfursat 5 Stelle*

**Ten lighter, modern-style Nebbiolo-based Piedmont wines**
Antichi Vigneti di Cantalupo *Ghemme Signore di Bayard*
Bricco Maiolica *Nebbiolo d'Alba Il Cumot*
Carema Cantina dei Produttori *Nebbiolo di Carema*
Michele Chiarlo *Langhe Barilot*
Aldo Conterno *Langhe Il Favot*
Nino Negri *Sassella Le Tense*
Pio Cesare *Nebbiolo d'Alba Il Nebbio*
Prunotto *Nebbiolo d'Alba Occhetti*
Luciano Sandrone *Nebbiolo d'Alba Valmaggiore*
G D Vajra *Langhe Nebbiolo*

**Five Nebbiolo wines from elsewhere**
Cavatappi *Red Willow Vineyards Maddalena (Washington)*
L A Cetto *Nebbiolo (Mexico)*
Il Podere dell'Olivos *Nebbiolo (California)*
Renwood *Nebbiolo (California)*
Viansa *Nebbiolo (California)*

*Delinquent or just misunderstood? Nebbiolo is proving to be one of the most frustrating grapes to grow outside its Piedmont heartland. Growers are examining the minutiae of soil and climate, but the grape still obstinately refuses to play ball.*

**Maturity charts**
Even Barolo is drinkable much earlier than it used to be. Tannins are softer these days, and the fruit is more to the fore.

**1997** Barolo (top Cru)

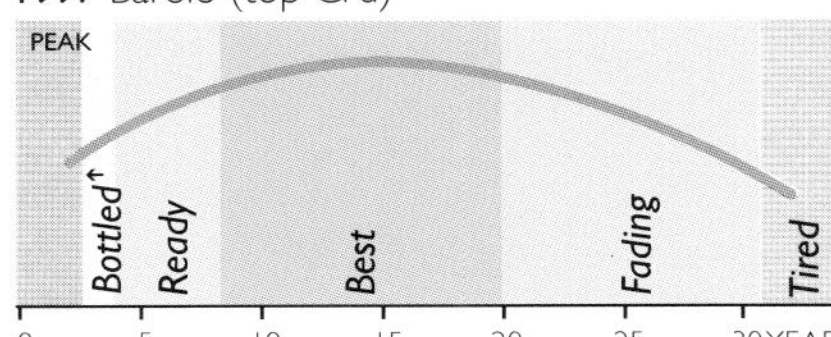


*1997 was an atypically lush vintage for Nebbiolo. The wines have great concentration and tannin, but soft fruit, fairly low acidity and high alcohol.*

**1997** Barbaresco

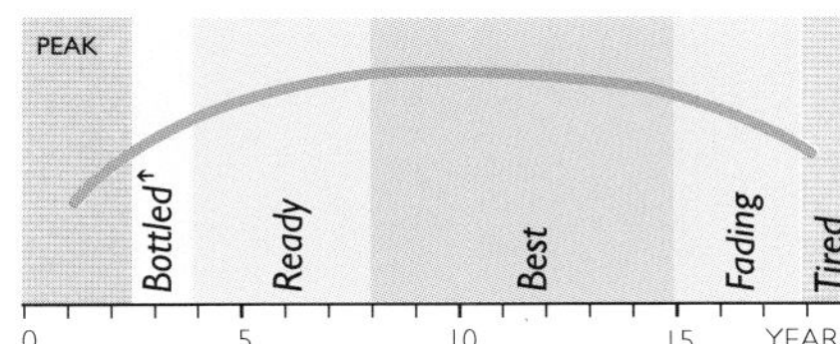


*Barbaresco too had a tremendously ripe vintage in 1997. Again, expect rich, dense but soft fruit and plenty of concentration and tannin.*

**1999** Nebbiolo d'Alba

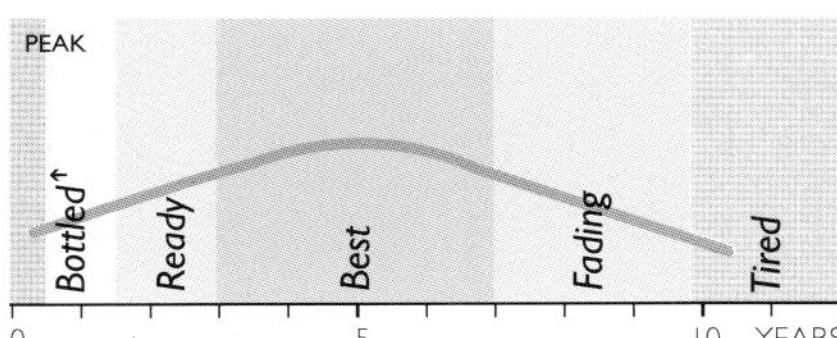


*A difficult year so stick to top producers only. Humidity and rot caused problems, but top growers made good, though not great, wines.*

## NEGOSKA

Soft, richly fruity grape found in northern Greece and blended in small quantities with the much tougher, more acidic Xynomavro (see page 285) to make Goumenissa. Best producer: (Greece) Boutari.

## NEGRA MOLE

This grape is probably not the same as the Tinta Negra Mole grown in Madeira. As Negra Mole it is found in Portugal's Algarve; and as Negramoll, in Spain's Canary Islands, where it makes soft reds for early drinking. In the Ribatejo it is known as Preto Martinho.

## NÉGRETTE

A grape of south-west France, found in the Côtes du Frontonnais appellation, where it must constitute 50 to 70 per cent of the blend. Négrette contributes a wonderfully silky texture and raspberry perfume to the best wines. It is also found in other wines from north of Toulouse. Common blending partners include Fer and Syrah, but inevitably there is a fair bit of Cabernet around as well, which tends to hijack the style of wines to which it is added. Admittedly Négrette's tendency to a slightly high level of volatile acidity means it sometimes needs a bit of stabilizing, but too often Cabernet ejects the silky Négrette baby with the bath water.

Négrette is usually best drunk young and without oak aging. The hot, dry climate of the region suits the grape well – it gets botrytis bunch rot rather easily in wet conditions. Best producers: (France) Bellevue-la-Forêt, la Colombière, Ferran, la Palme, le Roc.

CHÂTEAU BELLEVUE-LA-FORÊT

*Negrette may be the biggest single grape in this blend, but Cabernet Sauvignon comes next and it does give it that characteristic blackcurrant flavour. But the Negrette adds suppleness and good raspberry flavours.*

## NEGROAMARO

A rather odd-tasting grape found in the south of Italy, particular in Puglia. It could hardly be more Italian in its combination of perfume and bitterness – the name means 'black bitter' and the wine is indeed very dark in colour. It is sturdy in structure and can be both slightly farmyardy in flavour and have a distinctly medicinal edge. The combination takes a bit of getting used to but is certainly interesting.

It is far more easy to appreciate, however, if blended with a little of the far more scented, succulent Malvasia Nera (see page 122). Some of Puglia's best reds, notably Salice Salentino, feature this double act. Best producers: (Italy) Giuseppe Calò, Michele Calò, Candido, Leone de Castris, Pervini, Taurino, Vallone.

## NEGRU DE DRAGASANI

A new Romanian crossing which seems promising, and might well appear in vineyards (and even on wine labels) in years to come. One of its parents is Saperavi, one of Eastern Europe's best grapes (see page 207).

## NERELLO

A Sicilian grape that comes in two versions, Nerello Mascalese and the less commonly planted Nerello Cappuccio. Both contribute high alcohol to a blend, though they lack the concentration of Nero d'Avola, with which they are usually blended. Nero d'Avola is fast becoming fashionable, and Nerello may follow in its wake. Best producers: (Italy) Benanti, Palari, Duca di Salaparuta.

## NERO D'AVOLA

Sicilian grape that is attracting attention from winemakers in search of new grapes and flavours, and is a possibility for planting in Australia, especially in the warm Riverland region in South Australia. Nero d'Avola produces wines that are fashionably dark, soft yet robust, and age well, especially if given a short stint in oak. Best producers: (Italy) Abbazia Santa Anastasia, Calatrasi, COS, Donnafugata, Morgante, Planeta, Duca di Salaparuta, Spadafora, Tasca d'Almerita.

## NEUBURGER

Austria's Neuburger grape gives rich, spicy wines, sweet, semi-sweet or dry, that are not unlike a more intense version of Pinot Blanc; and indeed Pinot Blanc, or Weissburgunder is one of Neuburger's parents. But the other parent is Silvaner, and the broad, flat neutrality of Silvaner clearly influences Neuburger. It is happily susceptible to noble rot, but dry or off dry versions also age well in bottle, becoming nicely honeyed and nutty, but always lacking a little zing. Best producers: (Austria) Beck, Feiler-Artinger, Haider.

## NIAGARA

Vigorous, high-yielding hybrid grown especially in New York State, where its ability to withstand winter cold comes in useful. It was bred in 1872 from a crossing of Concord (see page 85) and Cassady, and has a strong dose of the so-called 'foxy' aromas that mark many *Vitis labrusca* grapes.

## NIELLUCCIO

This robust, tannic Corsican grape is said to be identical to the Sangiovese of Italy. It is generally blended with the less substantial though more complex Sciacarello: the two of them constitute the best of Corsica's red grapes. Other varieties on the island are often those, like Cinsaut and Carignan, that were planted by immigrants returning from North Africa in the 1960s. However, these last are on the decline, and Nielluccio and other indigenous grapes are on the increase. So, inevitably, are international varieties.

Nielluccio is more widely planted than Sciacarello, and has its heartland in Patrimonio in the North; the chalky clay here suits it particularly well. It is also grown in the South, around Porto Vecchio. Its acidity and tannins are good, though the tannins can be somewhat harsh and need careful handling. It also makes excellent rosés. Best producers: (France) de Alzipratu, Antoine Aréna, Orenga de Gaffory, Leccia.

## NOSIOLA

An unusually neutral grape found in Trentino-Alto Adige – unusual because most of the white vines in this part of north-eastern Italy have more character. Nosiola is crisp, light and ephemerally aromatic, and that's about it. It's a perfectly attractive wine, and is probably at its best in terms of personality when made into Vino Santo, the Trentino sweet wine made from dried grapes. It is grown only in small and diminishing quantities. Best producers: (Italy) Concilio Vini, Pojer & Sandri, Pravis.

## NURAGUS

Prolific Sardinian vine which makes light, neutral, perfectly attractive and perfectly unmemorable wine. Its main distinction seems to be that it was brought to Sardinia by the Phoenicians. Best producers: (Italy) Argiolas, Dolianova, Santadi, Trexenta.

## OJO DE LIEBRE

The Catalan name for Tempranillo (see pages 256–265). The name means 'hare's eye'.

Ripe Palomino Fino grapes growing in the almost pure chalk soil of Jerez. Palomino's neutrality becomes a virtue here – but only because sherry gains nearly all its flavour from the solera system.

## OLASZ RIZLING

The Hungarian name for Welschriesling (see page 284). For all its poor reputation in some countries, perhaps especially Britain, the grape can make very attractive wine if well handled. Britain's problem was that it imported the dirtier and cheaper examples from Hungary, thus comprehensively destroying the grape's reputation in the British market.

## ONDENC

Almost extinct grape found occasionally in south-west France. Nowadays it is unpopular for its low yields and susceptibility to rot. It used to be grown in Australia, but is almost extinct there, too, except in Victoria where it makes a very dry, cold-hearted base for sparkling wine. Best producers: (France) de Causses-Marines, Robert Plageoles.

## OPORTO

Hungarian name for Portugieser (see page 188), Kékoporto (see page 115), or Blue Oporto.

## OPTIMA

Mercifully declining German crossing, produced in 1930 from a Silvaner x Riesling vine crossed with Müller-Thurgau. Its virtue for the grower is that it ripens very early, and reaches high sugar levels; it is hard to think of any advantages it has for anyone else. Its wines are blowzy, dull and crudely sweet, but its late-budding habit endeared it in the past to growers in the Mosel. The Rheinhessen was its main field of operations; I have had one or two fairly rich dessert examples, but it is declining there as well. Best producer: (Canada) Quails' Gate.

## ORANGE MUSCAT

This should by rights be a form of Muscat, but it is believed to be unrelated. There is a little grown in California, of which the most famous example is made by Andrew Quady as a fortified wine called Essencia. He picks early for aroma and lightness, and stops the fermentation by adding grape spirit. The wine, which is elegant with a tangily sweet finish, has an intense orange-flower and orange peel aroma and flavour. Best producer: (USA) Quady.

## OREMUS

A Tokaj grape with good susceptibility to botrytis and good sugar levels at harvest. It is a cross between Furmint and Bouvier. There is also another Furmint x Bouvier grape in the region, called Zeta. Best producer: (Hungary) Disznókö.

## ORMEASCO

Ligurian name for Dolcetto (see pages 86–87).

## ORTEGA

A modern German crossing (Müller-Thurgau and Siegerrebe) which has little more than high sugar levels to recommend it. It has plenty of flavour but no complexity, which rather misses the point of the Spätlese, Auslese and higher Prädikat levels which it reaches with such ease. Acidity can be low, and the vine is prone to various diseases. In Germany it is mostly found in the Rheinhessen.

To be honest, the best Ortegas I've had have been from England, where the vine only just ripens in the cool conditions and gives wine with good grapefruit acidity and breezy elderflower scent with good potential for aging. Best producers: (England) Denbies, Throwley, Valley Vineyards.

## ORTRUGO

A grape found in the Colli Piacentini zone in Emilia, central Italy. The wine is deep in colour, with high alcohol and decent acidity, and may be blended with Malvasia. It also comes as frizzante. Best producers: (Italy) Gaetano Lusenti, Pernice.

## OZ

A white table grape found in Israel. Envigorating, juicy, refreshing, sure to please, just like this book's co-author.

## PAGADEBIT

Italian name for Bombino Bianco (see page 43).

## PAIEN

Gewürztraminer (see pages 102–111) may be called by this name in Switzerland.

## PAÍS

One of the earliest grapes to be planted in Chile by European settlers, it is still cultivated in large quantities in the south of the country. It is either the same as or very similar to the Mission of California (see page 138).

## PALOMINO FINO

One of the dullest grapes in the world, which just happens, when grown in a small area of white, chalky *albariza* soil in the south-west tip of Spain, and aged in a solera system, to produce wines of matchless complexity and pungency. The wine is, of course, sherry – a name that cannot, under EU law, be applied to the fortified wine of any other country or region. Palomino Fino is effectively the only grape grown in the sherry region and has replaced the lesser Palomino de Jerez or Palomino Basto.

The Pedro Ximénez that may be used for sweetening some wines nowadays largely comes from nearby Montilla-Moriles. A few bodegas opt for the rarer but fruitier Moscatel as a sweetening wine, but Moscatel's intense grapy perfume doesn't make for great sherry. Various ways of concentrating the sweetness of Palomino – leaving the grapes in the sun or concentrating the must – may also be used to producing sweetening wine, according to the desired quality and style of the final result.

The wine is always fermented to dryness, and fortified later. Considering the aridity of the climate and the lack of irrigation, yields are on the high side – between 75hl/ha and perhaps 150hl/ha. The received wisdom is that large companies restrict yields to improve quality, and smaller growers squeeze every last ounce of juice out of their grapes. Well, many small growers whose only objective is to cart as big a crop as possible to the local co-operative certainly do maximize yields. But many big companies make an awful lot of mediocre sherry. Are such companies really restricting their yields? Or perhaps they're buying juice from the very same co-operatives that the smaller growers are supplying. High yields are aided by the nature of the region's *albariza* soil, which has the capacity to retain 25 per cent of its volume in water. Not surprisingly vineyards on less good soils in the region have now been uprooted.

The vines are generally trained as low bushes, though training on wires is increasingly popular. Rot is less of a problem with wire-trained vines. In addition, wire training is necessary for mechanical harvesting, and while this has never so far been permitted in the region it is by no means impossible that it won't sometime be allowed. It was a ban imposed for social reasons, to provide seasonal work for unskilled labour, but nowadays the number of pickers looking for work is less each year.

However, despite my carping, high yields do not really seem to be a problem for quality. Some 90 per cent of the style and quality of sherry comes from the aging, and the position of the bodega and how the solera is managed are far more important than the base wine.

Sherry is divided into two basic styles: those that grow *flor*, and those that do not. Those that do are Fino and Manzanilla, and by extension Amontillado, since proper Amontillado is Fino or Manzanilla on which the flor has, after some years, died. Flor is a yeast, or rather several yeasts, and each bodega, indeed each solera, may have its own particular balance of yeast strains in its flor.

Flor grows on the surface of the wine in a porridgey white layer perhaps 1cm thick. It feeds on the wine, keeping it free of contact with oxygen (thus preventing oxidation while keeping it fresh-tasting) but it also changes the chemical make-up of the wine so that it develops a characteristic pungency.

Oloroso sherry does not grow flor. Wines for Olorosos are selected at an early stage – they may well come from the third pressing of the grapes, and thus be higher in phenols and coarser in flavour. (The most delicate wines are used for Finos and Manzanillas, and generally come from free-run juice.) Oloroso gains its dark colour and pungent flavour of nuts, prunes and coffee from long aging in solera, during which time the wine oxidizes and becomes more concentrated.

The solera system is a system of fractional blending in which wine is drawn off for bottling at one end, with the barrels in the last stage of the solera being replenished from those in the next stage back, and so on. New wine is added to the barrels in the first stage. Wine is drawn off for bottling perhaps four times a year, with perhaps a quarter of each barrel being taken at any time. All sherry is thus a blend of old and much younger wine; it is this that helps give it its balance of freshness and mature depth.

Palomino is also used for producing table wine in the three sherry towns of Jerez de la Frontera, Puerto de Santa María and Sanlucar de Barrameda. Many sherry bodegas now produce such a wine, and without exception they prove the utter dullness of the Palomino grape. The wine is low in acidity unless acidified in the winery, and neutral in flavour.

Palomino is also grown elsewhere in Spain, for table wines as well as for fortifieds and *rancio* styles: in Condado de Huelva, west of Jerez towards the Portuguese border, it is increasingly popular at the expense of Zalema, but in northern Spain, in Rueda and Galicia, it is fading in importance. It is a key variety in the Canary Islands, where its wine even manages to have a little character – I haven't quite worked out how.

Where it is drunk as a table wine the solution is sometimes to make it deliberately *rancio* by leaving it in glass demijohns in the sun to oxidize. It then acquires a taste reminiscent of sherry, even if not very good sherry. It is drunk this way in Rueda. And you wonder why Palomino is disappearing in Rueda.

It is found in decreasing quantities in western France, where it goes under the name of Listán or Listán de Jerez. It may be the same grape as the Perrum of Portugal's Alentejo region, and possibly the same as Portugal's Malvasia Rei.

Outside Europe it is found in South Africa where, as Fransdruif, it declined from 10.9 per cent of the total vineyard in 1985 to 2.4 per cent in 1999. There are relatively small plantings in California's San Joaquin Valley, where the variety used to be known as Golden Chasselas, and in Australia, where it makes sherry-style fortified wines. There are still smaller plantings in New Zealand and Cyprus, where it is used for fortified wines, and Argentina. Best producers: (Spain) Argüeso, Barbadillo, Delgado Zuleta, Díez Mérito, Domecq, Garvey, Gonzalez-Byass, Hidalgo, Lustau, Osborne, Valdespino.

## PANSA BLANCA

This is the name given to Xarel-lo (see page 285) in Spain's Alella region, on the coast north of Barcelona, where it seems to acquire some extra bite and aroma. It ripens to perhaps 14 per cent potential alcohol without losing its lime cordial aroma and flavour, and seems to suit the region's sandy *sauló* soil. Best producers: (Spain) Marqués de Alella, Parxet, Roura.

## PARDILLO

There's a lot of this neutral white grape in central Spain, and it's really very dull.

## PARELLADA

One of the three grape varieties grown for Cava, the Spanish name for Champagne-method fizz, Parellada contributes lemony freshness to the blend, providing that it is not overcropped: it tends to be high yielding in fertile soils. It is outplanted in the Catalan region of Penedès by the other Cava varieties, Macabeo and Xarel-lo. It also makes still wines, light, fresh and gently floral, with good acidity and, for Spain, lowish alcohol (between 9 and 11 per cent). Drink it as young as possible – and I mean young – while it still has the benefit of freshness.

It can be blended with other varieties, notably Chardonnay and Sauvignon Blanc, and is also found in the Costers del Segre region inland of Barcelona. Best producers: (Spain) Castellblanch, Codorníu, Freixenet, Juvé y Camps, Marqués de Monistrol, Parxet, Raimat.

## PEDRO GIMÉNEZ

This, logically, should be Pedro Ximénez (PX) under a different name, but the large plantings of Pedro Giménez in Argentina and Chile are believed to be a different variety from Spain's PX. In Argentina it makes everyday wines which never get exported; in Chile it is grown for pisco, the local brandy.

## PEDRO XIMÉNEZ (PX)

Along with Palomino Fino (see page 165), this used to be the other grape of the Jerez sherry region in southern Spain. It was used for sweetening purposes and sometimes for bottling on its own as a sweet, dark, dessert sherry. But the more disease-resistant Palomino has all but replaced it, and almost all the PX used in sherry these days is brought in from the nearby Montilla-Moriles area, where it is the main variety.

Traditionally it is picked and then left to dry and shrivel in the sun to concentrate its sweetness. It is also used for Málaga, and is found across Andalucía, Valencia and Extremadura. One legend attributes its introduction to a 17th-century Spanish soldier (called, presumably, Pedro Ximénez), who brought it from the Rhine on his return from the Netherlands. Why he thought that a vine that grew in the chilly Rhine would grow in Montilla – the hottest part of Spain – is not clear. Nor can one think of any current Rhine vine with much resemblance to PX, even allowing for the difference in climate. Still, it's a nice story.

Varietal PX sherry is one of the most immediately seductive of dessert wines. It has low acidity and a thick, silky, syrupy texture; its flavour is grapy and raisiny, with only the bare minimum of acidity to prevent it from cloying. In solera it becomes dark to the point of blackness, and can be very concentrated; examples of 60 years old or more, freshly bottled, nevertheless taste very similar to, though more intense than, younger examples. It is far less successful as a table wine grape, and gives flabby, dull-tasting wines with little acidity and no character.

It is grown in Australia, but in even smaller quantities than Palomino, with which it is lumped both in the statistics and, usually, in the blending vat. Best producers: (Spain) Alvear, Pérez Barquero, Domecq, Garvey, Gonzalez-Byass, Lustau, Toro Albalá, Valdespino.

GONZALEZ-BYASS

*Pedro Ximénez like this example has intense grapy flavours – but actually the wine develops very little over time, and while old versions are more concentrated the flavours are the same.*

*Tasting Jurançon Moelleux from a new oak barrique at Château Jolys in south-west France. The main grape used for sweet Jurançon is the intensely aromatic Petit Manseng, with its concentrated flavour of apricots and its high acidity.*

## PERIQUITA

The 'correct' name (if such a thing exists) for this soft, raspberry-flavoured Portuguese variety is Castelão (see page 58), though it is also known as João de Santarém and, in the Alentejo, as Trincadeira. Confusingly, this last synonym, in the Douro, applies to Tinta Amarela. Best producers: (Portugal) J M da Fonseca, J P Vinhos, Pegos Claros.

## PERLAN

One of the many names of Chasselas (see page 59). It is sometimes known by this name in the canton of Geneva in Switzerland, though Chasselas de Genève is more common.

## PERRICONE

Sicilian grape grown for deep-coloured, alcoholic red wines.

## PETIT COURBU

The other name for this grape is, rather unimaginatively, Courbu. It is one of the many old and obscure vines found in Gascony, in south-west France, where it is an ingredient in Pacherenc du Vic-Bilh, Béarn, Jurançon and Irouléguy. It adds richness and lemony, honeyed fruit to the blend. There is also a Courbu Noir vine in Béarn. Best producers: (France) Bru-Baché, Clos Thou, Clos Uroulat, de Souch.

## PETIT MANSENG

Petit Manseng is very similar to its cousin Gros Manseng (see page 113) and is found in the same places – namely between Gascony and the Pyrenees in south-west France. It has smaller berries, as its name might suggest, and is more extreme in its flavours, and more difficult to handle in the winery. It is very low yielding, sometimes giving less than 15hl/ha and like Gros Manseng, able to reach high levels of potential alcohol.

It is even more suitable for sweet wines than its Gros cousin, with the grapes regularly being left on the vine to become *passerillé*, or shrivelled in Jurançon; Gros Manseng, of which larger quantities are at present planted, may well be more suitable for dry wines than Petit Manseng. The latter has similar intense floral, spicy fruit and high acidity, and these flavours, together with its considerable finesse, are making it increasingly popular with growers in other regions. It is appearing more and more in Languedoc and in California, where its high acid profile fits neatly with the fashion for soft, scented Viognier.

Some authorities believe it to be the same as Albariño/Alvarinho (see pages 36–37). It behaves differently in the winery, but that may be simply because it is picked at a riper stage in France than Albariño is in north-west Iberia. Best producers: (France) Bru-Baché, Cauhapé, Clos Uroulat.

## PETIT ROUGE

An obscure variety from Italy's Valle d'Aosta region. It is usually blended. Best producer: (Italy) Les Crêtes.

## PETIT VERDOT

Petit Verdot may be planted only in small quantities in Bordeaux, but it is often highly valued for its colour, structure and lovely violet scent – and it is being regarded with increasing interest in California, Chile and Australia as a useful seasoning for Cabernet Sauvignon, and a wine in its own right.

It fell from favour in Bordeaux because it is so late ripening: it ripens even later than Cabernet Sauvignon, which makes it an impossibility for St-Émilion and Pomerol. It is, however, found in the Médoc and especially in Margaux, where the soils give lighter wines that need the extra tannin and colour provided by Petit Verdot. Château Margaux itself has 6 to 7 per cent Petit Verdot in the vineyard, and most of it goes into the Grand Vin; it has wonderful scent, but it supposedly lacks elegance, so winemaker Paul Pontallier says he would never want more than 10 per cent in the blend. In the 19th century, by contrast, 30 per cent of the vineyard at Château Margaux was planted with Petit Verdot. In Bordeaux generally it is reckoned to reach full ripeness in only one year in five, which makes châteaux in less favoured spots somewhat wary of planting it; in the 1980s this probably changed to one year in three, but the rainy 1990s have been less kind.

Pontallier describes his Petit Verdot as having a banana aroma when young, and developing violet aromas later. Violets are also a keynote elsewhere: in Tuscany, Spain (the Marqués de Griñón has some at Dominio de Valdepusa), Long Island, Chile (Errázuriz is enthusiastic about its potential) and Australia. In Australia's warm, irrigated Riverland Petit Verdot can give even better results than Cabernet Sauvignon, with better acid retention and fresher flavours. Even Australia, however, has areas too cool to ripen it.

There is an unrelated and less good variety known as Gros Verdot, which may still occasionally be found in Argentina. Best producers: (Italy) Castello dei Rampolla; (Spain) Marqués de Griñón; (USA) Araujo, Benziger, Cain Cellars, Jekel, Newton.

## PETITE ARVINE

A high quality grape of the Swiss Valais, which gives wines of elegance and finesse, with tense acidity and unusual minerally, leafy flavour. It demands, and gets, the best sites, and if the grapes are left on the vine until November or even December it can get noble rot. Otherwise sweet or medium-sweet wines can be made from shrivelled berries.

There are also dry versions; all age well in bottle. I should know. I've still got a bit of 1969 under my stairs. The vine is also found over the Italian border in the Valle d'Aosta. Best producers: (Switzerland) Charles Bonvin, Caves Imesch, du Mont d'Or.

## PETITE SIRAH

This grape is sometimes wrongly, and confusingly, spelled Petite Syrah – confusingly because some growers in the Rhône Valley refer to a small-berried version of their (true) Syrah as Petite Syrah.

Well, Petite Sirah is not real Syrah. It is, in fact, a seedling of a little grown French vine called Peloursin and Syrah, but the name of Petite Sirah has become attached to several different grape varieties all of which have been traditionally planted together in California, which is where it, or they, are mostly found today.

One of these varieties is true Syrah, now being grown under its own name; another is Peloursin, and another is Durif (see page 88), a grape that was thought to be identical to Petite Sirah until DNA fingerprinting established that it is a separate variety. In fact, Durif was only propagated in the 1880s in France, and in the early 1880s Petite Sirah was already being talked of in California wine literature. Since most of the 'Petite Sirah' in California is in fact Durif, the confusion must have happened at a later date.

Petite Sirah produces tannic wines which are even darker in colour than those of true Syrah; they have a savoury, almost meaty character and dense blackberry fruit. Its powerful style has long made it a useful blending wine, especially for Zinfandel: Paul Draper of Ridge Vineyards swears by an admixture of 10 to 15 per cent of Petite Sirah in his Zinfandels.

It can also be made as a varietal if the powerful tannins are handled well: it is intense and usually rustic, but good examples can be surprisingly supple and ageworthy, lasting up to 20 years in bottle. Mostly, though, it is a workhorse blending grape, and declined as varietals took over. In 1980 there were 11,000 acres (4450ha) of Petite Sirah in California; by 1999 this had fallen to 3208 acres (1298ha). It is at its best in Sonoma and Mendocino, where unirrigated vineyards of often very old vines produce wines of considerable depth, backbone and brutal power.

Mexico is another source of good Petite Sirah; Argentina and Brazil also have some. Best producers: (Mexico) L A Cetto; (USA) Carmen, Foppiano, Frick, La Jota, Ridge, Rockland, Stags' Leap Winery, Sean Thackray, Turley Cellars.

## PETITE SYRAH

Some Rhône producers like to use this name for a smaller-berried strain of Syrah, see pages 244–255. Petite Sirah (see left) is sometimes incorrectly spelled this way.

## PICOLIT

An overpriced north-eastern Italian variety which may, on occasion, be ravishing enough to justify the hype. But only on occasion, and prices are high.

*An experimental plot of Petit Verdot vines near Lago di Caldaro, Alto Adige, Italy. Varietal versions of the wine do appear from time to time, but usually it is a seasoning for other red varieties.*

The wines are sweet, with apricot and peach flavours and firm, sometimes apply acidity. There may be a flowery note, too, and *passito* wines are full of dried apricot and candied fruit flavours. Some may be aged in oak barriques.

The most intense versions are the *passito* wines, made from grapes picked late and then left to shrivel before being pressed and fermented. There are also late-harvest versions, made from grapes which are picked even later. Either way yields are tiny: the vine is prone to very poor fruit set, which is one reason why there is not much of it planted. It demands good sites, too, on hillsides with volcanic soil.

It has had periods of great popularity in the past, long before its current fashionability: in the 18th century it was exported to the courts of Tuscany, Austria, Holland, Russia, Britain, Saxony and France. Best producers: (Italy) Cà Ronesca, Dario Coos, Dorigo, Livio Felluga, Davino Meroi, Primosic, Bernarda Rocca, Paolo Rodaro, Ronchi di Cialla, Ronchi di Manzano, Ronco del Gnemiz, Le Vigne di Zamò.

## PICPOUL

Picpoul or Piquepoul is found in the Languedoc, usually in its white form, though there is also a Picpoul Noir and a Picpoul Gris. All are old varieties. Picpoul Noir is aromatic and reaches high levels of potential alcohol, but is pale in colour and doesn't age. It is almost extinct in France.

Picpoul Blanc is noted for its high acidity (*piquepoul* means 'lipstinger') and lemony fruit, and any grape with high acidity can be an asset in the hot Languedoc. It is not widely planted, but is useful for blending, and is found in AC form as Picpoul de Pinet near the coast north of Agde. It used to be widely grown for vermouth. Best producers: (France) Pinet co-op, de Gaujal, Genson.

## PIEDIROSSO

This grape, planted in Campania in southern Italy, has been in decline for some years but is part of the blend for the wines Lacryma Christi del Vesuvio and Falerno Rosso. Best producers: (Italy) Feudi di San Gregorio, Galardi, Cantine Grotte del Sole, Luigi Maffini, Mastroberardino, Ocone, Giovanni Struzziero, Villa Matilde.

## PIGATO

A grape found in the Riviera di Ponente zone in Italy's Liguria region which makes sturdy wine with plenty of fruit. All sorts of theories exist about its origins: it could be Greek, or brought by the Romans from central Italy, or be related to Arneis (see page 38) or Vermentino (see page 273); it is often confused with the latter.

It gains its name, which means 'spotted' from the appearance of the ripe grapes. Best producers: (Italy) Anfossi, Bruna, Maria Donata Bianchi, Cascina Feipu dei Massaretti, Parodi, Terre Rosse, La Vecchia Cantina, Claudio Vio.

## PIGNOLO

Pignolo is found in Friuli where it seems to be having something of a revival, albeit a fairly small and quiet one. It deserves to be revived: it balances good acidity with rich blackberry and plum flavours and silky tannins, and it seems to suit barrique aging. It is or is not (opinions vary) the same as Pignola, which is a black grape grown in the Valtellina region in Lombardy. Best producers: (Italy) Dorigo, Davide Moschioni, Le Vigne di Zamò.

## PINEAU D'AUNIS

A grape of the Loire Valley, now in decline but still blended into the reds and rosés of Anjou and Touraine. It has good fruit and adds a certain zesty pepperiness to blends, but other reds, notably Cabernet Franc, are more fashionable and have more commercial potential.

It is an old vine, and has been known since the Middle Ages; it takes its name from the Aunis priory near Saumur, and King Henry III of England imported its light red wine to England in 1246. Anyone who wanted darker coloured wine blended it with something else: the local Teinturier du Cher would have given the quickest results. Best producer: (France) des Hautes Vignes.

## PINOT BEUROT

Burgundian synonym for Pinot Gris (see pages 172–173). A little is still grown on the Côte de Beaune and Hautes-Côtes de Beaune, where it is blended with Chardonnay, and adds extra complexity to the wine. Odd patches survive in some pretty famous domaines, such as Simon Bize, Joseph Drouhin and Comte Senard. Best producers: (France) de l'Arlot, Simon Bize, Bruno Clair, Coche-Dury, Joseph Drouhin, Comte Senard.

## PINOT BIANCO

Italian name for Pinot Blanc (see pages 170–171). It is grown mostly in the North-East, and gives particularly good crisp, light wines in the Alto Adige, the Veneto and Friuli, and makes useful sparkling wines in Lombardy. Plantings of Pinot Bianco in Italy are much greater than those of Pinot Grigio (see below).

Until the mid-1980s the names Pinot Bianco and Chardonnay were considered interchangeable, and the two varieties are still interplanted in some vineyards. The grape name that then goes on the label might, in these circumstances, just possibly owe more to market forces than to strict accuracy. In which case, buy whichever wine is currently cheaper.

## PINOT BLANC

See pages 170–171.

## PINOT GRIGIO

Italian name for Pinot Gris (see pages 172–173). Pinot Grigio is often more highly regarded than Pinot Bianco (see above) in Italy, though the quality of the wine is at much the same level. It grows in the north-east of the country, and is picked for its acidity rather than for the plump richness that distinguishes the variety in Alsace; Italian wines from this grape are accordingly some of the lightest and crispest around, with delicate spice. If vineyard yields are too high, as they frequently are, this delicacy turns to blandness; however, this doesn't stop Pinot Grigio being archetypal Italian restaurant wine, and some of it is really very pleasant.

## PINOT GRIS

See pages 172–173.

## PINOT LIÉBAULT

A local variation of Pinot Noir found in Burgundy. See pages 174–185.

## PINOT MEUNIER

The most common synonym for Meunier (see page 138).

## PINOT NERO

Italian name for Pinot Noir (see pages 174–185). So far Italian examples lag behind Burgundy, California, New Zealand and the best of Australia; the Alto Adige seems to have the most potential.

## PINOT NOIR

See pages 174–185.

## PINOT ST GEORGE

Obscure Californian grape, included here because it is a synonym for Négrette (see page 164) and thus not a Pinot at all.

# Pinot Blanc

I CAN'T THINK OF A SINGLE REGION in the world where Pinot Blanc is regarded as a star grape. It's widely enough grown, but never plays the leading role. I'd say the nearest it gets are some parts of northern Italy, like the Alto Adige, where it makes very nice bright, fresh dry whites – and in Alsace, where, although the label on the bottle won't even mention its name, much of the best Crémant d'Alsace fizz is based on Pinot Blanc. But otherwise it is one of the wine world's genuine Cinderellas, and if there's going to be a ball to attend, Pinot Blanc's invitation hasn't arrived yet.

Yet it's got a fairly decent family tree. It's a mutation of Pinot Gris, which is itself a form of Pinot Noir, and its flavour closely resembles a mild Chardonnay, especially in northern Italy where the Chardonnay is often in any case light, and Pinot Blanc just tastes like an even lighter version. In Germany's Baden region it is called Weissburgunder and is often barrique-aged to quite good effect, and in Austria's Burgenland it produces excellent sweet botrytized wines. In Eastern Europe it is widely grown though with no great distinction.

So we're left with Alsace. But even here it plays at best fifth fiddle to Riesling, Gewurztraminer, Muscat and Pinot Gris, and is often blended with Auxerrois – again, the two grapes look very similar. The ampelographer Galet calls the more productive Alsace version of the vine Gros Pinot Blanc, to distinguish it from other Pinot Blancs.

The New World has not taken to Pinot Blanc in any great way, and clearly prefers the greater glamour of Chardonnay. California has some, though much of what was thought to be Pinot Blanc there turned out to be Melon de Bourgogne, better known as Muscadet. Californian views about how Pinot Blanc should taste vary, so that some wines are as fat as Chardonnay while others are lighter and less assertive. Oregon and Canada have shown a bit more idea of what to do with it, and British Columbia has definitely shown some promise.

So, to be honest, it's difficult to know how good Pinot Blanc could be. In a way its very lack of assertiveness, its mild but bright drinkability in a world awash with Chardonnay, is actually one of its most important characteristics.

### The taste of Pinot Blanc

Pinot Blanc in Alsace has a touch of spice to its round, creamy fruit, but not too much. It is not assertively spicy, or assertively aromatic. In Italy it becomes lighter and more minerally, sometimes with a reasonable pear and apple freshness. In Germany it takes submissively to new oak and makes a fair stab at a Chardonnay style with decent acidity and just enough body to cope with the wood.

*Pinot Blanc may not be allowed as a permitted variety in Alsace's Grand Cru vineyards, but that doesn't prevent Domaine Schoffit's Cuvée Caroline from having the intensity and ripe, honeyed flavours one might expect of a very favoured site. Old vines and low yields are the secret of its success.*

**Bergdolt**

*German examples can have higher acidity than most Pinot Blanc wines, especially when they are Trocken, or dry. This one is from the Pfalz and while the acidity is high it has good, creamy, nutty concentration.*

**Sumac Ridge**

*Sumac Ridge in British Columbia's Okanagan Valley gives this Pinot Blanc a spell in new oak for extra complexity and weight: that's one way of solving the grape's perennial personality problem.*

***Left:*** *Once it was seen that Pinot Noir did well in Oregon's Willamette Valley, local producers started to plant other members of the Pinot family including Pinot Blanc and Pinot Gris. This is the WillaKenzie Estate whose Pinot Blanc vines are planted on Willakenzie soil, known for its excellent drainage.* ***Above:*** *Pinot Blanc's greatest virtue is its subtlety; but the trouble is convincing wine lovers that subtlety on its own is a quality worth shelling out money for. I see their point. 'Subtlety' can so easily become a euphemism for dilution and blandness. Even so, Pinot Blanc is brilliant with food: there are plenty of dishes that don't require fireworks in the wine.*

## CONSUMER INFORMATION

### Synonyms & local names

Sometimes called Clevner or Klevner in Alsace. Known as Pinot Bianco in Italy and Weissburgunder or Weisser Burgunder in Germany and Austria. In Slovenia and the Balkans it may be called Beli (White) Pinot.

### Best producers

**FRANCE/Alsace** Adam, Blanck, Bott-Geyl, Boxler, Deiss, Dopff & Irion, Hugel, Josmeyer, Kreydenweiss, Kuentz-Bas, Meyer-Fonné, René Muré, Ostertag, Rolly Gassmann, Schaetzel, André Scherer, Schlumberger, Schoffit, Turckheim co-op, Weinbach, Zind-Humbrecht; **Burgundy** Maurice Écard, Jadot, Louis Lequin, Daniel Rion.
**GERMANY** Bercher, Bergdolt, Schlossgut Diel, Fürst, Dr Heger, Karl-Heinz Johner, Franz Keller, U Lützkendorf, Reinhartshausen, Sasbach co-op, Dr Werheim.
**AUSTRIA** Feiler-Artinger, Walter Glatzer, Hiedler, Lackner-Tinnacher, Hans Pittnauer, Fritz Salomon.
**ITALY** Buonamico, Colterenzio co-op, Marco Felluga, Jermann, Masùt da Rive, Ignaz Niedriest, Querciabella, Alessandro Princic, Puiatti, San Michele Appiano co-op, Schiopetto, Terlano co-op, Vallarom, Vignalta.
**USA/California** Au Bon Climat, Chalone, Laetitia, Saddleback, Steele, Tyee; **Oregon** Amity, Archery Summit, WillaKenzie.
**CANADA** Blue Mountain, CedarCreek, Inniskillen Okanagan, Mission Hill, Sumac Ridge.
**NEW ZEALAND** St Helena.
**SOUTH AFRICA** Nederburg.

## RECOMMENDED WINES TO TRY

### Ten dry or off-dry Alsace wines

**Blanck** *Alsace Pinot Blanc*
**Boxler** *Alsace Pinot Blanc*
**Josmeyer** *Alsace Pinot Blanc les Lutins*
**Kreydenweiss** *Alsace Pinot Blanc Kritt*
**Meyer-Fonné** *Alsace Pinot Blanc Vieilles Vignes*
**Réné Muré** *Alsace Pinot Blanc Tradition*
**Rolly Gassmann** *Alsace Pinot Blanc Auxerrois Moenchreben*
**Schlumberger** *Alsace Pinot Blanc*
**Schoffit** *Alsace Pinot Blanc Cuvée Caroline*
**Zind-Humbrecht** *Alsace Pinot d'Alsace*

### Five German/Austrian wines based on Weissburgunder

**Bercher** *Burkheimer Feuerberg Weissburgunder Spätlese Trocken Selektion*
**Fürst** *Franken Weissburgunder Spätlese Trocken*
**Dr Heger** *Achkarrer Schlossberg Weissburgunder Spätlese Trocken*
**Lackner-Tinnacher** *Südsteiermark Steinbach Weissburgunder*
**Fritz Salomon** *Weissburgunder*

### Five Italian Pinot Bianco-based wines

**Fattoria del Buonamico** *Vasario*
**Colterenzio co-op** *Alto Adige Pinot Bianco Weisshaus*
**Querciabella** *Batàr*
**San Michele Appiano co-op** *Alto Adige Pinot Bianco Schulthauser*
**Schiopetto** *Collio Pinot Bianco*

### Five US wines

**Au Bon Climat** *Santa Maria Valley Reserve (California)*
**Chalone Vineyard** *Chalone (California)*
**Saddleback Cellars** *Napa Valley (California)*
**Steele** *Santa Maria Valley Bien Nacido Vineyard (California)*
**WillaKenzie Estate** *Willamette Valley (Oregon)*

# PINOT GRIS

WHERE HAVE I HAD MY most memorable Pinot Gris? Was it in Alsace? Was it in Germany? Or in Italy, under its alter ego of Pinot Grigio? Well, these are the three most likely suspects. But might it have been from Switzerland? Might it have been from New Zealand, or Canada or Romania? Ever since I tasted a remarkable St Helena Pinot Gris '87 from New Zealand's South Island, I had a feeling Pinot Gris would shine there. Pinot Gris from both Ontario and British Columbia looks as though it could be the white variety that manages best to ride the 'hot short summer, freezing long winter' that makes Canada so intriguing yet challenging. And Romania? If you'd experienced the sumptuous, honeyed, hedonistic flavours of nobly rotten, sweet Pinot Gris from Murfatlar on the Black Sea coast you wouldn't be smirking behind your hand when I say Pinot Gris has the potential to be one of the world's great white grapes, in some of the world's most unlikely places.

But I still go back to Alsace to find what makes it tick. Here, it makes wine from gloriously sweet to dry but always mellow. And the reason no other region makes Pinot Gris like Alsace has a lot to do with climate. Alsace's dry autumns make long hang times possible; you can pick late for dry or off-dry wines, and even later for sweet ones.

In Alsace, and indeed in Baden, just over the Rhine, it needs ripeness – at least 12.5 per cent potential alcohol at picking – in order to have character. It needs low yields, too, to have quality: no more than 60hl/ha for good wine, or 40hl/ha for great.

In Germany, Pinot Gris does well in Baden and Pfalz and can be called Ruländer or Grauer Burgunder – or, indeed, Pinot Gris. It usually keeps the honeyed quality though rarely reaches Alsace levels. Italy grows a bit of this variety, though rarely wants to showcase its full potential, with the exception of a few growers in Collio in the North-East. Yet Switzerland, Hungary and Romania have all produced wonderful fiery spicy wines in the past, and hopefully will do so again. In the New World, especially in Canada, Oregon and New Zealand, Pinot Gris, because of its honeyed flavour, is seen as an excellent alternative to oaked Chardonnay – and it certainly is. Australia has some vines just getting into gear: it will be interesting to see if the Aussies really go for it full throttle. Let's hope so.

### The taste of Pinot Gris

I adore Pinot Gris' ability to produce fabulously honeyed wines, with the richness of brazil nut flesh and the merest suggestion of something slightly unwashed. (Don't flinch – many of the most memorable wines have something slightly 'incorrect' about them.) Alsace Pinot Gris revels in spicy, musky, honeyed and exotic flavours yet north-east Italy's copper-tinged, relentlessly popular Pinot Grigio with lightly spicy, minerally flavours makes a sort of superior glugger to go with the *fritto misto*. The Canadians and New Zealanders are too new at the game to court the dangerous flavours of decay, and Oregon, in the Pacific Northwest, is happy to exploit the honeyed quality of the grape, but is rapidly showing that a bright, almost spritzy style of white may actually prove to be the state's best wine. In Oregon and New Zealand, fruit flavours prevail – pear and apple, mango and spring flowers – but the best examples always have a flickery honeysuckle scent while Switzerland, Hungary and Romania still hold on to the richer, honeyed style.

*Domaine Weinbach makes some of the most characterful wines in Alsace. This Vendange Tardive or late-harvested version is elegant, concentrated and just medium-sweet, with beautifully balanced acidity.*

DRY RIVER

*New Zealand is rapidly developing an interesting Pinot Gris style based on pear and apple fruit and honeysuckle scent. Dry River, by contrast, is crisp and complex and will develop in bottle for a few years.*

BURROWING OWL

*Pinot Gris is fast becoming one of British Columbia's top white wine varieties. Vanilla, honey and pear flavours blend in this barrel-fermented example.*

***Left:*** *Olivier Humbrecht MW in his Pinot Gris vineyard on the Rotenberg hill at Wintzenheim in Alsace. Finesse is not the principal* raison d'être *of Alsace Pinot Gris, and his Pinot Gris is some of Alsace's weightiest.*
***Above:*** *Pinot Gris grapes, shown here in the early stages of* Botrytis cinerea, *are more susceptible to the fungus than any other noble Alsace variety. There are two strains of Pinot Gris in Alsace: the Gros Grains, which has pretty well taken over, gives much higher yields than the smaller-berried Petits Grains, which has almost disappeared. But Petits Grains gives stellar quality and the new breed of ambitious young growers won't let it disappear completely.*

## CONSUMER INFORMATION

### Synonyms & local names

French synonyms include Pinot Beurot or Burot (Burgundy), Malvoisie (Loire Valley and Savoie) and Tokay-Pinot Gris (Alsace). Called Pinot Grigio in Italy and in Germany Grauburgunder (for dry) and Ruländer (for sweet wines). Switzerland's Valais region calls it Malvoisie and Hungary Szürkebarát.

### Best Producers

**FRANCE/Alsace** Adam, Albrecht, Barmès-Buecher, Beyer, Blanck, Bott-Geyl, Boxler, Burn, Deiss, Dopff & Irion, Hugel, Josmeyer, Kientzler, Koehly, Kreydenweiss, Kuentz-Bas, Albert Mann, Ernest Meyer, Meyer-Fonné, Mittnacht- Klack, René Muré, Ostertag, Pfaffenheim co-op, Rolly Gassmann, Schaetzel, Schlumberger, Schoffit, Bruno Sorg, Trimbach, Turckheim co-op, Weinbach, Zind-Humbrecht; **Burgundy** de l'Arlot; **Loire** Henri Beurdin.
**GERMANY/dry** Bercher, Dr Heger, Karl-Heinz Johner, Müller-Catoir, Klaus Zimmerling; **sweet** Salwey.
**AUSTRIA** Feiler-Artinger, Gross, Schandl.
**ITALY** Avio co-op, Cesconi, Livio Felluga, Lageder, Le Monde, Ronco del Gelso, Russiz Superiore, Schiopetto/Poderi dei Blumeri, Vie di Romans, Villa Russiz. Elena Walch.
**USA/California** Long; **Oregon** Adelsheim, Archery Summit, Chehalem, Evesham Wood, Eyrie, King Estate, Ponzi, Rex Hill, Sokol Blosser, WillaKenzie.
**CANADA** Blue Mountain, Burrowing Owl.
**NEW ZEALAND** Dry River, Gibbston Valley, Kumeu River, St Helena, Seresin.
**SOUTH AFRICA** L'Ormarins, Van Loveren.
**AUSTRALIA** Mount Langhi Giran, T' Gallant.

## RECOMMENDED WINES TO TRY

### Ten dry or off-dry Alsace wines

Lucien Albrecht *Pfingstberg*
Léon Beyer *Comtes d'Eguisheim*
Ernest Burn *Goldert Clos St-Imer*
Kientzler *Vendange Tardive*
Kreydenweiss *Moenchberg Vendange Tardive*
Albert Mann *Vieilles Vignes*
Ostertag *Muenchberg Vendange Tardive*
Schoffit *Cuvée Alexandre*
Zind-Humbrecht *Clos Windsbuhl Vendange Tardive*

### Five Alsace Sélections de Grains Nobles

Bott-Geyl *Alsace Tokay-Pinot Gris*
Hugel *Alsace Tokay-Pinot Gris*
Kuentz-Bas *Alsace Tokay-Pinot Gris Cuvée Jeremy*
Weinbach *Alsace Tokay-Pinot Gris*
Zind-Humbrecht *Alsace Tokay-Pinot Gris Rangen Clos St-Urbain*

### Five German/Austrian Grauburgunders

Bercher *Burkheimer Feuerberg Spätlese Trocken Selektion*
Alois & Ulrike Gross *Südsteiermark*
Dr Heger *Ihringer Winklerberg Spätlese Trocken 3 Sterne*
Müller-Catoir *Haardter Herrenletten Spätlese Trocken*
Klaus Zimmerling *Landwein Trocken*

### Five top Italian Pinot Grigio

Livio Felluga *Colli Orientali del Fruili Pinot Grigio*
Lageder *Alto Adige Pinot Grigio Benefizium Porer*
Ronco del Gelso *Friuli Isonzo Pinot Grigio Sot Lis Rivis*
Schiopetto/Poderi dei Blumeri *Colli Orientali del Friuli Pinot Grigio*
Vie di Romans *Friuli Isonzo Pinot Grigio Desimis*

### Five New World wines

T' Gallant *Pinot Grigio (Australia)*
Gibbston Valley *Pinot Gris (New Zealand)*
Dry River *Pinot Gris (New Zealand)*
Evesham Wood *Pinot Gris Estate (Oregon)*
Burrowing Owl *Pinot Gris (Canada)*

E·R

# Pinot Noir

Winemakers the world over get themselves in an almighty tizz over how their Pinot Noirs should taste. And I'm afraid we wine enthusiasts don't help by endlessly criticizing rather delicious glasses of Pinot Noir wine on the grounds that they don't taste like Burgundy. But which of us can truly say what good Burgundy tastes like?

Is it pale, ethereal, with a scent as sweet and wistful as half-forgotten childhood summer memories hovering over the glass? Is it sensual or heady, swirling with the intoxicating excitement of the super-ripe cherries and strawberries and blackberries of the gardens of Paradise? Is it muscular, glowering, dark as blood, bitter sweet as black cherries and licorice and yet even within this brutish cave, invaded with an insidious, exotic scent?

*Seen through an arched Gothic window typical of those found at the Hospices de Beaune in Burgundy, Pinot Noir's homeland, is the château at Gevrey-Chambertin, one of the Côte de Nuits' best known wine villages. The moonlit scene is a tribute to the word 'Nuits'. On the windowsill are various references to Pinot Noir's many facets – the traditional silver Burgundian* tastevin *or tasting cup, a Champagne cork and wire cage, a cone from the Oregon pine and a Knave of Spades to symbolize Pinot Noir's capriciousness both in the vineyard and the wine cellar.*

And I can come up with a dozen more thoughts on what Burgundy tastes like. And then a dozen more again. And still some of the world's greatest Pinot Noirs would fail to resemble any of them. You know what we so-called experts should do with our Pinot Noirs. Lie back, open our mouths and enjoy them for what they are, not for their possible resemblance to the wines of France's most fickle area of greatness – Burgundy.

In the meantime, Pinot winemakers pursue what has now almost become a cliché – the Holy Grail. And we're back in never-never land. *What* is the Holy Grail? Burgundy? Which Burgundy...? And so it goes on. Well, I suppose one of the benefits of no-one really knowing quite how to coax the best flavours from Pinot Noir is that you never know when you're suddenly going to be faced with a delicious glass of Pinot from a completely unexpected quarter.

And another of the benefits is that whereas it's relatively simple to put a stylistic straitjacket around the two world favourites, Chardonnay and Cabernet, putting a stylistic harness on Pinot is as tricky as wrestling with an eel. While a host of talented but mainstream producers are perfectly happy to try to excel at Cabernet and Chardonnay in accepted styles, Pinot Noir attracts a much wilder bunch. A crowd who don't like being told what to do, a crowd who don't like a marketing manager to have more say in a wine than a winemaker. A self-indulgent crew of men and women who love flavour, who love perfume, who love the silky tactile experiences of a wine like Pinot, seductive, sultry, steamy, sinful if possible, but always solely there, solely made, to give pleasure.

Well, that's why I love Pinot Noir and enthusiastically stalk it round the world. And I've come to a few conclusions. First, some of the best Pinot Noir producers have given up trying to make 'Burgundy'; they've decided to let the grape have its head and see where it ends. The result is some lovely, exhilarating, unusual delights to please any hedonist. Second, it is just about possible to find good inexpensive Pinot Noir. Chile, for a start, is making something of a speciality of it. And third – stop arguing as to whether it tastes like it should. There's no 'should' with Pinot. Get it down you, and if it makes you smile, it's good.

**Pinot Noir: from Grape to Glass**

*Geography and History page 176; Viticulture and Vinification page 178; Pinot Noir around the World page 180; Enjoying Pinot Noir page 184*

# Geography and History

YOU'LL FIND MORE UNCERTAINTY among growers and winemakers about what they think they should be doing with Pinot Noir than with almost any other grape. How to trellis it, prune it, crop it; when to pick it, early or late; what style of wine to make, light or dark; and using which winemaking method, new or old. Doesn't anyone have the definitive method of producing great wine from this grape?

You only have to look at the myriad styles of wine produced in its homeland, Burgundy, and the regular changes of fashion every decade or so, to realize it's not just the newcomers who are confused. And the grape is a most finicky traveller, its inconsistency abroad an amplification of that which it displays at home, to the fascination and despair of those who grow it, vinify it and indeed drink it.

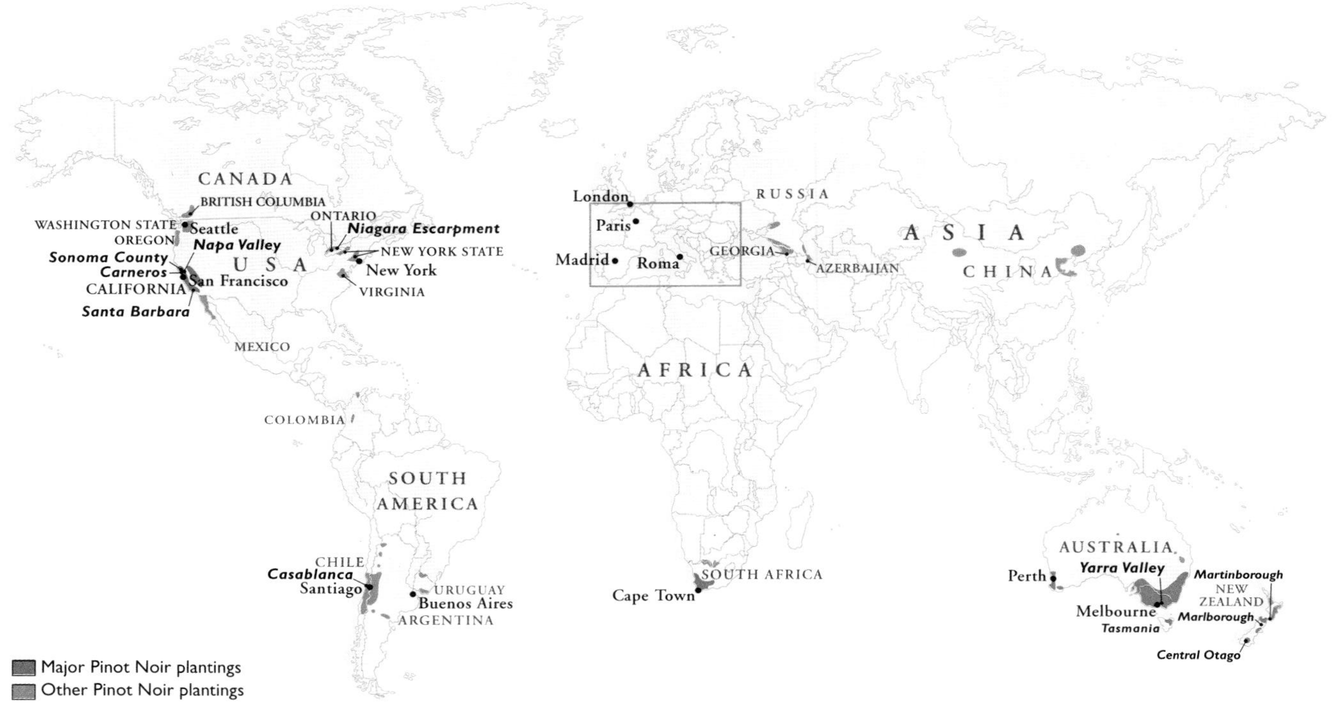

CANADA
BRITISH COLUMBIA
WASHINGTON STATE
Seattle
OREGON
Napa Valley
Sonoma County
Carneros
San Francisco
CALIFORNIA
Santa Barbara
U S A
ONTARIO
Niagara Escarpment
NEW YORK STATE
New York
VIRGINIA
MEXICO
COLOMBIA
SOUTH AMERICA
CHILE
Casablanca
Santiago
URUGUAY
Buenos Aires
ARGENTINA
London
Paris
Madrid
Roma
RUSSIA
GEORGIA
AZERBAIJAN
A S I A
C H I N A
A F R I C A
SOUTH AFRICA
Cape Town
AUSTRALIA
Perth
Yarra Valley
Melbourne
Tasmania
Martinborough
NEW ZEALAND
Marlborough
Central Otago
Major Pinot Noir plantings
Other Pinot Noir plantings

Major planting figures for Pinot Noir

AREA PLANTED (HECTARES)
25,000
20,000
15,000
10,000
5000
2000
France 25,871
Germany 8204
Moldova 8200
California (USA) 6316
Switzerland 4196
Italy 3538
South Africa 3438
Australia 2996
Romania 1480
New Zealand 1427

ENGLAND
NETHERLANDS
GERMANY
BELGIUM
Ahr
Rheingau
LUXEMBOURG
Montagne de Reims
Pfalz
Paris
Alsace
Franken
Champagne
Baden
CZECH REPUBLIC
SLOVAKIA
MOLDOVA
Loire
Burgundy
Côte d'Or
Jura
Côte Chalonnaise
SWITZERLAND
Burgenland
AUSTRIA
HUNGARY
Alto Adige
Trentino
Lombardy
ROMANIA
SLOVENIA
CROATIA
FRANCE
Piedmont
Emilia-Romagna
SERBIA
Languedoc-Roussillon
Douro
Somontano
Aragon
Cataluña
Tuscany
Marche
Umbria
Lazio
BULGARIA
PORTUGAL
Ribatejo
Madrid
SPAIN
Valencia
Roma
ITALY
Puglia
ALBANIA
GREECE
Mediterranean Sea

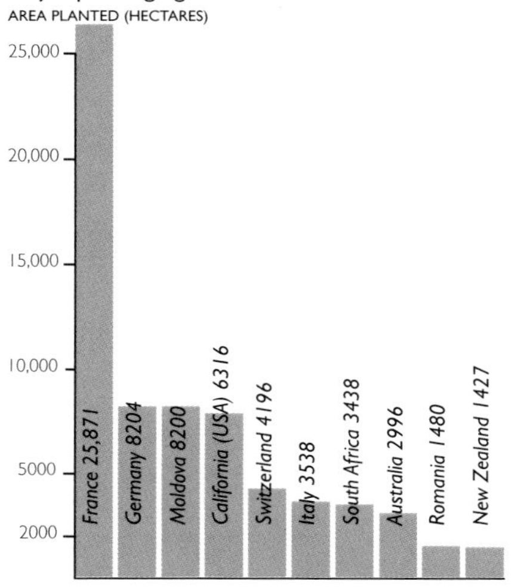

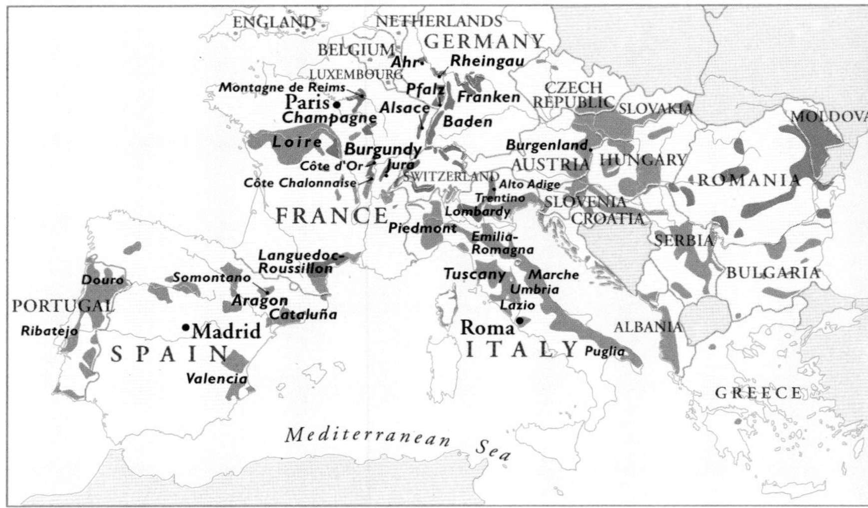

It claims to dislike hot climates, though as an Oregon grower says of Pinot Noir in the Margaret River, Australia, 'One has to rethink one's biases when one sees regions that have Shiraz and Pinot Noir growing next to each other, but producing world-class Pinot.' Too right. Certainly it is supposed to be a cool climate grape, yet some of its most exciting wines in Burgundy have come from seriously hot vintages. Everyone said that cool Oregon would be ideal for Pinot, but some of the USA's most exciting Pinots have come from warmer California. And so it goes on. Warm climate Pinot should be jammy and flat, but isn't always. Cool climate Pinot should be delicate and perfumed, but is equally often green and raw. This unpredictability ought to limit its spread around the globe. But you'd never think so from the map.

The truth is that everybody wants to make Pinot because, like true love, great Pinot is elusive but (perhaps) worth the effort. There is more promising Pinot in the world than there is really great Pinot – yet the 1990s saw a vast increase in the numbers of very good Pinots from the Old and, especially, the New World. True love is commoner than it was.

## Historical background

This is one of the oldest cultivated vines in existence. All cultivated vines owe their origins to wild vines; recent research into its genetic background suggests that Pinot Noir turned domestic some 2000 years ago. For a grape variety that's quite a pedigree.

There is some indication that Pinot was being grown in Burgundy as early as the 4th century AD; however, the first written record of it there dates only from 1375, and just 20 years later Philippe the Bold, Duke of Burgundy, issued an edict banning the inferior Gamay from the Côte d'Or in favour of Pinot.

Its age is the reason for its extreme genetic instability. The Pinot family (Pinot Noir, Pinot Blanc and Pinot Gris are, to ampelographers, all the same grape, because they share the same DNA) is prone to constant mutation. On the one hand this can be to the grower's advantage, since it means that vines will tend to adapt themselves to local conditions; on the other hand it makes it harder to retain the desirable characteristics in particular clones.

Its descendants are legion. Recent work at the University of California at Davis, in collaboration with French specialists, has determined that Pinot Noir is the ancestor of no fewer than 16 modern grape varieties, including Chardonnay, Gamay Noir, Aligoté, Melon de Bourgogne (alias Muscadet) and Auxerrois. Everyone is pretty pleased to have Pinot Noir as a parent. But the Pinot clearly married beneath itself: the other parent of this brood is the undistinguished Gouais Blanc, a grape so obscure that few wine lovers had ever heard of it until this research thrust it into the limelight. Gouais is banned from places like Burgundy and Champagne where Pinot Noir and Chardonnay dominate and, despite its newfound infamy, is likely to remain so. The grape known in California as Gamay Beaujolais is now thought to be a poor, pallid clone of Pinot Noir.

*In the foreground (above) is the Romanée-Conti vineyard in Vosne-Romanée; behind the wall is Romanée St-Vivant. The need to maintain the differences in taste given by such different but neighbouring terroirs has been a spur to the growth of biodynamism on the Côte d'Or.*

*Christophe Roumier of Domaine Georges Roumier, one of Burgundy's top winemakers, aims to coax his vines into perfect balance, so they produce an ideal yield for optimum ripeness.*

*Jacques Seysses tasting from barrel in the cellars of Domaine Dujac. He neither filters nor fines his wines, but says 'I might fine again one day; I might filter one day. I do what needs to be done.'*

# Viticulture and Vinification

THE AIM OF MOST PINOT NOIR PRODUCERS around the world is to make red wine which rivals that of Burgundy's Côte d'Or. That much is straightforward: there is, as yet, no other benchmark style than that of Burgundy, though both California and New Zealand are beginning to display distinctly non-Burgundian virtuosity.

It is less easy, however, to find agreement on what it is about the Côte d'Or, and about Burgundian winemaking, that needs to be imitated. Is it the soil? Is it the climate? And if so, which elements of either? And if you wanted to make great Pinot more consistently than Burgundy manages to, what should you try to improve?

Might it not be a good idea to forget about Burgundy altogether and just follow your instincts? Vanya Cullen of Margaret River said she began making her best Pinot Noir when she gave up doing anything Burgundian or particular to it and just let it stew in its own juice. An increasing number of New World Pinot-makers are indeed starting to follow their own instincts – and a number of New World Pinots are starting to taste very good indeed.

## Climate

Burgundy is cooler than Bordeaux, with a colder winter and a slightly cooler summer, and a larger fluctuation between day and night-time temperatures. There is a risk of frost in spring, abundant rain in May and June, and more rain in October which, if it falls just before or during the harvest, can spoil the quality of the wine. Sounds like an ill place to grow grapes? You're not wrong. Pinot Noir is an early-ripening variety, and sensitive to rot: slow ripening in a cool climate suits it, because its berries are particularly sensitive to heat; yet, paradoxically, it needs quite a bit of warmth to ripen properly. There are generally several years in each decade when Pinot Noir fails to ripen fully on the Côte d'Or, and many of the winemaking fashions there of the last few years have been aimed at extracting more colour and flavour from this not-very-strongly coloured and delicately scented grape.

In California, UCD for many years used to advise growers that there was nowhere in California cool enough for Pinot Noir. Now it has become clear that Carneros, Santa Barbara, Russian River and certain other regions can indeed produce smashing Pinot.

At the other extreme, there are now a few Pinot vineyards in Germany's cool Mosel Valley. To ripen at all here it must steal the best spots from Riesling; even in the warmer Pfalz growers are experimenting with must concentration to get more colour and extract into their wines. In Alsace, just to the north of the Pfalz, Pinot Noir tends to be pallid: perfectly nice as a dark rosé, but barely a red wine at all.

As Myron Redford of Amity Vineyards, Oregon, says, 'I still think the best Pinot Noir needs a cool climate, but my definition of "cool" has been expanded. You cannot grow great wine if the climate is unsuited to the grapes. Once you are in the right climate, however, terroir in the sense of soil, exposure, drainage and microclimate become the critical factors.'

*Lalou Bize-Leroy has embraced biodynamic viticulture at her estate, Domaine Leroy in Vosne-Romanée. She maintains she wants the taste of the terroir, not of the grape, in her Pinot Noir.*

## Soil

Pinot Noir in the Côte d'Or is planted on the more limy soils, though ample supplies of the clay mineral montmorillonite are found in the mid-slopes where the best vineyards are. Montmorillonite has a high cation-exchange capacity, a cation being an atom or molecule with a positive electrical charge; cation exchange is how a plant extracts nutrients from the soil. In addition, the topsoil in the Côte d'Or is shallow, seldom much more than 1–1.5m deep. This aids drainage and the fractured nature of the bedrock means that vines can generally find water even in drought conditions – but where the topsoil is thinnest there can be a greater concentration of virus-spreading nematodes. (It was the problem of viruses in Burgundy that led to the popularity of virus-free clones there; see below.) The high water table at the foot of the slope usually defines the lower boundary of the Grands Crus, the exceptions being Clos de Vougeot and Bâtard-Montrachet, both of which fail to stop in time, and which include land that is too waterlogged for top quality.

Elsewhere in the world shallow, well-drained, low-fertility soils are generally favoured for Pinot Noir. Some focus on clay to give depth to the wine: Hamilton Russell in South Africa believes that people pay too much attention to the limestone in Burgundy, and not enough to its clay. But then Josh Jensen, maker of some of California's greatest Pinot Noirs, says he couldn't do it without his limestone soils.

## Cultivation

The usual density of planting on the Côte d'Or is 1m by 1m, which means 10,000 vines per hectare. Some vineyards are planted even tighter, at 1m by 0.9m, although in the Mâconnais the density can be 8000 vines per hectare. Most of the vines are Guyot pruned, though there are some trials of spur pruning and cordon; short pruning, it is acknowledged, is the main weapon in the battle to keep yields down and concentration up. However, if you prune too short the vine ends up as a bush, with all the fruit tucked in where it won't get the sun. Dominique Lafon of Domaine des Comtes Lafon recommends leaving a longer cane and debudding one or two buds. 'It is better for the canopy and for aeration. You must think of the shape of the vine; shape is more important than the number of buds at pruning, because you need aeration to ward off botrytis.'

Crop thinning, generally done after *véraison* so that the green berries can be removed, is seen as a temporary solution to a problem, not a technique in itself. The ideal is to get the vine in balance, so that it is not overproducing in the first place; among the best growers, 35hl/ha is seen as the ideal yield for Pinot Noir. At Domaine de la Romanée-Conti average yields are even lower, at around 24hl/ha.

In New Zealand and Oregon around 2 tonnes/ha is a generally accepted figure for producing ripe Pinot. Compare that with yields for sparkling wine: in Champagne in the excellent 1990 vintage, yields averaged 110hl/ha. Houses that own their own vineyards generally use lower yields than growers who sell their grapes. Compare that, too, with Chile, where 8–10 tonnes/ha is considered low, and 12–13 tonnes/ha is not uncommon.

### Clones

Pinot Noir is notorious for its clonal variation. Some clones give pale-coloured, frail-tasting wine; others give robust, jammy flavours and dark colours; some give elegance, some chunkiness; some give large crops, some small. The 1970s were a bad time for Pinot Noir clones: high yields were the object, not high quality. New generations of clones, particularly the Dijon clones developed in Burgundy, are gradually improving the picture as they move into vineyards across the world. But change doesn't happen overnight, and even on the Côte d'Or there is likely to be a higher proportion of the high-yielding Pinot Droit than many Burgundians care to admit. (Lower-yielding strains are Pinot Fin or Pinot Tordu.) On the Côte d'Or alone over 1000 different clones have been noted.

Some Burgundian growers, having planted new-generation clones, found them too high-yielding and have moved to massal selection from their own vineyards, following the progress of each vine for at least five years before planting cuttings from it. Olivier Merlin in the Mâconnais has moved on again: he takes cuttings from the total population of his vineyards, omitting only diseased plants. 'Complexity from diversity' is his watchword. Those who disagree believe that complexity comes from the terroir, and that greater vine age, plus good vineyard management, tends to eradicate differences between clones, even clones of Pinot Noir.

### At the winery

While good winemaking is absolutely essential for Pinot Noir, it is a curious fact that what makes Pinot Noir interesting, rather than merely correct, is a series of low-level faults: aromas and flavours that are just slightly off squeaky clean. While Burgundians have generally reached this point by making their winemaking more correct, winemakers in some other countries perhaps have to become a little less correct if they are to make great Pinot.

The norm in Burgundy is to destem all or most of the crop, and ferment a high proportion of whole berries, without a pre-fermentation crush. Only a few estates (including, it should be said, some top ones) leave the stems in. Chaptalization is usual, and acidification not uncommon; it is illegal, though not unknown, for both to be done.

A pre-fermentation maceration (sometimes known as a cold soak) is often practised; it is now referred to as 'a traditional technique brought back to people's attention by Guy Accad', the fashionable consultant winemaker of the late 1980s who advocated an extreme form of this maceration. It usually lasts just a few days, while the wild yeasts (cultured yeast is little used in Burgundy) slowly get moving. The point of it is that at 15°C enzymes work on the cells of the skins to produce better aromas and tannins; it can be likened to tomatoes ripening on a windowsill. It gives more depth of flavour, and more detailed flavour. Fermentation temperatures usually rise to 30–32°C, and the total time in tank, including any post-fermentation maceration, varies from two to three weeks.

### Filtration

This does no harm to the wine if done gently and properly. However, the influence of the American critic Robert Parker, who is strongly opposed to filtration, means that many producers say they no longer do it.

#### THE CABERNISATION OF PINOT NOIR

How light should Pinot Noir be? How rich? Winemakers have been asking themselves this question ever since the early 1980s saw much lighter, perfumed, less weighty red Burgundies take over the market. Some consumers mourned the demise of the big, soupy wines of yore. (Go back far enough and the reason they were big and soupy was because they contained a good dose of red wine from elsewhere – usually the southern Rhône or North Africa.) Others relished the elegance and delicacy of the new wines, and saw them as authentic representatives of their terroirs.

But fashion in all other red wines currently focuses on big, rich flavours – like those of Cabernet Sauvignon. To be certain of selling in the international marketplace, wines must have huge but soft tannins, massive fruit, colossal intensity. Light reds are for wimps; they don't grab headlines, and they don't win blind tastings.

Accordingly more and more Pinot producers feel the pressure to make big, solid wines. And it's not just happening in California, where individual critics' opinions carry so much weight. In Burgundy too, Cabernisation is a factor. Take the 200 per cent new wood of *négociant* Dominique Laurent. Less than half new wood is usual in Burgundy, with the weightiest wines getting more than that, and the lighter wines less. About 18 months in barrel is normal. Laurent's policy is to age his wines for a year in new oak, and then rack them into more new barrels. His object, he says, is not to produce hugely tannic, oaky wines: instead he aims to produce dark, voluptuous, rich wines from perfectly balanced, perfectly ripe grapes; wines that reflect their terroir. New oak, he believes, gives better oxidation, better interaction between the lees and the wine, and slows down the aging process. It also gives deeper colours and fixes the aromas. Sure. And adds massively to the tannins. Only the biggest wines actually get 200 per cent new oak; he does not follow a precise recipe, and even his critics say that he uses oak with care and sensitivity.

Does Laurent's oak represent a fundamental shift of opinion? Will it, like Guy Accad's cold soak (see above), become 'a rediscovered traditional technique' and, in a modified form, enter the repertory? Will red Burgundy change its nature again and become a bigger, beefier wine? Only time will tell.

# Pinot Noir around the World

PINOT NOIR MAY HAVE a wide geographic spread, but its fussiness about soil and climate means that it is, in most places, in a minority. Even in California, where plantings are rising fast enough to make the state certain to become the world's biggest Pinot producer any moment now, it is in no danger of outpacing Cabernet Sauvignon.

## Burgundy

The Côte d'Or is the heart of Burgundy as far as Pinot Noir is concerned; the surrounding regions produce what is often referred to as 'affordable Burgundy', which is at least a tacit admission that it is neither cheap nor usually particularly good value compared to other wines. Such bottles can be very attractive but do not expect them to taste like the Pinots of the Côte d'Or.

In the Auxerrois to the north, Pinot is light and perfumed; in the Côte Chalonnaise to the south it is, at its best, richly fruity with good structure, but not much complexity. In the Mâconnais Gamay is the main red grape, and Pinot Noir vines are often poor clones.

In Burgundy, and especially in the Côte d'Or, growers' reverence is reserved not for the grape variety but for the terroir. Burgundians do not look first and foremost for varietal character in their Pinot; indeed, some claim not to want varietal character at all. The vine is merely a conduit for the terroir.

But how do you express that terroir? Biodynamists in Burgundy (a growing number) believe that the over-generous application of nitrogen and phosphorus fertilizers in the 1970s and beyond had the effect of masking the differences in terroir – a serious matter, since it is on terroir that the whole elaborate map of Premiers and Grands Crus is based. So it is not surprising that biodynamism is so influential here. It is helped by the small size of most Burgundian domaines, and by the high price of the wines, which means that the extra expense and manpower involved can be afforded and eventually recouped.

Ideally, then, each Côte d'Or vineyard should have its own recognizable flavour. To a large extent that is true – in Volnay, for example, Champans is more structured and dense than Santenots but the large number of owners in each vineyard, with the Grand Cru of Clos de Vougeot, for example, having over 50 different growers owning vines and making wines that bear the vineyard name, means that the terroir is seen through the grower as well as through the grape. When the producer is good this adds another layer of complexity; but, inevitably, all producers are not of equal quality. The unevenness of Pinot Noir grape quality is mirrored, and indeed exacerbated, by unevenness among producers. It is true, however, that standards of viticulture and winemaking have risen enormously in the last ten years.

The Côte de Nuits produces bigger, chewier, plummier, more solid reds than the Côte de Beaune. Gevrey-Chambertin is muscular, even pruney; Morey St-Denis, with a clutch of top-class growers, is lean yet savoury and complex; Chambolle-Musigny is all roses and violets. Vosne-Romanée is violets and cream, Nuits-St-Georges black cherry and chocolate. In the Côte de Beaune, Aloxe-Corton is savoury and perfumed, Savigny-les-Beaune scented with strawberries; Beaune is round and soft. Pommard is fatter, Volnay fragrant, and Chassagne-Montrachet quite chunky. Well, that's the idea. But for every example where you cry, 'Yippee, it tastes like it should', you'll find far more which leave you staring blankly into the middle distance fingering your grievously abused wallet.

## Champagne

This cold region might not, at first glance, seem to be ideal for the early-budding Pinot Noir. The grape is horribly susceptible to spring frosts here, and clonal research in Champagne is directed to producing more frost-resistant vines. Pinot Noir occupies some 37.5 per cent of the vineyards overall, and plantings are concentrated on the Montagne de Reims between Reims and Épernay – even, in this marginal climate, on certain north-facing slopes. Not very steep slopes, admittedly, but it is still surprising to find Grand Cru vineyards in such a position. Perhaps the chalk reflects enough heat to warm the air, and compensate?

Pinot Noir in Champagne never reaches what any other region would describe as ripeness. The warmest, south-facing sites are usually the best, but the undulating hills offer many different mesoclimates. The village of Ay, for example, is said by local growers never to get frost in spring – or perhaps hardly ever. The Marne valley broadens out on one side there, and the cold air can slip away – unlike at Épernay, a few kilometres away, where the valley retains the cold air. The river Marne

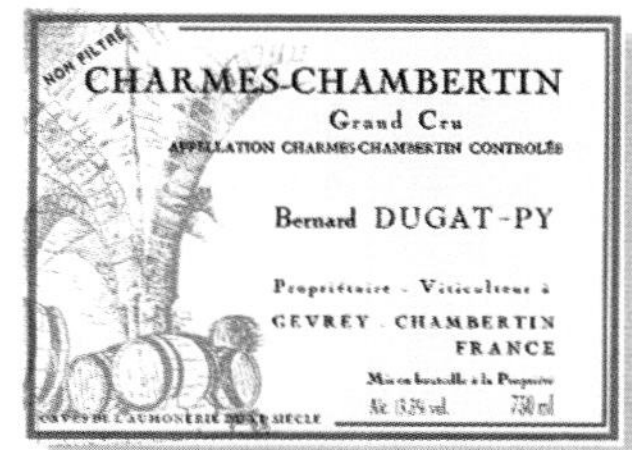


BERNARD DUGAT-PY

*Bernard Dugat-Py owns about 0.25ha of the Grand Cru Charmes-Chambertin and has been bottling his fresh, pure-fruited wines since 1989.*

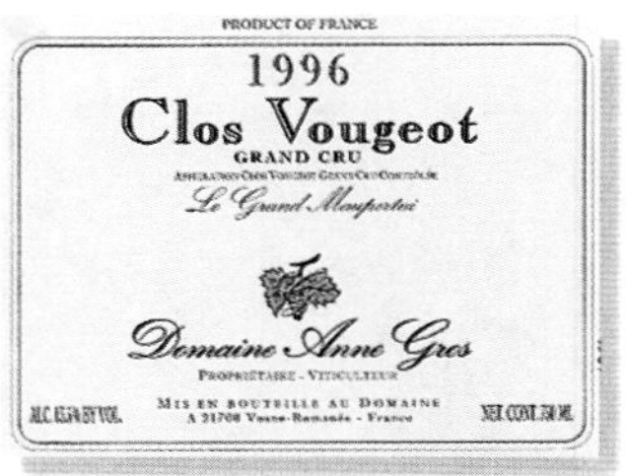


DOMAINE ANNE GROS

*Umpteen members of the Gros family make wine under their own labels. Anne's style is for deep-coloured wines which are balanced and elegant.*

DOMAINE ARMAND ROUSSEAU PÈRE ET FILS

*Domaine Armand Rousseau is one of the great names in Burgundy, making very traditional, long-aging wines of great individuality.*

*Pinot Noir vines on the Montagne de Reims above the Grand Cru village of Verzenay in Champagne. Pinot Noir is only possible here because Champagne needs base wines of aroma and finesse, not richness, tannin and deep colour, so barely ripe, pale Pinot Noir fits the bill nicely.*

runs right at the foot of the slope, too, which is the closest it gets to any of the Montagne vineyards: at Épernay, they say, there can be fog until 10 am when it's sunny in Ay.

Picking is by hand, and Pinot Noir must be pressed carefully and quickly to avoid too much colour getting into the wine. The still wine usually has a pale pink tint, but this rarely shows in the finished wine: the colouring matter combines with the dead yeasts in the bottle after fermentation but before the wine is disgorged. And of course Pinot Noir is usually blended with white Chardonnay. Only a few still red wines are produced, from Bouzy or Ay, where the wines are ripest and have the darkest colour, but they need a very warm year to shed their tartness. Blanc de Noirs – white wine made from black grapes – is thus Pinot Noir's best style in Champagne.

Varietal character is no more the object here than it is in Burgundy – in fact probably even less so since the grapes are less ripe. Much of the flavour and style of Champagne comes from the aging period on the lees; and while reflecting the terroir of an individual vineyard is seldom possible, the wine is certainly strongly marked by the terroir of the Champagne region as a whole.

## Other French Pinots

No other French regions get close to the ripeness and complexity of red Burgundy. There is some Pinot Noir in the Loire, most notably in Sancerre and Menetou-Salon, where in ripe years (these were frequent in the 1980s and 1990s) it can make wines with good fruit and structure. Most red Sancerre is, however, light and made to be drunk young; to make *vin de garde* here you have to dedicate your very best sites to Pinot Noir rather than to Sauvignon Blanc.

Alsace Pinot Noir is generally light – sometimes hardly more than dark rosé – but Pinot Noir here, often planted on light, fertile soils, has nearly doubled its area in the last ten years. Pinot Noir from the Savoie and Bugey regions is light and aromatic.

## Germany

The best Spätburgunders (as Pinot Noir is known here) tread a line between the traditional German pale, sweetish style and the trap of overoaking and overextraction. The latter is a natural reaction to the problem of producing fashionably ripe reds in a country with no history of making such a style, and with only a few places where black grapes will ripen sufficiently.

The most successful Spätburgunders come from Baden, the Pfalz and the Rheingau, but in the latter they have to compete for the best sites with Riesling. Rheingau producer Robert Weil has some Spätburgunder in a site too hot for Riesling, but usually a grower must choose between the two grapes. Some of the most attractive reds, particularly in Franken, are blends of Spätburgunder with other grapes such as Blauer Portugieser. You don't get much Spätburgunder character coming through, but they're nice, ripe, cherry- and raspberry-flavoured wines.

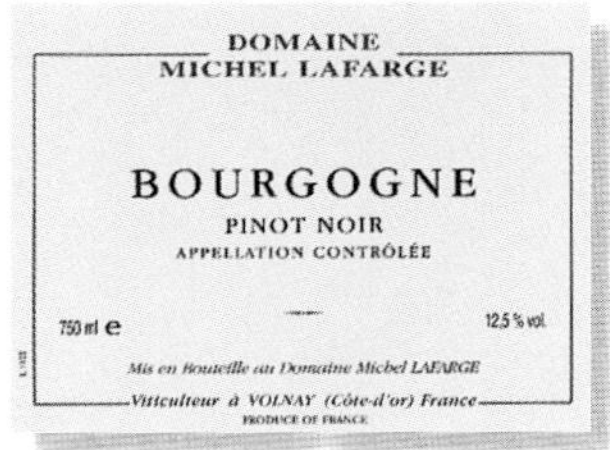


**DOMAINE MICHEL LAFARGE**

*Buying good Bourgogne Rouge can be just as tricky as buying Premier or Grand Cru Burgundy. This stylish example from Volnay has rich, ripe fruit.*

**PRODUCTEURS DE MAILLY**

*This co-operative group in the village of Mailly on the Pinot Noir-dominated Montagne de Reims produces consistently good, full-bodied Champagnes.*

**KARL-HEINZ JOHNER**

*Johner is as passionate about new oak as he is about his grapes and so produces international-style Pinots from his vineyards in Baden, Germany.*

## Austria

Blauer Burgunder (as Austria may call Pinot Noir) and other reds are on the increase in the warmer Burgenland, at the expense of less fashionable white varieties like Welschriesling. Many reds are blends of several varieties, such as Zweigelt, Blaufränkisch, Sankt Laurent and Cabernet Sauvignon. Overoaking and overextraction can be a problem, but hopefully only a temporary one as growers learn how to handle oak more deftly. Burgenland also has some oddities like Spätburgunder Trockenbeerenauslese: rich, white and sweet, they are rare and command high prices – but they're treacly, low acid oddities rather than masterpieces.

## Italy

Pinot Nero is hardly a new introduction here: it was first planted in Piedmont in 1825, though its pale colour and low acidity did not endear it to consumers, and its susceptibility to rot did little for its image among growers. Only in the last couple of decades have growers begun to look at better clones and better training systems (generally Guyot rather than pergola). It's very much a northern speciality, and quite a lot goes into sparkling wine, especially in Lombardy. On the still front, the west-facing slopes of Alto Adige east of the river seem to be particularly successful, helped by early morning shade and plenty of sun the rest of the day: one of the challenges has been finding spots with high fluctuations between day and night-time temperatures. In the North-West, blends with Barbera, such as Giacomo Bologna's Bacialè, are highly fashionable and often very good.

## Rest of Europe

Spain has one or two respectable Pinot Noirs from Somontano, and there are moves to get it added to the list of approved grape varieties in Rioja. In Switzerland it is mostly blended with Gamay to make Dôle, though varietal Pinot Noir can be light, strawberryish and quite good here. In Eastern Europe it is planted widely though not well, and few examples from Romania, Hungary, Slovakia or Bulgaria really deliver. Most are too soft and jammy, or too woody, or both, and merely prove that Pinot Noir is rarely both cheap and good.

## USA: California

Outside Burgundy the real Pinot Noir excitement is in the New World. So far California is ahead, with wines of richness and complexity, plus that elusive, fascinating fragrance and silkiness that are the most difficult to attain of all Pinot Noir's attributes. Lesser wines settle for charm and supple fruit. Most of the state is too hot for Pinot Noir, but Carneros, where the temperature is 3–5 degrees cooler than the northern part of the Napa Valley, and where the lack of rich humus in the soil produces less vigorous vines and lower yields, seems to be able to do the business.

Carneros wines have wild strawberry fruit flavours, without a great deal of weight in the middle palate; other parts of Napa can give earthier, leathery flavours. In Sonoma County, Russian River has the weight that Carneros can lack, and darker, blackberry flavours. Further south, in Santa Barbara County, Santa Maria can be weightier and more complex again, with black cherry and plum fruit.

It is in Santa Barbara, too, that the biggest increase in plantings of Pinot Noir – 57 per cent – has been seen in the last ten years.

**California Pinot Noir**

*Lane Tanner (right) is one of the best winemakers in California when it comes to this most subtle of grapes. California's success with Pinot Noir has come as a bit of a surprise: it has easily overtaken Oregon, which has long been hyped as the USA's answer to the Côte d'Or. Clearly, coolness of climate is not everything. But California has not established a distinct Californian style of Pinot. Does that mean it's doing a me-too? Yes and no. On the one hand, aiming at the top wines of the Côte d'Or is no bad thing – it's what California did with Chardonnay, after all. But with Chardonnay California ended up establishing more than one highly successful style. Pinot Noir seems to be different: it is, we know, far less obligingly malleable than Chardonnay. But it's worth noting, too, that stylistic differences between growers and areas in California are every bit as great as they are in Burgundy. As ever, great Pinot Noir is impossible to pigeonhole.*

CALERA

*In the early 1970s Josh Jensen planted Pinot Noir on limestone soil in California when the accepted wisdom was that it would never work there. He saw limestone as the key when everyone else there thought climate mattered most. He's been triumphantly vindicated.*

Pinot Noir is on the increase in every wine-growing county except Napa, which has a fairly constant 2400 acres/971ha: Sonoma is up 45 per cent to 4800 acres/1943ha, and the rugged Sonoma Coast, producing darker colour and brighter fruit, is attracting attention, even though the margin of error on this wild coast is small. And with large companies planting Pinot hundreds of acres at a time, the potential for both quality and quantity is very exciting for drinkers – though cult wineries may make Pinot Noir in smaller quantities even than Burgundy.

Some of the improvement in quality in California can be put down to better clones: some of the early clones had been developed for less warm and fertile places. Exposed to the Californian sun they produced wines that lacked the silkiness and finesse of great Burgundy. New clones, especially Dijon clones, seem to be better adapted, though site, vine age and grower are probably more important to the final result.

For sparkling wines the prime sites are cool ones with shallow soil and not too much sun: Schramsberg puts a premium on cool breezes to keep the temperature down. Russian River, Anderson Valley and Carneros all produce good sparklers.

### USA: Oregon

Charm and suppleness are generally the order of the day here, with a few producers making serious, structured, long-aging wine. But quality is still erratic. Low yields are essential for quality, and Pinot Noir tends to get the best, warmest sites. Favoured soils include a clay loam called Jory, which is well drained and has high cation-exchange capacity, and gives wines with bright cherry fruit; Nekia soil, which is fairly similar and known as Aiken in California; and a clay loam called Willakenzie, which needs both drainage and irrigation. Nekia and Willakenzie give black cherry fruit and bigger, fuller tannins. There are about 300,000 acres/121,410ha of Nekia and Jory in the foothills of Oregon, and overall some 3689 acres/1493ha of Pinot Noir in 1999.

### Canada

Canada is not a big player in the Pinot Noir stakes, but a new joint venture between Canadian company Vincor and Burgundian *négociant* Boisset to plant Pinot Noir on the Jordan Bench in the Niagara Escarpment might be promising. We'll have to wait until 2005, when the first wines are due to be released, to find out.

### Australia

Australia started off by planting Pinot Noir in sites that were too hot, where it produced baked, jammy wines. It's still only 1.5 per cent of production in Australia, but greater vine age (the average was still under 10 years in 2000), better understanding of yields and cultivation and, as winemaker Michael Hill Smith puts it, 'lots of expensive bottles of Burgundy drunk in the middle of the night', are all paying off. The general quality level is improved, with a few wines of exceptional quality and complexity. The areas around Melbourne are in the lead, especially the Yarra. Isolated producers elsewhere make good Pinot Noir. A lot is also used, blended with Chardonnay, for sparkling wine of high quality and often fuller fruit than Champagne.

### New Zealand

There is no shortage of cool climate regions here, and Martinborough, in the south of the North Island, has carved out a reputation as Pinot Noir capital of New Zealand, making elegant, fairly light styles from old river terraces above the flood plain. Canterbury gives equally good fruit though Marlborough has more body, particularly in Wairau Valley, where the soils retain water and ripening is slower and more even. Stony soils in much of Marlborough tend to advance ripening too quickly. Central Otago in the South Island also produces fine examples, including one or two of New Zealand's best. By 2001 1017ha of Pinot Noir are due to be planted, from a mere 234ha ten years before.

### Chile

Chile, despite seeming too hot for Pinot and being prone to excessive yields, has managed to capture Pinot Noir's silky texture surprisingly well. Flavour can be on the lighter side but the feeling of ripeness can be positively seductive. In terms of climate Casablanca, which is most producers' favourite Pinot region, is comparable to California's Santa Barbara. Other parts of the coastal hills could also be good. In 1998 589ha were planted.

### South Africa

South Africa is generally hotter for Pinot Noir than most competing New World wine regions. The grape has also suffered from leafroll virus, as do most reds here. The few producers who get it right produce round, rich Pinots of considerable Burgundian stylishness, but plantings comprise less than 1 per cent of the total.

**DOMAINE DROUHIN**

*Domaine Drouhin, the Oregon offshoot of the Burgundy négociant and grower, makes the most Burgundian wines in the Pacific North-West. Laurène is named after the first child of the winemaker, Veronique Drouhin.*

**GIACONDA**

*A tiny winery with cult status, Giaconda is in the far north-eastern corner of Victoria and produces only about 1000 cases each year.*

**ATA RANGI**

*The 1998 vintage at Ata Rangi saw a change to a more tannic, longer-lived style. Perhaps this is an indication of the direction in which the Martinborough region in general might evolve.*

**CONO SUR**

*The Chilean producer, Cono Sur has taken on a Burgundian winemaker as consultant: expect a more Burgundian style here from the 2000 vintage onwards and perhaps even more sensuous fruit.*

# Enjoying Pinot Noir

HOW WELL DOES PINOT NOIR AGE? It's a question that baffles many lovers of the grape. In Burgundy alone there are wines that will last up to a couple of decades and wines that seem hardly able to last the year. Admittedly the two types do not usually come from the same producers, though sometimes they do come from the same communes.

Almost all Pinot Noir is delicious young; some is delicious mature. In between there is a period of adolescence, when the wine can be closed, even gawky, and can be as awkward at table as human adolescents. Watch out for that, and when it happens tuck your wine away for a couple of years: it is the only stage at which the grape does not give enormous pleasure. Otherwise you can catch its youthful fruit and perfume by opening it within months of bottling or you can wait until it is grown up and full of savoury spice.

Lesser vintages on the Côte d'Or should often be drunk within three to eight years: if a vintage is described as 'early drinking', then don't delay. Only the bigger, oakier wines are a little unfriendly in youth, as the oak hides the fruit.

Most Pinot Noirs from outside the Côte d'Or should be drunk young, within five years or so. The exceptions are the weightiest wines from regions like California's Santa Barbara, where they last longer: a decade should not be a problem for these.

Champagne is different again: Blancs de Noirs can be sturdy wines, though their aging ability depends partly on how much Pinot Meunier the blend contains. Most focus on weight and breadth over long ageability – and almost all are non-vintage. So keep them for another 12 months or so after you buy them, and then get the glasses out.

## The taste of Pinot Noir

There's always something mouthwatering about good Pinot Noir; something that makes you want to go back to the glass again and again to pin down those elusive flavours. Because descriptions like strawberry, black cherry, game, leather, mushrooms, don't really tell you what it tastes like. Pinot is all these things, and none of them, sliding from one flavour to another. Words like 'complex', 'ethereal' or 'profound' may tell you even less about what it tastes like, but sometimes they are the only words that will do.

Simple Pinot wines are the easiest to describe. Less expensive fruit-first wines from the outlying parts of Burgundy, from Switzerland, from Carneros, from New Zealand, really do taste of strawberries. They are generally low in acidity, though the north of Burgundy can be an exception here, and low in tannin. Finer examples combine strawberries or black cherries with sensuous yet focused flavours, perhaps with a touch of incense and spice, sometimes with a pungent, gamy richness to balance the fruit.

Mature wines gain flavours of leather and woodsmoke, game and undergrowth, even a touch of rotting vegetables. Primary fruit is less important, though the wines should always taste fruity and perhaps slightly sweet. But labelling a mature Pinot with a particular type of fruit is usually quite impossible; every time you think you've caught the precise flavour, it's moved on to something else.

*Robert Chevillon's highly-rated Nuits-St-Georges domaine produces chewy, characterful wines from old vines. Half a world away, Dr Neil McCallum's Dry River is one of the leading names in Martinborough, New Zealand. Both estates succeed with Pinot Noir because of meticulous viticulture and winemaking.*

## Matching Pinot Noir and food

The great grape of Burgundy has taken its food-friendly complexity all over the wine world. However, nothing can beat the marriage of great wine with sublime local food that is Burgundy's heritage, and it is Burgundian dishes that spring to mind as perfect partners for Pinot Noir: coq au vin, chicken with tarragon, rabbit with mustard, braised ham, boeuf bourguignon... the list is endless.

Pinot Noir's subtle flavours make it a natural choice for complex meat dishes but it is also excellent with plain grills and roasts, and with most dishes based on mushrooms. Richer examples are the ideal match for roast or casseroled game birds, and in its lighter manifestations from, say, the Loire or Oregon, Pinot Noir is a good match for salmon or salmon trout.

## CONSUMER INFORMATION

### Synonyms & local names

Also known in France as Noirien or Pineau (and as Savagnin Noir in Jura), and as Spätburgunder or (less common) Blauburgunder or Blauer Burgunder in Germany, Pinot Nero in Italy, Blauburgunder or Blauer Spätburgunder in Austria, Burgundac Crni in Croatia and Serbia.

### Best producers

**FRANCE/Burgundy *(growers)*** Ambroise, d'Angerville, Comte Armand, Denis Bachelet, G Barthod, J-M Boillot, Chandon de Briailles, Charlopin, R Chevillon, Clair, J-J Confuron, Pierre Damoy, Dugat-Py, Dujac, Engel, Henri Gouges, Anne Gros, Grivot, Hudelot-Noëllat, Jacqueson, Lafarge, Lafon, des Lambrays, H Lignier, Méo-Camuzet, Montille, Denis Mortet, J-F Mugnier, Pavelot, Ponsot, Potel, Rion, Dom. de la Romanée-Conti, E Rouget, Roumier, Rousseau, Tollot-Beaut, de Vogüé; ***(merchants)*** Bouchard Père, Champy, Drouhin, Faiveley, Jadot, V Girardin, Labouré-Roi, D Laurent, Maison Leroy, Remoissenet, Rodet.
**GERMANY** Bercher, Rudolf Fürst, Karl-Heinz Johner, Rebholz.
**ITALY** Ca' del Bosco, Hofstätter, Marchesi Pancrazi, Castello della Sala.
**USA/California** Au Bon Climat, Calera, Chalone, Dehlinger, Etude, Gary Farrell, Fetzer, Hartford Court, Landmark, Littorai, Lane Tanner, J Rochioli Vineyards, Saintsbury, Sanford, Talley, Williams Selyem; **Oregon** Archery Summit, Beaux Frères, Bethel Heights, Cristom, Domaine Drouhin, Domaine Serene, Panther Creek, Ponzi, Rex Hill, Torii Mor, WillaKenzie, Willamette Valley Vineyards, Ken Wright.
**AUSTRALIA** Ashton Hills, Bannockburn, Bass Phillip, Coldstream Hills, Cullen, Diamond Valley, Freycinet, Giaconda, Lenswood Vineyards, Moorooduc, Paringa Estate, Plantagenet, Yarra Yering.
**NEW ZEALAND** Ata Rangi, Dry River, Felton Road, Isabel, Martinborough Vineyard, Neudorf, Palliser, Rippon, Seresin.
**SOUTH AFRICA** Bouchard Finlayson, Hamilton Russell.
**CHILE** Carmen, Cono Sur, Gracia, Valdivieso, Villard Estate.

## RECOMMENDED WINES TO TRY

### Ten top red Burgundies

Marquis d'Angerville *Volnay Clos des Ducs*
Joseph Drouhin *Beaune Clos des Mouches*
Faiveley *Chambertin Clos-de-Bèze*
Jean Grivot *Vosne-Romanée les Beaux Monts*
Louis Jadot *Gevrey-Chambertin Clos St-Jacques*
Domaine des Lambrays *Clos des Lambrays*
Maison Leroy *Richebourg*
Méo-Camuzet *Corton*
Domaine de la Romanée-Conti *La Romanée-Conti*
Comte Georges de Vogüé *le Musigny*

### Ten red Burgundies

Bouchard Père *Chorey-lès-Beaune*
Philippe Charlopin *Marsannay En Montchevenoy*
Faiveley *Mercurey Domaine de la Croix Jacquelet*
Vincent Girardin *Maranges Clos des Loyères*
Henri et Paul Jacqueson *Rully les Cloux*
Louis Jadot *Côte de Beaune-Villages*
Michel Lafarge *Bourgogne Rouge*
Jean-Marc Pavelot *Savigny-lès-Beaune la Dominode*
Nicolas Potel *Bourgogne Rouge*
Remoissenet Père et Fils *Santenay*

### Ten top New World Pinot Noirs

Ata Rangi *Pinot Noir (New Zealand)*
Au Bon Climat *Sanford & Benedict Vineyard Pinot Noir (California)*
Beaux Frères *Pinot Noir (Oregon)*
Calera *Jensen Pinot Noir (California)*
Felton Road *Pinot Noir (New Zealand)*
Giaconda *Pinot Noir (Australia)*
Hamilton Russell *Pinot Noir (South Africa)*
Lenswood *Pinot Noir (Australia)*
J Rochioli *West Block Reserve Pinot Noir (California)*
WillaKenzie *Pierre Leon Pinot Noir (Oregon)*

### Five good-value New World Pinot Noirs

Carmen *Reserve Pinot Noir (Chile)*
Fetzer *Pinot Noir (California)*
Martinborough Vineyard *Pinot Noir (New Zealand)*
Saintsbury *Garnet Pinot Noir (California)*
Valdivieso *Reserve Pinot Noir (Chile)*

### Five sparkling Blancs de Noirs

Ashton Hills *Salmon Brut Vintage (Australia)*
Edmond Barnaut *Champagne Blanc de Noir Brut Non-vintage (France)*
Egly-Ouriet *Champagne Blanc de Noirs Brut Vintage (France)*
Laurent-Perrier *Champagne Rosé Non-vintage (France)*
Schramsberg *Blanc de Noirs Vintage (California)*

*Pinot Noir continues to tantalize growers worldwide. Indeed, for many winemakers success with this grape represents the Holy Grail of winemaking as it is one of the most challenging of all the international varieties.*

### Maturity charts

Most Pinot Noir from anywhere in the world other than the Côte d'Or follows the Martinborough pattern of early drinkability.

**1996** Côte de Nuits red (Grand Cru)

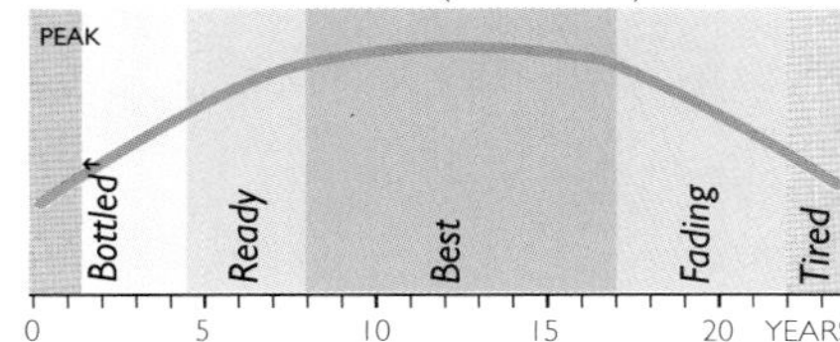


*A beautifully ripe vintage of seductive, silky Pinot Noir of generosity and good acidity. They will need three to five years in bottle before being opened.*

**1998** Top Santa Barbera Pinot Noir

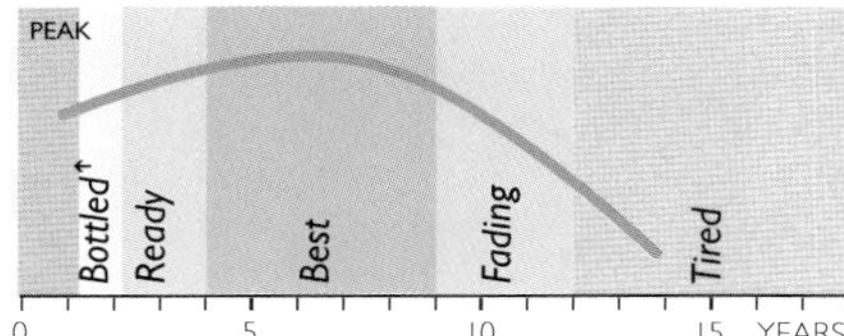


*Santa Barbara Pinot Noir varies in longevity. Only the most concentrated wines will improve beyond five or six years; many should be drunk earlier.*

**2000** Top Martinborough Pinot Noir

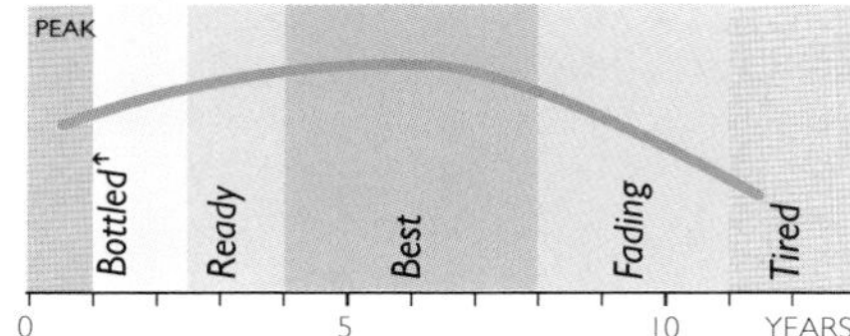


*These are wines of good intensity and balance that will be drinkable soon after release. Already fruity and perfumed, the best have concentration and depth.*

# PINOTAGE

DEAR, OH DEAR, I don't know any grape variety that arouses such fierce disagreement as Pinotage. It is potentially South Africa's national treasure, seeing that it was bred there – Professor Perold created it in 1925 at Stellenbosch University by crossing Pinot Noir and Cinsaut – and yet South Africans are some of its fiercest critics. You'd expect European wine coteries to disdain it because its wine is so startlingly, iconoclastically different to the European classics. But for half the South African establishment to bridle at the very thought of it needs some explanation.

The South African wine establishment has long been obsessed with being as European as possible – even to the extent that they didn't like being described as 'New World' at all – 'We've been making wine since 1659' and all that stuff. So they were obsessed with making wine that resembled as nearly as possible the European classics. And Pinotage is triumphantly different – capable of flavours no other traditional European vine possesses. Luckily a new wave of imaginative Cape winemakers has realized the fantastic possibilities of the South African flavours that Pinotage offers and we are now seeing a parade of deep, roaring reds, oaked and unoaked, with an astonishing mixture of mulberry, damson and blackberry fruit, mixed with the November scents of marshmallows toasted over a Thanksgiving bonfire.

Pinotage can be very high quality, but it needs low-yielding old bush vines, and very careful handling in the cellar to tame its wilder flavours and avoid the volatile esters which can mar it. It might be that super-ripeness is the way to curb its rusticity, and careful oaking (with French or American oak) suits the variety well. New clones seem able to avoid Pinotage's estery character.

At first it was trumpeted as offering the holy grail of Burgundy flavours from the Pinot Noir, and high yield and easy ripening from the Cinsaut (often called Hermitage in the Cape, hence the Pinotage name). In fact, it has never behaved or tasted remotely like Pinot Noir, or Cinsaut for that matter. So enjoy Pinotage for what it is – a true orginal – even if it is less than a century old.

One or two other countries have had a go with it – New Zealand, Zimbabwe, the USA and even Germany. None have had much luck so far, but now that South Africa is producing exciting examples, I would not be surprised to see a few New World tyros giving it a go.

**The taste of Pinotage**

Good Pinotage tastes and smells like no other wine – wonderful mulberry, blackberry and damson fruit, a flicker of lava and the deep, midwinter flavour of marshmallows toasted in front of the fire. There are two problems with Pinotage. One is that it is difficult to get the full flavours and yet control its rather aggressive tannins. And secondly those genuinely individualistic and exciting flavours do tread a knife edge and if the winemaking isn't really good, you can find a wine reeking of the spiritiness of gloss paint and a volatile acidity tottering towards raspberry vinegar.

*Beyers Truter is the winemaker behind Beyerskloof Pinotage – and it was Truter, with a small group of fellow Pinotage enthusiasts, who kept their faith in the grape during its most unfashionable period in the late 1970s and 1980s. It has been estimated that some 60 to 70 per cent of South Africa's Pinotage vines were uprooted during this period – including, inevitably, the sort of low-yielding, old bush vines that make the best wine. But some survived., and unirrigated old vines are the key to this wine's concentration and depth.*

SPICE ROUTE

*A dense, powerful example of Pinotage from the Swartland district north of Cape Town, where unirrigated bush vines produce small crops of ultra-ripe grapes.*

L'AVENIR

*Award-winning Pinotage is a speciality at this farm in the heart of Stellenbosch's red wine belt. New virus-free vines are giving good structure and more fruity generosity to the wine.*

**Left:** *Picking Pinotage on the Warwick Estate in the Stellenbosch district. The grape is early-ripening, beating Cabernet Sauvignon to the winery by about a fortnight.* **Above:** *Pinotage is high yielding – often too high for top quality. Many South African winemakers have yet to be convinced by it, and prefer Cabernet Sauvignon. But there is now a Pinotage Association in South Africa, which is dedicated to improving quality, and new clones, without the viruses that plagued the vine in the past, produce much better flavours. So long as the winemakers keep faith in Pinotage's fascinating personality and don't try to internationalize it, we should be in for some treats.*

## CONSUMER INFORMATION

### Synonyms & local names

The grape is known as Pinotage to one and all.

### Best producers

**SOUTH AFRICA** Ashanti, Avontuur, Backsberg, Graham Beck, Bellingham, Beyerskloof, Boplaas, Bouwland, Claridge, Clos Malverne, Fairview, Grangehurst, Groot Constantia, Jordan, Kaapzicht, Kanonkop, Laibach, L'Avenir, Longridge, Morgenhof, Neethlingshof, Robertson, Saxenburg, Simonsig, Spice Route, Stellenzicht, Swartland, Uiterwyk, Warwick, Wildekrans.

### RECOMMENDED WINES TO TRY

**Fifteen full-bodied Pinotage wines**

Graham Beck *The Old Road Pinotage*
Claridge *Trafalgar Bush Vine Pinotage*
Clos Malverne *Reserve Pinotage*
Kaapzicht *Steytler Pinotage*
Kanonkop Estate *Pinotage*
L'Avenir *Pinotage*
Laibach *Pinotage*
Longridge *Pinotage*
Morgenhof *Pinotage*
Neethlingshof *Lord Neethling Pinotage*
Saxenburg *Pinotage*
Spice Route *Flagship Pinotage*
Stellenzicht *Pinotage*
Uiterwyk Estate *Top of the Hill Pinotage*
Warwick Estate *Old Bush Vine Pinotage*

**Ten lighter-style Pinotage wines**

Ashanti Estate *Pinotage*
Avontuur *Pinotage*
Bellingham *Pinotage*
Beyerskloof *Pinotage*
Boplaas *Pinotage*
Fairview *Pinotage*
Groot Constantia *Pinotage*
Robertson Winery *Pinotage*
Simonsig Estate *Pinotage*
Swartland Winery *Pinotage*

UITERWYK ESTATE

*With lush texture and and ripe, fine tannins, Top of the Hill Pinotage reflects all owner/winemaker Daniel de Waal's red wine experience gleaned from working at the top St-Émilion estate of Château Angélus.*

## PLAVAC MALI

This grape, found on the Dalmatian coast and on the Adriatic islands off Croatia, seems to have some links with Zinfandel/Primitivo, though it appears that the two varieties are not identical, as was once thought. Its similarities do, however, provide clues as to the origins of Zinfandel (see pages 286–295 and Primitivo right). It may also be the case that several strains of Plavac Mali exist. The wine is robust, deep coloured and alcoholic, and can age in bottle; it is usually blended with other grapes, and plays a part in such wines as Dingac and Postup. There are a number of varieties in the region called Plavac 'Something-or-other'; 'Mali' means 'small' in Serbo-Croat.

## PORTUGIESER

Common in both Germany and Austria for reds that are too light, soft and sweetish to be of any interest except to the most red wine-deprived local. In Germany it is much planted in the Pfalz and Ahr regions. In Austria it goes by the name of Blauer Portugieser (see page 42).

In France it becomes called Portugais Bleu; in Hungary and Romania Kékoporto, which means the same thing, and in Croatia Portugizac Crni or Portugaljka. Another name is Oporto. In spite of all this determined association with Portugal, there is no proof that it came from that country, and many authorities reckon its origins are Austrian. Indeed, most Portuguese red varieties have considerably more character than Portugieser, so it's a bit of an affront. There is a pink Portugieser as well, and a white one; the name of the latter seems to be attached to several different vines. Neither white nor pink versions are related to the black one.

## POULSARD

An interesting pale red grape found in the Jura in eastern France. It lacks the colour to make a deep-coloured red on its own and is often blended with Trousseau and/or Pinot Noir. It is a delicate plant that buds early and so gets hit by frost; it also gets downy mildew, oidium and grey rot. It makes a good table grape, but its fragile skin makes transport difficult. Not surprisingly, it is not widely grown. It comes in white, pink and black musqué forms, as well as the light red-berried form. Best producers: (France) Jean Bourdy, Jacques Puffeney.

## PRESSAC

Malbec, usually known as Cot in Bordeaux, becomes known as Pressac when it is grown on Bordeaux's Right Bank (in the appellations of St-Émilion, Pomerol and Fronsac). The name is used to some extent in the nearby appellations of Bourg, Blaye and Entre-Deux-Mers, but is declining in popularity. See Malbec pages 118–119.

## PRETO MARTINHO

Portuguese grape found in the Ribatejo and Estremadura regions. It ripens early and gives deep-coloured wines with plenty of alcohol. In the Algarve it is called Negra Mole.

## PRIETO PICUDO

An interesting grape found in León in Spain. It has deep colour and an affinity for oak: it doesn't go solidly tannic in new oak, which Tempranillo can. It has slightly musky but not overt fruit. It does not appear in DO wines, which seems a loss, but to be more cultivated since León wines could do with beefing up. Best producer: (Spain) Bodegas de León-Ville.

## PRIMITIVO

Recently fashionable southern Italian grape making big, burly, alcoholic wines. DNA fingerprinting has established that Primitivo is the same as California's Zinfandel (see pages 286–295), but even fingerprinting cannot determine how the vine reached Italy, since it appears not to be native. It appears, too, to have arrived in Italy after it reached the USA. Maybe its relationship with the Plavac Mali of Croatia (see left) gives some idea as to where it comes from originally, but nothing has been proved so far.

It languished in obscurity in Puglia for many years, contributing colour and strength to blends and having a couple of DOCs of its own, but seldom being seen outside the country. The popularity of Zinfandel has changed all that, and now Primitivo is popping up all over the place, including Australia.

## PROCANICO

Sub-variety of Trebbiano (see page 270) grown in Umbria in central Italy. It has rather more character than most other Trebbiano sub-varieties.

## PROSECCO

Both the name of an Italian grape and of a wine made from it, Prosecco's most familiar incarnation is as a dry or off-dry sparkling wine with good acidity and a lightly creamy flavour. It is a favourite in Venice and the vicinity, and is the proper wine to use in a Bellini, the peach juice and sparkling wine cocktail served in Harry's Bar (and, inevitably, elsewhere). It is not high in alcohol or body, and is fresh and rather neutral rather than particularly aromatic – the ideal base for sparkling wine, in other words. There are occasional sweet versions, quite a lot of frizzante, or semi-sparkling, and a little still wine.

Virtually all Prosecco is grown in the Valdobbiadene and Conegliano zones, north of Venice, or in Colli Euganei near Padua. It ripens late, and it was this late-ripening that originally gave rise to the spumante tradition: the fermentation tended to stop in the autumn, leaving some carbon dioxide and perhaps some residual sugar in the wine, which would begin to ferment again in the spring – if it hadn't all been drunk by then. Most of the wine today is made by the Charmat method. Best producers: (Italy) Adami, Bisol, Bortolin, Carpenè Malvolti, Col Vetoraz, Le Colture, Nino Franco, Angelo Ruggeri, L & C Ruggeri, Tanoré.

## PRUGNOLO GENTILE

One of Sangiovese's many Tuscan synonyms (see pages 208–217). This one is used in the Vino Nobile di Montepulciano zone.

## PX

The usual abbreviation for Pedro Ximénez (see page 167).

## RABO DE OVELHA

Common Portuguese white grape that reaches high sugar levels but never has much finesse. The name means 'ewe's tail', and refers to the long clusters. An alternative name is Rabigato, or 'cat's tail', though this may be a different vine which in some places is known as Rabo de Ovelha. As usual in Portugal, sorting out vine synonyms is almost impossible. The Douro, Ribatejo, Dão, Alentejo and Bairrada regions all have vines called Rabo de Ovelha, and they are all, it seems, different. Rabo de Ovelha Tinto is an unrelated red grape found in Vinho Verde. Best producers: (Portugal) Quinta do Avelar, Quinta das Setencostas, Quinta do Carmo, Quinta da Romeira.

## RABOSO

A highly tannic, highly acidic red grape found in north-eastern Italy, where its austere wine lacks the charm of other acidic reds of the region such as Marzemino (see page 122). However, it makes a reasonable, rasping accompaniment to pasta dishes. The grape has

*The grey skies and windbreaks may be horribly reminiscent of English beach holidays, but these are Ramisco vines for Colares wine grown on the coastal sand dunes at Azenhas do Mar, Portugal.*

deep colour, not much sugar (so not much alcohol to add fat), and pretty lean fruit.

There are two vines called Raboso: one is Raboso Piave, which is also known as Raboso Friulara or Friularo, and the other is Raboso Veronese. They appear to take their name from the Raboso river, which is a tributary of the Piave. Both are planted less than they once were. Raboso is also found occasionally in Argentina where I would have thought it might add a certain zip to some of their juicy, warm-climate reds. Best producers: (Italy) Borletti, Cecchetto, Ivan Cescon.

## RAMISCO

An oddity in today's wine world, Ramisco is the grape of Colares, on Portugal's Atlantic coast, where it grows in the sand dunes. Since this sand has never been infected with phylloxera, these Ramisco vines are ungrafted. The grapes have thick skins and colossal tannins and need time to soften. In time good blackcurrant flavours are revealed although, while I've enjoyed the blackcurrant scent of one or two mature examples, the tannins have never softened enough for me to enjoy drinking it. But the Colares vineyard is shrinking and Ramisco has yet to catch the eye of winemakers from other regions. Best producers: (Portugal) Adegas Beira Mar, Tavares & Rodrigues.

## REFOSCO

A north-eastern Italian red grape that gives highly acidic, deeply coloured wines often with somewhat harsh, green tannins, but attractively berried, sometimes slightly grassy aromas. It is best young and unoaked but can develop good dark plum depth and black chocolate bitterness with a couple of years' aging. The grapes have good resistance to rot, but they ripen late, which is one reason for the high acidity and greenness. Despite all this talk of acidity and greenness I'm quite fond of Refosco in its angular way.

There are various sorts of Refosco: the ampelographer Galet describes Refosco dal Peduncolo Rosso, which is the best, and Refosco Nostrano; the author Nicolas Belfrage adds Refosco d'Istria and Refosco del Terrano, which seems to be the same as the Teran of Slovenia and the Cagnina of Romagna. Refosco is also grown in Greece, and recent plantings there seem to be the Pedunculo Rosso type. There is some disagreement over how much all these vines are actually related. But if you've read this far in the book you won't be surprised to hear that the unravelling of Italian and Portuguese grape varieties, in particular, won't be finished before I'm pushing up the daisies. Best producers: (Italy) La Castellada, Dorigo, Marco Felluga, Gravner, Jermann, Livon, Ronchi di Cialla, Ronchi di Manzano, Franco Terpin, Villa Russiz, Villanova, Volpe Pasini.

## REICHENSTEINER

A modern (1939) German crossing of no great quality, Reichensteiner has Müller-Thurgau (see page 142) for one parent and a crossing of Madeleine Angevine (see page 117) and Early Calabrese, which sounds like a cabbage but is actually an Italian grape, for the other. It is resistant to rot and reaches high sugar levels but has dull flavours which, though not exactly resembling a Calabrese cabbage, have swathed themselves in the neutrality of the Müller-Thurgau, though with even less aroma. Most of Germany's Reichensteiner is in the Rheinhessen; there is also some in New Zealand and some in England – none of it is memorable Best producers: (England) Carr Taylor, Chapel Down, Denbies, St George's, Valley Vineyards.

## RHEIN RIESLING

A synonym for Riesling (ee pages 190–201). An alternative spelling is Rhine Riesling.

## RHODITIS

An alternative spelling of Roditis (see page 202), a pink-skinned grape grown pretty much all over Greece.

## RIBOLLA GIALLA

An attractive Friulian grape from north-east Italy that combines good acidity with a certain nutty flavour but lacks much aroma or fullness. Quality is generally good, although there is not much planted. It grows on hillsides in the Collio and Colli Orientali zones, as well as in Slovenia across the border, where it is called Rebula. There is also a more neutral version called Ribolla Verde. In Greece, as Robola, it can give more lemony, flinty wines (see page 202). Best producers: (Italy) Primosic, Matjaz Tercic, Le Vigne di Zamò.

## RIBOLLA NERA

A synonym for Schioppettino (see page 229).

## RIESLANER

A 1921 crossing of Silvaner and Riesling. It is mostly grown in Franken in Germany, and is actually one of the better modern crosses, producing wines of good curranty flavour. But the grapes must be ripe, and since it is late ripening this can be a problem. Unripe Rieslaner is fearfully acidic. There is only a little planted, and it's not on the increase.

# RIESLING

I WONDER WHAT IT FEELS LIKE being the wine experts' favourite grape, yet failing to excite the palates of the vast majority of wine drinkers across the world? Does it feel as though you're the class swot cooed over by teacher for your straight 'A's in all your exams, yet when it comes to break-time you're left shunned and alone at the edge of the playground as your more raucous colleagues cavort and shout and delight in each other's company? It's no good that teacher comes by and tells you not to cry, that your time will come – because right now, no-one will play with you, however much you try to please them. And anyway, who says your time *will* come? Straight 'A's and teacher approval is no guarantee of a successful and happy life.

And I'm afraid that's how it is with Riesling. In the German-speaking world, Riesling is lauded as the globe's great white grape. A significant number of wine writers and experts elsewhere in Europe, and particularly in the UK, relentlessly repeat their view that Riesling is the greatest white grape in the world. Yet the world doesn't get the message. And there are a couple of good reasons for this.

*On the banks of the river Mosel stands the town of Bernkastel directly behind which climb steep Riesling vineyards, just like the ridged green stem of the traditional Mosel wine glass. The intricate ironwork shop signs typical of the Mosel region were surely originally inspired by coiling vine tendrils.*

First, Riesling is often used as a banner-waver for those who don't like the flavour of new oak in their wines. Fair enough. The few new-oaked Rieslings I've had have been pretty weird and completely unrecognizable as Riesling, so I'll certainly support the view that new oak and Riesling is an abomination – all the tingling acidity and orchard-fresh fruit that is Riesling's joy is lost. But the flavour of new oak on a lot of the other grape varieties is not an abomination – it's a delight. Chardonnay, Sémillon, Chenin Blanc, Verdelho, Viura and many others take to oak-aging with gay abandon and generally excellent results. And the world loves them for it.

And here's another thing. Did you notice that word in the last paragraph? Chardonnay? Hmm. In some people's eyes, Chardonnay is the great Satan; the ruthless colonizer and destroyer of the world's vineyards and the world's palates. Riesling lovers, in particular, bridle at the runaway success of the easy-going, crowd-pleaser Chardonnay, and chunter among themselves that Chardonnay is a little slip of a thing, a flibbertygibbet with no depth and no complexity.

Oh dear. Riesling lovers are going to have to stop all this Chardonnay envy. Most of the world loves Chardonnay and certainly isn't going to change its mind because of sniping and carping from the Riesling brigade. It's time for Riesling fanciers to forget about Chardonnay, forget about oak, and contentedly get on with appreciating the undoubted beauty of their own grape.

Riesling isn't a flavour that everyone likes, and it doesn't take kindly to dumbing down – as Chardonnay does. Riesling is something of an acquired taste. And in the present world that is mad for rich, broad, super-ripe – and alcoholic – flavours in its wines, Riesling, with its mouthwatering acidity, its icy cool perfume and its citrus fruit is a minority taste – and none the worse for that because, despite being grown all over the world, few areas really seem to have got the hang of it, and there's only just enough top quality wine available for those of us who crave it.

Yet there's no doubt Riesling *is* flexible. It's the only grape that can flourish in the unique cool climate conditions of the Mosel and the northern Rhine in Germany. But it can also produce wonderful results grown cheek by jowl with Shiraz and Cabernet in Australia's Clare Valley and Coonawarra, or South Africa's Constantia. It can make magnificently austere dry wines, delightfully delicate medium ones and thrilling super sweet ones that drench the palate and invigorate the mind. However, these results are only achieved in a few special places. Most of the Riesling grown in the world is dull, bland even, as wine companies try to force it to be a mass producer of crowd-pleasing wines and fail to give it the attention and commitment and imagination it requires. Ah, poor teacher's pet with your straight 'A's. No-one said everyone was going to like you.

**Riesling: From Grape to Glass**

*Geography and History page 192; Viticulture and Vinification page 194; Riesling around the World page 196; Enjoying Riesling page 200*

# Geography and History

In the final decades of the 20th century Riesling suffered a fall from grace of epic proportions. One hundred years earlier the greatest German Rieslings had commanded higher prices than even the finest red Bordeaux; now just a glimpse of a German label was enough to deter many consumers even from taking the bottle from the shelf. Germany had, by the mass production of ever cheaper, ever poorer quality wines, succeeded in getting the full force of wine snobbery turned on it. Riesling, Germany's finest grape, was in fact an innocent victim, because the dismal parade of flaccid Liebfraumilchs and Niersteiner Gutes Domtals rarely had a single Riesling berry in them. Merely to be German was, in some markets, enough to damn the grape.

Riesling (and especially German Riesling) has, however, benefited from its years in the wilderness. In Germany Riesling has got drier, been through a painful stage of skeletal

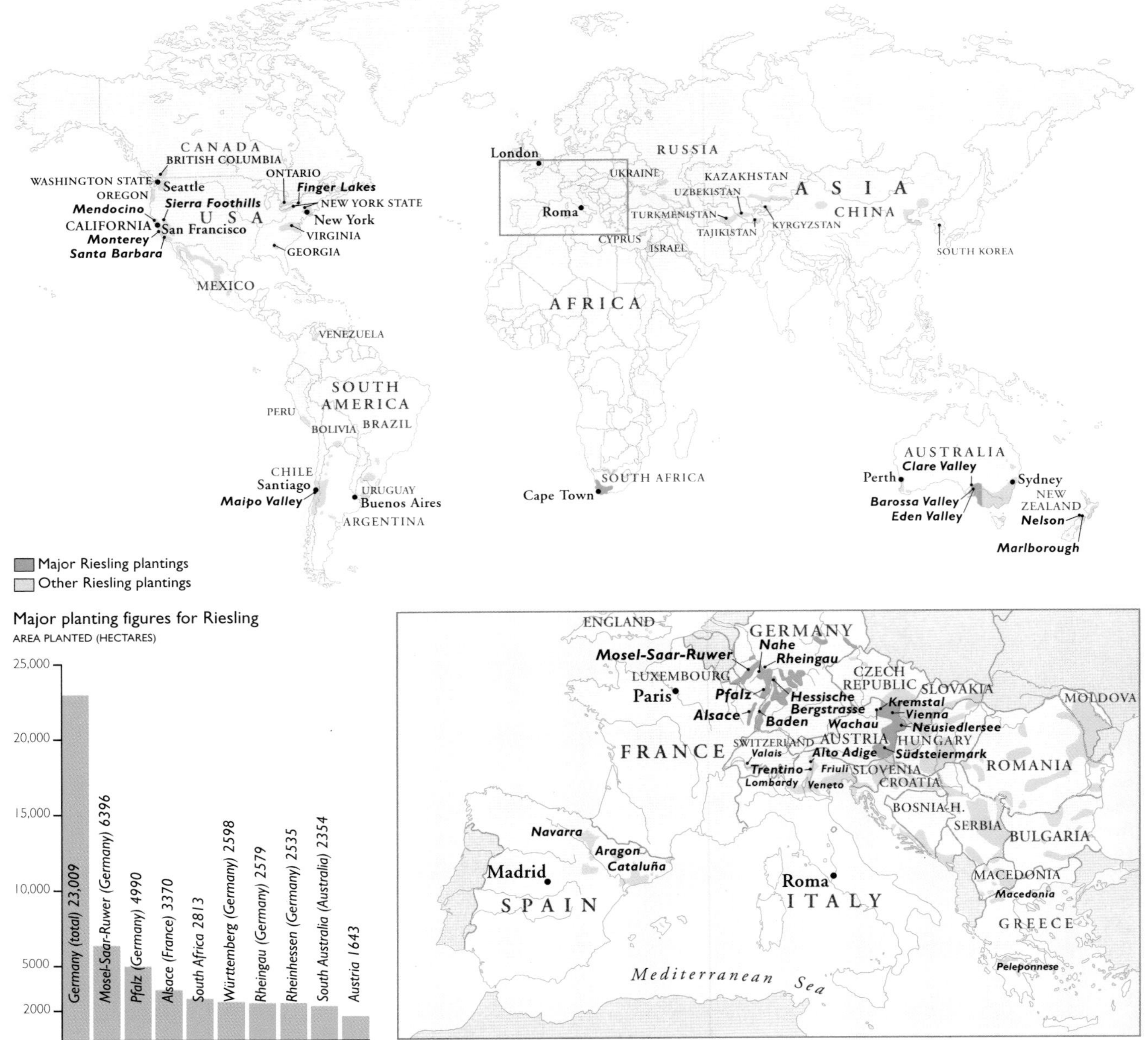

thinness, and emerged better balanced than before. The same is true of Austria.

In Australia the profound wines of the 1950s and 1960s gave way to lighter, less interesting ones as attention shifted to Chardonnay and winemakers ceased to regard Riesling as a serious grape variety. That is now changing again, and the Rieslings of the Eden and Clare Valleys rank as benchmark styles of the grape, along with those of Germany and Alsace.

The third benchmark style, that of Alsace, has both suffered and benefited from the lack of mass consumer adoration or revulsion. Styles have changed less here, which is great when it comes to those committed winemakers determined to produce fine wine. But co-operatives and merchant houses dominate proceedings in Alsace and too many of their offerings are lean, lemony – and not all that cheap. They should take note of the competition.

Elsewhere in the world, Riesling has not created many waves – largely because so many New World leaders are basically hot countries with conditions too warm for successful Riesling. But Australia has worked out how to do it, so expect exciting things from New Zealand, Chile and the Pacific Northwest in years to come.

**Historical background**

Riesling first appears in written records in Germany's Rhine Valley, and probably developed there from a wild vine some time during the Middle Ages. That first written reference is on an invoice of 1435, when the grape is spelled 'riesslingen' – the modern spelling doesn't appear until 1552, when it is so named in Hieronymus Bock's Latin herbal.

Its quality seems never to have been in dispute. It spread through the Rhine and Mosel Valleys from the 16th century onwards; in the 18th century it began to be planted as the sole variety in some vineyards – this was in itself a novel idea – in Germany, starting with Schloss Johannisberg in the Rheingau in 1720–21. Not until the expropriation of the Church's vineyards in the late 18th to early 19th centuries, however, was it planted on a large scale in the Mosel, and only in the late 19th century did it become the main variety there.

It seems to have been planted in Alsace from at least 1477, when Duke René of Lorraine praises it. By the end of the 19th century, however, with Alsace under German rule and reduced to the role of bulk winemaker, Riesling was little planted in Alsace. It was not until 1919, when the region returned to France, that Riesling began to increase there.

Once it reached Austria it was but a short step to the furthest reaches of the Austrian empire: regions that now include the Czech Republic, Slovenia and north-eastern Italy. German immigrants took it to the New World, and by 1857 it was established in California, by 1871 in Washington State, and by the early 1880s in South Australia. For a long time Riesling made many of Australia's finest whites before being eclipsed by the Chardonnay surge of the 1980s. Now it's on the way back.

*Autumnal Riesling vines in the Goldtröpfchen vineyard above Piesport in the Mosel Valley. Goldtröpfchen is a suntrap: the Mosel does one of its U-turns here, and provides a broad, sweeping slope where Riesling can ripen. However, the north-facing vineyards across the river have little hope of ripening their fruit and traditionally were usually planted to other crops.*

*Wilhelm Haag in the Brauneberger Juffer Sonnenuhr vineyard. Several south-facing Mosel vineyards have a sundial, or Sonnenuhr; an alternative to the church clock in the days before everyone had a watch.*

*Josef Leitz sampling wine from cask in his cellars at Rüdesheim. Large old oak casks like these, which give no oak flavour to the wine, are still favoured over stainless steel by many German producers.*

# Viticulture and Vinification

DESPITE BEING GEOGRAPHICALLY DIVERSE, Riesling is genetically rather stable. It does not mutate with the ease of the varieties in the Pinot family, and while there are some 60 different clones commercially available to growers in Germany (including the startlingly aromatic N90, grown by some innovative growers in the Pfalz and elsewhere) most of them are not hugely different from each other. The choice of clone is seldom a major factor in determining the style of the wine where Riesling is concerned.

Riesling wines can, however, exhibit a remarkable number of different flavours, from smoke to peach, from earthiness to petrol, from slate to spice to dried apricot. It reflects its vineyard more transparently than almost any other grape.

## Terroir

The ability to translate the vineyard into the glass through the medium of winemaker and vine is what makes Riesling so endlessly fascinating. Growers in the Mosel region of Germany speak of the minerally flavour of wines from the village of Wehlen; of the blackcurranty note that comes from Piesport clay, of the steeliness of Traben-Trarbach's blue slate. In the Nahe region, wines from the Traiser Bastei vineyard at the foot of the Rotenfels cliff can, in good years, express the fieriness of Riesling grown in red slate. In Alsace, where the whole Grand Cru system of 50 top vineyards only means anything if there is a difference imparted to the wine by different terroirs. The soils which Riesling prefers are principally sandy clay and loam. However, it seems to be able to adapt to most soil types, providing they are well drained and offer the sheltered, sunny position it likes.

To reflect its vineyard site to greatest advantage, and to produce wine with the best balance, the Riesling needs two things: a long, slow ripening season, and low yields.

*Harvesting Riesling in the Würzgarten vineyard at Urzig in the Mosel Valley. Riesling in the Mosel is trained in a particular and labour-intensive way, with one stake per vine. The stakes must be embedded in the solid rock beneath, itself no easy task.*

## Climate

Riesling is normally considered an early ripening variety: only in Germany, where crossings like Müller-Thurgau can beat it with even greater speed, is it considered late-ripening. In a warm climate it ripens too early and quickly to have much interest: it acquires none of its characteristic complexity and tastes rather dull and flat. The vine's hard wood makes it highly resistant to winter cold, which makes it useful in New York State and Canada, as well as in the chillier parts of Germany, and it buds late, which gives it some resistance to late frosts. Its frost resistance means that winter pruning can begin early.

In the Mosel-Saar-Ruwer it is nevertheless at its northern limit of cultivation in Europe, and must, if it is to ripen, have the sunniest, most sheltered sites; the angle to the sun of the steep vineyards is crucial if they are to benefit from every possible ray. In the Mosel it can find itself too enveloped in mists coming off the river if it is planted at the foot of the slopes, and too cold if it is planted above about 200m. In Australia the Clare and Eden Valleys, while they are on the face of it considerably warmer than the Mosel (Shiraz flourishes here, as well as Riesling), have such convoluted topography that many variations of soil, altitude and exposure are possible. Rainfall is low and nights are cool, and altitudes can rise as high as 400m.

## Yields

In Germany Riesling is considered low-yielding, simply because it appears so beside such cash-cow grapes as Müller-Thurgau. The latter can produce 300 hectolitres per hectare; the turning point at which Riesling on the Mosel begins to produce much lower quality is somewhere between 120 and 150hl/ha. By most standards that is extremely high, and while yields are certainly higher in the Mosel than, say, in the Rheingau, top growers will seldom admit to making more than 50–70 hl/ha. There seems to be little difference analytically in the wine from high and low yields but the taste is dramatically different.

Elsewhere, up to 60–70hl/ha is not uncommon, though winemakers in warmer climates may press less juice out of the grapes to avoid coarseness in the wine. Warmer climates give thicker skins, which can impart bitter-tasting tannins and phenols to the juice: Riesling from South Australia may have skins up to seven times as thick as Riesling from the Rheingau.

Riesling's compact clusters of small berries make it susceptible to *coulure* and grey rot, both of which are exacerbated by cold, wet, cloudy weather – and both of which can reduce yields.

### Winemaking

The main discussion here is over the relative merits of wood and stainless steel.

Stainless steel is favoured, both for fermentation and for aging, by those who seek a fresh, youthful briskness in their Riesling. When fermented and stored in steel Riesling has crystal clarity of fruit and almost antiseptic cleanliness.

When large, old wooden barrels are used the flavours are quite obviously different. The gentle oxygenation that occurs in old wooden barrels softens the edges of the wine and adds complexity. The choice is a stylistic one, with the steely style generally being seen as more modern.

### New oak versus old

Here there is less argument. A few producers use new oak barrels for fermenting and/or aging the wine, but only a minority. Riesling's acidity and floral perfume just don't go with new oak, and far from gaining complexity from the association, its beauty is smothered.

The only new-oak-aged Rieslings that are at all successful are those with plenty of weight. These tend to be from the Pfalz or from Baden, where warmer temperatures give greater alcohol and extract, and where the wines are likely to be fermented to dryness. (The more flowery style of Riesling made elsewhere in Germany loses everything in new oak.) In theory Alsace Rieslings also fit the dry, weighty template, but while there have been a few experiments, Alsace growers do not generally favour new oak for any of their wines. Experiments with new oak elsewhere in the world tend to prove them right.

### Varietal or blend?

Blended Rieslings are rare – not least because the tradition in both Germany and Alsace is of varietal wines. It is also true that Riesling is one of the few grape varieties that are complete in themselves, and need no improvement from others – to add another grape variety to Riesling even in small quantities means losing aroma, complexity and finesse.

Having said that, it can on occasion blend remarkably well, particularly with Pinot Blanc, for wines that combine finesse with weight. And it is in itself useful as a blender. Because of its high acidity, early picked Riesling can massively improve the balance of grapes like Gewürztraminer and Muscat grown in warm conditions. It is also suitable for use as a sparkling wine, but has fallen out of favour partly because of the dominance of the Champagne blend – Chardonnay and Pinot Noir – in fizz production, but also because of the poor quality of much German Riesling-based Sekt. Carefully made from rather underripe, sharp fruit, sparkling Riesling can be delicious.

*Harvesting Riesling for Icewine at Henry of Pelham estate in Ontario. Climate changes in Canada over the last decade mean that this harvest takes place about a month later than it used to.*

### RIESLING AS A SWEET WINE

This is one of the rare grapes that will make sweet wines that dazzle just as much as its dry ones. The reason is the grape's piercing acidity, which balances even the most intense sweetness so that a wine with 50 grams of residual sugar can taste light, refreshing and even delicate.

Rieslings may be made off-dry or sweet by simply stopping the fermentation (by adding sulphur dioxide, centrifuging the wine or chilling it) or (where the law allows) by adding Süssreserve – sweet unfermented grape juice. The method chosen affects the flavour: the sugar in a ripe grape is composed of roughly equal amounts of glucose and fructose. The glucose ferments before the fructose, so if you stop the fermentation the residual sweetness is fructose – and fructose tastes more fruity and refreshing than glucose. If you ferment the wine to dryness and then add Süssreserve, you don't gain that extra dimension.

Fine sweet Rieslings are made with the help of *Botrytis cinerea*, or noble rot, the fungus that shrivels the grapes and concentrates their sweetness and acidity. Botrytis-affected wines usually come from Germany, Austria, Alsace, New Zealand and California. For German growers they are a flagship, though not always a profitable one. An estimate from the Rheingau puts the cost of producing a bottle of Trockenbeerenauslese (TBA) at DM500 – around DM100 more than its selling price. Another estimate is that it takes one picker one day to pick the grapes for a single bottle of Beerenauslese, and one picker one day to pick the grapes for half a bottle of Trockenbeerenauslese.

Yet another sweet style is possible where Riesling thrives in cold climates. For Eiswein/Icewine the grapes are picked frozen solid, when the night-time temperature plunges to –6°C or lower. These low temperatures may not occur until January, but the wine nevertheless bears the vintage date of the previous year. When the grapes are pressed the water is left behind in the form of ice, and intensely sweet juice runs (very slowly) from the press. Some growers like to have some botrytized grapes in their Eiswein, while others prefer the purer flavours that come from having no botrytis-affected grapes.

Canada produces more Icewine than any other country – even more than Germany. In the late 1990s it made some 50,000 cases a year and bought Canada international attention. Oregon, Michigan and Luxembourg also make some.

# Riesling around the World

RIESLING PRODUCERS AROUND THE WORLD have two Old World prototypes to copy (or, alternatively, to rebel against). There is the flowery Germanic style, or the weightier, winier example from Alsace. So far only Australia has succeeded in establishing a home-grown style of equal stature, and its wines are more like those of Alsace than those of Germany. The Germanic style appears in watered-down versions elsewhere – showing, perhaps, how difficult it is to get right.

### Germany

Rieslings from Germany depend on acidity as taut as piano wire, yet they can be as seemingly delicate as a butterfly. The fruit can be intense yet ethereal, the residual sugar (if there is any) must be integrated and honeyed, and if the alcohol is low, the extract must be sufficient to give the wine balance.

That, admittedly, is not the pattern all over Germany. In the South – in the Pfalz and Baden – the wines are drier and weightier, with more substance. But they should still have that taut, knife-edge balance. The further north you go the more apparently ethereal the wines become. Wines from the Rheingau and Nahe have more weight than those from the Ahr or Mosel-Saar-Ruwer; and within this latter region, wines from the Saar and, especially, the Ruwer, seem ever more delicate. The paradox of German Rieslings is that, like a consumptive operatic heroine, they appear to be ready to collapse in the first scene, yet are in fact quite capable of lasting the full three Acts and singing half a dozen physically demanding arias into the bargain. These frail-seeming Rieslings are fit, lithe athletes – and potentially some of the longest-lived wines in the world.

### Concentration and yields

To fulfil their potential as the wine world's marathon winners, Rieslings need concentration. In the Rheingau the average yield is around 100–140hl/ha on the flatter land nearer the river, and perhaps 50–65hl/ha on the less fertile slopes. On a good soil, with good vineyard management, it is probably possible to raise the yield to 85–90hl/ha without quality problems – but that is in an ideal year, with sufficient sun and rain. The Rheingau doesn't always get enough sun. In the Mosel yields are generally higher, in spite of the exceptionally poor soil: 180hl/ha is not uncommon, and yields of this magnitude do not produce concentrated Riesling. Quality-conscious growers do not allow their vines to yield as generously as this: most quote an average yield of 50–70hl/ha, with the figure going down to 35hl/ha for very old vines. (The Rheingau equivalent might be an average yield of 45–50hl/ha from a top grower.)

More strictly selected wines like Auslese (made from selected clusters), Beerenauslese (from selected berries) and Trockenbeerenauslese (from selected berries that have been shrivelled by botrytis) are lower again.

Why should yields in the Mosel, where the soil in the best vineyards consists of nothing but flat shards of slate balanced on a steep hill, be so much higher? One reason is that Mosel Riesling sells for less in Germany than its counterpart from, say, the Pfalz so growers are not prepared to sacrifice any quantity. Yet the Mosel demands greater effort: one hectare of vineyard in the Mosel requires 1200 hours of work per year. In the Pfalz, 800 hours is average. In the Saar and Ruwer, where the climate is even less promising than that of the Mosel, yields are lower. The regions are close together, but even half a degree Centigrade can be the difference between losing part of your crop to frost and escaping, or between reaching, or failing to reach, an acceptable ripeness in your grapes.

### Acidity

Acidity is the key to German Riesling. It is also the key to understanding the differences between the major regions of Germany. Wines from Franken, where the summer is hotter, though shorter, are balanced at a much lower level of acidity than wines from the Mosel or Rheingau. Accordingly, they have less residual sugar.

In the warm Pfalz, residual sugar levels of as low as three grams per litre (very dry, in other words) are common, and acidity is likewise relatively low. In the Rheingau, a wine with 10g/l residual sugar, and acidity to balance, tastes only off-dry. That level of sugar in Franken would taste positively sweet. In the cold Saar, where acidity is higher, a wine with 40g/l sugar may taste as dry as a Rheingau with 10g/l.

Likewise, the type of acidity is important. Malic acid and tartaric acid taste totally different. Malic acid tastes raw and green like unripe cooking apples; tartaric tastes citrus, more intense, but riper. As a grape ripens,

DR LOOSEN

*The red slate of Erden gives earthy, minerally wines that are complex, racy and long-lived. Erdener Prälat Auslese should not be touched for 12 to 15 years.*

GUNDERLOCH

*In the Rheinhessen the Nackenheimer Rothenberg vineyard is also red slate; Gunderloch's TBA from here is one of the most thrilling from anywhere in Germany.*

MÜLLER-CATOIR

*The Pfalz producer Müller-Catoir makes a speciality of wines with piercingly pure fruit. yet marvellous biting acidity that ensures their longevity.*

*View over the town of Thann from the Clos St-Urbain vineyard on the hill of Rangen, Alsace's most southerly Grand Cru. Owned by Zind-Humbrecht, it is famous for Riesling. Here the grapes may be left on the vines until well into November. Thann is home to a local cult of St Urbain, whose litany prays for 'deliverance from the devastations of storm and frost'. The original chapel was built at the end of the 15th century, but destroyed in the revolution of 1789. The current chapel seen here was built in 1934.*

and its sugar content increases, so the total acidity in that grape decreases. But tartaric acid builds up while malic acid falls. At 40° Oechsle, the point at which the grapes begin to soften, Saar Riesling will have about 40 grams of acidity, all of which is malic. At 80° Oechsle – Kabinett level in the Saar – there will be about ten grams of acidity, of which five will be tartaric. At 90° Oechsle – Saar Spätlese level – there will be eight grams of acidity, of which six will be tartaric. But to get this much riper acidity, you have to pick late: in cool years Saar growers may wait until November to pick. Picking earlier can produce wines with rawer acidity.

In the sunnier, more sheltered Mosel, the grapes are riper and the acidity is more likely to be the riper-tasting tartaric acid: nine grams of acidity in the Mosel can taste very different from nine grams of acidity in the Saar. Levels of the different types of acidity can determine the picking date of Riesling in Germany as much as sugar levels.

## Clones

Although there is little clonal variation in Riesling compared to other vines, it is nevertheless true that the commercially available Riesling clones in the 1960s and 1970s were geared to high yields rather than high quality. The change to better, lower-yielding clones began in the early 1990s; by 2005 or 2010 it is estimated that about a third of German vineyards will have been replanted to better clones.

## Alsace

Riesling is on the increase here, and in the late 1990s covered some 20 per cent of the vineyard, but is choosy about where it is planted, preferring the hilliest, most sheltered sites. This might seem curious, given that Alsace is so much further south than much of Germany; but many of Alsace's vineyards are on the flat plains rather than on the steep, church-dotted hills beloved of photographers of the region. Soils on the plains are richer and more fertile, and while they should not be dismissed out of hand – good viticultural practice can produce expressive wines here – Riesling in less skilful hands can be thin and anaemic, or heavy and flat.

On the best hillside sites, particularly in the hillier Haut-Rhin in the south of Alsace, Riesling comes into its own. (There are 50 Grand Cru sites in Alsace; their names are a fair guide to the best sites, but not all the best wines are Grand Cru, and not all Grand Cru wines are worthy of the name.) The soils Riesling likes best here are sandy-clayey loams that warm up quickly in spring, although it is less fussy about soil than it is about aspect; key vineyards include Brand, Clos Ste-Hune, Elsbourg, Hengst, Kaefferkopf, Kastelberg, Kirchberg (Ribeauvillé), Kitterlé, Osterberg, Rangen, Schneckelsbourg, Schoenenbourg, Sporen and Zahnacker. But the geology of Alsace is so convoluted that soil types often change within the same vineyard: hence the habit of some top Alsace growers of picking such parcels separately and vinifying them as separate cuvées. Such cuvées may be blended

KOEHLER-RUPRECHT

*This estate makes weighty, dry wines: this Auslese Trocken has 12.5 per cent alcohol, and the structure and extract to go with both meat and fish dishes.*

DOMAINE OSTERTAG

*Yields are low at Domaine Ostertag, and consequently the wines have tremendous richness and depth. This Riesling from Muenchberg is especially aromatic.*

DOMAINE PAUL BLANCK

*Fastidious winemaking at this estate means that umpteen different Riesling cuvées are kept separate – with a detectable difference in taste.*

later out of commercial necessity, but tasting them from barrel shows that Riesling in Alsace reflects its terroir just as clearly as it does in Germany – as long as the yields are not too high.

Permitted yields in Alsace are over-generous: 100hl/ha for the basic AC wine and 55hl/ha for Grand Cru. Wines cropped at that level have little character and no concentration; serious growers quote around 50hl/ha for AC wine and less for Grand Cru.

Why does Alsace Riesling taste so different from German Riesling – different even from the Rieslings of the Pfalz, only just to the south? One reason is the soil: the predominantly calcareous, clayey soils of Alsace give a fuller character than does the slate of, say, the Middle Mosel. Another reason is their higher alcohol: Rieslings here commonly have over 12 per cent alcohol, and are chaptalized. They often spend longer in old barrels, too, which gives them greater roundness. But above all they are French wines, with the indefinable but recognizable French imprint. They are 'winier' wines than German Rieslings, lean and austere in youth but rich, honeyed and petrolly in maturity.

They are also generally dry, although Alsace growers are notoriously cavalier about residual sugar levels. It is impossible to predict from a label whether an Alsace wine will be bone dry or medium. Vendange Tardive, or late harvest, wines may be dry or semi-sweet: Rieslings in this category must attain 95° Oechsle (220g/l sugar at picking). The much rarer Sélection de Grains Nobles wines are always sweet, and must have reached 110° Oechsle (256g/l sugar) at harvest.

## Australia

Riesling is on a roll in Australia, although, since it only occupies some 3 per cent of total plantings, its current popularity may never amount to more than a mini-roll. Until Chardonnay overtook it in the early 1990s, Riesling was the most planted white grape in Australia, reflecting the belief current among many growers at the time that if you took a noble grape variety you could plant it anywhere and get great wine. It was (and still is) grown in many regions that were (and are) far too hot for such an early ripener; the resulting wines gave the variety a bad name.

Production in the Clare Valley, one of the key Riesling areas, is set to increase by a fifth over five years from the 1999 figure of 4000 tonnes of grapes; but Cabernet Sauvignon, Merlot and Shiraz between them are set to more than double over the same period. Investors in Clare, as elsewhere in Australia, are hypnotized by reds.

Clare is, on the face of it, a region far better suited to reds. Its climate is almost Mediterranean, but its topography makes it more adaptable than it would seem. Far from being a single valley, Clare is a series of gulleys, with hills rising to 400m to the east and west. This already gives significant variations in temperature. In addition, nights are cool and rainfall is low, and the free-draining soils – which include red soil over limestone and shaley slate – allow for great variations of wine style. Watervale is traditionally the finest part of Clare for Riesling, though the Polish Hill River gives wines with finesse and a certain mineral character. Clare's stylistic affinities are with Alsace.

In fact Clare may well be Australia's finest Riesling area. The great 1970s Rieslings from Leo Buring set a standard

### Another new classic from Australia

*Riesling growing in Pike's Polish Hill River vineyards in Clare Valley, South Australia (right). Polish Hill River Riesling has minerally finesse, but it is by no means the only part of Clare that has a name for great Riesling. The topography of the valley – a series of hills and valleys rather than just one valley – and its varied soils and climates offer great scope to the grower in search of individuality.*

*Riesling was first planted in South Australia by the Silesians who settled in the Barossa Valley, bringing with them their brass bands, Würst and Lutheran churches, all of which still flourish there. There is some Riesling grown in the Barossa itself, but most has moved to the cooler hills of the Eden Valley, just above the Barossa to the east, or to the Clare Valley, a couple of hours' drive to the north.*

*Pike is one of many producers who have made a name for themselves with Rieslings of appetizingly cool steely structure, yet refreshing fruit and scent.*

MOUNT HORROCKS

*This wine is yet more proof that the Clare Valley produces great Riesling. Mount Horrocks' example is typically limy and gets toasty with age.*

which is still inspiring winemakers three decades later. Such wines are lean, even apparently rather simple in youth: many are drunk in their youth because their crispness and freshness are attractive then and seem to promise nothing more. But with time the acidity softens, the palate deepens, and they develop flavours of toast and lime, a melting honey richness and a whiff of kerosene; they emerge into maturity as some of the richest, most complex dry white wines in the world. And all that without even a hint of new oak.

The Australian style of Riesling varies from light and delicate to more powerful, but all share winemaking techniques of no oxidation, low temperature fermentation in stainless steel, and early bottling. What you get is the unalloyed character of the grape.

In the Eden Valley the wines are more austere than in Clare; plantings have moved up here into the hills from their traditional place on the floor of the Barossa Valley. Some sparkling Riesling is also produced.

### Austria

The finest sites for Riesling are the granite, gneiss or mica-schist terraces of the Wachau, where the climate is cool and the soil free-draining, and irrigation is both necessary and permitted. However, it accounts for only about 10 per cent of total Wachau plantings, and has only attained even that figure since the Second World War. In Austria generally it has about 2.5 per cent of total plantings. It demands, and gets, the best spots: they include the Steinertal, Kellerberg, Schütt and Loibenberg in Loiben; the Tausendeimerberg, Singeriedl and Hochrain in Spitz; and Steinriegl, Achleiten and Klaus in Weissenkirchen. Alcohol levels are usually over 13 per cent and the wines are at their best after five years or so; Austrians, however, prefer to drink them much younger.

The styles, and soils, of the Wachau continue into the western part of Kremstal; elsewhere in Austria Riesling is more prevalent, and usually good, though less racy. The hilly region of South Styria, however, produces wine with good taut acidity, and Vienna itself makes some appetizing examples. Most Austrian Riesling is made dry: there is a little botrytis-affected Riesling made in that haven for botrytis, the shores of the Neusiedlersee, but it is vastly outplanted there by the totally different Welschriesling.

### New Zealand

Riesling was introduced to the country in the early 1970s. The cool climate is ideal for making wines of greater lightness and delicacy than are produced in Australia, and acidity levels are good, but all too often the wines inexplicably lack excitement. That is not the case, however, with the late-harvest sweet wines, which have great concentration and raciness. Marlborough is a leading area for both sweet and dry styles; Nelson also has a reputation for late-harvest Riesling.

### Canada

While dry Riesling wines here are of increasingly good quality and weight, it is the sweet Icewines for which Canada, and in particular Ontario, is most famous. They have greater breadth than German Eiswein, and balance slightly less finesse and complexity with greater weight and an impressively direct sweetness.

### USA

Riesling has been enthusiastically uprooted in California, though plantings still exist across the state. High-altitude regions like the Sierra Foothills produce the most elegant, racy wines, and Monterey, Mendocino and the cooler parts of Santa Barbara are also suitable. California has a short and intermittent history of sweet, botrytis-affected Rieslings, though only a few producers have sufficient understanding of botrytis to succeed.

Washington State, Oregon and the Finger Lakes of New York State all produce delicate, cool climate examples.

### Rest of the world

The vine is found widely in Europe, though west of Alsace only Spain has the occasional patch. Luxembourg makes dry, delicate versions and north-eastern Italy produces good quality examples that vary from light and aromatic in the Alto Adige to fuller but still dry in Friuli. There are substantial plantings in Slovenia and Croatia and, further north, in the Czech Republic. Riesling is also grown in Kazakhstan, Kyrgyzstan, Moldova, Russia, Tajikistan, Turkmenistan, the Ukraine and Uzbekistan.

It is scattered all over South America, though Chilean plantings have been mostly uprooted. In South America generally it is planted in sites that are too hot. South Africa suffers from the same problem, and unlike South America lacks the option of exploring ever more southerly sites.

**FRANZ HIRTZBERGER**

*Franz Hirtzberger is a leading light in the Wachau region of Austria, making wines of weight and piercing elegance. 'Smaragd' is a quality category unique to the Wachau.*

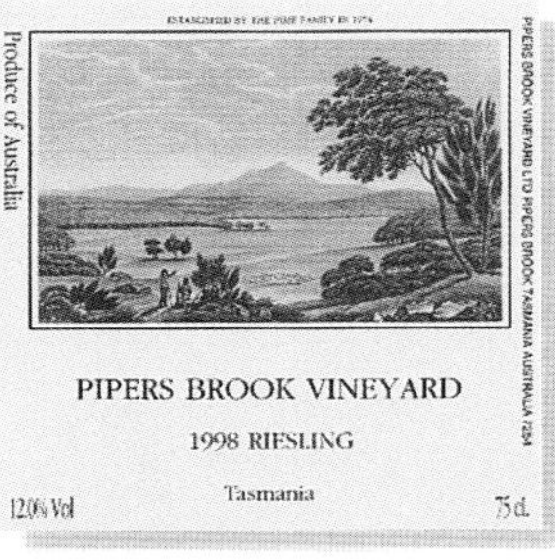


**PIPERS BROOK VINEYARD**

*Pipers Brook Rieslings, from cool Tasmania, have a floral, citrus delicacy that is very different from the fruit and steely structure of Clare Valley Rieslings.*

**CORBANS**

*Any New Zealand wine labelled 'Noble' will be sweet and nobly rotten. Some of New Zealand's best Rieslings are made in this style.*

**KIONA**

*Kiona is a small, family-owned company based in the sagebrush country of Washington State. Pure fruit flavours are the keynote of its wines. 'White Riesling' is a common synonym for Riesling.*

# Enjoying Riesling

HOW LONG CAN RIESLING AGE? Some Rieslings, it seems, can age almost indefinitely. It is possible to taste German Rieslings from the 1940s and 1950s that are still in fine fettle and, curiously, these are not always wines from the best vintages. Sometimes the high acidity levels that go hand-in-hand with cooler vintages have helped to preserve the wines into old age.

The question of aging Riesling is tied up with that of acidity. The acidity of many German Rieslings (in particular those of the cooler regions like the Saar) makes them taste lean and ungenerous in extreme youth. Taste them at this stage and you may well decide that Riesling is not for you. It is impossible to emphasize too strongly that good Rieslings must have bottle age: four to five years for a Kabinett, five to seven for a Spätlese, six to ten for an Auslese, and ten years plus for Beerenauslesen, Trockenbeerenauslesen and Eisweine. Simple QbA wines can be drunk a year or two after the vintage.

Dry (*trocken*) German Rieslings have a slightly different aging profile. They don't last as long as Rieslings with residual sugar, and become drinkable a little earlier.

Alsace Rieslings similarly need age – three or four years for simple AC wines, four or five years or more for Grands Crus, and at least five – ideally ten or more – for Vendange Tardive wines. Austrian Rieslings are drunk young in Austria, though the best wines of the Wachau can improve for six or eight years. Top Australian Rieslings from the Clare or Eden Valleys will improve for perhaps eight years, and last for 20.

### The taste of Riesling

The adjectives that can be used to describe the flavour of this grape are as varied as the terroirs in which it grows. Slate soil gives it a characteristic smoky tang; in other soils it may taste minerally, steely, tarry, earthy, flowery or slightly spicy. Peaches and green apples are common descriptors; quince is also found, as is citrus peel.

Riper wines may taste of apricot or even pineapple. Australian Riesling often tastes of ripe limes and toast.

With bottle age Riesling acquires a characteristic smell of petrol or kerosene; this may not sound appetizing but is delicious. Look also for honey and marzipan and uncooked buttery shortbread.

*Botrytis cinerea*, or noble rot, typically gives Riesling a dried apricot flavour, or one of honey, almonds or even raisins. Eiswein/Icewine, where it is made with unbotrytized grapes, has a distinctive icy smell literally like that of fresh snow; other aromas and flavours are suggestive of lemons, peaches, apricots, passionfruit, pineapples or baked apples.

*Gunter Künstler, who runs the family company of Franz Künstler at Hochheim in the Rheingau, describes winemaking as 'dancing on the edge of a knife', and his wines do indeed have the knife-edge balance and finesse that comes from intense attention to detail. 'I know all my barrels from the inside as well as from the outside,' he says. Jeff Grosset is based in South Australia's Clare Valley, and his Rieslings have equal intensity in a drier, limes-and-toast style that nevertheless gives off a seductive perfume when young, and keeps its citrus tang, young or old.*

### Matching Riesling and food

Good German Spätlese Rieslings have the acidity to counteract the richness of, say, goose or duck, but the endless permutations of sweetness, dryness and weight in German wines mean that you do have to think out your food and wine combinations rather carefully.

For example, a well-aged Mosel Kabinett or Spätlese will be perfect with trout, or with smoked fish pâté. The Rheingau Halbtrocken equivalent will be good with fish in creamy sauces, though a traditionally sweet Rheingau Spätlese may well be too sweet for most dishes. But only the weightiest Auslesen and upwards should be attempted with desserts, and then only with desserts that are not over-sweet. An Auslese from the Mosel will be too light for almost all desserts, and will be best drunk on its own.

Alsace Rieslings are far more food-friendly, and will partner everything from onion tart to spicy chicken dishes – and both Alsace and Australian Rieslings are perfect with Chinese and Thai food. Any dry Riesling with crisp apple or lime acidity will be a good match for salads.

## CONSUMER INFORMATION

**Synonyms & local names**
Also known as Johannisberger Riesling, Rhine Riesling or White Riesling and in Italy as Riesling Renano. Not to be confused with Laski Rizling, Olasz Rizling, Riesling Italico or Welschriesling.

**Best producers**
**DRY RIESLINGS**
**GERMANY** Bassermann-Jordan, Georg Breuer, Bürklin-Wolf, Dönnhoff, Gunderloch, Heyl zu Herrnsheim, Heymann-Löwenstein, Koehler-Ruprecht, Künstler, J Leitz, Müller-Catoir, St Antony, Horst Sauer, J L Wolf.
**AUSTRIA** Bründlmayer, Hirtzberger, Jamek, Knoll, Nigl, Nikolaihof, F-X Pichler, Rudi Pichler, Prager, Freie Weingärtner Wachau.
**FRANCE/Alsace** Beyer, Blanck, Boxler, Deiss, Dirler, Hugel, Kientzler, Kreydenweiss, Kuentz-Bas, Ostertag, Schoffit, Trimbach, Weinbach, Zind-Humbrecht.
**NON-DRY RIESLINGS**
**GERMANY** Georg Breuer, Bürklin-Wolf, J J Christoffel, Diel, Gunderloch, Haag, Heymann-Löwenstein, Jost, Karthäuserhof, Carl Loewen, von Kesselstatt, Künstler, Dr Loosen, Maximin Grünhaus, Markus Molitor, Müller-Catoir, Egon Müller, J J Prum, Richter, Horst Sauer, Willi Schaefer, Weil.
**AUSTRIA** Franz Prager.
**FRANCE/Alsace** Beyer, Deiss, Hugel, René Muré, Trimbach, Weinbach, Zind-Humbrecht.
**NEW WORLD RIESLINGS**
**AUSTRALIA** Tim Adams, Alkoomi, Leo Buring, Delatite, Grosset, Henschke, Howard Park, Leeuwin, Mitchell, Mitchelton, Mount Horrocks, Orlando, Petaluma, Pipers Brook, Plantagenet, Geoff Weaver, Wilson Vineyard, Wolf Blass.
**NEW ZEALAND** Cloudy Bay, Corbans, Dry River, Felton Road, Giesen, Millton, Neudorf, Ngatarawa, Villa Maria.
**SOUTH AFRICA** Neethlingshof, Thelema.
**USA/Oregon** Argyle; **Washington State** Covey Run, Kiona.
**CANADA** Gehringer Brothers, Reif Estate, Thirty Bench.

**RECOMMENDED WINES TO TRY**
**Ten classic European non-dry Rieslings**
Georg Breuer *Berg Rottland Auslese Gold Capsule (Germany)*
J J Christoffel *Urziger Würzgarten Auslese (Germany)*
Gunderloch *Nackenheimer Rothenberg Auslese Gold Capsule (Germany)*
Toni Jost *Bacharacher Hahn Auslese (Germany)*
Karthäuserhof *Eitelbacher Karthäuserhof Auslese Long Gold Capsule (Germany)*
Dr Loosen *Wehlener Sonnenuhr Auslese (Germany)*
René Muré *Vorbourg Clos St-Landelin Alsace Vendange Tardive (France)*
Willi Schaefer *Graacher Domprobst Auslese (Germany)*
Weinbach *Alsace Vendange Tardive (France)*
Zind-Humbrecht *Clos Windsbuhl Alsace Vendange Tardive (France)*

**Ten dry European Rieslings**
Paul Blanck *Furstentum Vieilles Vignes (France)*
Bründlmayer *Zöbinger Heiligenstein Alte Reben (Austria)*
Dönnhoff *Schlossbockelheimer Felsenberg Spätlese Trocken (Germany)*
Heyl zu Herrnsheim *Niersteiner Pettenthal Erstes Gewächs (Germany)*
Franz Hirtzberger *Spitzer Hochrain Smaragd (Austria)*
Koehler-Ruprecht *Kallstadter Saumagen Auslese Trocken (Germany)*
Emmerich Knoll *Dürnsteiner Kellerberg Smaragd (Austria)*
Franz Künstler *Hochheimer Hölle Auslese Trocken (Germany)*
Schoffit *Rangen Clos St-Théobald Alsace (France)*
Trimbach *Clos Ste-Hune Alsace (France)*

**Ten New World dry or off-dry Rieslings**
Tim Adams *Clare Valley Riesling (Australia)*
Covey Run *Late-Harvest White Riesling (Washington)*
Dry River *Craighall Riesling (New Zealand)*
Felton Road *Riesling Dry (New Zealand)*
Grosset *Polish Hill Riesling (Australia)*
Howard Park *Riesling (Australia)*
Millton *Opou Vineyard Riesling (New Zealand)*
Mitchelton *Blackwood Park Riesling (Australia)*
Orlando *Steingarten Riesling (Australia)*
Thelema *Riesling (South Africa)*

**Five sweet (dessert wine) Rieslings**
Neethlingshof *Noble Late Harvest (South Africa)*
Ngatarawa *Alwyn Reserve Noble Harvest Riesling (New Zealand)*
Franz Prager *Ried Achleiten Riesling TBA (Austria)*
Reif Estate *Riesling Icewine (Canada)*
Horst Sauer *Eschendorfer Lump Riesling TBA (Germany)*

*'You can't be a winemaker with Riesling,' says Mosel producer Johannes Selbach Oster. 'With Chardonnay you can have a recipe. With Riesling you decide what you are going to do and it turns out differently.'*

**Maturity charts**
Riesling is one of the longest-lived of all white grapes. Some Australian examples from Clare and Eden Valleys are made to be cellared far longer than this chart shows.

**1999** Mosel-Saar-Ruwer Riesling Auslese

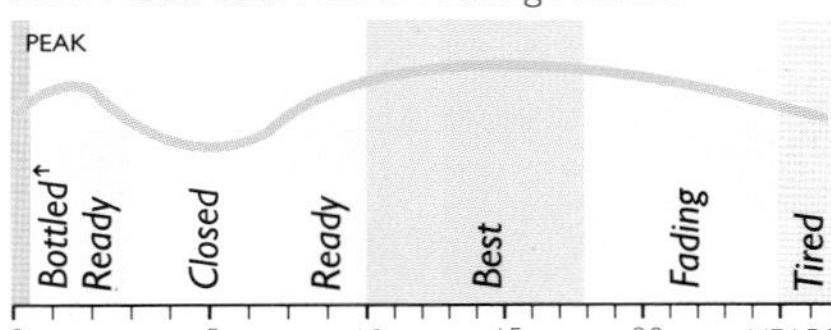


*Germany has had a run of very good vintages. Riesling Auslesen go through a closed period before emerging into maturity.*

**1998** Alsace Riesling Grand Cru

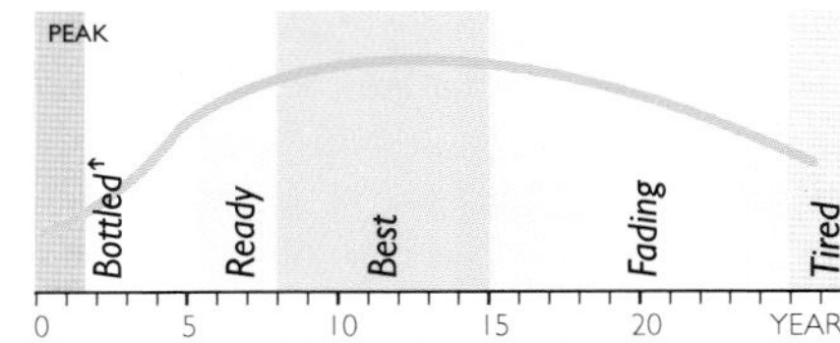


*A tricky year in Alsace, though with some superb Rieslings at the top level. Most will be at their best between five and 15 years old.*

**2000** Top Clare Valley Riesling

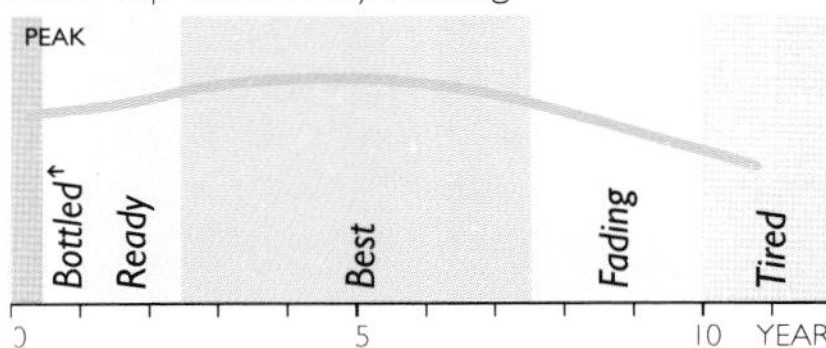


*Most Clare Rieslings are at their peak within their first five years. A few, however, are meant for longer aging and will go on improving beyond that time.*

## RIESLING ITALICO

A synonym for Welschriesling (see page 284), a lesser grape which is unrelated to the true Riesling of the Rhine and Mosel, and Hungary's Olasz Rizling (see page 165). It is known as Riesling Italico in Italy, where it is far more widely planted than Riesling, and is found mostly in Friuli and the Veneto and other parts of the North-East. It has even spread further south to the Colli Bolognesi in Emilia-Romagna where the Vallania family at Vigneto delle Terre Rosse make a delicious late-harvest wine as well as a flowery dry white from Riesling Italico blended with Riesling.

Riesling Italico wine is light, flowery or nutty, and quite crisp. It's not a patch on a true Riesling, of course, but it has its attractions as an everyday wine with moderate perfume. Best producers: (Italy) Mazzolino, Pieropan, Vigneto delle Terre Rosse.

VIGNETO DELLE TERRE ROSSE

*From Emilia-Romagna comes this unusual blend of Riesling Italico and Riesling. The wine is firm when young and improves with some age.*

## RIESLING RENANO

Italian synonym for Riesling (see pages 190–201) – the name is a translation of Rhein Riesling. It is outplanted in Italy by the lesser, and unrelated, Riesling Italico (see above) or Welschriesling (see page 284), and is mostly found, like Riesling Italico, in the North-East. The wines tend to be light and crisp, without much complexity. Best producers: (Italy) Paolo Caccese, Schiopetto, Le Vigne di San Petro, Villa Russiz.

## RIESLING-SYLVANER

A synonym for Müller-Thurgau (see page 142), dating from the days when the parents of this undistinguished grape were believed to be Riesling and Silvaner/Sylvaner. They were, in fact, Riesling and Chasselas.

Müller-Thurgau is known by this name in Switzerland, even though Dr Müller, who was responsible for the grape, came from the Swiss canton of Thurgau (the crossing was made in Germany). New Zealand has also sometimes preferred this name, and has periodically made some fairly decent wine from it.

## RIVANER

A synonym for Müller-Thurgau (see page 142) used in Luxembourg, where it is widely planted, and occasionally in England. It can produce pleasurable pale whites.

## RIZLING

Welschriesling (see page 284), under its various synonyms of Olasz Rizling and Laski Rizling, has been pressured to take this name to distinguish it from the superior Riesling.

## RIZLING ZILVANI

Synonym for Riesling-Sylvaner (see left), and thus for Müller-Thurgau (see page 142).

## RKATSITELI

Useful, all-purpose white grape found throughout the Russian Federation, especially in Georgia. It is resistant to winter cold and high in acidity; the latter quality helps to make it (relatively) resistant to bad winemaking. The two together help to explain why it could cover as much as 260,000ha in the Russian Federation. That is after President Gorbachev's vine pull scheme; before it, Rkatsiteli might even have outdone Spain's global leader Airén in numbers of hectares planted.

It is used for pretty well everything, including fortified wines and brandy; visiting Western winemakers have seen it as having the potential for reasonably good quality, though it doesn't seem to achieve more than a pleasant neutral flavour. It is also grown in Bulgaria, Romania and China, where it is called Baiyu, and to some extent in the USA, a surprising example of East-West *entente cordiale.* Best producer: (Moldova) Vitis Hincesti.

## ROBOLA

High-acid white grape which is the same as the Ribolla of north-eastern Italy (see page 189), and the Rebula of Slovenia. In Greece it is found in Cephalonia, where it grows ungrafted in limestone soils and gives lemony, flinty wines with plenty of character and weight. Best producer: (Greece) Calligas.

## RODITIS

A pink-skinned, Greek grape with good acidity, especially when planted on cooler, north-facing slopes rather than on flat land. It often forms part of the Retsina blend, along with Savatiano (see page 228). There are numerous different strains, including some redder-skinned ones, which produce more complex, interesting wine. The wine is best young, and has good acidity even in warm spots. Also spelt as Rhoditis. Best producers: (Greece) Achaia-Clauss, Kourtakis, Strofilia.

ACHAIA-CLAUSS

*The revival in quality of Greek wines has brought a whole raft of unfamiliar grapes, such as Roditis, to the fore. This wine comes from the northern Peloponnese.*

## ROLLE

Found in many blends in France's Languedoc-Roussillon, Rolle seems to mix happily with Roussanne, Marsanne, Viognier, Grenache Blanc and other local grapes. It makes characterful, aromatic wine with good acidity. It is the main grape of Bellet, a full-bodied Provençal wine that is hard to find in neighbouring Nice, never mind elsewhere. Rolle is believed to be the same as Italy's Vermentino (see page 273), and may or may not be the same as Liguria's obscure Rollo grape. Best producers: (France) de Bellet, Commanderie de Peyrassol, de la Courtade, de Crémat, Gavoty.

## ROMORANTIN

White grape used for Cour-Cheverny, a Loire Valley appellation created in 1993 along with that of Cheverny. White Cheverny is made from Chenin Blanc, Chardonnay and Sauvignon Blanc, but Romorantin, which is found nowhere else, has its own AC. Plantings are on the decline as the grape does not have any very distinctive character. When made traditionally it is a fierce, almost entirely unattractive wine, and when made in a modern, hygiene-conscious, stainless-steel style the examples I have so far tasted have seemed emasculated and forgettable. So if I had to choose between forgettably dull or memorably horrific, I think I'd choose the memorably horrific. Best producers: (France) le Chai des Vignerons, des Huards, Philippe Tessier.

## RONDINELLA

Rondinella is part of the Valpolicella blend, and usually the least characterful part. It lacks the elegance and aroma of Corvina, Valpolicella's main grape. It is disease-resistant and a reliably large cropper which is why growers like it. It doesn't reach enormously high sugar levels, but it dries very well so is a useful part of the *recioto* blend.

## RORIZ

Roriz, or Tinta Roriz, are the names by which Tempranillo (see pages 256–265) goes in Portugal's Douro Valley, where it is one of the five varieties recommended by the port authorities. It is so called after the Quinta do Roriz, the estate where it is said to have been first planted in Portugal in the 18th century. It gives large clusters of thick-skinned berries, and rich, tannic, intense wines with plenty of mulberry fruit and attractive floral fragrance. In the Alentejo it is called Aragonez and produces fairly rich, mulberry-flavoured wines for drinking young. Best producers: (Portugal) Quinta dos Carvalhais, Cortes de Cima, Quinta da Cortezia, Quinta do Côtto, Quinta do Crasto, Esporão, Casa Ferreirinha (Barca Velha), Niepoort, J P Ramos, Quinta dos Roques, Quinta de la Rosa, Casa Santos Lima, Quinta do Vale da Raposa.

## ROSENMUSKATELLER

Rose-scented, deep pink Muscat grape found in Trentino-Alto Adige in northern Italy. It's rare, but it does produce wines, both sweet and dry, with a heavenly tea-rose scent. See Muscat pages 144–153.

## ROSSESE

Probably Liguria's most characterful dark grape variety, Rossese can make wines either for early drinking or ones that will benefit from some aging. Since Ligurian wines are hard to come by outside the region, I've never absolutely been able to decide how Rossese should taste, but I've had a succession of bottles with good dark fruit and leaf and herb aroma that all tasted good – but different. So I can tell you I think the grape is good, but not precisely why. It has its own DOC in Dolceacqua at the western end of Liguria. Best producer: (Italy) Giuncheo.

## ROSSOLA NERA

Here we go again, trying to make head or tail of Italian grape varieties. The Rossola Nera grown to a small extent in Italy's Valtellina zone for the Nebbiolo-dominated blend might be the same as Mourvèdre (see pages 140–141), but the Rossola Nera grown in Corsica may be a different vine.

If it's not the same, then the latter might be the same as the dark-berried form of Ugni Blanc (see page 271). There is also a Rossola Bianca which is a synonym for Ugni Blanc. There is also an unrelated black grape called Rossolo grown in northern Italy, where it makes pretty inferior wine usually blended with Schiava. Got all that?

*The main street of the Austrian town of Gumpoldskirchen, south of Vienna. The town has given its name to a distinctive blend of Zierfandler and Rotgipfler, though the wine seems to be suffering from a dip in fashionability at the moment.*

## ROTGIPFLER

Pink-skinned grape found almost only in Austria's Thermenregion south of Vienna. Its wines are full, even sturdy, with high alcohol, and it has a big spicy bouquet. It is generally made off-dry or sweet, and is blended with the slightly superior Zierfandler (see page 285) for the long-lived, spicy white wine called Gumpoldskirchner, named after the eponymous town. Best producers: (Austria) Karl Alphart, Johann Stadlmann, Harald Zierer.

## ROUPEIRO

Southern Portuguese grape grown in the Alentejo region for gently aromatic, light white wine. It is perfectly pleasant if drunk young, meaning well before the year is out. It seems to fall apart and oxidize rather quickly. A synonym used in the Alentejo is Alva. The same grape is found in the Douro Valley, where it is called Códega. Best producers: (Portugal) Quinta do Carmo, Esporão.

*If you want to taste Rossese di Dolceacqua you may have to travel to Ventimiglia in Liguria and find the wine on its home ground. It's a characterful grape and makes leafy, herby wine.*

# Roussanne

Sometimes it's not such a bad thing to be given up as a lost cause. Ever since I can remember, I've been reading that, of the northern Rhône twins, Roussanne has far more style and elegance and class than Marsanne, but that it was a tricky, finicky, inconsistent beast of a vine and so was being uprooted in favour of the gutsier Marsanne. And it was. So the obituaries continued to flow from every vinous pen – Roussanne is wonderful: Roussanne is doomed.

Well, maybe Roussanne was pushed to the brink, but we're now seeing a typical fashion-led revival of interest in Roussanne – interestingly, at the same time people decided that Marsanne is a pretty decent vine as well. And now that we can taste some examples of Roussanne vinified by itself, it becomes clear that it does possess an almost herbal minerally perfume and surprisingly elegant texture for a warm climate wine. Yet it is easy to see why, from the grower's point of view, Marsanne rather than Roussanne dominates in Hermitage, Crozes-Hermitage, St-Joseph and St-Péray – because Roussanne yields irregularly, gets powdery mildew and rot at the drop of a hat, and doesn't like strong winds. It also ripens late, and is prone to oxidation in the cellar.

Only a few producers, notably the Perrins of Château de Beaucastel and Jaboulet, are still strongly in favour of it, even though new clones have alleviated some of its worst problems, and this is why Jaboulet's top white Hermitage and Crozes-Hermitage wines are usually so much brighter and more scented than the opposition. Those who persevere with it love the finesse it adds to blends and its ability to age: it is excellent young, within its first three or four years, but then can enter a dumb phase, from which it emerges at seven or eight with greater depth and complexity.

In Châteauneuf Marsanne isn't allowed, and Roussanne adds backbone and interest to Clairette, Bourboulenc and Grenache Blanc, but only occasionally is it a varietal wine. It likes a long growing season, and too much heat can send the alcohol up beyond 14 per cent and the wine out of balance. Some Languedoc-Roussillon and Provence examples are very exciting, particularly with a little judicious oaking. In Savoie Roussanne (known there as Bergeron) has an attractively glacial peppery, herby scent when properly made.

Rhône Rangers in California like the grape. There is also some grown in Liguria and Tuscany, and in Australia.

**The taste of Roussanne**

The flavour of Roussanne is intense but nevertheless elusive and intriguing. It is reminiscent of pears or aromatic herbal tea, floral in youth and nutty and winey with age. If picked underripe it has high acidity. However, wines picked at less than full ripeness do not age as well; and fully ripe Roussanne manages to be both low in acidity and long-lived. Some of the white Hermitages of old have aged even better than the reds.

*Château de Beaucastel's Châteauneuf-du-Pape Blanc Vieilles Vignes is, unusually for the appellation, made from 100% Roussanne. The normal bottling is about 80% Roussanne. Yields are low here at around 20hl/ha. The oak influence on this wine is extremely restrained – it is vinified half in stainless steel and half in one-year-old barrels and has superb aging potential.*

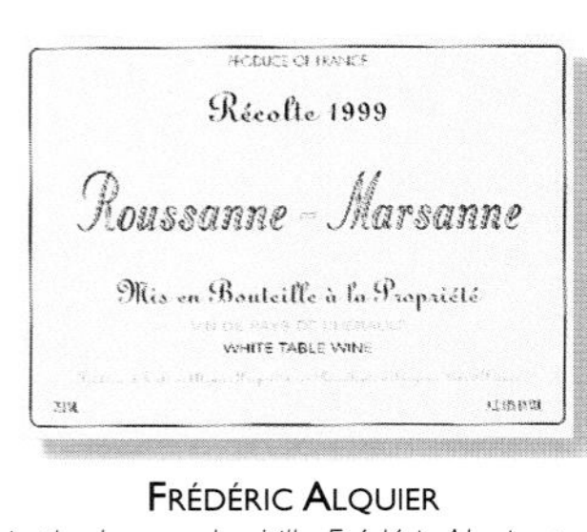


Frédéric Alquier

*Based in the Languedoc hills, Frédéric Alquier produces a Roussanne-Marsanne blend with good structure and character, overlaid with floral and honeysuckle notes.*

Mitchelton

*This Viognier-Roussanne blend would not be a common blend in the Rhône Valley, but here at Mitchelton in Australia's Goulburn Valley the 40% Roussanne contributes texture and backbone to the blend.*

***Left:*** *Randall Grahm of Bonny Doon in the South Central Coast is one of the leaders of California's Rhône Rangers, a self-styled group of producers mad about Rhône grape varieties. But matters do not always run smoothly: the Roussanne cuttings which Grahm imported from Châteauneuf-du-Pape turned out to be Viognier, and he found himself at the receiving end of legal action in the US, even though the confusion was not of his making.*
***Above:*** *Roussanne's popularity throughout the Rhône Valley rose dramatically in the 1990s.*

## CONSUMER INFORMATION

### Synonyms & local names

The grape grown in Provence as Roussanne du Var is unrelated, and although Roussanne is sometimes known as Roussette in the northern Rhône, it is not the same as the Altesse (alias Roussette) of Savoie. Also known as Bergeron in the Vin de Savoie cru village of Chignin.

### Best producers

**FRANCE/Rhône Valley** de Beaucastel, Belle, Chave, Clape, Clos des Papes, Yves Cuilleron, Delas, Florentin, Font de Michelle, la Gardine, Bernard Gripa, Guigal, Jaboulet, de la Janasse, Jean Lionnet, Pradelle, des Remizières, Marcel Ricaud, Sorrel, Tain l'Hermitage co-op; **Languedoc** Frédéric Alquier, de Cazeneuve, des Chênes, Clavel, des Estanilles, de Lascaux, Mas de Bressades, Mas Bruguière, de Nages, Prieuré de St-Jean de Bébian; **Provence** de Trévallon; **Savoie** Raymond Quénard.
**USA/California** Alban, Bonny Doon, Sobon Estate.
**ITALY** Bertelli.
**AUSTRALIA** Mitchelton.

## RECOMMENDED WINES TO TRY

### Ten Rhône Valley wines

**Ch. de Beaucastel** *Châteauneuf-du-Pape Blanc and Châteauneuf-du-Pape Blanc Vieilles Vignes*
**Domaine Belle** *Crozes-Hermitage Blanc*
**Chave** *Hermitage Blanc*
**Clos des Papes** *Châteauneuf-du-Pape Blanc*
**Yves Cuilleron** *St-Joseph Blanc Coteaux St-Pierre*
**Font de Michelle** *Châteauneuf-du-Pape Blanc Cuvée Etienne Gonnet*
**Ch. la Gardine** *Châteauneuf-du-Pape Blanc Vieilles Vignes*
**Domaine de la Janasse** *Châteauneuf-du-Pape Blanc*
**Marcel Richaud** *Côtes du Rhône les Garrigues Cairanne*

### Ten Languedoc-Roussillon and Provence wines

**Frédéric Alquier** *Vin de Pays d'Oc Roussanne/Marsanne*
**Ch. de Cazeneuve** *Coteaux du Languedoc Pic St-Loup Blanc and Coteaux du Languedoc Pic St-Loup Blanc Grande Cuvée*
**Domaine Clavel** *Coteaux du Languedoc Blanc*
**Ch. de Lascaux** *Coteaux du Languedoc Pierres d'Argent*
**Mas de Bressades** *Vin de Pays du Gard Roussanne/Viognier*
**Mas Bruguière** *Coteaux du Languedoc Blanc*
**Ch. de Nages** *Costières de Nîmes Réserve du Château Blanc*
**Prieuré de St-Jean de Bébian** *Coteaux du Languedoc Blanc*
**Domaine de Trévallon** *Vin de Pays des Bouches-du-Rhône Blanc*

### Five other Roussanne-based wines

**Alban Vineyards** *Edna Valley Roussanne (California)*
**Bertelli** *St-Marsan Bianco (Italy)*
**Bonny Doon** *Le Sophiste (California)*
**Mitchelton** *Viognier/Roussanne (Australia)*
**Sobon Estate** *Amador County Roussanne (California)*

*Snow-covered vineyards at Jongieux, Savoie in eastern France. The best parcels of vines in the village, as here, can use the Roussette de Savoie appellation.*

## ROUSSETTE

A very high quality grape found in Savoie in eastern France, which has strong similarities to Hungary's fine Furmint (see page 90), and seems to be pretty well indistinguishable in the vineyard. It is also known as Altesse. Like Furmint, it is late ripening, has high acidity, an unusual, pungent mineral, mountain herbs and lemon pith aroma, and ages extremely well. It is low yielding but resistant to rot.

As Roussette de Savoie it has several appellations of its own in the region (Frangy, Marastel, Monterminod and Monthoux), but if the grape name is not followed by an appellation name, then up to half the blend may be Chardonnay, making for a good but less distinctive wine. It may also be blended into the sparkling wines of the region to add aroma.

Roussette du Bugey, which comes from scattered vineyards halfway between Savoie and Lyon, is , confusingly, usually a blend of Altesse with a little Chardonnay. Best producers: (France) Maison Mollex, de Monterminod, Varichon & Clerc.

## RUBY CABERNET

A 1936 crossing of Carignan and Cabernet Sauvignon, produced by the University of California at Davis. The idea was to combine the heavy cropping of the former with the elegance and complexity of the latter. In this it was only partly successful, since whatever Ruby Cabernet's qualities may be, elegance and complexity are not among them. It was intended to be suitable for hot climates, and was heavily planted in California's Central Valley, though the few examples produced in cooler spots seem more successful. It is mostly blended, though varietal versions do exist.

It has declined in importance in California, but is found also in Australia, South America and South Africa. In both Australia and South Africa its ability to produce high yields of dark wine with a rather earthy flavour yet decent black fruit has made it popular for bulking up blends at a low price. Indeed, some heavily marketed, commercial Australian brands only manage to keep prices down and quality up by including Ruby in the blend. It also seems well suited to making fortified wines. Best producers: (South Africa) Longridge Winery, Van Loveren, Vredendal.

## RUCHÉ

Seldom-seen red Piedmontese variety making interesting, florally aromatic pale reds with good tannin structure and sometimes penetrating acidity. It has its own DOC in Castagnole Monferrato south of Asti. The name is also spelt as Rouchet. Best producers: (Italy) Bava, Biletta.

## RULÄNDER

Pinot Gris (see pages 172–173) in Germany may take either this name, or be called Grauburgunder. The former is more likely to be used for sweet wines, and the latter for dry, though there is no hard and fast rule.

The name Ruländer comes from Johann Seger Ruland, a wine merchant from the Pfalz region, who propagated the variety in the early 18th century. Like Pinot Blanc, it is genetically identical to Pinot Noir. In Germany today it is grown most heavily in the warmer, more southerly wine regions, especially Baden, and also the Pfalz and Rheinhessen.

In Baden it is usually known as Grauburgunder, and the wines are usually fermented dry. It produces quite weighty, fat wine; the aroma is mushroomy and earthy, with a modicum of runny honey. It's slightly spicy but less so than Alsace Pinot Gris.

In Austria, where it lives a similar double life, it makes rich, substantial wines in the Burgenland region and drier, lighter ones further south in Steiermark. In the more fashionable cellars of Austria and Germany it may be fermented in barrique, and this can result in big, ripe, dry honeyed wines. Best producers: (Austria) Feiler-Artinger, Peter Schandl; (Germany) Salwey.

**FEILER-ARTINGER**
*Ruländer forms part of the blend in this Pinot Cuvée: the other Pinots involved are Pinot Blanc, alias Weissburgunder and Neuburger, which, of course, isn't a Pinot at all. Never mind – it's a good name for the wine.*

## SACY

This French white grape is still found in the *département* of the Yonne in northern Burgundy, where it is grown not for Chablis but for Crémant de Bourgogne, the local Champagne-method fizz and white Bourgogne Grande Ordinaire. Its virtues are high productivity and decent acidity – fairly basic ones as virtues go.

That it is grown there at all is a legacy of the days, in the 18th and 19th centuries, when, before the advent of the railways, the region was an important producer of wine for Paris. High-yielding varieties were the most prized by the growers, and Chardonnay was confined to the Chablis hills. Even so, the authorities made vigorous attempts to get rid of Sacy, on the entirely reasonable grounds that its wine was dreadful and gave the region a bad name. Now Sacy is finally on its way out, having in the 20th century at last fallen victim to market forces. Those 18th-century bureaucrats must be rejoicing.

## SAGRANTINO

This Italian red grape, found around Perugia in Umbria, makes intense, tannic and strongly fruity wines with typically Italian cherry and smoke flavours, that used to be considered good for convalescents. It has its own DOCG in Montefalco, and its popularity is rising. The rarer *passito* wines from dried grapes are more powerful than the straight dry red versions. Some may be blended with Sangiovese.

Sagrantino used to be written off as being austerely tannic, but better vinification has helped round out the fruit, and it is clear that this is a grape with a good deal of personality. Sagrantino may soon be found growing in Australia, as the craze for Italian grapes takes hold there, and there's every reason to believe it could work in California and Argentina. It should also be more widely planted in Italy. Best producers: (Italy) Adanti, Antonelli, Arnaldo Caprai, Colpetrone, Rocca di Fabri.

## ST-ÉMILION

A synonym for Ugni Blanc (see page 271), especially in the Charente *département* in western France, where it is grown for distillation into Cognac.

## SÄMLING 88

Scheurebe (see page 228) is sometimes known by this name in Austria's Burgenland region, where it is used for high-quality sweet wines. Best producers: (Austria) Alois Kracher, Johann Münzenrieder.

## SANGIOVESE

See pages 208–217.

## SANGIOVETO

A Tuscan synonym for Sangiovese (see pages 208–217).

## SANKT LAURENT

An early-flowering Austrian grape that according to leading Austrian producer Axel Stiegelmar is a seedling of Pinot Noir – in other words both its parents were Pinot Noir, but, as happens with vines, it did not reproduce true to type. The ampelographer Galet says firmly that it is not a Pinot, and probably originated in the south of Alsace. But it is certainly like Pinot Noir – though perhaps even more like good Gamay – in its soft-centred, juicy cherry fruit. It is best drunk young, and doesn't age well – but so what? It's a complete delight when it's young. It is particularly popular in Thermenregion and in southern Burgenland. With Blaufränkisch it is a parent of Zweigelt.

Germany also has some and it is one of the few red varieties to find any success in the cool climate of the Czech Republic. I'd like to see some planted in areas like southern England and Canada. Best producers: (Austria) Paul Achs, Gernot Heinrich, Juris (G Stiegelmar), Willi Opitz, Hans Pittnauer, Joseph Umathum.

*Sagrantino vines at Montefalco, Umbria. Sagrantino is one of a number of Italian red varieties attracting attention in Australia and other parts of the New World.*

## SANTARÉN

Another name for the Portuguese grape João de Santarém, which is itself a synonym for Castelão Francês (see page 58).

## SAPERAVI

A dark-skinned, pink-fleshed, low-yielding variety found throughout the Russian Federation, where its resistance to winter cold no doubt comes in useful. It produces wines high in tannin, colour and acidity, which need time in bottle to soften to any degree of friendliness. It is late ripening and needs some warmth, or its acidity is too high for comfort; in such cases it is best blended. But I believe that as winemaking improves and they get their vineyards sorted out we are going to find that Saperavi is a superb grape, maybe even a classic grape just waiting for its moment in history to leap centre stage and do the jellyroll. Already I've seen 20- to 30-year-old examples from places like Moldova that had the style and perfume of a distinguished old Pauillac. Now that's high praise. It needs a few top class French and Australian vine and winemaking consultants to head eastwards, and some decent, sympathetic long-term finance to help solve vineyard and winery problems – all of which may take a while – and just you wait.

The Magaratch research institute in the Crimea has crossed Saperavi with Cabernet Sauvignon to produce Magaratch Ruby, which seems to have some potential; it has also crossed Saperavi with Bastardo (from Portugal) and produced Magaratch Bastardo. This latter is intended for fortified wine production. Which is all great fun. But I want to see Saperavi cherished and encouraged in all its own glory, and in the not too distant future, I'm sure I will. Best producer: (Moldova) Vitis Hincesti.

IOVIS

# SANGIOVESE

I'M A BIT OF A LATE CONVERT TO SANGIOVESE and its charms, but as any true 'born again' will tell you, the bug gets you far more potently if you come to it late. And I mean really late – like in the last couple of years. All through the 1970s and '80s when France was effortlessly the leading quality producer, and when the New World was still in its wine infancy, Italy as the globe's biggest wine producer did have a chance to wow the world with its most famous wine – Chianti, which is based on Sangiovese – but it failed to take the challenge seriously. Most people, me included, rarely came across Chianti except in raffia-covered flasks: you always ordered them in Italian *trattorie* and never anywhere else.

Although Chianti was supposed to be predominantly Sangiovese, antiquated wine laws meant that a substantial proportion was just as likely to be white Malvasia or Trebbiano. That didn't make for a very exciting red. There were some other local red grapes – Canaiolo is quite good – but the laws were so lax that much of the remaining blend was likely to be a great big soup of high strength, jammy this and that trucked in from Puglia or Sicily to add – well, I suppose to add colour, alcohol and flavour, however rough. When you've got badly cared-for vineyards of a late-ripening variety like Sangiovese, diluted by high acid Trebbiano – well, you're looking at a rather unsavoury, harsh rosé if you're not careful, and a dollop of wine soup from the South could hardly make it any worse.

*Sangiovese means literally the 'Blood of Jove'. Nice to think that this Roman god left off seducing mortals and dropping thunderbolts to give his name to this vine. Almost certainly of Tuscan origin, Sangiovese is still central Italy's most important grape. Florence, the heart of Tuscany, lies beyond and Keats's nightingale has strayed from Provence as this grape seems to give more truly 'a beaker full of the warm south'.*

My first moment of enlightenment came on the side of a country road south of Florence one spring time. We'd stopped off in a village to buy some wine. 'Chianti?' the man asked. We nodded. 'Where's your bottle?' Sorry, didn't know we had to bring our own bottle. 'Okay, okay.' He found us a reasonably clean litre vessel, went out the back – and squirted a jet of bright red fluid into it, charged us almost nothing – and off we went. My first mouthful of this fantastic, slightly spritzy, sweet-sour red wine, cherries and cranberries, redcurrants and a splash of fresh thyme, was a revelation. *This* was Sangiovese? If nothing else, it made supreme picnic wine.

Then during the 1980s, a few leading lights in Tuscany surveyed the rampant success of Bordeaux and the abject failure of their own Tuscan wines whose names – Chianti, Brunello di Montalcino, Vino Nobile di Montepulciano – were widely known but equally widely despised and set in train a massive movement for change. They were driven by high ambition and extreme seriousness. Bordeaux, with its glittering rank of Classed Growths and international favourites, was the target, but the grape would be Tuscany's own Sangiovese.

It has been a remarkable success. At the top end, the Tuscans have embraced modern winemaking and extensive investment in things like new oak barriques and low-yielding vineyards and they have produced a string of Italy's most challenging and ageworthy reds. Not easy-peasy stuff, not remotely New World in softness or richness – but real class, imbued with Florentine arrogance and an austere haughty beauty that makes no effort to seduce, demanding rather that you make the effort to understand. In which case, as Bordeaux becomes more and more homogenized in its flavours, perhaps Tuscany with its Sangiovese will become the new Mecca for those red wine lovers who demand a little pain with their pleasure.

**Sangiovese: from Grape to Glass**

*Geography and History page 210; Viticulture and Vinification page 212; Sangiovese around the World page 214; Enjoying Sangiovese page 216*

# Geography and History

IS THE 21ST CENTURY going to be the Italian century? Well, in grape terms, it is possible. The 20th century saw the spread of all the great French grapes around the globe and the startling success of many of them. Cabernet Sauvignon, Pinot Noir, Syrah/Shiraz and Merlot have had many breathtaking successes. And yet the world is still thirsting for more red varieties and of the many possibilities, Sangiovese from Tuscany is probably the most famous. It is not, however, the most readily adapted to travel. It is a sensitive grape at the best of times, and needs far more attention to site, clone and yield than does Cabernet Sauvignon: plant it carelessly and you will be rewarded with wine that bears no resemblance to the great Tuscan reds, and very little resemblance to anything you'd really want to drink.

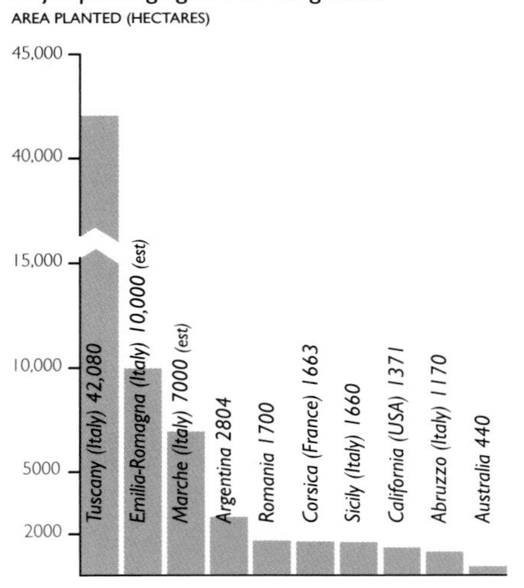

Yet the future could be bright. This is a grape that only started being taken seriously in its native land some 20 years ago: progress there has been dramatic. We are now at the beginning of a new phase, and new clones and more care over the hows and wheres of planting have the potential to produce some fascinating wines in California and Australia in the next decade.

Even so, Sangiovese is not a grape for everyone, nor everywhere. It likes warm climates, but its vigour needs to be controlled; and if it is not treated with sufficient sensitivity the flavours in the finished wine will be harsh and unattractive. It has naturally high acidity, high tannins, but somewhat fugitive colour, and it has not so far produced a single great wine outside Italy. Not yet, anyway. It is getting its chance to prove itself internationally because we are going through a period of enthusiasm for all things edible and drinkable from Italy. Somewhere, some impassioned genius surely will unlock the secret. If Sangiovese doesn't succeed, there's a queue of other red varieties – from Italy, but also from Portugal and Spain – awaiting their turn.

**Historical background**

Sangiovese probably had its origins in a wild *Vitis silvestris* vine which was caught and tamed at an early date, probably in Tuscany, though Emilia-Romagna has put in a claim to be its home. We may never know the answer to this.

Sangiovese means 'Blood of Jove', and the vine is certainly as changeable as its namesake. It is genetically as varied as Pinot Noir: so much so that the standard distinction between the small-berried Sangiovese Piccolo, or Sangioveto, and the large-berried Sangiovese Grosso has little validity. It is more useful to think of a broad spectrum of styles and qualities, with no clear dividing line between them.

The Sangiovese grown, as Brunello, in the Montalcino zone is not a separate clone: in fact around six different clones have been identified in Montalcino. (Montalcino seems to have gone off the idea that its Sangiovese was unique when it was pointed out that in that case the clone could be planted elsewhere, and there could be a Brunello di Puglia.)

Sangiovese goes by different names in other parts of Tuscany, and its personality is difficult to pin down: neither the Prugnolo Gentile of Montepulciano nor the Morellino of Scansano have much uniformity. Nor does that of Emilia-Romagna: Tuscan growers in the 1980s blamed their problems on the Sangiovese di Romagna that they had planted in the 1970s, but there is enormous clonal variation in Emilia-Romagna as well. Indeed some of the newest clones being selected and planted in Chianti originate there.

The difference is that now clones are being selected for colour, flavour and concentration of fruit, not high yields. Great vintages like 1997, but also the less-favoured years at the end of the century, showed the startling improvement already made with Sangiovese.

*Vineyards are often interspersed with olive groves in Chianti, and here in the south of the region, in the commune of Castelnuovo Berardenga, the land is more gently sloping than in other parts of the Classico zone. The local climate is also a touch warmer and the style of wine consequently a little richer.*

*Enologist Giacomo Tachis is one of the powers behind the Italian wine revolution: Tignanello and other super-Tuscans from Antinori were his babies.*

*Picking Sangiovese at the Castello di Volpaia estate in Chianti. Later picking for riper tannins is especially important in a high, cool vineyard like Volpaia.*

# Viticulture and Vinification

IN SOME WAYS SANGIOVESE is a most obliging grape: it will produce light, juicy wines or big, complex ones according to where it is grown and how it is cultivated. But in other ways it is both demanding and inconsistent: it is early budding but late ripening, so likes a warm growing season, and in the marginal regions where it produces its finest quality it will do so only in three or four years out of ten. The Italians would like to think of Sangiovese as their equivalent to Cabernet Sauvignon. Up to a point it is. The top examples age really well, and love to be aged in oak. And one more thing. Just as most top Cabernet wines are blended with something else, so far for every great varietal Sangiovese there are several great blends.

Work in the vineyard and winery is aimed at more focused fruit, increased intensity and colour, softer tannins and less aggressive acidity.

## Climate

You've got to have a fair amount of warmth for this late-ripening variety, but too much warmth does not produce the best Sangiovese. In Italy it won't ripen well north of Emilia-Romagna – though this is because of rain, not because of temperature – and in Emilia-Romagna earlier-ripening clones are necessary to dodge the October rains. In Tuscany, where so far all the finest Sangioveses have been made, it is less reliable a ripener in Chianti than in Montalcino, where nights are warmer and rainfall notably less. In Chianti it requires the best south or south-west facing slopes, and altitudes of between 150m and 550m, and the paucity of ideal sites means that only some ten per cent of the region is actually given over to vineyards. September rains, too, can spoil a vintage, and they do so several times a decade.

In Montalcino it will even ripen on north-facing slopes; these certainly produce lighter, more elegant wines than on Montalcino's south and south-west facing sites, but nevertheless the grapes ripen.

In the south of Tuscany, the Maremma, a lot of Sangiovese is being planted for the rich, broad character it acquires in the hotter climate and shorter growing season. But here too much alcohol and too little aroma can be a problem.

Trying to duplicate the extremely varied climates of Tuscany in California and Australia is proving something of a headache. Tuscany's climates are more markedly continental than those of Australia, but likely regions include Langhorne Creek, Strathalbyn and Port Lincoln in South Australia, and Karridale and Margaret River in Western Australia: these are as warm as Tuscany, though not as continental. Better matches of climate could include Canberra and Young in New South Wales, the western parts of the Great Dividing Range in Victoria, and Stanthorpe in Queensland.

The variety has yet to take California by storm, though many winemakers feel that, so far, Sangiovese has simply been planted in the wrong places and in the wrong way. Exposing the clusters to too much sun seems to give poor results; Marchese Piero Antinori, who realized Sangiovese's full potential in Tuscany in the 1970s (see caption, facing page) and now owns Atlas Peak Vineyards in the foothills of the Napa Valley (see page 215), points to the greater intensity of sunlight in California as a possible factor. Until California's great Sangiovese sites have been identified, Sangiovese wines are not going to rise above the 'Interesting; must try harder' standard.

## Soil

The soils of Tuscany are as varied as the climate and altitude. The heart of Chianti Classico is a highly desirable, friable, shaly clay called *galestro;* the lesser Chianti zones of Colli Senesi and Colli Aretini are clay; towards the coast the soils are lighter and sandier. Further south in Montalcino *galestro* alternates with limestone *alberese*. These two soils, in Tuscany, produce the best wines, with good body and flavour.

In the New World insufficient attention has been paid to soil. Leaving aside the vexed question of terroir, Sangiovese is a vigorous variety which needs a lot of work to keep it in balance, so you don't want too fertile a

*Sangiovese vines at the Montevertine estate at Radda in Chianti in the heart of the Chianti Classico zone. Planting densities must be high enough to keep Sangiovese's vigour in check but not too high: cram too many vines in and you get problems of fruit shading and consequent insufficient ripeness. The yield per vine must also be kept under control if the wines are to taste ripe.*

soil. When planting is done on fertile land, the temptation might be to increase dramatically the density of planting in the belief that this will keep the vine in check, but this approach may result in a positive jungle of foliage and no sunlight getting to the grapes at all. The vigour of the soil is crucial in determining the density of planting Sangiovese.

### Density and cultivation

In Tuscany planting densities have been rising steadily in recent years, with the Chianti 2000 research project advocating densities of 7000 vines per hectare or more. The traditional density in Chianti was 2700/ha: 'We planted vineyards for tractors, the opposite of what they did in France,' says consultant winemaker Dr Alberto Antonini. 'If you go to 5000/ha you get an improvement in quality at the same yield per hectare. I have experimented with densities of 10,000/ha, but I find no improvement in quality from 5000/ha upwards. But you need very poor soil for high density, or you get problems of canopy congestion, and shaded fruit. For Sangiovese you need open canopies with good filtered light and fruit distribution, and reasonably low yield per vine. Density must depend on the soil, the vigour and the rainfall. You can't generalize.'

The only generalization it is possible to make is that Sangiovese requires more attention in the vineyard – more cluster thinning, more selection, more careful canopy management – than Cabernet Sauvignon or Merlot of the same quality.

### Yields

Here, too, it is difficult to make worldwide generalizations, except to say that reasonably low yields produce better quality than high ones. But what counts as reasonably low varies with the site, the vigour of the soil and the climate.

In Chianti, most estimates put 1.5kg per vine as the maximum for quality (that is equivalent to 10.5 tonnes per hectare at 7000 vines per hectare). In Tuscany Sangiovese is not considered very prolific, largely because recent work has been aimed at reducing fertilizers, restricting yields and planting less vigorous clones and rootstocks. In California it is said to grow like a weed, even on poor soils; in southern Italy and in Emilia-Romagna, too, it is vigorous. It is certainly less self-regulating than Cabernet Sauvignon, and if yields exceed 10–12 tonnes/ha quality is likely to suffer. But the vine is so varied, and the conditions in which it grows are so equally varied, that what is true for one spot may not be true for another. Certainly Argentina manages to get a very pleasant fruity red with a slight bitter twist from yields far higher than those tolerated in Tuscany.

*When Marchese Piero Antinori launched Tignanello in the early 1970s, he set Tuscany on a new course. He blended Sangiovese with a small percentage of Cabernet instead of the usual Chianti grapes, and all the wine was aged in small oak barrels.*

### At the winery

Modern Tuscan winemaking is aimed at softening the tannins of Sangiovese, and at getting those tannins ripe in the first place. Picking dates are ten days to two weeks later than they used to be, which helps achieve better ripeness; and the length of the post-fermentation maceration on the skins, which had shortened to 7–12 days, has now lengthened again to three or four weeks. This gives greater polymerization of tannins – as does the (illegal) use of oak chips and the (legal) running of the wine into new oak barriques for the malolactic fermentation.

Barriques are not universal in Tuscany: the traditional aging of Sangiovese is in large oak *botti* of five or six hectolitres upwards in size. Most producers use a combination of different woods, sizes and ages of cask. Chestnut is often found in traditional cellars, but I've yet to see a new chestnut barrel. Sangiovese does seem to suck up the sweet vanilla of new oak with gay abandon, the resulting wine mellowing very attractively with age.

### BLENDING SANGIOVESE

Varietal Sangioveses can be superb: witness Flaccianello della Pieve. It is more traditional, and probably still more common, however, for Tuscan Sangiovese to be blended with something else: Canaiolo Nero, Cabernet Sauvignon, Syrah, Merlot, what you will. Primitivo, Montepulciano and Nero d'Avola are also said to be added, though not legally, since these are not Tuscan grapes, and their addition would involve trucking wine up from the South. But why add other grapes at all? Why can't Sangiovese stand on its own?

The answer lies partly in the climate of Chianti, partly in the character of the grape, and partly in the long gradual decline of Italian viticulture, which has only been arrested and reversed in the past three decades.

Chianti is relatively cool, and rain is likely to descend just as the grapes are nicely ripe. In such circumstances Sangiovese will benefit from the extra colour and flesh and softness provided by another variety. Its colour is also a factor: it is a bit short on a group of colour-giving substances called acylated anthocyanins, so here, too, other darker grapes can help.

The reason that Cabernet Sauvignon entered the equation was because of the low reputation of Sangiovese both at home and abroad in the 1970s. To make world class wines in Tuscany it seemed to be necessary to employ other grapes, and Cabernet Sauvignon had all the perfume, finesse and complexity that Sangiovese seemed to lack. It also had global renown as the great grape of the top red Bordeauxs. In the 1980s a raft of super-Tuscan vini da tavola appeared, blending Sangiovese with Cabernet in every possible percentage.

At the same time Tuscans were busy studying their vineyards and their winemaking. What they learnt about Sangiovese convinced them that it could be a great grape in its own right, and did not need to be dominated by Cabernet.

But while varietal Sangioveses will increase from the warmer parts of Tuscany, in Chianti the climate demands the option of blending. In that respect it is just like Bordeaux.

# Sangiovese around the World

WHAT SHOULD SANGIOVESE taste like when it is grown outside Italy? If New World producers shun the bitter-cherry and tea-scented styles of Tuscany, what will they replace them with to make the wine distinctly different from Cabernet and Merlot? Well, they're not really sure in most cases, so Sangiovese flavours worldwide are pretty haphazard and top Chianti flavours still seem to be the goal of many experimenters.

## Tuscany

Sangiovese in Tuscany has evolved over the past three decades. For all the excitement about Cabernet Sauvignon, and the worry that Italian flavours were being drowned in a rush of Francophilia, Sangiovese has emerged more serious, better understood and more polished, with a distinctive Italian bittersweet savoriness to the fruit. It has also emerged confusingly varied in taste.

The most traditional styles emphasize the herb and bitter cherry flavours we have always associated with Chianti and other Sangiovese-based reds; the most international styles stress plum and mulberry flavours, and use new oak barriques for extra richness and spice. In between these extremes every possible permutation is being made.

The blend of grapes also varies: Chianti may now be made with 100 per cent Sangiovese, or it may include an admixture of other grapes, which may be Cabernet Sauvignon, Merlot, Syrah, Canaiolo Nero or other softening varieties. (Chianti has no fewer than seven sub-zones, of which Chianti Classico and Chianti Rufina produce the best, most substantial wines. Colli Fiorentini wines are lighter and fresher, Colli Senesi wines can be solid and rustic, and wines from the three remaining zones, Colli Aretini, Colline Pisane and Montalbano, have no distinctive character.) In Carmignano a percentage of Cabernet is called for in the DOCG regulations; for Vino Nobile di Montepulciano, where quality has still not quite caught up with Chianti and Montalcino, 20 per cent of Canaiolo is supposed to be added, and perhaps sometimes even is. Only in Brunello di Montalcino must the DOCG wine be made entirely of Sangiovese.

For the consumer, therefore, predicting the likely style of a Tuscan red is increasingly difficult. The raft of super-Tuscan vini da tavola which emerged in the 1980s did little to help the confusion: each had its own blend, its own fantasy name and its own style, even if that style was similar to that of several other super-Tuscans.

But it is unreasonable to complain of confusion when it is the mania for experimentation by producers that has given us the current massive leap in quality. The Chianti 2000 research project, which involves the universities of Pisa and Florence and the Chianti Classico *consorzio*, has been very valuable, but it has also been outstripped by the work of individual estates, who have been doing their own work on clonal selection, planting density, cultivation methods, rootstocks and soil selection. And the improvement we have seen so far is primarily the result of better viticulture and better selection in old vineyards: good selected clones have only been available in the past ten years: as Dr Alberto Antonini puts it, 'The word 'clone' didn't exist in Italy 20 years ago'. He believes that 50 per cent of the possible improvement in Tuscan Sangiovese is yet to come.

## Rest of Italy

Most of the Sangiovese in Italy is concentrated in the centre; it is officially recommended in 53 provinces from Piedmont southwards, and authorized in a further 13, but it plays a progressively smaller part the further one strays from Emilia-Romagna and Tuscany.

In central Italy it is the workhorse red grape, producing everyday wines as well as world class ones, and it may be made into *rosato* wines, sweet *passito* ones and even into Vin Santo. Umbria's finest examples include Torgiano, where Sangiovese may be blended with Canaiolo, and Montefalco Rosso, a blend with the local Sagrantino grape; in the Marche there is Rosso Piceno, blended with the Montepulciano grape. In Romagna all varietals come under the umbrella of Sangiovese di Romagna, which covers all conceivable qualities, up to and including some slick international barrique-aged versions. In the South it is mostly blended.

## USA

Cal-Ital is the word here: it is being used by a group (a self-styled *consorzio*, indeed) of Californian growers to describe the wines they make from Italian grape varieties. So far many are good, but are notable more for

FONTODI

*Fontodi makes exemplary Chianti Classico, but Flaccianello is 100% Sangiovese from a single vineyard of old vines, aged in new oak.*

FATTORIA PETROLO

*Torrione is a pure Sangiovese, from the theoretically humble Chianti Colli Aretini zone, and is proof that great vineyard sites are not always confined to established DOCGs.*

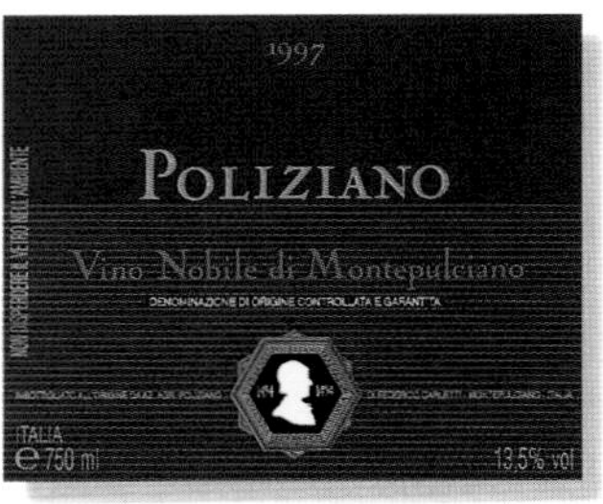


POLIZIANO

*Poliziano is a leading light in the rejuvenated Vino Nobile zone. This is his regular bottling of Vino Nobile, and there is also a single-vineyard wine called Asinone.*

*Vineyards at Atlas Peak in Napa County. When Piero Antinori became the sole owner of this winery in 1993 he brought Sangiovese cuttings from his vineyards in Tuscany. Early vintages failed to show the finesse and complexity of good Tuscan Sangiovese, even though Atlas Peak is in a relatively cool region for California, with low rainfall, chilly nights, and a reasonably long growing season. Lately quality has looked far more promising.*

their ambitious pricing than their quality to justify it.

By 1998 there were over 2900 acres/ 1175ha of Sangiovese planted in California, much of it in the wrong places. Until the 1980s the only Sangiovese in the state was a small patch in Alexander Valley left over from pre-Prohibition days: the arrival of Tuscan innovator Piero Antinori (see caption, page 213) at Napa Valley's Atlas Peak in 1986 was the catalyst that set Sangiovese on its current somewhat rocky path. Antinori points out that Sangiovese is not easy to grow even in Tuscany; many Californian growers would ruefully agree that it requires far more work in the vineyard than they had bargained for.

There is far from being a definitive style of California Sangiovese. Most of these wines have attractively bright fruit, cherryish and spicy, but they can be rustic or, alternatively, thin and lean. A touch of Cabernet or Merlot in the blend often helps the flavour, but too many California Sangioveses are already being made to taste too much like overoaked Cabernet.

Napa is generally hotter than Tuscany, and Tuscan-based consultant winemaker Dr Alberto Antonini believes that it is essential to extend the ripening period for as long as possible so that the tannins get fully ripe: sugar ripeness is no problem, and these wines often have 14.5 per cent alcohol compared to Tuscany's 12–12.5 per cent. Site selection and vineyard management will be crucial to the improvement he expects to see over the next decade. Washington State could also have possibilities in the future. Hopefully they've learnt from California's problems how important site selection is.

### Australia

If interest in Sangiovese has failed to gain momentum in California, it is on the rise in Australia. Growers here seem to be taking a more planned approach to the grape – principally through better site selection – than was sometimes the case in California. Both large and small companies are taking Sangiovese seriously: the next five years could see some interesting wines.

### Central and South America

Mexico has some Sangiovese, though it hasn't produced any outstanding examples yet. Some Sangiovese is being planted in Chile, though experience elsewhere suggests that producers there may be only at the beginning of a learning curve, starting with the question of which clone to plant and, even more importantly, where.

Argentina has a reasonable amount since waves of Italian immigrants often brought their Italian vine varieties with them. There haven't been any exciting top level, oak-aged wines yet – largely because the quality revolution here is so recent that other more international grapes offer better rewards for less effort and headscratching. But some very nice juicy young reds are now appearing and they do manage to offer that characteristic bitter cherry kernel twist at the finish.

**BIONDI-SANTI**

*Biondi-Santi claims to have created Brunello di Montalcino, but it has not always been regarded as the best producer there. New investment is helping.*

**SEGHESIO**

*The Seghesio family are no newcomers to California: they've been growing grapes there for over a century. These Sangiovese vines were planted in 1910.*

**PIZZINI**

*Victoria's King Valley is strongly influenced by Italian farmers who settled there after the Second World War. Even so, Sangiovese is a minority grape in the region.*

# Enjoying Sangiovese

SANGIOVESE CAN AGE, and age well, but most of its wines are intended to be drunk within a year or two of the harvest; they quickly develop an almost tomato-like leanness if left hanging around for too long. In fact with so many different styles and qualities being made in Italy and with many of these being very recent innovations, it's pretty tricky to generalize about aging. In most regions there are a few producers trying to do something smart with Sangiovese that could merit aging, but even they themselves couldn't guarantee you a successful conclusion.

The longest-lived wines are Brunello di Montalcino and the finest Sangiovese-based super-Tuscans. (Originally these all held the lowly status of vino da tavola, but now they may opt to move up the ranks of the Italian wine classification system, to DOC or IGT – Indicazione Geografica Tipica). In a good vintage these wines may well keep for 20 years, but most can start to be drunk after about five.

Vino Nobile di Montepulciano and the lighter Rosso di Montalcino can also be broached after five years, or sometimes less; all these should be drunk within eight to ten years. Carmignano has roughly the same lifespan.

Chianti is even more variable. Most basic Chianti should be drunk within three or four years of the vintage; Riserva may need a year or two more before it is ready. The very top wines, but only these, may last up to 15 years or more.

Other Italian Sangioveses should be drunk young unless they come from a producer who is specifically making a wine for aging. The same goes for New World Sangioveses: nearly all should be drunk within three or four years, and examples from Argentina are good within a year of the vintage.

### The taste of Sangiovese

The traditional flavours of Tuscan Sangiovese are of bitter cherries and violets, with a certain tomatoey savoriness to the fruit, a definite rasp of herbs and a tea-like finish. Acidity is high and so is tannin: upfront fruit flavours are not the be-all and end-all of traditional Tuscan reds.

That has partly, though by no means entirely, changed. Those traditional flavours are likely nowadays to be richer and more concentrated, with better textures and finer tannins; the acidity is still there, but the greater concentration of flavour makes it less obvious.

The most international styles have a seasoning of vanilla and spice from new oak barriques, and the fruit leans more towards black cherry, plums and mulberries; where Cabernet Sauvignon joins the blend it is likely to be disproportionately dominant, with blackcurrant and plum flavours. These may become less noticeable as the wine ages.

Less ripe and concentrated Sangiovese can be stringy and rustic; that from warm climates can be heavier, broader and more alcoholic, tasting rather stewy and soupy and lacking the finesse of Tuscany's finest.

New World flavours vary from oaky, plummy Cabernet lookalikes to attractively bright, cherry-fruited bottles to those with high alcohol but unripe tannins. None so far really have the poise and finesse of good Montalcino. Not for the first time, emulators have found it surprisingly difficult to capture the quintessence of Italy.

*Fattoria di Felsina's Chianti Classico Riserva is the result of some serious quality-oriented thinking. It is a blend of Sangiovese and Canaiolo from a single vineyard, Vigneto Rancia. Back in the 1980s Paolo de Marchi of Isole e Olena, also in the Chianti Classico zone, planted Syrah to pep up his Sangiovese; and paradoxically in so doing realized that with more careful viticulture the Sangiovese could be just as good on its own. Cepparello was the result.*

### Matching Sangiovese and food

Tuscany is where Sangiovese best expresses the qualities that can lead it, in the right circumstances, to be numbered among the great grapes of the world. And Tuscany is very much food-with-wine territory. Sangiovese-based wines such as Chianti, Rosso di Montalcino or Montepulciano, Vino Nobile and the biggest of them all, Brunello, positively demand to be drunk with food – such as *bistecca alla fiorentina* (succulent grilled T-bone steak), roast meats and game, calves' liver, *porcini* mushrooms, casseroles, pizza, hearty pasta dishes and almost anything in a tomato sauce (Sangiovese's acidity helps here), and tangy Pecorino cheese.

## CONSUMER INFORMATION

**Synonyms & local names**
Also known (especially in Tuscany) as Sangioveto, Brunello or Prugnolo Gentile and Morellino. Corsica calls it Nielluccio.

**Best producers**
**TUSCANY/Brunello di Montalcino** Argiano, Banfi, Barbi, Biondi-Santi, Campogiovanni, Caparzo, Casanova di Neri, Ciacci Piccolomini, Col d'Orcia, Donatella Cinelli Colombini, R Cosimi, Costanti, Fuligni, Gorelli, Lambardi, Lisini, Mastrojanni, Montalcino co-op, Pertimali, Pieve Santa Restituta, Poggio Antico, Poggio Salvi, Il Poggione, Siro Pacenti, Soldera (Case Basse), Valdicava, Val di Suga; **Carmignano** Ambra, Capezzana, Pierazzuoli; **Chianti** Ama, Antinori, Badia a Coltibuono, Basciano, Brolio, Carobbio, Casaloste, Castellare, Cennatoio, Collelungo, Casa Emma, Felsina, di Fonterutoli, Fontodi, Frescobaldi, Isole e Olena, La Massa, Monsanto, Monte Bernardi, Ormanni, Paneretta, Poggerino, Poggio al Sole, Poggiopiano, Querciabella, Rampolla, Riecine, Rietine, Rocca di Castagnoli, Rocca di Montegrossi, Ruffino, San Giusto a Rentennano, Selvapiana, Terrabianca, Valtellina, Vecchie Terre di Montefili, Vicchiomaggio, Villa Cafaggio, Volpaia; **Morellino di Scansano** Erik Banti, Moris Farms, Le Pupille; **Super-Tuscan Sangiovese blends** Altesino, Ama, Antinori, Badia a Coltibuono, Basciano, Biondi-Santi, Capaccia, Castellare, Felsina, Fonterutoli, Fontodi, Frescobaldi-Mondavi, Gagliole, Isole e Olena, Lilliano, Monsanto, Monte Bernardi, Montepelosio, Montevertine, Moris Farms, Paneretta, Petrolo, Poggerino, Poggio Scalette, Poggiopiano, Querceto, Riecine, Rocca della Macie, Rocca di Montegrossi, Ruffino, San Felice, San Giusto a Rentennano, Michele Satta, Valtellina, Vecchie Terre di Montefili, Villa Cafaggio, Volpaia; **Vino Nobile di Montepulciano** Avignonesi, Bindella, Boscarelli, La Braccesca, Le Casalte, Contucci, Dei, Del Cerro, Il Macchione, Poliziano, Valdipiatta.
**Other central Italy** Boccadigabbia, Castelluccio, La Carraia, Lungarotti, Zerbina.
**USA/California** Altamura, Atlas Peak, Dalla Valle, Ferrari-Carano, Plumpjack, Saddleback, Seghesio, Shafer, Staglin Family, Swanson; **Washington State** Leonetti.
**AUSTRALIA** Coriole, Pizzini.

## RECOMMENDED WINES TO TRY

**20 classic Tuscan reds (from Sangiovese and other Tuscan varieties)**
Biondi-Santi *Sassoalloro*
Castello di Brolio *Chianti Classico*
Castellare *I Sodi di San Niccolò*
Fattoria di Felsina *Fontalloro*
Castello di Fonterutoli *Chianti Classico Riserva*
Fontodi *Flaccianello*
Frescobaldi *Chianti Rufina Montesodi*
Isole e Olena *Cepparello*
Castello di Lilliano *Anagallis*
La Massa *Chianti Classico Giorgio Prima*
Monte Bernardi *Sa'etta*
Montevertine *Le Pergole Torte*
Castello della Paneretta *Quattrocentenario*
Poggio Scalette *Il Carbonaione*
Fattoria Petrolo *Torrione*
Riecine *La Gioia*
Rocca di Montegrossi *Geremia*
San Giusto a Rentennano *Percarlo*
Selvapiana *Chianti Rufina Riserva Vigneto Bucerchiale*
Castello di Volpaia *Coltassala*

**Five Sangiovese-Cabernet blends**
Antinori *Tignanello*
Gagliole *Gagliole Rosso*
Moris Farms *Avvoltore*
Querciabella *Camartina*
Ruffino *Cabreo Il Borgo*

**Five Sangiovese-Merlot blends**
Castello di Ama *Chianti Classico La Casuccia*
Antinori/Badia a Passignano *Chianti Classico Riserva*
Castello di Fonterutoli *Siepi*
Frescobaldi-Mondavi *Luce*
Poggerino *Primamateria*

**Five non-Tuscan Sangiovese wines**
Boccadigabbia *Marche Sangiovese Saltapicchio*
Castelluccio *Ronco delle Ginestre*
La Carraia *Umbria Sangiovese*
Lungarotti *Torgiano Rubesco Riserva Vigna Monticchio*
Zerbina *Sangiovese di Romagna Superiore Riserva Pietramora*

**Five New World Sangiovese wines**
Atlas Peak *Reserve Sangiovese (California)*
Coriole *McLaren Vale Sangiovese (Australia)*
Dalle Valle *Napa Valley Pietre Rosse (California)*
Leonetti *Walla Walla Valley Sangiovese (Washington)*
Shafer *Firebreak (California)*

*Intensive research is pushing the quality of Sangiovese further and further forward. We have probably not yet seen the best it can do – and if that applies to Tuscany, which it does, it applies even more to the New World.*

**Maturity charts**
Simple wines can and should be drunk within a few years; more concentrated, tannic versions need longer.

**1997** Chianti Classico/Rufina (both Riserva)

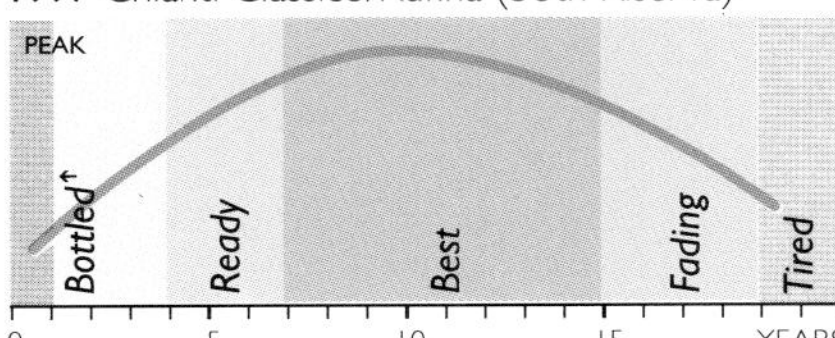


*1997 was a super-ripe vintage in Tuscany, and Chianti from this year displays lush, opulent fruit and relatively low acidity and tannin.*

**1997** Brunello di Montalcino

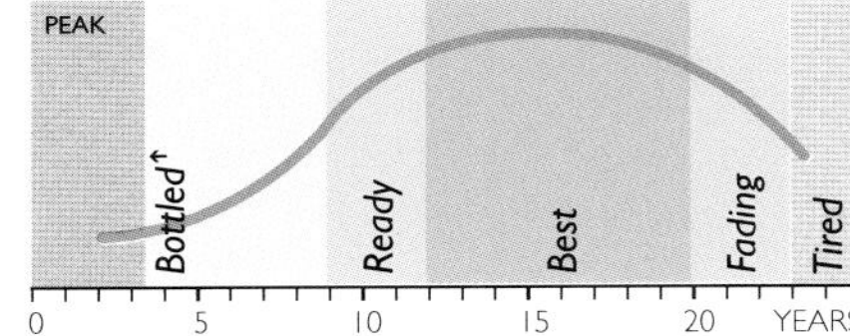


*An exceptionally rich vintage, in which the Sangiovese reached high levels of ripeness. The wines have seductive fruit and lowish acidity.*

**1999** Chianti (Normale)

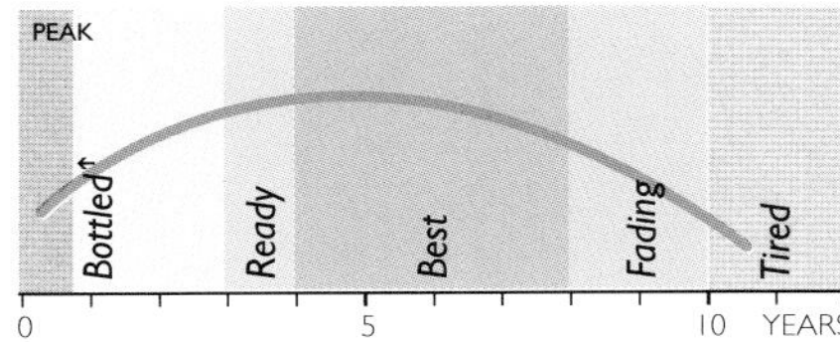


*Tuscany did better than some more northerly regions in 1999. Basic Chianti is good to very good in this vintage.*

# Sauvignon Blanc

Sauvignon Blanc is the most useful grape variety in the world. To me, that is. Not to everybody, I admit. But I love it and I *need* it. And the reason is simple. Every year I travel the country holding tastings for people – not wine buffs, connoisseurs – but nice straightforward people who drink wine but don't pretend they're experts. And if I'm going to persuade them that it's worth making the effort to learn a bit more about wine, maybe pay a bit more for their bottles – I have to have wines that sing, that shout, that roar their personalities. This is my one chance to win them over. Many of my audience have never been asked to tell the difference between two wines before. When I say 'this wine tastes of nettles, or capsicum, or gooseberries' it *has* to do just that, no shillyshallying about subtlety, elegance, finesse – they're nuances that can come later. And when I say 'let's have a "how to wine taste" lesson – you know, look at it, smell it, taste it' I need a wine that is totally reliable, that will always give me a painter's palette of flavours anyone can understand as soon as they smell and taste the wine. Sauvignon Blanc. That's my baby.

*A piece of the trellis from the potager garden at the château of Villandry illustrates the Loire Valley's long association with Sauvignon Blanc. The flavours of Sauvignon Blanc are linked to myriad fruit and vegetables but none more so than gooseberries. Behind the trellis stretch the vineyards and hills of the Marlborough region in New Zealand's South Island, the world's new classic area for Sauvignon Blanc.*

Well, not just any Sauvignon Blanc. The Sauvignon Blanc that I mean is New Zealand Sauvignon, and if you ask me why I don't use one of the so-called classic Sauvignon Blancs of Bordeaux or the Loire Valley in France – I say – *this* is the classic. South Island New Zealand Sauvignon Blanc is the classic. I know the first example was only made 25 years ago, but there wasn't a classic Sauvignon before then. Oh, sure, there was the occasional good Sancerre or Pouilly-Fumé – the Loire Valley's top Sauvignon wines – but growers there weren't thinking about the flavour of the grape. They were – and are – obsessed with the flavour of their terroirs, the places where the grapes grow rather than the actual grapes themselves.

It took the South Island of New Zealand, and the region of Marlborough in particular, to show the astonishing flavours of the grape itself, largely because there wasn't any terroir talk in Marlborough because no-one had ever planted grapes there before to create any terroir. Nowadays it's different and you do find people talking about this or that bit of land, but to begin with no-one cared or knew. What they did realize, immediately, was that this brilliant, pungent, aggressively green yet exotically ripe style of wine was unlike anything the world had ever seen before. There had never been a wine with such outspoken, cut-glass purity of flavours, and I've been a devoted fan ever since – partly because I do love drinking wine that is so refreshing, and partly because no wine has ever been such a brilliant teaching aid, before or since.

I think it's fair to say that New Zealand Sauvignon Blanc changed the wine world by changing our ideas of what wine could be like, just as the Chardonnays of California and Australia did. And for a long time these two wines occupied seats in the opposite corners of the White Wine boxing ring: the pungent, green, aggressive Sauvignon, bare and unoaked, and the warm, round, soft, creamy, spicy, tropical Chardonnay with oak a fundamental part of its attraction. Now, many other varieties are filling in the gaps between the two, but in the early years of the Wine Revolution, when we looked for leaders we found two whites – sexy Chardonnay, swathed in oak, and Sauvignon, naked as nature intended – and with attitude.

**Sauvignon Blanc: from Grape to Glass**

*Geography and History page 220; Viticulture and Vinification page 222; Sauvignon Blanc around the World page 224; Enjoying Sauvignon Blanc page 226*

# Geography and History

I'M TEMPTED TO SAY there's no geography and no history of Sauvignon Blanc before 1973. That's when the first Sauvignon vines were planted in the Marlborough region of New Zealand's South Island. Within a few years they'd produced a wine of such shocking, tongue-tingling pungency that the world of wine was never the same again. Well, I'm tempted, but in fact Sauvignon Blanc had been around elsewhere for donkey's years, except that it never gave a wine with half so much excitement as New Zealand's offerings, and you virtually never saw the name Sauvignon on the label, so in any case you probably didn't know you were drinking it.

But look at the map – it's all over the place. New Zealand, sure, Australia, South Africa, Chile, northern Italy, Hungary, the Loire Valley, Bordeaux – almost everywhere you look

Major Sauvignon Blanc plantings
Other Sauvignon Blanc plantings

Major planting figures for Sauvignon Blanc
AREA PLANTED (HECTARES)

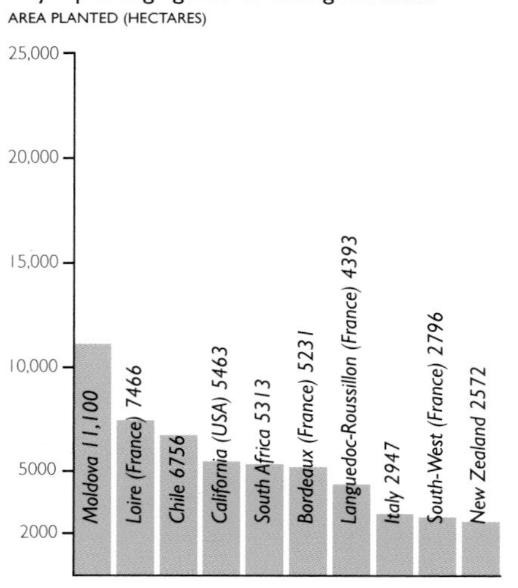


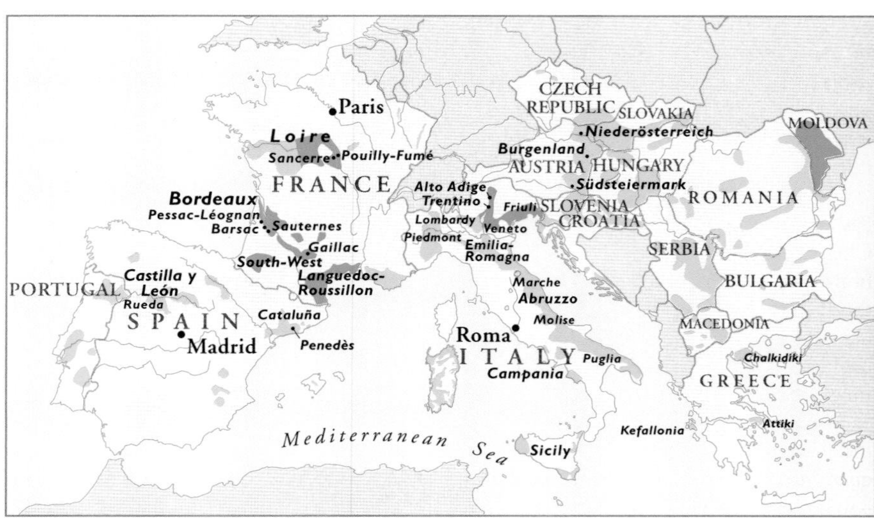

you'll find Sauvignon planted. Which might imply it's a marvellously adaptable grape. But in fact it isn't. Sauvignon Blanc is widely planted largely because growers in new areas couldn't think what else to plant. They'd planted the great Bordeaux red grapes and Burgundy's red and white grapes – Pinot Noir and Chardonnay. So what next? Well, Bordeaux's white grapes might be an idea – and so, regardless of how suitable the variety was, that's what got planted.

It didn't always work. Sauvignon seems to be indigenous to the Loire Valley and Bordeaux in France, and it can make good wine in both regions if looked after carefully. But although Sauvignon likes sunshine, too much heat quickly destroys its whole purpose in life – its fantastic, memorably shrill aroma and tang. So neither Australia nor California have found it easy though Australia does now have some good, cool-climate versions. California has barrel-fermented a few good ones. Parts of South Africa have succeeded and there's a fair amount of successful Chilean Sauvignon. Hungary is producing good stuff, as is northern Italy. And then there's New Zealand South Island. Which is where we came in.

## Historical background

Both south-west France and the Loire Valley claim Sauvignon Blanc as an indigenous grape. Wherever its birthplace, at some time in or before the 18th century, and presumably in Bordeaux, it got together with Cabernet Franc to produce the seedling that became known as Cabernet Sauvignon, and if it had done nothing else in its existence, wine lovers would have to thank it for that.

Its current fame in the South-West, however, is of quite recent date. Until the late 1980s it languished behind Sémillon and Ugni Blanc in terms of the numbers of hectares planted and its wine was generally rather raw and earthy. If it did indeed originate in the South-West and spread from there to the Loire (the opposite seems more likely) then we have a rare instance of the sort of happy accident by which a vine that produces generally indifferent wine at home (and frankly 'indifferent' is high praise for most of the Bordeaux Sauvignon that was creeping out into the critical spotlight until the 1990s) suddenly excels elsewhere.

The Sauvignon Blanc planted in Bordeaux in the 19th century must have been mixed up with Sauvignon Vert, alias Sauvigonasse, a pretty uninspired poor relation of the real Sauvignon. Since Chile got its Sauvignon from Bordeaux cuttings before phylloxera and the two are mixed up there, it is reasonable to assume its field blend was inherited.

There is also a pink mutation of Sauvignon Blanc, known as Sauvignon Gris: Chile has a bit of this, and there is also some in Bordeaux. It gives 20 per cent lower yields than Sauvignon Blanc, one degree more alcohol and a less pungent but spicier aroma. The berries are more deeply coloured than those of Sauvignon Blanc. However, Sauvignon Gris seems not to be the same as the Sauvignon Rosé mutation found in small quantities in the Loire. And modern history – well, that started in New Zealand, in 1973.

*Vineyards in the rolling hills above Bué and Venoize in the Sancerre appellation in the eastern Loire Valley. There are 14 different communes in the appellation, so wine styles are necessarily heterogeneous. In fact, the much-vaunted differences between Sancerre and Pouilly-Fumé are far less than the differences to be found within the Sancerre appellation, with its varying soil types and its hills that rise to 400m.*

*The chalky soils of Bué produce some of the most highly regarded Sancerres of all, with minerally fruit and balance.*

*Picking Sauvignon in Didier Dagueneau's vineyard at les Berthiers, Pouilly-Fumé. Dagueneau's most famous wine is probably his Cuvée Silex, grown on flinty soil.*

# Viticulture and Vinification

IF SAUVIGNON BLANC DOESN'T ASSAULT THE TASTE BUDS with a great whoosh of flavour, there's no real point to it. Sauvignon ripened so much that the acidity falls is Sauvignon flattened and eviscerated. Which unfortunately means – since Sauvignon is a late budder and early ripener – that there is a lot of pretty dull Sauvignon around from warmer regions. And yet warm countries that really put their mind to it are beginning to make good Sauvignon.

South African Sauvignon gets better every year. Australia's baking hot Riverland region is beginning to produce full-flavoured but tangy examples. The south of France and northern Spain are showing they can do it too. The key is arranging your harvest according to the taste of the grapes – picking them when they are full of aroma and streaked with green acidity – rather than going by supposedly suitable sugar ripeness – and temperature-controlled fermentation.

## Climate

To get that too-often-elusive balance between sugar ripeness, acidity and aroma, the right climate is crucial. Once you've got it then the soil can certainly influence the flavour of Sauvignon; but the climate must be the first priority for anyone thinking of planting Sauvignon Blanc.

It's both a late budder and an early ripener: it doesn't therefore need enormous heat. In France it flourishes in both the maritime climate of Bordeaux and the more continental climate of Sancerre in the Loire Valley; New Zealand's climates, too, are maritime. Slow ripening gives better flavour development, but optimum aroma occurs just before optimum sugar ripeness. This is important. The greatest intensity of aroma is found just before the ideal balance of sugar and acidity; choosing the picking date means a slight compromise in one direction or the other. Ideally, it should, of course, be as slight as possible. Part of the improvement in quality in Sancerre in the last decade has been produced by better-judged picking dates; and while logically these should be later, since the wines clearly taste riper, better balanced and more interesting than they did, most growers, when questioned, claim they pick earlier. Global warming, they say, has advanced the maturity of the grapes. I'd say awareness of the success of Sauvignon from New Zealand's South Island is an equally likely answer.

In New Zealand, different systems of canopy management have been aimed at producing riper grapes in a cool but sunny climate and with soils of high potential vigour. I think they're overdoing it. New Zealand Sauvignon has begun to lose much of its pungency and become fatter and more tropical. Since New Zealand Sauvignon was a world standard of pungent citrus fruit, this is surely barmy. Why change a world standard?

*Sorting Sauvignon Blanc at Domaine de Chevalier. Work on the variety by Professor Denis Dubourdieu and others has enabled Bordeaux to produce Sauvignon that is both riper and more aromatic than in the past. Careful selection, as here, is a vital part of the process.*

In warmer climates growers generally pick early to keep acidity in their grapes, but in so doing seldom get the best aromas. California is a classic example: few of its Sauvignon Blanc wines have any character, so much so that the belief that Americans don't like Sauvignon Blanc becomes a self-fulfilling prophecy.

In Chile, winemaker Ignacio Recabarren picks Sauvignon Blanc for one of his best wines at several different times: unripe for grapes with high malic acid; 20 per cent riper for red and green pepper flavours; 40 per cent at perfect ripeness, and 10 per cent at overripeness. Where did he learn such habits? Why, the South Island of New Zealand, which he luckily visited before New Zealand started to damp down the tongue-tickling fire of its Sauvignon Blanc.

## Soil

The question of soil and Sauvignon Blanc is really confined to France, with a glance at New Zealand; no other countries pay such attention to the issue.

In the Loire Valley soils in Sancerre and neighbouring Pouilly vary from chalk over Kimmeridgean marl – this produces the best balanced wines, with richness and complexity – to the compact chalk, or *caillotte*, found at the base of the hills – this gives finesse and perfume – to flint, or silex, which gives wines with a certain gunflint sparkiness and vigour. There are also warm terraces of sandy or gravelly soil near the river which give spicy, floral flavours to wines which are the earliest maturing of all. Marl gives wines that age better, but the longest lasting wines come from silex.

In Bordeaux the soils usually allotted to Sauvignon are more alluvial and produce high yields, which accounts for the lesser pungency of the wine here.

In New Zealand's Marlborough region the soils are particularly varied, though not all are well suited to vines. The best soils are stony or sandy over shingle: these have poor fertility and good drainage.

In the Sauvignon Blanc heartland of the flood plain of the Wairau Valley soil types run in bands that go from east to west. This means that if vines are planted from north to south, then a single row can contain vigorous vines with large canopies and feeble ones with small canopies. Interestingly, this mix of weak and strong, of tropical and citrus, unripe and ripe, can produce pretty interesting, pungent stuff. The heavier soils tend to be later ripening, and give more herbaceous flavours; stony soils are warmer and thus earlier ripening, and give riper, lusher flavours. Mix them cleverly and you've got pungency plus ripeness. That's what made New Zealand famous. That's what the best producers still try to do.

*Sauvignon Blanc vines at Montana's Brancott Estate near Blenheim, Marlborough. Blenheim gets more sunshine hours than any other town in New Zealand, thanks to the Southern Alps, which provide a handy rain shadow. Yet it's still quite cool in viticultural terms, so Sauvignon Blanc doesn't race to overripeness and flabbiness. Because it ripens fairly early the vine can handle cooler summers than othe varieties.*

## Yields

If what you're after with Sauvignon Blanc is aromatic, crisp, fresh wine for early drinking, then yields are not an enormous problem. The maximum yield in Sancerre is 68hl/ha, including the *plafond limite de classement*, by which growers are allowed an increase in yield in prolific years, and actually Sauvignon Blanc tolerates these sort of levels reasonably well. If you want more serious, fatter, weightier wines, ones that will last and improve for several years, then 50hl/ha is enough. At 40hl/ha the wine may well be very serious, with extract and staying power, but it will not be most people's idea of Sancerre. Yields are lower than this at good Graves and Pessac-Léognan estates in Bordeaux: under 40hl/ha at a few estates, notably Château Pape-Clément, but more usually between 40hl/ha and 55hl/ha. Since most top Bordeaux whites are barrel-fermented, these lower yields are crucial for a deep, balanced, ageworthy wine.

In the New World yields are higher again: at least six tons per acre (108 hl/ha) in California, and around 8–12 tonnes per hectare (58–87hl/ha) even in Chile's relatively low-yielding Casablanca Valley. In the Central Valley yields can top 15 tonnes per hectare (109hl/ha).

## At the winery

Fermentation temperatures are a major point of difference between producing Sauvignon in the Loire Valley and the New World. In the Loire, fermentation (either in steel or wood) is at around 16–18°C, in order to avoid the tropical fruit aromas that the New World seeks with its cooler temperatures. These relatively warm fermentations produce wines that are minerally rather than exuberantly fruity: Loire winemakers prefer their wines to reflect their terroir rather than the grape variety.

Denis Dubourdieu, Professor of Enology at Bordeaux University and high priest of Sauvignon, whose work has lifted Bordeaux Sauvignon to a level unthought of 20 years ago, points out that it is the peaks of temperature during barrel fermentation – these peaks can touch 25°C – that give richness and varietal aroma to the wine. Fermentation in stainless steel, with tightly controlled temperatures, gives more monochrome wines with fewer characteristics of the terroir and the variety.

But there are far more ways of tinkering with Sauvignon in the winery than merely adjusting fermentation temperatures. California winemakers have largely abandoned pre-fermentation skin contact, believing that it produces wines that age poorly, but some Loire producers favour a short period of skin contact for more explosive fruit flavours. But only for a small part of the wine, and not for so long that harsh flavours are extracted.

When South Island New Zealand Sauvignon first hurled itself on to our unsuspecting palate with thrilling flavours of gooseberry, green pepper, asparagus and passion fruit, one reason for the wine's intensity was an unavoidable period of skin contact. Why? Well: there were no wineries in the South Island and grapes had to travel by truck and ferry all the way to Auckland up through North Island. At the height of harvest time that could mean grapes and juice sloshing around together for up to 24 hours. A wine of particular pungency could always be blamed on a traffic jam at the ferry.

Both the gooseberry/green pepper flavours and the tinned asparagus flavours of Sauvignon come from a group of flavour compounds called methoxypyrazines, which are found at higher levels in Cabernet Sauvignon; their concentration is greater in grapes from cooler climates.

Aging on the fine lees works well: the lees protect against oxidation, and if the wine is in barrel they prevent it from becoming too oaky. *Bâtonnage* is used to increase weight; there is even some malolactic fermentation being practised in the Loire, though I don't really think it has an exciting effect; it seems better suited to the less pungent wine of Bordeaux's barrel-fermented wines.

There is also some new oak creeping into the Loire, though again the style is atypical. New oak is more usually thought of as belonging to Bordeaux and the New World: the term Fumé Blanc, which has no legal meaning, may be applied to New World Sauvignons (or indeed wines from other white grapes) with new oak aging. I suppose the idea for the name came from Pouilly Blanc Fumé in the Loire, but the flavours have nothing whatever to do with traditional Loire Sauvignon Blanc.

# Sauvignon Blanc around the World

FRANKLY, I GENERALLY PREFER my Sauvignon Blanc as a pure thrilling blast of all the grape has to offer – unashamed, unblended, love it or loathe it. But some faint hearts prefer to temper the wine's fierceness by blending – and Sémillon is usually the grape of choice. Indeed, in Pessac-Léognan, I actually prefer the Sémillon-Sauvignon Blanc blend.

## France: Loire Valley

The finest appellations are Sancerre and Pouilly-Fumé; and in spite of the tradition that Pouilly-Fumé is marked by a characteristic whiff of gunflint, the difference between the two wines is less than differences found between Sancerre's different soils and villages. The 'Fumé' was appended to Pouilly to distinguish the area's Sauvignon from its Chasselas, which has the AC of Pouilly-sur-Loire.

At the moment Sancerre is a more go-ahead, quality-conscious AC than Pouilly-Fumé. The older style of grassy, gooseberry-ish fruit is being replaced by richer, more peach and melon notes. The growers of the Loire have certainly been spurred into doing better by the success of New Zealand's South Island. New Zealand Sauvignon is not made in a style they admire. They want to achieve something different.

Lesser, but attractive and distinctive wines are produced nearby in Quincy, Reuilly and Menetou-Salon. These are snappy, fresh wines, but lack the depth of a good Sancerre or Pouilly-Fumé: Quincy is the most intense and gooseberryish, sometimes a bit too green if the grapes are picked too early. Reuilly is more like a gentle version of Sancerre. Menetou-Salon has the Sancerre snap and nettly tang; because it is less popular, costs less too.

## France: Bordeaux

Denis Dubourdieu has been the greatest single influence behind the enormous improvement in quality of Sauvignon Blanc at all levels. Winemaking, viticulture and clones have all improved, particularly in the last ten years; even basic Bordeaux Blanc (which may be 100 per cent Sauvignon or Sémillon, or a blend of Sauvignon and Sémillon with or without a little Muscadelle) is now pretty reliably fresh. It is always less pungent than Sancerre; the dry wines of the Graves and Pessac-Léognan are often fermented and aged in new oak.

In Sauternes and Barsac, and in the outlying sweet wine appellations of Loupiac, Cérons, Cadillac and Ste-Croix-du-Mont, the proportion of Sauvignon in the vineyards varies between 10 and 40 per cent. Its thin skin makes it highly susceptible to botrytis, and its acidity adds freshness to the blend. See Sémillon (pages 234–235) for more detail.

## Rest of France

Sauvignon is widely planted for white wines throughout the South-West; in the Languedoc it is fairly popular for Vin de Pays wines, usually made in a slightly fat Chilean style. The warm climate and high yields combine to make it difficult to produce Sauvignon with Loire-style aroma and freshness but there are one or two standouts.

## Rest of Europe

Spain's most notable Sauvignon comes from the Rueda DO in Castilla y León, where it was introduced in the early 1980s. Styles are ripe, but the peachy fruit is balanced by a nettly acidity, usually retained by picking the grapes as early as August. Some new oak may be used. It is also an authorized variety for the La Mancha DO. Austrian versions are nettly or spicy, with the ripest wines coming from Niederösterreich and Burgenland, and restrained, understated wines from Sudsteiermark.

It is not known when Sauvignon was first planted in Italy, but it seems, in its early days at least, to have been grown alongside Sauvignonasse. Its first port of call was Piedmont, although such Sauvignon Blanc as is made there today (there isn't much, and Gaja's is the best) was planted in the 1980s and 1990s. It is a grape of the North, producing its most typical varietal aromas in Collio, Friuli and Alto Adige. Italian producers may make Sauvignon as a varietal, or they may blend it with anything and everything: Chardonnay, Müller-Thurgau, Ribollo, Picolit, Vermentino, Inzolia, Tocai, Malvazia Istriana, Pinot Bianco and Erbamatt.

The grape has potential for good quality in many countries of central and eastern Europe; there are large plantings in Romania and Moldova; and the Czech Republic and parts of the former Yugoslavia, notably Slovenia, can make promising styles.

## New Zealand

The first Sauvignon Blanc was not planted in Marlborough, in the South Island, until the

LUCIEN CROCHET

*Crochet's domaine-bottled Sancerres are textbook examples, all fresh minerally fruit and ripe acidity. This wine is for drinking young.*

CHÂTEAU COUHINS-LURTON

*This property produces a wonderfully rich and intense barrel-fermented Sauvignon Blanc which, unlike most Pessac-Léognan wines, contains no Sémillon.*

CLOUDY BAY

*The most famous New Zealand Sauvignon of them all, Cloudy Bay achieved instant cult status with its first vintage in 1985.*

1970s, when Montana made the inspired decision to plant a trial plot. The main attraction of this then little-known region was its cheapness, and its possible suitability for Müller-Thurgau, then a far more important vine than Sauvignon in New Zealand. Montana's first Marlborough Sauvignon was made in 1980; by 2001 there were 1923 ha of Sauvignon planted in Marlborough and it is now the leading region for the grape. (By contrast, Müller-Thurgau across the country is down to less than a quarter of its 1991 figure.) The typical Marlborough style – the climate is cool, dry and sunny – has a startling clarity and purity of fruit, made more complex, but no less distinctive, when techniques such as *bâtonnage*, malolactic fermentation, new oak fermentation, the blending in of a little Semillon, and the leaving of a few grams of residual sugar to balance the acidity, are used to season and temper this taste.

Wines from Hawkes Bay are slightly less distinctive, from Nelson a little softer, and Wairarapa makes several styles, some of which can equal most Marlborough examples for pungency and fruit.

### Chile and South America

Just how much of Chile's Sauvignon Blanc is in fact Sauvignonasse is not clear: a distinction between the two varieties was only made in the early 1990s. Only plantings after about 1995 are of Sauvignon Blanc proper, and nearly all wines at the moment are still blends of the two. As a result many wines lack aroma, and fade quickly. But even pure Sauvignon Blanc from Chile is lower in acidity than that of New Zealand, and does not need the latter's residual sugar to achieve balance. Yields are also high. Chile clearly has enormous potential with Sauvignon and styles are likely to improve dramatically.

*An oil burning stove in Chile's Casablanca Valley; such stoves are effective ways of keeping frost away, but may be forbidden under new anti-pollution laws. Casablanca is close to ultra-polluted Santiago.*

The leading region, though not the largest, is Casablanca, where plantings began in about 1990, and where Sauvignon takes second place to Chardonnay. Only about 5 per cent of Chile's Sauvignon is in Casablanca, and nearly all the rest languishes in much hotter climates. Casablanca, with its Mediterranean climate, has similar daily temperature variation to Marlborough, but is closer to the Equator than comparable regions like Carneros, Burgundy and indeed Marlborough. This means that autumn days are longer and warmer, and picking can be later. Casablanca is Chile's coolest wine region to date: to put it into a Chilean context, Sauvignon is picked in Casablanca five to six weeks later than it is in warmer Curico.

Sauvignon Blanc is also found in Mexico, Argentina, Uruguay and Bolivia. In Brazil, according to Galet, the vine called Sauvignon Blanc is really Seyval Blanc.

### Australia

A distinctive Australian style seems to be emerging: riper than that of New Zealand, with white peach and lime juice flavours rather than gooseberry and herbaceous ones. It is not minerally like the Loire, but has good acidity. If too alcoholic and overripe these wines become oily and heavy; they can also be too soft and undefined. But when they get it right they're extremely promising. Adelaide Hills, Padthaway and the Southern Vales are leading areas, and Margaret River makes some excellent blends with Semillon.

### USA

The situation in California is complicated by the relatively low price that Sauvignon fetches – about half that of Chardonnay. Growers are therefore encouraged to allow the vine to crop heavily, which depresses quality. There are two main styles: one, New Zealand-influenced, is bright-fruited, with passionfruit and citrus flavours; the other is rounder and more melony, and often oaked. Many winemakers fear that consumers won't like the grape's assertive nature, and so tone it down.

There is some Sauvignon in Washington State, but the acreage is declining.

### South Africa

Nettles, asparagus and gooseberries are common flavours here; the wines are vivid and brisk and less tropical than in Australia or California. Stellenbosch has the most planted and Elgin wines are noteworthy. In 1999 Sauvignon covered 5.1 per cent of the total vineyard to Chardonnay's 5.7 per cent.

**Viña Casablanca**
*Casablanca is both the name of the company and of the region here; the wine has marvellously aggressive grassy, herby and gooseberry fruit.*

**Lenswood**
*Tim and Annie Knappstein of Lenswood make wines with great balance and finesse, combined with considerable complexity, in cool Adelaide Hills vineyards.*

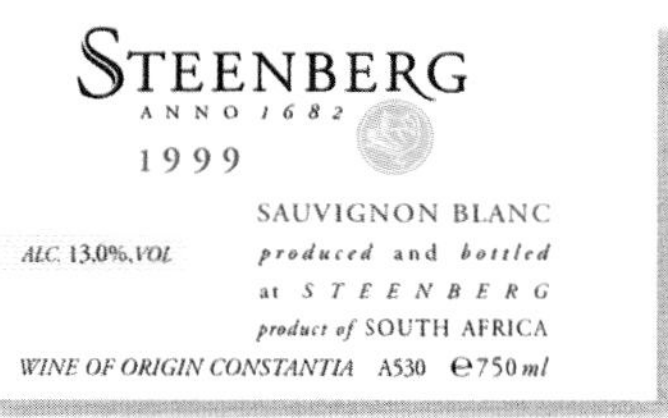


**Steenberg**
*Wine was first made at Steenberg in 1695, though the vineyards have been replanted since the 1990s and today's Sauvignon is tangy and super-fresh.*

# Enjoying Sauvignon Blanc

GENERALLY SPEAKING, SAUVIGNON BLANC is not a wine made to last. Its attraction is its youthful freshness and zest, and the fact that it can be drunk immediately, in the spring following the harvest. Most Sauvignon Blancs, if kept longer than a couple of years, fade rapidly and lose their aroma.

Those that will improve in bottle come not just from particular areas but from particular growers who take a decision to make a wine for keeping. This means, first and foremost, restricting yields: serious, ageworthy wines do not come from the same generous crops as light, early-drinking ones.

New Zealand's best known Sauvignon Blanc, Cloudy Bay, improves in bottle for between five and ten years, depending on the vintage. Top Sancerres and Pouilly-Fumés can develop in bottle for eight to ten years, developing flavours of honey and toast to replace those of nettly fruit, but always keeping a mineral streak; top white Graves and Pessac-Léognan change dramatically with age. They often start out with nettly acidity, bright nectarine fruit and gentle custardy oak and change over ten to 15 years to magnificently deep nutty, creamy wines. Domaine de Chevalier, Smith-Haut-Lafitte, Haut-Brion and Laville-Haut-Brion can last even longer, and may well stay in excellent shape for 20 years or even longer.

Sweet wines are made from blends of Sauvignon and Sémillon in Sauternes and Barsac, and sometimes from pure Sauvignon in California and New Zealand. Classed growth Sauternes reach maturity after a decade or so, but are nevertheless delicious younger; California and New Zealand versions vary in their ageability, but most will improve for up to five years. Top Australian examples are beautiful at two to three years, but should age for a decade or so.

### The taste of Sauvignon Blanc

When I'm trying to describe why I absolutely love the flavour of Sauvignon Blanc, I have to accept it's one of those grapes some wine people simply can't stand. That's okay. They don't have to drink it. All I ask is that they don't try to change it into something bland and soft or they'll find me challenging them to a furious fistfight. I love its taste of gooseberries, its taste of green peppers sliced with a silver knife, passion fruit and kiwi scattered with lime zest, nettles crushed up with blackcurrant leaves. These are the kind of flavours that make Sauvignon for me irresistibly refreshing.

If you like riper tastes, well, Sauvignon develops a spectrum of white peach, nectarine and melon masking any excess acidity; wines with a touch of botrytis may have a whiff of apricot. Sancerre and Pouilly Fumé often have a minerally streak, particularly if they are grown on flinty silex soil: generally clay gives more richness, chalk lightness and perfume.

Lower fermentation temperatures produce a range of tropical fruit flavours: pineapple, banana and guava – dangerous unless balanced by good acidity.

New oak aging will give the wines a vanilla sheen; malolactic fermentation may add butter to the palate. With bottle age Sauvignon takes on tastes of honey and toast and quince, less obviously fruity, but rich and complex.

Botrytized sweet wines have flavours of pineapple and marzipan, oranges and apricots, with often piercing acidity to cut through the richness.

*At Château Smith-Haut-Lafitte in Bordeaux about 5 per cent of Sauvignon Gris is included in what is otherwise a pure (and very long-lived) Sauvignon Blanc wine; this mutation was discovered in the vineyards around 40 years ago. Mulderbosch was one of the first South African estates to show how thrilling Cape Sauvignon can be.*

### Matching Sauvignon Blanc and food

This grape makes wines with enough bite and sharpness to accompany quite rich fish dishes as well as being an obvious choice for seafood. The characteristic acid intensity makes a brilliant match with dishes made with tomato but the best match of all is white Sancerre or Pouilly-Fumé with the local sharp *crottin* goats' cheese of the Upper Loire Valley. With their strong, gooseberry-fresh taste Sauvignons make good thirst-quenching apéritifs.

## CONSUMER INFORMATION

### Synonyms & local names

Sometimes called Blanc Fumé in the central Loire; there are variations called Jaune, Noir, Rose or Gris and Violet according to the berry colour; it is called Muskat-Silvaner or Muskat-Sylvaner in Germany and Austria (though Steiermark often uses the name Sauvignon Blanc); oaked versions are also known as Fumé Blanc in California and Australia. The variety Sauvignon Vert or Sauvignonasse is unrelated.

### Best producers

**FRANCE/Pouilly-Fumé** Henri Bourgeois, Alain Cailbourdin, Jean-Claude Chatelain, Didier Dagueneau, Serge Dagueneau, de Ladoucette, Alphonse Mellot, de Tracy; **Sancerre** F & J Bailly, Bailly-Reverdy, Pierre Boulay, Henri Bourgeois, Francis & Paul Cotat, Lucien Crochet, Pierre & Alain Dezat, Gitton, Pascal Jolivet, Alphonse Mellot, Joseph Mellot, Millérioux, Henry Pellé, Vincent Pinard, Bernard Reverdy, Jean-Max Roger, Lucien Thomas, Vacheron, André Vatan; **Bordeaux/Pessac-Léognan** Carbonnieux, Domaine de Chevalier, Couhins-Lurton, de Cruzeau, de Fieuzal, Haut-Brion, la Garde, la Louvière, Malartic-Lagravière, Pape-Clément, de Rochemorin, Smith-Haut-Lafitte.
**NEW ZEALAND** Babich, Cairnbrae, Clifford Bay Estate, Cloudy Bay, Corbans (Stoneleigh, Cottage Block), Craggy Range, Kim Crawford, Delegat, Giesen, Goldwater (Dog Point), Grove Mill, Hunter's, Isabel Estate, Jackson Estate, Lawson's Dry Hills, Matua Valley, Montana, Mount Riley, Nautilus, Neudorf, Nobilo, Ponder Estate, Palliser, St Clair, Allan Scott, Seifried, Selaks, Seresin, Vavasour, Villa Maria, Wither Hills.
**ITALY** Abbazia di Novacella co-op, Peter Dipoli, Gravner, Haderburg, Inama, Edi Kante, Lageder, Nalles-Magré Niclara, Tenuta dell'Ornellaia, San Michele Appiano co-op, Terlano co-op, Vie di Romans, Villa Russiz, Baron Widmann.
**AUSTRIA** Erich & Walter Polz, Gross, Lackner-Tinnacher, E & M Tement.
**SPAIN** Hermanos Lurton, Marqués de Riscal, Torres.
**USA/California** Araujo, Babcock, Caymus, Kenwood, Matanzas Creek, Murphy-Goode, Navarro, Quivira, J Rochioli Vineyards.
**AUSTRALIA** Alkoomi, Bridgewater Mill, Brookland Valley, Chain of Ponds, Hanging Rock, Karina, Katnook, Lenswood Vineyards, Ravenswood Lane, Shaw & Smith, Geoff Weaver.
**CHILE** Casa Lapostolle, Villard, Viña Casablanca.
**SOUTH AFRICA** Buitenverwachtung, Graham Beck, Kanu, Klein Constantia, Morgenhof, Mulderbosch, Rustenberg, Saxenburg, Springfield, Steenberg, Vergelegen, Villiera, Waterford.

### RECOMMENDED WINES TO TRY

**Ten classic Loire Sauvignon wines**

Henri Bourgeois *Sancerre le MD de Bourgeois*
Jean-Claude Chatelain *Pouilly-Fumé Cuvée Prestige*
Francis & Paul Cotat *Sancerre Chavignol la Grande Côte*
Lucien Crochet *Sancerre Cuvée Prestige*
Didier Dagueneau *Pouilly-Fumé Pur Sang*
Gitton Père et Fils *Sancerre la Mijonnette*
Alphonse Mellot *Sancerre Cuvée Edmond*
Henry Pellé *Menetou-Salon Morogues Clos des Blanchais*
Vincent Pinard *Sancerre Florès*
Ch. de Tracy *Pouilly-Fumé*

**Five classic Sauvignon-dominated dry white Bordeaux wines**

Domaine de Chevalier *Pessac-Léognan*
Ch. Malartic-Lagravière *Pessac-Léognan*
Ch. Pape-Clément *Pessac-Léognan*
Ch. Margaux *Bordeaux Pavillon*
Ch. Smith-Haut-Lafitte *Pessac-Léognan*

**Five other European Sauvignon wines**

Peter Dipoli *Alto Adige Sauvignon Voglar (Italy)*
Marqués de Riscal *Rueda Sauvignon (Spain)*
Tenuta dell'Ornellaia *Poggio alle Gazze (Italy)*
Vie di Romans *Friuli Isonzo Sauvignon Vieris (Italy)*
Villa Russiz *Collio Sauvignon de la Tour (Italy)*

**Ten New Zealand Sauvignon wines**

Craggy Range *Marlborough Old Renwick*
Grove Mill *Marlborough*
Hunter's *Marlborough*
Isabel Estate *Marlborough*
Lawson's Dry Hills *Marlborough*
Palliser Estate *Martinborough*
Seresin *Marlborough*
Vavasour *Marlborough Single Vineyard*
Villa Maria *Marlborough Clifford Bay Reserve*
Wither Hills *Marlborough*

*The exciting thing about Sauvignon Blanc is its wonderful, unabashed fruit salad bowlful of flavours, all tumbling over one another. These tastes are most obvious when the wine is young, but, especially when barrel-fermented and blended with Sémillon, Sauvignon can produce deep, complex, long-lasting wines.*

### Maturity charts

Sauvignon is usually made for early drinking, though some Sancerres and Pouilly-Fumés will keep and improve for much longer.

**1999** Sancerre/Pouilly-Fumé

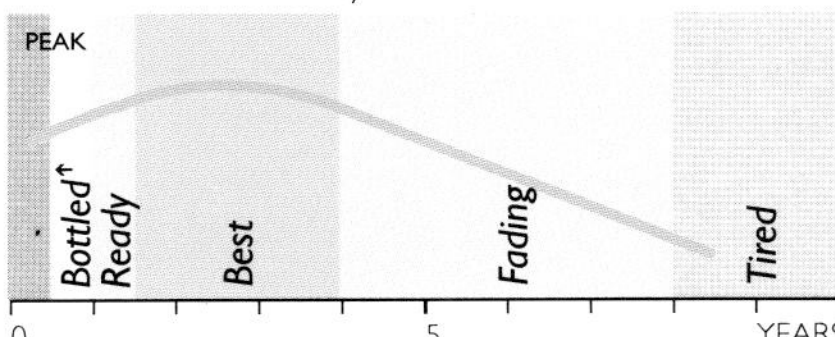


*A few producers make Sancerre that will age in bottle, but generally it is a light, fresh wine intended for early drinking.*

**1999** Pessac-Léognan Cru Classé

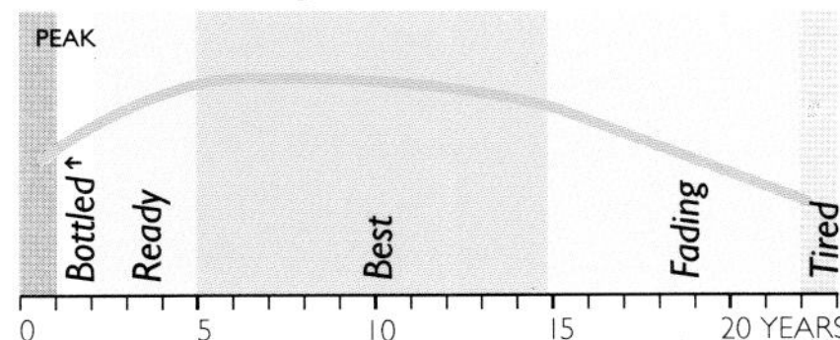


*An excellent vintage at the top level. The slowest maturing wines, like Domaine de Chevalier, need seven or eight years; others are ready earlier.*

**2000** New Zealand Marlborough Sauvignon Blanc

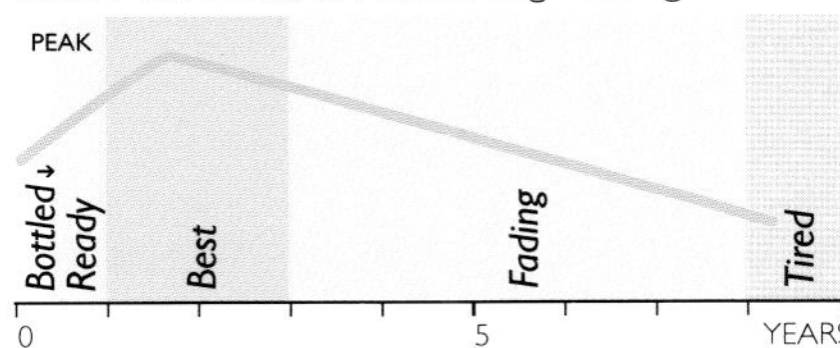


*A couple of years in bottle is generally enough for even the top wines; more everyday bottles are best drunk within the year of the vintage.*

## SAUVIGNONASSE

It was only in the early 1990s that a distinction was made in Chilean vineyards between Sauvignon Blanc, which was what producers thought they had, and Sauvignonasse, which was what they actually had much of the time. The two varieties look similar: it is an easy mistake to make. And since the original cuttings of 'Sauvignon Blanc' that populated Chile's vineyards came from Bordeaux in the 19th century, the Chilean field mix merely reflected the 19th century Bordelais field mix.

Sauvignonasse is also known as Sauvignon Vert, but unlike Sauvignon Gris (also found in small amounts in Chile) it is not a mutation of the more pungent Sauvignon Blanc. Sauvignonasse and Sauvignon Blanc may prove to be related, but the former lacks the assertive nose, the acidity and the staying power of the latter. When very young it can have quite good aroma and flavour, but it is an aroma of green apples rather than the blackcurrant leaf and gooseberries of Sauvignon Blanc. It has little character if picked underripe, and reaches high levels of alcohol – up to 14.5 per cent – very easily. Acidity, however, drops rapidly, and at over 13 cent alcohol the wine can taste dull and featureless.

The large amount of Sauvignonasse that is mixed with Sauvignon Blanc in Chilean vineyards, planted before the distinction was made clear, means that much Chilean 'Sauvignon Blanc' fades quickly. Newer plantings are of pure Sauvignon Blanc and have more character. It will, however, be a few years before older vineyards are superseded.

Opinions are divided as to whether Sauvignonasse is identical to Italy's Tocai Friulano, but there are certainly strong similarities in the glass. If it is the same, it would be interesting to see if Chile could produce versions as impressive as those of the best growers of Collio and Colli Orientali del Friuli by dramatically lowering yields and taking their winemaking a little more seriously. See also Sauvignon Blanc pages 218–227.

## SAUVIGNON GRIS

An alternative name for Sauvignon Rosé, which is a pink-skinned version of Sauvignon Blanc (see pages 218–227). It is much less aromatic than Sauvignon Blanc, but makes elegant, rather interesting wines. Chile has some, as does Bordeaux: Château Smith-Haut-Lafitte in Pessac-Léognan, for example, adds around 5 per cent Sauvignon Gris to its dry white which is otherwise entirely Sauvignon Blanc. Best producers: (France) Carsin, de Courteillac.

*Savagnin vines growing on the limestone and marl soils of the Château Chalon appellation, in the Jura region of eastern France. The Vin Jaune from here is one of the very few flor-growing wines in the world. I'm not convinced that it's as good as good sherry, but it has its moments.*

## SAUVIGNON VERT

An alternative name for Sauvignonasse (see left).

## SAVAGNIN

This is the only grape used for Vin Jaune, the curious flor-aged wine of the Jura region in eastern France. It ripens late, being picked at 13 to 15 per cent potential alcohol in November or even December. The flor-like yeast covering, here called *voile* or veil, grows more slowly and more thinly than the flor of Jerez (see Palomino Fino page 165), and dies earlier. There is no solera system used in the Jura; in Jerez and the other sherry towns it is the constant refreshing of wine in solera that keeps the flor alive. In addition, the much cooler temperatures of the Jura do not encourage such lavish growth. The wine is left in cask for six years and three months, by the end of which time it has developed a pungent, oxidized, nutty flavour with piercing acidity. Once bottled it is said to be able to last 50 years or more.

Savagnin is grown throughout the Jura region, and while it is a permitted addition to any white wine there, it is usually kept for Vin Jaune. It is the same as the Heida found in the Valais in Switzerland.

There is also a Savagnin Rose which is the same as Traminer (see page 270); the *musqué* form of Savagnin Rose is the more aromatic Gewürztraminer (see pages 102–111). Savagnin Noir is not a dark form of Savagnin; instead, it is a name given to Pinot Noir in the Jura. Best producers: (France) d'Arlay, Jean Bourdy, Hubert Clavelin, Jean-Marie Courbet, Durand-Perron, de l'Étoile, Henri Maire, de Montbourgeau.

## SAVAGNIN NOIR

A name occasionally used in the Jura for Pinot Noir (see pages 174–185). It should not be confused with the local grape Savagnin (see left).

## SAVATIANO

Greece's workhorse grape, usually neutral and low in acidity, and used for inexpensive branded wines and for retsina. More acidic grapes, especially Assyrtiko and Roditis, are often added to retsina to give balance. The pine resin is added in pieces to the must, and removed only when the wine is racked. If grown on good sites and picked slightly earlier Savatiano can produce surprisingly well-structured wines. Best producers: (Greece) Achaia-Clauss, Kourtakis, Semeli Winery, Skouras Winery, Strofilia.

## SCHEUREBE

The purpose of propagating this German crossing in 1915 was, for once, not to try and produce an improved version of Riesling.

Instead, according to the late Dr Helmut Becker of Geisenheim, Georg Scheu was aiming at a better Silvaner, with more perfume, greater frost resistance and greater resistance to chlorosis, the ailment suffered by many vines when planted on limestone. Limestone soils are common in the Rheinhessen and some other parts of Germany.

To this end Scheu crossed Silvaner with Riesling – or at least, he is supposed to have done. DNA fingerprinting suggests that the parentage traditionally attributed to Scheurebe may be as erroneous as the supposed Riesling x Silvaner parentage of Müller-Thurgau.

Scheurebe's wine bears more resemblance to that of Riesling than to that of Silvaner. It lacks the taut elegance of good Riesling and even at its best tends to be clumsier, but it is complex and rich at high Prädikat levels, making powerful sweet wines with a fantastic flavour of ripe pink grapefruit swathed in honey which age quite well, though not for as long as equivalent Rieslings. It ripens to higher sugar levels, is high-yielding, and seems to produce its most exciting wines in the Pfalz. When made dry there is a danger of catty, white grapefruit flavours if the grapes are underripe; Scheurebe like this can be raw and aggressive. It is however far and away the most successful of the modern German crossings, and the only one highly regarded by serious winemakers.

In Austria it is known as Sämling 88 (see page 207) – seedling number 88 was the seedling selected from all those propagated by Scheu in 1915. Best producers: (Germany) Andreas Laible, Lingenfelder, Müller-Catoir, Hans Wirsching, Wolff-Metternich.

**Alois Kracher**
*Scheurebe comes into its own at high Prädikat levels. In Austria the grape is often called Sämling 88, but Alois Kracher clearly can't be bothered with such a workaday name.*

### SCHIAVA

This Italian red grape produces perfectly pleasant everyday wines in Trentino-Alto Adige, but doesn't seem capable of anything of real excitement. It yields generously, and gives wines of light smoky strawberry fruit and a mildly creamy texture. Concentration, depth and complexity, however, are generally lacking. It is declining as growers see the greater commercial opportunities of weightier reds, but still covers a substantial area.

Its Italian name, meaning 'little slave' is generally taken as evidence of Slavic origin, though the author Nicolas Belfrage points to use of its German name, Vernatsch, in the Süd-Tirol (Alto Adige) as indicating that it has long been thought of there as a local variety. The name it is given in Germany itself, Trollinger, also suggests a link with the Tyrol. Records of the vine in Trentino-Alto Adige go back to the 13th century. (See Vernatsch page 273 and Trollinger page 270.)

In any case, there are lots of different Schiavas found in Trentino-Alto Adige. There is Schiava Grigia, or Grauvernatsch; Schiava Gentile, or Kleinvernatsch; and Schiava Grossa, or Grossvernatsch. The last is the least distinguished, but particularly high yielding so inevitably it is the most planted.

It is the main grape in DOC Santa Maddalena, where it may be given more character by Lagrein or some other red grape and I have to say I've had some really lovely gentle, fresh picnic wines sitting among the vines high above the city of Bolzano. It is also found in numerous other DOC and non-DOC wines in the area. Declining Schiava may be, but in the high mountains of Alto Adige in early summer, well, find me a more delightful red than one of these. Best producers: (Italy) Cornaiano co-op, Cortaccia co-op, Franz Gojer, Gries co-op, Lageder, Josephus Mayr, Thomas Mayr, Niedermayr, Georg Ramoser, Hans Rottensteiner, Heinrich Rottensteiner, San Michele Appiano co-op, Santa Maddalena co-op, Termeno co-op.

### SCHIOPPETTINO

A fairly characterful north-eastern Italian variety currently being rescued from terminal decline. It is native to Friuli, and a grape of this name has been grown there since at least the 13th century. Its flavour is peppery and raspberryish, fairly light in body and alcohol, and with high acidity. There is also a young and fizzy version drunk locally. Best producers: (Italy) Dorigo, Davide Moschioni, Petrussa, Ronchi di Cialla, Ronco del Gnemiz, La Viarte.

### SCHÖNBURGER

A German crossing (Pinot Noir with a crossing of Chasselas Rose and Muscat Hamburg) now more grown in England than in Germany, where it is concentrated in the Rheinhessen and Pfalz. It ripens easily, yields well and is disease-resistant. Its berries are pink but it is used for making white wine. Its perfume is heavy and somewhat Muscatty which is very attractive in a light English wine, but can be rather cloying in wine from warmer climes. Best producers: (Canada) Gehringer; (England) Carr Taylor, Denbies, Valley Vineyards, Wootton.

### SCHWARZRIESLING

German name for Meunier (see page 138); it is found mostly in Württemberg. Best producers: (Germany) Dautel, Drautz-Able, Fürst zu Hohenlohe-Öhringen, von Neipperg.

### SCIACARELLO

A grape that seems to be unique to Corsica, but is on the decline relative to Nielluccio (see page 164). It is at its best in the South-West of the island, around Sartène and Ajaccio, and while its wines are light in colour and not particularly tannic, they have a lively herby pepperiness. With age they develop hints of woodsmoke and tobacco. The name means 'the grape that bursts under the teeth' and – guess what – the grapes have tough skins and lots of juice. Best producers: (France) Albertini Frères, Clos Capitoro, Clos Laudry, Martini, Peraldi, de Torraccia.

### SCUPPERNONG

A vine found in the south-western states of the USA and in Mexico, Scuppernong is a *Vitis rotundifolia* vine, and belongs to the genus Muscadinia. The thick-skinned berries grow in small clusters and are low in sugar; chaptalization is normal. Pressing can be difficult, too, because of the thick, fleshy pulp. The flavour is strong and musky and the wines are usually made sweet. Virginia Dare, a North Carolina wine that enjoyed great popularity in the early years of the 20th century, was made from Scuppernong and named after the first child born in the American colonies to English settlers.

### SEIBEL

A group of French hybrids produced by Albert Seibel (1844-1936). Seibel 4643, also known by the somewhat ambitious name of Roi des Noirs, used to be widely planted in western France, though is so no longer: its wine was rustic and dark. Other Seibels include 7053, otherwise known as Chancellor; and 5279, or Aurore.

# SÉMILLON

SÉMILLON IS A GRAPE THAT DOESN'T LIKE to do it the easy way. Give it a nice normal vineyard in a nice warm climate to ripen its grapes – and Sémillon blows it. Give it a reliable springtime to bud in, a warm dry summer to fruit and ripen in and a dry autumn to gather in the harvest – and Sémillon says thank you by filling your vats with dull, tasteless juice that is as good for distilling into brandy as it is for anything else. And I'm not kidding. Various countries have tried their hand at Sémillon over the years. South Africa was awash with it in the 19th century. Chile was overrun with it for much of the 20th. Argentina and the rest of South America also gave it its head, and Sémillon rewarded them by making wines that were a by-word for dullness and which revelled in sharing with Chenin Blanc the dubious distinction of attracting sulphur like carrion attracts crows.

*Seen here in the honeyed autumnal light evocative of its precious golden wine, Château d'Yquem is the supreme example of the majestic sweet wines of Sauternes. Pickers will go through the vineyards four times or more during the harvest, picking fully botrytized grapes or even green ones, according to the constantly changing instructions from the winery.*

So why's it here as a classic grape in this book? You may well ask. A couple of quirks of nature is all I can offer by way of explanation – and the most famous quirk is Sauternes, just on the edge of the Bordeaux region in south-west France. Bordeaux is most famous for its red wines, and besides craving a decent summer, every grower's last words before he or she goes to sleep at night are 'Please God, let's have a fine autumn to pick in'. Rot is the curse of the red crop. Yet rot is the making of the crop in Sauternes. But not any old rot. Noble rot, so called because it concentrates the sugar in the grapes without turning the juice sour and undrinkable, needs very special conditions to flourish, and in Sauternes' little patch of land it gets them. The local rivers Ciron and Garonne get very foggy in the autumn mornings. If that fog is compounded by rain, we're in trouble. All kinds of rot will threaten, none of them noble. But if the sun fills the sky and burns off the morning fog, the whole vineyard becomes muggy and humid – and hot. And in these conditions noble rot sets to work thinning the skins of the Sémillon grape and intensifying the sugar.

It seems likely that Sémillon naturally evolved in the Sauternes district – so we can thank nature for arranging matters sufficiently well to create a world-class wine. But nature wasn't trying to be helpful in the other place that has made Sémillon famous – Australia's Hunter Valley, north of Sydney.

Nature knows that no sane person would try to grow grapes there, and ever since the first magnificent obsessives decided to have a go in the 1830s nature has done her best to flout their efforts. She's washed away the decent soils by a never-ending succession of tropical storms. She's arranged sub-tropical heat during the summer, brackish bore water unfit for irrigation, and frequent winter droughts, just in case you were thinking you might build a few dams to store water to help your vines survive. And to make quite sure you got the message that grape-growing is doomed in the Hunter, she arranges cyclones to sweep down the coast and batter the valley just before the hapless grapes are ripe – just so you know who's really in charge. And without me attempting to seek refuge in rhyme or reason, it is precisely these woeful conditions that have created the classic Hunter Valley Semillon.

In the occasional perfect summer, with the grapes fully ripe and the harvest safely in, Hunter Semillon is full and fat, a bit blowzy – good grog, but quick to flower and fade. But in the years when nature does its worst, when the grape's alcohol level sometimes barely reaches 9 or 10 per cent alcohol before the fruit must be ripped off the vine to save the crop – the result – if you wait ten years for it to mature – is one of the world's great classic whites.

**Sémillon: from Grape to Glass**

*Geography and History page 232; Viticulture and Vinification page 234; Sémillon around the World page 236; Enjoying Sémillon page 238*

# Geography and History

WHERE HAS ALL THE SÉMILLON GONE? Look at the map and it's just isolated patches: South-West France has the most, and there's some in Australia, some in Chile and other South American countries, and a bit in South Africa.

Yet 50 years ago some three-quarters of Chile's white vines were Semillon (it's usually seen without the accent in New World winemaking countries). It smothered South Africa – in 1822 it covered 93 per cent of the vineyard area, and was referred to with perfect logic simply as Wyndruif, or wine grape.

Chile now has much less. South Africa has much, much less. First of all, Sémillon gave way to Chardonnay; now white grapes are in their turn giving way to red. And Sémillon, planted all over the world for its disease resistance and

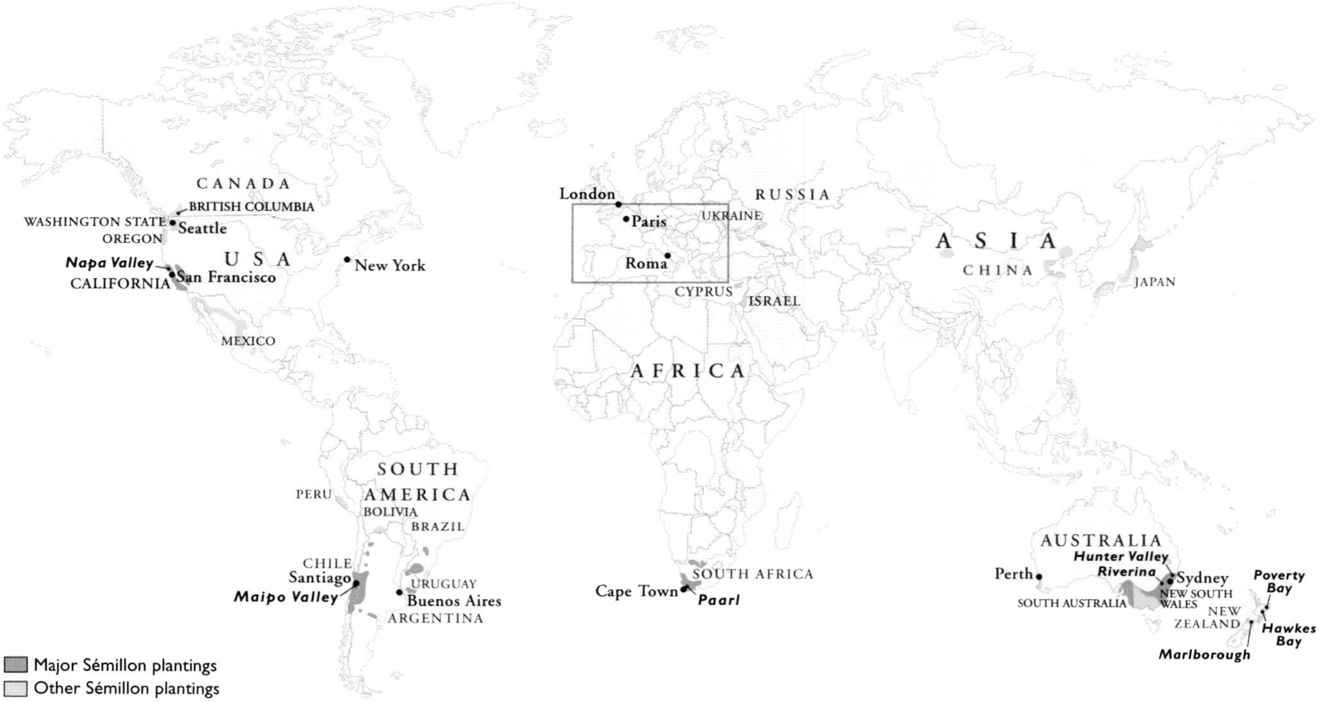

Major Sémillon plantings
Other Sémillon plantings

Major planting figures for Sémillon
AREA PLANTED (HECTARES)

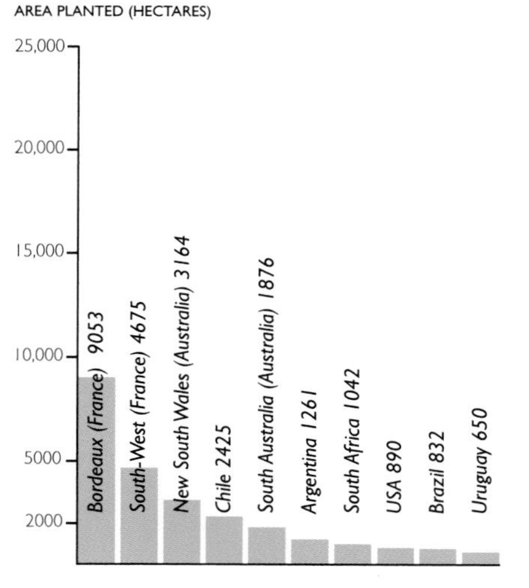

its ability to produce large quantities of grapes, might have had its day, were it not for two facts. One is that it produces outstanding sweet wine in Sauternes and Barsac on the left bank of the Garonne in Bordeaux. The other is that it produces outstanding dry wine in Pessac-Léognan and Graves, next door to Sauternes, and in Australia's Hunter Valley north of Sydney. And that's just enough of a CV to keep Sémillon being planted around the world as growers wonder whether they can ever emulate Sémillon's performance in Sauternes and the Graves.

The thing is, Sémillon was planted as an all-purpose grape – you could make dry or sweet wine from it, sherry, brandy, whatever. Yet it isn't an all-purpose grape. At high yields it is dilute and thin; when underripe it is green and stringy. Even its sweet wines usually need a touch of Sauvignon Blanc to brighten them up. And then, in misty, humid Sauternes and the subtropical Hunter Valley where the conditions are hardly suitable for grape growing at all, it produces world class wine. So why does it make so few great wines in much easier conditions? Let's have a look.

**Historical background**

The ampelographer Galet thinks that Sémillon probably originated in Sauternes and spread from there to the rest of the Gironde. It was found in St-Émilion by the 18th century, and indeed St-Émilion used to be one of its local synonyms; nowadays this is a synonym for the Ugni Blanc variety, which Sémillon does not remotely resemble. It was still planted in St-Émilion and indeed in the Médoc, along with Sauvignon Blanc and Muscadelle, for white wines well into the 20th century.

When was Sauternes first made sweet? The usual date given, for want of concrete evidence of an earlier date, is the mid-19th century. To suppose that it was not made sweet before, however, requires a suspension of disbelief: Tokaj had been famous for its botrytis-affected sweet wines since the late 17th century, so the technique of making sweet wines from botrytized grapes was well known; and *Botrytis cinerea* occurred in Sauternes and Barsac then just as it does now. Picking in the region in the 18th century was not until late November; in Cadillac – on the right bank of the Garonne – the Abbé Bellet, who kept records of every vintage from 1717 to 1736, confirms that by October the grapes were affected by noble rot, and were picked in selective *tries.* He does not confirm that only the rotten ones were used, but if the non-rotten ones had been the most desired, and had been picked in the first *trie,* it is hard to see the point of further pickings. Clearly the selective trips through the rows of vines were to pick out the grapes affected by noble rot.And he would have found it pretty hard to make ordinary dry wine from such grapes.

If you judged a grape solely on its appearance, it's hard to see why anyone would make wine from ugly, squishy, noble rotted grapes. But squeeze the gooey syrup from the grapes and lick your fingers and the fabulous rich flavour will persuade you in a flash.

*A golden autumn day at Château Suduiraut at Preignac in the Sauternes appellation. This is just the sort of weather that favours the development of noble rot: humid, foggy nights are essential, but if humidity persists throughout the day then the rot will turn grey and ignoble and there will be little chance of rich, sweet wine.*

*A carved coat of arms at Château d'Yquem, Sauternes. The Yquem family owned the property from the Middle Ages, and married with the Lur-Saluces in 1785.*

*A gooey mess of nobly rotten Sémillon grapes. Grapes in this state are highly desirable but difficult to press and slow to ferment.*

# Viticulture and Vinification

I CAN SEE WHY SÉMILLON was such a popular variety in the old days. It will grow just about anywhere and will recklessly produce gigantic crops of grapes. Which taste of...? Er, nothing really. Just vatsful of juice of no discernible personality but enough sugar to ferment into whatever kind of concoction you fancy. From dry white to sweet, from sherry to brandy. But in one or two little corners of the globe, if you really restrict its crop and vinify it skilfully, it has for centuries produced world class dry and sweet whites. It buds slightly later than Sauvignon Blanc but ripens earlier, and is actually less subject to noble rot than Sauvignon. It is resistant to most other diseases but rainy years, however, can produce attacks of grey rot. This is is caused by the same fungus as noble rot, but has different results.

## Climate

It is an unarguable fact that if you were to match your vine varieties with your sites on a rigid degree-day basis you would never plant Semillon in the Hunter Valley. It is, by this reckoning, far too warm there to produce dry Semillon, and indeed far too humid, especially when the humidity takes the form of tropical downpours at harvest time. But Semillon thrives there, and even more curiously, it is rainy years that often produce the best Hunter Semillon, years which are so poor in accepted terms that the alcohol level hardly reaches 10 per cent. There's got to be a reason. Up to a point, there is. So here's the science. It still won't all add up, because Hunter Semillon is one of the wine world's great enigmas. But here goes.

That the region produces such good Semillon is partly due to its cloud cover: the temperature may rise to 42°C during the day, but the sunlight is muted. The Hunter has a relatively low ratio of hours of bright sunshine to effective day degrees, and this helps to keep sugar levels down. (It also helps to mitigate tannins and astringency in the red wines of the region.) The humidity also helps: high afternoon relative humidity is associated with higher acidity levels in grapes, and low-acidity Semillon needs all the help it can get in this department.

The climate in the Hunter Valley could hardly be less like that of Bordeaux, except that autumn humidity here is essential for the growth of *Botrytis cinerea* in Sauternes. Early morning mists, caused by the confluence of the ice-cold river Ciron and the warmer Garonne, spread back up the Ciron valley and encourage the growth of botrytis on the ripe grapes. The botrytis has the effect of concentrating both sugar and acidity, again giving the grape's low acidity levels a helping hand.

Interestingly, Bordeaux's relative humidity and sunshine hours compare with those of the Upper Hunter, though of course the latter is much, much hotter. The Lower Hunter however has less relative humidity than Bordeaux, and slightly fewer sunshine hours.

## Soil

In the Hunter Valley the usual soils for growing vines are light, sandy ones, simply because heavy soils become quagmires after heavy rain. In a way, suitable doesn't come into it in the Hunter. There are some soils that will support vines – whatever variety – and there are poor clays that won't support anything at all. Period. The best-drained soils in the Lower Hunter Valley include the friable loam and friable red soils. In the Upper Hunter black silty loams over dark clay loam are successful; the red-brown duplex soils, which resemble those of the Lower Hunter, are better drained, which is not automatically an advantage in this hotter, drier region where irrigation is essential.

The soil in Sauternes is sandy gravel in varying depths over calcareous clay; Barsac is much flatter, and lacks the gravel of Sauternes. Barsac has well-drained, sandy, limey soil, and vine stress can be a problem in very dry years. In Sauternes, by contrast, the clay subsoil can be poorly drained where the topsoil is very thin: Château d'Yquem put in some 10km of drainage in the 19th century to correct this. Because Sémillon ripens earlier than Sauvignon Blanc it may be planted

*Picking grapes is never easy work – it's backbreaking at the best of times. But when you have to search each cluster for berries of the right degree of noble rot, as is happening here at Château d'Yquem in Sauternes, then it requires great concentration. Only skilled pickers can be trusted with such a task.*

on more clayey soil in Sauternes; it is sometimes suggested that clay can favour the development of botrytis. Sauvignon Blanc may be planted on the gravel. In the Graves, however, where a different style of wine is desirable, Sémillon gets the warmer soils and the better exposed sites.

The AC rules state that the vines must be eight years old for Sauternes; some châteaux maintain that they don't get good levels of botrytis until they are ten. Root depth seems to be a factor in determining whether rot turns noble or grey; shallowly rooted vines tend to get grey rot.

### Cultivation and yields

In Bordeaux growers may leave a long cane in order to have some spare buds in case of spring frost; what is liable to happen then is that the buds at the end of the cane grow vigorously and impede the development of those further down. The cane may be trained in an *arcure* to balance this.

Yields of Sémillon must be low if quality is the aim. Sauternes is the lowest: the legal limit here is 25hl/ha, and most leading properties make much less. Yquem famously makes just one glass of wine per vine, or 9hl/ha. In Monbazillac legal maximum yields have been cut from an over-generous 40hl/ha to a sensible 27hl/ha. In the lesser regions of Bordeaux yields may reach 80–100hl/ha. Australian yields are around 3.5 tons per acre and 8 to 9 tonnes per hectare (roughly 60hl/ha); in New Zealand it produces thinner, grassier wine at 10–17 tonnes/ha.

### At the winery

The big question with Sémillon is to oak or not to oak? It certainly has a great affinity with oak, in particular with new oak, and the increased proportion of new oak used by the Sauternes châteaux since the mid-1980s has been a factor in the improvement of the region's wines. In the Graves, too, it is commonly fermented and aged in new oak; in youth these blends of Sémillon and Sauvignon Blanc can seem too intensely oaky but they age remarkably well. New oak here is being used with a more delicate hand than it was a few years ago, as is sulphur in Sauternes: Sémillon oxidizes easily, and one of the effects of noble rot is to make it require extra sulphur to protect it against oxidation. It takes a strong nerve for a grower to hold back. Many Sauternes are still noticeably sulphurous in youth, though this should fade with age.

*These are what partly nobly rotten clusters look like: there are some shrivelled grapes and some healthy ones. The healthy ones will be left for a later picking, by which time they will have shrivelled.*

The alternative style in Bordeaux and the South-West is the stainless steel-fermented one of crisp fruit and youthful acidity. Such wines are not intended to age, and a large proportion of Sauvignon Blanc is essential if the wine is to have sufficient flavour; young, unoaked Sémillon can taste lemony and grassy, but that's about all. If it doesn't have Sauvignon to help it along, then it needs oak. Sometimes, as in the great Sémillon-based dry whites of the Graves, it gets both.

In Australia, unoaked Semillon is a classic style of the Hunter. These wines, neutral – indeed positively unattractive and flavourless – in youth, mature (after a decade or so) into rich, honeyed toastiness – tasting for all the world as though they had spent their infancy in new oak barrels, although they haven't. Australian wine buffs love to bemuse visiting Brits in blind tastings with these unoaked beauties that we unerringly pronounce to be mature French Burgundy. The Hunter has toyed with oak-aging and now mostly rejected it; Barossa Semillons seem better suited to oak, and those from the Clare Valley can be handled either way. For the consumer, and perhaps for the producer, oak-aging means that the wine has more complexity at a young age: unoaked Hunter ones are beloved by those who know them, but they are about as far removed from wines for instant gratification as it is possible for a white wine to get.

### SEMILLON AND BOTRYTIS

What makes Sémillon and *Botrytis cinerea* so suited to each other? And what makes some rot noble and delicious and other rot merely grey and unpleasant-tasting?

Noble rot and grey bunch rot are the same fungus: both are *Botrytis cinerea*. The fungus can develop in either direction, depending on circumstances; and it is by no means unknown to have grey rot and noble rot together, even on the same cluster.

For botrytis to turn noble, fluctuating humidity is necessary: just the sort of damp, foggy nights and early mornings, followed by warm dry days, that are typical of a Sauternais autumn. Under these conditions the grapes will be infected with botrytis, but fungal growth will be limited, and its metabolism modified. Constantly high humidity favours grey rot, as does berry splitting in heavy rain; Sémillon's thick skin helps to protect against this.

The flavour of botrytized Sémillon is not solely due to the concentration of sugars and acidity, though the drying and shrivelling of the grapes (which helps to protect against invasion by other bacteria that could produce off-flavours) is an important part. Noble rot metabolizes grape acids, especially tartaric, which falls dramatically in infected grapes. The concentration of the remaining acidity, however, means that the pH of botrytized Sémillon grapes is slightly, but only slightly, higher than that of healthy ones. It also metabolizes sugar, and total sugar content drops by 35–45 per cent; this is compensated by water loss as the grapes shrivel, and the remaining sugar becomes more concentrated. It produces glycerol, which contributes to the silky mouthfeel of botrytized wines, and it both degrades the esters that give white wines their fruitiness, and destroys the terpenes that give varietal aroma. Sémillon has so little varietal aroma and so few fruity flavours to start with that this is no great loss; and in their place the wine gains great aromatic complexity. (Muscat, by contrast, loses more than it gains, which is why most sweet Muscats are made sweet by the fortification method or by shrivelling.)

Sweet Sémillon's new aromas come from over 20 aromatic compounds synthesized by noble rot. Sotolon is one: in conjunction with others, it helps to give nobly rotten wines their distinctive honeyed aroma.

# Sémillon around the World

SÉMILLON'S CLASSIC STYLES – Sauternes, white Pessac-Léognan and Graves, and the Hunter Valley – are so unlike each other that if growers elsewhere wanted to make serious Sémillon they would be forgiven for not knowing where to aim. Sémillon also seems to be its own worst enemy: unless severely regulated it produces huge amounts of insipid wine.

## Sweet white Bordeaux

Sémillon is the main grape throughout Sauternes, Barsac, Monbazillac and the lesser sweet and semi-sweet white regions of Cadillac, Ste-Croix-du-Mont, Loupiac and Cérons. It accounts for between 60 and 90 per cent of most vineyards, the balance being Sauvignon Blanc (of which yields are slightly higher, so the percentage of Sauvignon in the wine may be greater than in the vineyard); and sometimes a few per cent of Muscadelle. Sauvignon adds freshness, lightness and acidity and Muscadelle aroma.

All these regions follow, more or less closely, the practices of Sauternes and Barsac, and the great improvements in quality in the latter regions since the 1980s have spread, with encouraging results. Monbazillac, for example, used to permit mechanical harvesters – a complete contradiction of the style of the wine, since botrytized wines depend on selection, and the hand and eye of the picker. Since 1994 these harvesters have been phased out, and minimum must weights have been increased from 13 per cent potential alcohol to 14.5 per cent. In fact careful selection here can regularly produce potential alcohol levels of 18–19 per cent, and more in top years.

The rules in Sauternes are stricter. The alcohol in the finished wine must be at least 13 per cent, and is usually 14 per cent; the residual sugar usually amounts to another four to seven per cent potential alcohol. Château d'Yquem, for example, picks at between 20 and 22 per cent potential alcohol. It might be necessary to send the pickers through the vineyards for up to ten separate passages to attain these levels, though three or four times is more usual. Much depends on the year: noble rot arrives more years than not, but it is unpredictable and often patchy. Years in which it blankets the vineyards, like 1990, are few and far between. Balance can also be a problem in Sauternes: Yquem sends constant instructions to its pickers to select more botrytized grapes or more healthy ones, as the balance of the must requires.

Chaptalization is permitted, though the best properties claim only to use it in poor years; cryoextraction is also permitted. This is an expensive technique which involves leaving the picked grapes for about 20 hours in a cold room to freeze their water content so that just the sweet juice runs from the press. It can be useful in a rainy year, but it is no substitute for careful selection.

There are, inevitably, years in which noble rot fails to appear. In these years wine will be made from overripe and shrivelled grapes only: it will be sweet and concentrated and pretty tasty, but will lack the characteristic flavours of Sauternes.

## Dry white Bordeaux

The lighter, sandier soils of Pessac-Léognan (in the whole appellation reds outnumber whites four to one) are the ones usually given over to white wines, of which at least 25 per cent of the blend must be Sauvignon Blanc. Usually Sauvignon's figure is much higher, and may even be 100 per cent, though Sémillon is valued for the richness it brings to the blend, and for its affinity for the new oak in which the wine is usually fermented and aged. Flavours can be remarkably exotic, with apricot, nectarine and mango, nuts and buttered toast, and the wines can age for many years.

Dry whites from the rest of Bordeaux and the South-West are seldom intended for aging. The proportion of Sémillon in the blend may range from 100 per cent to nil. It is being outplanted not just by Sauvignon, but by more fashionable red varieties.

## Australia

Hunter producer Bruce Tyrrell's comment that 'Semillon has a structure and acidity like Riesling; it's like Riesling in everything except flavour' might come as a shock to anyone comparing a high-acid Saar Riesling with a low-acid Bordeaux Sémillon. But Semillon in the Hunter Valley does have much higher acidity than it manages in France. The ripeness at which it is picked is a factor: for unoaked Semillon, which needs acidity to age for a decade or more in bottle, picking is often at 10–10.5 Baumé. For Semillon destined for oak-aging more ripeness and substance is required, and picking is at 12–12.5 Baumé, to prevent the wine

CHÂTEAU D'YQUEM
*This most famous of all Sauternes properties is now owned by the multinational company LVMH, which makes luggage, perfume and Champagne.*

CHÂTEAU CLIMENS
*Climens produces one of the richest wines in Barsac, where the rules allow wines to be labelled either as Barsac or as Sauternes.*

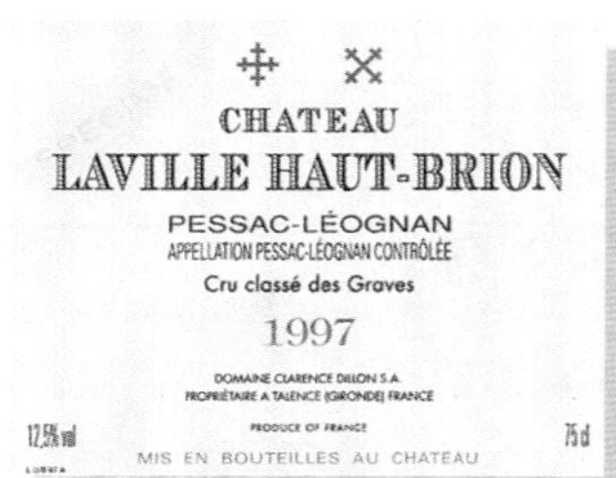


CHÂTEAU LAVILLE-HAUT-BRION
*This is the white wine of Château la Mission-Haut-Brion. Slow to open out, good vintages become deeper and more complex over 20 to 30 years.*

*The Lindemans Ben Ean winery in the Lower Hunter Valley. This is one of the great historic wineries of an historic region: vines have been grown in the Lower Hunter since the 19th century. The unoaked Semillons for which it is known are a unique world wine style and predate by many years the birth of modern Australian wine. Difficult to appreciate when young, they were nearly washed away by a tide of fashionable Chardonnay – nearly but luckily not quite.*

being swamped by the wood. If you pick at that ripeness for unoaked Semillon, the fruit loses its freshness and gets broad and blowzy after five or six years. The best vintages for unoaked Hunter Semillon are often the rainy ones: in 1971, says Tyrrell, it rained for three months, 'and the wine is still going strong'. A bit of botrytis – say 10–20 per cent – is also useful for the complexity and aroma it adds.

Semillon from the rest of Australia is usually oaked; only that from Hilltops in New South Wales seems to share the Hunter structure. The warm irrigated regions produce large amounts of fairly basic but pleasant dry white that is often blended with Chardonnay and seasoned with oak chips. The Riverina area of New South Wales gets natural botrytis infection and has produced some astonishing Sauternes-like sweet wines.

### USA

Plantings in California are falling rapidly, and what little there is is usually blended with Sauvignon Blanc. Clones and yields may be at fault, but the main problem is probably that few care enough to treat the vine seriously. Clos du Val makes a good varietal, though it is not as long lived as a Graves or Hunter version; Semillon in California is often more successful as a sweet late-harvest wine. There were experiments in the 1950s and 1960s by Myron and Alice Nightingale with spraying picked grapes with botrytis spores; Beringer made some wines commercially this way into the 1990s. The results were astonishingly intense, but lacked finesse. Results from botrytis infection in the vineyard have generally been better and there are a few world class examples from top Napa wineries. Both Oregon and Washington State make a certain amount of grassy, Sauvignon-ish wine.

### New Zealand

Marlborough, Poverty Bay and Hawkes Bay are the main regions for Semillon in New Zealand. There were 237ha planted in 2000. The usual clone here – UCD2 – has loose knit clusters like the Semillon in California, but unlike the compact clusters of Bordeaux. It is often blended with Sauvignon Blanc, and can improve the longevity of the latter. One or two interesting sweet wines have been made from Semillon.

### Rest of the world

Chile had 2425ha of Semillon in 1998 – which seems a lot, until compared with 1985's figure of 6195ha. Plantings are still falling slowly, but the vine's high productivity makes it a useful supplier of bulk wine. A few producers, notably Morande and Casa Silva, see it as having potential, and there are some old vines capable of interesting quality. South Africa had 1042ha in 1999, most of which went through co-operative cellars. Here, too, it is seen as a bulk variety by all but a few growers. There is some in Croatia and other parts of Eastern Europe, but little interest has been shown in Sémillon there.

CHÂTEAU TIRECUL-LA-GRAVIÈRE

*Monbazillac lies about 80km north-east of Sauternes and its Sémillon vines also get noble rot. Single estates such as this one can make good, rich wine.*

DE BORTOLI

*The wine that has shown that Australia, too, can produce outstanding nobly rotten Semillon. Intense and sweet, it compares favourably with top Sauternes.*

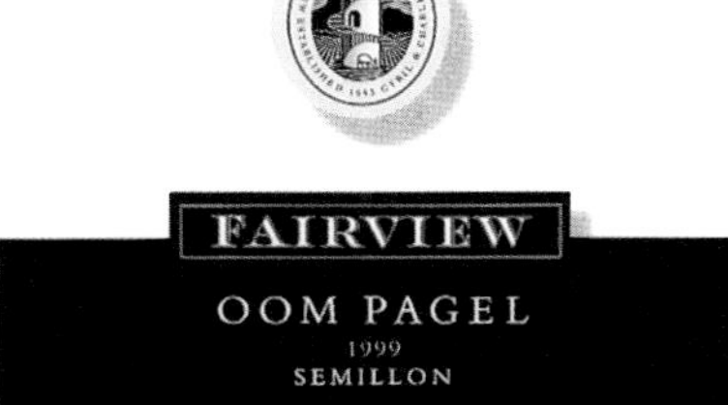


FAIRVIEW

*This bold, rich style of South African Semillon augmented by new French oak spice comes from Paarl and benefits from a year or two of aging.*

# Enjoying Sémillon

ASK MOST OF THE WORLD'S WINEMAKERS how Sémillon ages, and you'll be told that it doesn't. In most countries it's a bland, bulk wine for drinking young.

Australia and France know differently. This is where Sémillon from a serious producer can last 20 years – and if botrytis is brought into the equation, wines from top châteaux can still be drinking a century later. Most Australian and French Sémillons, happily for those of us with neither the cellar nor the expected lifespan to keep wines for a century, are less extreme in their longevity.

Simple Bordeaux Blanc should be drunk within a year of two of the vintage; even decent quality white Pessac-Léognan and Graves should mostly be drunk within five years. Classed growth wines can last longer, from ten to 20 years in a good vintage, and are good at almost any age; the exceptions, and the longest-lived wines, are the magically rich yet dry Haut-Brion and Laville-Haut-Brion, which are seldom at their best before their fifth birthday but which positively demand 20 years in a good vintage.

Much is written about Sauternes needing time to mature. Of course it does improve in bottle, but it is also utterly delicious young, and while Classed Growth wines should probably not be touched for at least ten years, lesser wines, and those from Monbazillac, can be drunk with great enjoyment earlier.

Unoaked Hunter Semillon is, however, less expressive in youth. It needs five or six years to show much character, but the rainiest, most acidic vintages (the best ones, in other words) can last 20 years or more. Upper Hunter Semillon develops faster than Lower Hunter wine. Oaked Semillon can sometimes last three or four years, but the point of the oak is to make the grape more complex, and therefore drinkable, at a younger age. It is always drinkable immediately upon release, though many improve with a little extra aging.

### The taste of Sémillon

'Battery acid' is how winemaker Michael Hill Smith succinctly describes the flavour of young, unoaked Hunter Valley Semillon. The same could be said of much German Riesling; and like Riesling, Hunter Semillon develops astonishing flavours of honey and toast. After six or ten years in bottle it does not have clear fruit flavours, although in youth it is citrus and fresh; maturity subsumes its fruit into a rich, silky wininess.

Oaked Australian Semillon is different: being picked riper it has richer fruit flavours, of greengages and apricots and mangoes, all mixed with the custardy vanilla of the wood. It is richer, broader and fuller than good white Graves or Pessac-Léognans, which usually have a substantial addition of Sauvignon Blanc for acidity and freshness; they are generally tighter and more subtle, and elegance, finesse and a complex nuttiness are more to the fore, particularly with maturity. In young Graves wines, if they are not dominated by new oak, the Sémillon adds creaminess and nectarines to the Sauvignon, which in any case is less grassy in Bordeaux than it is in New Zealand.

The flavours of Sauternes are of marzipan, apricots, mangoes, honey, nuts, toast, pineapple, peach, orange, honeysuckle, beeswax, barley sugar and coconut, all wrapped up in a creamy, silky, unctuous texture, enthralling and rich. If that seems like an awful lot of flavours, get yourself a mature bottle of top Sauternes to check. You may end up needing more words, not fewer.

*Like all Sauternes, Château Lafaurie-Peyraguey is usually delicious when young – so much so that it can require great willpower to lock the wine away until it reaches its full, complex maturity at 15 or more years. From the Hunter Valley, Tyrrell's Vat 1 Semillon, far right, is powerful and immensely long-lived, taking on honeyed, waxy aromas with maturity, but, unlike Sauternes, is not nearly so interesting in youth.*

### Matching Sémillon and food

Dry Bordeaux Blancs are excellent with fish and shellfish; fuller, riper New World Semillons are equal to spicy food and rich sauces, often going even better with meat than with fish. Sweet Sémillons can partner many puddings, especially rich, creamy ones. Sémillon also goes well with many cheeses, and Sauternes with Roquefort or other blue cheese is a classic combination.

## CONSUMER INFORMATION

### Synonyms & local names

Known in 19th-century South Africa as Wyndruif ('winegrape') or 'groen' ('green') grape and in Australia's Hunter Valley as Hunter Valley Riesling. Called Boal in the Douro Valley, Portugal.

### Best producers

**FRANCE/Bordeaux/Graves and Pessac-Léognan** Ardennes, Brondelle, de Chantegrive, Domaine de Chevalier, Clos Floridène, de Fieuzal, Domaine la Grave, Haut-Brion, de Landiras, Laville-Haut-Brion, Latour-Martillac, Pape-Clément, Rahoul, Respide-Médeville, de Roquetaillade-la-Grange, du Seuil, Vieux Château Gaubert, Villa Bel Air; **Sauternes/Barsac** d'Arche, Barréjats, Bastor-Lamontagne, Bonnet, Caillou, Climens, Clos Haut-Peyraguey, Coutet, Doisy-Daëne, Doisy-Dubroca, Doisy-Védrines, de Fargues, Filhot, Gilette, Guiraud, Haut-Bergeron, les Justices, Lafaurie-Peyraguey, Lamothe-Guignard, Liot, de Malle, Moulin-de-Launay, de Myrat, Nairac, Rabaud-Promis, Raymond-Lafon, Rayne-Vigneau, Rieussec, Romer-du-Hayot, St-Amand, Sigalas-Rabaud, Suau, Suduiraut, la Tour-Blanche, d'Yquem; **other Bordeaux** Birot, Carsin, Cayla, du Cros, Fayau, Grand Enclos du Château de Cérons, Lagarosse, Loubens, Loupiac-Gaudiet, les Miaudoux, du Noble, du Pavillon, la Rame, Reynon, de Ricaud, de Sours, de Toutigeac, Turcaud; **Bergerac** le Raz, Tour des Gendres; **Monbazillac** l'Ancienne Cure, Bélingard, Bellevue, la Borderie, Grande-Maison, les Hébras, Theulet, Tirecul-la-Gravière, Treuil-de-Nailhac.

**AUSTRALIA** Tim Adams, Allandale, Ashbrook, Bethany, Brokenwood, Leo Buring, Cape Mentelle, Chateau Xanadu, De Bortoli, Huntington Estate, Peter Lehmann, Lindemans, McWilliams, Moss Wood, Nepenthe, Petersons, Reynolds Yarraman, Rothbury Estate, St Hallett, Tyrrell's, Vasse Felix, Yalumba.

**USA/California** Carmenet, Duckhorn Vineyards, Far Niente, Matanzas Creek, Murrieta's Well, Robert Pepi, Preston, Signorello, Simi, Spottswoode, Swanson; **Washington State** Columbia, Hogue Cellars, L'Ecole No 41, Matthew Cellars, Woodward Canyon.

**NEW ZEALAND** Pegasus Bay, Selaks, Seleni, Seresin.

**SOUTH AFRICA** Boekenhoutskloof, Fairview, Neethlingshof, Steenberg, Stellenzicht.

### RECOMMENDED WINES TO TRY

**Classic sweet white Bordeaux**

See Best producers for Sauternes/Barsac left.

**Ten other sweet white Bordeaux**

Ch. d'Arche *Sauternes*
Ch. Bastor-Lamontagne *Sauternes*
Ch. Barréjats *Sauternes*
Grand Enclos du Château de Cérons *Cérons*
Ch. Loubens *Ste-Croix-du-Mont*
Ch. les Miaudoux *Saussignac*
Ch. du Noble *Loupiac*
Ch. la Rame *Ste-Croix-du-Mont Réserve du Château*
Ch. Theulet *Monbazillac Cuvée Prestige*
Ch. Tirecul-la-Gravière *Monbazillac*

**Ten French Sémillon-dominated dry white wines**

Ch. Carsin *Bordeaux Blanc Cuvée Prestige*
Ch. de Chantegrive *Graves Cuvée Caroline*
Clos Floridène *Graves*
Domaine la Grave *Graves*
Ch. Haut-Brion *Pessac-Léognan*
Ch. Laville-Haut-Brion *Pessac-Léognan*
Ch. le Raz *Montravel Cuvée Grande Chêne*
Ch. Reynon *Bordeaux Blanc Vieilles Vignes*
Ch. Tour des Gendres *Bergerac Cuvée des Conti*

**Ten dry New World Semillon wines**

Allandale *Hunter Valley (Australia)*
Bethany *Barossa Valley Wood Aged (Australia)*
Boekenhoutskloof *Franschhoek (South Africa)*
L'Ecole No 41 *Columbia Valley Barrel-Fermented (Washington)*
McWilliams *Hunter Valley Mount Pleasant Elizabeth (Australia)*
Moss Wood *Margaret River (Australia)*
Nepenthe *Adelaide Hills (Australia)*
Seleni Estate *Hawkes Bay (New Zealand)*
Stellenzicht *Stellenbosch Reserve (South Africa)*
Tyrrell's *Hunter Valley Vat 1 (Australia)*

**Five sweet New World Semillon wines**

Tim Adams *Clare Valley Botrytis (Australia)*
Chateau Xanadu *Margaret River Noble Semillon (Australia)*
De Bortoli *Noble One (Australia)*
Swanson *Napa Valley Late Harvest Semillon (California)*
Yalumba *Eden Valley Botrytis (Australia)*

*It seems perverse of Sémillon to be thick-skinned and actually not that susceptible to botrytis, when much of its survival in Bordeaux depends on it becoming nobly rotten, but those thick skins help protect it against grey rot in less than perfect years.*

### Maturity charts

Sauternes and Pessac-Léognan can be drunk young, but greatly improve with age while unoaked Hunter Semillon demands age.

**1997** Sauternes Premier Cru Classé

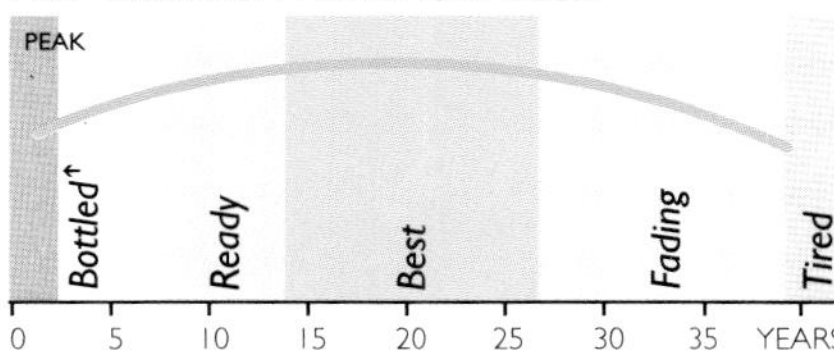


*1997 produced probably the most complex wines of the trio of good years of 1996, 1997 and 1998. They will be delicious young, but will also last well.*

**1999** Pessac-Léognan Cru Classé

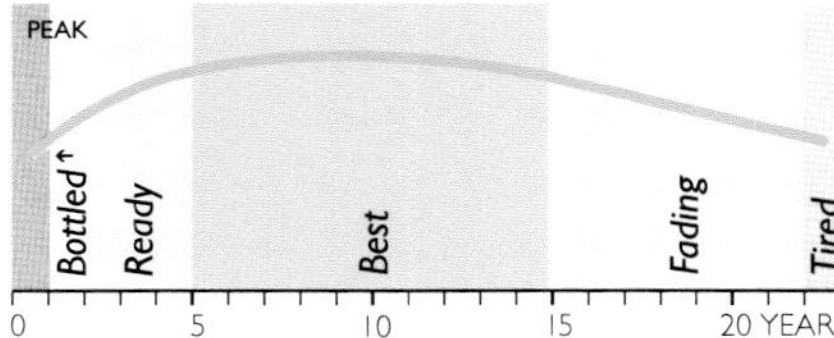


*One of the longest lived white wines, Château Laville-Haut-Brion contains around 70% Sémillon – an unusually high proportion for a Pessac-Léognan white.*

**1998** Hunter Valley Semillon (premium unoaked)

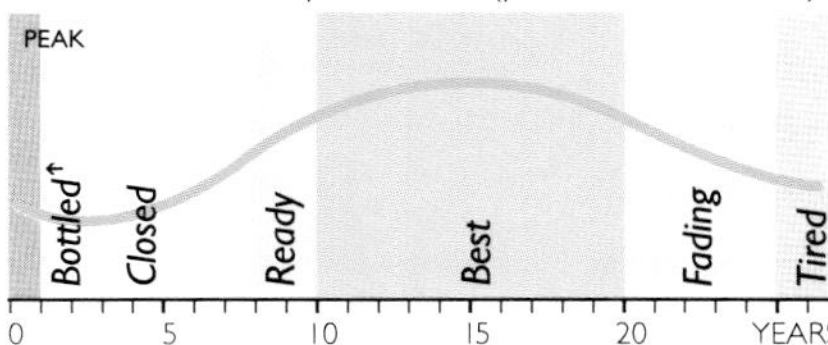


*A vintage of big, rich wines. The acidity is nevertheless good, and the Semillons should be excellent when they emerge from their closed, ungenerous period.*

## SERCIAL

One of the noble grape varieties of Madeira, now much reduced in area and really only found around Câmara de Lobos in the south of the island, and around Porto Moniz in the North-West. It is often grown at high altitude, and being in any case late ripening, it is the last variety to be picked.

It is high in acidity and makes the driest and lightest of all Madeira styles. Since 1993 any Madeira with the word 'Sercial' on the label must contain at least 85 per cent of that grape. Modernizing efforts by the island's largest producer, the Madeira Wine Company, mean that its Sercial Madeiras are slowly becoming less austere in style, and retaining more fruit flavour. This is partly achieved by maturing the wine not by the short, sharp shock of a warm *estufa* (which in any case is only used for the cheaper wines of Madeira), but by the gentler *canteiro* method, in vats stored under the roof. Since the brilliant, shocking tang of Sercial, and its astonishing longevity were two of the reasons for Sercial wine's greatness, I hope modernization doesn't mean 'dumbing down'.

Sercial is the same grape as the Esgana Cão, or 'Dog strangler', of the Portuguese mainland (see page 89). Best producers: (Portugal) Barros e Sousa, Blandy, Cossart Gordon, Henriques & Henriques, Leacock.

*The northern coast of Madeira near the village of Seixal is one of the best spots for growing Sercial, but as always in Madeira, other vines – here, notably, the hybrid Jacquet – outnumber the noble variety.*

## SEYVAL BLANC

A French hybrid vine that was produced by crossing two Seibel hybrids – Seibel (see page 229) being the name of a group of French hybrids. Seyval Blanc is a so-called Seyve-Villard hybrid, one of a group of about 100 French hybrids bred by Bertille Seyve and Victor Villard. As it is not pure *vinifera* it is forbidden by EU law from being used in Quality Wine. Ironically, it is one of the most successful varieties of all in England, producing well-structured wines which, at their best, have a strong, Chablis-like austerity to start with and attain a complex honeyed richness with bottle age. Seyval Blanc may be coarse and dull in warmer France, but every grape has a day of glory, and low-cropped Seyval in cool England can make a smashing dry white. The fact that the EU would like to stop England using it is merely proof, for some, of a France-inspired conspiracy against poor old Blighty.

Elsewhere it is found in Canada, and the eastern states of the USA. Best producers: (England) Breaky Bottom, Chapel Down, Valley Vineyards; (USA) Benmarl, Wagner.

BREAKY BOTTOM

*Breaky Bottom makes exemplary, dry, weighty and grassy Seyval Blanc, which becomes honeyed and Burgundian after three or four years' aging.*

## SHIRAZ

The Australian name for Syrah. But it's a lot more than that. To many drinkers, the very word Shiraz evokes images of rich, dark, spicy red wines that perfume the brain and fill the nostrils with billowing waves of flavours. Many of these drinkers would not know what Syrah was, but Australian Shiraz is their nirvana.

Nowadays you can often tell what style a winemaker is after in the rest of the world by whether he or she labels the wine Syrah – for a more austere, less ebullient style, or Shiraz – for a richer, riper, more hedonistic experience. And consequently you may find Shiraz or Syrah in such countries as South Africa, Argentina, Chile, California, Spain or even Australia and France.

To generalize, the Australian Shiraz style is less austere than Rhône Syrah, with broader, sweeter fruit, and more flavours of chocolate and berries instead of the smoke and minerals of the northern Rhône. Syrah is generally thought to be the original name of the grape – but is it? The town of Shiraz in ancient Persia is one of the first recorded centres of winemaking. The vine, though, is now thought to have originated in the Rhône Valley. See Syrah/Shiraz pages 244–255.

## SILVANER

See pages 242–243.

## SOUSÃO

Portuguese black grape found in the Douro Valley. It is not one of the five varieties recommended for planting by the port authorities, but plays a part in one of the greatest of all ports, Quinta do Noval's Nacional as well as in many others. It is normally thought of as a rather rustic grape that contributes youthful colour and a certain raisiny fruit to the blend, but which does not age particularly well. It may be the same grape as Vinhão (see page 273), which is the main grape grown for red Vinho Verde. It appears in Galicia as Sousón. In South Africa it is regarded as one of the better fortified wine varieties for its high sugar levels and deep colour. Best producers: (Portugal) Quinta do Noval; (South Africa) Boplaas, Die Krans, KWV, Overgaauw.

## SPANNA

Nebbiolo is known by this name in the Vercelli and Novara hills of Piedmont. It is found under this name in the DOCs of Gattinara, Ghemme and several others, usually blended with other grapes. The wines have a reputation for being somewhat tough and stringy, but improved winemaking is yielding better results in the form of rounder fruit and riper tannins. See Nebbiolo pages 154–163.

## SPÄTBURGUNDER

German name for Pinot Noir (see pages 174–185). Styles of this wine have changed out of all recognition in Germany, with the pale, sweet wines of yore being superseded by dark, dry, sometimes overoaked international versions. The best wines come from the Pfalz and Baden regions. Best producers: (Germany) Bercher, Rudolf Fürst, Karl-Heinz Johner, Rebholz.

## STEEN

Traditional South African name for Chenin Blanc (see pages 74–83), although the latter name is increasingly used. It covers nearly a quarter of the total vineyard area there, making it easily the most planted variety, and it is the national workhorse, producing everything from dry table wine, through sweet and sparkling wines, to fortified wines and brandy. The handful of producers determined to improve the quality and image of the grape in South Africa tend to prefer the name of Chenin Blanc. Best producers: (South Africa) Cederberg, Glen Carlou, Hartenberg, L'Avenir, Stellenzicht, Villiera.

## SYLVANER

Silvaner (see pages 242–243) is spelled with a 'y' in all countries except Germany. In Austria it may also be known as Grüner Sylvaner.

## SYRAH

See pages 244–255.

## SZÜRKEBARÁT

Hungarian name for Pinot Gris (see pages 172–173). At their best the wines are lively, with good earthy, spicy fruit, but winemaking and viticulture are of variable standards and a vaguely grubby flat flavour is still unfortunately more common than the bright, minerally, honeyed style it can achieve.

## TAMÎIOASA

Romanian name for Muscat. Tamîioasa Alba or Tamîioasa Românească are names for Muscat Blanc à Petits Grains, while Tamîioasa Ottonel is Muscat Ottonel and Tamîioasa Hamburg or Neagra is Muscat of Hamburg.

Tamîioasa Românească is used, unusually, for nobly rotten wines in Romania, and the name may appear on the label as the only grape in such wines, although in the sweet wine region of Cotnari a blend of several grapes, including Grasa, Feteasca Alba and Frîncusa as well as Tamîioasa Românească is supposed to be traditional. Noble-rotted versions of Tamîioasa lack the characteristic Muscat Blanc à Petits Grains aroma, and have instead a more typically botrytis spice and marzipan note. These are difficult times for Romanian wines but there are still occasionally lovely examples of Tamîioasa, young and old, to make you hope that, one day, Romania will begin to fulfill its massive potential once more. Best producers: (Romania) Cotnari Cellars, Dealul Mare Winery.

*Tannat, here seen in the region of Madiran where the vine is believed to have originated, needs softening with other grapes in France – but in Uruguay it could have found its perfect home.*

## TANNAT

A grape of the Basque region of south-west France, seldom seen on wine labels there, but now making its mark as one of the more interesting grapes grown in South America.

In France its wines are most notable for their high tannin levels, and in Madiran the grape benefits from being blended with Cabernets Sauvignon and Franc as well as Fer, to make it more approachable. It also plays a part in Côtes de St-Mont, Irouléguy, Tursan and Béarn. All these reds have been or are being improved by modern winemaking methods, which emphasize fruit and soften tannins, but all can be tough to the point of astringency if not made well. The grape gives deep colours and takes well to oak aging. It is generally in decline in France.

In Uruguay Tannat is increasing in quality every year, and is marked by its fine, ripe tannins and elegant blackberry fruit. These wines seem far more European in style than most from South America, and it's a matter of opinion whether they could actually use more of the juicy, ripe fruit that is so typically South American, or whether they should jealously guard their rather more severe European style.

Growers in Uruguay report a difference in character between their old vines, which are descendents of the original cuttings brought from south-west France, and the newest clones imported from France. The newer clones tend to give more powerful but more simple wines, with an extra degree of alcohol –13–13.5 per cent instead of 12–12.5 per cent – but less acidity. The best solution, if complexity and depth are the aims, might be to produce virus-free clones from the old vine stocks – but, of course, the removal of viruses can itself change the character of the wine. Best producers: (France) d'Aydie, Berthoumieu, Brana, du Cayrou, la Chapelle Lenclos, du Crampilh, Montus; (Uruguay) Establecimiento Juanico, Castel Pujol, Hector Stagnari.

## TARRANGO

A recent Australian crossing, bred in 1965 from Touriga x Sultana to give wines of good colour and acidity but low tannin – Beaujolais-style, in other words. Quality is pretty good and prices are low. Tarrango needs hot climates to ripen properly and is at its best in the torrid, irrigated Riverland region where producing a bright, breezy, juicy Beaujoulais lookalike in such semi-desert conditions is quite an achievement. Best producer: (Australia) Brown Brothers.

# SILVANER

I USED TO LOOK FORWARD TO SILVANER (or Sylvaner as the French and most other people spell it) wines in blind tastings for the simple reason that it seemed to have almost no taste of its own, except for a certain green apple peel acidity when it came from a decent Alsace producer, and consequently, you could always taste the terroir. Well, the vineyard anyway. Actually what I really mean is the mud, the earth, and, if the wine was mature, it developed a strange taste of tomatoes. So I suppose I'm saying it never exactly tasted clean.

But that doesn't mean it was necessarily unattractive because, in either Alsace or Germany, surrounded by wines that were either highly aromatic or definitely on the sweet side, this broody, earthy quality actually made it rather appetizing. However, having earthiness as your main taste characteristic is not going to endear you greatly to producers in new wine regions looking for new vines to plant. Despite having once been widespread throughout Central Europe, Silvaner is now in decline and has made virtually no mark at all on the New World, though California has a few vines of no recognizable virtue, and one or two Australians have had a sniff of it.

It was originally planted in Germany – in Franken, to be precise – coming from further east, perhaps from Austria. Its first German appearance was at Castell in Franken in 1659, during the replanting that followed the Thirty Years War.

It's still Franken's main speciality. It has that dry (usually less than 4 grams per litre of residual sugar), slightly earthy fruit, good acidity and a lightness that can veer towards the insubstantial, but at low yields it reflects the subtleties of its terroir and can last a few years, though never very many. In the Rheinhessen it also occupies some good sites. I still find that earthy undertow in the wines, but it can make good apple-crisp dry whites (often labelled RS – Rheinhessen Silvaner) as well as surprisingly long-lived, fat, sweet styles – I've just tried a 1971 Oppenheimer Spätlese which was excellent.

In Alsace planting figures of Sylvaner have remained stable for many years, but its job as the workhorse grape for still wines and fizz has largely been taken over by Pinot Blanc. You can still find a little Sylvaner in Central and Eastern Europe – Hungary, Slovenia, Russia and the Czech Republic have some, and Austria has a few plots. But the only other places where it performs remotely memorably are Italy's Alto Adige where there are some good, snappy examples, and Switzerland (it's called Johannisberg there) where in some of the warmer sites in the Valais it can produce surprisingly deep, golden, minerally wine.

**The taste of Silvaner**

Even at its best, Silvaner seems to have more of a style than an actual flavour. In Germany its wines are dry, light and gently earthy, but never pungently so, and never at all aromatic. In Alsace it takes on the region's prevailing smoky spice, together with greater breadth, but still has little in the way of identifiable flavour. Alto Adige makes a fairly bright acid style and Switzerland can add a little earthy honey. But most examples won't age and will start to taste of tomatoes if you do try to cellar them.

**MARTIN SCHAETZEL**

*Sylvaner is not one of the 'noble' grapes permitted for planting in Grand Cru sites in Alsace, but Vieilles Vignes examples can be surprisingly concentrated.*

*The Juliusspital winery, right in the centre of Würzburg, is one of the many old religious and medical foundations of Germany, and has 121ha of vineyards, as well as farming and timber interests. The Stein vineyard in Würzburg is probably the best site in Franken for Silvaner: its deep limestone soil and southern exposure bring the grapes to high levels of ripeness and give the wines a rich, vanilla earthiness. The bottle is the traditional Bocksbeutel of the Franken region.*

**HORST SAUER**

*The Lump vineyard at Escherndorf in Franken has excellent south-east exposure: Silvaner gets very ripe here. Generally speaking the vine doesn't like the very hottest sites but it's not complaining chez Herr Sauer.*

***Left:*** *As is so often the case in Germany, the village – in this case, Escherndorf in Franken – is on the flat land beside the river, while vines are planted on the steep slopes above, benefiting from the sunny aspect. Here, the bends in the meandering river Main create several excellent vineyard sites.*
***Above:*** *Growers in Germany like Silvaner for its early ripening and consistently high yields – higher than Riesling. In Franken it likes warm sites with deep, rich soil and with good moisture retention. Its character is subtle rather than strong, but it can make wines of all levels of sweetness, up to Trockenbeerenauslese.*

## CONSUMER INFORMATION

### Synonyms & local names

The grape is spelt Silvaner in Germany and Sylvaner in France and Austria. The Swiss have many synonyms for it, including Johannisberg in the Valais.

### Best producers

**FRANCE/Alsace** Blanck, Bott-Geyl, Albert Boxler, Dirler, Keintzler, Seppi Landmann, René Muré, Ostertag, Pfaffenheim co-op, Martin Schaetzel, Bruno Sorg, Weinbach, Zind-Humbrecht.
**GERMANY** Rudolf Fürst, Fürstlich Castell'sches Domänenamt, Freiherr Heyl zu Herrnsheim, Staatliche Hofkeller, Juliusspital, U Lützkendorf, Horst Sauer, Schmitt's Kinder.
**ITALY** Abbazia di Novacella co-op, Köfererhof, Peter Pliger/Kuenhof, Valle Isarco co-op.
**SWITZERLAND** Robert Gilliard, Caves Imesch.
**AUSTRALIA** Ballandean Estate.
**SOUTH AFRICA** Overgaauw.

## RECOMMENDED WINES TO TRY

### Ten German wines

**Rudolf Fürst** *Bürgstadter Centgrafenberg Silvaner Kabinett*
**Fürstlich Castell'sches Domänenamt** *Casteller Hohnart Silvaner Kabinett Trocken*
**Freiherr Heyl zu Herrnsheim** *Niersteiner Rosenberg Silvaner QbA Trocken*
**Staatliche Hofkeller** *Würzburger Stein Silvaner Spätlese Trocken*
**Weingut Juliusspital** *Iphofer Julius-Echter-Berg Silvaner Spätlese Trocken* and *Würzburger Stein Silvaner Spätlese Trocken*
**U Lützkendorf** *Pfortener Köppelberg Silvaner Trocken*
**Horst Sauer** *Eschendorfer Lump Silvaner Auslese* and *Eschendorfer Lump Silvaner Spätlese Trocken*
**Schmitt's Kinder** *Randsackerer Sonnenstuhl Silvaner Spätlese Trocken*

### Ten Alsace wines

**Blanck** *Alsace Sylvaner Vieilles Vignes*
**Dirler** *Alsace Sylvaner Vieilles Vignes*
**Kientzler** *Alsace Sylvaner*
**Seppi Landmann** *Alsace Sylvaner Vallée Noble*
**René Muré** *Alsace Sylvaner Clos St-Landelin Cuvée Oscar*
**Domaine Ostertag** *Alsace Sylvaner Vieilles Vignes*
**Pfaffenheim co-op** *Alsace Sylvaner Vieilles Vignes*
**Martin Schaetzel** *Alsace Sylvaner Vieilles Vignes*
**Bruno Sorg** *Alsace Sylvaner*
**Weinbach** *Alsace Sylvaner Réserve*

### Eight other Sylvaner wines

**Abbazia di Novacella co-op** *Alto Adige Valle Isarco Sylvaner (Italy)*
**Ballandean Estate** *Late Harvest Sylvaner (Australia)*
**Robert Gilliard** *Johannisberg du Valais Porte de Novembre (Switzerland)*
**Caves Imesch** *Johannisberg du Valais Sylvaner (Switzerland)*
**Köfererhof** *Alto Adige Valle Isarco Sylvaner (Italy)*
**Overgaauw** *Stellenbosch Sylvaner (South Africa)*
**Peter Pliger/Kuenhof** *Alto Adige Valle Isarco Sylvaner (Italy)*
**Valle Isarco co-op** *Alto Adige Valle Isarco Sylvaner Dominus (Italy)*

# SYRAH/SHIRAZ

I CONDUCT A LOT OF TASTINGS around the country each year. These aren't tastings for wine connoisseurs – quite often the people attending have never been to a wine tasting before and very few of them have ever been subjected to the sort of 'This is how you taste wine' session they get from me. So I have to use wines with loads of flavour. For instance, I start out with Sauvignon Blanc from New Zealand because you'd have to have the palate of a lamp post not to appreciate and recognize some of the powerful fruit flavours there. And after roaming through the world of white and red, I need a grand finale: I need a great big walloping wine packed with personality and passion that will make any doubters cry – yes, this wine-tasting lark's for me. I need a wine to put a smile on people's faces and start a jitterbug in their hearts. And I always choose Australian Shiraz, as Syrah is called there – from the Barossa Valley or McLaren Vale. But as the cult of Shiraz spreads I can see myself using Californian, South African, Argentinian, Spanish or Italian examples. Or, of course, French – where they'd call it Syrah, if they put the name of the grape on the label at all, that is. But right now, I stick with Australian Shiraz. And when I ask people what their favourite wine is – they chorus 'Shiraz'. And if I'd asked them what the greatest wine grape in the world was – they'd scream 'Shiraz'. And perhaps they're right. Perhaps it is.

*Whether it originated in the northern Rhône, its European heartland, in Syracuse in Sicily or was brought back from Shiraz in Persia, there is something exotic about this grape. It's a nice idea but probably untrue that Syrah reached Europe at roughly the same time as the damask rose, and along the same route. In the far distance is the chapel on the Hermitage hill high above the river Rhône and in the foreground some of the wild violets that seem to perfume the wine itself. The parrot is an Adelaide Rosella whose habitat includes the Barossa Valley where Shiraz excels.*

You wouldn't have found many supporters for that view even ten years ago. The name Syrah was largely unknown to anyone who wasn't a Rhône red fanatic. In Australia the name Shiraz was applied to a style of rich, ripe, ebullient red wine in Australia that was generally thought of as a rather old-fashioned alternative to Cabernet Sauvignon.

Ah, yes. Cabernet Sauvignon. The grape that has enjoyed a century and maybe more revelling in its reputation as the world's great red wine grape. And until the 1980s at the very earliest, there wasn't a suggestion of any rival to its position. Bordeaux reds were thought of as the world's greatest. Anyone in the New World wishing to be taken seriously in the wine game would plant their best land with Cabernet Sauvignon and pray the wine turned out something like Bordeaux.

But why wasn't anyone doing the same with Syrah? Well, the fact is that there was a large amount of good quality Bordeaux red available and it had first been exported to northern Europe and then around the world as long as 800 years ago. The great red wines of the Rhône, however, were very scarce, hardly written about, rarely exported and consequently few of the New World pioneers had ever tasted them, so why on earth would they even think of planting Syrah? Only Australia, pursuing its own merry path thousands of miles away in the southern Pacific, had significant plantings, used mainly for port styles, because it was one of the first varieties taken out to the country in the early 19th century.

Well, luckily the Rhône began to get its share of the limelight during the 1990s – just at a time when Bordeaux was starting to look very expensive and, after a string of poor vintages, not very good value. And, what a surprise, Australia suddenly realized what a fantastic grape Shiraz was and how lucky they were to have so much of it. The 1990s were also a decade when warm climate flavours began to be appreciated more than cool climate ones, and if you have to look for France's greatest warm-climate variety, Syrah is unquestionably the star. And if you have to decide on the most opinionated, bumptious, most irrepressibly self-confident of the warm climate nations, Australia takes the crown without breaking sweat. If the world is going to continue its love affair with warm-climate wines, well, I wouldn't argue too hard against Shiraz or Syrah – call it what you will – waltzing off with the title of world's best red grape in the none-too-distant future.

**Syrah/Shiraz: from Grape to Glass**

*Geography and History page 246; Viticulture and Vinification page 248; Syrah/Shiraz around the World page 250; Enjoying Syrah/Shiraz page 254*

# Geography and History

A MAP SHOWING THE SPREAD of Syrah (or Shiraz, the name it takes in the New World) would have appeared very different 20 years ago. Then it would have looked like a grape in decline. And in any case, there would be only two patches on the map with any significant plantings at all – Australia and southern France. In Australia its vineyards were shrinking under the pressure to plant fashionable Cabernet Sauvignon, and indeed Chardonnay. Only in its homeland of the Rhône Valley – in particular Hermitage – and in the Midi, where it has long been regarded as an 'improver' vine, ideal for bolstering the aroma and flavour of otherwise tough reds, was it recovering from a long period of drab stagnation.

Back in the first half of the 19th century the picture would have looked different again. Then Bordeaux would

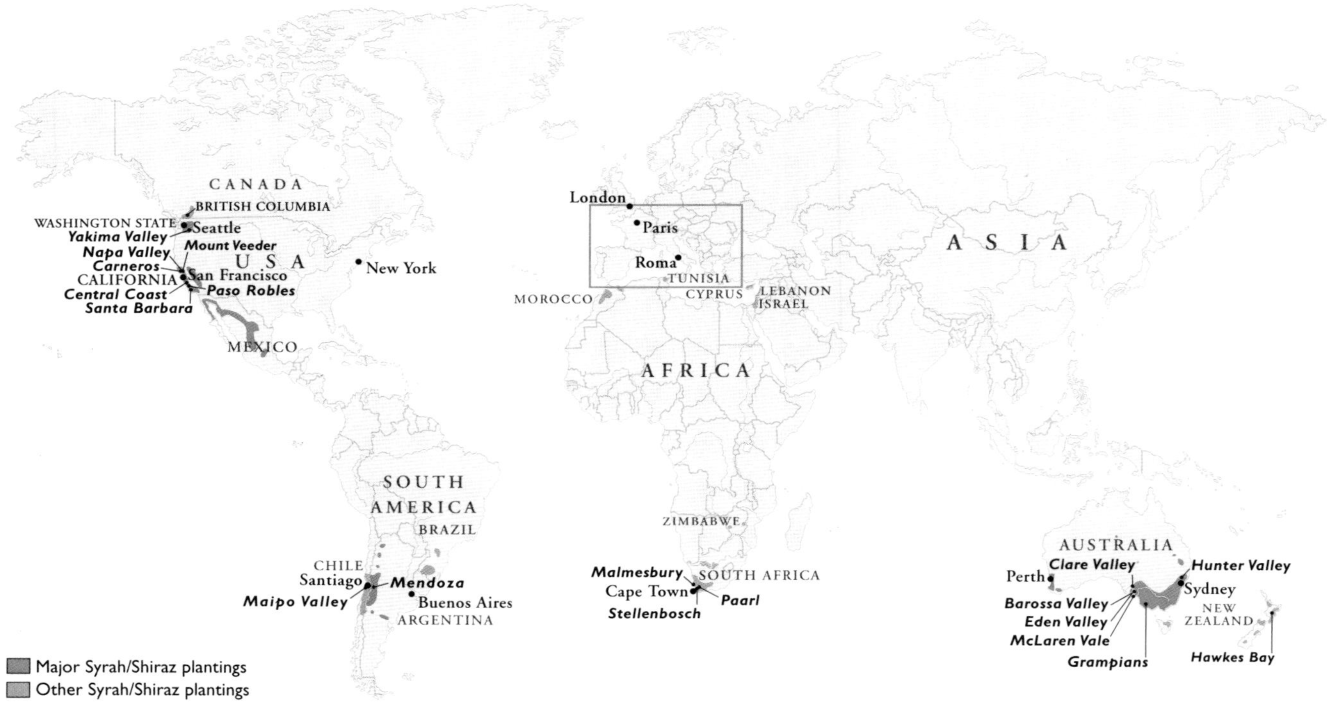

CANADA
BRITISH COLUMBIA
WASHINGTON STATE
Seattle
Yakima Valley
Napa Valley
Mount Veeder
Carneros
U S A
CALIFORNIA
San Francisco
Central Coast
Paso Robles
Santa Barbara
New York
MEXICO
London
Paris
Roma
TUNISIA
MOROCCO
CYPRUS
LEBANON
ISRAEL
ASIA
AFRICA
SOUTH AMERICA
BRAZIL
CHILE
Santiago
Mendoza
Maipo Valley
Buenos Aires
ARGENTINA
ZIMBABWE
Malmesbury
SOUTH AFRICA
Cape Town
Paarl
Stellenbosch
AUSTRALIA
Perth
Clare Valley
Hunter Valley
Sydney
Barossa Valley
Eden Valley
McLaren Vale
Grampians
NEW ZEALAND
Hawkes Bay
Major Syrah/Shiraz plantings
Other Syrah/Shiraz plantings

Major planting figures for Syrah/Shiraz

AREA PLANTED (HECTARES)
25,000
20,000
15,000
10,000
5000
2000
Languedoc-Roussillon (France) 27,992
Australia 25,596
Rhône (France) 13,615
Argentina 4282
Provence (France) 4195
California (USA) 4168
Mexico 1744
Gaillac (France) 898
Washington State (USA) 607
Corsica (France) 347

Paris
FRANCE
SWITZERLAND
Valais
AUSTRIA
Burgenland
Côte-Rôtie
St-Joseph
Hermitage
Valle d'Aosta
SLOVENIA
Gaillac
Rhône
South-West
Languedoc-Roussillon
Provence
Navarra
Tuscany
PORTUGAL
SPAIN
Aragon
Cataluña
Banyuls/Collioure
Corsica
Madrid
Roma
Lazio
Ribatejo
Toledo
ITALY
Macedonia
Alentejo
GREECE
Mediterranean Sea
Sicily

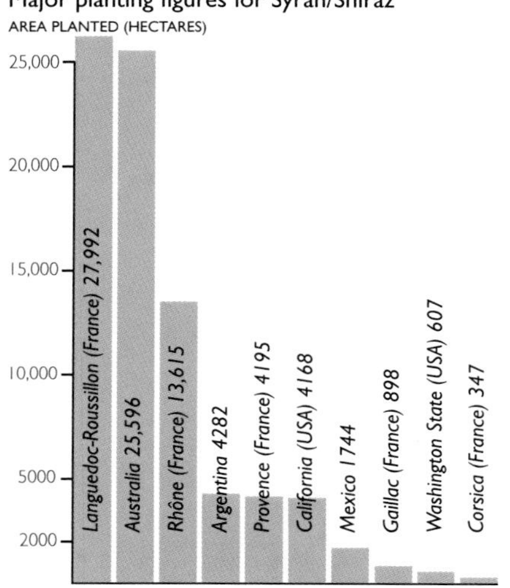

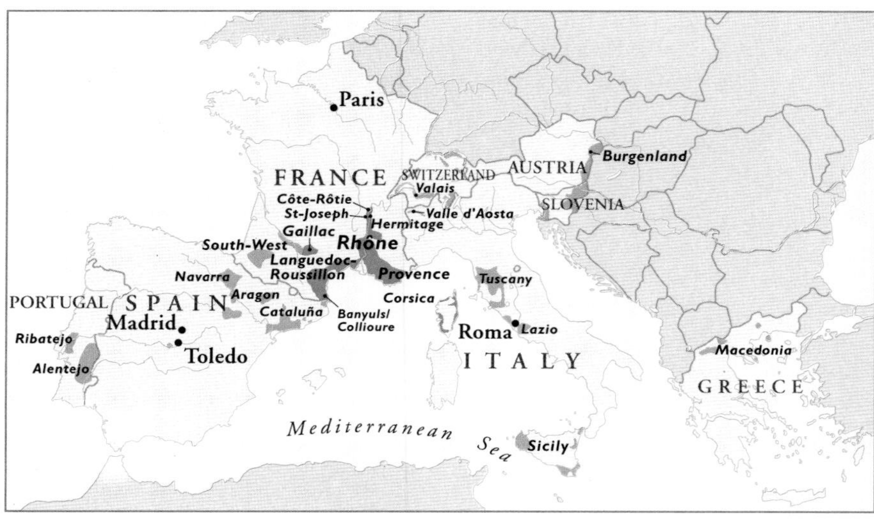

have been highlighted on the map. There were small but significant amounts planted here, at châteaux as eminent as Cos d'Estournel, Lafite and Latour.

Syrah was also grown and vinified in the Rhône, to be blended with top-level red Bordeaux to add colour and structure; the resulting 'improved' wines were described as 'Hermitagé'. As merchant Nathaniel Johnston wrote to his partner Monsieur Guestier in Bordeaux in the early 19th century, 'The Lafitte of 1795, which was made up with Hermitage, was the best liked wine of any of that year.'

But while Syrah blends like a dream with other varieties and also makes some of the finest, most exotically aromatic of red wines on its own, it is quite fussy about where it grows. Now that it is fashionable again it is spreading rapidly across the world – into California, Spain, Switzerland and Chile, Argentina, South Africa and New Zealand, and even, experimentally, into the vineyards of Austria, Italy and Portugal. But its choosiness means that it is unlikely ever to become as widespread in every wine region of the world as, say, Cabernet Sauvignon.

**Historical background**

It is somehow appropriate that a grape so startlingly perfumed should come laden with legend. Were the first vines on the hill of Hermitage in the Rhône planted by St Patrick as he made his way to the monastery of Lérins by the sea? Were they planted by a returning Crusader, Gaspard de Stérimberg? Did the Syrah vine originate in Persia and take one of its names from the city of Shiraz? Did it originate in Egypt, and travel to the Rhône via Syracuse in Sicily, gaining its name on the way? Or did it simply originate in the northern Rhône and stay there?

Sadly for romantics, it looks as though the last theory is the correct one, though mere reality won't stop me preferring all the other stories. A vine known as Allobrogica was being cultivated in the northern Rhône by the Gallic tribe of that name, the Allobroges, during the Roman Empire, and seems to have been selected from vines growing wild in the locality. The wine rapidly gained a reputation for fine quality and an unusual tarry flavour.

At some point it seems to have developed into the vine we know as Syrah. DNA fingerprinting at the University of California at Davis suggests that Syrah's parents could be red Dureza and Mondeuse Blanche, an old Savoie vine. However, much-quoted resemblances between the larger-berried form of Syrah and Mondeuse Noir have yet to be supported by genetic evidence. Syrah in the northern Rhône does nevertheless have enormous genetic variation, which would be expected if this was indeed its birthplace.

Syrah's introduction to Australia is less of a mystery. Scottish-born James Busby (1801–71), often described as the father of Australian viticulture, settled in New South Wales in 1824. In 1832, during a four-month visit to Europe, he collected more than 400 vine cuttings; Syrah was one of the most successful at adapting to the hotter, drier conditions of its new homeland.

*A little chapel dedicated to St Christopher sits on the peak of the hill of Hermitage in the Rhône. The hill is said to have got its name from Gaspard de Stérimberg who lived there as a hermit after his return from the 13th-century Crusades, though the name of Hermitage was not used for wine until the 16th century.*

*Vines have been grown on the hill of Hermitage (seen here across the Rhône) at least since the Romans occupied the area. The town at the foot of the hill is Tain, which the Romans knew as Tegna: Pliny and Martial both mention its wines by name.*

*Vines on the Côte Blonde sector of Côte-Rotie. The traditional method of training vines here is up chestnut stakes arranged in a wigwam, to protect the vines from damage by the Mistral, the cold wind that blows down the Rhône Valley.*

# Viticulture and Vinification

SYRAH'S REPUTATION IS FOR WINES that seem positively to ooze sunshine and warmth, especially if you take the marvellous monsters of South Australia as your starting point. But if you start from the perfumed delicacy of, say, a Côte-Rôtie from the northern Rhône, you would think you were dealing with a much cooler climate vine. Syrah/Shiraz will lose its heavenly floral fragrance if the climate gets too hot – though if it gets too cold, as in a poor vintage on the Rhône, the wine will taste too much like root vegetables for comfort. And in the heat of Australia's Barossa Valley or McLaren Vale, it can go from ripe to overripe in as little as a day.

It grows vigorously and gives its finest, most fascinating flavours on poor soils; on fertile soils the canopy must be open and well spread out. You can't overcrop it and produce excellent wine, though pleasant light wine is possible at high yields. And as for aging it in new oak – well, it has so much personality it doesn't need much new oak. Too much and it will taste like Barossa Shiraz – wherever it comes from.

### Climate

Syrah has a low tolerance both of too much heat and too little. It buds late and ripens early to mid-season, and in too hot a climate will rush to overripeness. It is naturally resistant to disease, though grey rot and bunch rot can be a problem.

In terms of climate the Rhône Valley is marginal. The vine needs sheltered, south-south-east- or south-west-facing sites here, where it is protected from the full strength of the Mistral, which originates further east and rips down the valley at speeds of up to 145km/h. The only virtue of such winds for the vine grower is that they help to dry the grapes after often violent rainstorms.

Hermitage has a mean temperature a couple of degrees higher than that of Côte-Rôtie, and Cornas can be hotter again, set as it is in a south-facing amphitheatre, protected from the wind. Growers in the northern Rhône seek the warmest spots for Syrah, and thus the greatest ripeness; in Australia, where Shiraz is frequently planted in regions considerably hotter than the Rhône, the last few decades have seen first a crisis of confidence, and then renewed enthusiasm for the heavyweight styles that result.

In the Barossa, source of Australia's benchmark Shiraz, drought is a problem on the hot valley floor. The best Shiraz comes from old, old vines whose roots are deep enough to find what little subterranean water there is. A lot of the top Shiraz from the Barossa is dry-farmed; irrigated vines here can require five litres of water per vine per day during the summer.

In Australia in the late 1980s a lot of effort went into seeking out cooler climates, in an attempt to mimic the climate of the Rhône. But too many of these new sites, which produced wines that were euphemistically described as 'elegant', were too cool, and the wines were peppery and lean rather than focused and complex. Even so, the search continues in Australia for sites that will give something of the Barossa's ripeness with something of Côte-Rôtie's perfume.

### Soil

This vigorous vine needs shallow, rocky, well-drained soils if it is to produce its most intense flavours. The best Rhône sites have soils derived from primary rock, especially granite, which retains the heat. Côte-Rôtie's Côte Blonde vineyard has some gneiss (a form of granite), and the Côte Brune has mica schist (a heat-retaining schistose rock that is rich in potassium, magnesium and iron; mica weathers to clay); the granite of the southern slopes of Côte-Rôtie gives softer, more aromatic wines, while the schistose of the northern slopes gives greater tannic structure. Parts of Hermitage, Crozes-Hermitage and St-Joseph are planted on alluvial terraces, and Hermitage also has granitic soil in the west.

In the New World site selection has, as we've seen, traditionally been by climate rather than by soil, and many different soils

*Breakfast time – on the flat road, not on the precipitous slopes – during the harvest at Côte-Rôtie. Picking has to be by hand on these slopes: much of Côte-Rôtie is terraced, making any sort of mechanization impossible. Spraying against rot, however, is often done by helicopter, which saves man-hours.*

are planted with Shiraz. In Australia's McLaren Vale alone, major soil types include thin, shaly soil over limestone in the north, with next to no underground water; deep sandy soils to the south-east; clay on limestone, sand on marly limestone clay and grey loam on clay to the west and south-west; grey clay over limestone and red earth over limestone in Willunga Flats near the sea; and heavy red loam and shale in the foothills of the Southern Mount Lofty Ranges. These five major terroirs produce, respectively, concentrated, spicy, bold wines; fleshy, soft ones; wines with peppery, spicy, dark plum characters; wines with firm tannic structure; and potentially heavily cropped wines with a danger of lack of intensity.

### Cultivation

When I think of the Rhône, I immediately think of stubby bush vines, gobelet-trained with one stake per vine, as in Hermitage, or Guyot simple-trained with four stakes per vine, as in Côte-Rôtie, for maximum protection against the wind. However, even within the Rhône Valley Syrah is subject to many variations in training, with training on wires increasingly popular on flatter ground where some mechanization is possible. It grows floppily, and where it is particularly vigorous it needs careful trellising to spread out its canopy: training methods elsewhere vary from bush, as in the oldest vines of the Barossa, to new systems of canopy management like Lyre and Smart-Dyson: this last is especially favoured by the Marqués de Griñon in Spain for his silky, complex Syrah, grown near Toledo. It gives, he says, some 40 per cent extra leaf surface.

### Yields

As usual, concentration goes hand in hand with excitement. The usual maximum permitted yield in the appellations of the northern Rhône is 40hl/ha, and while average yields have sometimes risen, particularly in Cornas, the steep hillside vineyards usually get less than this. Rhône producers maintain that at over 45hl/ha quality suffers; Australian viticulturalist Richard Smart, however, believes that it is possible to increase the crop to 10 tons/acre – that's about 180hl/ha – with no loss of quality, provided the vigour, vegetation and fruit production are in balance. (I'm waiting to taste the results before I decide.) This balance must be achieved by a combination of canopy management, correct pruning and water stress. Barossa producer Charles Melton, by contrast, puts the limit for quality at 2 tons/acre (36hl/ha).

### Glucosyl-glucoses

These are flavour precursors, chemical compounds found in the grapes. Current research in Australia indicates that they are unique to Syrah/Shiraz. Their measurement could prove a useful forecast of quality.

### At the winery

A grape as rich in colour-producing anthocyanins as Syrah can take hot fermentation temperatures of 30–35°C, and a maceration of anything from a few days to three weeks to maximize extraction. It used to be traditional in the Rhône to include all the stems, which made for some relentlessly tannic wines; now total or partial destemming is more common, with some growers taking up an extreme anti-stem position. Everything we want from the grape is in the skins, they say: add the stalks and you're adding harsh tannins. I'm inclined to agree.

In the 1970s, open fermenting tuns made of wood were the rule in the northern Rhône, and can still be found there, but now closed vats and stainless steel are also part of the picture. In Australia some growers are returning to shallow, open fermenters with excellent results.

Later picking of riper grapes in France means that the wines are better suited to longer macerations (it also means, incidentally, that the stems, if used, are likely to be riper and less green), and to the small new oak barrels that are increasingly used for aging the wine. Côte-Rôtie has the biggest concentration of small new oak barrels; Hermitage has fewer. Cornas, under the influence of enologist Jean-Luc Colombo, has been looking more kindly on new oak in recent years but I'm not convinced northern Rhônes benefit from much new oak. The oak here is French, more subtle in its flavours than American; Australian Shiraz is often run into new small American oak barrels before the fermentation is complete. Finishing the fermentation in new oak gives better integration of oak flavours, and helps to fix the colour. New oak must, however, be handled with care: it can easily muffle the excitement of Syrah. Grange, the archetypal Australian Shiraz, has the concentration to withstand lashings of American oak; not all would-be Granges do.

### SYRAH AS A BLEND

We've become so accustomed to varietal Syrahs from the Rhône and the south of France, California and (in the form of Shiraz) from Australia that it's easy to forget that this is also a superlative blending grape. That was for long its role in the south of France, where it was being added as a *cépage améliorateur* ('improver' variety) to the appellation wines of Provence and Languedoc-Roussillon long before Australian-trained winemakers starting bottling it as a varietal vin de pays. In the southern Rhône, too, it plays its part in the blend along with such grapes as Grenache and Mourvèdre.

Even in the northern Rhône it has commonly been blended – and, more remarkably, with white grapes. Up to 20 per cent Viognier may be added in Côte-Rôtie – though with only five or six per cent of Viognier grown in the whole appellation, the figure is much less than that, and often zero. In Hermitage, the blending grapes are Marsanne and Roussanne: 15 per cent may be added, though again, it is increasingly rare to add any. The reasoning was that white grapes helped to soften Syrah's high tannins; a function performed nowadays by picking at greater ripeness and better control of fermentation and maceration. New oak, too, can paradoxically help to polymerize and thus soften tannins, although one might think its only role would be to add more.

So-called Rhône Ranger producers in California – lovers of all things Rhône-ish – produce some impressive blends of Syrah with the likes of Grenache and Mourvèdre; in the Napa Valley, in particular, temperatures can be on the hot side for Syrah, making it less successful as a varietal. Areas like Santa Barbera further south are more successful.

Until the late 1990s the typical Australian red blend was Shiraz and Cabernet Sauvignon, mimicking an old 19th-century Bordeaux style. Cabernet/Merlot blends are now more popular, and Shiraz is being planted for varietal wines as fast as growers can get it in the ground. But blends with Viognier are not unknown here, too, for the extra aroma and complexity it contributes. In Tuscany, Syrah blends well with Sangiovese, and in Sicily with Nero d'Avola; in Spain with Garnacha, Cabernet Sauvignon, Cariñena and Tempranillo, even all together.

# Syrah/Shiraz around the World

YOU CAN TELL A WINEMAKER'S allegiance from the name he chooses for the grape. In France's Rhône Valley it's always Syrah, but then this is the archetype of the minerally, smoky herbs style. In the Languedoc it may sometimes change its name to Shiraz, reflecting brand loyalty to the Australian style of sweeter, more chocolaty fruit. In Australia it's almost always Shiraz. In other countries like Spain, South Africa, Argentina and the USA you get both names, according to what role model the winemaker wants to adopt – French Hermitage or Australian Grange.

**France: Rhône Valley**

Until the 1970s, the Syrah vineyards of the northern Rhône were struggling for survival. The terraces on the granite slopes overlooking the river had been replanted after phylloxera, but what was the point in working them? The wine sold for a pittance. Most wasn't domaine bottled: it was either sold by the *pichet* in local bars, or taken away by the *négociant* houses – in fact, it is thanks to the latter that many of these vineyards survived at all. What changed was not the wine or the grape, but the market. People at home and abroad woke up to these wines: and the decades since have seen rapid improvements in winemaking and the expansion, not always for the best, of the vineyards.

Terroir matters here for Syrah just as much as it does for Pinot Noir in Burgundy, and the steepness of the slopes (up to 55 degrees in Côte-Rôtie) gives maximum exposure to the sun, which is crucial in a climate that is marginal for the grape. Unlike Pinot Noir in Burgundy, however, the best northern Rhône Syrahs are nearly always blends of several different terroirs.

Syrah in the northern Rhône tends to be peppery at lower ripeness levels (10–11 per cent potential alcohol), but fruity and perfumed at 12.5–13.5 per cent potential alcohol. Côte-Rôtie can be paler in colour than Hermitage and more aromatic, with floral and roasted characters on the nose. Hermitage is firm, minerally and tannic; Cornas is the darkest and most robust of the lot, but lacks the thrill of Hermitage. Crozes-Hermitages and St-Joseph are lighter versions, more peppery, sometimes coffeeish, but increasingly good. The reasons there are such differences in flavour between Syrahs grown close to each other are to do with aspect, temperature and soil. The Hermitage hill's daunting slopes shouldn't and don't give the same flavours as the flatter vineyards of Crozes-Hermitage.

Let's take the two main appellations of the northern Rhône as examples. The Côtes Brune and Blonde in Côte-Rôtie are both larger sectors and specific *lieux-dits* (named plots of land) within those sectors. The Côte Brune, the broad northern sector, has clay- and iron-rich soil and produces the denser, more long-lived wine; but Viognier, the aromatic white grape which is permitted to be blended with Syrah here, does better on the more limestone soils of the Côte Blonde. However, since it ripens before Syrah yet must be picked at the same time (the grapes must be added at fermentation) it is generally picked overripe and quantities used in the wine, if any, are smaller than its 5 per cent share of the total vineyard would suggest.

Other key *lieux-dits* in Côte-Rôtie include La Mouline, which makes rich, balanced wines which fade slightly earlier than some others; La Landonne, which makes solid, backward wines which take longer to come round and which last for decades. La Turque is firm, concentrated and elegant. These are generalizations because the chief producer is Guigal whose wines are a) very oaky, and b) very, very expensive. Consequently, comparisons are rare.

At Hermitage, the granitic les Bessards vineyard makes dark, tough, concentrated wines that are the backbone of many blends. The loess soil of l'Hermite makes softer, more supple wine; les Beaumes, with limestone and ferruginous clay, gives scented, complex wines with fairly low tannins; chalky le Méal gives supple wines; the brown limestone of les Greffieux also gives supple styles, but is even better for the inclusion of white grapes, Marsanne and Roussanne.

Turning back to the vine itself for a moment, there is a much-quoted difference between the smaller-berried, finer Petite Syrah and the larger-berried Grosse Syrah, which is sometimes identified as the Mondeuse of Savoie. Certainly smaller-berried Syrah vines have a greater concentration of phenolics, but the clonal variation of Syrah in the northern Rhône is very great, and the two types, rather than constituting separate sub-species, have evolved by selection over the years.

**GUIGAL**
*The grapes for Guigal's La Turque come from one hectare on the Côte Brune, just at the boundary with the Côte Blonde. The wine contains about 5% Viognier.*

**CHAPOUTIER**
*Chapoutier's Ermitage Le Pavillon uses Syrah vines that are on average 65 years old, giving a concentrated, perfumed wine.*

**LES VIGNERONS DU VAL D'ORBIEU**
*Syrah is only a part of the blend in La Cuvée Mythique from Languedoc-Roussillon; the other grapes are Mourvèdre, Carignan and Grenache.*

In the southern Rhône Syrah loses its monopoly of the red vineyards. The North is part of what the French label the Septentrionale zone; the South is in the Méridionale. The North's climate is Continental, with cool winters and warm summers; the South is hotter and drier. Too hot, often, for Syrah, which can race to over-ripeness, and good producers give it the cooler, north-facing sites to slow it down a little. Grenache gets first refusal of the best sites in the South, though there are some high-altitude (up to 550m) spots where the harvest can be two weeks later than the norm; it will be interesting to see if the vogue for Syrah produces some new plantations here.

## France: the South

Syrah plays an increasing part both in Provence and in the Midi. It is mandatory in some appellations and optional in others, and along with Mourvèdre it is one of the main *cépages améliorateurs*. For varietal Syrah here one must look to vin de pays, and in particular to Vin de Pays d'Oc, where it may even be rechristened Shiraz in deference to the allegiance of the winemaker.

*These Shiraz vines, in Henschke's Hill of Grace vineyard, are over 100 years old. That makes them older than just about all the Syrah in the northern Rhône. The Henschkes call them 'the grandfathers', and the Barossa vines from which they were taken as cuttings some time in the 19th century were themselves brought to Australia from Hermitage before phylloxera struck the French vineyards. Hill of Grace Shiraz is now one of Australia's most expensive wines.*

## Australia

If the Australian settlers who first planted Shiraz in their new, untried country had been searching for a place that mimicked Hermitage, they would never have touched the Barossa or Hunter Valleys.

But of course they weren't looking for ideal sites; they had a hot climate and wanted to plant a grape that could cope. For them Shiraz was a dream come true. It was first planted in the Hunter in the 1830s and in the Barossa in the 1840s, and from the 1860s until the 1970s and 1980s, when the vogue for Cabernet Sauvignon took over, Shiraz was the national workhorse grape. It could do anything: it could produce light, soft, jammy flavours and big, beefy reds. It could make sparkling wines (one producer used to strip the colour out with carbon) and it could make fortified wines. Not surprisingly, it was taken less than seriously and Cabernet Sauvignon seemed to be the grape of the future.

In 1986 a government vine-pull scheme was the opportunity growers had been waiting for: they could get rid of their Shiraz, and be paid for doing so. Those who hung on could find themselves having to sell their grapes for Shiraz raisin muffins.

Some of the biggest companies, however, did hang on. Penfolds, Lindemans, Wynns, Tyrrells and other major names continued to grow Shiraz, often blending it with the more fashionable Cabernet. What might have happened if the supply of virus-free Cabernet vines had been able to equal demand is another question; we should just be grateful that Australians came round in the nick of time to appreciating what they had in their own backyard.

Even after the vine-pull scheme had done its worst, there remained far more old, ungrafted Shiraz vines – often planted in the middle of the 19th century – in the Barossa

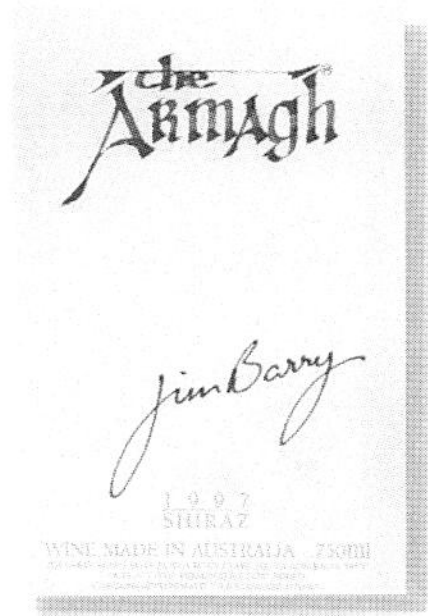


JIM BARRY

*Jim Barry's The Armagh is about as big as Australian Shiraz can get – and that's pretty big. American oak is part of the secret.*

BROKENWOOD

*The Graveyard vineyard is in the Lower Hunter Valley, and is marked as a cemetery on 1882 plans. It now produces an exciting, intense, well-structured Shiraz.*

ROCKFORD

*Rocky O'Callaghan acquired his antique basket press when other Australian wineries were throwing such things out.. He's made sumptuous Shiraz ever since.*

and Hunter Valleys than in the whole of the Rhône. These vines had originally come as cuttings from the Rhône, and so form a sort of living museum of pre-phylloxera France.

The knowledge of how to handle Shiraz had come from French expertise at around the same time, and had been handed down the generations even during the years when cheap fortified Shiraz was the rule. By the time Shiraz was resurrected in public esteem the knowledge still existed, though growers had forgotten its source. Barossa producer Robert O'Callaghan remembers meeting Rhône producers like Chapoutier and Jaboulet and being astounded that they knew the same things that he did.

Shiraz is now Australia's most significant red wine, and after some ill-starred forays into too-cool climates, styles seem pretty established. Barossa makes dark chocolate-tasting wines to McLaren Vale's milk chocolate; Eden Valley wines have more focused black fruit. In Victoria the Grampians produces wines that have a touch of Rhône spice. Geelong and Sunbury can have problems ripening, but make elegant, restrained styles, and the Hunter makes extracted, leathery wines, even though the famous 'sweaty saddles' flavour of Hunter Shiraz is now regarded as a winemaking fault – sadly for those who love a whiff of the old larrikins' leather now and then.

Fashionability has not put an end to Shiraz's versatility in Australia. It continues to make port styles of high quality, and sparkling wines that have attained a small but devoted following abroad. There are two styles of these: those made with young base wines are simple and sweetly blackberryish; those made with aged base wine are remarkable – they taste like serious Shiraz – but with bubbles. Wicked! Shiraz could also have a future as a base grape for a fuller, richer, white sparkling wine with very different flavours to those of Champagne.

Plantings of Shiraz are rocketing. In the Clare Valley alone, plantings of Shiraz, along with Cabernet and Merlot, are set to more than double by 2005.

The Shiraz planted in Australia is of the small-berried Petite Syrah type. Clonal variation in the Barossa, and in any vineyards planted with massal selections from old vineyards, reflects that in the Rhône. The commercially available clones, however, do not vary much and are probably less exciting.

### USA: California

The first Syrah was planted here in the 1970s, but it took a loose-knit group of winemakers known as Rhône Rangers to show that it could be an alternative to the ubiquitous Cabernet: deeply coloured but less tannic, well-flavoured and food-friendly.

Now fashion is on their side, and Californians are planting Shiraz as fast as they can. Between 1998 and 1999 plantings increased by one-third, to 10,298 acres (4168ha), and are scattered across the state. With such a young style – and with such young vines – it is too early to say which are the most promising regions. Mountain regions like Mount Veeder, which give such intense, brooding Cabernets, clearly show great potential, as do Santa Barbara and Paso Robles. Even cooler Carneros is interesting.

### The Grange story

*It was 1949 when the late Max Schubert (right) fell in love with the great reds of Bordeaux; he returned to Australia determined to recreate them in the Penfolds winery where he worked as winemaker. But there was hardly any Cabernet in Australia, which is why Grange is made from Shiraz – there was loads of that planted, and at the time it was generally used to make sweet 'port'. Schubert named his wine Grange Hermitage in honour of the famous Rhône red; in the 1980s the European Union insisted that Hermitage be dropped. The first commercial vintage was 1952; and in 1957 Schubert was told to stop making the wine . Nobody liked it; nobody wanted it. Did he obey? Well, no. And just as well: come 1960, those first vintages were tasting so superb that Penfolds revoked its order. Moral: never obey orders.*

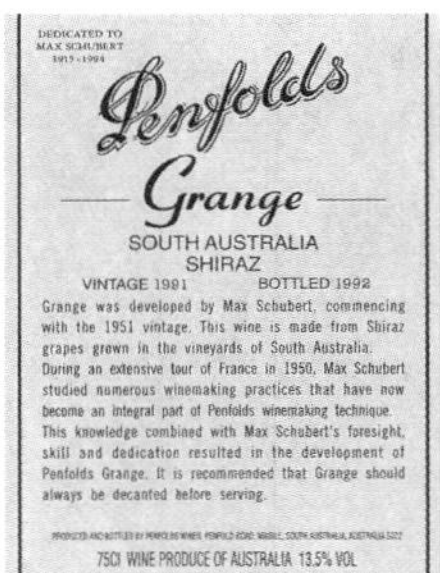


PENFOLDS

*Grange (then known as Grange Hermitage) first found public acclaim in Australia in 1962 – after a decade of criticism. These days you'll pay more for a single bottle of Grange than a whole ton of Shiraz grapes cost back in the wine's early days.*

The Australian wine giant Southcorp has bought 750 acres (303ha) in the Central Coast, and much of this land is being planted with Syrah. And there's French interest, too: the Perrins of Château de Beaucastel in Châteauneuf-du-Pape have a joint venture with their US importer, Robert Haas, to grow Syrah in 120 acres (48ha) west of Paso Robles.

As in Australia, there is a search for climate and style. The hottest parts of the Napa can be too hot, and Syrah is then best blended into Châteauneuf-du-Pape-style reds with Mourvèdre and Grenache. Here, as elsewhere, Syrah is a most obliging blender. Varietal Syrahs declare their allegiance on the label, with those calling themselves by the grape's French name being generally spicier and meatier than the fruit-first styles that go under the name of Shiraz. These 'Shiraz' are a more recent development in California, and they generally denote some Australian or New Zealand influence in the winery.

Syrah in California (which is usually larger berried than in Australia, either because it is a different clone or because of the climate) should not be confused with Petite Sirah. What California calls Petite Sirah is usually Durif; according to Dr Carole Meredith of UCD, 'Durif/Petite Sirah is a seedling of Syrah. It was known that Durif was "derived" from Peloursin, but how so was not known. What we know now, from DNA research, is that Durif grew from a seedling resulting from a cross between Peloursin (mother) and Syrah (father). We think this cross was not deliberate because Dr Durif would certainly have reported the pollen source as Syrah if he had known it. It was probably just a chance pollination in his research plot.'

## USA: Washington State

There is so much Syrah being planted here that by 2005 it will be as important in terms of area as Merlot. The first plantings were by David Lake MW of Columbia Winery in 1985; it began to boom in the late 1990s. The hottest, earliest ripening vineyards give blackberry, cassis and plum flavours; cooler sites in the Yakima Valley are more mulberry and black cherry with some bacon fat. McCrea Cellars, for one, blends a little Viognier into its best cuvée.

## South America

Argentina has most, with 2500ha, mostly in Mendoza. Growers like it for its large crops of 12–14 tonnes/ha, but this cropping level stops it excelling so far. Even at this level, quality is respectable if canopy management is good, but these yields will need to be halved at least if we're going to see how good Argentine Syrah could be. Its favoured climates tend to be like the southern Rhône rather than like the hotter temperatures of Australia: in hotter spots it loses structure and acidity.

In Chile, where plantings are still small but are increasing fast, the grape shows considerable promise. It can be highly productive, so some new plantings are on very rocky soil, which ought to restrict vigour somewhat. It can taste flowery when young, and seems to need a couple of years to develop recognizable flavours, though this may just be a stage on the learning curve of how to handle it, rather than a defining characteristic of the country.

## Rest of the world

As Shiraz it is being planted more widely in South Africa, helped by the availability of better clones. In 1999 40 per cent of new plantings were of Shiraz and Cabernet Sauvignon. Paarl and Stellenbosch are the regions in vogue at the moment, and Malmesbury for fortified versions. The aim is usually Australian-style fruit rather than Rhône-style complexity.

In New Zealand it needs the warmest sites, but has proved successful for a handful of producers. In Austria's Burgenland it is being planted in small quantities at the expense of lesser white varieties like Welschriesling. In Switzerland's Valais, where it has been planted in small quantities since 1920, it can make good, deep-coloured wines if the yields are kept down to 40hl/ha and the cap of skins and pips is well punched down during fermentation to maximize extraction. It is the first vine to bud in the spring, and one of the latest to ripen. Work is currently being done on selecting the best clones for the Valais. So far Swiss Syrah doesn't have great structure, but it is attractive young – and it does taste like Syrah.

Syrah is scattered in small quantities all over Italy, and blends well with local varieties like Sangiovese and Nero d'Avola as well as being made as a varietal. In Rioja there are moves to get it added to the approved list of varieties, and it could have interesting possibilities in Portugal and in Greece. It plays a small part in the blend at Chateau Musar in the Lebanon, and there is some, rather less distinguished, in Morocco and Tunisia.

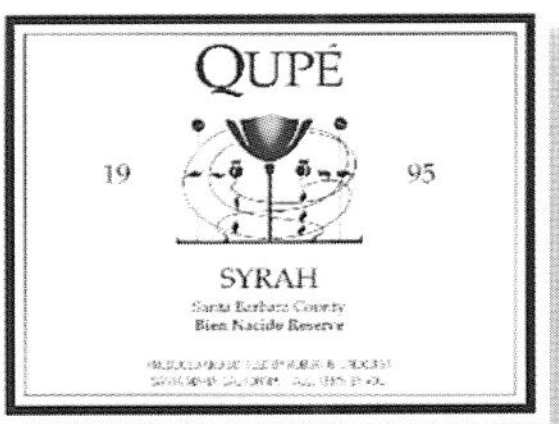


**QUPÉ**

*Rhône Ranger time: California producer Qupé favours French oak, and not too much of it new. The result is a lovely, dark, savory wine with deep, chewy fruit.*

**EDMUNDS ST JOHN**

*Another producer with a passion for the Rhône. The name of the grape on the label – Syrah rather than Shiraz – shows he's a Francophile.*

**MCCREA CELLARS**

*Doug McCrea is getting rather keen on Rhône grapes: his Syrah is one of Washington's best, and he's also planted some Grenache and Viognier.*

**VIÑA TARAPACÁ**

*Syrah has always been a great blending grape – it was much prized in 19th-century Bordeaux. This Chilean blend of Cabernet Sauvignon, Merlot and Syrah might mirror such 19th-century classics.*

**ISOLE E OLENA**

*Paolo de Marchi first planted Syrah in Chianti to beef up his Sangiovese. Now it's bottled alone and is a fantastic wine – Syrah in fruit flavour but unmistakably Tuscan.*

# Enjoying Syrah/Shiraz

OUR KNOWLEDGE OF HOW SYRAH AGES might, at first glance, seem rather limited. Until the 1960s Cornas, for example, was seldom domaine bottled, and though Hermitage and Côte-Rôtie were domaine bottled earlier than that it was common until as late as the 1970s to bottle wines on demand rather than when the development of the wine in cask called for it. Wines could vary dramatically according to whether they were bottled early or late; happily, such cavalier attitudes are a thing of the past.

Hermitage is the longest living and slowest maturing wine in the Rhône. Good vintages need ten or more years to loosen up, though the palate can remain tight and tannic even after the bouquet has developed plummy, spicy characters. Lesser vintages may be ready in five or six years. Top Côte-Rôties take nearly as long, though the Côte Blonde is earlier maturing than the Côte Brune, and top Cornas can need nearly a decade. The influence of enologist Jean-Luc Colombo is producing Cornas that is much lusher, but beware of the dumb phase Cornas goes through between two and eight. Crozes-Hermitage and St-Joseph are often delicious after about three years though they can age well.

Southern Rhônes need less time than the northerners: even the very best Châteauneuf-du-Pape is ready by eight, and most can be drunk by five.

Most Australian Shirazes are made to be drunk within five years or so, though top old-vine versions, and especially Grange, may need up to a decade. They'll keep easily for 20 years, and Grange for much, much longer.

### The taste of Syrah/Shiraz

Young Syrah often smells, surprisingly for a red, of exotically scented flowers like carnations and violets. To those add woodsmoke, perhaps with a few sprigs of rosemary thrown on to the fire, plus raspberries, blackberries and blackcurrants. Côte-Rôtie is even smokier, but also more perfumed, than Hermitage, which in turn is more blackcurranty. Hermitage, too, has a centre of tremendous succulence and richness – though if you open a bottle too young you may wonder where this is – and it keeps this richness longer than any comparable French red wine. With age it gains gamy, leathery smells, an almost chocolaty character, and a whiff of violets and tobacco.

In Australia, flavours are creamier and more chocolaty, with blackcurrant and black cherry fruit. Hunter Valley Shiraz is leathery, earthy and extracted; Barossa versions are all black cherry and spice. Clare wines have purer fruit and fine structure and Coonawarra fruit is bright and peppery. The Victoria style of lean, peppery fruit might be on the way out; fleshier, richer wines have won the day. Nearly all Aussie Shiraz gains a thick dark taste of licorice, prunes and black chocolate as it ages.

Much of the sweet, vanilla spice of Australian versions comes from the new American oak in which it is aged. This is now being handled more sensitively than it was; and earlier bottling means less leatheriness, more fruit.

*Syrah on the left, Shiraz on the right. Domaine de Thalabert is Jaboulet's top Crozes-Hermitage, needing up to ten years to show its best. Peter Lehmann, Barossa producer extraordinaire, bought the Stonewell plot in 1994, after using the fruit for several years in his 'Stonewell' label. The wine gets 100 per cent new American oak, which tests the fruit to its limit, but now that Lehmann controls the vineyard quality, the results are spectacular.*

### Matching Syrah/Shiraz and food

Whether from France (in the northern Rhône), Australia, California or South Africa, this grape needs food with plenty of flavour. Northern Rhône Syrah goes well with game, beef and venison; the same goes for the more concentrated, leathery or smoky Australian Shiraz. Australian and South African versions are perfect for picking up the smoky flavours of barbecued food.

Lighter, more berryish wines are good with turkey, guinea fowl, lamb casseroles and even chicken, providing the chicken has some flavour. Liver is a good partner, too. Light Shiraz can be a good match for the subtle spices of Indian dishes, but Chinese or Thai flavours are seldom successful with this grape. Hard English cheeses like Cheddar are good.

## CONSUMER INFORMATION

### Synonyms & local names

It's Syrah in France and Shiraz in Australia. Most producers in other regions use either name depending on the style of wine being made. Also called Petite Syrah in the northern Rhône. Not to be confused with Petite Sirah, a grape grown mainly in California and Central and South America.

### Best producers

**FRANCE/Rhône Valley** Allemand, Balthazar, Gilles Barge, P Barge, Belle, Bonserine, Burgaud, Chapoutier, Jean-Louis Chave, Auguste Clape, Clusel-Roch, Colombo, du Colombier, Courbis, Coursodon, Yves Cuilleron, Delas Frères, Durand, B Faurie, Pierre Gaillard, J-M Gérin, A Graillot, J-L Grippat, Guigal, Jaboulet, Jamet, Jasmin, du Monteillet, E Pochon, des Remizières, René Rostaing, St-Désirat co-op, M Sorrel, Tardieu-Laurent, Verset, François Villard; **Languedoc-Roussillon** l'Aiguelière, des Estanilles, Gauby, l'Hortus, Mas Blanc, du Peyre-Rose, les Vignerons du Val d'Orbieu.
**ITALY** Bertelli, D'Alessandro, Fontodi, Isole e Olena, Le Macchiole, Villa Pillo.
**SPAIN** Marqués de Griñón.
**AUSTRALIA** Tim Adams, Banrock Station, Jim Barry, Basedow, Best's, Bowen, Brokenwood, Grant Burge, Chapel Hill, Charles Cimicky, Clarendon Hills, Coriole, Craiglee, Dalwhinnie, D'Arenberg, Fox Creek, Hardys, Haselgrove, Henschke, Jasper Hill, Leasingham, Leconfield, Peter Lehmann, Lindemans, McWilliams, Charles Melton, Mitchelton, Mount Horrocks, Mount Langi Ghiran, Penfolds, Plantagenet, Redbank, Rockford, Rosemount, Rothbury, St Hallett, Seppelt, Seville Estate, Taltarni, Taylors, Tyrrell's, Vasse Felix, Wendouree, Wirra Wirra, Wynns, Yarra Yering, Zema.
**USA/California** Alban, Araujo, Dehlinger, Edmunds St John, Jade Mountain, Ojai, Joseph Phelps, Qupé, Swanson, Sean Thackrey, Truchard, Zaca Mesa; **Washington State** McCrea Cellars.
**NEW ZEALAND** Stonecroft, Te Mata.
**SOUTH AFRICA** Boekenhoutskloof, Fairview, Graham Beck, Saxenburg, Spice Route, Stellenzicht.
**ARGENTINA** Luigi Bosca/Leoncio Arizu, Finca El Retiro.
**CHILE** Carmen, Errázuriz, Montes Alpha, Viña Tarapacá.

## RECOMMENDED WINES TO TRY

### Ten classic northern Rhône Syrah

Chapoutier *Ermitage le Pavillon*
Jean-Louis Chave *Hermitage*
Auguste Clape *Cornas*
Clusel-Roch *Côte-Rôtie les Grandes Places*
Delas Frères *Hermitage les Bessards*
Pierre Gaillard *St-Joseph Clos de Cuminaille*
Alain Graillot *Crozes-Hermitage la Guiraude*
Guigal *Château d'Ampuis Côte-Rôtie la Turque*
Jaboulet *Hermitage la Chapelle*
Réne Rostaing *Côte-Rôtie Côte Blonde*

### Ten other classic Shiraz/Syrah from around the world

Tim Adams *Aberfeldy Shiraz (Australia)*
Brokenwood *Graveyard Vineyard Shiraz (Australia)*
Hardys *Eileen Hardy Shiraz (Australia)*
Henschke *Hill of Grace Shiraz (Australia)*
Mount Langhi Ghiran *Shiraz (Australia)*
Penfolds *Grange (Australia)*
Qupé *Bien Nacido Reserve Syrah (California)*
Stellenzicht *Syrah (South Africa)*
Stonecroft *Syrah (New Zealand)*
Sean Thackrey *Orion Old Vines Rossi Vineyard (California)*

### Five good-value northern Rhône Syrah

Domaine du Colombier *Crozes-Hermitage Cuvée Gaby*
Yves Cuilleron *St-Joseph*
E Pochon *Crozes-Hermitage Domaine Curson*
Domaine des Remizières *Crozes-Hermitage*
St-Désirat co-op *St-Joseph*

### Ten other good-value Shiraz/Syrah

Basedow *Shiraz (Australia)*
Best's *Great Western Shiraz Bin 'O' (Australia)*
Errázuriz *Reserve Syrah (Chile)*
Ch. des Estanilles *Faugères Cuvée Syrah (France)*
Fairview *Shiraz (South Africa)*
Finca El Retiro *Syrah (Argentina)*
Leasingham *Shiraz Bin 61 (Australia)*
Peter Lehmann *Shiraz (Australia)*
Lindemans *Padthaway Shiraz (Australia)*
Domaine Peyre Rose *Coteaux du Languedoc Syrah Léone (France)*

### Five sparkling Shiraz wines

Banrock Station *Sparkling Shiraz (Australia)*
Fox Creek *Vixen Sparkling Shiraz Non-vintage (Australia)*
Charles Melton *Sparkling Shiraz (Australia)*
Rockford *Black Shiraz Non-vintage (Australia)*
Seppelt *Show Sparkling Shiraz Vintage (Australia)*

*Shiraz is the comeback kid: one moment growers were pulling it up as fast as they could and the next the whole world was crying out for old-vine Shiraz. Or failing that, any Shiraz at all. And there's a massive difference between old-vine Shiraz and young, high-yielding Shiraz.*

### Maturity charts

This grape is so new to most regions except the Rhône and Australia that identifying national styles is tricky.

**1999** Hermitage (Red)

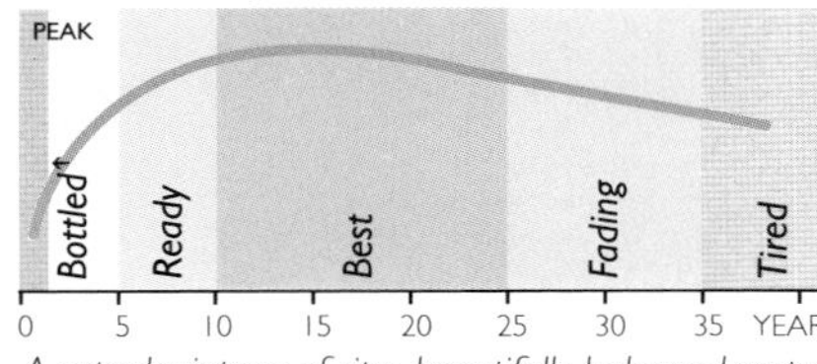


*A superb vintage of ripe, beautifully balanced grapes to end the century. They have sufficient fruit to be drinkable young, but will age magnificently.*

**1998** Barossa Valley Old Vine Shiraz

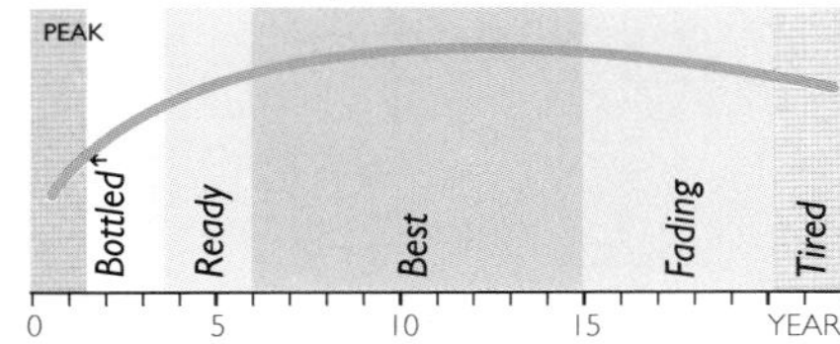


*Styles here vary, but the most intense wines can combine early drinkability with the ability to age for a couple of decades or more.*

**1999** California Syrah

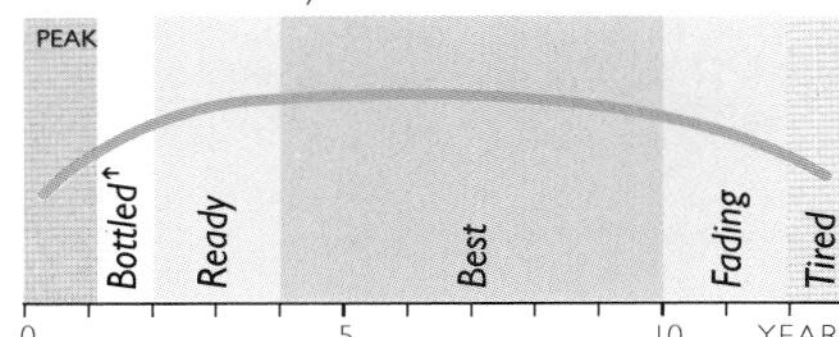


*A typical California Syrah or Shiraz does not yet exist. Many wines are made to be drunk within three or four years; gutsier ones will, however, last longer.*

# TEMPRANILLO

THIS BOOK IS FULL OF FASCINATING details, facts and figures – I hope you will agree. But they don't come much more eye-popping than the prediction that the volume of Tempranillo in Australia will have increased by 1500 per cent between 1999 and 2004. Now, we're not talking shedloads of grapes here – only 439 tonnes projected for 2004 – but that is a significant amount of fruit. And there's no doubt at all that the Aussies will be making some good Tempranillo wines by then. Well, I'm glad those boys Down Under are taking this grape seriously, because it would seem to have everything the New World wants – good colour, relatively low acid, low tannin when you ripen it fully (which isn't a problem in most New World regions), bright, upfront fruit and an eager-to-please affinity with oak barrels and their soft spicy characters. It sounds as though it was just made for the wine tyros of California, Chile, South Africa, Australia, and, well – Argentina, maybe New Zealand, definitely the Mediterranean basin. And yet of all these, only Argentina – where it's been for a long time – and Australia, where they have a very good nose for good grapes and a very good knack for spotting the Next Big Thing, seem to be interested in it. I am certain things will change, but why has Tempranillo been so slow in getting off the ground?

*The history and grandeur of Tempranillo is conveyed by references to two Spanish artists of the 17th century: first, Diego Velázquez, the great Spanish court painter whose paintings of the royal family, including the Infanta Maria Teresa depicted here, were much admired at the time, and second, Juan Sánchez Cotán who is best known for his still lifes with their detailed realism.*

The main point is that maybe nobody knew it was there. Or at least nobody knew that what they were tasting was Tempranillo, despite it being the main grape for Spain's two most famous reds, Rioja and Ribera del Duero, and a crucial part of Portugal's most famous wine – port. Aye, there's the rub. It was indeed called Tempranillo in Rioja, though it's only in relatively recent times that Rioja has been accorded any great fame and respect outside Spain. And remember that when the Californians and Australians and South Africans and the rest were looking for grape varieties to transplant back to their distant lands they were obsessed by the French classics – they wanted to make Pauillac lookalikes and never cast a glance south of the Pyrenees. But if they had glanced at Ribera del Duero – and they could have done since Vega Sicilia has been Spain's most famous and expensive red wine for a century – they'd have discovered that the main grape was called – well, Tinto Fino, actually. What about port then, over the border in Portugal? Ah, you must be thinking of Tinta Roriz. Or Aragonez. Not Tempranillo? Hmm.

And back in Spain, it is the major grape in most of the other top red wines. But Toro's best grape is Tinto de Toro. Valdepeñas and La Mancha boast about their Cencibel. Penedès swears by the Ull de Llebre – and so it goes on, through province after province. Always an excellent red grape, always called something different. But always actually Tempranillo. I can't think of any red grape that so completely dominates a country's quality wines as Tempranillo does Spain's – and which hides under so many aliases, with Portugal willingly adding a few of its own.

Since the New World winemakers are the ones who have made grape varieties famous by putting their names on the labels, you have to attract these trailblazers if your grape is to achieve any renown. Tempranillo was in the wrong place – nobody wanted to copy Spanish wines. And Tempranillo was never on the label, anyway. Spain and Portugal followed the French appellation idea of naming their wines after areas, not grapes. And since table winemaking in Iberia was dreadfully old-fashioned, who would care to spend time and money finding out if their grapes were actually any good?

It was only when a few 'flying winemakers' turned up in Iberia – and they were there to try to transform hopelessly old-fashioned wines into something saleable – that the vineyards actually got a close look. Such winemakers have turned Iberia's wine into some of the most exciting in Europe. And time and again, they've discovered that the local grape which they've transformed into the best grog was – the Tempranillo. Which is why Australia's crop of Tempranillo will have increased by 1500 per cent by 2004. I look forward to tasting it.

**Tempranillo: from Grape to Glass**

*Geography and History page 258; Viticulture and Vinification page 260; Tempranillo around the World page 262; Enjoying Tempranillo page 264*

# Geography and History

SUDDENLY, EVERYONE'S TALKING TEMPRANILLO. If you're one of those who's not, let me explain. Tempranillo is the great Spanish grape that is at the heart of Rioja and Ribera del Duero. In turns up almost everywhere else in Spain under a variety of names (see below) like Cencibel, Tinta de Toro and the rest, and could rightly be said to express the fundamental character of Spanish red wines from the cool high vineyards of the North-East to the arid plains way south of Madrid. As such, we've all drunk Tempranillo wines, even if we've not seen the name on the label. But Spain's explosion of quality in the last few years has meant that its grape varieties have suddenly come under examination by growers worldwide keen to find commercial alternatives to Cabernet Sauvignon and Merlot.

London
Madrid
Roma
ASIA
USA
New York
San Francisco
CALIFORNIA
MOROCCO
MEXICO
AFRICA
SOUTH AMERICA
Mendoza
Buenos Aires
ARGENTINA
Cape Town
SOUTH AFRICA
AUSTRALIA
Sydney

Major Tempranillo plantings
Other Tempranillo plantings

Major planting figures for Tempranillo
AREA PLANTED (HECTARES)

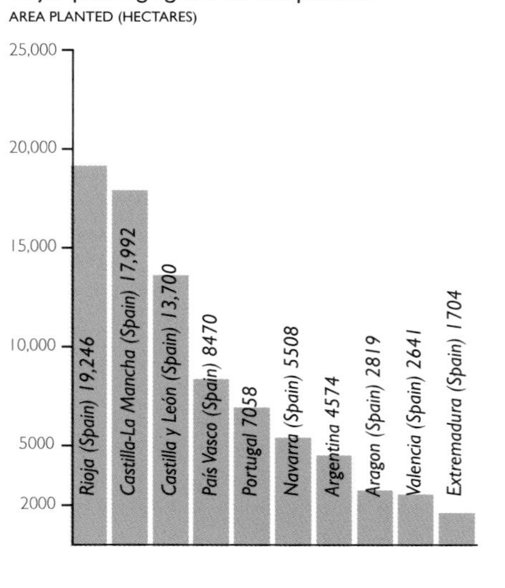


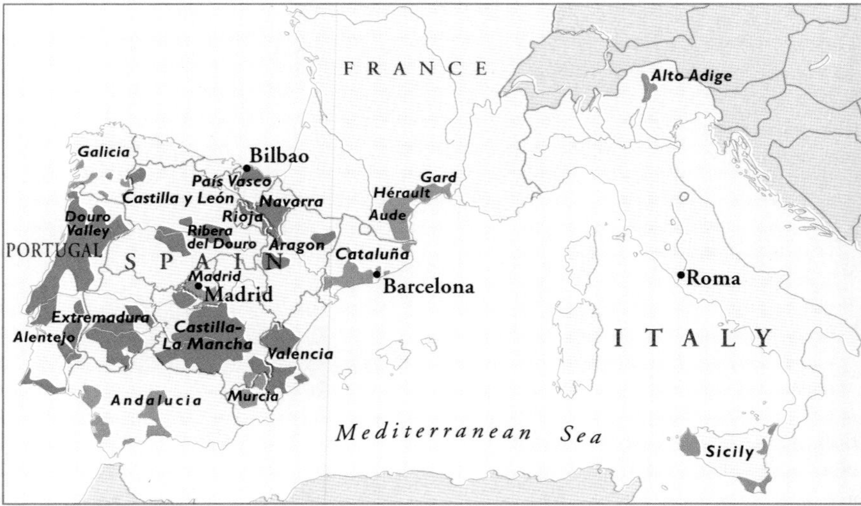

Southern France has had experimental plantings for quite a while. Portugal has a good deal of it under the name of Tinta Roriz, and Argentina has fairly wide plantings. But it is the current wave of interest from such places as Australia and , belatedly, California and Italy that is really going to make the name Tempranillo internationally famous at last.

What's the attraction? The world is running mad for red wine – but it wants its red wine to be both reliable and different. It wants lush textures and appealing fruit, low tannin and not too much acidity. And it seems to want a grape name to be, in effect, a brand, tasting recognizable no matter what the geographical source. Tempranillo can do all this. It has the right exuberant flavour, and what it lacks in perfume can be supplied by other grapes. It takes to new oak like a dream. It lacks acidity, but that too can be rectified by canny blending. It will grow obligingly in warm climates, and it will produce anything from light young wines to older, oak aged ones. And all the time it will taste of itself. It doesn't reflect its terroir in the way that Pinot does, and it doesn't have much complexity. But it is utterly, utterly seductive.

**Historical background**

Tempranillo probably originated in Spain, and probably in Rioja, although according to one story it was brought to Spain by monks from Burgundy on their way to Santiago: the idea is that they would have given vine cuttings to the Spanish monasteries of their orders. There is certainly a similarity of flavour between mature Pinot Noir and mature Tempranillo, and it has been suggested that the two could share a common ancestor. In Toro it is said to have been planted after the Moors left.

From Spain it spread to Portugal and to Argentina and Mexico. Its arrival in the Douro Valley seems to date from the 18th century, when it was planted at Quinta de Roriz by the quinta's Scottish owner, Robert Archibald. Archibald is also thought to have introduced Tinta Francisca to the Douro.

Further south in Portugal, its Spanish origins are reflected in its local name of Aragonez. In the Setúbal Peninsula it is known as Tinto de Santiago, again pointing to a Spanish source.

In Spain, local names include the Catalan Ull de Llebre (Ojo de Liebre in Spanish) in Penedès; Tinto Fino in Ribera del Duero; Cencibel in Valdepeñas; and Tinto Madrid, Tinto de la Rioja, Tinta del País and Tinta de Toro in other parts. All these regions tend to claim that their Tempranillo is a local clone that gives a unique character to their wines; the truth is that genetic differences between Tinto Fino, Cencibel and the rest are smaller than those found among Tempranillo vines in any one region. Variations include, for example, a Tempranillo Peludo, or Hairy Tempranillo, which has more down on the back of the leaf.

Its internationally accepted name of Tempranillo comes from the Spanish word *temprano*, meaning early, and reflects its early ripening nature; it also buds late, and needs only quite a short growing season. In other words, it's a grape grower's delight.

*The house of the Rodriguez family on the beautiful Remelluri estate up in the hills above Labastida in Rioja Alavesa. For many years Remelluri, a single-estate wine, was a rarity in Rioja, a region where bodegas have traditionally bought grapes from many different spots. Now there is an increasing focus on individual terroirs.*

*A traditionally bush-trained Tempranillo vine in Rioja: the local name for this form of training is* en vaso. *More and more growers, succumbing to the delights of mechanization, are training their vines on wires.*

*Isaac Muga of Bodegas Muga checking wine as it is being racked from one barrel to another. This most traditional of bodegas eschews stainless steel: everything is aged in oak.*

# Viticulture and Vinification

CLIMATE AND WINEMAKING seem to be the main influences on the flavour of Tempranillo – in fact until quite recently the flavour most associated with it was the vanilla softness of American oak. This was because the only widely known Tempranillo-based wine was Rioja, whose reputation during the 1970s and 1980s had been built almost entirely on the felicitous marriage of vanilla-scented oak and strawberry-flavoured Tempranillo. There is no long tradition of single-vineyard wines in Spain, and little emphasis until recently on the importance of different terroirs to the styles of individual wines. Most Tempranillo was and is a blend of different soils and climates; more emphasis on individuality might lead to greater understanding of how the vine reacts to different terroirs. The most likely place for this to occur is the top-quality region of Ribera del Duero, though single-vineyard Riojas are now beginning to make their presence felt. At the moment complexity comes from winemaking; in the future certain vineyard sites are sure to shine.

## Climate

To get elegance and acidity out of Tempranillo, you need a cool climate. But to get high sugar levels and the thick skins that give deep colour you need heat. In Spain these two opposites are best reconciled in the continental climate but high altitude of Ribera del Duero. Here, at up to 850m, summer daytime temperatures may hit 40°C, which gives sugar and colour, but fall dramatically by up to 20° at night, thus conserving acidity. Summers are short here and winters long and hard: there may be as few as a hundred frost-free days a year.

Nearby Toro also has a short, hot ripening season, though at 230 days it is significantly longer than Ribera del Duero's 171 to 198 days. Vines are planted here at between 600 and 750m, Rioja Alta is appreciably lower at 500–600m, and Rioja in general is less continental in climate: producers point to the temperate, Atlantic-influenced west wind that blows across their vineyards, while the Sierra de Cantabria mountains give shelter from the colder north wind. The result is lower alcohol in Rioja – typically 12–12.5 per cent, compared to 13.5 per cent in Toro – and less pronounced structure and fruit than in Toro and Ribera del Duero. In Portugal's part of the Douro Valley, vineyards are planted at almost all altitudes, with the highest ones giving wines with the greatest acidity, and ripening two to three weeks after the lowest ones, but even late-ripening sites can usually produce grapes with pronounced perfume and fruit.

*Peter Sisseck with 60-year-old Tinto Fino (Tempranillo) vines at Dominio de Pingus in Ribera del Duero, Castilla y León. 'I used to believe that Tempranillo was a second-rate grape,' he says. 'I didn't realize it was just a matter of vinifying it in a way more Burgundian than Bordelais to get the flavours out.'*

## Soil

Soils in Rioja are relatively homogeneous compared to some regions: in the North they are mostly clay-based, and in Rioja Alta and Rioja Alavesa there are patches of chalky or iron-rich subsoil. The best wines, as is so often the case, come from the chalky clays: chalk gives acidity and elegance, clay gives body. But the ferruginous clays are full of trace elements that can add complexity.

In Toro most of the soils are alluvial, and there is some limestone in the sub-zone of Morales; an admixture of clay, because it holds water, enables the vines to survive the hot summers.

Both in Ribera del Duero and in the Douro Valley further west, the river Duero/Douro flows through schist. At the Spanish end there is also a lot of chalk: limestone and chalk comprise a third of the soils in the west of Ribera del Duero, and over half the soils in the east. Limestone and chalk, here again, give acidity. Aragonez from the Alentejo in south-east Portugal lacks the acidity of Tinta Roriz from the Douro, but can be a delicious, juicy mouthful.

## Cultivation

The traditional pruning method for Tempranillo in Spain is the same as for so many Spanish grapes: *en vaso*, or gobelet. Three or four branches are left, with about 20 fruiting buds in all. Mechanization is not possible with this system, and so training on wires, usually with double cordon, is on the increase. Indeed, it seems to suit the vigour and upright habit of the vine. Densities for mechanical harvesting are lower: about 2200 vines per hectare, compared to the more traditional figure of 2500–3000 per ha. In the Douro Valley training on wires is the norm.

Tempranillo crops generously: sometimes too generously for quality because, like Pinot Noir, it is considered a very yield-sensitive vine. In the Alentejo, 1.5 to 2 kilos per vine is about the limit for quality: go to twice that and the wine will be dilute. In the Douro, seven tonnes per hectare (equivalent to about 49hl/ha) is considered quite high; in Argentina, however, 12 tonnes per hectare is considered quite low. In Argentina's warmer, heavily irrigated spots the vine will give 30 tonnes per hectare (or about 200hl/ha) of dilute wine, but as a simple soft glugger sold at a low price – well, it still has enough flavour to perform that task.

In Rioja the legal limit is 45hl/ha (49hl/ha in Ribera del Duero), but while many vineyards produce less than this, there's no doubt much of the region overproduces – particularly on newer valley-floor vineyards. Indeed, dilution through overcropping has been one of Rioja's major problems in recent years. Training on wires increases yields, and in addition to this, irrigation was introduced in the late 1990s.

### At the winery

Ribera del Duero is leading the way in Spain, producing the sort of dark, rich wines that fashion requires. Rioja is following, with longer maceration times, shorter oak aging and more use of French oak in place of the American oak that used to be the norm; some of the newest and priciest Riojas are unrecognizable as Rioja, if you take the traditional style as the benchmark. But that is because Rioja's traditional style is one that comes from long oak aging rather than from terroir or particular fruit characters. Now the flavour profile of Tempranillo is shifting more towards plums and black cherries, and away from strawberries, coffee and spice.

Winemakers are just beginning to understand what Tempranillo can do, and so far what it seems to be best at is wines in the full but soft Merlot mould. This might point to a bright future in both California and Australia.

Maceration times vary enormously, though too much extraction can give oily, rancid flavours, since the pips are less hard than those of Cabernet Sauvignon, and need to be handled gently. For the more modern wines, the malolactic fermentation is done in barrel to fix the colour and tannins. But the crucial difference between old-style and modern Rioja winemaking is in the length of wood aging. Tempranillo is particularly resistant to oxidation, so is able to take a lot more wood aging than many grape varieties – even when the crop is on the high side and the colour and structure are light. But such wines, even if they're technically healthy, aren't a great mouthful because the creamy oak will completely dominate the mild fruit. A lot of Rioja is still made in this style, particularly at Reserva level, but more modern wines are usually vinified to maximize fruit and bottled earlier, to preserve freshness.

*Putting finishing touches to barrels in the cooperage of Bodegas Vega Sicilia at Ribera del Duero. After a period of maturation in small barrels like these, Vega Sicilia continues the aging process in larger barrels.*

Carbonic maceration is also very popular in many regions, either for a part of or the whole of the blend, for early-drinking wines, and the Viño Joven – young wine – made from Tempranillo is very successful: deep, soft and bursting with fruit.

Much Tempranillo is blended. It may only need a seasoning of other varieties – Garnacha, Mazuelo, Graciano, Cabernet Sauvignon, Merlot, Syrah and so forth – to add perfume, acidity, flesh, or whatever the particular lack might be. But that seasoning can make all the difference. Although Tempranillo resists oxidation successfully, many unblended Tempranillo wines seem to gain little with extended aging. A seasoning with other grapes adds the spark which makes for a complex, interesting maturity.

#### TEMPRANILLO AND OAK

The taste of oak, which has long been associated with Rioja and thus with Tempranillo, traditionally comes from 225 litre *barricas bordelesas* of American oak. These were introduced by French merchants who came prospecting for wine during France's phylloxera crisis in the 1860s and '70s. (Phylloxera did not hit Rioja until slightly later, in 1901.)

Tempranillo and American oak get on like a house on fire: the rich vanilla flavour of the oak suits the ebullient fruit of the grape. But new oak flavours are not the ones generally found in Rioja: instead the wine gets its character from long aging in old barrels. The wine acquires a mature character over the years from all its exposure to air, and needs no further aging in bottle – although it nevertheless seems to be able to survive in bottle – albeit in what can sometimes seem a mummified form – for many, many more years.

These are the rules in Rioja: Crianza and Reserva reds must spend at least a year in oak; Gran Reservas must spend at least two years. Joven wines are unoaked. There are also legal minima for the time wines must spend in bottle or tank before release. Traditional Rioja may be given much longer in old oak barrels than the law demands. Not surprisingly, there are an awful lot of barrels in Rioja: over 600,000 at the last count.

However, the trend in Spain is towards shorter oak aging, and a greater use of new oak – and French oak at that. The lead was given by Ribera del Duero, which pioneered a style of more youthful fruit and more pronounced new oak. More and more Rioja bodegas are now busy making wines in the same style.

But alongside the move towards new oak – seen not just in these classic regions, but in every part of Spain where Tempranillo is grown – there is an equal move towards unoaked, juicily-fruited Tempranillo. Given the amount of Tempranillo growing all over Spain, even in less well-known regions, this is an encouraging trend. Wines aged for many years in old oak are no longer the norm: in fact they are rapidly becoming a niche market – even, just possibly, an endangered one.

# Tempranillo around the World

IS TEMPRANILLO A GREAT GRAPE? The jury is still out on this one. Just as Italy's Nebbiolo has only excelled in Piedmont thus far, so it's only Spain that has produced world class Tempranillo – as yet. But Portugal's new wave winemakers have made some lovely examples, and since it is basically a forgiving grape I'd expect it quickly to find fans among New World producers.

## Rioja

Rioja is flourishing. Prices have risen (though grape prices fell back for the 2000 vintage), and the Spanish market seems to have an insatiable demand for super-premium wines – wines made to higher standards than normal, and sold for prices that are pulled out of the air: even the producers admit that such prices have no relationship to costs. Not surprisingly, producers are eyeing every patch of half-decent land not yet planted with grapes. Obtaining planting rights is a problem, particularly for bodegas – the authorities prefer to give planting rights to growers – but the price of such potential vineyard land is already rising. Klondike? Gold rush? Yup. Every few years one vineyard region or another gets the nod and opens its gates to a flood of would-be vineyard owners and wine producers waving massive wads of banknotes in their grubby paws. Rioja has been in this foment of investment and exploitation for several years, and I do hope the authorities don't just casually grant planting rights to growers, wherever they're situated. To be honest, almost all the top vineyards were in production 20 years ago, and the number of overpriced but characterless Riojas on the market today is frequently because the grapes are from newly planted inferior land. Of course, I suppose you could argue that if more planting rights are not granted then growers will simply overproduce even more to meet demand. Depressing, but true. And already production has shot up: in 1970, around 40,000ha were under vine in Rioja. By 2000 it had risen to 55,000ha – but production had doubled.

In Rioja Alavesa, the only part of the region in the Basque country, and thus under different local government, there is a policy of encouraging growers to become bodegas. In the long run this must surely produce greater individuality in Rioja: until recently the picture has been one of disappointing uniformity, with bodegas making their wine to a standard house style, and generally blending wines from all parts of the region. Unless the bodega is a really serious outfit, lowest common denominator often rules. Single-vineyard wines, along with special cuvées, are becoming more popular, but Tempranillo does not seem to reflect its terroir as clearly as some other grapes. So a single-vineyard wine will not necessarily be more interesting than a carefully created blend.

There are, however, differences between the three parts of Rioja. Rioja Alavesa produces the most delicate, scented wines; those from Rioja Alta, where about half the region's Tempranillo is planted, are firmer, darker, richer. Overall just over half Rioja's vineyards are planted with Tempranillo, but in the hotter Rioja Baja Garnacha takes over as the main grape and the wines are frequently rather thick and stewy.

## Ribera del Duero

This region, where Tempranillo covers some 85 per cent of the vineyard, owes its fame in the first place to Vega Sicilia, which was making world-class wine here more than 100 years ago. More recently a raft of new bodegas has been established: in the late 1990s there were fewer than 60 bodegas, but by 2000 there were more than twice that number. The surface area of the vineyard has not, however, increased by the same amount; not surprisingly, grape prices have shot up.

Many of the wines are sold as Jovenes, full of blackberry and mulberry fruit; Crianzas and Reservas have one year's oak aging, and Gran Reservas have two years. Reservas and Gran Reservas must have additional aging in tank or bottle before release. At least 75 per cent of the wines must be Tinto Fino (the local name for Tempranillo); the rest may be Cabernet Sauvignon, Garnacha, Malbec, Merlot or Albillo. The most complex wines are generally the blends.

## Other Spanish Tempranillos

There are few places in Spain where Tempranillo does not pop up: in fact Spanish growers are planting more Tempranillo than Cabernet Sauvignon, Merlot and Syrah put together. In Navarra it can be silky and voluptuous; in La Mancha anything from light, pale, fresh and

ARTADI
*This is the way Rioja is heading: towards special cuvées like Artadi's Grandes Añadas, made only in the best years, in very limited quantities.*

MARQUÉS DE MURRIETA
*Even at ultra-traditional Murrieta, where the Castillo Ygay label is used only for Gran Reservas, the winemaking is being updated.*

BODEGAS VEGA SICILIA
*Unico is Vega Sicilia's top wine. Tinto Fino (Tempranillo) makes up 65–80% of the blend, with the rest Cabernet Sauvignon, along with a touch of Merlot and Malbec.*

*Quinta do Cotto vineyards in the Douro Valley, Portugal. This property is typical of most in the Douro: there is Tinto Roriz (Tempranillo) in the vineyards but not a huge percentage. Blending a number of varieties together is the usual game in the Douro. Roriz is now the second most planted variety in the Douro, and it's increasing both for table wine and for port.*

blended with white grapes, to surprisingly intense and juicy; in Cataluña red-fruited and generally blended; in Somontano perhaps a little green; in Toro dark, savory, solid, low in acidity and high in alcohol; in Costers del Segre well balanced and savory.

### Portugal

In the Douro Valley, Tinta Roriz is one of the five vines recommended for port. It flourishes in all parts of the region, but favours soils that are rich in minerals. The wine can be almost astringently tannic, but new wave winemakers are finding ways of taming this harshness and emphasizing its raspberry and mulberry fruit and surprising floral fragrance. It has less colour than some other port grapes, but its ability to resist oxidation means that it keeps its colour well; this lighter tint makes it highly suitable for tawnies.

Further south, in the Alentejo, if traditionally made, Aragonez (as it is known here) has less tannin and acidity than it does in the Douro and this has in the past consigned it to a subsidiary role since it quickly loses colour and freshness. But new wave winemakers are finding ways to capitalize on its lower tannin and acidity and we are now seeing gorgeous, juicy, plummy, spicy reds, ready to guzzle as soon as they are bottled. Will they age? No idea. Time will tell.

In Dão the Sogrape company is persuading its growers to graft over to Aragonez, in an attempt to add some perfume and fruit to what is traditionally a rather lean wine. Plantings of Aragonez in Portugal multiplied by three between 1997 and 1999.

### Australia

Large and small companies alike are interested in Tempranillo: in 1999 29 tonnes were crushed. Doesn't sound much, but this figure is projected to rise to 439 tonnes in 2004: the largest percentage growth of any grape variety in Australia. Will it be any good? I'd put serious money on saying yes.

### South America

Tempranilla, as the grape is sometimes called in Argentina, is used as a workhorse grape, and until recently was never thought of as capable of producing good wine. But that's largely because the Argentinians grossly overproduce it. With lower yields, companies like Finca El Retiro have made some really serious wines – and you bet there'll be more to come.

### Rest of the world

I've had quite nice examples from Mexico and it has been grown in the south of France, particularly the Aude, for over 20 years. In both France and Italy it was one of the most planted grapes in 1999–2000.

In California it is probably the same as the variety known as Valdepeñas, a grape in decline and little regarded. But no-one's ever asked it to perform. Expect young tyros to take it more seriously in the next few years, expecially since it can make wines similar in softness and texture to Merlot. Globetrotting viticulturist Richard Smart feels it has potential in Oregon.

**BODEGAS ALEJANDRO FERNANDEZ**
*One of the most hyped wines in Spain, Pesquera is 100% Tempranillo. Gran Reservas are made only in the best years.*

**JOÃO PORTUGAL RAMOS**
*100% Tempranillo under its Alentejo name of Aragonez from João Portugal Ramos, who is one of Portugal's top enologists.*

**ANUBIS**
*This is a new Argentine venture between leading Argentine winemaker Susanna Balbo, and roving Italian enologist Alberto Antonini. The first vintage was 1999.*

# Enjoying Tempranillo

STRAIGHT TEMPRANILLO CAN BE absolutely gorgeous young – crunchy, juicy, herb-scented, irresistible. But its reputation, established in places like Rioja and Ribera del Duero, is as a grape that ages. Well, pure Tempranillo rarely evolves much with age; it cries out for a slug of something else – Graciano, Mazuelo, Cabernet Sauvignon – and in wines destined for aging it usually gets it. Although Rioja was the first region to make a name for itself, Ribera del Duero is now the most exciting exponent of what Tempranillo is capable of – and it's interesting that the soil (a lot of limestone) and the climate (massive temperature differences between day and night) encourage acid retention and development of perfume and fruit. Here the finest Gran Reservas can last up to perhaps 30 years, and even lesser wines can last a decade quite happily. But wines labelled Joven or Crianza, here as everywhere in Spain, are not intended to be aged: Joven wines should be drunk within a year or so, and most Crianza wines within a couple of years.

The best Riojas are only slightly less age-worthy. But traditionally made Rioja, which has been aged for several years in old oak barrels, follows a different aging pattern to more modern wines that are bottled earlier. Traditional Rioja seems to age relatively fast in its first few years in bottle, before reaching a plateau which continues for a couple of decades before gradually fading. Is this a form of mummification? The wines seem to evolve little in this period; they couldn't be said to improve. But they certainly last.

Tempranillo and Tempranillo blends from elsewhere in Spain age according to their concentration, the quality of the vintage, and the nature of the blend: five to ten years is probably the limit for the most age-worthy.

In Portugal, new- wave Roriz from the Douro Valley and Aragonez from the Alentejo should be drunk young, but port lasts for donkey's years and Tempranillo/Roriz plays an important part in the blend. For more details see Touriga Nacional pages 268–269.

### The taste of Tempranillo

Think of a cross between Cabernet Sauvignon and Pinot Noir, and you have the flavour of Tempranillo. Well, sort of. It has the deep colour and rich flavour of the one, plus the strawberry fruit of the other – yet the complexity of neither at its best. But complexity is not the point of Tempranillo: its attractions are its lush texture and its supple, exuberant fruit, all blackberries and black cherries, mulberries and raspberries. In Ribera del Duero and Toro these flavours have a sensational savoury butter and blackcurrant slant when young and move towards tobacco, plums, prunes and cocoa with age.

In lightweight Tempranillo intended for early drinking the taste is more of strawberries and plum jam. Overripe Tempranillo is figgy and sweet; with long oak aging the flavours become savory and strawberryish, with a touch of coffee bean and dried fruits.

Acidity can vary from low to quite high, and tannins are generally soft and ripe, though Toro and Ribera del Duero are sometimes a bit tough, as can be Roriz in the Douro. In lesser Ribera del Duero wines the acidity and oak between them can outpace the fruit in a poor year. Both tannins and acidity are, however, essential if the wine is to improve for more than a year or two.

*The Marqués de Griñón, alias Carlos Falcó, studied winemaking at the University of California at Davis and planted his family estate near Madrid, convinced that he could make a success of both Tempranillo and the Bordeaux red varieties on previously unplanted land. He later moved into Rioja. Dominio de Pingus, from Ribera del Duero, is made by another newcomer: Peter Sisseck is a Dane who takes fruit from two old Tempranillo vineyards, destems them by hand, does the malolactic fermentation in barrel and ages the wines for 20 months in new French oak. All the sorts of techniques, in other words, that produce top-of-the-range international wines. The price is sky-high.*

### Matching Tempranillo and food

Spain's best red native grape makes aromatic wines for drinking young, and matures well to a rich and usually oaky flavour. North Spanish cuisine is well suited to the gentle oaky flavours of Navarra and Rioja, as well as the more intense fruit of Ribera del Duero. Tempranillo is good with game, local cured smoked hams and sausages, especially spicy *chorizo*, casseroles and meat grilled with herbs; it is particularly good with roast lamb. It can partner some Indian dishes and goes well with soft cheeses such as ripe Brie.

## CONSUMER INFORMATION

**Synonyms & local names**
Tempranillo has many synonyms in Spain – the Ribera del Duero uses Tinto Fino and the La Mancha region, especially Valdepeñas, uses the name Cencibel. Other synonyms include Tinto del País, Tinto de Toro, Tinto de Madrid and (in Penedès) Ull de Llebre (Catalan) and Ojo de Liebre (Spanish); in Portugal it is known as (Tinta) Roriz or (Tinta) Aragonez/Aragonês. Sometimes called Tempranilla in Argentina.

**Best producers**
**SPAIN/Castilla-La Mancha** Dehesa del Carrizal, Uribes Madero, Manuel Manzaneque, Marqués de Griñón; **Rioja** Allende, Artadi, Bretón, Campillo, Campo Viejo, Contino, CVNE, Faustino Martínez, Viña Ijalba, López de Heredia, Marqués de Cáceras, Marqués de Griñón, Marques de Murrieta, Marqués de Riscal, Marqués de Vargas, Martínez Bujanda, Montecillo, Muga, Palacio, Remelluri, Fernando Remírez de Ganuza, la Rioja Alta, Bodegas Riojanas, Roda, Señorío de San Vicente, Sierra Cantabria; **Ribera del Duero** Abadía Retuerta, Alión, Ismael Arroyo, Arzuaga, Balbás, Hijos de Antonio Barceló, Del Campo, Alejandro Fernández, Fuentespina, Condado de Haza, Hermanos Cuadrado Garcia, Grandes Bodegas, Hacienda Monasterio, Matarromera, Bodegas Mauro, Emilio Moro, Pago de Carraovejas, Pedrosa, Dominio de Pingus, Protos, Teófilo Reyes, Rodero, Hermanos Sastre, Valduero, Valtravieso, Vega Sicilia, Viñedos y Bodegas, Winner Wines; **Navarra** Chivite, Guelbenzu, Ochoa, Orvalaiz; **other Spanish wines** Albet I Noya, Pirineos, Romero Almonazar, San Isidro, Schenk, Viñas del Vero.
**PORTUGAL** Quinta dos Carvalhais, Cortes de Cima, Esporão, João Portugal Ramos, Quinta dos Roques, Quinta de la Rosa, Quinta do Vale da Raposa.
**ARGENTINA** Anubis, Finca El Retiro, Salentein.

**RECOMMENDED WINES TO TRY**
**Ten classic Ribera del Duero (or equivalent) wines**
Abadía Retuerta *El Campanario*
Bodegas Alión
Ismael Arroyo *Val Sotillo Gran Reserva*
Alejandro Fernández *Pesquera Janus*
Condado de Haza *Alenza*
Bodegas Mauro *Vendimia Seleccionada*
Pedrosa *Gran Reserva*
Dominio de Pingus *Pingus*
Teófilo Reyes *Crianza*
Vega Sicilia *Unico*

**Ten classic Rioja wines**
Finca Allende *Rioja Aurus*
Artadi *Rioja Reserva Pagos Viejos*
Bodegas Bretón *Rioja Alba de Bretón*
Viña Ijalba *Rioja Reserva Especial*
Marqués de Griñon *Rioja Coleccion Personal*
Marqués de Murrieta *Rioja Reserva Dalmau*
Muga *Rioja Reserva Especial Torre Muga*
Remelluri *Rioja Gran Reserva*
Bodegas Roda *Rioja Reserva Roda II*
Señorío de San Vicente *Rioja Tempranillo*

**Five good-value Ribera del Duero and Rioja wines**
Abadía Retuerta *Primicia*
Fuentespina *Ribera del Duero Crianza*
Matarromera *Ribera del Duero Crianza*
Palacio *Cosme Palacio y Hermanos Rioja*
Martínez Bujanda *Conde de Valdemar Rioja Reserva*

**Ten other Spanish Tempranillo wines**
Albet I Noya *Penedès Tempranillo Collecció*
Chivite *Navarra Gran Feudo Viñas Viejas Reserva*
Guelbenzu *Navarra Tinto*
Ochoa *Navarra Tempranillo Crianza*
Orvalaiz *Navarra Crianza*
Pirineos *Señorio de Lazán Reserva*
Romero Almonazar *Ribera del Guadiana Crianza*
San Isidro *Castillo de Maluenda Calatayud*
Schenk *Las Lomas Utiel-Requena Reserva Especial*
Viñas del Vero *Somontano Tempranillo*

**Ten non-Spanish Tempranillo wines**
Anubis *Tempranillo (Argentina)*
Quinta dos Carvalhais *Dão Tinta Roriz (Portugal)*
Cortes de Cima *Alentejo Aragonez (Portugal)*
Esporão *Alentejo Aragonês (Portugal)*
Finca El Retiro *Tempranillo (Argentina)*
João Portugal Ramos *Alentejo Aragonês (Portugal)*
Quinta do Vale da Raposa *Douro Tinta Roriz (Portugal)*
Quinta dos Roques *Dão Tinta Roriz (Portugal)*
Quinta de la Rosa *Tinta Roriz (Portugal)*
Salentein *Tempranillo (Argentina)*

*We can expect a flood of Tempranillo from all corners of the earth in the next few years. Australian growers are planting it as fast as they can get it in the ground, and it could be terrific in South America. Only California has yet to spot its potential.*

**Maturity charts**
The attraction of most Tempranillo is that it can be drunk quite early. Only a few top Ribera del Dueros need extended aging.

1998 Ribera del Duero Crianza

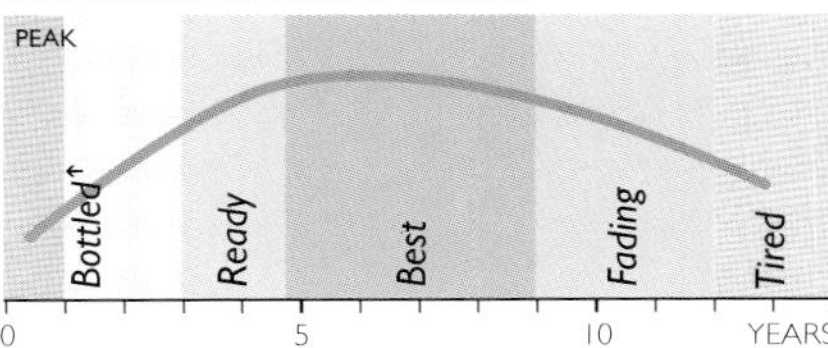


*The best growers in Ribera del Duero were able to salvage an excellent crop after a difficult growing season, producing wines of good concentration.*

1998 Rioja Reserva

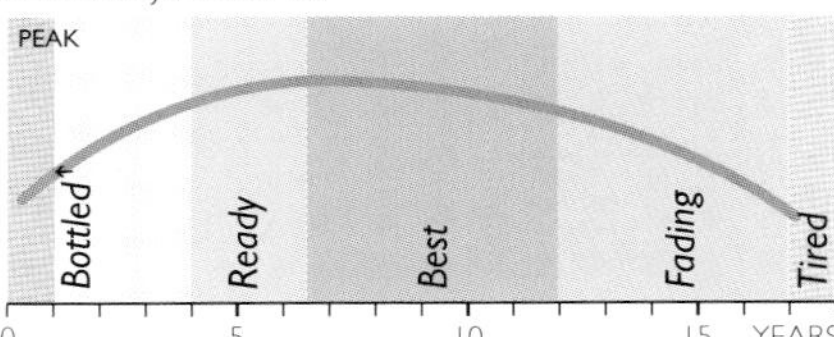


*A middling to good vintage in Rioja, despite some rain at harvest. Quality is irregular, but the wines at their best have rich tannins and plenty of depth.*

1999 Toro

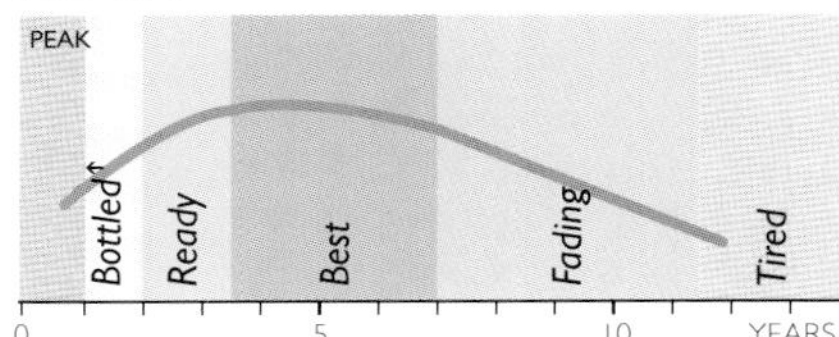


*Toro seldom gains complexity in bottle, and is at its best within four or five years of the vintage. With more investment going into the region, quality should rise in the next few years.*

## TEROLDEGO

Teroldego may at the moment be a relatively little-known north-eastern Italian grape, but it has a devoted fan club, and could have great potential in parts of the New World – New Zealand, for example. Already Californian producer Jim Clendenen of Au Bon Climat is on the record as saying that California should have planted Teroldego and Barbera instead of Merlot – and that if it replaced Syrah on the slopes of Crozes-Hermitage in the northern Rhône the wine would be greatly improved.

Not everybody would go that far – particularly those with a soft spot for the Syrah grape – but there is no doubt that Teroldego is an extremely interesting grape. In Trentino, which is at present almost its only home, it is often overcropped, and produces perfectly pleasant red with a rough-cut leafy earthiness that is best drunk young. But if the yield is restricted, the tannins allowed to get really ripe, and a bit of barrique aging added into the scheme of things the wine gains enormously in complexity and depth. Made like this it retains all its typically Italian bitter-cherry fruit but adds smoke and plums and mulberries; the acidity is in balance, though high (this is Italy, after all) and the tannins give a good firm backbone.

Its only DOC is Teroldego Rotaliano, and the grape itself is sometimes also called by this longer name. Best producers: (Italy) Barone de Cles, Marco Donati, Dorigati, Endrizzi, Foradori, Conti Martini, Mezzacorona, Cantina Rotaliana, G Sebastiani, A & R Zeni.

## TERRANTEZ

Almost extinct Madeira grape of very good quality. It is very low yielding. Tastings of old bottlings reveal a wine of great complexity and length. The Terrantez of the Portuguese mainland is, according to the ampelographer Truel, a different vine. Best producers: (Portugal) Barbeito, Barros e Sousa, Blandy.

## TERRET

It is tempting to see Terret – or the Terrets, since it is a grape that comes in different colours – as a relic of the old wine world. Terret Gris and Terret Blanc, both of which make white wine, used to be standbys of the vermouth industry in the south of France; as that industry fades, there is less and less need for grapes that will produce up to 150hl/ha of wine that is fresh, light and dry but without much else to say for itself.

Terret Noir has a bit more to say for itself, but it too is in terminal decline. It is part of the permitted blend in Châteauneuf-du-Pape, Corbières, Minervois and other wines of the South, but it is light in colour and body. It adds perfume to the blend and marries quite well with sturdier wines like Mourvèdre and Grenache. That, though, is not enough to justify its presence in most modern vineyards. It is so prone to mutation that it is possible to find the same plant giving black grapes and white grapes. Best producer: (France) Mas Jullien.

## TINTA AMARELA

A Portuguese variety grown for port in the Douro valley and for table wines further south, especially in the Dão and Alentejo. In the Douro it is less favoured than it once was because of its susceptibility to rot; it also has to be picked at exactly the right moment, and the window of opportunity between under-ripeness and overripeness is very short. On the plus side it has an intriguing tea-like perfume and good colour. It also yields well. In more southerly parts of Portugal it gives good quality reds of intensity and depth. Its other names include Trincadeira. Best producers: (Portugal) Quinta das Maias, Valle Pradinhos.

## TINTA ARAGONEZ

Like Aragonez, a Portuguese synonym for Tempranillo (see pages 256–265). It takes these names in the Alentejo. It gives wines with attractive blackberry aromas, but yields must be kept severely in check. Best producers: (Portugal) Quinta do Carmo, Cortes de Cima, Esporão, J M da Fonseca, J P Ramos, Reguengos de Monsaraz co-op.

## TINTA BARROCA

A useful part of the port blend, favoured in the Douro Valley because it will produce good deep colours even on north-facing slopes and high up, and is unworried by drought. In fact, it even prefers north-facing sites, because its thin-skinned berries do not like too much strong sunlight. Hotter sites do, however, give it the tannin that it can lack; a soft, even slack texture despite being fairly full-bodied is its main drawback. It does not always age well. There is now the occasional varietal wine.

It is highly regarded for fortified wines in South Africa, but plantings of it there are small. It also produces sturdy, earthy varietal table wines there. Best producers: (Portugal) Quinta da Estação, Ferreira, Ramos Pinto.

## TINTA CAIADA

This is probably not, as is sometimes thought, a synonym for the Portuguese grape Bastardo (see page 42). Instead, it is a grape found in the Alentejo region, where it is attracting some attention from go-ahead winemakers.

## TINTA FRANCISCA

A port grape, though not one of the five officially recommended varieties. Its wines are high in sugar but light in body and flavour.

## TINTA MIÚDA

Portuguese grape found mainly in Estremadura and Ribatejo. Yields are low and the wine is sturdy and full-bodied with a flowery aroma. It is thought to be the same grape as Rioja's Graciano (see page 112), and the Morrastel of the Languedoc. Best producers: (Portugal) Arruda co-op, Casa Santos Lima.

## TINTA NEGRA MOLE

A high-yielding vine grown on the Portuguese island of Madeira which has all but replaced the four noble varieties of Sercial, Verdelho, Bual and Malvasia or Malmsey. Efforts to revive the production of these four continue, and they have increased to around 12 or 14 per cent of the vineyard. Since hybrids are believed to constitute about 20 per cent of the Madeiran vineyard, that gives Tinta Negra Mole about 68 per cent of the total.

It is the basis of most Madeira. Only since 1993 has it been required by law that any bottle bearing the name of one of the noble varieties should contain at least 85 per cent of that grape; nowadays, if the label does not mention a grape name, it will have been made from Tinta Negra Mole.

Wine from the four noble varieties is superior in quality to that from Tinta Negra Mole but if Tinta Negra Mole were cropped at the 4 tonnes per hectare that is usual for the noble varieties, instead of at the 10 tonnes per hectare or more (sometimes much more) that it is currently encouraged to give, its quality might be much improved. It is picked early, at 9.5-10 per cent potential alcohol, because viticultural practices on Madeira tend to encourage rot. Tinta Negra Mole is, however, pretty disease-resistant, and indeed this was the reason it became so widely planted on Madeira after phylloxera. The Madeira shippers are trying hard to get the law changed so that they may put the name of Tinta Negra Mole on Madeira labels. It is permitted rather than recommended for table wines.

Its origins are unclear. Some believe it to be a crossing of Pinot Noir and Grenache, but its presence on Madeira has been traced back to the early 19th century. It is not the same as the Negra Mole of the mainland (see page 164).

## TINTA RORIZ

A northern Portuguese synonym for Tempranillo (see pages 256–265). It is more often known just as Roriz (see page 203).

## TINTO CÃO

High quality but low-yielding port variety grown in the Douro Valley. The name means 'Red dog', in keeping with the Portuguese habit of naming vines after animals whenever possible. In what way Tinto Cão resembles a dog of any colour is, however, not clear.

It is believed to have appeared in Portugal in the 17th century, and possibly originated in the Douro Valley. Its colour is not particularly deep, and the wine oxidizes easily, but it has good aroma – spicy when grown in warm sites, floral in the cooler sites to which it is best suited – and when young seems inferior to the other four recommended port varieties. After five years, however, it develops extra finesse. It is attracting some attention in California and Australia. Best producers: (Portugal) Quinta do Vale da Raposa; (USA) St Amant, Quady.

## TINTO FINO

The name given to Tempranillo in Spain's Ribera del Duero region (see pages 256–265). It is sometimes claimed to be a particular clone of Tempranillo, but this seems not to be the case: instead, it is simply an instance of a vine producing better balanced and structured wines of more complexity in a different climate and in different soils.

The vines are grown at about 750m above sea level, which makes Ribera del Duero the highest wine region in Spain, and summers are hot, winters cold and springs and autumns short. Nights are cool: in August and September there can be a difference of 20°C between daytime and nighttime temperatures. This produces more intensely flavoured wines with good acidity and powerful blackcurrant fruit.

There is a rather short vegetative period between the last spring frost and the first autumn frost, too, so it is very important that no ripening time is lost through drought. In the 15-20 days of summer drought when the vines shut down and ripening ceases, judicious irrigation can make all the difference to quality. Best producers: (Spain) Arroyo, Alejandro Fernández, Mauro, Viña Pedrosa, Peñalba López, Teófilo Reyes, Valduero, Vega Sicilia.

## TINTO DEL PAÍS

Spanish synonym for Tempranillo (see pages 256–265).

## TOCAI FRIULANO

It seems very likely that Tocai Friulano is the same grape as the Sauvignonasse grown in Chile (see page 228). In retrospect it is hard to see how the Chileans could have confused the flavour of this light, delicately floral or appley wine with the pungent gooseberry or white peach-flavoured Sauvignon Blanc, but the plants do look very similar.

In Italy's Friuli it can make wine of high though subtle quality, with good structure and balance and some depth. It can also turn out very large quantities of unexciting everyday wine rather as it is prone to do in Chile. Either way, it is best drunk young. It is the main white grape in the DOC zones of Collio, Colli Orientali, Friuli Grave and Friuli Isonzo, accounting for around a fifth of the vineyard area here. There is some Tocai Friulano (or at least, a vine of that name) grown in Argentina.

It is probably the same as the Tocai Italico of the Veneto region (though is probably not, according to the author Nicolas Belfrage, the same as the Tocai Italico found in the Breganze DOC), but has nothing whatever to do with Tokay d'Alsace, a now-redundant name for Pinot Gris in Alsace. Best producers: (Italy) Rocco Bernarda, Borgo Conventi, Borgo San Daniele, Borgo del Tiglio, Drius, Livio Felluga, Marco Felluga, Edi Keber, Miani, P Pecorari, Princic, Paolo Rodaro, Ronchi di Manzano, Russiz Superiore, Russolo, Schiopetto, La Viarte, Vie di Romans, Villa Russiz, Volpe Pasini, Le Vigne de Zamò.

## TOKAY

Australian name for Muscadelle (see page 142), making classic sweet wines called Liqueur Tokay at Rutherglen and Glenrowan in north-eastern Victoria.

## TOKAY D'ALSACE

Pinot Gris was for many years known by this name in Alsace, but the bureaucrats of Europe have decreed that it should be phased out to avoid any confusion, however unlikely, with the Tokaj wine of Hungary. Tokay-Pinot Gris was a medium-term compromise; nearly all Alsace labels now call it simply Pinot Gris (see pages 172–173).

## TORBATO

A characterful Italian white grape found on Sardinia, where it was heading for extinction until rescued by the company of Sella & Mosca in the late 20th century. It produces well-structured dry wines with good body and buttery richness. Best producer: (Italy) Sella & Mosca.

## TORRONTÉS

Argentina's white speciality, Torrontés is, in fact, several varieties, not all of which have the full Muscat-like aroma for which the grape is known. The aroma of Torrontés can also be reminiscent of Gewürztraminer; it also brings to mind the smell of air freshener. It is floral, soapy and sometimes lightly spicy; when not overcropped and when carefully vinified it can be positively heady and beautifully refreshing. I've never come across a Torrontés wine that has aged well beyond a couple of years. Less than a year is usually best.

The most planted and most aromatic Torrontés in Argentina is Torrontés Riojano; the name comes from the Argentine province of La Rioja, not the Spanish region of the same name. The less aromatic and less widely planted Torrontés Sanjuanino likewise takes its name from San Juan province. The least aromatic variety of all is Torrontés Mendocino or Mendozino, which is found in Rio Negro in the South, at the northern end of Patagonia. There is some Torrontés planted in Chile – mostly Torrontés Riojano, which is used for pisco, the local brandy.

Spain also has a vine called Torrontés and logically it should be the same as at least one of Argentina's versions. However, no definite relationship has been proved. Spain's Torrontés is found in Ribeiro, around Madrid, in Montilla-Moriles and in the Canaries. Best producers: (Argentina) Etchart, Norton, Michel Torino; (Spain) Bodega Alanis, Viña Meín, Vitivinícola del Ribeiro, Emilio Roja, Vilerma.

## TOURIGA

More of an abbreviation than a synonym, this is what Australians call Touriga Nacional (see pages 268–269), though it seems to have been used erroneously in the past for several other beefy red vines as well. California's Touriga is probably Touriga Franca (see below). Best producers: (Australia) Brown Brothers, St Hallett.

## TOURIGA FRANCA

Called Touriga Francesa until 2001, this is a robustly flavoured port grape, one of the five varieties officially recommended by the port authorities, with good colour and tannins, though without the quality of Touriga Nacional (see pages 268–269), Tinta Barroca (see left) or Tinta Roriz (see above). Its great advantage is its powerful aroma of mulberries and roses, which adds an exotic note to the blend. It yields well but needs hot sites to ripen properly. It is also found in the Trás-os-Montes region north of the Douro Valley.

# Touriga Nacional

PORTUGAL'S WINE REVOLUTION is so recent and the changes for the better in winemaking and wine styles are so dramatic that it is difficult to say which of a marvellous array of individuals is Portugal's best grape variety. But it may be, just may be, Touriga Nacional. I've no doubt that this variety can produce thrilling flavours – dark, damsony fruit, violet and new leather perfumes – but it also has a fairly powerful tannic attack which means it often does best when blended with other varieties – and luckily Portugal has a whole fistful just waiting for this call: Touriga Franca, Tinta Roriz (a.k.a. Tempranillo), Tinto Cão, Periquita, and a heap of others. In this, Touriga Nacional is like Cabernet Sauvignon – a grape of tremendous personality but aggressive tendencies that is usually much better when softened by something else.

However, Touriga Nacional is also the star grape in the great fortified Port wines of the Douro Valley, even though as recently as the 1980s, some port houses, including Ferreira, opted to omit Touriga Nacional from new plantings because of its low yields, which are half or even less than those of the other top grapes. Its yields are low because its berries are so tiny; this is, of course, the source of its intense flavour and its powerful tannins. Though recent clonal research is making some progress towards easing the aggression, there are in fact hundreds of mutations of Touriga Nacional already, producing anything from dense, dark, heady reds in small quantities to large crops of pale, insipid juice. It's going to take a few years to sort out the good from the bad, and meanwhile, Touriga Nacional won't spread as widely as its potential deserves.

In Dão it used to cover 90 per cent of the vineyards until phylloxera; as so often, post-phylloxera replanting favoured more productive varieties and it is only now once more on the increase; the regulations demand a 20 per cent minimum in the blend. It is also increasing in Bairrada, Estremadura, Beiras and the Alentejo, though opinions differ on whether the great heat of Portugal's far South is good for it.

In the New World, only Australia, with its great 'port' tradition, has shown much interest in Touriga Nacional, but producers there have so far found it tricky to isolate clones that balance productivity with quality. Certainly it should be looked at by red table wine producers in such warm places as Chile, Argentina, California and South Africa when they want to expand their range.

### The taste of Touriga Nacional

Young Touriga Nacional has a deep and intense aroma that is reminiscent of young Cabernet Sauvignon, even to the extent of mixing dark, sweet fruit with a leafy freshness and hint of violets. As it matures as a port it gains wonderfully rich mulberry and blackberry richness without losing its black peppery attack and its seductive scent of flowers. As a table wine, it takes a while to lose its tannin, unless blended with other varieties, but the fruit gets sweeter and richer with age.

*Quinta do Crasto's revival as a wine estate dates from the 1980s, when its current owner returned from a sojourn in Brazil – he'd gone there in 1975 after the Portuguese revolution. Now Crasto makes both port and table wine from the same grapes. Like most port estates, it has both old mixed plantings and newer ones planted by variety; it also has a fantastic situation, perched right on top of a mountain overlooking the river Douro. As so often in Europe, the Romans were here first: the name comes from the Latin word 'castrum', or fort.*

TAYLOR

*Quinta de Vargellas is the backbone of Taylor's vintage port, and also makes a single-quinta wine. Apart from Touriga Nacional the blend includes Touriga Francesa, Tinta Barroca, Tinta Roriz and Tinta Cão, and the wine has a characteristic floral note.*

QUINTA DO VALE DA RAPOSA

*Touriga Nacional can certainly make successful varietal wines, as here, but winemakers are divided over whether or not it is generally improved by the addition of other grapes.*

**Left:** *Terraces at Quinta do Noval in the Douro Valley with parcels of the ungrafted Nacional vines on the left. Nacional is a field blend, but as individual vines die they are usually replaced with Touriga Nacional.*
**Above:** *It is not clear where Touriga Nacional originated, but to judge from the amount of clonal variation in the Dão and Douro regions it has been in both for a very long time. There is a village called Tourigo in the Dão, and nearby is another village called Mortágua, which is another of Touriga Nacional's synonyms.*

## CONSUMER INFORMATION

### Synonyms & local names

Sometimes called Mortágua in Portugal. Known as Touriga in Australia. California's Touriga is probably Touriga Franca.

### Best producers

**PORTUGAL/Dão** Boãs Quintas, Quinta dos Carvalhais, Quinta da Pellada, Quinta dos Roques; **Douro** Aliança, Quinta do Cachão, Calem, Churchill, Cockburn, Quinta do Côtto, Quinta do Crasto, Croft, Ferreira, Fonseca, Quinta da Gaivosa, Graham, Niepoort, Quinta do Noval, Ramos Pinto, Sandeman, Sogrape, Taylor, Quinta do Vale da Raposa, Warre; **rest of Portugal** Quinta da Cortezia, Esporão, Quinta de Pancas, Caves Primavera, Casa Santos Lima.
**SOUTH AFRICA** Boplaas.

### RECOMMENDED WINES TO TRY

**Ten premium ports containing Touriga Nacional**

Churchill *Vintage Port*
Cockburn *Vintage Port*
Croft *Quinta da Roêda Vintage Port*
Fonseca *Guimaraens Vintage Port*
Graham *Malvedos Vintage Port*
Niepoort *Vintage Port*
Quinta do Noval *Vintage Port*
Sandeman *Vau Vintage Port*
Taylor *Quinta de Vargellas Vintage Port*
Warre *Traditional Late-Bottled Vintage Port*

**Ten Douro reds made wholly or predominantly from Touriga Nacional**

Aliança *Foral Grande Escolha*
Calem *Lagarde Sá Touriga Nacional*
Quinta do Côtto *Grande Escolha*
Quinta do Crasto *Touriga Nacional*
Ferreira *Quinta da Leda Touriga Nacional*
Quinta da Gaivosa
Quinta do Noval *Corucho*
Ramos Pinto *Duas Quintas Reserva*
Sogrape *Reserva*
Quinta do Vale da Raposa *Touriga Nacional*

**Eight other Portuguese reds**

Boãs Quintas *Dão Touriga Nacional*
Quinta dos Carvalhais *Dão Touriga Nacional*
Quinta da Cortezia *Estremadura Touriga Nacional*
Esporão *Alentejo Touriga Nacional*
Quinta da Pellada *Dão Touriga Nacional*
Caves Primavera *Beiras Touriga Nacional*
Quinta dos Roques *Dão Touriga Nacional*
Casa Santos Lima *Estremadura Touriga Nacional*

QUINTA DOS ROQUES
*Varietal Touriga Nacional wines like this one from Quinta dos Roques are still a relative rarity in the Dão. This example is smoky and intense, and will age for three to five years.*

QUINTA DA CORTEZIA
*A medium-bodied, supple wine from Quinta da Cortezia in Estremadura. The wines here are made by arch-modernist winemakers Caves Aliança.*

## TRAJADURA

One of the grapes grown in northern Portugal for Vinho Verde, Trajadura is less aromatic than Loureiro or Alvarinho, and tends to be blended with one of them, usually the former. It is early ripening and needs to be picked before full ripeness if it is to keep its acidity. It contributes lemony, peppery fruit to the blend. Over the border in Spain it becomes Treixadura (see right). Best producers: (Portugal) Quinta da Aveleda, Quinta da Franqueira, Casa de Sezim.

## TRAMINER

This name should really only be used for the less aromatic form of Gewürztraminer (see pages 102–111). In practice, the two names are often used interchangeably. Traminer is found throughout central and eastern Europe, in Austria, Germany and Romania, Italy, and in Australia where the name is usually used for Gewürztraminer. Best producers: (Canada) Hainle; (Germany) Andreas Laible, U Lützkendorf, Fritz Salomon, Wolff-Metternich, Klaus Zimmerling.

## TREBBIANO

Trebbiano is Italy's least avoidable grape: there are very few regions where you do not run the risk of bumping into it. It is permitted in around 80 DOCs and probably produces about a third of all Italy's DOC white wine. Total monotony is avoided by the fact that Trebbiano is a collection of vines rather than one single one; nevertheless, boredom is the state most often induced by Trebbiano.

All Trebbiano is distinguished, if that is the word, by its high acidity and neutral flavour. The better sorts can add some leafy fruit and even some waxy depth, and just now and then Trebbiano comes up with some pretty pleasant wines.

CA' DEI FRATI

*Surprisingly good Lugana, this – in fact, it's about as good as Lugana gets. It has complexity and depth, and can even age for a few years, and you don't get that with most Trebbiano.*

The best Trebbiano sub-variety is Trebbiano di Soave, which is sometimes known as Trebbiano di Lugana or Trebbiano Veronese. It is also the least planted, and forms only a minor part of the Soave blend. Even here the lesser Trebbiano Toscano outplants it and outyields it. Trebbiano di Soave/di Lugana is however the only grape used for Lugana, from the southern end of Lake Garda, a wine of significantly more weight and character than most Soave. The grape may in fact be a version of Verdicchio, which would certainly explain its greater interest. Umbria's Procanico, also a form of Trebbiano, is another version with character.

Trebbiano Toscano is the most heavily planted Trebbiano of all, and certainly the most bland. It is followed in terms of area planted by Trebbiano Romagnolo, Trebbiano d'Abruzzo (DOC Trebbiano d'Abruzzo is made from Bombino Bianco, and not from Trebbiano at all, though Trebbiano Campolese and Trebbiano d'Abruzzo – the ampelographer Galet reckons these are separate varieties – are among its synonyms), and Trebbiano Giallo. There is also a Trebbiano della Fiamma, with slightly pinkish berries, found in Emilia-Romagna.

Trebbiano's most famous misuse in recent decades was as part of the (red) Chianti blend: an example of the law compromising on wine quality in order to accommodate growers who had Trebbiano in their vineyards and were disinclined to uproot it for such a facile reason as improving the flavour of the wine. (However, the addition of white grapes to the Sangiovese fermentation vats might have had the effect of fixing the always elusive colour of Sangiovese: see co-pigmentation theory page 31.) The addition of Trebbiano to the blend was mandatory, even if, as time went on, fewer and fewer producers put any in. After a lot of wrangling and arm twisting by the more progressive producers, Trebbiano is no longer required in Chianti – and about time too. Wines like Galestro sprang up in order to mop up the surplus Trebbiano; however, with plantings of Trebbiano generally in decline now in Tuscany, production of Galestro is already becoming less important.

It is the same grape as France's Ugni Blanc, and under one name or another is present in Bulgaria, Greece, Russia, Portugal (as Thalia), Mexico (for brandy), Brazil, Argentina, Uruguay, South Africa, California (a bit) and Australia. Best producers: (Italy) Antinori, Barberani, Ca' dei Frati, Falesco, Valentini; (USA) Iván Tamás, Viansa.

## TREIXADURA

The Trajadura of Portugal (see left) takes this name when it crosses the border into Spain. In Galicia it performs much the same role as it does in Vinho Verde, adding light lemony fruit to the blend. In Ribeiro it is the main grape, and is blended with Torrontés and Lado; in Rías Baixas, Albariño and Loureira are its partners. Best producers: (Spain) Gargalo, Viña Meín, Vitivinícola del Ribeiro, Emilio Rojo, Vilerma.

## TRESALLIER

The traditional variety of white St-Pourçain, an obscure VDQS from the centre of France. These days no more than half the blend may be Tresallier, with the balance being Chardonnay and/or Sauvignon Blanc. Tresallier is believed to be a local variant of Sacy (see page 206), and gives light, acidic wines. Sacy is itself on the decline and Tresallier won't be far behind. Best producer: (France) St-Pourçain co-op.

## TRINCADEIRA

This Portuguese grape has many aliases: Rabo de Ovelha Tinto, Tinta Amarela, Espadeiro, Crato Preto, Mortágua and Murteira. But an awful lot of Portuguese vines share the same synonyms: Trincadeira is the same vine as Tinta Amarela, but that's probably all. It is found all over Portugal, but especially in the Alentejo. Its wines are deep coloured with good tannins, good blackberry-jelly fruit and a touch of herbaceousness when young. It is early ripening, high yielding and prone to rot. Best producers: (Portugal) Fundação Eugenio de Almeida, Cortes de Cima, Esporão, J P Ramos, Tapada do Chaves.

## TROLLINGER

The name given in Germany to Italy's Schiava (see page 229), or Vernatsch (see page 273). The name seems to point to an origin in the Tyrol region, and in Germany it is grown almost exclusively in Württemberg. German producers also prefer the higher-yielding Grossvernatsch or Schiava Grossa clone – almost inevitably the one with the least quality – and use it to satisfy the local taste for sweetish, reddish wine. The style is thus very different to the light, dry, mildly acidic wine of Trentino-Alto Adige. The vine may also be known as Blauer Trollinger; it also doubles as a table grape.

## TROUSSEAU

A now little planted grape in the Jura region of eastern France which has trouble ripening unless planted on the warmest gravel soils. It makes sturdy, dark wines, 1.3 times as dark as those of Cinsaut, but yields irregularly, and Pinot Noir has tended to replace it in many vineyards. It is often blended with Poulsard, which contributes finesse. It is believed by some authorities to be the same as Portugal's Bastardo and the Cabernet Gros of Australia, as well as the Malvoisie Noir of France's Lot *département*. There is also a paler mutation, Trousseau Gris, which may be the same as the rare variety called Gray Riesling in California. Best producers: (France) Jacques Forêt, Henri Maire, Rolet Père et Fils, Tissot.

## UGNI BLANC

This neutral-tasting, acidic variety is the same as the Trebbiano of Italy (see left), and probably arrived in France along with the establishment of the Papal court at Avignon in the 14th century, a kind of vinous Trojan Horse from the Vatican. It is still found in Provence, and in the Midi, where its wines tend to be less acid, but its main French stronghold is in the Cognac and Armagnac regions of the South-West. Here, at the limit of its cultivation, it produces wines of much lower alcohol – 9 per cent or lower – than in the Midi, where it easily reaches 11 or 12 per cent, but its acidity is searingly high – good for making brandy, but requiring considerable skill to make it into attractive table wine.

Although good Charente white is hard to come by, Vin de Pays des Côtes de Gascogne from the Armagnac region has been a considerable success, especially when the Ugni is mixed with a bit of Colombard. Yields are high everywhere: up to 150hl/ha. Ugni Blanc features in many blends of the Midi, though usually in a minor (and decreasing) way.

It is grown in Mexico, Brazil, Argentina and Uruguay; South Africa, California and Australia also have some, under either one name or another, as do Bulgaria, Romania and Portugal, where it is called Thalia. Other French synonyms include Clairette Ronde in Provence, Rossola in Corsica and St-Émilion in the Charentes. Best producers: (France) Meste-Duran, Producteurs Plaimont, le Puts, du Tariquet.

## ULL DE LLEBRE

The name given to Tempranillo in Cataluña in north-east Spain. See pages 256–265.

## UVA RARA

This may be a synonym for Bonarda – true Bonarda, that is, as found in the Novara/Vercelli region of Piedmont. Bonarda also takes this name in Lombardy and Emilia. Or it may be a different grape. Either way, it is blended with the local Nebbiolo (here called Spanna) to make the latter's ferocious tannins more approachable. Best producers: (Italy) Antichi Vigneti di Cantalupo, Antoniolo, Nervi, Travaglini.

*Picking Verdejo grapes in Rueda, Castilla-León. Verdejo is one of Spain's best white vines, and is seldom improved by an admixture of dull Viura. The best Rueda has little or none of the latter.*

## UVA DI TROIA

Found in central Puglia in Italy, Uva di Troia can be complex and can age well if properly handled. It is part of the blend in several DOCs, and brings deep colour and substantial alcohol to the blend, but its potential for quality is outweighed, as far as most growers are concerned, by its low yields. Best producers: (Italy) Rivera, Santa Lucia, Torrevento.

## VALDIGUIÉ

Now hardly grown in France, Valdiguié was once a workhorse grape of the South-West, producing huge quantities of poor wine. It can also occasionally be found still in Languedoc and in Provence.

In 1980 the French ampelographer Galet identified the variety known in California as Napa Gamay as being merely Valdiguié. It had already been exported from California to Australia, Brazil and Uruguay under the name of Napa Gamay. Beringer in California makes a varietal under the name of Valdiguié, and the wine shows that, while throwing the full force of New World winemaking at the grape can certainly bring out lots of juicy fruit, it cannot impart any character. Best producers: (USA) Hop Kiln, J Lohr.

## VELTLINER

Another name for Austria's Grüner Veltliner (see page 114) – and also for the earthy-tasting red Roter Veltliner, Brauner Veltliner and reddish Frühroter Veltliner.

## VERDEJO

A good quality Spanish grape making greengage- and pear-flavoured wine that is one of the mainstays of Rueda, one of Spain's top DOs for white wines. The blend must include at least 50 per cent Verdejo. The other native grape in the blend is the rather monochrome Viura (see page 284). Sauvignon Blanc, first introduced here in the 1980s for varietal wines, can also be used.

The grape seems to have originated in Rueda, and was widely planted there until phylloxera. Afterwards the growers opted for the higher yields of Viura and it was the Marqués de Riscal and the Marqués de Griñon who in the 1980s rescued it from obscurity. Early examples tended to be clean but too neutral, no doubt in reaction to the grape's tendency to oxidize easily. Wine from Verdejo has the structure and balance to age well, and becomes nutty and honeyed with a few years in bottle. Its high glycerol content gives it good roundness. Pre-fermentation skin maceration and some barrel fermentation and aging can help the complexity of the wine, though the oak can easily be overdone.

It is also found in the Toro and Cigales DOs. Best producers: (Spain) Alvarez y Díez, Belondrade y Lurton, Angel Lorenzo Cachazo, Marqués de Griñón, Marqués de Riscal, Vega de la Reina, Angel Rodríguez Vidal, Castilla la Vieja.

## VERDELHO

A grape or several grapes found in Portugal. On the island of Madeira Verdelho is both a grape and a style; since 1993 any Madeira wine calling itself Verdelho must be made with at least 85 per cent of that grape, which was not the case before.

As a style of Madeira, Verdelho comes between Sercial and Bual, less dry than the former (though Sercial styles are becoming less dry) but less rich than the latter. It has high, piercing acidity but when young and unaged has more fruit than the other noble Madeira varieties and is the most 'complete'. It is now also being used for table wines in Madeira, sometimes with an admixture of Arnsburger to soften the acidity. These table wines do not at the moment seem to offer great quality potential, partly perhaps because of the difficulty of persuading growers on the island to wait until their grapes are ripe before picking them.

In the Douro Valley Verdelho takes the name of Gouveio. Here it reaches high sugar levels and is used for white port. It seems to be the same as the Godello (alias Verdello) found across the border in Spain's Galicia region (see page 112).

Verdelho is also proving very successful in Australia, where plantings are relatively small but quality is high. The wines have intense flavours of lime cordial and honeysuckle, tending to oiliness at the riper end. The Swan District of Western Australia, Cowra in New South Wales and Langhorne Creek in South Australia all have some Verdelho. Best producers: (Australia) Bleasdale, Chapel Hill, Fox Creek, Moondah Brook, Rothbury; (Portugal) Barros e Sousa, Blandy, Cossart Gordon, Henriques & Henriques, Leacock.

## VERDICCHIO

The best and most characterful white grape of Italy's Marche region, Verdicchio has benefited in recent years from lower yields and better balance and structure. The wine is not aromatic but has a good flavour of nuts and lemons. It is typically central Italian in that it relies on subtlety rather than flamboyance for its effects – and it goes with food extremely well.

There are two DOCs: Castelli dei Jesi, which is by far the larger, and di Matelica, which insists on lower yields from hillside vineyards and is therefore often the more substantial wine. There is also some spumante. Best producers: (Italy) Belisario, Bisci, Brunori, Bucci, Colonnara, Coroncino, Fazi Battaglia, Garofoli, Mecella, La Monacesca, Santa Barbara, Sartarelli, Tavignano, Umani Ronchi, Vallerosa-Bonci, Fratelli Zaccagnini.

## VERDUZZO

Found mainly in Friuli and Veneto in northeastern Italy, wine from Verduzzo varies from the pleasantly crisp to the interestingly sweet. Much depends on whether it comes from the Verduzzo Trevigiano grape or from the apparently unrelated Verduzzo Friuliano.

The latter is finer, and has been in Friuli longer, since at least the early 19th century. The former seems to have appeared in the early 20th century and, since it yields both generously and reliably, has inevitably taken a larger share of the vineyard for itself. The DOCs in which it features most heavily are Lison Pramaggiore and Piave north of Venice. Other DOCs are Friuli Aquileia, Friuli Grave, Colli Orientali, Friuli Isonzo and Friuli Latisana. Colli Orientali is generally the best: hillside vineyards suit the vine better than ones down on the plains, and besides, the Verduzzo grown here is more likely to be Verduzzo Giallo, which seems to be a superior subvariety of Verduzzo Friuliano. The other subvariety, Verduzzo Verde, is less interesting and more likely to be found in the plains.

Verduzzo Giallo makes good sweet wines, honeyed and quite rich though not with enormous complexity. Even better sweet wines come from Verduzzo Rascie, which is billed as yet another sub-variety, this time with looser clusters. Such details matter: looser clusters are less likely to succumb to rot while the grapes linger on the vine late into the autumn.

Sweet Verduzzo is usually made from late-picked grapes; an alternative is to pick the berries and leave them to dry and shrivel before fermenting them. The wines may then be aged in barrique. The DOC of Ramandolo has the reputation of producing the best sweet Verduzzo. Best producers: (Italy) Dario Coos,

*Verdicchio growing in Italy's Marche region. The wine used to taste largely of glue, but these days is far fresher and more attractive, though colossal flavour is not one of its virtues.*

Dorigo, Giovanni Dri, Filiputti, Lis Neris-Pecorari, Livon, Paolo Rodaro, Ronco del Gnemiz, Ronchi di Manzano, Torre Rosazza.

### VERMENTINO

A protean vine found the length of Italy, from Liguria in the North-West to Sardinia. It appears to have originated as an import from Spain in the Middle Ages and was possibly a sub-variety of Malvasia; in the intervening centuries it has mutated enthusiastically, so that around 40 versions of Vermentino can now be found in Italy.

The best wines come from Tuscany and Sardinia, where they have a weight and breadth not found elsewhere. The wines have a typically Italian aroma of lemons and nuts and leaves, combined with racy acidity and robust structure. There is no point in aging them, but when young they have considerable character.

The grape is also grown in Languedoc-Roussillon and is believed by many authorities to be the same as Provence's Rolle (see page 202). It is widely planted on Corsica, where it sometimes takes the name of Malvoisie de Corse. It could have possibilities in the New World if producers were not currently so fixated on reds. Best producers: (Italy) Argiolas, Capichera, Giovanni Cherchi, Gallura co-op, Le Macchiole, Piero Mancini, Santadi co-op, Sella & Mosca.

### VERNACCIA

Vernaccia wines are found all over Italy, but to try and relate them to each other is often a waste of time. A number of different vines, mostly white, but sometimes red go under this name. The racy Vernaccia grown in San Gimignano for the DOC of the same name is probably unrelated to the Vernaccia grown on Sardinia for the sherry-like Vernaccia di Oristano, for example; and the name, translated directly into German, becomes Vernatsch, which is the name of a red grape grown in Trentino-Alto Adige.

The reason for this abundance is that the name has the same root as the word 'vernacular': it simply indicates a local grape. There is thus absolutely no reason whatever why a Vernaccia from the south of Italy should bear any resemblance to one from the North. In the Marche region, for example, there is sparkling red Vernaccia di Serrapetrona.

Vernaccia di San Gimignano is, nevertheless, the best known Vernaccia, not least because it is the local wine of one of Tuscany's most touristy towns. It is usually good and reliable, with crisp acidity and ripe, leafy, citrus fruit. It is one of the better, more characterful whites produced in Tuscany, though given the local prevalence of Trebbiano, that is not in itself much of a distinction. Vernaccia di San Gimignano is never more than a good everyday drink, but that, after a hard day slogging round the sights, is often very welcome. Best producers: (Italy) Le Calcinaire, Vincenzo Cesani, Attilio Contini, La Lastra, Melini, Montenidoli, Giovanni Panizzi, Paradiso, Pietrafitta, Guicciardini Strozzi, Teruzzi & Puthod, Vagnoni.

TERUZZI & PUTHOD

*This refined, barrique-aged Vernaccia di San Gimignano shows subtle use of oak and good length. The name of the wine, Terre di tufi, refers to the type of soil – tufa – on which the grapes are grown.*

### VERNATSCH

Like Vernaccia (see left), this grape takes its German name from the same root as the word 'vernacular', showing that in Italy's Alto Adige it has long been regarded as home grown. Its Italian name there, Schiava (see page 229), translates as 'slave' or possibly 'Slav', but the name given to it in Germany, Trollinger (see page 270), points to a Tyrolean origin. All this is rather more interesting than the wine itself, which, at best, is a perfectly pleasant light red with fair acidity and a flavour of strawberries and smoked ham.

### VESPAIOLO

Make Vespaiolo dry and you get attractive, quite characterful white wine, light and startlingly acidic but with no particular aroma. Make it sweet, however, from *passito* or noble-rotted grapes, and you have one of Italy's finest sweet wines. Indeed, the name is said to derive from the wasps, or *vespe*, which are attracted to the grapes as they hang on the vines getting sweeter and sweeter. It is found in the Veneto in the Breganze DOC. *Passito* sweet wines come under the heading of Torcolato, for which Vespaiola is blended with Garganega and Tocai. The grapes are left to dry until January and the wine may be aged in new barriques. The acidity is still there in the wine, balancing the intense sweetness, but the flavour is of dried grapes and apricots, honeysuckle and spice. Alternative names for the grape are Vespaiola or Vesparolo. Best producer: (Italy) Maculan.

### VESPOLINA

Logically this name should derive from *vespe*, or wasps, in the same way as Vespaiolo (see above). However, such a derivation is never mentioned, which makes one wonder about the authenticity of such stories. Vespolina is a red grape grown in Piedmont and Lombardy. It is blended with Nebbiolo and sometimes with Bonarda in wines from Gattinara and Ghemme. In Lombardy it is known as Ughetta. Best producer: (Italy) Antichi Vigneti di Cantalupo.

### VIDAL

A French hybrid grown in Canada and New York State, where it is able to withstand the cold winters. Its main fame is as one of the grapes used for Canada's Icewine. Vidal Icewines do not have the elegance or longevity of Riesling Icewines, but they have good concentration and intensity and a certain four-square appeal. Best producers: (Canada) Inniskillin Okanagan, Marynissen, Vineland Estate.

### VINHÃO

The main grape grown for red Vinho Verde, where it constitutes some 80 per cent of red plantings. It is currently enjoying a revival as demand for red Vinho Verde increases in the wake of the worldwide fashion for red wine; however, the swing to red is not as great in Vinho Verde as the swing to white was 15 or 20 years ago. There is an obvious reason for this – red Vinho Verde is a shocking, highly acidic, rip-roaring wine not at all in tune with the current New World vogue for soft ripe fruit and oak spice. Plantings of red and white grapes are now divided fairly evenly in Vinho Verde, with perhaps a slight bias towards white. Vinhão is thought to be the same as the Sousão of the Douro. Best producers: (Portugal) Domingos Alves de Sousa, Sogrape.

# VIOGNIER

WE SHOULD BE EXTREMELY GRATEFUL to Viognier for coming along when it did. But then, Viognier should be extremely grateful we came along when *we* did. We as wine drinkers desperately needed a new taste experience that wasn't oaked Chardonnay or green-streaked snappy Sauvignon Blanc. Viognier was in danger of virtual extinction except for a few precious vines at Condrieu in the northern Rhône Valley.

Luckily for Viognier, the vast hordes of wine lovers who had lapped up so much soft, creamy, oaky Chardonnay were beginning to tire of it. They wanted a new white wine experience not dependent on the creaminess of oak, yet many of them weren't at all fond of the high acid style of Sauvignon or Riesling. They wanted the weight and softness and suppleness of an oaky Chardonnay – but without the oak.

Other grapes that might give weight were too dependent on oak – Chenin Blanc and Sémillon, for instance, demanded oak to give a round, toasty softness to what were usually rather raw, dull wines. On the other hand, wines like Gewürztraminer or Muscat were too overpoweringly fruity and aromatic: people often thought they were sweet, even when they weren't. Wasn't there a middle way?

Well, there was. But it was a difficult route involving a grape that was notoriously difficult to grow and certainly difficult to persuade to give a regular crop – Viognier. But, wow! If you wanted serious, swooning wine, with texture as soft and thick as apricot juice, perfume as optimistic and uplifting as mayblossom, and a savoury sour creamy richness like a dollop of crème fraîche straight from the ladle of a smiling farmer's wife – in other words, a wine which just oozed sex and sensuality – Condrieu, from the Viognier grape, was it.

But Condrieu is a tiny place. The fame of the wine spread and prices soared as new-found enthusiasts fought for the precious bottles. Suddenly the original swooner was being properly valued, and like any good sex goddess she was being far from liberal with her favours and charging mightily for the swish of a sequinned hem, the fleeting caress of her glistening crimson lips. Viognier was in danger of becoming the silver screen sex symbol that many wanted, but few could have.

*If a wine can be described as pretty without insulting it, then Viognier is pretty. It tastes deliciously of the apricots that grow along with Viognier on the banks of the Rhône at Condrieu. There is a theory that the vine arrived by boat, along with Syrah, during the Roman occupation, found a landing place at Condrieu and then remained there as a tiny island of distinctive taste until its recent explosion around the world. Château-Grillet, a tiny estate within Condrieu, can be glimpsed in the picture behind the apricot fruit and blossom.*

But haven't you ever heard of lookalikes, the ones that make you double-take and say – did I see what I think I saw? Well, you didn't. But did it give you a thrill? Did your heart leap for just a second? Sure it did. And what did it cost you? A diamond ring? A champagne dinner for two at the Ritz? Nope. Maybe just what it costs to get your glasses repaired after you walked into a lamp post. Maybe a new round of drinks for someone as your swivelling elbow knocked over the waiter's drinks tray.

Well, after a small band of well-heeled wine fanatics discovered Viognier and all its sumptuous charms, the Condrieu lookalike business was born. It probably started in California where there was a tiny band of wine growers who didn't want to march down the Cabernet and Chardonnay highway. Find a red wine maker keen on Pinot Noir and Syrah and you should find someone getting all wide-eyed and wobbly at the knees at the prospect of Viognier too.

Indeed, all the new-wave countries had a few passionate wine producers who were inspired by making a Condrieu lookalike that could fool all of us devotees until she was so close we could smell the cologne on her neck – Australia, South Africa, Italy and Spain, Argentina, even Greece – have joined the fray. But the country that has flooded the wine world with lookalikes – some very good, some mere pale photocopies of a publicity postcard, but almost all affordable – is France itself.

After all, where do you find the most movie-star lookalikes? Hollywood and LA, that's for sure.

**Viognier: from Grape to Glass**

*Geography and History page 276; Viticulture and Vinification page 278; Viognier around the World page 280; Enjoying Viognier page 282*

# Geography and History

WE'RE LUCKY TO HAVE ANY VIOGNIER left in the world at all. And yet here it is, at the beginning of the 21st century, one of the most fashionable grapes in the universe. But a mere decade ago few wine enthusiasts had ever heard of the grape and even fewer had tasted its wine. Indeed, back in 1965, Condrieu – the tiny Rhône appellation which virtually single-handedly kept Viognier alive – was down to a mere eight hectares of measly, impoverished Viognier vines. One vintage produced precisely 1,900 litres of wine. 1,900 litres of Viognier to satisfy the world. Of course, then, the world didn't care. But a couple of people were taking note. Josh Jensen of Calera in California decided Viognier was a great grape and he planted some in his mountain-top vineyard. And Georges Duboeuf, the king of Beaujolais, was desperate to expand out

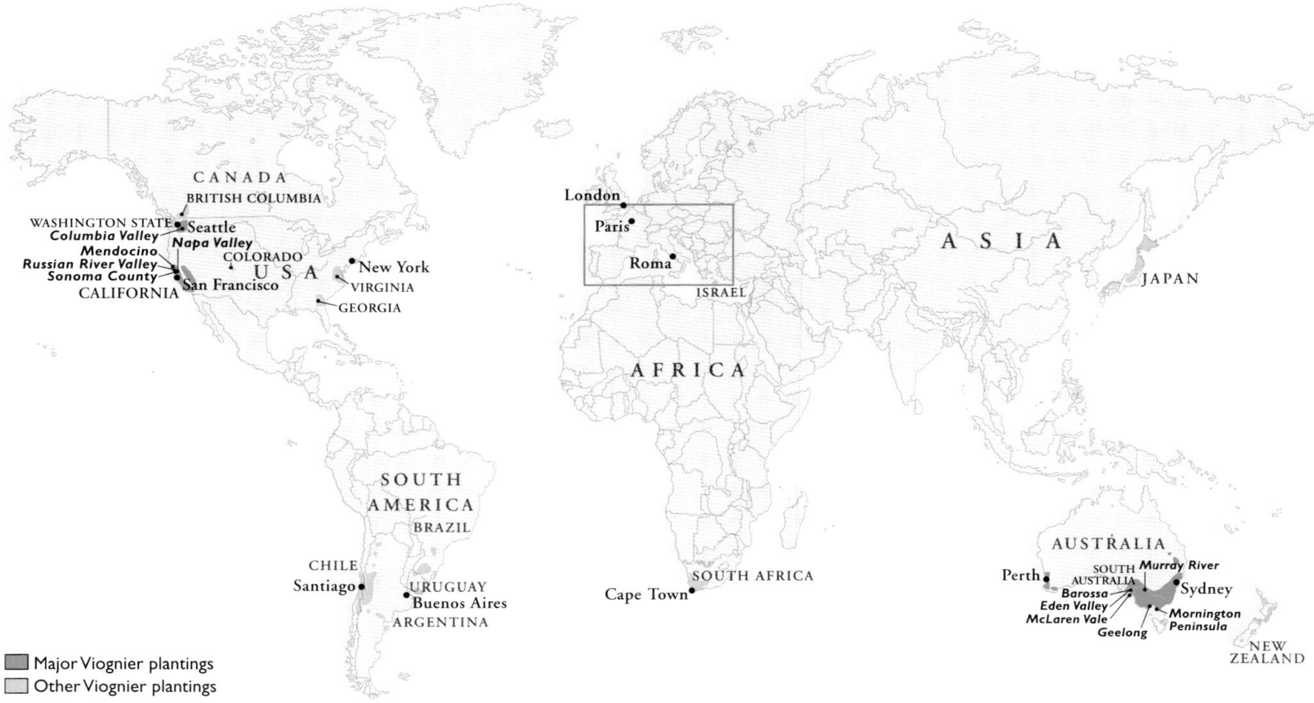

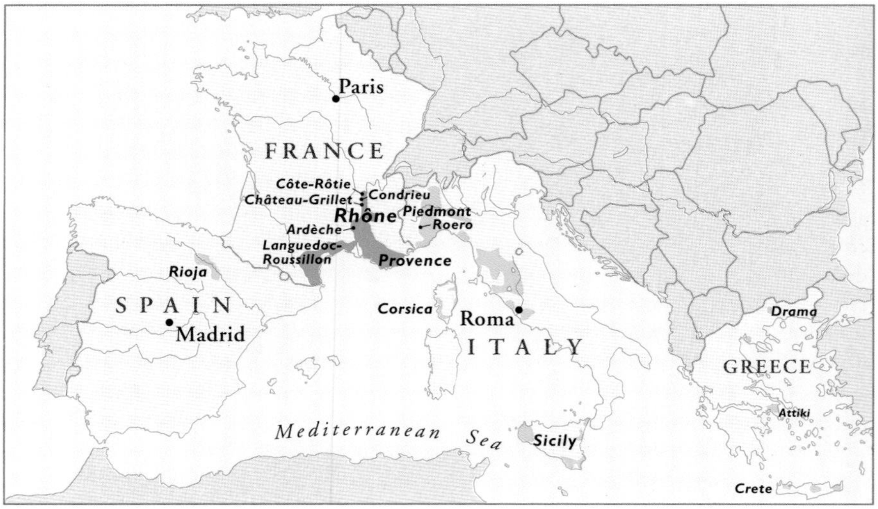

Major Viognier plantings
Other Viognier plantings

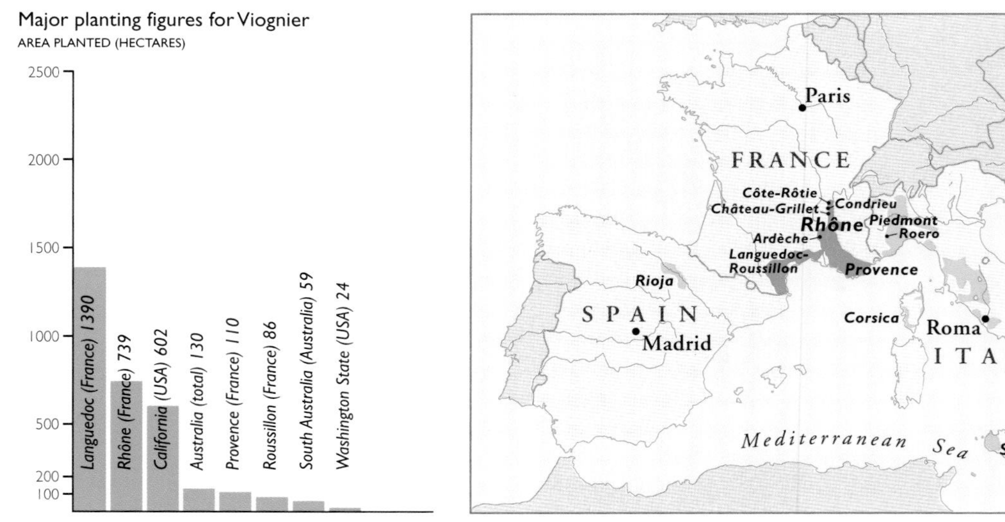

of the Beaujolais Nouveau straitjacket. He planted a great swathe of Viognier in the Ardèche in central southern France. By the mid-1990s, not only was Viognier being planted with great enthusiasm in all the New World countries, but, led by Georges Duboeuf, Viognier was appearing as a relatively low-priced vin de pays all over the south of France. And just a generation before, the entire world production of one vintage had been precisely 1900 litres.

It hadn't always been this bad. In 1928 the Condrieu vineyards covered 146ha. But the slopes are incredibly steep here and difficult to work. A market that was indifferent, and the lure of easier, better-paying jobs in Lyon and Vienne, meant that all but the hardiest wine-growers abandoned their terraces. The fightback from 1965 was painfully slow, but by 1990 plantings had risen to 40ha; by 1995 they were up to 80ha; the current figure is 102ha. These figures are for Viognier that is old enough to make AC wine; plantings of under three years are not included. The total plantable area of the appellation is 250ha, so less than half is planted, even now. But most of the best dirt is.

### Historical background

The origins of Viognier are obscure: the myths that have Syrah travelling up the Rhône in the baggage of Roman legionaries often include Viognier cuttings in the same luggage. Dalmatia is the region most named as the source of the grape, and it is said to be the Emperor Probus who brought the vine to the Rhône in AD 281. A galley of cargo for the Beaujolais is said to have been captured in the Rhône, conveniently close to Condrieu, and its contents looted by local bandits called *culs de piaux*, after the leather patches on their breeches. It's a suitably inventive story. One can only wonder how Viognier would have fared on the granite hills of Beaujolais.

What is true is that the nearby city of Vienne was a major Roman centre. It is said to take its name from *via Gehennae*, the road to hell, Gehenna also being the name given to the main public rubbish tip in Jerusalem. Anyone stuck in holiday traffic on the A7 *autoroute* south of Vienne might perhaps draw a crumb of comfort from this.

Wherever Viognier came from, it has probably been in the Rhône for a couple of thousand years or so. In more recent times a new mutant strain has appeared, according to Remington Norman in *Rhône Renaissance*. This strain looks and tastes quite different and, most worryingly, doesn't seem to possess the fabulous scent of the original. It is found mostly, though not entirely, in the southern Côtes du Rhône. It may have found its way, as a Samsonite clone (see page 19), to other countries, and that could explain why, despite Herculean efforts by really serious New World producers, an awful lot of the resulting wines lack the magic and mystery of true Viognier.

*The best Condrieu vineyards are perched on terraces above the Rhône, where some of the slopes are as steep as 70%. Even so, slopes at Hermitage, Côte-Rôtie or Cornas elsewhere in the northern Rhône are even steeper. This vineyard is La Maladière, just south of Condrieu itself.*

*Viognier vines trained one vine to a stake on the steep slopes of Château-Grillet. This is the traditional method of training Viognier in the Rhône.*

*Georges Vernay in his Coteau de Vernon site. The 40-year-old vines here go to make Vernay's top cuvée, which gets up to 18 months' barrel age.*

# Viticulture and Vinification

IF YOU'RE A VINE GROWER and you want an easy life, the last grape you should choose is Viognier. It gives yields that are both low and unpredictable, and in order to get the seductive, exotic flavours of apricots and candied peel, spice and flowers that characterize the wine at its best, you have to pick at precisely the right moment. Pick too early and you will get sugar ripeness but not flavour development; pick too late and all the beauty and perfume is lost as the wine descends into heavy, oily mediocrity. Acidity is generally low, but picking early to retain acidity doesn't work either. Viognier develops its perfume and richness very late. Until tantalizingly close to ripeness, the grape gives bland acid wine, unrecognizable as Viognier. And while some oak can be welcome, too many producers are loading on the new oak – which is pointless. Viognier is brimming full of flavour and perfume. Why swamp it?

## Climate

The revival of Viognier in the northern Rhône, and its worldwide spread, are both so recent that it is tempting to look upon almost all its vineyards as experimental. Experience so far suggests that Viognier must be allowed to get very ripe in order to allow its flavour precursors to develop, and that these develop late, after sugar ripeness. It needs a good 13 per cent alcohol to achieve that fine balance between weight and aroma, and it is generally one of the last varieties to be picked in any region. So it needs warmth: this will hardly come as a surprise to anyone who has visited the northern Rhône in summer.

Condrieu's vineyards tend to be angled towards the South and South-East, where they get plenty of sun in the morning, but are protected from the full strength of the hot evening sun. These steep terraces get as cold in winter as they get hot in summer; the Mediterranean climate does not reach much further north than Valence. The Mistral, too, adds its influence, racing down from the North at speeds of up to 140km an hour whenever there is high pressure over northern France and low pressure over the western Mediterranean.

Across the world, however, Viognier is grown in a range of climates from cool to warm. It generally produces some varietal characteristics, providing it is well cultivated, but few examples match the finesse of a good Condrieu. Perhaps it is a question of soil; perhaps it is a question of accumulated experience, and subtlety of touch.

*At Mas de Daumas Gassac, Aimé Guibert blends Viognier with Chardonnay, Petit Manseng, Muscat, Marsanne and Roussanne into an extraordinarily complex cocktail. Few winemakers would be brave enough to blend so many highly aromatic grapes together: the result could so easily be a discordant mess. But Guibert pulls it off brilliantly.*

## Soil

Condrieu's steep terraces are cut out of heat-retaining granite, with a deep sandy topsoil known as *arzelle*. Erosion can be a terrible problem, in spite of the retaining walls, and since the *arzelle* contributes much to the perfume of the wine, if it washes down the hill the *vigneron* must go down and fetch it back up.

In the South, where plantings of Viognier are on the increase in the Languedoc, Roussillon and Ardèche, soils are extremely varied, as they are in the Viognier vineyards of the New World. It seems clear that heavy soils can result in a loss of freshness and zest. In Australia's Eden Valley, where Yalumba produces some of its best Viognier, the soil is principally grey loam over clay, tightly structured and with low amounts of readily available water. It's very different from Condrieu, but does seem to produce fragrant, sensual wine.

## Cultivation

The exotic flavours of Viognier can prove elusive, so it is hardly a shock that at high yields they disappear altogether. Maximum yields in Condrieu are set at 35hl/ha, and unusually there is no *plafond limite de classement*, the legal dodge that elsewhere raises the maximum yield in prolific years. Condrieu's top yield used to be 30hl/ha: it was put up to 37hl/ha during the 1980s, when the appellation rules were revised to take account of the sudden popularity of the vine, only to be lowered again in 1991 by the local Syndicat. There have been discussions since about raising the maximum to 40hl/ha in years when the quality is good enough, provided there is a retrospective reduction for the preceding vintage. But most of the people arguing for an increase are the new johnny-come-latelys up on the wind-swept plateau above Condrieu, where Viognier doesn't ripen properly whatever the yield. Since these fertile orchard soils can't ripen Viognier, I hope the plateau producers' self-interested plans for an increase are regularly defeated – but when did questions like quality and reputation ever influence the considerations of local politicians?

At the moment, yields are usually under 30hl/ha, and there's no doubt that over 35–40hl/ha the wine loses the gorgeous, sexy lusciousness that makes Condrieu fans get all hot and bothered. In Australia, Yalumba says that yields vary according to the region, the style of wine sought and the price it sells for. Yields for The Virgilius are less than five tonnes per hectare (around 35hl/ha); for Eden Valley Viognier, less than eight tonnes per hectare; for Oxford Landing ten to 12 tonnes. Remarkably, the Viognier from Oxford Landing is usually succulent and good. In California, yields for the better wines seem to be between three and five tons per acre (approximately 50–90hl/ha).

Trellising systems vary, as well. Some favour split canopies like Geneva Double Curtain; in Condrieu a single cane is usual, and this is the most widely adopted system elsewhere. The basal buds on Viognier are not very fruitful, so it must be allowed to set fruit higher up the cane.

Poor fruit set is an endemic problem, even if the flowering goes well. The fruit also tends to develop unevenly, leading to what is known as 'hen and chicken' – normal sized berries and tiny, undeveloped ones on the same bunch. Leaf roll virus is another headache. The first Viognier to be imported into Australia was chosen by Alan Antcliff of the Commonwealth Scientific and Industrial Research Organisation (CSIRO) from the Institut National de Recherche Agronomique (INRA) collection, and came from Montpellier; it was supposedly free of leaf roll. However, later studies have shown that this clone (Montpellier 1968; also known as 642) has leaf roll type one, and more work needs to be done to clean it up. CSIRO has developed a leaf roll-free clone called HT Koolong; this should be available in Australia within a year or so. Yalumba is also doing its own work on producing virus-free clones; it says its aim is to redevelop the genetic pool in Australia, as well as looking at massal selection from elsewhere.

*The terraces of Château-Grillet in its amphitheatre above the Rhône. Ripeness should not be a problem in a sun trap like this, but the key factor with Viognier is to pick at precisely the right time to get all that fat apricot fruit into the wine, yet not lose its haunting, seductive scent.*

The age of the vines is crucial to quality: they really need to be at least 15 years old, and preferably 20, in order to produce good wine. Given that the explosion of interest in Viognier has happened well within the last 20 years, and most plantings are less than a decade old, the quality around the world is likely to improve as the vines reach maturity. At the moment only in the Rhône are there very old vines of 70 years or more.

### At the winery

Viognier is made above all in the vineyard: its winemaking is generally characterized by simplicity. In warm climates it is often picked early in the morning, because cool grapes mean clearer juice; the juice may also be allowed to settle for up to 24 hours to fall clear. A few hours of skin contact are sometimes tried, though many winemakers are nervous of getting too many oily, phenolic characters into the wine. Ripe Viognier has soft skins, and phenols are easily extracted.

If the wine is put through the malolactic fermentation it is to increase weight rather than decrease perceived acidity, since the variety's low acidity deters many people from doing the malolactic. In the New World acidification is a possiblity. *Bâtonnage*, or lees stirring, is quite common, with the new wine being left on the fine lees until bottling, generally in the spring or summer following the vintage.

### SWEETNESS – AND OAK

Sweet Condrieu is a traditional style which a few growers continue to make on a small scale. The grapes are usually overripe or *passerillé* rather than nobly rotten, though botrytized Viogniers are by no means unknown, either in Condrieu or in other countries. The fermentation may be stopped by the addition of sulphur, or it may be allowed to continue until it has practically finished of its own accord. Chilling and sterile filtering may then be necessary to ensure the wine is stable. One argument is that the latter method – allowing the fermentation to finish – gives better aromas, since these are formed towards the end of fermentation.

The leading late-harvest Condrieu is Yves Cuilleron's Récolte Tardive – he wasn't allowed to call it Vendange Tardive because of possible confusion with, of all things, Alsace. He picks the grapes in late October or early November; he does not, however, pick them *à l'assiette*, which was the traditional method. A plate was held underneath the cluster, and the cluster was shaken to make the overripe grapes drop on to the plate. The quantities, one might suspect, were not very large, but I can just see one of the more switched on Condrieu producers using this *assiette* method in the future, labelling the wine as such – and then selling the handful of bottles for a fortune. I hope it wasn't me that gave them the idea.

Oaked Viogniers are a commoner proposition than sweet ones. Some oak, even some new oak, can add complexity and structure, but the perfume of Viognier is easily overwhelmed by too great a dose of vanilla. Fermentation may be in stainless steel, cement or oak but, as with other grapes, greater integration of oak flavours is achieved if the wine is fermented in oak rather than run into oak after fermentation.

Few producers use more than ten to 30 per cent new oak, with the balance being of various ages. Old oak is just as desirable, since it gives oxygenation and good fruit expression without obvious oaky flavours. Extreme lightness of touch is required in this, as in all aspects of making Viognier. The lighter your touch the more you'll be rewarded by lushness of texture and perfume.

# Viognier around the World

AH, THE SCENT OF MAYBLOSSOM and hothouse apricots, the texture of crème fraîche festooned with apricot and peach and floral fragrance – these are the glories of Condrieu and these are the sensations that inspire Viognier growers all around the world. But few achieve such heights – around the world, or, for that matter, in Condrieu itself.

## Condrieu, the Rhône and the South

Condrieu comprises seven communes strung out along a 22km stretch of the Rhône. They are the three original communes named in 1940 when the appellation was created (Condrieu, Chavanay and St-Michel-sur-Rhône) plus Vérin, St-Pierre-de-Boeuf, Malleval and Limony. The soil is relatively homogeneous, but differences do occur and the wines may have aroma or more power according to the balance of *arzelle* and clay. Land just outside the AC boundaries – above 300m altitude, on clay plateaux or on the valley floor – may be used to grow vin de pays with reasonable success.

Château-Grillet has its own very small appellation (3.08ha). Its granite soil is marked by a high proportion of decomposed mica, and the wines have higher acidity and are slower to mature than Condrieu. In quality Château-Grillet currently lags behind the best of Condrieu. The vineyard has been under single ownership since 1840 and has expanded since 1965 from 1.75ha to its current size; the oldest vines are about 80 years old. Much of the criticism hangs on the perceived diluteness of the wine (though yields are only slightly higher than in Condrieu), and relatively early picking, lack of new wood, and late bottling, combining to give a lack of concentration.

The point is not that the wine is poor, but that it is poor value compared to what is obtainable in Condrieu. However, I had a bottle last year which showed much more style and perfume than any I've had for a generation. And at Château-Grillet's prices, it's normally only once a generation that I try it.

*Josh Jensen of the Calera Wine Company in San Benito, California, seems to have a knack of managing difficult grapes. Not only is his Pinot Noir one of the best in the West, but his immensely rich, honeyed and powerful Viognier is California's best so far.*

In Côte-Rôtie, up to 20 per cent Viognier is allowed in the red wine under the AC rules. This may have evolved as a way of adding softness and perfume to the Syrah, or the presence of Viognier might have been an accident. In any case, many growers add no Viognier at all and few add more than 5 per cent. The Viognier must be added to the vat at fermentation. This is tricky, since Viognier ripens earlier than Syrah, and is overripe by the time the Syrah is ripe. So what might start out as 5 per cent in the vineyard is very much less by the time the wine is made. Nevertheless, its effect is much greater than just an increase in aroma or finesse. The point of adding Viognier at fermentation is a process known as co-pigmentation: the different phenolic compounds added by the white grapes help to fix the colour of the red. Most Côte-Rôtie Viognier is on the Côte Blonde, where the limestone-rich soil suits it better than the clay of the Côte Brune.

Viognier is also grown in the southern Rhône, often for blending with Marsanne, Roussanne, Bourboulenc, Clairette and others. It doesn't stand heat as well as Chardonnay, and in the Midi, where it is often made as a varietal vin de pays, it is best grown in the higher spots of the Languedoc rather than in Roussillon, where temperatures are an average of three or four degrees higher. Provence and the Ardèche also have large plantings.

## Australia

Yalumba's first commercial Viognier was planted in 1979, and by the following year the company had propagated enough to plant 1.2ha in its Vaughan vineyard at Angaston. This vineyard is now 8ha in size; the oldest vines here are the oldest Viognier vines in Australia. In all there are now about 50ha of Viognier in Australia, scattered between Murray River, McLaren Vale,

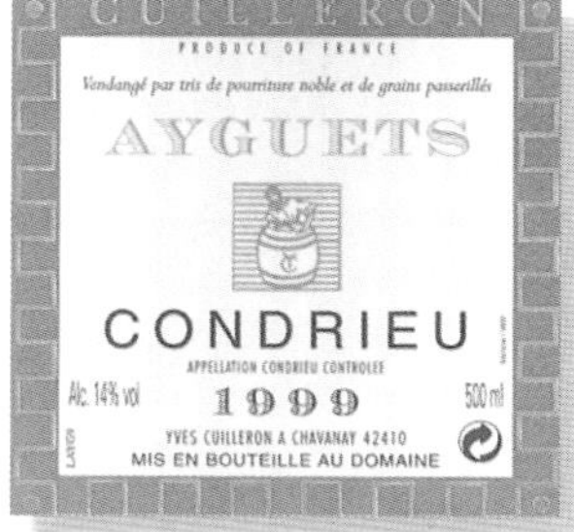


**YVES CUILLERON**

*Cuilleron is an up-and-coming star of the appellation; this wine, made from late-picked grapes is concentrated and unctuous.*

**E GUIGAL**

*Guigal ferments one-third of his Condrieu in new oak to give it the structure he feels the grape lacks. He also has some Viognier in his Côte-Rôtie vineyards.*

**CHÂTEAU PECH-CÉLEYRAN**

*Pech-Céleyran made its name as a producer of exciting, herb-scented Languedoc reds, but also makes a rich, intense Viognier.*

*Yalumba's Heggies Vineyard in South Australia. Viognier often gets a welcome touch of botrytis here, which adds complexity. Yalumba is fine-tuning its viticulture and hopes to find a way of getting aroma at lower sugar levels, so keeping alcohol levels down. I'm not sure they'll succeed: Viognier is a succulent, sybaritic creature.*

Geelong, Central Victoria, Mornington Peninsula, Barossa, Eden Valley and Adelaide Hills; in the next few years that figure is set to double at least.

Quality varies hugely: the best wines are glorious, the worst dull and lacking in aroma. Says Louisa Rose, senior winemaker at Yalumba, 'The most frustrating thing about Viognier during the ripening season is wondering when the hell it is going to get some flavour. The sugar in the berries may be high, but this is no indication that the flavours are ripe. You have to wait and wait – and then all of a sudden it changes from being dull and boring, and there is a massive explosion of musky apricot character.' Getting the fruit flavours ripe determines the moment of picking; the alcohol level normally turns out between 13 and 14 per cent, but it can go higher and retain its perfume.

### California

With over 1100 acres (445ha) planted, and more going in all the time, California is in love with Viognier, largely because people were searching for a white alternative to Chardonnay that could be made as cult wine and sold for a high price. Viognier fitted the bill perfectly. It was first planted in the early 1980s by Josh Jensen at Calera; Mendocino now has the most, followed by Napa. The same issues apply here as everywhere: yields, ripeness, skin contact and oak. With plantings rising so fast it is inevitable that many will turn out to be in unsuitable places, but the best California Viogniers are superb.

However, it is not set to take over Chardonnay's crown. Viognier is far more demanding about site, and far less easy-going in the winery. Only the most dedicated wineries will go to the lengths of Renaissance and carve terraces out of rocky hillsides. And while Viognier can never be a cheap variety to grow, the high prices being charged in California do not always bear a discernible relationship to quality. And even with the best treatment in the world, Viognier is not a versatile variety. Even the most dedicated Rhône Rangers are beginning to wonder if there might not be more future in Marsanne and Roussanne.

Other states with Viognier plantings include Colorado and Virginia, but the original California plantings came from France via a nursery in New York State.

### Rest of the world

Central Italy is having considerable success with Viognier, and there are also some good examples from Piedmont. In Spain it is grown experimentally in Rioja, and there are also plantings in Greece and the wines, while lacking a little perfume, are fat and full and sensuous in the mouth. In the New World, there is some in Brazil, Argentina, Uruguay and some, recently planted, in Chile. South Africa, New Zealand and Japan have small plantings. Everyone wants to do it. But just as in Condrieu, not everyone will know how.

**ALBAN VINEYARDS**

*John Alban is fanatical about Rhône varieties – in fact he grows nothing else at his Californian winery. His first vintage of this peach- and pear-laden wine was 1991.*

**KUNDE**

*A supple, oaky example made by a rapidly expanding operation in Sonoma County, California. Like most Viognier, it is best drunk young.*

**GIACOMO ASCHERI**

*Piedmontese producer Giacomo Ascheri may be a Barolo specialist, but he grows Viognier in his Roero vineyards by way of experiment.*

# Enjoying Viognier

THERE ARE EXCEPTIONS TO EVERY RULE, but the general rule for Viognier is that it does not improve with age. The point of the wine is its head-spinning, intoxicating perfume, and, frankly, this is often at its most sexy and seductive at not much more than a year old. Well, I'd say even less. I'd say, if you wheedle your way into a Condrieu cellar, you need to pour yourself a glass straight from the barrel. Viognier is never better than this. If you must age it – well, it can age. I've had ten-year-old examples – deep, brooding, rich and rather exciting, but with none of the eye-spinning thrill of the young model. Four or five years old and a good Viognier will be at the halfway point between self-indulgent youth and corruptly delicious middle age. This youth fetish makes Condrieu an anomaly. High prices usually go hand in hand with ability to age. But the money here flows after youth, making Condrieu the most expensive early-drinking wine in the world. It also seems to change rapidly in bottle according to unknown rules of its own. Certainly if you age it each bottle can turn out wildly different from all the others.

The wines that do age are as likely to come from Australia and California as from Condrieu. But beware. Do you really want to give up the exuberant flirtatious perfumes of youth in the half-hope the wine will gain a beeswax rich majesty with age? Are you sure you want to risk it?

## The taste of Viognier

Think of every aromatic flower and fruit you can, and throw them together into a glass. Viognier can remind you of honeysuckle, jasmine or primroses, apricots and peaches, candied peel, musk and spice. But that in itself is not sufficient: all these aromas could become top-heavy and wearying without finesse and subtlety on the one hand, and a remarkable texture half way to crème fraîche on the other.

Acidity is usually low, and if growers pick early for acidity they miss out on perfume. But it must still be a weighty wine, of at least 13 per cent alcohol, and sometimes 15 per cent. The danger then is of the flavours turning oily and phenolic and lumpen.

Simpler versions – not necessarily inexpensive, because the enormous demand for Viognier means that it is hardly ever very inexpensive – may lack complexity and have a perfectly attractive, though hardly thrilling, aroma of apricots. In that case one might wonder what all the fuss is about. Well, a cheap vin de pays Chardonnay is hardly going to resemble a Corton-Charlemagne, so why should cheap Viognier resemble Condrieu? As it happens, quite a few do have at least a passing similarity – at a fraction of the price. It is a difficult grape to grow and vinify, requiring the same balance as Chardonnay but not possessing any of Chardonnay's obliging nature, and achieving that subtlety of flavour is extremely difficult. It also requires mature vines, which is another reason why only a minority of Viogniers throughout the world are as good as they could be. Or will be: what Viognier mostly needs now, worldwide, is time and experience. As each year passes, it gets a bit more of both.

*From the northern Rhône, Yves Cuilleron's Condrieu les Chaillets Vieilles Vignes is one of the most complex, concentrated wines in the appellation; it's a wine of irresistible perfumed personality yet finesse; it even breaks the 'drink immediately' rule. It won't last forever, but a good vintage will last five or six years. Calera Mount Harlan Viognier from California can't quite hit those heights – yet – but it does have tremendous apricot and crème fraîche succulence and a honeyed scent.*

### Matching Viognier and food

Fresh, young Viognier is at its best drunk as an apéritif. Its subtle flavours have an affinity with such aromatic herbs and spices as rosemary and saffron, and it is also a good partner for mildly spiced Indian dishes, such as chicken korma. In fact, it is not a bad match with any chicken in a creamy sauce, or with rich-tasting seafood such as crab, lobster and scallops. The apricot aroma that typifies even inexpensive Viognier suggests another good pairing: pork or chicken with an apricot stuffing.

## CONSUMER INFORMATION

### Synonyms & local names
There are none to worry about.

### Best producers
**FRANCE/Rhône Valley** Gilles Barge, P & C Bonnefond, Burgaud, Chapoutier, Château-Grillet, Louis Chèze, Yves Cuilleron, Delas Frères, Pierre Dumazet, Christian Facchin, Phillipe Faury, Gilles Flacher, Font de Michelle, Pierre Gaillard, Jean-Michel Gérin, les Goubert, Guigal, Jaboulet, Jamet, Jasmin, du Monteillet, Robert Niero, Niero-Pinchon, Alain Paret, André Perret, Christophe Pichon, Phillipe Pichon, René Rostaing, Ste-Anne, Ets L de Vallouit, Georges Vernay, Vidal-Fleury, Gérard Villano, François Villard; **Languedoc** de l'Arjolle, Ch. Cazal-Viel, de Clovallon, Mas de Daumas Gassac, Pech-Céleyran, Skalli/ Fortant de France; **Provence** Ch. Routas.
**ITALY/Piedmont** Giacomo Ascheri; **Tuscany** D'Alessandro/Manzano.
**USA/California** Alban Vineyards, Araujo, Arrowood, Calera, Caymus, Cayuse, Jade Mountain, Kunde, La Jota, McDowell Valley, Qupé, Zaca Mesa; **Washington State** McCrea Cellars; **Virginia** Horton.
**AUSTRALIA** Clonakilla, Haselgrove, Heggies, Yalumba.
**SOUTH AFRICA** Fairview.

### RECOMMENDED WINES TO TRY

**Fifteen classic Rhône wines**
Guigal *Château d'Ampuis Condrieu la Doriane*
Gilles Barge *Condrieu*
Château-Grillet
Louis Chèze *Condrieu Coteau de Brèze*
Cuilleron *Condrieu les Chaillets Vieilles Vignes*
Delas Frères *Condrieu Clos Bondes*
Pierre Dumazet *Condrieu*
Pierre Gaillard *Condrieu*
Jean-Michel Gerin *Condrieu Coteau de la Loye*
Guigal *Condrieu*
Domaine du Monteillet *Condrieu*
André Perret *Condrieu Coteau de Chéry*
René Rostaing *Condrieu la Bonette*
Vernay *Condrieu Coteau de Vernon*
François Villard *Condrieu les Terrasses du Palat*

**Ten other French Viognier-dominated whites**
Domaine de l'Arjolle *Vin de Pays des Côtes de Thongue Equinoxe*
Ch. Cazal-Viel *Vin de Pays d'Oc Viognier*
Domaine de Clovallon *Vin de Pays d'Oc Viognier*
Font de Michelle *Côtes du Rhône Cépage Viognier*
Domaine les Goubert *Côtes du Rhône Viognier*
Ch. Pech-Céleyran *Vin de Pays d'Oc Viognier*
Ch. Routas *Coteaux Varois Cuvée Coquelicot*
Domaine Ste-Anne *Côtes du Rhône Viognier*
Skalli/Fortant de France Collection *Vin de Pays d'Oc Viognier*
Georges Vernay *Vin de Pays des Collines Rhodaniennes Viognier*

**Ten New World Viognier wines**
Alban Vineyards *Edna Valley Viognier (California)*
Arrowood Vineyards *Russian River Valley Viognier (California)*
Calera *Mt Harlan Viognier (California)*
Fairview *Paarl Viognier (South Africa)*
Haselgrove *McLaren Vale 'H' Viognier (Australia)*
Horton Vineyards *Viognier (Virginia)*
Jade Mountain *Mount Veeder Viognier (California)*
Kunde Estate *Russian River Valley Viognier (California)*
McCrea Cellars *Columbia Valley Viognier (Washington)*
Yalumba *The Virgilius (Australia)*

*Lovers of Condrieu shouldn't despair at the small quantities available: there are another 145ha in the appellation which could be planted with Viognier. However, the vines do need to be properly mature to give expressive flavours.*

### Maturity charts
Occasional Viogniers will age in bottle, even for ten years or more. The vast majority, however, should be drunk within a few years.

1999 Condrieu

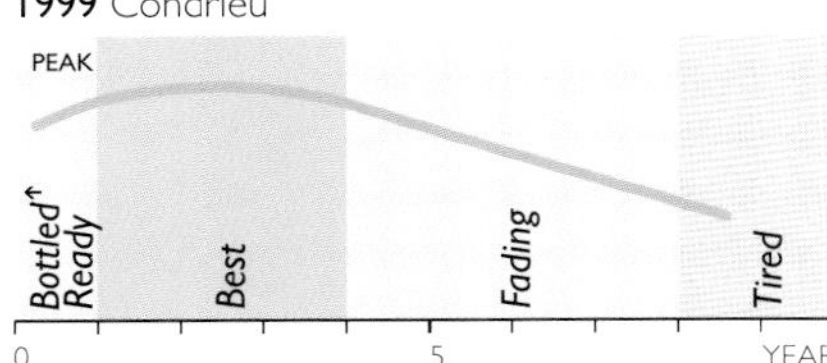


*Condrieu is one of the few high-priced whites which really does not need bottle age. Some commentators maintain it is at its best straight from the barrel.*

2000 Languedoc Vin de Pays

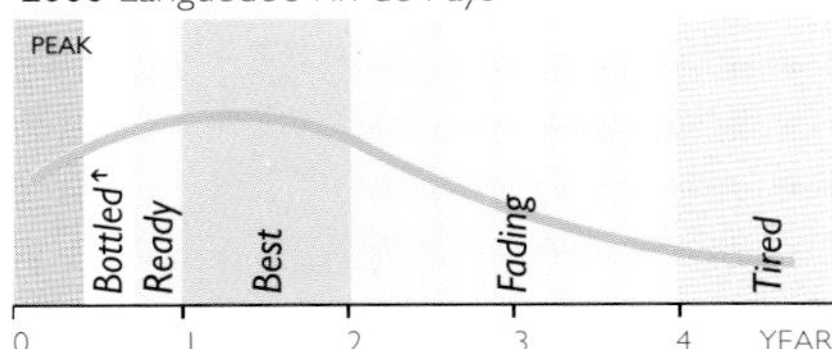


*Viognier from Languedoc-Roussillon should be drunk within a couple of years of the harvest. It will not improve in bottle.*

1999 Top California

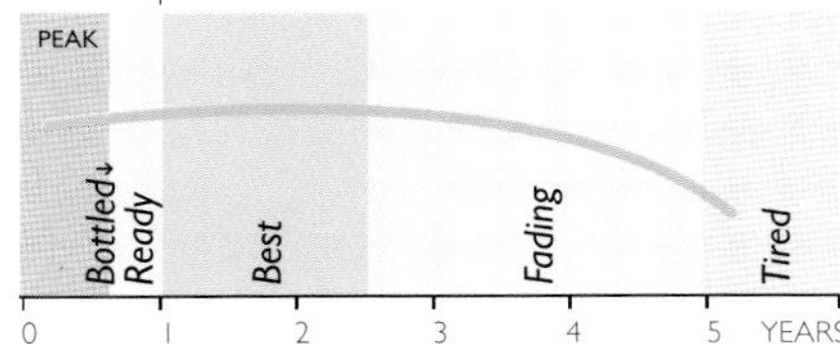


*California Viognier is no exception to the 'drink young' rule. Left in bottle too long it loses its perfume and gains nothing in return.*

## VIOSINHO

A northern Portuguese grape, grown in the Douro and Tras-os-Montes regions – and, as is the way these days, anywhere else an enterprising winemaker decides to try it out. It gives structure and flavour to a blend, and these are valuable qualities in Portugal, which has a wealth of characterful red grapes, but is slightly less well off for good white ones. Best producers: (Portugal) Domingos Alves de Sousa, Sogrape.

## VITAL

This is the name given in the Estremadura region of Portugal to a grape that may or may not be the same as the Malvasia Fina grape of the Douro Valley further north (see pages 120–121), though the leaves are different. Generally reckoned to be somewhat bland, it can give interesting, minerally wine with clever winemaking and a touch of oak. In the Douro Malvasia Fina is grown at high altitudes, to retain its acidity.

## VIURA

The name by which Macabeo (see page 117) is known in Rioja in northern Spain, where it probably arrived around the turn of the 20th century. It is a grape which obstinately defies nearly all attempts to turn it into world-class wine.

It used to have a role as a softening grape in red Rioja: many regions employed such strategems in the past, if tannins and acidity were likely to be too high for fun. Nowadays higher yields from the vines have softened up the reds, though diluted their character too, and better viticulture and winemaking are also producing softer results. This has left some of the world's more neutral white grapes in search of a job. (Trebbiano in Chianti is in the same boat as Viura – see page 270.)

Viura-based white Rioja is seldom exciting, though as always there are exceptions: the best white Rioja usually gains character from Malvasia, not Viura.

It had its day in the white wine boom of the 1980s, when bodegas promoted young, aggressive, grapefruit-scented wines. Viura's neutrality is the reason that much white Rioja tastes of oak, youthful fruit and little else: it takes obligingly to barrique fermentation and lees aging, and if picked early has good acidity. However, early picking means that it tends to lose flavour, so you pays your money and you takes your choice.

Nowadays everybody is calling for red, and the last thing that current fashion demands is lighter reds diluted with white so growers in Rioja are tending to replace Viura with Tempranillo.

At the moment Rioja plantings of Viura (about 7500ha) constitute some 15 per cent of its plantings in Spain but new plantings of Viura are being passed over in favour of red varieties, especially Tempranillo; in Navarra, too, where it forms some 8 per cent of the vineyard and is the principal white grape, red wines are in the ascendant at the expense of white. Best producers: (Spain) Artadi, Bodegas Bretón, Martínez Bujanda, CVNE, Enomar, López de Heredia, Marqués de Caceres, Marqués de Murrieta, Montecillo, La Rioja Alta.

*Harvesting Xarel-lo grapes in Cataluña, Spain. The wine, whether sparkling or still, will be neutral and clean – not exactly exciting, but not offensive, either.*

## WEISSBURGUNDER

Sometimes written as Weisser Burgunder, this is a common German and Austrian synonym for Pinot Blanc (see pages 170–171). Best producers: (Germany) Bercher, Bergdolt, Schlossgut Diel, Fürst, Dr Heger, Karl-Heinz Johner, Franz Keller, U Lützkendorf, Müller-Catoir, Rebholz; (Austria) Feiler-Artinger, Walter Glatzer, Hiedler, Lackner-Tinnacher, Hans Pittnauer, Fritz Salomon.

## WEISSER RIESLING

Anything called Weisser Riesling or White Riesling is proper Riesling (see pages 190–201), as opposed to the lesser Welschriesling (see below).

## WELSCHRIESLING

A grape that is not entitled to its suffix of Riesling, since it is unrelated (so far as is known) to the genuine Riesling of Germany. However, one can perhaps see how the name arose. 'Welsch' means 'foreign' in Germanic tongues, and since it is known by the name of Welschriesling in Austria, where it is widely grown, one can surmise that it did not originate there. (It is not grown in Germany.) But the name might possibly derive from Wallachia in Romania: the Slav name for Wallachia is Vlaska, and an alternative name

**VELICH**

*Roland and Heinz Velich, in Austria's Neusiedlersee region, make both dry and sweet Welschriesling wines. This is their dry one – stylish and well structured.*

for the grape, used in Slovenia and Vojvodina, is Laski Rizling. Other names include Olasz Rizling in Hungary and Rizling Vlassky in the Czech Republic and Slovakia. The Italians prefer to call it Riesling Italico, and use the name of Riesling Renano for true, or Rhine Riesling.

In northern Europe it has long had a reputation for extremely poor quality, but this reputation is based on the large quantities of Yugoslav Laski Rizling exported until the end of communism (and indeed until the break-up of Yugoslavia). The low quality of this wine – its taste was more reminiscent of drains than of wine – was the result of poor winemaking, poor storage and, perhaps, the determination of the shippers to meet certain price points; certainly the same state-owned winery that produced the worst wine also produced some perfectly creditable wine from the same grape.

Good Welschriesling, as made in Austria and Slovenia, is non-aromatic, nutty and quite weighty, with good roundness and relatively low acidity. It makes excellent noble-rotted sweet wines in Burgenland. However, the fashion for reds means that it is being uprooted to make way for Merlot, Pinot Noir and Austrian red varieties like Blaufränkisch. Northern Italian versions are lighter and crisper.

It yields generously, and yields must be controlled if the wine is to have the concentration it needs. In future times it may regain its own reputation as a good, if not superlative grape. But for now it is still blighted by its notoriety as the grape that has contributed so cynically to the entirely blameless real German Riesling's loss of popularity and respect in much of the wine world. Best producers: (Austria) Feiler-Artinger, Alois Kracher, Velich.

## WHITE RIESLING

Another name for proper Riesling; the Riesling of Germany, or Rhine Riesling (see pages 190–201).

## WILDBACHER

An Austrian grape found in the region of Weststeiermark, where it is almost all made into the local pink speciality, Schilcher. It may also be called Blauer Wildbacher. Schilcher is notable both for its startlingly high acidity, and for the speed with which it sells out, to locals and visitors alike. It has a light redcurrant flavour and is best drunk young. There are some examples of red wine from the grape, too.

## XAREL-LO

A high-yielding Spanish grape found especially in Cataluña, where it is one of the mainstays of the Cava blend. Parellada (see page 166) and Macabeo (see page 117) are its traditional blending partners and the addition of a little Chardonnay can do wonders for Cava's flavour and quality. Xarel-lo oxidizes easily, and until recently gave Cava its characteristic earthy, rooty flavour. That problem seems to have been largely solved now, and Cava is much cleaner-tasting than of yore, though more neutral – no bad thing in most cases.

Xarel-lo seems to have more character in the Alella region, north of Barcelona, where it takes the name of Pansa Blanca (see page 166) and gains an attractive lime cordial aroma and flavour. Best producers: (Spain) Albet i Noya, Castellblanch, Codorníu, Freixenet, Augustí Torelló.

## XYNISTERI

White grape grown on Cyprus. If one is to judge its quality by that of the table wines emerging from the island one would not give it the time of day, but there are signs that better winemakers are taking an interest in what Cyprus has to offer. It may well be capable of making perfectly good table wines, especially if carefully blended with international varieties like Chardonnay or Sémillon and vinified with New World care. It is the best grape (better and more subtle than the black Mavro, that is) for making Commandaria, the dark, sweet, often fortified wine made from raisined grapes, that is Cyprus's claim to vinous fame. Best producer: (Cyprus) KEO.

## XYNOMAVRO

A highly acidic, dark-coloured Greek grape, the name of which means, oddly enough, 'acidic and black'. All that tannin and acidity does mean that the wines age well – between 5 and 12 years. It is the backbone of rich, full-bodied spicy reds from Naoussa and Goumenissa in the heart of the Macedonian region, and is grown widely all over northern Greece. It ripens late, and may be planted at high altitudes for sparkling wine. Best producers: (Greece) Boutari, Tsantalis.

## ZIBIBBO

Sicilian name for the less aromatic and less seductive Muscat of Alexandria (see pages 144–153). Best producers: (Italy) Colosi, D'Ancona, De Bartoli, Donnafugata, Pellegrino, Salvatore Murana.

## ZIERFANDLER

Austrian grape found in the Thermenregion area south of Vienna, where it is the traditional partner for Rotgipfler (see page 203) in the white wine called Gumpoldskirchner, though sometimes it is also made as a varietal. It makes big, spicy, long-lived wines of substance, especially when not fully dry. When dry the high alcohol level can seem too much of a good thing. It makes excellent sweet wines. Best producers: (Austria) Karl Alphart, Gottfried Schellmann, Johann Stadlmann, Richard Thiel.

## ZINFANDEL

See pages 286–295.

## ZWEIGELT

Austria's most planted red grape, Zweigelt is high-yielding and easy to grow. It buds late and ripens early, so avoiding bad weather at both ends of the growing season and, perhaps as a result of its high yields, is often treated as a workhorse grape. It can, however, produce quite rich, cherry-fruited wines with an attractive pepper tingle which take well to oak ageing if they are sufficiently concentrated. The best can even age for a few years in bottle. Most, however, are soft, jammy, but not at all innocuous.

The vine was produced in 1922 by a Dr Zweigelt, who crossed Sankt Laurent (see page 207) with Blaufränkisch (see page 42); alternative names are Blauer Zweigelt or Zweigeltrebe. It has potential for other cool climates, notably England and parts of Germany. Best producers: (Austria) Feiler-Artinger, Gernot Heinrich, Hans Pitnauer; (Hungary) Akos Kamocsay.

And that – unless anyone can find us any grapes starting Zx, Zy or Zz – is IT.

**GERNOT HEINRICH**

*Gernot Heinrich is one of the most go-ahead producers in the Neusiedlersee region of Austria. His Zweigelt is a simple everyday wine: a perfect light summer red.*

EL DORADO
OF THE
OF AMERICA
E·R

# ZINFANDEL

THE WAY SOME CALIFORNIAN WINE marketeers go on, you might think that Cabernet Sauvignon was a Californian grape. Merlot too, and Chardonnay as well. But they're not. They're French classics that the Californians have adopted with a revivalist fervour that threatens to eclipse their origins in Bordeaux and Burgundy.

But the Californians *do* have a grape that they just might claim as their own – indeed a lot of them do just that. And that's the Zinfandel. As Californian as the Hollywood sign, Big Sur and the Golden Gate Bridge. Except that now some smart alec techno-wizards have proved that Zin ain't Zin. Zinfandel is Primitivo from the south of Italy. And if that idea isn't hard enough for Zin fanatics to swallow, there are some people who say that Zin might actually be the Plavac Mali from the Croatian coast. Or if not Plavac Mali itself, something very similar from the same area.

*Zinfandel, now known to be the same as southern Italy's Primitivo grape, may have come to the eastern seabord of the USA with Italian immigrants. From there it made the journey west to California with the Gold Rush – hence the gold pan, the poster advertising the Gold Fields – the new El Dorado – and the nugget of gold. The popularity of the grape has proved to be a more lasting national treasure than the seams of gold which fast ran out.*

Now, let's get one thing straight. Zin was nothin', but *nothin'* before the Californians came along. So, maybe it *is* the same as Primitivo, but in the last 1000 years, did you ever read a single word saying how great Primitivo from the heel of Italy was? How many syllables can you remember being expended on the merits of Croatian Plavac Mali in the world's library of wine wisdom? So no matter what the evidence I'm brazenly going to wave the Californian flag for the wine that above all others symbolizes the spirit of the Golden West – red, pink or white, heavy or light, sweet or dry – California Zin.

Yes, Italy's Primitivo probably did stray across the Adriatic sea from Croatia. But vines need prophets and producers of genius if we are ever to see how good they are – and California had two. Both very different, with very different objectives, but both of vital importance. First there was Bob Trinchero of Sutter Home Winery. Now, if you don't live in the USA, the 'white' Zinfandel craze may have passed you by, but during the 1980s and early 1990s this pale pink, sweetish winey glugger was one of the most popular wine styles in America. Bob Trinchero invented it. He says he saved Zinfandel from extinction as California went crazy for Cabernet and Chardonnay, uprooting all its old vines – many of them Zin – with mindless determination. Perhaps he did. In which case – here's to you, Bob.

But the true prophet, the bearded, bespectacled genius of Zinfandel is Paul Draper of Ridge Vineyards, high in the earthquake-rattled hills above San José. With some grapes you can point towards a bunch of growers and say – these producers saved the grape, or showed us how great its wine could be. But with Zinfandel, if you're looking for the one person who has raised its wine to an art form, a heady, swirling, perfumed riot of richness and ripeness that never somehow topples over into the banality of stewed fruit and jam, it is the scholarly, amiable Paul Draper. For me, he's the true king pin of Zin. And from round the world I hear faint but lusty cries of support – from Cape Mentelle in Western Australia, from Fairview Estate in South Africa, from Mexico and Chile, from Puglia in Italy and even, at the last, a hoarse but hearty shout from the Croatian coast itself.

**Zinfandel: from Grape to Glass**

*Geography and History page 288; Viticulture and Vinification page 290; Zinfandel around the World page 292; Enjoying Zinfandel page 294*

# Geography and History

To listen to some of the Californian Zin producers talk you would think that Zinfandel only grows in California, only ever could grow in California, only ever did grow in California and that anyone who dares protest that they too can grow Zinfandel in any less favoured area than America's Golden West is cruisin' for a bruisin' (or mushin' for a crushin' depending on your preferred winemaking style).

Well, folks, it ain't so. Let's start with the USA: 14 States of the Union grow Zin, and some of it's pretty good. And there's more. Zin isn't Californian; it isn't even American. It's Italian. They call it Primitivo there – but it's Italian.

Now, I can feel the hurt, because there's a powerful movement that declares Zinfandel as California's indigenous grape. But California didn't *have* any indigenous *Vitis*

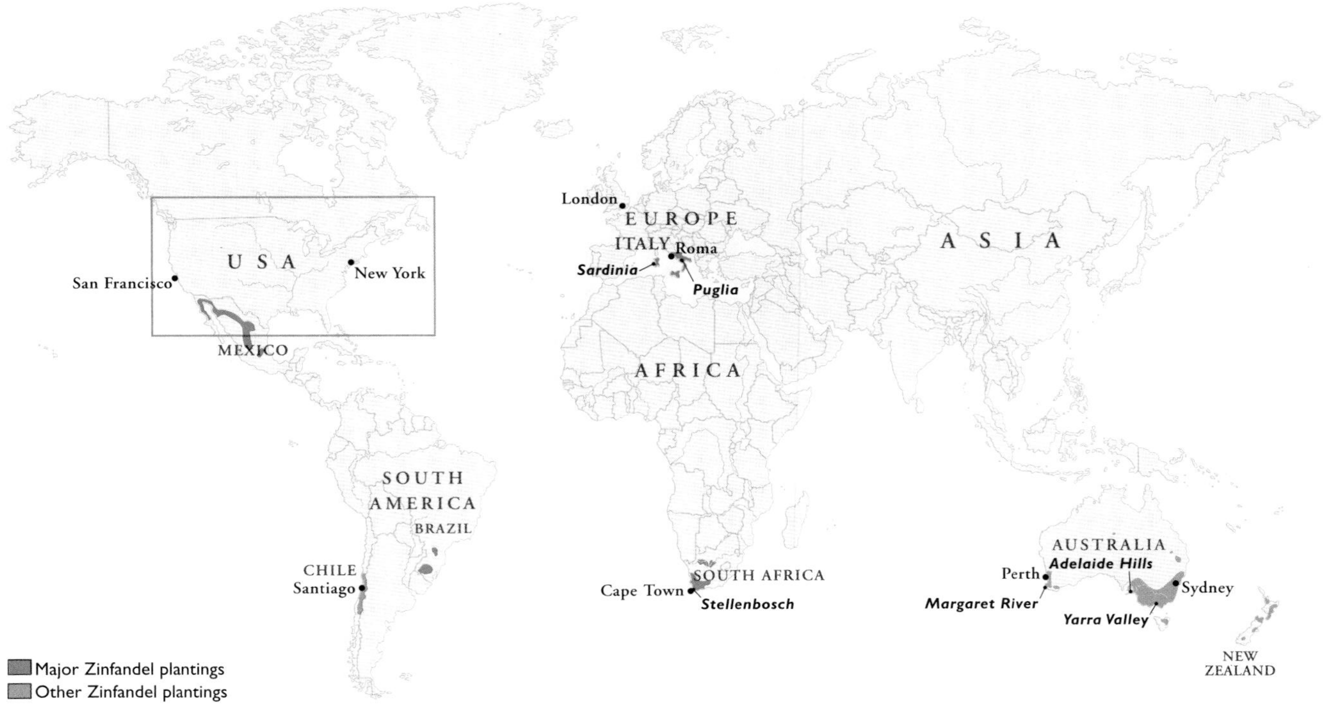

Major planting figures for Zinfandel
AREA PLANTED (HECTARES)

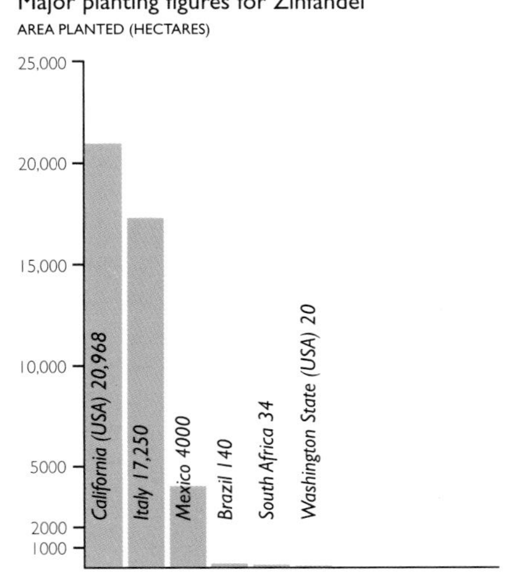

*vinifera* – the grape had to come from somewhere and it was California's own University of California at Davis that proved the grape was the same as southern Italy's Primitivo.

But this is going to salve the hurt of those who proclaim Zin as California's own. It flourished in California in the gold rush, providing the forty-niners with cheap hooch, and home winemakers turned to it with enthusiasm during Prohibition. California's first wine boom, in the 1880s, was based on Zinfandel. Then the wine was red. A century later, another California wine boom galloped along on the broad shoulders of Zin – but this time it was pink. It can be pink or deepest black, sweet or dry; a properly white mutation even exists. In its protean nature it is the epitome of the American dream, constantly able to reinvent itself to suit the Zeitgeist. When Americans talk about Zinfandel, they are often, it seems, talking about themselves. But perhaps they are talking about themselves more than they know. Modern America is based on waves of immigrants adapting deftly to the unique conditions of the continent. Zinfandel, the Italian immigrant, has achieved that par excellence.

## Historical background

Knowing that Zinfandel and Italy's Primitivo are the same vine – and thanks to DNA fingerprinting by the University of California at Davis (UCD) we are now certain of that – is only part of the solution to the problem of where Zinfandel comes from.

The earliest mention of Zinfandel in the USA is in the 1820s, when George Gibbs, a nurseryman from Long Island, brought cuttings from the Imperial botanical collection in Vienna. By 1832 'Zinfendal' vines were being sold by a Boston nursery; in subsequent years it was cultivated as a table grape in the north-eastern states. From there, some time in the 1840s, it is believed to have been taken to California by another nurseryman, Frederick Macondray.

Its name was probably acquired because of confusion with an Austrian vine called Zierfandler; and it probably arrived in Vienna in the first place from Croatia. For a while it was thought that Zinfandel was identical to a Croatian vine called Plavac Mali, though in 1998 UCD's DNA studies revealed that while the two grapes are related, they are not the same.

It has also been suggested that Zinfandel was taken to Italy from the USA by returning emigrants: some 20 per cent of Italian emigrants returned to their homeland. In which case, of course, there should be no problem with Italian Primitivos labelling themselves as Zinfandels, should there? But do you really think that returning emigrés would bring vines back to somewhere like Italy? And anyway, there are records in Puglia of Primitivo arriving there from Dalmatia (now Croatia) in the 18th century.

So, Zin moved from Sicily to California – but it hasn't stopped moving yet. Central and South America are doing well with it. Australia and South Africa make some exciting examples. There's even a Zin vineyard on the hallowed Syrah domain of the hill of Hermitage (for amusement only, I must add).

*Early morning mist in the Lytton Springs Vineyard on the eastern side of Dry Creek Valley, Sonoma, California. Zinfandel was first brought to the valley by Italian immigrants in the 1870s. Part of the secret of this region's great Zinfandel is unique, old gravelly soil, known as Dry Creek Conglomerate.*

*100-year-old vines in Lytton Springs, Dry Creek Valley. Over half the Zinfandel acreage in California is reckoned to be over 50 years old.*

*Benny Dusi of Dusi Ranch in Paso Robles, San Luis Obispo County, believes these bush-trained Zinfandel vines were planted in 1923.*

# Viticulture and Vinification

FOLKS DO GET STUCK ON ZIN. Pinot Noir devotees work themselves into paroxysms of distress over their beloved grape's many shortcomings. But Zin worshippers are a more laid-back crew. Zin does have shortcomings, but these guys seem to welcome them.

It will produce obligingly large crops on poor soil in dry conditions, but it ripens so unevenly that it is not uncommon to find green, ripe and raisined berries all on the same cluster. This, in any other variety, would be regarded with horror for the unbalanced flavours it produces; Zinfandel aficionados seem to treat it with the indulgence of a parent for a spoilt child. Green, ripe and raisined? Why, that sounds pretty balanced to me. And so it often is. If Zin rarely achieves the heights of Shiraz or Cabernet Sauvignon, producers who understand it make some magnificent, mouth-warming, heart-warming stuff.

## Climate

What Zinfandel likes is warmth, but a long growing season as well. To get the full flavours of the grape you seem to need 14 per cent alcohol: Paul Draper of Ridge Vineyards says, 'You must go to 101 per cent ripeness to get the true fruit. Whereas an overripe Cabernet does not work, an overripe Zinfandel does not exist.'

Ripeness carried to the ultimate can mean 17 per cent alcohol, and such wines are not everyone's idea of a balanced, appealing drink (and Draper's wines do not go to these levels). Such monster Zins had their greatest vogue during the 1970s, but they are still around – and some are magnificent.

The vine performs best in a Mediterranean climate with plenty of sun; but let it get too hot and the wine will taste jammy and baked. Dry Creek Valley, California's prime Zinfandel region, points to its hot days (95° to 100°F) and its relatively cool nights (45° to 50°F) as factors that produce good ripeness and preserve acidity at the same time. Lodi, where a lot is planted for sweetish 'blush' (in reality pale pink) Zin, is also very hot, but without the redeeming cool nights.

In the south of Italy, where the quality of Primitivo has yet to rival that of Californian Zinfandel, much of the wine hits 16 per cent alcohol. A few quality-oriented producers are looking at ways of slowing the maturation of this grape – not easy given that it's a variety that used to be famous for ripening before any other grape in Europe.

Much of the style of the wine depends on when the grapes are picked. Earlier picking gives flavours in the cranberry/raspberry range; later picking gives black cherry, blackberry and plum flavours, right through to prune, dates and raisin. Up to 23 Brix (the standard US measurement of fermentable sugars in the grape juice, see page 307) strawberry flavours dominate; cherry comes into the picture between 23 and 24, and at 25 the flavours turn blackberryish. Raisined grapes will be even higher in sugar, about 27 Brix; most growers will accept ten per cent or so raisined grapes in the vat. Much blush Zin is picked at 20 Brix: varietal character is not at a premium for such wines, but, interestingly, pink (or so-called white) Zin does have a very recognizable apple flesh and tobacco flavour, dappled with raisin.

*Paul Draper, the 'King of Zin', arrived at Ridge Vineyards as winemaker in 1969; Ridge Zinfandels, made with grapes from various sources, are famed for their great intensity, concentration and long life.*

## Soil

Italian settlers, planting Zinfandel in the 19th century, usually chose hillsides and river benchlands, and soils that were too poor for other crops. Such practices reflected their habits back home; and it turned out that they suited Zinfandel very well. Poor, well-drained but mineral-rich soils produce good results, but good wines also come from the valley floors. As is usual in California, choice of soil is secondary to choice of climate. And in any case, even the now-revered Dry Creek Valley was originally planted for bulk wine. That it turned out to produce excellent quality was as happy a chance as those that founded many a European wine region.

## Cultivation

California abounds in Zinfandel vines of 50 years old or more; these are one of the reasons why the USA regards the variety as a national treasure. Such vines are bush-trained and head-pruned, just as they are in the south of Italy, but bush-trained vines are not necessarily naturally low-yielding: 3 to 5 tons per acre is about the maximum for top quality; 6 tons is all right, but even bush-trained vines can give up to 8 tons per acre – at a low density of perhaps 512 vines per acre – unless they are carefully managed. (The oldest vines may not manage more than 2 tons per acre.) Clusters need to be thinned, and the second-crop clusters, which Zinfandel is adept at producing because of its uneven flowering, need to be removed.

More recent plantings, including those of the higher-yielding, virus-free clones produced by the University of California at Davis (UCD) in the 1970s, are trained on wires, and mechanically cultivated and picked. At the other end of the scale, some producers try to achieve more consistent ripeness by hand-picking in successive *tries*.

UCD has been busy preserving the diversity of Zinfandel in California. In its Heritage Vineyard at Oakville, Napa Valley, it has gathered together massal selections from vineyards planted before 1930 from all over California: from Sonoma, Mendocino, Napa, Contra Costa, Sierra Foothills, San Luis Obispo, San Joaquin, Lake, Amador, El Dorado, Calaveras, Alameda, Santa Cruz, San Bernardino and Riverside Counties.

*Autumnal vineyards of Chateau Potelle on the slopes of Mount Veeder AVA in Napa, California. Much of the Zinfandel planted in the 1990s is trained on wires such as these ones. There is also a debate over whether hillside vineyards like this one, or vineyards in the valleys, produce the longest-lived wines. Mount Veeder reds, from whatever variety, are some of the deepest and longest lived in California.*

There are 90 selections from 60 vineyards, all grafted on to the same St George rootstock, head trained and spur pruned, and planted in 9ft by 8ft spacing. Such variables as cluster weight, berry size and cluster tightness will be studied, to see which ones are genetic and which are derived from site and circumstances. Some of these vines may be cloned and made commercially available to growers in the future, hopefully with the emphasis on flavour; in any case, such a collection will provide an invaluable research resource.

### At the winery

Zinfandel can be treated in almost any way the producer likes, once it gets to the winery. For a start, you can make it like a white wine: run the juice off quickly, ferment it nice and cool. Leave a fair amount of sugar in it and – hey presto – you've got a white Zin, or, more accurately, a Blush Zin since there'll be a shy pink bloom to the wine.

If you're making reds, whole cluster fermentation will give fruitier, strawberryish flavours; a long maceration with the skins will give more colour and tannin and that'll set the wine up for some oak aging. Traditionally, old oak barrels or even just great big redwood vats were used for softening the wine. In the right hands, this produced some blockbuster stuff. But nowadays the risk of oxidation and infection isn't seen as acceptable, so small, frequently new barrels are used. I still prefer it when the oak is kept in check, but the rich vanilla and spice flavours of American oak do seem to marry pretty well with the deep date and raisin richness of Zin. Some French oak is now being used, though I can't see the point for such an over-the-top grape as Zin. Anyway, a good winemaker using well-coopered barrels will make a wine where you don't notice the oak at all.

Late-harvest Zins, made from completely raisined grapes, may be fermented out to dryness, which gives somewhat intimidating alcohol levels of over 17 per cent; Paul Draper at Ridge has also made sweet Zinfandel Essence from botrytized grapes. Remarkably, the wine kept its dark colour: most botrytized reds end up a nondescript pink, since one of the effects of *Botrytis cinerea* is to destroy colour.

### BLENDING ZINFANDEL

If you go back far enough, traditional Zinfandel came from vineyards in which the vine was interplanted with other varieties such as Petite Sirah, Carignan, Grenache, Mourvèdre, Mission and even Muscat: a hotchpotch of warm-climate grapes which, blended together, gave extra complexity to the exotic flavours of Zinfandel. Carignan would give tartness to such a blend; Petite Sirah colour and tannin; Zinfandel would have provided the richness. It's what's known as a field blend, or *mista nera* (dark blend), and although such plantings are rare today, they are reputed to survive in some of California's oldest vineyards.

If a single variety is to be blended with Zinfandel, it is usually Petite Sirah. Paul Draper of Ridge Vineyards is particularly keen on this mix, and all his Zins have 10 to 15 per cent of Petite Sirah included in the blend. J Lohr adds a touch of Riesling to its white Zinfandel, to help the aroma. Since Riesling is such a despised grape in California, it's nice to see it being put to good use for once.

# Zinfandel around the World

WE'RE LIVING IN AN AGE when great big gobfuls of super-ripe juicy red fruit win a lot of the plaudits in the wine world. Well, step forward Zin. That's what Zin does best, and that explains a rash of plantings around the world, as well as a great increase in respect and enthusiasm for Zinfandel as Primitivo in its homeland of southern Italy.

## USA: California

Zinfandel can be found all over California, from the Sierra Foothills to the coastal hills of Paso Robles. And plantings are still rising: 1986's figure of 25,979 acres (10,513ha) had risen to 51,811 acres (20,968ha) in 1999. It is now California's most widely planted red grape, outdoing even Cabernet Sauvignon. Zinfandel Advocates and Producers (ZAP), a group formed in 1992 by 22 wineries to promote the vine and its wines, now has 229 members. And that the vine has been proved to be identical to Primitivo has not altered the opinion of the Bureau of Alcohol, Tobacco and Firearms (BATF) that Zinfandel is unique to the US, and that the two wines must be labelled differently when sold in the USA.

Dry Creek Valley in Sonoma County is the region associated above all with the grape. Here it is warm, but not excessively so: it is cooler than Alexander Valley, where the vine also flourishes, though warmer than Russian River Valley, where Zinfandel really needs warm years to excel. Typical Zin flavours from Dry Creek are juicy blackberry and pepper; bright flavours with good acidity. Russian River examples also show acidity – too much so in cool vintages when the grapes don't ripen fully. Sonoma generally gives this character, with relatively low tannin. Santa Cruz wines have complexity and depth – but is that Santa Cruz character, or the skill of Paul Draper at Ridge Vineyards? Its advocates say that Zinfandel reflects its site more than, say, Cabernet, but it reflects its producer as well, and the diversity of styles make it difficult definitively to assign particular characters to particular regions – even if the regions were homogeneous, which they are not. But single-vineyard Zins, often made in tiny quantities, are enjoying a vogue.

Paso Robles Zinfandels often have soft, round flavours without the acidity of Sonoma, and in Napa, where the vine has generally given way to Cabernet Sauvignon, the flavours can be plummy and intense. Contra Costa wines can be dusty and earthy; Lodi produces fleshy, approachable wines, and a great deal of blush.

There is a distinct difference in flavour when the vines are old – and with California so rich in vines of half a century and more, there is not the same devaluation of the term 'old vines' that one sees in most countries. 'Old Vines' on the label still means something here. Old vine Zinfandels are denser, more compact and have a much more thrilling, brooding intensity. Their balance is partly because of their lower yields and partly because they were planted on poorer soils and on hillside sites where maturation is slower and flavours have longer to develop.

Blush, or white, Zinfandel was the great commercial success of the 1980s, although the grape had been used for flavoursome rosé in the 19th century. Bob Trinchero of Sutter Home is the man credited with commercializing the first modern blush Zin: the red wine one year was pallid, so he took the skins from one vat and added them to another, leaving him with some properly deep coloured wine, and some pale pink wine.

In the late 1990s this bland, sweetish style of wine was still outselling red Zinfandel. It is often said that it was blush Zin that saved the vine in California: sales had been falling until then and vineyards were being uprooted. But the vines that produce blush wine are not the same as those that produce top quality red: blush is nearly always a commodity wine, produced from high yielding, wire-trained, mechanically harvested vines. Old bush-trained vines cannot be mechanically harvested.

Nevertheless, blush Zin did revive interest in the variety. If nothing else, it paved the way for Zinfandel's current cult status.

## Rest of the USA

Other states that cultivate Zinfandel are Arizona, Colorado, Illinois, Indiana, Iowa, Massachusetts, Nevada, New Mexico, North Carolina, Ohio, Oregon, Tennessee, Texas and Washington. They rarely achieve the ripeness of California, but, as I write, I've just tasted a New Mexico example that was unmistakably Zinfandel and deliciously different.

## Italy

Primitivo was a widely planted but deeply obscure grape from the south of Italy until its

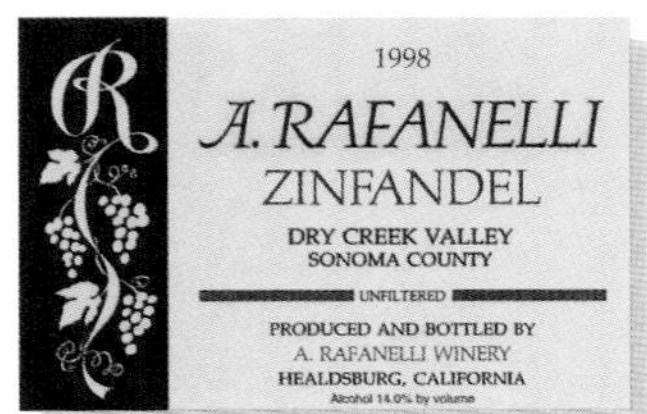


A RAFANELLI

*There's typically brambly Dry Creek fruit in this unfiltered, old vine example. The vineyards are unirrigated, and yields are low.*

DE LOACH

*The best wines from this Russian River Valley winery are labelled 'OFS'. Zinfandel, from selected lots of old vines between 70 and 100 years old, is the star here.*

MARTINELLI

*The Martinelli family have been growing grapes in the Russian River Valley for over a century now but they started their own winery only about a decade ago.*

*Vineyards in the Primitivo di Manduria DOC in Puglia, southern Italy. Primitivo is reckoned to be the original Zinfandel, though it was basically just used for blending until southern Italy beame trendy in the 1990s. So far the American versions are still better, but there's no reason why Italy's can't match them in the future.*

resemblance to Zinfandel was noted in the 1970s. I don't mean to imply that Primitivo wasn't popular – it was. It was tremendously popular in the North, with the red wine producers of Tuscany and Piedmont. Whoops. Wash my mouth out with soap and water for even entertaining such an idea. But, as a very early ripener, with loads of flavour and alcohol, it could be fermented and finished down in the South before the North had even begun picking. And in poor years, a good slug of Primitivo will have turned many an insipid thin northern red into something rather nice. So Primitivo was popular. But, understandably, it was obscure. After all, you don't go round praising a grape's name when it's been blended with something entirely different hundreds of miles away from where it was grown.

Anyway, as soon as the similarity between Zinfandel and Primitivo was noted, obscurity was out. The University of California at Davis proved that the two are identical in 1994, and now not only are bottlings of Primitivo on the increase (the bulk still goes to the North for blending), but some is being renamed Zinfandel – though not if it is to be sold in the USA.

Varietal DOCs are Primitivo di Manduria and Primitivo di Gioia, both of which are in Puglia. Most wines are somewhat rustic and solid, and the Manduria DOC allows for sweet and fortified versions as well; alcohol in even the unfortified versions commonly tops 16 per cent. However, young and ambitious producers are busy producing imitations of the California style, with 14 per cent alcohol and oak (often American oak) aging.

### Rest of the world

South Africa is beginning to look at the possibilities of Zinfandel, though plantings there so far account for less than 1 per cent of the total vineyard acreage. Even so, producers generally have to blend it with something like Cinsaut or Carignan to bring the alcohol level down below 16 per cent. It's pretty beefy, brawny stuff.

In Chile, too, quantities are still small, but rising and the flavour is absolutely spot on. The producer Cono Sur suggests that the colour of Chilean Zin is lighter than that of California – but then produces a sample that is black as night. Mexico and Brazil also have some, and I'd expect Argentina to give it a try.

In Australia, its most successful plantings so far are in Western Australia's Margaret River region where the flavours are tremendous, even if the alcohol heads unerringly towards 16 per cent most years. Adelaide Hills is less successful, and so is Yarra Valley. Well, I'm not surprised – they're supposed to be cool areas. As for New Zealand – well, they've got some too. In a super hot vintage like 1998 the wine could be interesting, but doubt if it'll do much in a normal year. Even so, the New Zealanders got so excited by the super hot 1998 vintage that some guys couldn't resist planting a raft of warm climate varieties. When – if – they have another hot year I'll be able to tell you whether they're any good or not.

**AMADOR FOOTHILL WINERY**

*A bold, supple, blackberry-fruited wine from unirrigated vineyards. The producers reckon it goes well with the flavour of sage – hence the picture on the label.*

**PERVINI**

*This is an example of the new-wave Primitivo wines from Puglia in southern Italy which have been inspired by Californian Zin.*

**CAPE MENTELLE**

*Founder David Hohnen was trained in California; no wonder when he went back to Australia he wanted to grow Zin. This is chewy, spicy – and rich.*

# Enjoying Zinfandel

I'VE NOT HAD A GREAT DEAL OF LUCK in aging Zin, and it's a lot to do with the nature of the grape. Ripe Zinfandel, even when young, seems to have the seeds of its own maturity lurking in its flavours. Flavours like prune, dates and raisins do friendly battle with the pepper and blackberry of youth, and within a year or two they've usually gained the upper hand as the blackberry browns and the pepper is soothed away.

So why age them? Why indeed. Many Zins are made to be drunk young anyway – and the grape is ideally suited for it. Especially when grown in high-yielding vineyards in hot areas like California's Central Valley, the low tannin and soft raspberry and blackberry fruit is perfect for a young glugger. And even from old vines in a top area like Dry Creek – well, the low tannin and burly brawny fruit makes them so easy to drink young, and few undergo any great sea-change in flavour, just a gradual loss of sweetness and fruit, eventually becoming almost tarry with time. If you like that style, well, fine. Many Zinophiles (as American lovers of the grape call themselves) do like it. Others feel that the point of Zinfandel is the young, spicy, forward fruit, the exotic perfume, the exuberance of its style, and when that fades the best of the wine is gone. Few Zinfandels have much tannic structure or complexity, and neither attribute can be magically produced by bottle age if it's not there in the first place.

Paul Draper of Ridge Vineyards maintains that it goes through a dumb phase between losing its primary fruit after eight or ten years, and becoming more interesting after 12 or 15 years. But remember, he is a genius winemaker – and he always includes a proportion of grapes like Petite Sirah to add backbone and muscle.

**The taste of Zinfandel**

Zinophiles probably already know about the Zinfandel Aroma Wheel produced by ZAP; they may also know about the website (www.zinfandel.org) from which the Aroma Wheel can be downloaded, and on which they can add to the fund of knowledge via the 'submit-a-smell' option.

It already includes pretty well all the aromas of which red wine is capable, including faults like TCA (cork taint) and horsiness. Spicy flavours include black pepper, clove, cinnamon and oregano; floral flavours are violets and roses; fruits range from cranberries, strawberries and raspberries (these are found on lighter wines) through to the blackcurrant, black cherry, plums, prunes and raisins of increasing ripeness. Unripe wines taste of green beans, artichokes, green bell peppers, mint and eucalyptus.

Zinfandel can taste nutty or chocolaty, cedary or of green oak – the latter, one assumes, because of winemaker error. The malolactic fermentation gives creamy tastes; oak gives toast, spice and vanilla flavours.

So that's that then. Zinfandel is the protean grape, capable of just about any flavour in the world. Well, no. It's just that these Zin freaks get a bit evangelical if you let them. In fact, you're only likely to find such wealth of flavours in wines from great winemakers like Paul Draper of Ridge and from superb, low-yielding, old vines. Otherwise, Zinfandel is blackberryish, raspberryish, peppery when young, with the flavours of dates, raisins, prunes and herbs never far from the surface. And that's not a bad bunch of flavours.

*Paul Draper (see page 290) is the man behind Ridge Vineyards, and he makes the most complex, long-lived Zins around: they're concentrated, but they're a great deal more than just blockbusters – they're some of the world's greatest red wines. The 1974 vintage of this wine was still fresh and rich – and sensationally good – when I tasted it in 2000. Callaghan Vineyards Zinfandel, from southern Arizona, can spring some surprises, too. In this very hot state Callaghan demonstrates that high-altitude planting (1460m/4800ft), a big difference between day and high temperatures, and low yields, can give powerful, robust but balanced flavours.*

**Matching Zinfandel and food**

California's most versatile grape is used for a bewildering variety of wine styles from bland, slightly sweet pinks to rich, elegant fruity reds. And the good red Zinfandels themselves vary greatly in style. If they aren't too oaky they are good with barbecued meats, venison and roast chicken. The hefty, old-style wines are a great match with the spicy, mouthfilling San Francisco cuisine. The pale blush style of Zin goes well with tomato-based dishes, such as pizza and pasta, as well as with hamburgers.

## CONSUMER INFORMATION

**Synonyms & local names**
It has now been established by DNA fingerprinting that Zinfandel and Primitivo are one and the same variety.

**Best producers**
**USA/California** Alderbrook, Amador Foothill Winery, Beringer, Robert Biale, Blockheadia Ringnosii, Boeger, Burgess Cellars, Cakebread Cellars, Chateau Potelle, Cline Cellars, David Coffaro, De Loach, Deux Amis, Dickerson Vineyard, Dry Creek, Eberle, Edmunds St John, Elyse, Edmeades (Kendall-Jackson), Gary Farrell, Fetzer, Frog's Leap, Gallo Sonoma, Green and Red Vineyard, Hartford Court, Hidden Cellars, Hop Kiln, Kendall-Jackson, Kenwood, Kunde, Limerick Lane, Martinelli, Louis M Martini, Nalle, Newlan, Niebaum-Coppola, Norman Vineyards, Peachy Canyon, Pedroncelli, Preston, Quivira, Rabbit Ridge, A Rafanelli, Ravenswood, Renwood, Ridge, Rosenblum, Saddleback Cellars, St Francis, Sausal, Scherrer, Seghesio, Steele, Storybook Mountain, Sutter Home, Joseph Swan, The Terraces, Topolos, Trentadue, Turley Cellars, Villa Mt Eden, Wellington, Williams Selyem; **Washington State** Sineann; **Arizona** Callaghan Vineyards.
**ITALY** A Mano, Leone De Castris, Felline, Masseria Pepe, Pervini, Pichierri/Vinicola Savese, Sava co-op, Sinfarosa, Torrevento.
**AUSTRALIA** Cape Mentelle, Kangarilla Road, Lenwsood, Nepenthe.
**SOUTH AFRICA** Blaauwklippen, Fairview, Hartenberg.
**CHILE** MontGras.

## RECOMMENDED WINES TO TRY

**Fifteen top California Zinfandel wines**
De Loach *OFS Russian River Valley Zinfandel*
Gary Farrell *Russian River Valley Zinfandel*
Hartford Court *Russian River Valley Fanucchi Wood Vineyard Zinfandel*
Kunde *Sonoma Valley Century Vines Zinfandel*
Martinelli *Russian River Valley Jackass Vineyard Zinfandel*
Niebaum-Coppola *Napa Valley Edizione Pennino*
Peachy Canyon *Paso Robles Dusi Ranch Zinfandel*
A Rafanelli *Dry Creek Valley Unfiltered Zinfandel*
Ravenswood *Sonoma Valley Monte Rosso Vineyard Zinfandel*
Renwood *Shenandoah Valley Grandpère Vineyard Zinfandel*
Ridge *Lytton Springs Dry Creek Valley Zinfandel*
Rosenblum *Napa Valley George Hendry Vineyard Zinfandel*
Saddleback Cellars *Napa Valley Old Vines Zinfandel*
Scherrer *Alexander Valley Old & Mature Vines Zinfandel*
Turley Cellars *Napa Valley Moore Vineyard Zinfandel*

**Ten good-value California Zinfandels**
Amador Foothill Winery *Shenandoah Valley Ferrero Vineyard Zinfandel*
Boeger *El Dorado Zinfandel*
Dry Creek Vineyards *Zinfandel*
Fetzer *Mendocino County Barrel Select Zinfandel*
Gallo Sonoma *Frei Ranch Vineyard Zinfandel*
Kendall-Jackson *Grande Reserve Zinfandel*
Ravenswood *North Coast Vintners Blend Zinfandel*
Sausal *Alexander Valley Zinfandel*
Sutter Home *Reserve Zinfandel*
Villa Mt Eden *Cellar Select Zinfandel*

**Ten southern Italian Primitivo wines**
Leone De Castris *Primitivo di Manduria Santera*
Felline *Primitivo di Manduria*
A Mano *Primitivo di Puglia*
Masseria Pepe *Primitivo di Manduria Dunico*
Pervini *Primitivo di Manduria Archidamo and Primitivo del Tarantino I Monili*
Pichierri/Vinicola Savese *Primitivo di Manduria Mamma Teresa*
Sava co-op *Primitivo di Manduria Terra di Miele*
Sinfarosa *Primitivo di Manduria Zinfandel*
Torrevento *Primitivo del Tarantino I Pastini*

**Five other Zinfandel wines**
Cape Mentelle *Margaret River Zinfandel (Australia)*
Hartenberg *Stellenbosch Zinfandel (South Africa)*
MontGras *Single Vineyard Zinfandel (Chile)*
Nepenthe *Zinfandel (Australia)*
Sineann *Columbia Valley Old Vine Zinfandel (Washington)*

*When Zinfandel grapes are really ripe they look like this: not just black but starting to shrivel. Growers of many other red varieties would take fright at this point, and think their grapes overripe. Zinfandel growers just love them like this. These grapes come from very old vines at Summit Lake Vineyard in the Howell Mountain AVA in Napa.*

**Maturity charts**
Whether you think that old vine Zin actually improves in bottle or merely changes depends on your taste.

**1998** California Zinfandel (full-bodied, old vine)

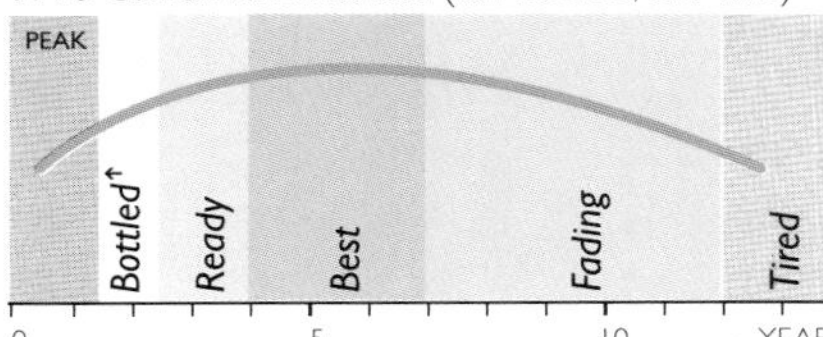


*A better vintage than 1997, producing plump, low-acid wines that will be immediately delicious, and won't dramatically improve though they will last quite well.*

**1999** California Zinfandel (fruity red)

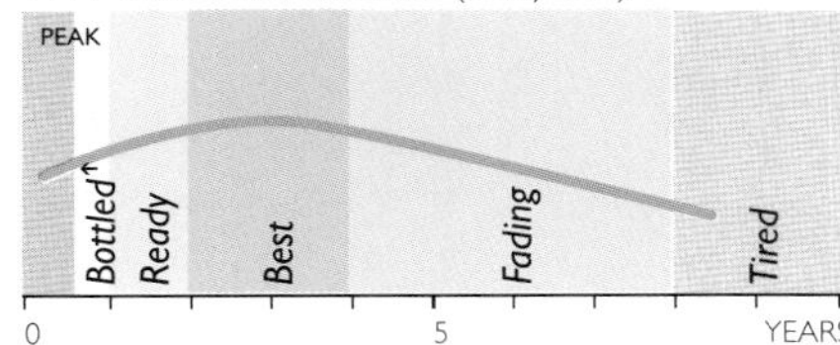


*Simple Zinfandel is best drunk young, while it has all its delicious blackberry flavours. Aging this style in bottle does it no favours.*

**1999** Primitivo di Manduria

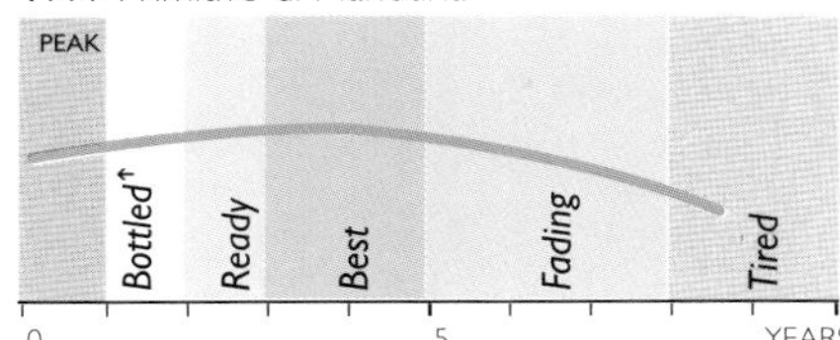


*This DOC from southern Italy makes medium-bodied wines with ripe, spicy black fruit and typical southern warmth and smoothness.*

# KEY TO THE CLASSIC GRAPE PAINTINGS

In each painting the artist Lizzie Riches has combined an accurate rendition of the grape in question with images associated with myths and legends surrounding the grape and its most famous wine styles. In so doing she has taken considerable licence with reality – for example, in the painting of Sauvignon Blanc the gooseberry and the grape should not fruit in the same season, nor do the hills of the Marlborough region in New Zealand's South Island lie directly behind the château of Villandry in the Loire Valley. The exact origin of these classic grape varieties is often conjectural, but the myths that remain behind are more evocative than prosaic fact. Here follows a brief explanation of the images chosen to convey the nature and quality of 17 of the world's classic grape varieties and the wonderful wines they can produce.

**LIZZIE RICHES**

Lizzie Riches, who lives at Norwich in the UK, has been a mainstay of London's Portal Gallery for many years. Largely self-taught, for figurative art was not much in vogue at art school in the 1960s, she has always been drawn to rich detail, exotic subjects and obscure references. Her taste in art is broad, but she has an especial fondness for the glittering surfaces of Veronese and wishes she could paint a feather as well as Bogdani. She was one of the artists selected for the 'Art in the Underground' series and has recently completed a cycle of 16 paintings for P&O. She has had exhibitions in London, Paris and Chicago, in Holland and Germany and is represented in collections worldwide.

**CABERNET SAUVIGNON (PAGE 46)**
Aristocratic and magnificent, Cabernet Sauvignon is represented here by the sunburst, the emblem of France's king Louis XIV, also known as Le Roi Soleil or the Sun King. His brilliant court at the Palace of Versailles was filled with images of Louis' glory. The painting captures Cabernet Sauvignon's self-importance and regal position in the world of wine.

**CHARDONNAY (PAGE 62)**
There seem to be more flavours associated with Chardonnay than with any other grape and it also has a wonderful affinity with new oak barrels. So here carved in the fresh new oak are many of these flavours, including cloves, hazelnuts, warm brioche and a host of different fruits.

**CHENIN BLANC (PAGE 74)**
Floating on air and water, the château of Chenonceau stretches across the river Cher in Touraine, with the river Loire beyond. Chenin Blanc was first planted here in the heart of the Loire Valley in the 15th century and Anjou and Touraine are still where it produces its most exciting wines, whether sweet or dry, sparkling or still.

**GARNACHA TINTA/GRENACHE NOIR (PAGE 92)**
High on the skyline loom the gaunt ruins of the castle built by the popes during their sojourn at Avignon in the southern Rhône, instead of Rome. This was their new castle, their *château neuf* – and the vineyards spread around the castle walls are those of Châteauneuf-du-Pape. Gnarled old Grenache vines grow in a soil covered in large white pebbles or *galets roulés* that retain the heat of the southern sun long into the night. The pebbles make good homes for lizards, too.

**GEWÜRZTRAMINER (PAGE 102)**
The Gothic-style, carved wooden spice cabinet, stork, and buttery, knot-shaped sweet pretzels suggest Alsace where the grape achieves its highest fame. The pomander, cloves, nutmeg and spice mortar are references to the word 'Gewürz' which means spice in German but Gewürztraminer is a complex story. The wines are far more than just being fat and spicy, famous only for their heady perfume.

**MERLOT (PAGE 126)**
Merlot's name is said to be derived from *merle*, French for blackbird, which apparently loves its sweet, early-ripening fruit. Being planted like fury around the world, Merlot has done particularly well in the Napa Valley where the California poppy, as orange as a blackbird's bill, grows wild.

**MUSCAT (PAGE 144)**
The wine-dark sea around the island of Samos, a home of the Muscat grape, lies in the distance. Was this the wine drunk by Dionysus, the Greek god of wine? The ancient Greeks clearly loved wine and all its paraphernalia. Here is a mixing jar, a wine jug and a beautiful *kylix*, or two-handled wine cup. And goodness me, some hooligan has been scrawling his name on the side of the cup. Disgraceful.

**NEBBIOLO (PAGE 154)**
Dusty purple grapes and bright red leaves seen against a background of snow and fog – this is the Piedmontese town of Alba, famous for red wine and white truffles, in autumn. Against a backdrop of snow-capped Alps, the town's medieval towers emerge from the late autumn fogs or *nebbie* that have given the Nebbiolo grape its name. In late autumn truffle hunters and their specially trained dogs set off into the oak forests bordering the vineyards in search of the revered delicacy.

**PINOT NOIR (PAGE 174)**
Seen through an arched Gothic window typical of those found at the Hospices de Beaune in Burgundy, Pinot Noir's homeland, is the château at Gevrey-Chambertin, one of the Côte de Nuits' best known wine villages. The moonlit scene is a tribute to the word 'Nuits'. On the windowsill are various references to Pinot Noir's many facets – the traditional silver Burgundian *tastevin* or tasting cup, a Champagne cork and wire cage, a cone from the Oregon pine and a Knave of Spades to illustrate Pinot Noir's capriciousness both in the vineyard and the wine cellar.

**RIESLING (PAGE 190)**
On the banks of the river Mosel stands the town of Bernkastel directly behind which climb steep Riesling vineyards, just like the ridged green stem of the traditional Mosel wine glass. The intricate ironwork shop signs typical of the Mosel region were surely originally inspired by coiling vine tendrils.

**SANGIOVESE (PAGE 208)**
Sangiovese means literally the 'Blood of Jove'. Nice to think that this Roman god left off seducing mortals and dropping thunderbolts to give his name to this vine. Almost certainly of Tuscan origin, Sangiovese is still central Italy's most important grape. Florence, the heart of Tuscany, lies beyond and Keats's nightingale has strayed from Provence as this grape seems to give more truly 'a beaker full of the warm south'.

**SAUVIGNON BLANC (PAGE 218)**
A piece of the trellis from the *potager* garden at the château of Villandry illustrates the Loire Valley's long association with Sauvignon Blanc. The flavours of Sauvignon Blanc are linked to myriad fruit and vegetables but none more so than gooseberries. Behind the trellis stretch the vineyards and hills of the Marlborough region in New Zealand's South Island, the world's new classic area for Sauvignon Blanc.

**SÉMILLON (PAGE 230)**
Seen here in the honeyed autumnal light evocative of its precious golden wine, Château d'Yquem is the supreme example of the majestic sweet wines of Sauternes. Pickers will go through the vineyards four times or more during the harvest, picking fully botrytized grapes or even green ones, according to the constantly changing instructions from the winery.

**SYRAH/SHIRAZ (PAGE 244)**
Whether it originated in the northern Rhône, its European heartland, in Syracuse in Sicily or was brought back from Shiraz in Persia, there is something exotic about this grape. It's a nice idea but probably untrue that Syrah reached Europe at roughly the same time as the damask rose, and along the same route. In the far distance is the chapel on the Hermitage hill high above the Rhône and in the foreground some of the wild violets that seem to perfume the wine. The parrot is an Adelaide Rosella whose habitat includes the Barossa Valley where Shiraz excels.

**TEMPRANILLO (PAGE 256)**
The history and grandeur of Tempranillo is conveyed by references to two Spanish artists of the 17th century: first, Diego Velázquez, the great Spanish court painter whose paintings of the royal family, including the Infanta Maria Teresa depicted here, were much admired at the time, and second, Juan Sánchez Cotán who is best known for his still lifes with their detailed realism.

**VIOGNIER (PAGE 274)**
If a wine can be described as pretty without insulting it, then Viognier is pretty. It tastes deliciously of the apricots that grow along with Viognier on the banks of the Rhône at Condrieu. There is a theory that the vine arrived by boat, along with Syrah, during the Roman occupation, found a landing place at Condrieu and then remained there as a tiny island of distinctive taste until its recent explosion around the world. Château-Grillet, a tiny estate in Condrieu, can be glimpsed in the picture behind the apricot fruit and blossom.

**ZINFANDEL (PAGE 286)**
Zinfandel, now known to be the same as southern Italy's Primitivo grape, may have come to the eastern seabord of the USA with Italian immigrants. From there it made the journey west to California with the Gold Rush – hence the gold pan, the poster advertising the Gold Fields – the new El Dorado – and the nugget of gold. The popularity of the grape has proved to be a more lasting national treasure than the seams of gold which fast ran out.

# WHICH GRAPES MAKE WHICH WINES

Most European wines are named according to where they come from – the appellation – rather than according to the grape variety or varieties in them. This list of the main European appellations and their authorized grape varieties (with local names in brackets) will help you find the grape you want in the A–Z section (pages 32–295). Grape names listed in *italic* are approved minor varieties that are either seldom used nowadays or allowed only in small percentages, sometimes as experimental. Virtually all the grapes listed here have an entry in the A–Z. A few, however, although they are still listed in the regulations do not in practice appear in the vineyards any more and have, therefore, been omitted from the A–Z for reasons of space.

The list does not include varietal wines which state their main grape on the label and also appellations named after their sole or principal grape, for example Muscat de Beaumes-de-Venise and Vernaccia di San Gimignano. If you cannot find the name you are looking for try the Index of Grape Names and their Synonyms on page 308.

**Grapes allowed for the following wines:**

Red wine
White wine
Rosé wine
Sparkling white wine
Sparkling rosé wine
Sparkling red wine

## AUSTRIA

**Gumpoldskirchner** Rotgipfler, Zierfandler
**Steiermark Schilcher** Blauer Wildbacher

## CYPRUS

**Commandaria** Mavro, Xynisteri

## FRANCE

**Ajaccio** Sciacarello, Barbarossa, Nielluccio, Vermentino, *Carignan, Cinsaut, Grenache Noir*; Vermentino, Ugni Blanc
**Aloxe Corton** Pinot Noir, *Pinot Gris (Pinot Beurot), Pinot Liébault, Pinot Blanc, Chardonnay*; Chardonnay
**Anjou, Anjou Gamay, Anjou Pétillant** Cabernet Franc, Cabernet Sauvignon, Pineau d'Aunis; Cabernet Franc, Cabernet Sauvignon, Pineau d'Aunis, Gamay, Cot, Groslot; Chenin Blanc (Pineau de la Loire), *Chardonnay, Sauvignon Blanc*; **Anjou Coteaux de la Loire** Chenin Blanc (Pineau de la Loire); **Anjou Mousseux** Chenin Blanc (Pineau de la Loire), *Cabernet Sauvignon, Cabernet Franc, Cot, Gamay, Groslot, Pineau d'Aunis*; **Anjou Mousseux** Cabernet Sauvignon, Cabernet Franc, Cot, Gamay, Groslot, Pineau d'Aunis; **Anjou-Villages** Cabernet Franc, Cabernet Sauvignon
**Arbois** Poulsard Noir (Ploussard), Trousseau, Pinot Noir (Gros Noirien, Pinot Gris); Savagnin Blanc (Naturé), Chardonnay (Melon d'Arbois, Gamay Blanc), Pinot Blanc; **Arbois Vin de Paille** Poulsard Noir (Ploussard), Trousseau, Chardonnay (Melon d'Arbois, Gamay Blanc), Savagnin Blanc (Naturé)
**Auxey-Duresses** Pinot Noir, *Pinot Gris (Pinot Beurot), Pinot Liébault, Pinot Blanc, Chardonnay*; Chardonnay, *Pinot Blanc*
**Bandol** Mourvèdre, Grenache Noir, Cinsaut, *Syrah, Carignan*; Mourvèdre, Grenache Noir, Cinsaut, *Syrah, Carignan plus any of the following white varieties*; Bourboulenc, Clairette, Ugni Blanc, *Sauvignon Blanc*
**Banyuls, Banyuls Grand Cru VDN** Grenache Noir, Grenache Gris, Grenache Blanc, Maccabéo, Tourbat (Malvoisie du Roussillon), Muscat Blanc à Petits Grains, Muscat of Alexandria (Muscat Romain), *Carignan Noir, Cinsaut, Syrah*
**Barsac** Sémillon, Sauvignon Blanc and Muscadelle
**Les Baux-de-Provence** Grenache Noir, Syrah, Mourvèdre, *Cinsaut, Counoise, Carignan, Cabernet Sauvignon*; Grenache Noir, Syrah, Cinsaut, *Mourvèdre, Counoise, Carignan, Cabernet Sauvignon*
**Béarn** Tannat, Cabernet Franc (Bouchy), Cabernet Sauvignon, Fer (Pinenc), Manseng Noir, Courbu Noir; Petit Manseng, Gros Manseng, Courbu, Lauzet, Camaralet, Raffiat, Sauvignon Blanc
**Beaujolais, Beaujolais Supérieur, Beaujolais-Villages** Gamay, *Pinot Noir, Pinot Gris, Chardonnay, Aligoté, Melon de Bourgogne*; Chardonnay, Aligoté
**Beaune** Pinot Noir, *Pinot Gris (Pinot Beurot), Pinot Liébault, Pinot Blanc, Chardonnay*; Chardonnay, *Pinot Blanc*
**Bellet** Braquet, Folle Noir (Fuella), Cinsaut, *Grenache Noir plus any of the following white varieties*; Braquet, Folle Noir (Fuella), Cinsaut, *Grenache Noir plus any of the following white varieties except Chardonnay and Muscat Blanc à Petits Grains*; Rolle, Roussanne, Spagnol (Mayorquin), *Clairette, Bourboulenc, Chardonnay, Pignerol, Muscat Blanc à Petits Grains*
**Bergerac, Bergerac Sec** Sémillon, Sauvignon Blanc, Muscadelle, Ondenc, Chenin Blanc, *Ugni Blanc*; **Bergerac, Côtes de Bergerac** Cabernet Sauvignon, Cabernet Franc, Merlot, Malbec (Cot), Fer Servadou, Merille (Périgord)
**Blagny** Pinot Noir, *Pinot Gris (Pinot Beurot), Pinot Liébault, Pinot Blanc, Chardonnay*; Chardonnay, *Pinot Blanc*
**Blanquette de Limoux** Mauzac, Chardonnay, Chenin Blanc; **Blanquette Méthode Ancestrale Mousseux** Mauzac
**Blaye** Cabernet Sauvignon, Cabernet Franc, Merlot, Malbec, *Prelongeau (Bouchalès), Cahors, Béguignol, Petit Verdot*; Ugni Blanc, *Colombard, Sémillon, Sauvignon Blanc, Muscadelle, Chenin Blanc (Pinot de la Loire)*
**Bonnes-Mares** Pinot Noir, *Pinot Liébault, Pinot Blanc, Pinot Gris, Chardonnay*
**Bonnezeaux** Chenin Blanc (Pineau de la Loire)
**Bordeaux, Bordeaux Clairet, Bordeaux Rosé, Bordeaux Supérieur** Cabernet Sauvignon, Cabernet Franc, Merlot, *Carmenère, Malbec, Petit Verdot*; **Bordeaux, Bordeaux Blanc Sec** Sémillon, Sauvignon Blanc, Muscadelle, *Merlot Blanc, Colombard, Mauzac, Ondenc, Ugni Blanc*;
**Bordeaux-Côtes de Francs** Cabernet Franc (Bouchet), Cabernet Sauvignon, Merlot, Malbec (Pressac); Sauvignon Blanc, Sémillon, Muscadelle; **Bordeaux Haut-Benauge** Sémillon, Sauvignon Blanc, Muscadelle;
**Bordeaux Mousseux** Sémillon, Sauvignon Blanc, Muscadelle, Cabernet Sauvignon, Cabernet Franc, Carmenère, Merlot, Malbec, Petit Verdot, *Ugni Blanc, Merlot Blanc, Colombard, Mauzac, Ondenc*; Cabernet Sauvignon, Cabernet Franc, *Carmenère, Merlot, Malbec, Petit Verdot*
**Bourg, Côtes de Bourg, Bourgeais** Cabernet Sauvignon, Cabernet Franc, Merlot, Malbec; Sauvignon Blanc, Sémillon, Muscadelle, Merlot Blanc, Colombard
**Bourgogne, Bourgogne-Hautes Côtes de Beaune, Bourgogne-Hautes Côtes de Nuits, Bourgogne-Côte Chalonnaise** Pinot Noir, *Pinot Gris (Pinot Beurot), Pinot Liébault, Pinot Blanc, Chardonnay*, plus César and Tressot in the Yonne, plus Gamay in Beaujolais; Chardonnay (Beaunois, Aubaine), Pinot Blanc; **Bourgogne Passetoutgrains** Gamay, Pinot Noir, *Pinot Blanc, Pinot Gris, Chardonnay*
**Bourgueil** Cabernet Franc (Breton), *Cabernet Sauvignon*
**Bouzeron** Aligoté
**Brouilly** Gamay, *Chardonnay, Aligoté, Melon de Bourgogne*
**Buzet** Merlot, Cabernet Sauvignon, Cabernet Franc, Malbec (Cot); Sémillon, Sauvignon Blanc, Muscadelle
**Cabardès** Grenache Noir, Syrah, Cabernet Sauvignon, Merlot, Cabernet Franc, Malbec (Cot), Fer, *Cinsaut*
**Cabernet d'Anjou, Cabernet de Saumur** Cabernet Sauvignon, Cabernet Franc
**Cadillac** Sémillon, Sauvignon Blanc, Muscadelle
**Cahors** Malbec (Cot, Auxerrois), Merlot, Tannat
**Canon, Canon-Fronsac** Merlot, Cabernet Franc (Bouchet), Cabernet Sauvignon, Malbec (Pressac)
**Cassis** Grenache Noir, Mourvèdre, Carignan, Cinsaut, Barbaroux, *Terret Noir*; Clairette, Marsanne, *Doucillon, Pascal Blanc, Sauvignon Blanc, Terret, Ugni Blanc*
**Cérons** Sémillon, Sauvignon Blanc, Muscadelle
**Chablis, Chablis Premier Cru, Chablis Grand Cru, Petit Chablis** Chardonnay (Beaunois)
**Chambertin, Chambertin Clos-de-Bèze, Chapelle-Chambertin, Charmes-Chambertin, Griotte-Chambertin, Latricières-Chambertin, Mazis-Chambertin, Mazoyères-Chambertin and Ruchottes-Chambertin** Pinot Noir, *Pinot Gris (Pinot Beurot), Pinot Liébault, Pinot Blanc, Chardonnay*
**Chambolle-Musigny** Pinot Noir, *Pinot Liébault, Pinot Gris (Pinot Beurot), Pinot Blanc, Chardonnay*
**Champagne** Pinot Noir, Pinot Meunier, Chardonnay
**Charlemagne** Chardonnay, *Aligoté*

**Chassagne-Montrachet** 🍷 Pinot Noir, *Pinot Gris (Pinot Beurot), Pinot Liébault, Pinot Blanc, Chardonnay*; 🍷 Chardonnay, *Pinot Blanc*
**Château-Chalon** 🍷 Savagnin Blanc
**Château-Grillet** 🍷 Viognier
**Châteaumeillant** 🍷 Gamay, Pinot Gris, Pinot Noir
**Châteauneuf-du-Pape** 🍷🍷 Grenache Noir, Syrah, Mourvèdre, Picpoul, Terret Noir, Counoise, Muscardin, Vaccarèse, Picardan, Cinsaut, Clairette, Roussanne, Bourboulenc
**Châtillon-en-Diois** 🍷🍷 Gamay, *Syrah, Pinot Noir*; 🍷 Aligoté, Chardonnay
**Chénas** 🍷 Gamay, *Chardonnay, Aligoté, Melon de Bourgogne*
**Cheverny** 🍷 Gamay, Pinot Noir, *Cabernet Franc, Cot*; 🍷 Gamay, Pinot Noir, *Cabernet Franc, Cot, Pineau d'Aunis*; 🍷 Sauvignon Blanc, *Chardonnay, Arbois (Menu Pineau), Chenin Blanc*
**Chinon** 🍷🍷 Cabernet Franc (Breton), *Cabernet Sauvignon*; 🍷 Chenin Blanc (Pineau de la Loire)
**Chiroubles** 🍷 Gamay, *Chardonnay, Aligoté, Melon de Bourgogne*
**Chorey-lès-Beaune** 🍷 Pinot Noir, *Pinot Gris (Pinot Beurot), Pinot Liébault, Pinot Blanc, Chardonnay*; 🍷 Chardonnay, *Pinot Blanc*
**Clairette de Die** 🍷 Muscat Blanc à Petits Grains, Clairette Blanche
**Clos des Lambrays** 🍷 Pinot Noir, *Pinot Gris (Pinot Beurot), Pinot Liébault, Pinot Blanc, Chardonnay*
**Clos de la Roche** 🍷 Pinot Noir, *Pinot Liébault, Pinot Blanc, Pinot Gris, Chardonnay*
**Clos St-Denis** 🍷 Pinot Noir, *Pinot Gris (Pinot Beurot), Pinot Liébault, Pinot Blanc, Chardonnay*
**Clos de Tart** 🍷 Pinot Noir, *Pinot Gris (Pinot Beurot), Pinot Liébault, Pinot Blanc, Chardonnay*
**Clos de Vougeot** 🍷 Pinot Noir, *Pinot Gris (Pinot Beurot), Pinot Liébault, Pinot Blanc, Chardonnay*
**Collioure** 🍷 Grenache Noir, Mourvèdre, Syrah, *Carignan, Cinsaut*; 🍷 Grenache Noir, Mourvèdre, Syrah, *Carignan, Cinsaut, Grenache Gris*
**Condrieu** 🍷 Viognier
**Corbières** 🍷🍷 Carignan, Grenache Noir, Lladoner Pelut, Mourvèdre, Piquepoul Noir, Terret Noir, Syrah, *Cinsaut, Maccabeu, Bourboulenc*; 🍷 Bourboulenc (Malvoisie), Maccabeu, Grenache Blanc, *Clairette Blanche, Muscat Blanc à Petits Grains, Piquepoul Blanc, Terret Blanc, Marsanne, Roussanne, Vermentino Blanc*
**Cornas** 🍷 Syrah
**Corton** 🍷 Pinot Noir, *Pinot Gris (Pinot Beurot), Pinot Liébault, Pinot Blanc, Chardonnay*; 🍷 Chardonnay
**Corton-Charlemagne** 🍷 Chardonnay
**Costières de Nîmes** 🍷🍷 Carignan, Grenache Noir, Mourvèdre, Syrah, Cinsaut; 🍷 Carignan, Grenache Noir, Mourvèdre, Syrah, Cinsaut *plus any of the following white varieties*; 🍷 Clairette Blanche, Grenache Blanc, Bourboulenc Blanc, Ugni Blanc, Roussanne, Rolle, Maccabéo, Marsanne
**Côte de Beaune, Côte de Beaune-Villages** 🍷 Pinot Noir, *Pinot Gris (Pinot Beurot), Pinot Liébault, Pinot Blanc, Chardonnay*; 🍷 Chardonnay, *Pinot Blanc*
**Côte de Brouilly** 🍷 Gamay, *Chardonnay, Aligoté, Melon de Bourgogne, Pinot Noir, Pinot Gris*
**Côte de Nuits-Villages** 🍷 Pinot Noir, *Pinot Liébault, Chardonnay, Pinot Blanc, Pinot Gris*; 🍷 Chardonnay, *Pinot Blanc*
**Côte Roannaise** 🍷🍷 Gamay
**Côte-Rôtie** 🍷 Syrah, *Viognier*
**Coteaux d'Aix-en-Provence** 🍷🍷 Cinsaut, Counoise, Grenache Noir, Mourvèdre, Syrah, *Cabernet Sauvignon, Carignan, plus any of the following white varieties*; 🍷 Bourboulenc, Clairette, Grenache Blanc, Vermentino Blanc, *Ugni Blanc, Sauvignon Blanc, Sémillon*
**Coteaux de l'Aubance** 🍷 Chenin Blanc (Pineau de la Loire)
**Coteaux Champenois** 🍷🍷🍷 Pinot Noir, Pinot Meunier, Chardonnay
**Coteaux de Die** 🍷 Clairette Blanche
**Coteaux du Giennois** 🍷🍷 Gamay, Pinot Noir; 🍷 Sauvignon Blanc
**Coteaux du Languedoc** 🍷 Carignan, Grenache Noir, Lladoner Pelut, Cinsaut, Mourvèdre, Syrah, *Counoise Noir (Aubun), Grenache Rosé, Terret Noir, Piquepoul*; 🍷 Carignan, Grenache Noir, Lladoner Pelut, Cinsaut, Mourvèdre, Syrah, *Counoise Noir (Aubun), Grenache Rosé, Terret Noir, Piquepoul plus any of the following white varieties*; 🍷 Grenache Blanc, Clairette Blanche, Bourboulenc, Piquepoul, Marsanne, Roussanne, Rolle, *Maccabéo, Terret Blanc, Carignan Blanc, Ugni Blanc*
**Coteaux du Layon** 🍷 Chenin Blanc (Pineau de la Loire)
**Coteaux du Loir** 🍷 Chenin Blanc (Pineau de la Loire); 🍷 Pineau d'Aunis, Cabernet Franc, Cabernet Sauvignon, Gamay, Cot; 🍷 Pineau d'Aunis, Cabernet Franc, Cabernet Sauvignon, Gamay, Cot, *Groslot*
**Coteaux du Lyonnais** 🍷🍷 Gamay; 🍷 Chardonnay, Aligoté
**Coteaux de Pierrevert** 🍷 Grenache Noir, Syrah, *Carignan, Cinsaut, Mourvèdre*; 🍷 Grenache Noir, Syrah, Carignan, Cinsaut *plus any of the following white varieties*; 🍷 Grenache Blanc, Vermentino, Ugni Blanc, Clairette, Roussanne, Marsanne, Piquepoul
**Coteaux de Saumur** 🍷 Chenin Blanc (Pineau de la Loire)
**Coteaux du Tricastin** 🍷🍷 Grenache Noir, Cinsaut, Mourvèdre, Syrah, Picpoul Noir, *Carignan plus any of the following white varieties*; 🍷 Grenache Blanc, Clairette Blanche, Picpoul Blanc, Bourboulenc, *Ugni Blanc, Marsanne, Roussanne, Viognier*
**Coteaux Varois** 🍷 Grenache Noir, Syrah, Mourvèdre, *Carignan, Cinsaut, Cabernet Sauvignon*; 🍷 Grenache Noir, Cinsaut, *Syrah, Mourvèdre, Carignan, Tibouren plus any of the following white varieties*; 🍷 Clairette, Grenache Blanc, Rolle, *Sémillon, Ugni Blanc*
**Coteaux du Vendomois** 🍷 Chenin Blanc, Chardonnay; 🍷 Pineau d'Aunis, Gamay, Pinot Noir, Cabernet Franc, Cabernet Sauvignon; 🍷 Pineau d'Aunis, Gamay
**Côtes d'Auvergne** 🍷 Gamay, Pinot Noir; 🍷 Chardonnay
**Côtes de Blaye** 🍷 Colombard, *Sémillon, Sauvignon Blanc, Muscadelle, Merlot Blanc, Folle Blanche, Chenin Blanc (Pinot de la Loire)*
**Côtes de Bordeaux-St-Macaire** 🍷 Sémillon, Sauvignon Blanc, Muscadelle
**Côtes de Bourg, Bourg, Bourgeais** 🍷 Cabernet Franc (Bouchet), Cabernet Sauvignon, Merlot, Malbec (Pressac); 🍷 Sauvignon Blanc, Sémillon, Muscadelle, *Merlot Blanc, Colombard*
**Côtes de Castillon** 🍷 Cabernet Franc (Bouchet), Cabernet Sauvignon, Merlot, Malbec (Pressac)
**Côtes de Duras** 🍷 Sauvignon Blanc, Sémillon, Muscadelle, Mauzac, Chenin Blanc (Rouchelin, Pineau de la Loire), Ondenc, *Ugni Blanc*; 🍷🍷 Cabernet Sauvignon, Cabernet Franc, Merlot, Malbec (Cot)
**Côtes du Forez** 🍷 Gamay
**Côtes du Frontonnais** 🍷🍷 Négrette, *Malbec (Cot), Mérille, Fer, Syrah, Cabernet Franc, Cabernet Sauvignon, Gamay, Cinsaut, Mauzac*
**Côtes de Gien, Coteaux du Giennois** 🍷 Gamay, Pinot Noir; 🍷 Chardonnay
**Côtes du Jura** 🍷 Poulsard Noir (Ploussard), Trousseau, Pinot Noir (Gros Noirien, Pinot Gris); 🍷 Savagnin Blanc (Naturé), Chardonnay (Melon d'Arbois, Gamay Blanc); **Mousseux** 🍷 Poulsard Noir (Ploussard), Trousseau, Chardonnay (Melon d'Arbois, Gamay Blanc), Savagnin Blanc (Naturé)
**Côtes du Jura Vin Jaune** 🍷 Savagnin Blanc
**Côtes du Lubéron** 🍷 Grenache Noir, Syrah, Mourvèdre, *Carignan, Cinsaut, Counoise Noir, Picpoul Noir, Gamay, Pinot Noir*; 🍷 Grenache Noir, Syrah, Mourvèdre, *Carignan, Cinsaut, Counoise Noir, Picpoul Noir, Gamay, Pinot Noir plus any of the following white varieties*; 🍷 Grenache Blanc, Clairette Blanche, Bourboulenc, Ugni Blanc, Vermentino (Rolle), *Roussanne, Marsanne*
**Côtes du Marmandais** 🍷🍷 Cabernet Franc, Cabernet Sauvignon, Merlot, *Abouriou, Merlot (Cot), Fer, Gamay, Syrah*; 🍷 Sauvignon Blanc, *Muscadelle, Ugni Blanc, Sémillon*
**Côtes de Montravel** 🍷 Sémillon, Sauvignon Blanc, Muscadelle
**Côtes de Nuits-Villages** 🍷 Pinot Noir, *Pinot Gris (Pinot Beurot), Pinot Liébault, Chardonnay, Pinot Blanc*; 🍷 Chardonnay, Pinot Blanc
**Côtes de Provence** 🍷🍷 Cinsaut, Grenache Noir, Mourvèdre, Syrah, Tibouren, *Cabernet Sauvignon, Carignan plus any of the following white varieties*; 🍷 Clairette, Sémillon, Ugni Blanc, Vermentino Blanc (Rolle)
**Côtes du Rhône** 🍷🍷 Grenache Noir, Syrah, Mourvèdre, Terret Noir, *Carignan, Cinsaut, Counoise, Muscardin, Camarèse, Vaccarèse, Picpoul Noir, Terret Noir, Grenache Gris, Clairette Rose plus any of the following white varieties*; 🍷 Grenache Blanc, Clairette Blanche, Marsanne, Roussanne, Bourboulenc, Viognier, *Ugni Blanc, Picpoul Blanc*
**Côtes du Rhône-Villages** 🍷 Grenache Noir, Syrah, Mourvèdre, Cinsaut *plus any of the other red varieties for Côtes du Rhône*; 🍷 Grenache Noir, *Camarèse, Cinsaut, Carignan plus any of the other red varieties for Côtes du Rhône*; 🍷 Clairette, Roussanne, Bourboulenc, *Grenache Blanc plus any of the other white varieties for Côtes du Rhône*
**Côtes du Roussillon** 🍷🍷 Carignan, Cinsaut, Grenache Noir, Lladoner Pelut Noir, Syrah, Mourvèdre, *Maccabeu Blanc*; 🍷 Grenache Blanc, Maccabeu Blanc, Tourbat Blanc (Malvoisie du

Roussillon), Marsanne, Roussanne, Vermentino

**Côtes du Roussillon-Villages** Carignan, Grenache Noir, Lladoner Pelut Noir, Syrah, Mourvèdre, *Maccabéo*

**Côtes du Ventoux** Grenache Noir, Syrah, Cinsaut, Mourvèdre, Carignan, *Picpoul Noir, Counoise, Clairette, Bourboulenc, Grenache Blanc, Roussanne*; Clairette, Bourboulenc, Grenache Blanc, *Roussanne*

**Côtes du Vivarais** Grenache Noir, Syrah, Cinsaut, Carignan; Grenache Noir, Cinsaut, Syrah Clairette Blanche, Grenache Blanc, Marsanne

**Cour-Cheverny** Romorantin Blanc

**Crémant d'Alsace** Pinot Blanc, Pinot Noir, Pinot Gris, Riesling, Muscat, Sylvaner, Chasselas; Pinot Noir

**Crémant de Bordeaux** Sémillon, Sauvignon Blanc, Muscadelle, Cabernet Sauvignon, Cabernet Franc, Merlot, Carmenère, Malbec, Petit Verdot, *Colombard, Ugni Blanc*; Cabernet Sauvignon, Cabernet Franc, Merlot, Carmenère, Malbec, Petit Verdot

**Crémant de Bourgogne** Pinot Noir, Pinot Gris, Pinot Blanc, Chardonnay, *Gamay, Aligoté, Melon, Sacy*

**Crémant de Die** Clairette Blanche

**Crémant du Jura** Poulsard (Ploussard), Pinot Noir (Gros Noirien), Pinot Gris, Trousseau, Savagnin (Naturé), Chardonnay (Melon d'Arbois, Gamay Blanc)

**Crémant de Limoux** Mauzac, Chardonnay, Chenin Blanc

**Crémant de Loire** Chenin Blanc, Cabernet Franc, Cabernet Sauvignon, Pineau d'Aunis, Pinot Noir, Chardonnay, Menu Pineau, *Grolleau Noir, Grolleau Gris*

**Crépy** Chasselas Rouge, Chasselas Vert

**Crozes-Hermitage** Syrah, *Marsanne, Roussanne*; Marsanne, Roussanne

**Echézeaux** Pinot Noir, *Pinot Gris (Pinot Beurot), Pinot Liébault, Pinot Blanc, Chardonnay*

**Entre-Deux-Mers, Entre-Deux-Mers Haut-Benauge** Sémillon, Sauvignon Blanc, Muscadelle, *Merlot Blanc, Colombard, Mauzac, Ugni Blanc*

**L'Étoile** , **L'Étoile Mousseux** Chardonnay, Poulsard (Ploussard), Savagnin; **L'Étoile Vin Jaune** Savagnin

**Faugères** Carignan, Cinsaut, Grenache Noir, Lladoner Pelut Noir, *Mourvèdre, Syrah*

**Fitou** Carignan, Grenache Noir, Lladoner Pelut Noir, Mourvèdre, Syrah, *Cinsaut, Maccabéo Blanc, Terret Noir*

**Fixin** Pinot Noir, *Pinot Gris (Pinot Beurot), Pinot Liébault, Pinot Blanc, Chardonnay*; Chardonnay (Beaunois, Aubaine), Pinot Blanc

**Fleurie** Gamay, *Chardonnay, Aligoté, Melon de Bourgogne*

**Fronsac** Merlot, Cabernet Franc (Bouchet), Cabernet Sauvignon, Malbec (Pressac)

**Gaillac** Duras, Fer Servadou, Syrah, *Cabernet Sauvignon, Cabernet Franc, Merlot, Gamay*; **Gaillac, Gaillac Premières Côtes** , **Gaillac Mousseux** Len de l'El, Mauzac, Mauzac Rosé, Muscadelle, Ondenc, Sauvignon Blanc, Sémillon

**Gevrey-Chambertin** Pinot Noir, *Pinot Gris (Pinot Beurot), Pinot Liébault, Pinot Blanc, Chardonnay*

**Gigondas** Grenache Noir, *Syrah, Mourvèdre, plus any of the varieties for Côtes du Rhône, except Carignan*; Grenache Noir *plus any of the varieties for Côtes du Rhône, except Carignan*

**Givry** Pinot Noir, *Pinot Gris (Pinot Beurot), Pinot Liébault, Chardonnay*; Chardonnay, *Pinot Blanc*

**La Grande Rue** Pinot Noir, *Pinot Blanc, Pinot Gris, Chardonnay*

**Grands Echézeaux** Pinot Noir, *Pinot Gris (Pinot Beurot), Pinot Liébault, Pinot Blanc, Chardonnay*

**Graves** Cabernet Sauvignon, Cabernet Franc, Merlot, *Malbec, Petit Verdot*; **Graves, Graves Supérieures** Sauvignon Blanc, Sémillon, Muscadelle; **Graves de Vayres** Cabernet Sauvignon, Cabernet Franc, Merlot, *Carmenère, Malbec, Petit Verdot*; Sémillon, Sauvignon Blanc, Muscadelle, *Merlot Blanc*

**Haut-Médoc** Cabernet Sauvignon, Cabernet Franc, Merlot, *Carmenère, Malbec, Petit Verdot*

**Haut-Montravel** Sémillon, Sauvignon Blanc, Muscadelle

**Hermitage** Syrah, *Marsanne, Roussanne*; Marsanne, Roussanne

**Irancy** Pinot Noir, César

**Irouléguy** Cabernet Sauvignon, Cabernet Franc, Tannat; Courbu, Manseng

**Jasnières** Chenin Blanc (Pineau de la Loire)

**Juliénas** Gamay, *Chardonnay, Aligoté, Melon de Bourgogne*

**Jurançon, Jurançon Sec** Petit Manseng, Gros Manseng, Courbu, *Camaralet, Lauzet*

**Ladoix** Pinot Noir, *Pinot Gris (Pinot Beurot), Pinot Liébault, Pinot Blanc, Chardonnay*; Chardonnay, *Pinot Blanc*

**Lalande-de-Pomerol** Merlot, Cabernet Franc (Bouchet), Cabernet Sauvignon, *Malbec (Pressac)*

**Limoux** Mauzac, Chardonnay, Chenin Blanc

**Lirac** Grenache Noir, Syrah, Mourvèdre, Syrah, Cinsaut, *Carignan*; Grenache Noir, Syrah, Mourvèdre, Syrah, Cinsaut, *Carignan plus any of the following white varieties*; Clairette Blanche, Grenache Blanc, Bourboulenc, *Ugni Blanc, Picpoul, Marsanne, Roussanne, Viognier*

**Listrac-Médoc** Cabernet Sauvignon, Cabernet Franc, Merlot, *Carmenère, Malbec, Petit Verdot*

**Loupiac** Sémillon, Sauvignon Blanc, Muscadelle

**Lussac-St-Émilion** Cabernet Franc (Bouchet), Cabernet Sauvignon, Merlot, *Malbec (Pressac)*

**Mâcon, Mâcon Supérieur, Mâcon-Villages or Mâcon followed by a commune name e.g. Mâcon-Lugny** Chardonnay, Pinot Blanc; **Mâcon, Mâcon Supérieur or Mâcon followed by a commune name e.g. Mâcon-Lugny** Gamay, Pinot Noir, Pinot Gris, *Chardonnay, Aligoté, Gamay Blanc (Melon)*

**Madiran** Tannat, Cabernet Sauvignon, Cabernet Franc (Bouchy), Fer (Pinenc)

**Maranges** Pinot Noir, *Pinot Gris (Pinot Beurot), Pinot Liébault, Pinot Blanc, Chardonnay*; Chardonnay

**Marcillac** Fer Servadou, *Cabernet Sauvignon, Cabernet Franc, Merlot*

**Margaux** Cabernet Sauvignon, Cabernet Franc, Merlot, *Carmenère, Malbec, Petit Verdot*

**Marsannay** Pinot Noir, *Pinot Gris (Beurot), Chardonnay, Pinot Blanc*; Chardonnay, *Pinot Blanc*

**Maury VDN** Grenache Noir, *Grenache Gris, Grenache Blanc, Maccabeu, Carignan Noir, Syrah*; Grenache Blanc, Grenache Gris, Maccabeu, Tourbat (Malvoisie du Roussillon), *Muscat of Alexandria, Muscat Blanc à Petits Grains*

**Médoc** Cabernet Sauvignon, Cabernet Franc, Merlot, *Carmenère, Petit Verdot, Malbec*

**Menetou-Salon** Sauvignon Blanc; Pinot Noir

**Mercurey** Pinot Noir, *Pinot Gris (Pinot Beurot), Pinot Liébault, Chardonnay*; Chardonnay

**Meursault** Pinot Noir, *Pinot Gris (Pinot Beurot), Pinot Liébault, Pinot Blanc, Chardonnay*; Chardonnay, *Pinot Blanc*

**Minervois** Grenache Noir, Syrah, Mourvèdre, Lladoner Pelut Noir, Carignan, *Cinsaut, Picpoul Noir, Terret Noir, Aspiran Noir*; Grenache Noir, Syrah, Mourvèdre, Lladoner Pelut Noir, Carignan, *Cinsaut, Picpoul Noir, Terret Noir, Aspiran Noir plus any of the following white varieties*; Grenache Blanc, Bourboulenc (Malvoisie), Maccabeu Blanc, Marsanne, Roussanne, Vermentino (Rolle), *Picpoul Blanc, Clairette Blanche, Terret Blanc, Muscat Blancs à Petits Grains*

**Monbazillac** Sémillon, Sauvignon Blanc, Muscadelle

**Montagne-St-Émilion** Cabernet Franc (Bouchet), Cabernet Sauvignon, Merlot, *Malbec (Pressac)*

**Montagny** Chardonnay (Beaunois, Aubaine)

**Monthélie** Pinot Noir, *Pinot Gris (Pinot Beurot), Pinot Liébault, Pinot Blanc, Chardonnay*; Chardonnay, *Pinot Blanc*

**Montlouis** Chenin Blanc (Pineau de la Loire)

**Montrachet, Bâtard-Montrachet, Bienvenues-Bâtard-Montrachet, Chevalier-Montrachet, Criots-Bâtard-Montrachet** Chardonnay (Beaunois, Aubaine)

**Montravel** Sémillon, Sauvignon Blanc, Muscadelle, Ondenc, Chenin Blanc, *Ugni Blanc*

**Morey St-Denis** Pinot Noir, *Pinot Gris (Pinot Beurot), Pinot Liébault, Pinot Blanc, Chardonnay*; Chardonnay (Beaunois, Aubaine), *Pinot Blanc*

**Morgon** Gamay, *Chardonnay, Aligoté, Melon de Bourgogne*

**Moulin-à-Vent** Gamay, *Chardonnay, Aligoté, Melon de Bourgogne*

**Moulis, Moulis-en-Médoc** Cabernet Sauvignon, Cabernet Franc, Merlot, *Carmenère, Malbec, Petit Verdot*

**Muscadet, Muscadet-Côtes de Grandlieu, Muscadet-Coteaux de la Loire, Muscadet-Sèvre et Maine** Melon de Bourgogne

**Musigny** Pinot Noir, *Pinot Gris (Pinot Beurot), Pinot Liébault, Pinot Blanc, Chardonnay*; Chardonnay

**Nuits-St-Georges** Pinot Noir, *Pinot Gris (Pinot Beurot), Pinot Liébault, Pinot Blanc, Chardonnay*; Chardonnay, *Pinot Blanc*

**Pacherenc du Vic Bilh** Gros Manseng,

Arrufiac, Courbu, Petit Manseng, *Sauvignon Blanc, Sémillon*
**Palette** Mourvèdre, Grenache Noir, Cinsaut (Plant d'Arles), *Manosquin (Téoulier), Durif, Muscat Noir, Carignan, Syrah, Castets, Brun-Fourcat, Terret Gris, Petit-Brun, Tibourenc, Cabernet Sauvignon plus any of the following white varieties*; Clairette (various forms), *Ugni Blanc, Ugni Rosé, Grenache Blanc, Muscat, Pascal, Terret-Bourret, Piquepoul, Aragnan, Colombard, Tokay*
**Patrimonio** Nielluccio, *Grenache Noir, Sciacarello, Vermentino*; Vermentino, *Ugni Blanc*
**Pauillac** Cabernet Sauvignon, Cabernet Franc, Merlot, *Carmenère, Malbec, Petit Verdot*
**Pécharmant** Cabernet Sauvignon, Cabernet Franc, Merlot, Malbec (Cot)
**Pernand-Vergelesses** Pinot Noir, *Pinot Gris (Pinot Beurot), Pinot Liébault, Pinot Blanc, Chardonnay*; Chardonnay, *Pinot Blanc*
**Pessac-Léognan** Cabernet Sauvignon, Cabernet Franc, Merlot, *Carmenère, Malbec (Cot), Petit Verdot*; Sauvignon Blanc, Sémillon, Muscadelle
**Petit Chablis** Chardonnay (Beaunois)
**Pomerol** Merlot, Cabernet Franc (Bouchet), Cabernet Sauvignon, *Malbec (Pressac)*
**Pommard** Pinot Noir, *Pinot Gris (Pinot Beurot) Pinot Liébault, Chardonnay, Pinot Blanc*
**Pouilly-Fuissé**, **Pouilly-Vinzelles**, **Pouilly-Loché** Chardonnay (Beaunois)
**Pouilly-Fumé**, **Blanc Fumé de Pouilly** Sauvignon Blanc (Blanc Fumé)
**Pouilly-sur-Loire** Chasselas, Sauvignon Blanc (Blanc Fumé)
**Premières Côtes de Blaye** Cabernet Sauvignon, Cabernet Franc, Merlot, *Malbec (Cot)*; Sémillon, Sauvignon Blanc, Muscadelle, *Merlot Blanc, Colombard, Ugni Blanc*
**Premières Côtes de Bordeaux** Cabernet Sauvignon, Cabernet Franc, Merlot, *Carmenère, Malbec (Cot), Petit Verdot*; Sauvignon Blanc, Sémillon, Muscadelle
**Puisseguin-St-Émilion** Cabernet Franc (Bouchet), Cabernet Sauvignon, Merlot, *Malbec (Pressac)*
**Puligny-Montrachet** Pinot Noir, *Pinot Gris (Pinot Beurot), Pinot Liébault, Pinot Blanc, Chardonnay*; Chardonnay, *Pinot Blanc*
**Quarts de Chaume** Chenin Blanc (Pineau de la Loire)
**Quincy** Sauvignon Blanc
**Rasteau VDN** Grenache Noir, Grenache Gris, Grenache Blanc
**Régnié** Gamay
**Reuilly** Sauvignon Blanc; Pinot Noir, Pinot Gris
**Richebourg** Pinot Noir, *Pinot Gris (Pinot Beurot), Pinot Liébault, Pinot Blanc, Chardonnay*
**Rivesaltes VDN** Grenache Noir, Grenache Gris, Grenache Blanc, Maccabeo, Tourbat (Malvoisie du Roussillon), Carignan, Cinsaut, Syrah, Listan; Grenache Gris, Grenache Blanc, Maccabéo, Tourbat (Malvoisie du Roussillon), *Muscat Blanc à Petits Grains, Muscat of Alexandria (Muscat Romain)*
**La Romanée**, **Romanée-Conti**, **Romanée-St-Vivant** Pinot Noir, *Pinot Gris (Pinot Beurot) Pinot Liébault, Chardonnay, Pinot Blanc*
**Rosé d'Anjou** Cabernet Franc, Cabernet Sauvignon, Pineau d'Aunis, Gamay, Cot, Groslot
**Rosé de Loire** Cabernet Franc, Cabernet Sauvignon, Pineau d'Aunis, Pinot Noir, Gamay, Grolleau
**Rosé des Riceys** Pinot Noir
**Rosette** Sémillon, Sauvignon Blanc, Muscadelle
**Roussette de Savoie** Altesse
**Rully** Pinot Noir, *Pinot Gris (Pinot Beurot), Pinot Liébault, Chardonnay*; Chardonnay
**St-Amour** Gamay, *Chardonnay, Aligoté, Melon de Bourgogne*
**St-Aubin** Pinot Noir, *Pinot Gris (Pinot Beurot), Pinot Liébault, Pinot Blanc, Chardonnay*; Chardonnay, *Pinot Blanc*
**St-Chinian** Grenache Noir, Lladoner Pelut Noir, Mourvèdre, Syrah, Carignan, Cinsaut
**St-Émilion**, **St-Émilion Grand Cru** Merlot, Cabernet Franc (Bouchet), Cabernet Sauvignon, *Carmenère, Malbec (Pressac, Cot)*
**St-Estèphe** Cabernet Sauvignon, Cabernet Franc, Merlot, *Carmenère, Malbec, Petit Verdot*
**St-Georges-St-Émilion** Cabernet Franc (Bouchet), Cabernet Sauvignon, Merlot, *Malbec (Pressac)*
**St-Joseph** Syrah, *Marsanne, Roussanne*; Marsanne, Roussanne
**St-Julien** Cabernet Sauvignon, Cabernet Franc, Merlot, *Carmenère, Malbec, Petit Verdot*
**St Nicolas-de-Bourgueil** Cabernet Franc (Breton), Cabernet Sauvignon
**St-Péray** Roussanne (Roussette), Marsanne
**St-Pourçain** Gamay, Pinot Noir, Gamay Teinturiers; Tressallier, St-Pierre-Doré, Aligoté, Chardonnay, Sauvignon Blanc
**St-Romain** Pinot Noir, *Pinot Gris (Pinot Beurot), Pinot Liébault, Pinot Blanc, Chardonnay*; Chardonnay, *Pinot Blanc*
**St-Véran** Chardonnay
**Ste-Croix-du-Mont** Sémillon, Sauvignon Blanc, Muscadelle
**Ste-Foy-Bordeaux** Cabernet Sauvignon, Cabernet Franc, Merlot, *Malbec, Petit Verdot*; Sémillon, Sauvignon Blanc, Muscadelle, *Merlot Blanc, Colombard, Mauzac, Ugni Blanc*
**Sancerre** Sauvignon Blanc; Pinot Noir
**Santenay** Pinot Noir, *Pinot Gris (Pinot Beurot), Pinot Liébault, Pinot Blanc, Chardonnay*; Chardonnay, *Pinot Blanc*
**Saumur Blanc** Chenin Blanc, Chardonnay, Sauvignon Blanc; **Saumur-Champigny** Cabernet Franc, Cabernet Sauvignon, Pineau d'Aunis; **Saumur Mousseux** Chenin Blanc, Cabernet Franc, Cabernet Sauvignon, Cot, Gamay, Grolleau, Pineau d'Aunis, Pinot Noir, *Chardonnay, Sauvignon Blanc*; **Saumur Mousseux** Cabernet Franc, Cabernet Sauvignon, Cot, Gamay, Grolleau, Pineau d'Aunis, Pinot Noir; **Saumur Rouge** Cabernet Franc, Cabernet Sauvignon, Pineau d'Aunis
**Saussignac** Sémillon, Sauvignon Blanc, Muscadelle, Chenin Blanc
**Sauternes** Sémillon, Sauvignon Blanc, Muscadelle
**Savennières** Chenin Blanc (Pineau de la Loire)
**Savigny-lès-Beaune** Pinot Noir, *Pinot Gris (Pinot Beurot), Pinot Liébault, Pinot Blanc, Chardonnay*; Chardonnay, *Pinot Blanc*
**Seyssel** Roussette; Molette, Chasselas (Bon Blanc, Fendant), Roussette
**La Tâche** Pinot Noir, *Pinot Gris (Pinot Beurot), Pinot Liébault, Pinot Blanc, Chardonnay*
**Tavel** Grenache Noir, Cinsaut, Clairette Blanche, Clairette Rose, Picpoul, Calitor, Bourboulenc, Mourvèdre, Syrah, *Carignan*
**Touraine** Chenin Blanc (Pineau de la Loire), Arbois (Menu Pineau), Sauvignon Blanc, *Chardonnay*; Cabernet Franc (Breton), Cabernet Sauvignon, Cot, Pinot Noir, Pinot Meunier, Pinot Gris, Gamay, Pineau d'Aunis; Cabernet Franc (Breton), Cabernet Sauvignon, Cot, Pinot Noir, Pinot Meunier, Pinot Gris, Gamay, Pineau d'Aunis, Grolleau; Chenin Blanc (Pineau de la Loire), Arbois (Menu Pineau), *Chardonnay, Cabernet Franc (Breton), Cabernet Sauvignon, Pinot Noir, Pinot Gris, Pinot Meunier, Pineau d'Aunis, Cot, Grolleau*; Cabernet Franc (Breton), Cot, Noble, Gamay, Grolleau; Cabernet Franc (Breton)
**Vacqueyras** Grenache Noir, Syrah, Mourvèdre *plus any of the varieties for Côtes du Rhône, except Carignan*; Grenache Noir, Mourvèdre, Cinsaut *plus any of the varieties for Côtes du Rhône, except Carignan*; Grenache Blanc, Clairette Blanche, Bourboulenc, *Marsanne, Roussanne, Viognier*
**Vin de Corse** Nielluccio, Sciacarello, Grenache Noir, *Cinsaut, Mourvèdre, Barbarossa, Syrah, Carignan, Vermentino (Malvoisie de Corse)*; Vermentino (Malvoisie de Corse), *Ugni Blanc (Rossola)*
**Vin de Savoie** Gamay, Mondeuse, Pinot Noir, *Persan, Cabernet Sauvignon, Cabernet Franc, Étraire de la Dui, Servanin, Joubertin*; Aligoté, Altesse, Jacquère, Chardonnay, Velteliner Rouge Précoce, Mondeuse Blanche, *Gringet, Roussette d'Ayze, Chasselas, Marsanne, Verdesse*; the above white grapes plus Gamay, Pinot Noir, Mondeuse, Molette
**Viré-Clessé** Chardonnay
**Volnay**, **Volnay-Santenots** Pinot Noir, *Pinot Gris (Pinot Beurot), Pinot Liébault, Pinot Blanc, Chardonnay*
**Vosne-Romanée** Pinot Noir, *Pinot Gris (Pinot Beurot), Pinot Liébault, Pinot Blanc, Chardonnay*
**Vougeot** Pinot Noir, *Pinot Gris (Pinot Beurot), Pinot Liébault, Pinot Blanc, Chardonnay*; Chardonnay (Beaunois. Aubaine), *Pinot Blanc*
**Vouvray** Chenin Blanc (Pineau de la Loire, Gros Pinot), Petit Pinot (Menu Pinot)

## GERMANY

**Liebfraumilch** Müller-Thurgau, Silvaner, Kerner, and/or most unusually Riesling must constitute 51 per cent.

## GREECE

**Amindeo** Xynomavro
**Anhialos** Roditis, Savatiano
**Archanes** Kotsifali, Mandelaria
**Côtes de Meliton** Limnio, Cabernet Sauvignon, Cabernet Franc; Athiri, Roditis, Assyrtico
**Dafnes** Liatiko
**Goumenissa** Xynomavro, Negoska
**Limnos** Muscat of Alexandria

**Mantinia** Moscophilero, Asproudos
**Mavrodaphne of Cephalonia** Mavrodaphne
**Mavrodaphne of Patras** Mavrodaphne
**Messenikola** Messenikola, Carignan, Syrah
**Muscat of Cephalonia** Muscat Blanc à Petits Grains
**Muscat of Limnos** Muscat of Alexandria
**Muscat of Patros** Muscat Blanc à Petits Grains
**Muscat of Rhodes** Muscat Blanc à Petits Grains, Muscat Trani
**Muscat of Rio of Patros** Muscat Blanc à Petits Grains
**Naoussa** Xynomavro
**Nemea** Aghiorghitiko
**Paros** Monemvassia Malvaria, Mandelaria
**Patras** Roditis
**Peza** Kotsifali, Mandelaria; Vilana
**Rapsani** Xynomavro, Krassato, Stavroto
**Rhodes** Mandelaria; Athiri
**Samos** Muscat Blanc à Petits Grains
**Santorini** Assyrtiko, Aidini
**Sitia** Liatiko
**Zitsa** Debina

## HUNGARY

**Bulls Blood of Eger, Egri Bikaver** Kékfrankos, Merlot, Kadarka
**Tokaji Aszú** Furmint, Hárslevelü, Muscat Lunel

## ITALY

**Alcamo** Catarratto, Ansonica, Inzolia, Grillo, Grecanino, Chardonnay, Müller-Thurgau, Sauvignon Blanc *and others*; **Alcamo Classico** Catarratto *and others*; Nerello Mascalese, Calabrese, Nero d'Avola, Sangiovese, Frappato, Perricone, Cabernet Sauvignon, Merlot, Syrah; Calabrese, Nero d'Avola, Frappato, Sangiovese, Perricone, Cabernet Sauvignon, Merlot, Syrah *and others*
**Aleatico di Puglia** Aleatico, *Negroamaro, Malvasia Nera, Primitivo*
**Alto Adige** mostly varietal wines; Pinot Bianco, Pinot Nero, Chardonnay
**Asti** Moscato Bianco
**Barbaresco** Nebbiolo
**Barbera d'Alba** Barbera; **Barbera d'Asti, Barbera del Monferrato** Barbera, *Freisa, Grignolino, Dolcetto*
**Bardolino** Corvina, Rondinella, Molinara, Negrara, *Rossignola, Barbera, Sangiovese, Garganega*
**Barolo** Nebbiolo
**Bianco di Custoza** Trebbiano Toscano (Castelli Romani), Garganega, Tocai Friulano, Cortese (Bianca Fernanda), Malvasia Toscana, Riesling Italico, Pinot Bianco, Chardonnay
**Bolgheri** Trebbiano Toscano, Vermentino, Sauvignon Blanc *and others*; Cabernet Sauvignon, Merlot, Sangiovese *and others*;
**Bolgheri Sassicaia** Cabernet Sauvignon, *Cabernet Franc and others*
**Breganze** Tocai Friulano, *Pinot Bianco, Pinot Grigio, Riesling Italico, Sauvignon Blanc, Vespaiolo*; Merlot, *Groppello Gentile, Cabernet Franc, Cabernet Sauvignon, Pinot Nero, Freisa*
**Brunello di Montalcino** Sangiovese
**Caldaro (Kalterer)/Lago di Caldaro (Kalterersee)** Schiava, *Pinot Nero, Lagrein*
**Cannonau di Sardegna** Garnacha Tinta (Cannonau)
**Carema** Nebbiolo *and others*
**Carmignano, Barco Reale di Carmignano** Sangiovese, Canaiolo Nero, Cabernet Franc, Cabernet Sauvignon, *Trebbiano Toscano, Canaiolo Bianco, Malvasia del Chianti and others*
**Castel del Monte** Pampanuto, Chardonnay, Bombino Bianco *and others*; Uva di Troia, Aglianico, Montepulciano *and others*; Bombino Nero, Aglianico, Uva di Troia *and others*
**Chianti, Chianti Classico** Sangiovese, *Canaiolo Nero, Trebbiano Toscano, Malvasia del Chianti, Cabernet Sauvignon, Merlot and others*
**Cirò** Gaglioppo, *Trebbiano Toscano, Greco Bianco*; Greco Bianco, *Trebbiano Toscano*
**Colli Berici** mostly varietal wines; Pinot Bianco, Pinot Grigio, Chardonnay, Sauvignon Blanc
**Colli Bolognesi** mostly varietal wines; **Colli Bolognesi Bianco** Albana, Trebbiano Romagnolo
**Colli Euganei** Garganega, Prosecco, Tocai Friulano, Sauvignon Blanc, *Pinella, Pinot Bianco, Riesling Italico, Chardonnay*; Merlot, Cabernet Franc, Cabernet Sauvignon, Barbera, Raboso Veronese
**Colli Piacentini Gutturnio** Barbera, Croatina (Bonarda)
**Est! Est!! Est!!! di Montefiascone** Trebbiano Toscano (Procanico), Malvasia Bianca Toscana, *Rossetto (Trebbiano Giallo)*
**Falerno del Massico** Falanghina; Aglianico, Piedirosso, Primitivo, Barbera
**Fiano di Avellino** Fiano, *Greco, Coda di Volpe Bianca, Trebbiano Toscano*
**Franciacorta** Pinot Bianco, Chardonnay, Pinot Nero
**Frascati** Malvasia Bianca di Candia, Trebbiano Toscano, Greco, Malvasia di Lazio *and others*
**Garda** Riesling Italico, Riesling Renano; Groppello, Sangiovese, Marzemino, Barbera
**Gattinara** Nebbiolo (Spanna), *Vespolina, Bonarda di Novarese (Uva Rara)*
**Gavi, Cortese di Gavi** Cortese
**Ghemme** Nebbiolo (Spanna), Vespolino, Bonarda Novarese (Uva Rara)
**Greco di Tufo** Greco, *Coda di Volpe Bianca*
**Lambrusco di Sorbara** Lambrusco di Sorbara, Lambrusco Salamino
**Locorontondo** Verdeca, Bianco d'Alessano, *Fiano, Bombino, Malvasia Toscano*
**Lugana** Trebbiano di Soave (Trebbiano di Lugana) *and others*
**Malvasia delle Lipari** Malvasia di Lipari, Corinto Nero
**Marino** Malvasia Bianca di Candia (Malvasia Rossa), Trebbiano Toscano, Trebbiano Romagnolo, Trebbiano Giallo, Trebbiano di Soave, Malvasia del Lazio (Malvasia Puntinata) *and others*
**Marsala** Grillo, Catarratto, Pignatello, Calabrese, Nerello Mascalese, Inzolia, Nero d'Avola, Damaschino
**Montecarlo** Trebbiano Toscano, Sémillon, Pinot Grigio, Pinot Bianco, Vermentino, Sauvignon Blanc, Roussanne; Sangiovese, Canaiolo Nero, Ciliegiolo, Colorino, Malvasia Nera, Syrah, Cabernet Franc, Cabernet Sauvignon, Merlot *and others*
**Montefalco** Sangiovese, Sagrantino *and others*; Grechetto, Trebbiano Toscano *and others*
**Montepulciano d'Abruzzo** Montepulciano *and others*
**Montescudaio** Trebbiano Toscano, Malvasia del Chianti, Vermentino *and others*; Sangiovese, Trebbiano Toscano, Malvasia del Chianti *and others*
**Morellino di Scansano** Sangiovese, *Canaiolo, Malvasia Nera*
**Moscato d'Asti** Moscato Bianco
**Moscato di Pantelleria** Muscat of Alexandria (Zibibbo)
**Oltrepò Pavese** Barbera, Croatina, Uva Rara, Ughetta (Vespolina), Pinot Nero
**Orvieto** Trebbiano Toscano (Procanico), Verdello, Grechetto, Canaiolo Bianco (Drupeggio), Malvasia Toscana
**Piemonte** varietal wines; Chardonnay, Pinot Bianco, Pinot Grigio, Pinot Nero
**Pomino** Sangiovese, Canaiolo, Cabernet Sauvignon, Cabernet Franc, Merlot *and others*; Pinot Bianco, Chardonnay, Trebbiano Toscano *and others*
**Prosecco di Conegliano-Valdobbiadene/ Prosecco di Conegliano/Prosecco de Valdobbiadene** Prosecco, *Verdiso*
**Riviera del Garda** Gropello, Sangiovese, Barbera, Berzamino (Marzemino)
**Riviera del Garda Bresciano** Groppello, Sangiovese, Barbera, Marzemino (Berzemino), Barbera *and others*; Riesling Italico, Riesling Renano *and others*
**Riviera Ligure di Ponente Ormeasco** Dolcetto *and others*
**Roero** Nebbiolo, *Arneis and others*; Arneis
**Rosso Conero** Montepulciano, Sangiovese
**Rosso di Montalcino** Sangiovese
**Rosso di Montepulciano** Sangiovese (Prugnolo Gentile), *Canaiolo Nero, Mammolo, Cabernet Sauvignon, Merlot, Syrah*
**Rosso Piceno** Montepulciano, Sangiovese *and others*
**Salice Salentino** Negroamaro, *Malvasia Nera*; *Chardonnay*
**San Gimignano** Sangiovese *and others*; Sangiovese, Canaiolo Nero, *Trebbiano Toscano, Malvasia del Chianti, Vernaccia di San Gimignano and others*
**Sangiovese di Romagna** Sangiovese
**Soave** Garganega, Pinot Bianco, Chardonnay, Trebbiano di Soave (Nostrano) *and other Trebbiano varieties*
**Taurasi** Aglianico *and others*
**Teroldego Rotaliano** Teroldego
**Terre di Franciacorta** Chardonnay, Pinot Bianco, Pinot Nero; Cabernet Sauvignon, Cabernet Franc, *Barbera, Nebbiolo, Merlot and others*
**Torcolato** Vespaiolo
**Torgiano** Trebbiano Toscano, Grechetto *and others* ; Sangiovese, Canaiolo, Trebbiano

Toscano *and others*; Chardonnay, Pinot Nero; **Torgiano Rosso Riserva** Sangiovese, Canaiolo, *Trebbiano Toscano, Ciliegiolo, Montepulciano*
**Trebbiano d'Abruzzo** Trebbiano d'Abruzzo (Bombino Bianco), Trebbiano Toscano *and others*
**Trebbiano di Romagna** Trebbiano Romagnolo *and others*
**Trentino** mostly varietal wines; Chardonnay, Pinot Bianco, Sauvignon Blanc, Müller-Thurgau, Incrocio Manzoni; Cabernet Franc, Cabernet Sauvignon, Merlot; Enantio, Schiava, Teroldego, Lagrein
**Trento** Chardonnay, Pinot Bianco, Pinot Nero, Pinot Meunier
**Val di Cornia** Trebbiano Toscano, Vermentino *and others*; Sangiovese, Cabernet Sauvignon, Merlot *and others*
**Valdadige** Pinot Bianco, Pinot Grigio, Riesling Italico, Müller-Thurgau, Chardonnay, *Bianchetta Trevigiana, Trebbiano Toscano, Nosiola, Vernaccia, Garganega*; Schiava, Lambrusco, *Merlot, Pinot Nero, Lagrein, Teroldego, Negrara*
**Valle d'Aosta Chambave** Petit Rouge, Dolcetto, Gamay, Pinot Nero *and others*; **Valle d'Aosta Donnas** Nebbiolo (Picoutener), *Freisa, Neyret*
**Valpolicella, Amarone, Recioto** Corvina Veronese, Corvinone, Rondinella, Molinara, *Croatina, Dindarella, Oseletta, Rossignola, Negrara Trentina, Terodola, Cabernet Sauvignon, Cabernet Franc, Merlot*
**Valtellina** Nebbiolo (Chiavennasca), *Pinot Nero, Merlot, Rossola, Pignola Valtellinese*
**Vino Nobile di Montepulciano** Sangiovese (Prugnolo Gentile), *Canaiolo Nero, Mammolo, Cabernet Sauvignon, Merlot, Syrah*
**Vin Santo Occhio di Pernice** Sangiovese

## PORTUGAL

**Alenquer** Camarate, Trincadeira, Periquita, Preto Martinho, Tinta Miúda; Vital, Jampal, Arinto, Fernão Pires
**Bairrada** Baga, Camarate, Castelão Nacional, Bastardo, Jaén, Alfrocheiro Preto, Touriga Nacional, Trincadeira Preta, Rufete; Arinto, Bical, Cercial Branco, Rabo de Ovelha, Fernão Pires
**Bucelas** Arinto, Esgana Cão, Rabo de Ovelha
**Colares** Ramisco; Malvasia
**Dão** Alfrocheiro Preto, Bastardo, Jaén, Tinta Amarela, Tinta Roriz, Touriga Nacional, Cabernet Sauvignon; Encruzado, Assario Branco, Borrado das Moscas, Cerceal, Verdelho, Arinto, Rabo de Ovelha, Tamarez, Malvasia Fina, Rabo de Ovelha, Terrantez, Uva Cão
**Douro** Touriga Nacional, Touriga Franca, Tinta Roriz, Tinto Cão, Tinta Barroca, *Tinta Amarela, Mourisco Tinto, Bastardo, Periquita, Rufete, Tinta da Barca, Tinta Francisca*; Couveio, Viosinho, Rabigato, Malvasia Fina, Donzelinho, Esgaña Cão, Folgazao
**Madeira** varietals from Sercial, Verdelho, Boal or Malvasia; *minor grapes include Tinta Negra Mole, Bastardo, Malvasia Roxa, Verdelho Tinto, Terrantez*
**Port** Touriga Franca, Touriga Nacional, Tinta Roriz, Tinto Cão, Tinta Barroca, *Bastardo, Mourisco Tinto, Tinta Amarela, Cornifesto, Donzelinho, Malvasia, Periquita, Rufete, Tinta de Barca*; Gouveio Verdelho, Malvasia Fina, Rabigato Rabo di Ovelha, Viosinho, Donzelinho, Códega, Arinto, Boal, Cerceal, Esgaña Cão, Folgasão, Moscatel de Bago Miúdo, Samarrinho, Vital
**Setúbal** Moscatel de Setúbal (Muscat of Alexandria); Moscatel Roxo
**Vinho Verde** Vinhão Sousão, Espadeiro, Padeiro Basto, Rabo de Ovelha, Borraçal, Brancelho, Pedral; Loureiro, Trjadura, Padernã, Azal, Batoca, Alvarinho

## ROMANIA

**Cotnari** Grasa, Tamîioasa, Francusa, Feteasca Alba

## SPAIN

**Alella** Garnacha Tinta, Tempranillo (Ull de Llebre), *Merlot, Cabernet Sauvignon, Pansá Rosada, Garnacha Peluda, Garnacha Negra, Pinot Noir*; Xarel-lo (Pansá Blanca), Garnacha Blanca, *Chardonnay, Macabeo, Parellada, Picapoll, Chenin Blanc, Sauvignon Blanc*
**Bierzo** Mencía, *Garnacha Tintorera*; Godello, Doña Blanca, *Palomino (Jerez), Malvasía*
**Calatayud** Garnacha Tinta, Tempranillo, Cariñena (Mazuelo), *Monastrell, Cabernet Sauvignon, Merlot, Syrah*; Macabeo (Viura), Malvasía, *Garnacha Blanca, Muscat of Alexandria (Moscatel Blanco/Moscatel Romano), Chardonnay*
**Campo de Borja** Garnacha Tinta, Tempranillo, *Macabeo, Cariñena (Mazuelo), Cabernet Sauvignon, Merlot, Syrah*; Macabeo, *Muscat of Alexandria (Moscatel Romano)*
**Cariñena** Garnacha Tinta, Tempranillo, Cariñena (Mazuela), *Cabernet Sauvignon, Monastrell, Merlot, Moristel (Juan Ibáñez)*; Macabeo (Viura), *Garnacha Blanca, Muscat of Alexandria (Moscatel Romano), Parellada*
**Cava** Macabeo (Viura), Parellada, Xarel-lo, *Chardonnay, Subirat Parent (Malvasía Riojana), Pinot Noir*; Macabeo (Viura), Parellada, Xarel-lo, *Chardonnay, Garnacha Tinta, Monastrell, Pinot Noir*
**Conca de Barberá** Macabeo, Parellada, *Chardonnay, Viognier*; Trepat, Garnacha, Tempranillo (Ull de Llebre), *Cabernet Sauvignon, Merlot*
**Costers del Segre** Macabeo, Parellada, Chardonnay, Xarel-lo, Garnacha Blanca; Garnacha Tinta, Cabernet Sauvignon, Tempranillo (Ull de Llebre/Gotim Bru), Pinot Noir, Trepat, Monastrell, Merlot, Cariñena
**Jerez y Manzanilla** Palomino Fino (Listán), Pedro Ximénez, Muscat of Alexandria (Moscatel Romano)
**Málaga** Pedro Ximénez, Muscat of Alexandria (Moscatel Romano)
**La Mancha** Tempranillo (Cencibel), Garnacha Tinta, Cabernet Sauvignon, Moravia, Merlot; Airén, Macabeo, Pardilla, *Chardonnay*
**Montilla-Moriles** Pedro Ximénez, Moscatel Romano, Airén (Lairén), Baladí, Torrontés
**Navarra** Garnacha Tinta, Tempranillo, Cabernet Sauvignon, Mazuelo, Graciano, *Merlot and other experimental varieties*; Macabeo (Viura), Muscat Blanc à Petits Grains (Moscatel de Frontignan, Moscatel de Grano Menudo), *Chardonnay, Malvasía, Garnacha Blanca and other experimental varieties*
**Penedès** Tempranillo (Ull de Llebre), Garnacha Tinta, Cabernet Sauvignon, Cariñena, Monastrell, Samsó, *Cabernet Franc, Pinot Noir, Merlot*; Xarel-lo (Pansá Blanca), Macabeo, Parellada, Subirat Parent (Malvasía Riojana), *Sauvignon Blanc, Riesling, Chardonnay, Muscat d'Alsace, Gewürztraminer, Chenin Blanc*
**Priorat** Garnacha Tinta, *Cariñena, Garnacha Peluda, Cabernet Sauvignon, Syrah, Merlot*; Garnacha Blanca, Macabeo, Pedro Ximénez, *Parellada, Chenin Blanc*
**Rías Baixas** Albariño, Loureiro Blanco (Marqués), Treixadura, Caiño Blanco, Torrontés; Caiño Tinto, Brancellao, Espadeiro, Mencía, Loureira Tinta, Sousón
**Ribeiro** Treixadura, Godello, Lado, Loureiro, Albariño, *Palomino (Jerez), Torrontés, Macabeo, Albillo*; Caiño, Ferrón, Sousón, Brancellao, *Garnacha Tintorera (Alicante), Tempranillo, Mencía*
**Ribera del Duero** Tempranillo (Tinto Fino,Tinto del País), *Garnacha Tinta (Tinto Aragonés), Cabernet Sauvignon, Merlot, Malbec, Albillo*
**Rioja** Tempranillo, Garnacha Tinta (Garnacha), Cariñena (Mazuelo), Graciano, *Cabernet Sauvignon, Merlot*; Macabeo (Viura), Malvasía Riojana, Garnacha Blanca, *Chardonnay, Sauvignon Blanc*
**Rueda** Verdejo, Sauvignon Blanc, Macabeo (Viura), Palomino Fino, *Chardonnay*
**Somontano** Moristel, Tempranillo, Cabernet Sauvignon, Garnacha Tinta, Parreleta, *Merlot, Pinot Noir and others*; Macabeo, Garnacha Blanca, Chardonnay, Alcañón, *Gewürztraminer and others*
**Tarragona** Garnacha Tinta, Cariñena (Mazuela), Tempranillo (Ull de Llebre), *Cabernet Sauvignon, Merlot, Syrah*; Macabeo, Parellada, Xarel-lo, Garnacha Blanca, *Chardonnay, Muscat of Alexandria (Moscatel Romano)*
**Toro** Tempranillo (Tinto de Toro), *Garnacha Tinta, Cabernet Sauvignon*; Garnacha Tinta; Malvasía, Verdejo Blanco
**Valdeorras** Mencía, *Garnacha Tintorera (Alicante), Gran Negro, Merenzao (Bastardo), Cabernet Sauvignon, Tempranillo, Merlot*; Godello, *Doña Blanca (Valenciana), Palomino (Jerez)*
**Valdepeñas** Tempranillo (Cencibel), *Garnacha Tinta, Cabernet Sauvignon, Merlot, Pinot Noir*; Airén, *Macabeo, Chardonnay*

## SWITZERLAND

**Dôle** Pinot Noir, Gamay
**L'Oeil-de-Perdrix de Neuchâtel** Pinot Noir

# GLOSSARY OF TECHNICAL TERMS

Words in SMALL CAPITALS have their own entries elsewhere in the glossary.

**Acidity** Naturally present in grapes; gives red wine an appetizing 'grip' and whites a refreshing tang. Too much can make a wine seem sharp, but too little and it will be flabby. See also MALIC ACID, TARTARIC ACID.

**Aging** Essential for many fine wines and for some everyday ones. May take place in vat, barrel or bottle, and may last for months or years. It has a mellowing effect on a wine, but if the wine has too long in storage it may lose its fruit.

**Alcoholic content** Alcoholic strength, sometimes expressed in degrees, equivalent to the percentage of alcohol in the total volume.

**Alcoholic fermentation** Biochemical process whereby yeasts convert the grape sugars into alcohol and carbon dioxide, transforming grape juice into wine. It normally stops when all the sugar has been converted or when the alcohol level reaches about 15 per cent.

**American Viticultural Area (AVA)** American appellation system introduced in the 1980s. AVA status requires that 85 per cent of grapes in a wine come from a specified region. It does not guarantee any standard of quality.

**Ampelography** The study of grape varieties.

**Appellation d'Origine Contrôlée (AC or AOC)** Official designation in France guaranteeing a wine by geographical origin, grape variety and production method.

**Assemblage** French term for blending of wines.

**Auslese** German and Austrian category for wines made from selected bunches of grapes. The wines will be generally sweet and sometimes touched by noble rot or BOTRYTIS. Sometimes they are fermented dry, making rich, powerful wines.

**Barrel aging** Time spent maturing in wood, normally oak, during which the wines may take on flavours from the wood if new barrels are used. The gentle oxygenation caused by gaseous penetration through the pores of the wood is important to the aging process.

**Barrel fermentation** Oak barrels may be used for FERMENTATION instead of stainless steel. If new barrels are used for this, the integration of oak flavour is better than if the wine is merely put into barrels after fermentation.

**Barrique** The *barrique bordelaise* is the traditional Bordeaux oak barrel of 225 litres (50 gallons) capacity, used for aging and sometimes for fermenting wine.

**Bâtonnage** A traditional Burgundian practice of stirring the LEES of fine white wines, now increasingly taken up by producers around the world. It is occasionally used for red wines.

**Baumé** Hydrometric MUST weight scale which determines the sugar content of grape juice by measuring its density. This indicates the potential ALCOHOLIC CONTENT. Commonly used in France and Australia. See table page 307.

**Beerenauslese** German and Austrian QMP category for wines made from individually selected berries. Almost always affected by noble rot or BOTRYTIS fungus. The wines are sweet to very sweet. Beerenauslese from new, non-Riesling grapes can be dull but Riesling and many a Scheurebe and occasionally a Silvaner will be astonishing.

**Bereich** German wine regions, of which there are 13, are subdivided into Bereiche.

**Bin number** Australian system used by wine companies to identify batches of wine. Bin numbers are often used as brand names.

**Biodynamic viticulture** This approach works with the movement of the planets and cosmic forces to achieve health and balance in the soil and in the vine. Vines are treated with infusions of mineral, animal and plant materials, applied in homeopathic quantities according to the position of the planets.

**Blanc de blancs** White wine, especially Champagne, made only from white grapes. Blanc de Noirs is white wine from black grapes.

**Blending** The art of mixing together wines of different origin, styles or age, often to balance out ACIDITY, weight etc.

**Bodega** Spanish winery or wine firm.

**Bordeaux mixture** Copper sulphate, slaked lime and water, sprayed on to vines throughout the growing season to prevent downy mildew. It is one of the few chemical treatments permissable in organic viticulture.

**Botrytis** Botrytis is rot – a fungus which attacks grapes. Often rot is bad news; but under certain circumstances (see page 27) the fungus concerned, *Botrytis cinerea*, has beneficial effects. It is often known as noble rot and is responsible for many of the world's great sweet wines.

**Brix** A scale used in the USA and New Zealand to measure sugar levels in grape juice. See table page 307.

**Brut** Term for 'dry', usually seen on Champagne labels and sparkling wines in the New World. In Champagne the term 'Extra Dry' is, in fact, slightly sweeter.

**Cal-Ital** A group of Californian wineries that promotes Italian grape varieties grown in California. Also refers to the wines.

**Canopy** The above-ground part of the vine, including stem, leaves and fruit.

**Canopy management** Term that includes pruning, training and everything that is done to control or alter the shape of the vine's vegetation. It is aimed at regulating the position of the fruit within the canopy, and at controlling the amount of sunlight and shade on the fruit, to produce a crop of optimum size and ripeness.

**Carbonic maceration** Winemaking method traditional to Beaujolais and now often used elsewhere. Bunches of uncrushed grapes are fermented whole in closed containers to give well-coloured, fruity wine for early drinking.

**Cava** Spanish fizz made by the traditional method, as used in Champagne.

**Cask** Wooden (usually oak) barrel used for AGING and storing wine. Known in France as *foudres* and Italy as *botti*.

**Cépage** French for 'grape variety'. Often followed on labels by the name of a single variety, such as Merlot. In the southern Rhône and the Midi the expression *cépage améliorateur* or 'improving variety' refers to the better quality local grapes: Syrah, Grenache or Mourvèdre.

**Chai** French term for the building in which wine is stored.

**Champagne method** Traditional way of making sparkling wine by inducing a second FERMENTATION in the bottle in which the wine will be sold. Now known as the traditional method. The term Champagne method is, ludicrously, no longer legal currency.

**Chaptalization** Addition of sugar during FERMENTATION to raise a wine's alcoholic strength. More necessary in cool climates where lack of sun may mean insufficient natural sugar in the grapes.

**Château** A wine-producing estate, especially in Bordeaux. Applied to all sizes of property.

**Claret** English term for red Bordeaux wine.

**Clarification** Term covering any winemaking process (such as FILTERING or FINING) that involves the removal of solid matter either from the MUST or the wine.

**Classico** Italian term for the heartland of a wine zone where its best wines are produced.

**Climat** French term for a specifically defined area of vineyard, often very small.

**Clone** Propagating vines by taking cuttings produces clones of the original plant. However, the term is more usually taken to mean laboratory-produced, virus-free clones, selected to produce higher or lower quantity or quality, or selected for resistance to frost or disease.

**Clos** Term for a vineyard that is (or was) wall-enclosed; traditional to Burgundy.

**Cold fermentation** Long, slow FERMENTATION at low temperature to produce wines of maximum freshness. Crucial for whites in hot climates.

**Consorzio** Italian for consortium or association, especially of wine producers. Each DOC has a Consorzio which lays down the rules for its region. The French equivalent is Comité Interprofessionel and in Spain Consejo Regulador.

**Corked/corky** Wine fault derived from a cork that has become contaminated, usually with Trichloranisole or TCA. Nothing to do with pieces of cork in the wine. The mouldy, stale smell is unmistakable.

**Commune** A French village and its surrounding area or parish.

**Cosecha** Spanish for 'vintage'.

**Côtes/Coteaux** French for 'slopes'. Hillside vineyards often produce better wine than low-lying ones.

**Coulure** Failure of the fruit to set after flowering, often accentuated by cold, wet or windy weather.

**Crémant** Traditional-method sparkling wine from French regions other than Champagne, e.g. Crémant de Bourgogne.

**Crianza** Spanish term used to describe both the process of AGING a wine and the youngest official

category of matured wine. A Crianza wine is aged in barrel, tank and/or bottle for at least two years.
**Cross, Crossing** Grape bred from two *Vitis vinifera* varieties.
**Cru** French for 'growth'. Used to describe a wine from a single vineyard.
**Cru Bourgeois** In Bordeaux, a quality ranking immediately below CRU CLASSÉ.
**Cru Classé** Literally 'Classed Growth', indicating that a vineyard is included in the official ranking system of its region.
**Cryoextraction** Technique of freezing grapes to remove excess water and increase concentration, adopted in Sauternes in the 1980s to improve lesser vintages.
**Cultivar** Term mainly used in South Africa for a single grape variety.
**Cuve close or Charmat method** A bulk process used to make sparkling wines. The second FERMENTATION, which produces the bubbles, takes place in tank rather than in the bottle (as in the superior but more costly TRADITIONAL METHOD).
**Cuvée** The term usually indicates a blend, which may mean different grape varieties or simply putting together the best barrels of wine.
**Degree** The alcoholic strength of wine, usually expressed in degrees equivalent to the percentage of alcohol in the total volume. It is used in a broad-brush way to classify regions by the warmth of their climate and determine which vines might succeed.
**Degree days** A system devised to measure the growth potential of vines in a specific area in terms of the climate.
**Demi-sec** Confusingly, it means medium tending to sweet, rather than medium-dry.
**Denominación de Origen (DO)** The main quality classification for Spanish wine. Rules specify each region's boundaries, grape varieties, vine-growing and winemaking methods.
**Denominação de Origem Controlada (DOC)** Portugal's top quality classification. Rules specify each region's boundaries, grapes, vine-growing and winemaking methods.
**Denominacíon de Origen Calificada (DOC)** New Spanish quality wine category, one step up from DO. So far only the Rioja DO has been promoted to DOC.
**Denominazione di Origine Controllata (DOC)** Italian quality wine classification for wines of controlled origin, grape varieties and style.
**Denominazione di Origine Controllata e Garantita (DOCG)** Top Italian quality wine classification meant to be one notch above DOC, with tighter restrictions on grape varieties, yields and a tasting panel. It is supposed to give recognition to particularly good vineyard sites.
**Density** The number of vines planted per hectare or acre.
**DNA** DNA fingerprinting techniques are being used to identify vine varieties. They can also identify the ancestry of vine varieties. Cabernet Sauvignon, for example, is now known to be the offspring of Cabernet Franc and Sauvignon Blanc.
**Domaine** Estate, especially in Burgundy.
**Downy mildew** Common vine fungus, also called peronospera, which destroys leaves and shrivels fruit. It can reach epidemic proportions in heavy rain.
**Einzellage** German for an individual vineyard. The name of the vineyard is generally preceded by that of the village, e.g. Wehlener Sonnenuhr is the Sonnenuhr vineyard in the village of Wehlen.
**Eiswein** Rare German and Austrian wine made from grapes harvested and pressed while still frozen to remove the slimy, sweet concentrate. The water, in its icy state, stays separate. Known as icewine in Canada.
**Élevage** French term covering all wine-making stages between FERMENTATION and bottling.
**Embotellado de/en Origen** Spanish term for ESTATE-BOTTLED.
**Engarrafado na Origem** Portuguese term for ESTATE-BOTTLED.
**Enologist/Oenologist** Winemaker. The role has become increasingly high profile in recent years.
**Espumoso** Spanish for 'sparkling'.
**Estate-bottled** Wine made from grapes grown on the estate's vineyards and then bottled where it has been made. In France, this is indicated on the label as *mis en bouteilles* followed by *au domaine*, *au château*.
**Fermentation** See ALCOHOLIC FERMENTATION, MALOLACTIC FERMENTATION.
**Filtering** Removal of YEASTS, solids and any impurities from a wine before bottling.
**Fining** Method of clarifying wine by adding coagulants, traditionally egg whites, to the surface. As these fall through the wine they collect solids. They are removed and leave nothing of themselves behind.
**Flavour or aroma compounds** Substances in wine that can be smelled or tasted.
**Flor** Film of YEAST that grows on the surface of certain wines when in barrel, especially sherry. Protects the wine from OXIDATION, and imparts a unique taste.
**Flying winemaker** Term coined in the late 1980s to describe ENOLOGISTS, many of them Australian-trained, brought in to improve quality in many of the world's under-performing wine regions.
**Fortified wine** Wine which has high-alcohol grape spirit added, either before or after the ALCOHOLIC FERMENTATION is completed.
**Foxy** Used to describe the very distinctive, perfumed character of many American native and HYBRID grapes, especially red ones. Usually a pejorative term.
**Frizzante** Italian term for lightly sparkling wine.
**Garrafeira** Portuguese term for high-quality wine with at least half a per cent of alcohol higher than the required minimum, that has had at least three years' AGING for reds, and at least one year for whites.
**Geographical Indication (GI)** Australian term to indicate the origin of a wine.
**Grafting** Since phylloxera the only sure method of growing grape vines. It involves grafting a cutting of *Vitis vinifera* on to a phylloxera-resistant American rootstock.
**Gran Reserva** Top-quality, mature Spanish wine from an especially good vintage, with at least five years' aging (cask and bottle) for reds, and four for whites.
**Grand Cru** 'Great growth'; the top quality classification in Burgundy, Alsace and Champagne. The Grands Crus of Bordeaux may be subdivided into different ranks, according to the region.
**Grand vin** Term used in Bordeaux to indicate a producer's top wine. Usually bears a CHÂTEAU name.
**Hectolitres** 100 litres; 22 imperial gallons or 133 standard 75-cl bottles. See Measurements page 305.
**Hybrid** Grape bred from an interspecific crossing of an American vine species and European *Vitis vinifera*.
**Indicação de Proveniência Regulamentada (IPR)** Official Portuguese category for wine regions aspiring to DOC status.
**Indicazione Geografica Tipica (IGT)** A quality level for Italian wines (roughly equivalent to French vin de pays) in between VINO DA TAVOLA and DOC.
**Institut National des Appellations d'Origine des Vins et des Eaux-de-Vie (INAO)** The organization in charge of administering the French appellation contrôlée system.
**Kabinett** Lowest level of German QMP wines. Made from ripe grapes, usually lighter in alcohol than ordinary QBA and often delicious. In Austria Kabinett is a sub-division of Qualitätswein.
**KMW** The Austrians use KMW or Klosterneuburger Mostwaage as a scale to determine the MUST weight or original sugar in freshly picked grapes. Like OECHSLE degrees in Germany, each quality category of wine sets a minimum number of KMW. See table page 307.
**Late harvest** Late-harvested grapes contain more sugar and concentrated flavours; the term is often used for sweetish New World wines. Vendange Tardive is the French term.
**Lees** Sediment – dead YEASTS etc – thrown by wine in a CASK and left behind after RACKING. Some wines stay on the fine lees for as long as possible to take on extra flavour. See BÂTONNAGE.
**Lieu-dit** Burgundian term for a single vineyard below the rank of PREMIER CRU that may nevertheless be named on the label.
**Liquoroso** Italian term for wines high in alcohol, often – but not always – fortified.
**Maceration** An important winemaking process whereby colour, flavour and/or TANNIN are extracted from grape skins before, during or after FERMENTATION.
**Macroclimate** Refers to the climate of a region. See also MESOCLIMATE, MICROCLIMATE.
**Maderization** A form of OXIDATION in white wines caused by heating, usually over a period of time. It takes its name from Madeira, which is the epitome of the style. Unintentional maderization, for example in a light white wine, is a fault.
**Malic acid** One of the two principal acids found in grapes (the other is TARTARIC). Levels of malic acid are significantly higher in cooler climates.
**Malolactic fermentation** Secondary fermentation whereby sharp, appley-tasting MALIC ACID is converted into riper-tasting lactic acid and

carbon dioxide; usually occurs after ALCOHOLIC FERMENTATION. It is encouraged in red wines, softening them and reducing their acidity, but often prevented in whites to preserve a fresh taste, especially in wines made in warm regions, where natural acidity will be lower.

**Maturation** The beneficial AGING of wine.

**Meritage** American, primarily Californian, term for red or white wines made from Bordeaux grape varieties.

**Mesoclimate** Describes the climate of a specific geographical area, be it one vineyard or simply a hillside or valley. See also MACROCLIMATE, MICROCLIMATE.

**Microclimate** Describes the immediate physical environment of a vine. Often confused with MESOCLIMATE. See also MACROCLIMATE.

**Mildew** See DOWNY MILDEW and OIDIUM.

**Millerandage** The failure of some young grapes to develop normally on an otherwise normal bunch of grapes.

**Mousseux** French term for sparkling wine not made by the TRADITIONAL METHOD.

**Musqué** A French term meaning both musky and Muscat-like. Some grape varieties, for example Chardonnay, have a Musqué mutation which is particularly aromatic.

**Must** The mixture of grape juice, skins, pips and pulp produced after crushing (but prior to completion of FERMENTATION), which will eventually become wine.

**Must weight** An indicator of the sugar content of juice – and therefore the ripeness of grapes. See BAUMÉ, BRIX, KMW, OECHSLE and table page 307.

**Négoçiant** French term for merchant or shipper who buys in wine from growers, then matures, maybe blends and bottles it for sale.

**Noble rot** See BOTRYTIS.

**Nouveau, novello** French and Italian terms for new wine. Wine for drinking very young, from November in year of vintage.

**Oak** The most common wood for wine CASKS. During AGING or fermenting it gives flavours, such as vanilla and TANNIN, to the wines. The newer the wood, the greater its impact. French oak is subtler in flavour than American.

**Oechsle** In Germany, MUST WEIGHT is measured in degrees Oechsle; in effect, it indicates the level of sweetness in the juice. Each quality category has a minimum required Oechsle degree. See table page 307.

**Oidium** Common fungal disease, also called powdery mildew, attacking vine leaves, tendrils and shoots.

**Organic** There is no such thing, properly speaking, as organic wine, only organic viticulture. Term applied to an increasing number of wines which have been subjected to restrictive viticulture and winemaking practices. There are as yet no universally agreed regulations. Unlike BIODYNAMISM, the concept is simply to eliminate the use of chemical fertilizers and pesticides.

**Oxidation** Over-exposure of wine to air, causing loss of fruit and flavour.

**Passito** Italian term for strong, sweet wine made from dried or semi-dried grapes.

**Pétillant** French for semi-sparkling.

**Phenolics** Chemical compounds found in the pips, stalks, skins, juice and pulp of grapes, especially in red ones. Also known as polyphenols, they include tannins, colour-giving anthocyanins and flavour compounds.

**Phylloxera** Vine aphid (*Phylloxera vastatrix*) which devastated viticulture worldwide in the late 19th century onwards. Since then, the vulnerable European *Vitis vinifera* has been grafted on to phylloxera-resistant American ROOTSTOCKS. Phylloxera has never reached Chile and parts of Australia, so vines there are ungrafted and can live up to twice as long – if allowed to.

**Plafond limité de classement (PLC)** A French system whereby the maximum YIELD permitted within an APPELLATION CONTRÔLÉE is increased in abundant years.

**Prädikat** One of the six German QUALITÄTSWEIN MIT PRÄDIKAT or QmP categories of wine.

**Premier Cru** 'First growth'; the top quality classification in parts of Bordeaux, but second to GRAND CRU in Burgundy. Used in Champagne to designate vineyards just below Grand Cru.

**Prohibition** 18th Amendment to the US Constitution, passed in 1920, banning the manufacture, sale and transportation of alcoholic beverages; the measure ruined many wineries, but some survived making grape juice, communion and medicinal wines. Repealed in 1933.

**Pruning** Method of trimming the vine which takes place mainly in the dormant winter months. Also the primary means of controlling the YIELD.

**Pulp** The flesh of the grape.

**Pumping over** The process, called *remontage* in French, whereby the fermenting must is drawn over the cap of skins in the vat. Essential for red wines. It helps to extract colour and tannin.

**Qualitätswein bestimmter Anbaugebiete (QbA)** German wine classification for 'quality wine from designated regions' – the German equivalent, in EU terms, of French AOC and Italian DOC. Most QbA wines are distinctly ordinary in quality; only buy QbA wines from good growers.

**Qualitätswein mit Prädikat (QmP)** German wine classification for 'quality wine with distinction'. There are six categories, in order of increasing ripeness of the grapes: KABINETT, SPÄTLESE, AUSLESE, BEERENAUSLESE, TROCKENBEERENAUSLESE and EISWEIN. Some wines (usually Kabinett or Spätlese) may be Trocken (dry) or Halbtrocken (half-dry). Austria has a similar system but regards Kabinett as a subdivision of simple Qualitätswein. Other Austrian Prädikats are Ausbruch (between Beerenauslese and Trockenbeerenauslese) and Strohwein.

**Quinta** Portuguese farm or wine estate.

**Racking** The transferring of wine from one barrel to another, leaving the LEES or sediment behind. Racking also produces aeration necessary for the AGING process and softens TANNINS.

**Rancio** Style of wine that is deliberately oxidized; either naturally strong or fortified, it is aged in the sun in glass bottles, earthenware jars or wooden barrels.

**Récoltant** French for 'grower'. They may make their own wine or sell the grapes to a merchant or *NÉGOCIANT*.

**Reserva** In Spain, quality wine from a good vintage with at least three years' AGING (cask and bottle) for reds, and two for whites. In Portugal it designates wine that has an alcohol level at least half a per cent higher than the minimum for the region.

**Reserve** Many New World producers use this term, or similar ones such as Private Reserve and Special Selection, freely on their wine labels to indicate different wine styles or a special selection rather than a better wine. It has no legal meaning.

**Reverse osmosis** Method of MUST concentration whereby the wine or juice to be concentrated passes through a filter, leaving the water behind. Other methods of MUST concentration include evaporation in a vacuum and CRYOEXTRACTION.

**Rhône Ranger** A phrase coined in the 1980s in California to describe local winemakers fascinated with traditional Rhône grape varieties, especially Syrah and Viognier.

**Ripasso** Valpolicella wine refermented on the LEES of Amarone della Valpolicella to give extra richness.

**Riserva** Italian term for wines aged for a specific number of years according to DOC(G) laws.

**Rootstock** The root of the vine on to which the fruiting branches are grafted. Most rootstocks are from PHYLLOXERA-resistant American vines.

**Rosado, rosato** Spanish, Portuguese and Italian for pink wine or rosé.

**Sec** French for 'dry'. When applied to Champagne, it actually means medium-dry.

**Second wine** A CUVÉE put together from wines selected out of a producer's main wine. It may come from young vines or from less favoured parts of the vineyard. Usually lighter and quicker-maturing than the main wine.

**Sekt** German term for sparkling wine.

**Solera** Blending system used for sherry and some other FORTIFIED wines. When mature wine is run off a cask for bottling, only a quarter or so of the volume is taken, and the space is filled with similar but younger wine from another cask, which in turn is topped up from an even younger cask, and so on.

**Spätlese** German QMP category for wines made from 'late-picked' (therefore riper) grapes. Often moderately sweet, though there are now dry versions.

**Spumante** Italian for sparkling.

**Sugar** Naturally present in grapes. Transformed during FERMENTATION into alcohol and carbon dioxide.

**Sulphur** Commonly used during VINIFICATION as a disinfectant for equipment; with fresh grapes and wine as an anti-oxidant; and added as sulphur dioxide to the MUST to arrest or delay FERMENTATION.

**Supérieur** French term for wines with a higher alcohol content and made according to slightly stricter rules than the basic AC.

**Superiore** Italian term for wines with higher alcohol, maybe more AGING too.

**Super-Tuscan** English term for high-quality, non-DOC Tuscan wine.

**Sur lie** French for 'on the LEES', meaning wine bottled direct from the FERMENTATION vat or cask to gain extra flavour from the lees. Muscadet is the most famous example.
**Tafelwein** German for 'table wine', the most basic quality designation.
**Tannin** Harsh, bitter element in red wine, derived from grape skins, pips, stems and from AGING in oak barrels; softens with time and is essential for a wine's long-term aging. Producing wines with ripe TANNINS that may be drunk earlier is a priority of red winemaking these days.
**Tartaric acid** One of the two most important acids naturally present in grapes, the other being MALIC. It tends to be the predominant acid in warm areas. It may be added during FERMENTATION to correct low ACIDITY.
**Teinturier** Black *vinifera* grapes with red pulp. All other *vinifera* grapes have colourless pulp. Teinturier grapes are usually inferior in quality.
**Terroir** A French term used to denote the combination of soil, climate and exposure to the sun that makes each vineyard and region unique. It is the basis of the French *APPELLATION D'ORIGINE CONTRÔLÉE* or AC/AOC system.
**Traditional method** The accepted term for what used to be called the CHAMPAGNE METHOD. The Champenois have succeeded in preventing other regions from using the term.
**Training** Method of vine management using a permanent vine structure, either free-standing, up stakes, along wires or onto a trellis or training system, which will determine the type of PRUNING.
**Trocken** German for 'dry'.
**Trockenbeerenauslese (TBA)** German quality wine category for wines made from individually picked single grapes, shrivelled by noble rot – often the highest level of sweetness.
**University of California at Davis (UCD)** The leading US viticultural research institute and college for aspiring wine-growers and winemakers.
**Varietal** The character of wine derived from the grape; also wine made from, and named after, a single or dominant grape variety and usually containing at least 75 per cent of that variety. The minimum percentage varies slightly between countries and, in the USA, between states.
**Vendange tardive** See LATE HARVEST.
**Vieilles vignes** Wine from old vines. The term has no legal weight, and 'old' may mean 25 or 100 years.
**Vigneron** French for 'wine-grower'.
**Vigour** The growth rate of the vine. Vigorous vineyards are often, but not always, associated with high yields. Soils, too, may be regarded as having low or high potential vigour according to the growth of vines planted on them.
**Vin Délimité de Qualité Supérieure (VDQS)** Second category of French wines, below AC, abbreviated to VDQS.
**Vin de garage** Wines made on so small a scale they could be made in one's garage. Particularly applies to some wines from Bordeaux's St-Émilion area. Such wines may be made from vineyards of a couple of hectares or less, and are often of extreme concentration. *Vins de garage* are highly fashionable and sell for high prices.
**Vin de paille** Wine made by drying the grapes on straw (*paille*) before FERMENTATION. This concentrates the sugar in the grapes: the resulting wines are sweet. Mostly from the Jura region of France. Similar wines may be made in other countries, for example Italy and Austria.
**Vin de pays** French for 'country wine'. Although it is the third category in the official classification of French wines, it includes some first-class wines which don't follow local AC rules.
**Vin doux naturel (VDN)** French sweet wine fortified with grape spirit. Mostly from Languedoc-Roussillon.
**Vin Santo** Historic but extremely variable sweet white wine from Tuscany and Italy. Called Vino Santo in Trentino.
**Viña** Spanish for 'vineyard'.
**Vinification** The process of turning grapes into wine.
**Vino da tavola** Italian for 'table wine'. Quality may be basic or exceptional, although many of these latter wines are now being reclassified as INDICAZIONE GEOGRAFICA TIPICA or IGT.
**Vino tipico** New Italian category for VINO DA TAVOLA with some regional characteristics.
**Vintage** The year's grape harvest, also used to describe the wine of a single year.
**Viticulture** Vine-growing and vineyard management.
**Vitis vinifera** The species of vine, native to Europe and Central Asia, responsible for all the world's fine wine, as opposed to other species such as the native American *Vitis labrusca*, which is still used in the eastern USA to make grape juice and sweetish wines but which is more suited to juice and jelly manufacture.
**Wine of Origin (WO)** South African system of controlled appellations which certifies the wine's area of origin, grape variety/varieties and vintage.
**Yeast** Organism which, in the wine process, causes grape juice to ferment. In the New World it is common to start FERMENTATION with cultured yeasts, rather than rely on the natural yeasts, known as ambient yeasts, present in the winery.
**Yield** The amount of fruit, and ultimately wine, produced from a vineyard. Measured in hectolitres per hectare (hl/ha) in most of Europe and in the New World as tons/acre or tonnes/hectare. Such figures on their own are, however, meaningless unless taken in conjunction with the density of planting. It is the yield per vine that is important, and a yield of 50hl/ha will obviously give very different wine at 2000 vines per hectare than at 10,000 vines per hectare. High yields are traditionally associated with lesser quality, because the larger the quantity of grapes per vine the less chance the vine has of ripening them successfully. However, reducing the yield per vine below a certain point will not improve quality. Modern viticultural techniques such as the production of virus-free clones tend to increase yields, and modern methods of canopy management may enable larger crops to be ripened successfully. Ascertaining the optimum crop for a vineyard so that the ideal size of crop may be brought to ideal ripeness is the aim of many viticulturalists. This is rather overturning the traditional idea that small yields are invariably better than high.

### Measurements and conversions

**Mass**
1 metric tonne = 0.9842 imperial ton
1 imperial ton = 1.016 metric tonne
**Surface area**
1 hectare (ha) = 10,000 square metres = 2.471 acres
**Temperature**
To convert Celsius into Fahrenheit, multiply by 1.8 and add 32. To convert Fahrenheit into Celsius, subtract 32 and multiply by 0.5555.
**Volume/capacity**
1 hectolitre (hl) = 100 litres = 22 gallons (British) or 26 gallons (USA)
1 USA gallon = 3.78 litres
**Yields**
In Europe these are measured in hl/ha. To convert to tons/acre (the system used in the New World) use the figure of 18hl/ha = 1 ton/acre. This is necessarily an approximate conversion.

### EQUIVALENT MUST WEIGHTS

These are the systems used in different parts of the world to measure the ripeness of grapes or their must weight. Equivalents are necessarily approximate.

| Baumé (degrees) | Brix (degrees) | Oeschle (degrees) | KMW (degrees) | Potential alcohol (%) |
|---|---|---|---|---|
| 10 | 18 | 74 | 14.8 | 10 |
| 10.55 | 19 | 78 | 15.6 | 10.55 |
| 11 | 19.8 | 83 | 16.6 | 11 |
| 11.55 | 20.8 | 88 | 17.6 | 11.55 |
| 12 | 21.7 | 92 | 18.4 | 12 |
| 12.55 | 22.6 | 96 | 19.2 | 12.55 |
| 13 | 23.4 | 100 | 20 | 13 |
| 13.55 | 24.4 | 107 | 21.4 | 13.55 |
| 14 | 25.2 | 108 | 21.6 | 14 |
| 14.55 | 26.2 | 112 | 22.4 | 14.55 |

# INDEX OF GRAPE NAMES AND THEIR SYNONYMS

Page numbers in **bold** refer to main references in the A-Z.


## A

Abouriou **34**
Acolon 88
Agiorgitiko (Mavro Nemeas) **34**
Aglianico 19, **34**, 53
Aidani **34**, 38
Aidani Mavro 34
Airen (Lairén) **34**
Albana (Greco; Greco di Ancona) **34-5**, 68
Albana Gentile di Bertinoro 34
Albariño (Albarin Blanco; Alvarinho; Cainho Branco) **36-7**, 112, 117, 123, 167, 270
Albarola **35**
Albillo **35**, 262
Aleatico **35**, 147, 150
Alfrocheiro Preto (Pé de Rato) **35**, 115
Alicante Bouschet (*see also* Garnacha Tintorera) **35**, 38, 134, 139
Aligoté **35**, 112, 177
Altesse see Roussette
Alva see Roupeiro
Alvarinho see Albariño
Amigne **35**
Ancellotta **35**, 116
Ansonica see Inzolia
Antão Vaz **35**
Anzonica see Inzolia
Aragón/Aragonés see Garnacha Tinta
Aragonez see Tempranillo
Aramon 35, **38**, 84
Arbois **38**
Arinto (Pedernã; see *also* Assario Branco) **38**, 89
Arinto Cachuda 38
Arinto do Dão 38
Arinto Galego 38
Arinto Miudo 38
Arneis **38**, 159, 169
Arnsburger **38**, 272
Arrufiac **38**
Assario Branco (Arinto; Malvazia Fina; see *also* Boal Cachudo) **38**
Assyrtico **38**, 228
Athiri **38**
Aubun **38**, 85
Aubun Vert 112
Aurore 229
Auxerrois (see *also* Malbec) **38**, 65, 112, 170, 177
Avesso **38-9**
Azal Branco **39**
Azal Tinto **39**

## B

Bacchus **39**
Bachet Noir 112
Baco **39**
Badener see Blauer Portugieser
Baga (Bago de Louro; Poerininha; Tinta Bairrada; Tinta de Baga; Tinta Fina) **39**
Baiyu see Rkatsiteli
Barbarossa (Barberoux) **39**
Barbera **40-1**, 52, 122, 160
Barbera d'Asti 41
Barbera Dolce 41
Barbera Fine 41
Barbera Forte 41
Barbera Grossa 41
Barbera Riccia 41
Barbera Vera 41
Barbaroux see Barbarossa
Baroque (Barroque) **42**
Bartolomeu see Castelão Francès
Bastardo (María Ardona; Merenzao; see *also* Trousseau) **42**, 207
Bastardo Espanhol see Castelão Francès
Beaunoir 112
Beli Pinot see Pinot Blanc
Bergeron see Roussanne
Bical (see *also* Borrada das Moscas) **42**
Bidure see Cabernet Sauvignon
Black Muscat (see *also* Muscat of Hamburg) **42**
Blanc de Morgex **42**
Blanc Fumé (see *also* Sauvignon Blanc) **42**
Blauburger **42**
Blauburgunder (see *also* Pinot Noir) **42**
Blauer Portugieser (Badener; Vöslauer; see *also* Kékoporto; Portugieser) **42**, 181
Blauer Spätburgunder see Pinot Noir
Blauer Trollinger see Trollinger
Blauer Wildbacher see Wildbacher
Blauer Zweigelt see Zweigelt
Blaufränkisch (Franconia; Frankovka; see *also* Kékfrancos; Lemberger; Limberger) **42**, 91, 182, 207, 284, 285
Blue Oporto 165
Boal (see *also* Sémillon) **42**
Boal Bagudo 42
Boal Bonifacio 42
Boal Branco 42
Boal Cachudo (see *also* Assario Branco) 42
Boal Carrasquenho 42
Boal de Alicante 42
Boal Espinho 42
Bobal **42-3**
Bogazkere **43**
Bombino Bianco (Pagadebit; Straccia Cambiale; Trebbiano bruzo) **43**, 270
Bonarda (see *also* Croatina; Uva Rara) **43**, 59, 159, 273
Bonarda di Cavaglia 43
Bonarda di Gattinara 43
Bonarda Grossa 43
Bonarda Novarese 43
Bonarda Piccola 43
Bonarda Piedmontese 43
Bonarda Pignola 43
Bordo see Cabernet Franc
Borrado das Moscas (see *also* Bical) **43**
Bouchet see Cabernet Franc
Bourboulenc (see *also* Clairette) **43**, 122, 204, 280
Bouvier **43**, 165
Brachetto **43**
Braquet 43
Brauner Veltliner 114, 271
Brown Muscat **43**, 149, 151, 153
Brunello see Sangiovese
Bual **43**, 117
Bukettraube **43**
Burgundac Crni see Pinot Noir

## C

Cabernet Dorio 88
Cabernet Dorsa 88
Cabernet Franc (Bordo; Bouchet; Cabernet Frank) **44-5**, 49, 52, 54, 114, 131, 132, 160, 169, 221
Cabernet Sauvignon (Bidure; Petite Vidure) 9, 15, 17, 21, 30, **46-57**, 84, 88, 98, 130, 132, 134, 182, 198, 206, 207, 213, 214, 221, 249
Cagnina see Refosco del Terrano
Cainho Branco see Albariño
Calabrese 90
Calabrese Nero see Nero d'Avola
Caladoc **58**
Camarate (see *also* Fernão Pires) **58**
Canaiolo **58**
Canaiolo Bianco (Drupeggio) 58
Canaiolo Nero 213, 214
Cannonau (see *also* Garnacha Tinta) 53, **58**, 97, 99
Cape Riesling see Crouchen
Carignan (Carignano; Cariñena; see *also* Mazuelo) 19, 52, 53, **58**, 97, 132, 164, 206, 291
Carmenère (Grande Vidure) 11, 51, **60-1**, 129, 131, 134, 135
Cassady 164
Castelão (Bartolomeu; Bastardo Espanhol; Mortágua de Vide Branca; Tinta Merousa; Trincadeira; see *also* João de Santarém; Periquita) **58**
Castelão Nacional (Castelão Portugues) 58
Catarratto **58**, 69, 114
Catawba (Francher Kello White; Mammoth Catawba) **58**
Cencibel see Tempranillo
Cerceal see Sercial
Cesanese **59**
César **59**
Chambourcin **59**
Chancellor 229
Charbonneau 59
Charbono (see *also* Dolcetto) 43, **59**
Chardonnay (see *also* Morillon; Pinot Chardonnay) 9, 14, 15, 16, 19, 27, 30, 54, **62-73**, 107, 112, 114, 138, 166, 169, 177, 181, 195, 198, 203, 206
Chardonnay Musqué 67
Chardonnay Rose 67
Chasan **59**
Chasselas (Fendant; Moster; Perlan; Wälscher; see *also* Gutedel) **59**, 147, 202
Chasselas de Courtillier 142
Chasselas de Genève 167
Chasselas Rose 229
Chenin Blanc (Pineau, Pineau de la Loire; Pinot Blanco; see *also* Steen) 9, 26, 27, **74-83**, 84, 203
Chiavennasca see Nebbiolo
Cienna **84**
Cinsaut **84**, 95, 97, 164
Clairette (see *also* Bourboulenc; Ugni Blanc) **84**, 122, 204, 280
Clairette Gris 84
Clairette Ronde 84
Clare Riesling see Crouchen
Clevner see Klevner
Códega see Roupeiro
Colombard **84**
Complexa **84**
Concord **85**, 164
Cortese 68, **85**
Corvina **85**, 203
Corvinone **85**
Cot see Malbec
Counoise **85**
Courbu see Petit Courbu
Courbu Noir 167
Crato Preto see Trincadeira Preta
Criolla Chica see Mission
Criolla Grande **85**, 138
Croatina (see *also* Bonarda) **85**, 159
Crouchen (Cape Riesling; Clare Riesling; Paarl Riesling; South African Riesling) **85**
Cygne Blanc **85**

## D

Delaware **85**
Dimiat **85**, 138
Dolcetto (Douce Noir; Ormeasco; see *also* Charbono) 19, 43, **86-7**
Domina 181
Doña Blanca (Doña Branca) **88**, 112
Dornfelder **88**
Douce Noire see Dolcetto
Drumin see Gewürztraminer
Drupeggio see Canaiolo Bianco
Duras **88**
Dureza 138, 247
Durif **88**, 168, 253
Dusty Miller see Meunier
Dutchess **88**

## E

Early Calabrese 189
Ehrenfelser **88**
Elbling **88**
Encruzado **88-9**
Erbaluce 68, **89**
Ermitage see Marsanne
Esgana Cão (see *also* Sercial) 38, **89**
Espadeiro (see *also* Trincadeira Preta) **89**
Esparte see Mourvèdre

## F

Faber (Faberrebe) **89**
Falanghina **89**
Farineux see Meunier
Favorita 68, **89**
Fendant see Chasselas
Fer (Fer Servadou) **89**
Fernão Pires (see *also* Camarate; Maria Gomes) **89**
Feteasca **89**
Feteasca Alba (see *also* Leányka) 89, 241
Feteasca Neagra 89
Feteasca Regala (Királileányka) 89
Fiano **89**
Folle Blanche (Gros Plant; Piquepoul) 39, 42, **89-90**, 114
Folle Noire (Jurançon; see *also* Négrette) 90
Franc Noir de la Haute-Saône 112
Francher Kello White see Catawba
Franconia see Blaufränkisch
Frankovka see Blaufränkisch
Fransdruif see Palomino Fino
Frappato Nero **90**
Freisa **90**
Freisa di Chieri 90
Freisa Grossa 90
Freisa Piccola 90
French Colombard 84
Frîncusa 241
Frontignac (see *also* Muscat) 51, 153
Frühroter Veltliner 114, 271
Furmint (Mosler; Zapfner) **90**, 112, 165, 206

## G

Gaglioppo 53, **91**
Gamay **91**
Gamay Beaujolais 91
Gamay Blanc Gloriod 112
Gamay de Bouze 91
Gamay de Castille 91
Gamay de Chaudenay 91
Gamay Fréaux 91
Gamay Mourot 91
Gamay Noir 112, 177
Gamza see Kadarka
Garganega **91**, 273
Garnacha Blanca (see *also* Grenache Blanc) **91**, 95, 117
Garnacha Peluda see Lladoner Pelut
Garnacha Tinta (Aragón; Aragonés; Garnacho Tinto; Garnaxta; Granaccia; Granacha; Grenache Noir; Tinto Aragonés; see also Cannonau) 9, **92-101**, 123, 249, 262
Garnacha Tintorera (see *also* Alicante Bouschet) **91**, 95
Garnacho Tinto see Garnacha Tinta
Garnaxta see Garnacha Tinta
Gelber Muskateller **91**, 143, 150
Gewürztraminer (Drumin; Heida; Heiden; Liwora; Mala Dinka; Païen; Pinat Cervena; Princ; Rotclevner; Roter Traminer; Rusa; Termano Aromatico; Traminac; Tramini; see *also* Klevner; Traminer) 9, 15, **102-11**, 195, 228
Godello (Verdello; see *also* Verdelho) **112**, 123
Golden Chasselas see Palomino Fino
Goldmuskateller (see *also* Moscato Giallo) **112**, 150, 153
Gouais Blanc 65, 84, **112**, 177
Gouveio (see *also* Verdelho) **112**
Graciano (Morrastel; see *also* Tinta Miúda) 98, **112**
Granaccia see Garnacha Tinta
Granacha see Garnacha Tinta
Grande Vidure see Carmenère
Grasa 89, 90, **112**, 241
Grasevina see Welschriesling
Grauburgunder (see *also* Pinot Gris; Rülander) **112**

Grauvernatsch *see* Schiava Grigia
Gray Riesling 271
Grecanico Dorato 69, 91, **112**
Grechetto **112**
Greco *see* Albana
Greco di Ancona *see* Albana
Greco Bianco **112**
Greco Nero **112**
Greco di Tufo 34, 112
Grenache Blanc (*see also* Garnacha Blanca) 97, **113**, 202, 204
Grenache Gris 95, 97
Grenache Noir *see* Garnacha Tinta
Grenache Poilu/Velu *see* Lladoner Pelut
Grenache Rose 95
Grignolino **113**
Grillo **113**
Grolleau (Groslot) **113**
Grolleau Gris 113
Groppello 122
Gros Manseng 38, **113**, 167
Gros Pinot Blanc 170
Gros Plant *see* Folle Blanche
Gros Syrah 138, 250
Gros Verdot 168
Groslot *see* Grolleau
Grosse Roussette *see* Marsanne
Grossvernatsch *see* Schiava Gross
Grüner Sylvaner *see* Silvaner
Grüner Veltliner (*see also* Veltliner) **114**
Gutedel (*see also* Chasselas) **114**

# H

Hanepoot (*see also* Muscat) 151, 153
Harriague **114**
Hárslevelü **114**
Heida *see* Gewürztraminer *and* Savagnin
Heiden *see* Gewürztraminer
Heiligensteiner Klevener *see* Savagnin Rose
Helfensteiner 88
Heroldrebe 88
Hondarrabi Beltza **114**
Hondarrabi Zuri **114**
Humagne Blanche **114**
Humagne Rouge **114**
Hunter Valley Riesling *see* Sémillon
Huxelrebe **114**

# I/J/K

Incrocio 69
Inzolia (Ansonica; Anzonica) **115**
Irsay Oliver (Irsai Olivér) **115**
Izsáki (White Kadarka) 115
Jacquère **115**
Jaen (*see also* Mencìa) **115**
Jaen Blanco 38
Joao de Santarém (Santarén; *see also* Castelão Francès) **115**
Johannisberg (Silvaner) **115**, 242
Johannisberg Riesling (*see also* Riesling) **115**
Jurançon *see* Folle Noire
Kadarka (Gamza) **115**
Kékfrankos (*see also* Blaufränkisch) **115**
Kéknyelu **115**
Kékoporto (Oporto; *see also* Blauer Portugeiser; Portugeiser) **115**
Kerner **115**
Királileányka *see* Feteasca Regala
Kleinvernatsch *see* Schiava Gentile
Klevner (Clevner; *see also* Gewürztraminer; Pinot Blanc *and* Pinot Noir) **115**
Knipperlé 112
Kotsifali **115**

# L

Lado 270
Lagrein **115**
Lairén *see* Airén
Lambrusco **116**
Lambrusco Grasparossa 116
Lambrusco Salomino 116
Lambrusco Sorbara 116
Laski Rizling (*see also* Welschriesling) **116**
Leányka (*see also* Feteasca Alba **116**
Lemberger (*see also* Blaufränkisch; Limberger) 88, **116**
Len de L'El **116**, 122
Lexia 147, 151
Limberger (*see also* Blaufränkisch; Lemberger) **116**
Limnio **116-17**
Listán (*see also* Palomino Fino) **117**
Listán Negro **117**
Liwora *see* Gewürztraminer
Lladoner Pelut (Garnacha Peluda; Grenache Poilu/Velu) **117**
Loureiro **117**, 123, 270

# M

Macabeo (Maccabéo; Maccabeu; *see also* Viura) **117**, 122, 166, 285
Madeleine Angevine **117**, 189
Magaratch Bastardo 207
Magaratch Ruby 207
Mala Dinka *see* Gewürztraminer
Malagousia (Malagoussia) 38, **117**
Malbec (Cot; Malbeck; *see also* Auxerrois; Pressac) 51, 54, 55, 89, **118-19**, 262
Malbeck *see* Malbec
Malmsey **117**
Malvasia 112, 117, **120-1**
Malvasia di Bosa 120
Malvasia di Cagliari 120
Malvasia di Candia 120
Malvasia Fina 120, 284
Malvasia di Grottaferrata 120
Malvasia Istriana 120
Malvasia di Lazio 120
Malvasia delle Lipari 120
Malvasia Nera 120, **122**, 164
Malvasia di Planargia 120
Malvasia Puntinata 120
Malvasia Rei 120, 166
Malvasia Rosada 117
Malvasia de S. Jorge 117
Malvasia di Toscana 120
Malvazia Fina *see* Assario Branco
Malvoisie (*see also* Pinot Gris) 121, **122**
Malvoisie de Corse *see* Vermentino
Mammolo **122**
Mammoth Catawba *see* Catawba
Mandelari **122**
Manzoni 69
Maréchal Foch **122**
Margarita 122
María Ardona *see* Bastardo
Maria Gomes (*see also* Fernão Pires) **122**
Marsanne (Ermitage; Grosse Roussette) **124-5**, 202, 204, 249, 250, 280, 281
Marzemino **122**, 189
Mataro (*see also* Mourvèdre) **122**
Mauzac 84, **122-3**
Mauzac Noir 123
Mavro **123**, 285
Mavro Nemeas *see* Agiorgitiko
Mavrodaphne **123**
Mavrud **123**
Mazuelo/a (*see also* Carignan) 98, **123**
Médoc Noir *see* Merlot
Melnik **123**
Melon de Bourgogne (*see also* Muscadet) 112, **123**, 177
Mencía (*see also* Jaen) **123**
Mendoza 107
Merenzao *see* Bastardo
Merlot (Médoc Noir; Merlau; Merlot Noir) 9, 15, 17, 52, 53, 54, 60, 98, 122, **126-37**, 159, 160, 198, 213, 214, 249, 262, 284
Merlot Blanc 137
Merlot Noir *see* Merlot
Meunier (Dusty Miller; Farineux; Noirin Enfariné; Müller-Traube; Pinot Meunier; *see also* Müllerrebe; Schwarzriesling) **138**
Misket **138**
Mission (Criolla Chica; *see also* País) **138**, 139, 291
Molette **138**
Molinara **138**
Monastrell (*see also* Mourvèdre) **138**
Mondeuse Blanche 138, 247, 250
Mondeuse Noire **138-9**, 247
Monica **139**
Montepulciano **139**, 213, 214
Morastell *see* Mourvèdre
Morellino (*see also* Sangiovese) **139**
Morillon (*see also* Chardonnay) **139**
Morio-Muskat **139**
Moristel **139**
Morrastel *see* Graciano and Mourvèdre
Mortágua *see* Touriga Nacional and Trincadeira Preta
Mortágua de Vide Branca *see* Castelão Francès
Moscadello **139**, 153
Moscatel **139**, 153
Moscatel de Alejandria 150, 153
Moscatel de España 153
Moscatel de Frano Menudo 150, 153
Moscatel de Frontignan 153
Moscatel Gordo Blanco 153
Moscatel de Málaga 150, 153
Moscatel de Setúbal 150, 153
Moscato 153
Moscato di Alexandria **139**, 153
Moscato d'Asti 153
Moscato Bianco **139**, 153
Moscato Canelli 153
Moscato Giallo (*see also* Goldmuskateller) **139**
Moscato Rosa (*see also* Rosenmuskateller) **139**
Moschofilero **139**
Mosler *see* Furmint
Moster *see* Chasselas
Mourisco **139**
Mourvèdre (Esparte; Morastell; Morrastel; *see also* Mataro; Monastrell) 15, 95, 97, 99, **140-1**, 203, 251, 291
Müller-Thurgau (*see also* Riesling-Sylvaner; Rivaner) 11, 23, 81, 89, **142**, 165, 189, 194, 225
Müller-Traube *see* Meunier
Müllerebe (*see also* Meunier) **142**
Murteira *see* Trincadeira Preta
Muscadel 151, 153
Muscadelle (*see also* Tokay) **142-3**, 151, 233, 236
Muscadet (*see also* Melon de Bourgogne) **143**
Muscardin **143**
Muscat (*see also* Tamîioasa) 9, 15, 143, **144-53**, 195, 291
Muscat of Alexandria **143**, 144-53
Muscat Blanc à Petits Grains **143**, 144-53
Muscat d'Alsace 153
Muscat Canelli **143**, 151, 153
Muscat de Colmar 147, 150
Muscat Dr Hogg 151
Muscat de Frontignan **143**, 153
Muscat Gordo Blanco **143**, 151, 153
Muscat of Hamburg (see also Black Muscat) **143**, 149, 151
Muscat Lunel 150, 153
Muscat Ottonel **143**, 146-53
Muscat de Rivesaltes 149
Muscat Romain 153
Muscat de Saumur 147
Muskadel **143**, 153
Muskat-Ottonel 153
Muskat-Silvaner/Sylvaner (*see also* Sauvignon Blanc) **143**
Muskateller **143**, 147, 153
Muskotály **143**, 153

# N

Napa Gamay (*see also* Valdiguié) 91
Nebbiolo (Chiavennasca; Picutener; *see also* Spanna) 9, 38, 52, 69, 85, 89, **154-63**, 271, 273
Negoska **164**
Negra Corriente 138
Negra Mole (Negramoll; *see also* Preto Martinho) **164**
Negra Peruana *see* País
Negramoll *see* Negra Mole
Négrette (Pinot St George; *see also* Folle Noire) **164**
Negroamaro 122, **164**
Negru de Dragasani **164**
Nerello 90, **164**
Nerello Cappuccio 164
Nerello Mascalese 164
Nero d'Avola (Calabrese Nero) 53, **164**, 213, 249, 253
Neuburger **164**
Niagara **164**
Nielluccio (*see also* Sangiovese) **164**
Nocera 90
Noirien *see* Pinot Noir
Noirin Enfariné *see* Meunier
Nosiola **164**
Nuragus 69, **164**

# O

Oeillade 84
Ojo de Liebre *see* Tempranillo
Olasz Rizling (*see also* Welschriesling) **165**
Ondenc **165**
Oporto *see* Kékoporto *and* Portugieser
Optima **165**
Orange Muscat 151, **165**
Oremus **165**
Ormeasco *see* Dolcetto
Ortega **165**
Ortrugo **165**
Oz **165**

# P

Paarl Riesling *see* Crouchen
Paderña 117
Pagadebit *see* Bombino Bianco
Païen *see* Gewürztraminer
País (Negra Peruana; *see also* Mission) 139, **165**
Palomino Basto 165
Palomino Fino (Fransdruif; Golden Chasselas; *see also* Listán) 38, 59, 112, **165-6**
Palomino de Jerez 165
Pansa Blanca (*see also* Xarel-lo) **166**
Pardillo **166**
Parellada 117, **166**, 285
Pé de Rato *see* Alfrocheiro Preto
Pedernã *see* Arinto
Pedro Giménez **166**
Pedro Ximénez (PX) **167**
Peloursin 88, 168, 253
Periquita (*see also* Castelão Francès) **167**, 268
Perlan *see* Chasselas
Perricone **167**
Perrum 166
Petit Bouschet 35
Petit Courbu (Courbu) 38, **167**
Petit Manseng 38, **167**
Petit Rouge **167**
Petit Verdot 51, 54, **167-8**
Petite Arvine **168**
Petite Sirah 88, **168**, 253, 291
Petite Syrah (*see also* Syrah) **168**, 250, 252
Petite Vidure *see* Cabernet Sauvignon
Peurion 112
Picolit **168-9**
Picpoul (Piquepoul) **169**
Picutener *see* Nebbiolo
Piedirosso **169**
Pigato **169**
Pignola 159, 169
Pignolo **169**
Pinat Cervena *see* Gewürztraminer
Pineau *see* Chenin Blanc *and* Pinot Noir
Pineau d'Aunis **169**
Pineau de la Loire *see* Chenin Blanc
Pinot Beurot (*see also* Pinot Gris) **169**
Pinot Bianco (*see also* Pinot Blanc) **169**
Pinot Blanc (Beli Pinot; *see also* Klevner; Pinot Bianco; Weissburgunder) 9, 65, 68, 89, 123, 164, 177, 195, **170-1**
Pinot Blanco *see* Chenin Blanc
Pinot Chardonnay (*see also* Chardonnay) 11
Pinot Droit 179
Pinot Fin 179
Pinot Grigio (*see also* Pinot Gris) **169**
Pinot Gris (Tokay-Pinot Gris; see

*also* Grauburgunder; Malvoisie; Pinot Beurot; Pinot Grigio; Rülander; Szürkebarát; Tokay d'Alsace) 10, 122, 147, **172-3**, 177
Pinot Liébault **169**
Pinot Meunier see Meunier
Pinot Nero (*see also* Pinot Noir) 159, **169**
Pinot Noir (Blauer Spätburgunder; Burgundac Crni; Noirien; Pineau; Savagnin Noir; *see also* Blauburgunder; Klevner; Pinot Nero; Spätburgunder) 9, 10, 12, 14, 15, 18, 19, 21, 65, 88, 112, 138, **174-85**, 188, 195, 207, 284
Pinot St George see Négrette
Pinot Tordu 179
Pinotage 11, 84, **186-7**
Piquepoul see Folle Blanche *and* Picpoul
Plavac Mali **188**, 289
Poerininha see Baga
Portugieser (Oporto; Portugais Bleu; Portugaljka; Portugizac Crni; *see also* Blauer Portugieser; Kékoporto) 42, **188**
Posip 90
Poulsard **188**
Pressac (*see also* Malbec) **188**
Preto Martinho (*see also* Negra Mole) **188**
Prieto Picudo **188**
Primitivo (*see also* Zinfandel) **188**, 213
Princ see Gewürztraminer
Procanico 69, **188**, 270
Prosecco **188**
Prugnolo Gentile (*see also* Sangiovese) 159, **188**
PX see Pedro Ximénez

## R

Rabigato see Rabo de Ovelha
Rabo de Anho 188
Rabo de Lobo 188
Rabo de Ovelha (Rabigato) **188-9**
Rabo de Ovelha Tinto (*see also* Trincadeira Preta) 188
Rabo de Porco 188
Rabo de Vaca 188
Raboso **189**
Raboso Piave (Raboso Friulara) 189
Raboso Veronese 189
Ramisco **189**
Rebula see Ribolla Gialla *and* Robola
Refosco 138, **189**
Refosco d'Istria 189
Refosco Nostrano 189
Refosco del Peduncolo 189
Refosco del Terrano (Cagnina; Teran) 189
Reichensteiner **189**
Rhein Riesling see Riesling
Rhine Riesling see Riesling
Rhoditis (*see also* Roditis) **189**, 202
Ribolla 68
Ribolla Gialla (Rebula; *see also* Robola) **189**
Ribolla Nera see Schioppettino
Rieslaner **189**
Riesling (Rhein Riesling; Rhine Riesling; Weisser Riesling; White Riesling; *see also* Johannisberg Riesling; Riesling Renano) 9, 11, 15, 16, 19, 23, 85, 88, 106, 114, 115, 138, 142, 229, 291, **190-201**
Riesling Italico (*see also* Welschriesling) **202**
Riesling Renano (*see also* Riesling) **202**
Riesling-Sylvaner (Rizling Zilvani; *see also* Müller-Thurgau) **202**
Rivaner (*see also* Müller-Thurgau) **202**
Rizling **202**
Rizling Vlassky see Welschriesling
Rizling Zilvani see Riesling-Sylvaner
Rkatsiteli (Baiyu) **202**
Robola (Rebula; *see also* Ribolla Gialla) **202**
Roditis (*see also* Rhoditis) **202**, 228
Roi des Noirs 229
Rolle **202**, 273
Romorantin **203**
Rondinella **203**
Roriz (*see also* Tempranillo) **203**, 268
Rosenmuskateller (*see also* Moscato Rosa) 149, 150, 153, **203**
Rossese **203**
Rossola Bianca see Ugni Blanc
Rossola Nera 159, **203**
Rossolo 203
Rotclevner see Gewürztraminer
Roter Gutedel 114
Roter Muskateller 143
Roter Traminer see Gewürztraminer
Roter Veltliner 114, 271
Rotgipfler **203**, 285
Roublot 112
Rouchet see Ruché
Roupeiro (Alva; Códega) **203**
Roussanne (Bergeron; Roussette) 124, 202, **204-5**, 249, 250, 280, 281
Roussanne du Var 205
Roussette (Altesse; *see also* Roussanne) **206**
Ruby Cabernet **206**
Ruché (Rouchet) **206**
Ruländer (*see also* Grauburgunder; Pinot Gris) **206**
Rusa see Gewürztraminer

## S

Sacy 112, **206-7**, 270
Sagrantino **207**, 214
St-Émilion see Ugni Blanc
Sämling (*see also* Scheurebe) **207**
Samtrot 138
Sangiovese (Brunello; Sangioveto; *see also* Morellino; Nielluccio; Prugnolo Gentile) 9, 51, 53, 58, **208-17**, 122, 133, 207, 249, 253
Sangioveto see Sangiovese
Sankt Laurent 182, **207**, 285
Santarén see Joao de Santarém
Saperavi 164, **207**
Sauvignonasse (Sauvignon Vert) 11, 134, 221, 224, 225, 227, **228**, 267
Sauvignon Blanc (*see also* Blanc Fumé; Muskat-Silvaner/ Sylvaner) 9, 16, 24, 42, 49, 54, 106, 114, 166, 203, **218-27**, 228, 233, 235, 236, 271
Sauvignon Gris 221, **228**
Sauvignon Rosé 221, 228
Sauvignon Vert see Sauvignonasse
Savagnin (Heida) **228**
Savagnin Noir see Pinot Noir
Savagnin Rose (Heiligensteiner Klevener; Traminer) 105, 115
Savatiano 38, 189, 202, **228**
Scheurebe (*see also* Sämling) 11, **228-29**
Schiava (*see also* Trollinger; Vernatsch) **229**
Schiava Gentile (Kleinvernatsch) 229
Schiava Grigia (Grauvernatsch) 229
Schiava Grossa (Grossvernatsch) 229, 270
Schioppettino (Ribolla Nera) **229**
Schönburger **229**
Schwarzriesling (*see also* Meunier) **229**
Sciacarello 164, **229**
Scuppernong **229**
Seibel **229**, 240
Sémillon (Boal; Hunter Valley Riesling; Wyndruif) 9, 224, **230-9**
Sercial (Cerceal; *see also* Esgana Cão) 117, **240**
Seyval Blanc 9, 225, **240**
Shiraz (*see also* Syrah) 19, 31, 54, 55, 97, 99, 198, **240**
Siegerrebe 165
Silvaner (Grüner Sylvaner; Johannisberg; Sylvaner) 88, 89, 139, 142, 164, 229, **242-3**
Sipon 90
Sousão **240-1**, 273
South African Riesling see Crouchen
Spagna 139
Spanna (*see also* Nebbiolo) **241**
Spätburgunder (*see also* Pinot Noir) **241**
Steen (*see also* Chenin Blanc) **241**
Straccia Cambiale see Bombino Bianco
Sultana 241
Sumoll 84
Sylvaner see Johannisberg *and* Silvaner
Syrah (*see also* Petite Syrah; Shiraz) 9, 15, 19, 31, 51, 88, 95, 98, 138, 168, 213, 214, **244-55**, 280
Szürkebarát (*see also* Pinot Gris) **241**

## T

Tamîioasa (*see also* Muscat) **241**
Tamîioasa Româneasca 112
Tannat **241**
Tarrango **241**
Teinteurier du Cher 35, 169
Tempranillo (Aragonez; Cencibel; Ojo de Liebre; Tempranilla; Tinta Roriz; Tinto de Madrid; Tinto del País; Tinto de la Rioja; Tinto de Santiago; Tinto de Toro; Ull de Llebre; *see also* Roriz; Tinta Aragonez; Tinto Fino) 9, 95, 97, 98, 249, **256-65**, 284
Tempranillo Peludo 259
Teran see Refosco del Terrano
Termeno Aromatico see Gewürztraminer
Teroldego 53, **266**
Terrantez (Truel) **266**
Terret **266**
Thalia see Trebbiano *and* Ugni Blanc
Tinta Amarela (*see also* Trincadeira Preta) 89, **266**
Tinta Aragonez (*see also* Tempranillo) **266**
Tinta Bairrada see Baga
Tinta Barroca **266**, 267
Tinta Caiada **266**
Tinta de Baga see Baga
Tinta Fina see Baga
Tinta Francisca 259, **266**
Tinta Merousa see Castelão Francès
Tinta Miúda (*see also* Graciano) **266**
Tinta Negra Mole 84, 117, 164, **266**
Tinta Roriz see Roriz *and* Tempranillo
Tintilla 117
Tinto Aragonés see Garnacha Tinta
Tinto Cão **267**, 268
Tinto Fino (*see also* Tempranillo) **267**
Tinto de Madrid see Tempranillo
Tinto del País see Tempranillo
Tinto de la Rioja see Tempranillo
Tinto de Santiago see Tempranillo
Tinto de Toro see Tempranillo
Tocai 273
Tocai Friulano 228, **267**
Tocai Italico 267
Tokay (*see also* Muscadelle) **267**
Tokay d'Alsace (*see also* Pinot Gris) **267**
Tokay-Pinot Gris see Pinot Gris
Torbato 122, **267**
Torrontés **267**, 270
Torrontés Mendocino 267
Torrontés Riojano 267
Torrontés Sanjuanino 267
Touriga (*see also* Touriga Nacional) 241, **267**
Touriga Franca **267**, 268
Touriga Nacional (Mortágua; *see also* Touriga) 15, 39, 115, **268-9**
Trajadura (*see also* Treixadura) 117, **270**
Traminac see Gewürztraminer
Traminer (*see also* Gewürztraminer; Savagnin Rose) **270**
Tramini see Gewürztraminer
Trebbiano (Thalia; *see also* Ugni Blanc) 68, 112, 114, 188, **270**
Trebbiano d'Abruzzo see Bombino Bianco
Trebbiano Campolese 270
Trebbiano della Fiamma 270
Trebbiano Gallio 270
Trebbiano di Lugana 270
Trebbiano Romagnolo 270
Trebbiano di Soave 270
Trebbiano Toscano 270
Trebbiano Veronese 270
Treixadura (*see also* Trajadura) 112, 117, 123, **270**
Tresallier **270**
Trincadeira (Crato Preto; Mortágua; Murteira; *see also* Espadeiro; Rabo de Ovelha Tinto; Tinta Amarela) **270**
Trollinger (Blauer Trollinger; *see also* Schiava; Vernatsch) 115, 116, **270**
Trousseau (*see also* Bastardo) 188, **270-1**
Trousseau Gris 271
Truel see Terrantez

## U/V

Ughetta see Vespolina
Ugni Blanc (Rossola Bianca; St-Émilion; Thalia; *see also* Clairette; Trebbiano) 39, 84, 203, 233, **271**
Ull de Llebre see Tempranillo
Uva Rara (*see also* Bonarda) **271**
Uva di Troia **271**
Valdepeñas 263
Valdiguié (*see also* Napa Gamay) **271**
Valenciana 112
Valpolicella 53
Veltlin Zelene 114
Veltliner (*see also* Grüner Veltliner) **271**
Veltlinkske Zelené 114
Verdeca 69
Verdejo **271-2**
Verdelho (*see also* Godello; Gouveio) 77, 117, **272**
Verdello see Godello
Verdicchio **272**
Verduzzo **272**
Verduzzo Friulano 272
Vermentino (Malvoisie de Corse) 68, 122, 169, 203, **273**
Vernaccia **273**
Vernatsch (*see also* Schiava; Trollinger) **273**
Vespaiolo (Vespaiola; Vesparolo) **273**
Vespolina (Ughetta) 159, **273**
Vidal **273**
Vinhão 240, **273**
Viognier 9, 31, 69, 124, 202, 249, **274-83**
Viosinho **284**
Vital 120, **284**
Viura (*see also* Macabeo) 271, **284**
Vöslauer see Blauer Portugieser

## W/Y/Z

Wälscher see Chasselas
Weissburgunder (Weisser Burgunder; *see also* Pinot Blanc) 89, 139, **284**
Weisser Gutedel see Chasselas
Weisser Riesling see Riesling
Welschriesling (Grasevina; Rizling Vlassky; *see also* Laski Rizling; Olasz Rizling; Riesling Italico) 43, 253, **284**
White Kadarka see Izsáki
White Riesling see Riesling
Wildbacher (Blauer Wildbacher) **285**
Wyndruif see Sémillon
Xarel-lo (*see also* Pansa Blanca) 117, 166, **285**
Xynisteri **285**
Xynomavro 164, **285**
Zapfner see Furmint
Zeta 165
Zibibbo 150, 153, **285**
Zierfandler 203, 289, **285**
Zinfandel (*see also* Primitivo) 9, 19, 20, 134, 188, **286-95**
Zöldveltelini 114
Zweigelt (Blauer Zweigelt; Zweigeltrebe) 182, 207, **285**

# GENERAL INDEX

**Page numbers in *italic* refer to illustration captions.**


## A
A Mano 295
Abadía Retuerta 265
Abbazia di Novacella 111, 227, 243
Abbazia Santa Anastasia 164
Abbona, Marziano & Enrico 87
Abrigada, Quinta da 35, 58
Accordini 85, 203
Accornero, Giulio & Figli 41, 113
Achaïa-Clauss 123, *202*, 202, 228
Achard-Vincent 84, 153
Achiary, D 101
Achs, Paul 207
acidification see chaptalization and acidification
Acquese, Viticoltori dell' 41, 43, 87
Adam, J-B 111, 153, 171, 173
Adami 188
Adams, Tim *99*, 101, 201, 239, 255
Adanti 207
Adelmann, Graf 88, 116
Adelsheim 173
Adler Fels 111
Agapito Rico 138, 141
Agramont 117
Agricola, La 43, 119
Agro de Bazán *36*, 37
Aguilas, Pierre 83
Aigle, Dom. de l' 83, 123
Aiguelière, L' 255
Aires Hautes, Dom. des 119
Alamos Ridge 119
Alanis, Bodega 267
Alario, Claudio 87, 163
Alary, D & D 101
Alaveses, Cosecheros 117
Alban Vineyards 101, 205, 255, *281*, 283
Albertini Frères 229
Albet i Noya 265, 285
Albrecht, Lucien 111, 153, 173
Alcântara Agricola 38
Aldea de Abaixo 37
Alderbrook 295
Alderiz, Quinta de 37
Alella, Marqués de 166
Além, Quinta de 37
Alessandria, Gianfranco 41, 163
Alfonso, Bodegas 141
Aliança, Caves 35, 39, 42, 122, 269
Alión, Bodegas 265
Alkoomi 201, 227
All Saints 43, 125, 143, 153
Allandale 239
Allegrini 85, 203
Allemand 255
Allende, Finca 265
Alliet, Philippe 45
Allimant-Laugner 111
Almeida, Fundaçáo Eugenio de 270
Almonazar, Romero 265
Almondo, Giovanni 38
Alpha Domus 57
Alphart, Karl 203, 285
Alquier, Frédéric 125, *204*, 205
Alquier, Jean-Michel 141
Alta Vista 119
Altamura 217
Altare, Elio *40*, 41, 87, *160*, 163
Altesino 139, 217
altitude, vineyard 16
Alto Adige, Viticoltori 115, 116
Altos de Medrano 119
Altos de Temporada 119
Alvarez y Díez 272
Alvear 167
Alves e Sousa, Domingos 273, 284
Alzipratu, de 164
Ama, Castello di *133*, 137, 217
Amador Foothill Winery *293*, 295
Amarine, Ch. de l' 84, 91
Ambra 217
Ambroise, Maison Bertrand 185
Amézola de la Mora 123
Amiot-Bonfils, Guy 73
Amity Vineyards 111, 171, 178
Amouriers, Dom. des 101
Ancienne Cure, Dom. de l' 239
Anderson, S 138
Andrew Will 57, *133*, 137, 153
Anfossi 169
Angélus, Ch. 19, *136*, 137
Angerville, Marquis d' 185
Angludet, Ch. d' 57
Annereaux, Ch. 137
Anselmi 91
Antichi Vigneti di Cantalupo 163, 271
Antinori, Marchesi *52*, 57, 73, 122, 212, *213*, 215, 217, 270
Antonelli 112, 207
Antoniolo 89, 271
Antonopoulos 34
Anubis 43, 119, *263*, 265
Apaltagua 61
Araldica Vini Piemontesi 38, 41
Araujo 57, 168, 227, 255, 283
Archanes co-op 115, 122
Arche, Ch. d' 239
Archery Summit 171, 173, 185
Ardennes, Ch. d' 239
Aréna, Antoine Dom. 153, 164
Argentina *24*, *26*, 40, 41, 43, 53, 55, 89, 118, 119, 131, 135, 138, 161, 166, 168, 189, 215, 225, 240, 253, 261, 263, 265, 267, 271, 281
  Mendoza 85, 118, *119*, 253
  Tupungato 71
Argiano 217
Argiolas 58, 101, 139, 164, 273
Argüeso 166
Arizu, Leoncio 255
Arjolle, Dom. de l' 283
Arlay, Ch. d' 228
Arlot, Dom. de l' 169, 173
Armand, Comte 185
Arrowood Vineyards 73, 137, 283
Arroyo, Ismael 265, 267
Arruda co-op 266
Artadi 112, *262*, 265, 284
Arzuaga 265
Ascheri, Giacomo 87, 90, *281*, 283
Ashanti Estate 187
Ashbrook 239
Ashton Hills 185
Ata Rangi *183*, 185
Atlas Peak 212, *215*, 217
Attems, Conti 169
Au Bon Climat 73, *161*, 171, 185
Aubineau, Jean 84
Aubuisières, Dom. des 83
Aucoeur, Dom. 91
Aujoux, Jean-Marc 91
Ausone, Ch. 45, 137
Australia 23, *26*, 41, 45, 57, 59, 73, 88, 101, 106, 109, 111, 119, 125, 131, 137, 141, 153, 173, 185, 201, 205, 215, 217, 227, 239, 243, 255, 263, 270, 271, 283, 295
  Adelaide Hills 70, 225, 281, 293
  Barossa Valley 19, 54, 99, 235, 248, 249, 251, 281
  Clare Valley *19*, 54, 194, 198, 235, 252
  Coonawarra 50, *51*, 54, 84
  Cowra 272
  Eden Valley 70, 194, 199, 278, 279, 281
  Geelong 70, 281
  Glenrowan 43, 148, 151, 267
  Hunter Valley 66, 70, 84, 234, 235, 236-7, 251
  King Valley 160
  Langhorne Creek 272
  Macedon Ranges 70
  Margaret River 51, 54, 70, 81, 161, 225, 293
  McLaren Vale 99, 249, 280
  Mornington Peninsula 161, 281
  Murray River 280
  Padthaway 70, 225
  Riverina 237
  Riverland 70, 164, 241
  Rutherglen 43, 112, 148, *149*, 149, 151, 267
  Southern Mount Lofty Ranges 249
  Southern Vales 225
  Swan Valley 81, 85, 272
  Tasmania 66, 70, 109
  Willunga Flats 249
  Yarra Valley 54, 70, 135, 183, 293
Austria 42, 53, 69, 73, 108, 111, 131, 139, 143, 153, 164, 171, 173, 188, 201, 206, 224, 227, 229, 271, 284
  Burgenland 42, 43, 90, 133, 150, 161, 170, 182, 207, 253
  Kamptal 114
  Neusiedlersee 43, 115, 150, 199
  Thermenregion 203, 207, 285
  Vienna 199
  Wachau 25, 114, 150, 199
Avelar, Quinta do 38, 188
Aveleda, Quinta da 37, 117, 270
Avignonesi 35, 112, 137, 217
Avio co-op 173
Avontuur 187
Aydie, Ch. d' 114, 241
Ayuso 34
Azelia 41, 163

## B
Babcock 227
Babich 227
Bachelet, Denis 185
Backsberg 187
Baden, Markgraf von 39, 142
Badia a Coltibuono 217
Badoux, Henri 59
Bágeiras, Quinta das 39
Baguinha, Quinta da 37
Baileys 143, 153
Bailly, F & J 227
Bailly-Reverdy 227
Balatonboglar Winery 115
Balbás 265
Balbi, Bodegas 119
Balcona, Bodega 141
Ballandean Estate 243
Balthazar 255
Banfi, Castello 57, 73, 85, 217
Banfi Strevi 43
Bannockburn 185
Banrock Station 255
Banti, Erik 139, 217
Barbadillo 166
Barbaresco, Produttori del 163
Barbatella, La 41
Barbeito 121, 266
Barberani-Vallesanta 112
Barbero, Pietro 41
Barbi 217
Barbier, René *97*
Barbieri, Casimiro 116
Barceló, Antonio, Hijos de 265
Barge, Gilles 255, 283
Barge, P 255
Barkham Manor 39
Barmès-Buecher 111, 173
Barnaut, Edmond 185
Barone de Cles co-op 115
Barraud, Dom. Daniel 73
Barréjats, Ch. 239
barrels 30-1
Barros e Sousa 42, 117, 121, 240, 266, 272
Barry, Jim 119, *251*, 255
Barthod, Ghislaine 185
Basciano 217
Basedow 255
Bass Phillip 185
Bassermann-Jordan 201
Bastide Blanche, Dom. de la 141
Bastor-Lamontagne, Ch. 239
Battistotti 122
Baudoin, Patrick *80*, 83
Baudry, Bernard 45
Baumard, Dom. des 78, 83
Baume, Dom. de la *132*, 137
Bava 41, 206
Beau-Séjour Becot, Ch. 137
Beaucastel, Ch. de 85, 113, 125, 141, *204*, 205, 253
Beauregard, Ch. 45
Beaurenard, Dom. de 101
Beaux Frères 185
Bechtold 111
Beck (Austria) 164
Beck, Graham (South Africa) 84, 187, 227, 255
Becker 153
Bedell Cellars *135*, 137
Beira Mar, Adegas 189
Belair, Ch. 45
Belingard, Dom du 239
Belisario 272
Bellavista 73
Belle, Dom. 205, 125, 255
Bellei, Francesco 116
Bellerive 83
Bellet, Ch. de *203*
Bellevue, Ch. 239
Bellevue-la-Forêt, Ch. *164*, 164
Bellingham 45, 187
Belondrade y Lurton 272
Benanti 164
Benmarl 240
Benziger 168
Bera, Fratelli 153
Berberana 123
Bercher 112, 153, 171, 173, 185, 241, 284
Bergdolt *170*, 171, 284
Berger, Frères 83
Bergerie, Dom. de la 83
Beringer 57, 73, *135*, 137, 295
Bernarda, Rocco 267
Bernardins, Dom. des 153
Bernhard-Reibel 111
Berrod, Dom. 91
Bertani 85, 203
Bertelli 41, 125, 205, 255
Berthoumieu, Dom. 241
Bertolotto 43
Best's 255
Bethany 239
Bethel Heights 185
Beurdin, Henri 173
Beyer, Léon 111, 173, 201
Beyerskloof *186*, 187
Biale, Robert 295
Biddenden 114
Biletta 206
Billecart-Salmon *68*, 138
Bindella 217
biodynamic viticulture 22
Biondi-Santi *215*, 217
Birot, Ch. 239
Bisci 272
Bisol e Figli, Desiderio 188
Bisquertt 61, 137
Bize, Simon 169
Blaauwklippen 295
Blanca, Casa 101
Blanck, Dom. Paul 111, 171, 173, *197*, 201, 243
Blankenhorn 114
Blandy 42, 117, 121, 240, 266, 272
Blard et Fils 115
Bleasdale 272
blending 31
  Cabernet Sauvignon 51
  Nebbiolo 159
  Riesling 195
  Sangiovese 213
  Syrah/Shiraz 249
  Zinfandel 291
Blin et Fils, R 138
Blockheadia Ringnosii 295
Blue Mountain 171, 173
Blumeri, Poderi dei 173
Boãs, Quintas 269
Boavista, Quinta do 38, 89
Boccadigabbia 139, 217
Bocopa co-op 141, 153
Boeger 295
Boekenhoutskloof 57, 239, 255
Boesch 111
Boglietti, Enzo 41, 87, 163
Boillot, Jean-Marc 73, 185
Bolivia 225
Bolognani, Nilo 142
Bon Pasteur, Ch. le 137
Bongiovanni 41, 87
Boniface, Pierre 115
Bonneau, Henri 101
Bonneau du Martray 73
Bonnefond, P & C 283
Bonnet, Ch. 239
Bonnevaux, Dom. de 45
Bonny Doon 41, 101, 121, 141, 153, *205*, 205
Bonserine, Dom. de 255
Bonvin, Charles 168
Boplaas 187, 241, 269
Borderie, Ch. la 239
Borges, H M 121
Borgo Conventi 267

Borgo San Daniele 267
Borgo del Tiglio 121, 267
Borlettí 189
Borrajo, Dominguez 37
Borruel, Bodegas 139
Borsao, Bodegas 101
Bortolin, Fratelli 188
Bosca, Luigi 119, 255
Boscaini 85, 203
Boscarelli 122, 217
Boschis, Francesco 87
Bosquet des Papes, Dom. 101
Botromagno 112
*Botrytis cinerea* 27, 195, 235
Bott-Geyl 111, 153, 171, 173, 243
Bouchaine 111
Bouchard Finlayson 185
Bouchard Père 185
Boulay, Pierre 227
Bourdy, Jean 188, 228
Bourgeois, Henri 227
Bourillon-Dorléans 83
Boussière, Dom. la 101
Boutari 34, 164, 285
Bouvet-Ladubay 45, 113
Bouwland 187
Bouzereau, Michel 73
Bovard, Louis 59
Bowen 255
Boxler, Albert 111, 153, 171, 173, 201, 243
Boyar, Dom. 115, 123, 138
Braćcesca, La 217
Braida *40*, 41, 43, 113, 153
Brana, Etienne 114, 241
Brancott Estate *223*
Brazil 85, 168, 225, 270, 271, 281
Breaky Bottom *240*
Brédif, Marc 83
Brema 87
Bretón, Bodegas 265, 284
Breuer, Georg 201
Breuil, Ch. du 83
Bricco Maiolica 87, *162*, 163
Bridgeview 111
Bridgewater Mill 227
Broglia, Gian Piero 85
Brokenwood 239, *251*, 255
Brolio, Castello di 217
Brondelle, Ch. 239
Brookfields 137
Brookland Valley 137, 227
Brotons, Bodegas 141
Brovia 38, 41, 87, 163
Brown Brothers 41, 153, 241, 267
Bru-Baché, Dom. 167
Brulesécaille, Ch. 137
Bruna 169
Bründlmayer, Willi *114*, 201
Brunelli 85
Brunori 272
Brusset, Dom. 101
Bryant family 57
Buçaco Palace Hotel 39
Bucci 272
Buitenverwachtung 227
Bulgaria 35, 53, 69, 85, 91, 123, 133, 138, 151, 182, 202, 270, 271
Buonamico, Fattoria del 171
Burgaud, Dom. Bernard 255, 283
Burgess Cellars 295
Buring, Leo 201, 239
Bürklin-Wolf 153, 201
Burn, Ernest 111, 153, 173
Burrowing Owl *172*, 173
Bussola, Tommaso 85, 203

C

Ca' del Bosco 45, 57, *69*, 73, 169, 185
Ca' dei Frati *270*
Ca' Ronesca 121, 169
Ca' Rugate 91
Ca'Viola 87
Caccese, Paolo 121, 202
Cachão, Quinta do 269
Cachazo, Angel Lorenzo 272
Cadalora, La 122
Cady, Dom. Philippe 83
Cafaro 137
Caggiano, Antonio 34
Cailbourdin, Alain 227
Cailleau, Pascal 83
Caillou, Ch. 239
Cailloux, les 101
Cain Cellars 168
CAIR 122
Cairnbrae 227
Cakebread Cellars 295
Calatrasi 58, 164
Calcinaire, Le 273
Caldaro co-op 111, 142
Calem 269
Calera 73, *182*, 185, 276, *280*, 281, *282*, 283
California 15, 16, *18*, 19, *20*, 23, *25*, 35, 41, 57, 59, 73, 83, 86, 87, 88, 101, 111, 121, 137, 138, 141, 143, 153, 163, 165, 169, 171, 173, 178, 185, 205, 217, 222, 223, 225, 227, 239, 255, 263, 270, 271, 279, 281, 283, 295
  Bay Area 58, 290, 291, 292
  Central Valley 69, 81, 84, 96, 99, 130, 151, 161
  Mendocino County 54, 109, 133, 168, 199
  Monterey County 54, 69, 109, 134, 199
  Napa Valley 54, 69, 91, 131, 133, 182, 212, 214-5, 237, 290
  San Luis Obispo County 54, 252, 253, 292
  Santa Cruz 54
  Santa Maria Valley 161, 182
  Sierra Foothills 199
  Sonoma Valley 50, 54, 69, 109, 133-4, 168, 182, 290, 292
Caliterra *61*, 137
Callaghan Vineyards 115, *294*, 295
Calligas 202
Calò, Giuseppe 164
Calò, Michele 164
Cambas 139
Campbells 43, 88, 143, *152*, 153
Campillo 112, 265
Campo Viejo 265
Campogiovanni 217
Can Ràfols dels Caus 137
Canada 39, 45, 85, 111, 122, 171, 173, 183, 195, 229, 270, 273
  British Columbia 70, *106*, 170, 173
  Ontario 70, 199; Pelee Island 45
  Quebec 70
Canale, Humberto 119
Candido, Francesco *35*, 35, 122, 164
Canet-Valette 141
Canon, Ch. 45
Canon-de-Brem, Ch. 137
Canon-la-Gaffelière, Ch. *7*, 45, 51, 131, 137
Capaccia 217
Caparra & Siciliana 91
Caparzo 217
Capçanes, Celler de 91, 101
Cape Mentelle 57, 239, *293*, 295
Capezzana 121, 217
Capichera 273
Caprai, Arnaldo 112, 207
Caraguilhes, Ch. de 43, 84, 113
Carballo 121
carbonic maceration 29-30
Carbonnieux, Ch. 227
Carema, Cantina dei Produttori 163
Carillon et Fils, Dom Louis 73
Carletti 139
Carmen 57, 61, 168, 185, 255, 137
Carmenet 239
Carmo, Quinta do 35, 89, 189, 203, 266
Carobbio 217
Carpenè 188
Carr Taylor 114, 189, 229
Carraia, La 217
Carras, Dom. 38, 117
Carretta 38
Carsin, Ch. 137, 228, 239
Carvalhais, Quinta dos 203, 265, 269
Carvalhinho, Quinta do 39
Casa Lapostolle 61, 73, *136*, 137, 227
Casa Silva *60*, 61, 137
Casal Branco, Quinta do 58, 89
Casale della Ioria 59
Casale Marchese 59
Casali 116
Casalinho, Caves do 39
Casaloste 217
Casalte, Le 217
Casanova di Neri 217
Casarta Monfort 142
Cascabel 141
Casenove, Dom. la 91, 117
Caslot-Galbrun 45
Casòn Hirschprunn 125
Cassan, Dom. de 101
Castaño 138, 141
Castelgiocondo 137
Castellada, La 189
Castellare 217
Castellblanch 117, 166, 285
Castellina, La 183
Castelluccio 217
Castilla, Camilo 153
Castilla, Vinicola de 34
Castilla la Vieja 272
Castlèt, Cascina 41
Catena 57, 73, *118*, 119
Cattin, Théo 153
Caudrina 90, 153
Cauhapé, Dom. 114, 167
Causses-Marines, Dom. de 116, 165
Cavas Hill 117
Cavas de Weinert 119
Cavatappi 163
Cavicchioli 116
Cavit 122
Cayla 239
Caymus 57, 227, 283
Cayron, Dom. de 101
Cayrou, Ch. du 119, 241
Cayuse 283
Cazal-Viel, Ch. 283
Cazeneuve, Ch. de 205
Cazes, Dom. 101, 137, 153
Ceccetto 189
Cecchi 139
Cedar Creek 171
Cederberg 241
Cèdre, Ch. du *118*, 119
Celli 35
Cellier Le Brun *71*
Cellier des Samsons 91
Cellier des Templiers *99*, 101
Cennatoio 217
Cephalonia 202
Ceretto 38, 163
Cérons, Grand Enclos du Ch. de 239
Cesani, Vincenzo 273
Cesari, Umberto 35
Cescon, Ivan 189
Cesconi 111, 173
Cetto, L A 163, 168
Chai des Vignerons, le 203
Chain of Ponds 227
Chalone *69*, 73, 83, 171, 185
Chambers 143, 153
Champalou, Didier 83
Champy, Maison 185
Chandon de Briailles, Dom. 185
Chantegrive, Ch. de 239
Chapel Down 39, 88, 142, 189, 240, 272
Chapel Hill 255
Chapelle Lenclos, Dom. la 241
Chapoutier 101, *125*, *250*, 252, 255, 283
Chappellet 83
chaptalization and acidification 29
Charbonnière, Dom. de la 101
Charlopin, Philippe 185
Charmes, Ch. de 45
Charvet, Fernand 91
Chasse-Spleen, Ch. 57
Château-Grillet *275*, 277, *279*, 280, 283
Chateau Musar 53, *84*, 253
Chateau Potelle *291*
Chateau Reynella 28
Chateau St Jean 115, 137
Chateau Ste Michelle 73, 137
Chateau Tahbilk *124*, 125
Chateau Xanadu 239
Chatelain, Jean-Claude 227
Chatsfield 45
Chave, Jean-Louis 125, 205, 255
Chehalem 173
Chênes, Dom. des 101, 205
Chenonceau, Ch. de *75*
Cherchi, Giovanni 273
Chéreau-Carré, Dom. 123, 143
Cheval-Blanc, Ch. *44*, 45
Chevalier, Dom. de *30*, 57, *222*, 226, 227, 239
Chevillon, Robert, Dom. *184*, 185
Chèze, Louis 283
Chiarlo, Michele 41, 85, *162*, 163
Chico, Cascina 38
Chidaine, Yves 83
Chile 19, 23, 35, 45, 53, 60, 61, 73, 106, 111, 113, 119, 134, 137, 161, 165, 166, 179, 183, 185, 215, 222, 227, 228, 237, 253, 255, 267, 281, 293, 295
  Aconcagua Valley 55, 135
  Bío-Bío 138
  Casablanca Valley 71, 223, *225*, 225, 227
  Central Valley 16
  Colchagua Valley 55, *61*, 135
  Maipo Valley *50*, 55, 71, 135
  Maule 55, 138
China 202
Chinook 45
Chionetti, Quinto *86*, 87
Chivite, Julian *98*, 265
Choblet , Luc 123, 143
Christoffel, J J 201
Churchill 112, 269
Ciacci Piccolimini 217
Cigliuti, Fratelli 41, 87, 163
Cimicky, Charles 101, 137, 255
Clair, Bruno 169, 185
Clairette de Die co-op 153
Clape, Auguste 205, 255
Clarendon Hills *100*, 101, 137, 255
Claridge 187
Clavel, Dom. 113, 205
Clavelin, Hubert 228
Clendenen, Jim 161
Clerico, Domenico 41, 87, 163
Cles, Barone de 115, 266
Clifford Bay Estate 227
climate
  and vineyards 14-15, 16
  *for climatic influences on specific varieties see under* viticulture, varietal
Climens, Ch. *236*, 239
Cline Cellars 58, 141
Clonakilla 45, 283
clones and cloning 10-11, 19
  Chardonnay 67
  Garnacha Tinta/Grenache Noir 97
  Gewürztraminer 107
  Merlot 131
  Pinot Noir 179
  Riesling 197
  Viognier 279
Clos, Les (Chablis) *65*
Clos Baudoin 83
Clos du Bois 137
Clos du Caillou 101
Clos Capitoro 229
Clos Carreyrès 119
Clos des Cazaux, Dom. le 101
Clos Centeilles 58, 84
Clos de Coulaine 83
Clos l' Église 45
Clos Erasmus 101
Clos Floridène 239
Clos Fourtet 137
Clos de Gamot 119
Clos Haut-Peyraguey, Ch. 239
Clos des Jacobins 45
Clos du Joncuas 101
Clos Lapeyre 114
Clos Laudry 229
Clos Malverne 187
Clos des Marronniers 45
Clos de Maurières 83
Clos Mogador *95*, *97*, 101
Clos du Mont Olivet 101
Clos Naudin, Dom. du *80*, 83
Clos Nicrosi 153
Clos de l'Oratoire *129*
Clos des Papes 101, 205
Clos des Paulilles 101
Clos Rességuier 119
Clos Rougeard 45, *81*, 83
Clos St-Urbain *197*
Clos Ste-Magdelaine 125
Clos & Terrasses *97*
Clos Thou 167
Clos Triguedina 119
Clos Uroulat 114, 167
Clos du Val 237

Closel, Dom. du 78, 79, *80*, 83
Cloudy Bay 18, 73, 201, *224*, 226, 227
Clovallon, Dom de 283
Clusel-Roch, Dom. 255
Coche-Dury, Dom. Jean-François 35, *66*, 73, 169
Cockburn 269
Codorníu 117, 166, 285
Coffaro, David 295
Coffele 91
Cognard, Max 45
Cogno, Elvio 41, 163
Col d'Orcia 57, 139, 153, 217
Col Vetoruz 188
Coldstream Hills 134, 185
Colin, Marc 73
Colla, Podere 90
Colli di Lapio 89
Colline, Ch. de la 137
Collards 83
Collelungo 217
Collina, La 87
Colombier, Dom. du 255
Colombière, Ch. la 164
Colombini, Donatella Cinelli 217
Colombo, Jean-Luc 101, 255
Colonnara 272
Colosi 115, *120*, 121, 285
Colpetrone 207
Colterenzio co-op 73, 111, 115, 153, 169, 171
Colture, Le 188
Columbia Crest 137
Columbia Winery 137, 239, 253
Commanderie de Peyrassol 203
Compostela, Casa de 37
Concha y Toro *50*, 61, 111
Concilio Vini 122, 164
Conestoga 58
Confuron, Jean-Jacques 185
Cono Sur 137, *183*, 185, 293
Conseillante, Ch. la 45, 137
Constanti 217
Conterno, Aldo 41, *87*, 90, *157*, 163
Conterno, Giacomo 41, 87, 90, *160*, 163
Conterno, Paolo 163
Conterno-Fantino 41, 87, 163
Contero 43
Conti, Leone 35
Contini, Annalysa Rossi 87
Contini, Attilio 273
Contino 112, 265
Contratto, Giuseppe 41, 153
Contucci 122, 217
coopering 30, *31*
Coos, Dario 169, 272
co-pigmentation 31
Coppo, Luigi 41, 90
Corbans 111, 137, *199*, 201, 227
Corbin-Michotte, Ch. 45
Corino 41, 87
Coriole 101, 217, 255
Cormons co-op 121
Cornacchia 139
Cornaiano co-op 111
Coroncino 272
Correggia, Matteo 38, 41, 43, *161*, 163
Cortaccia co-op 142, 229
Cortes de Cima 203, 265, 266, 270
Cortezia, Quinta da *269*, 269
COS 164
Cos d'Estournel, Ch. 26, 57
Cosentino 137
Cosimi, R 217
Cossart Gordon 42, 117, 121, 240, 272
Costers del Siurana 91, *97*, 101
Cotat, Francis & Paul 227
Cotnari Cellars 89, 112, 241
Côtto, Quinta do 203, *263*, 269
Coudraye, Dom. de la 45
Couhins-Lurton, Ch. *224*, 227
Coulée-de-Serrant, Ch. de la *77*, *78*
Coulée-de-Serrant, Clos de la 83
Couly-Dutheil 45
Courbet, Jean-Marie 228
Courbis, Dom. 255
Couroulu, Dom. de 101
Coursodon, Pierre 125, 255
Courtade, Dom. de la 141, 203
Courteillac, Dom. de 228
Couselo, Quinta de 37
Coutale, Clos la 119
Coutet, Ch. 239
Covela, Quinta de 39
Covey Run 116, 201
Craggy Range 227
Craiglee 255
Crampilh, Dom. du 241
Cranswick Estate 125
Crasto, Quinta do 203, *268*, 269
Crawford, Kim 73, 137, 227
Crémat, Ch. de 203
Crêtes, Les 167
Cristom 185
Crittenden, Garry 41
Croatia 90, 112, 188, 199, 237
Crochet, Lucien *224*, 227
Croft 269
Cros, Ch. du 239
Cros, Dom. du 89
Cros de la Mûre 101
Crotta di Vegneron, La 153
Crown Estates 90
Cruzeau, Ch. de 227
Cuadrado Garcia, Hermanos 265
Cuilleron, Yves 125, 205, 255, 279, *280*, *282*, 283
Cullen 57, 73, 185
Curicó co-op 61
CVNE 112, 123, 265, 284
Cyprus 123, 151, 285
Czech Republic 42, 114, 199, 207, 224, 242, 284

## D

Dagueneau, Dom. Didier *221*, 227
Dagueneau et Fils, Dom. Serge 59, 227
Daheuiller 45
Dal Forno 85
D'Alessandro 73, 255, 283
D'Alfonso del Sordo, Giovanni 43
Dalla Valle 57, 217
Dalwhinnie 255
Damoy, Pierre 185
D'Ancona 285
D'Angelo 34
D'Antiche Terre 34
D'Arenberg 101, 141, 255
Darting, Kurt 114
Dassault 45
Dautel 229
Dauvissat, René et Vincent *68*, 73
De Bartoli, Marco 113, 115, *151*, 153, 285
De Bortoli *237*, 239
De Castris, Leone 295
De Loach *292*, 295
De Martino 61
De Tarczal 122
Dealul Mare Winery 241
Dehesa de los Canónigos 35
Dehesa del Carrizal 265
Dehlinger 185, 255
Dei 122, 217
Deiss, Marcel, Dom. *108*, 111, 153, 171, 173, 201
Del Campo 265
Del Cerro 217
Del Golfo, Roda 122
Del Tufo, Antonio 87
Delas Frères 101, 125, 205, 255, 283
Delatite 111, 201
Delegat 227
Delesvaux, Philippe 83
Deletang, Dom. 83
Deletto 38
Delgado Zuleta 166
Delheim 111, 153
Denbies *88*, 165, 189, 229
Deutz 73
Deux Amis 295
Devalle 87
Dezat, Pierre & Alain 227
D F J Vinhos 37, 39
Di Majo Norante 34, 89
Diamond Creek 57
Diamond Valley 185
Dickerson Vineyard 295
Die co-op (France) 84
Die Krans 241
Diel, Schlossgut 112, 171, 201, 284
Díez Mérito 166
Dipoli, Peter 227
Dirler, Jean-Pierre 111, *150*, 153, 201, 243
diseases of vines 10, 20
Disznókö 90, 114, 165
Doisy-Daëne, Ch. 239
Doisy-Dubroca, Ch. 239
Doisy-Védrines, Ch. 239
Dolianova co-op 139, 164
Domaine Chandon 73
Domaine Drouhin *183*, 185
Domaine Paul Bruno 26, *134*
Domaine Serene 185
Domecq 166, 167
Dominus 57
Donata Bianchi, Maria 169
Donati, Marco 266
Donnafugata 164, 285
Dönnhoff 201
Donoso, Casa 61
Dopff & Irion 111, 171, 173
Dopff au Moulin 111
Dörflinger 114
Dorigati 266
Dorigo, Girolamo 169, 189, 229, 273
Dournie, Ch. Etienne 58
Draper, Paul 287, *290*, 290, 291, 292
Drautz-Able 88, 112, 116, 142, 229
Dri, Giovanni 273
Drius 169, 267
Droin, Dom. Jean-Paul 73
Drouhin, Joseph 73, 169, 185
Druet, Pierre-Jacques 45
Druis, Sergio & Mauro 121
Dry Creek Vineyards 83, 295
Dry River 107, 111, *172*, 173, *184*, 185, 201
Duboeuf 91, 276, 277
Dubois, Les Frères 59
Dubourdieu, Denis 223
Duckhorn Vineyards 137, 239
Ducru-Beaucaillou, Ch. 57
Dugat-Py, Bernard *180*, 185
Dujac, Dom. 19, *177*, 185
Dumazet, Pierre 283
Dunn 57
Durand, Joël & Noël 255
Durand-Perron, Dom. 228
Durban, Dom. de *150*, 153
Dusi Ranch *289*
Duxoup 59, 87

## E

Eberle 295
Écard, Maurice 171
Ecu, Dom. de l' 123, 143
Edmeades 295
Edmunds St John 141, *253*, 255, 295
Edwards, Luis Felipe 61
Egli, Augusto 43
Église-Clinet, Ch. l' 137
Egly-Ouriet 185
Eguisheim co-op 111
Egypt 8
Einaudi, Luigi 41, 87, 163
Eiswein (Icewine) 26, 195, 199, 273
Eizaguirre 114
Ellwanger, Jürgen 115
Elyse 295
Emma, Casa 217
Enate 73, 111
Endrizzi 266
Engel, Dom. René 185
England 39, 88, 117, 142, 165, 189, 229, 240
Enomar 284
Entrefaux, des 125
Eola Hills 111
Épiré, Ch. d' 79, 83
Ermita, Clos l' *95*, *97*
Errázuriz 57, 73, 137, 255
Erste & Neue 111
Esclans, Ch. d' 141
Esk Valley 45, 57, 83, 119, *135*, 137
Espiers, Dom. des 101
Esporão 38, 57, 203, 265, 266, 269, 270
Establecimiento Juanico 241
Estaçáo, Quinta da 266
Estanilles, Ch. des 125, 205, 255
Esteves Ferreira, António 37, 266
Etchart 119, 267
Étoile, Ch. de l' (France/Jura) 228
Étoile, l' (co-op) (France/ Roussillon) 101
Etude 185
Eugénie, Ch. 119
Evangile, Ch. l' 45, 137
Evans & Tate 73
Evesham Wood 173
Eyrie 173

## F

Fabas Augustin, Ch. 125
Fabre Montmayou *118*, 119
Facchin, Christian 283
Fairview 141, 187, *237*, 239, 255, 283, 295
Faiveley 185
Falesco 112, 137, 270
Far Niente 239
Fargues, Ch. de 239
Farina, Stefano 163
Farrell, Gary 185, 295
Faure 84
Faurie, Bernard 255
Faury, Phillipe 283
Fay 163
Fayau, Ch. 239
Feiler-Artinger 42, 164, 171, 173, *206*, 284, 285
Felgueiras co-op 37
Feipu dei Massaretti, Cascina 169
Felline *293*
Felluga, Livio 137, 169, 173, 267
Felluga, Marco 45, 169, 171, 189, 267
Felsina, Fattoria di *216*, 217
Felton Road 185, 201
Ferme Blanche, Dom. de la 125
fermentation temperatures 28-9
Fernández, Alejandro, Bodegas *263*, 265, 267
Ferran, Ch. 164
Ferrando, Luigi 89, 163
Ferrari-Carano 73, 137, 217
Ferraris, Roberto 41
Ferreira 266, 269
Ferret, Dom. J-A 73
Ferrière, Ch. 57
Ferrucci, Stefano 35
Fesles, Ch. de 79, 83
Fessy, Henry 91
Fessy, Sylvain 91
Fetzer 185, 295
Feudi di San Gregorio 34, *89*, 112, 169
Fieuzal, de 227, 239
Fife 59
Figeac, Ch. 45
Filhot, Ch. 239
Filliatreau, Dom. 45
Filiputti 273
Fitz-Ritter 111
Flacher, Gilles 283
Flagstone 83
flavour
  effect of oak barrels 30, 195, 235
  effect of vineyard latitude 12-13
flavours, varietal
  Albariño 36
  Barbera 40
  Cabernet Franc 44
  Cabernet Sauvignon 56
  Carmenère 60
  Chardonnay *63*, 72
  Chenin Blanc 82
  Dolcetto 86
  Garnacha Tinta/Grenache Noir 100
  Gewürztraminer 110
  Malbec 118
  Malvasia 120
  Marsanne 124
  Merlot 136
  Mourvèdre 140
  Muscat 152
  Nebbiolo 162
  Pinot Blanc 170
  Pinot Gris 172
  Pinot Noir 184
  Pinotage 186
  Riesling 200
  Roussanne 204
  Sangiovese 216
  Sauvignon Blanc 226
  Sémillon 238
  Silvaner 242
  Syrah/Shiraz 254
  Tempranillo 264
  Touriga Nacional 268
  Viognier 282
  Zinfandel 294

Flein, W G 115
Fleur-Pétrus, Ch. la 137
Florentin 125, 205
Florio 113, 115
Fogarty, Thomas 111
Fonda, Cascina 153
Fonseca, José Maria da 35, *36*, 37, 38, 58, 89, *151*, 153, 167, 266, 269
Font Cause, Dom. 79
Font de Michelle, Dom. 101, 113, 205, 283
Font-Sane, Dom. 101
Fontainerie, Dom. de la, 83
Fontana, Graziano 142
Fontanabianco 163
Fontanafredda 87
Fontenil, Ch. 137
Fonterutoli, Castello di 217
Fontodi *214*, 217, 255
food and wine matching
  Cabernet Sauvignon 56
  Chardonnay 72
  Chenin Blanc 82
  Garnacha Tinta/Grenache Noir 100
  Gewürztraminer 110
  Merlot 136
  Muscat 152
  Nebbiolo 162
  Pinot Noir 184
  Riesling 200
  Sangiovese 216
  Sauvignon Blanc 226
  Sémillon 238
  Syrah/Shiraz 254
  Tempranillo 264
  Viognier 282
  Zinfandel 294
Foppiano 168
Foradori 266
Foreau, Philippe 79
Fores, Bàrbara 91
Forêt, Jacques 271
Forges, Dom. des 83
Foris 111, 137
Forman 137
Formentini, Conti 169
Forte Rigoni 121
Forteto della Luja 153
Fortia 101
fortified wines 97
Fourmone, Dom. de la 101
Fox Creek 255, 272
Foz de Arouce, Quinta de 39
Fracce, La 43
France 59, 270
  Alsace 23, 68, 105, 106, 107, 108, 111, 115, 148, 150, 153, 169, 170, 171, 172, 173, 181, 194, 197-8, 201, 242, 243, 267
  Bordeaux *17*, 17, 23, *31*, 44, 45, 50-1, 52, 57, 84, 118, 130, 131, 133, 137, 142, 167, 188, 222, 223, 227, 228, 234-5, 236, 239
  Burgundy 16, 23, 35, 59, 66, 67, 68, 73, 91, 123, 169, 172, 173, 178, 179, 180, 185, 206
  Champagne 17, 66, 67, 68, 138, 180-1, *181*
  Corsica 35, 39, 84, 164, 229
  Jura 68, 188, 228, 270
  Languedoc-Roussillon (Midi) 57, 58, 67, 84, 85, 98-9, 117, 125, 131, 141, 169, 202, 205, 251, 271
  Loire Valley 27, 35, 42, 44, 45, 59, 68, 75, 77, 78, 79, 80, 83, 90, 91, 113, 132, 143, 169, 173, 181, 203, *221*, 222, 224, 227
  Provence 85, 125, 132, 141, 205, 251, 271, 283
  Rhône Valley 35, 43, 58, 85, 96, 98-9, 101, 124, 125, 141, 143, 150, 153, 168, 204, 205, *247*, *248*, 248, 249, 251-2, 255, *277*, 278, *279*, 280, 283
  Savoie 68, 86, 115, 124, 132, 138, 181, 204, 205, *206*, 206
  South-West 35, 38, 42, 88, 89, 113, 116, 122, 132, 164, 165, 167, 241, 271
Franco Españolas 117
Franco, Nino 188
Frankland Estate 45
Franqueira, Quinta da 37, 117, 270
Frégate, Dom. de 141
Freixenet 117, 166, 285
Frescobaldi 217
Frescobaldi-Mondavi 217
Fresne, Ch. du 83
Freycinet 185
Frick (USA) 168
Frick, Pierre (France/Alsace) 111
Frog's Leap 295
Fuentes 101
Fuentespina 265
Fuissé, Ch. 73, 91
Fuligni 217
Fürst, Rudolf 171, 185, 241, 243, 284
Fürstlich Castell'sches Domänenamt 243
Frühwirth 139

# G

Gaffelière, Ch. la 45
Gaffory, Orenga de 164
Gagliardo, Gianni 89
Gagliole 217
Gagnard, Jean-Noël *68*, 73
Gagnard-Delagrange 73
Gaillard, Pierre *124*, 125, 255, 283
Gainey 115
Gaivosa, Quinta da 269
Gaja, Angelo *41*, 57, 73, *158*, 163, 224
Galantin, Dom. 141
Galardi 34, 169
Galegas, Adegas 37
Galet des Papes, Dom. du 101
Gallant, T' 173
Gallo, Silvano 169
Gallo Sonoma 295
Gallura co-op 273
Gancia 153
Gandía 43, 139
Garde, Ch. la 227
Gardine, Ch. de la 101, 113, 205
Garetto 41
Gargalo 270
Garofoli 139, 272
Garvey 166, 167
Gasparini 45
Gastaldi 87
Gatti, Piero 43
Gauby 255
Gaudou, de 119
Gaudrelle, Ch. 83
Gaujal, de 169
Gaussen, Jean-Pierre 141
Gautoul, Ch. 119
Gavoty, Dom. 141, 203
Gazin, Ch. 137
Gehringer Brothers 201, 229
Geil 114, 115
genetic modification of vines 11
genetics of vines 10
Genson 169
Gentile, Dom. 153
geographical influences on vines 12-13
geology *see* soils, vineyard
Gérin, Jean-Michel 255, 283
Germanier, Bon Père 35
Germano, Ettore 87, 163
Germany 73, 111, 142, 153, 173, 185, 196-7, 201, 243
  Ahr 188, 196
  Baden 106, 109, 114, 138, 170, 171, 172, 181, 196, 206, 241
  Franken 39, 138, 181, 202, 242, *243*
  Mosel-Saar-Ruwer 16, *17*, 19, 39, 88, 178, *193*, *194*, 194, 196
  Nahe 194, 196
  Pfalz 88, 106, 109, 114, 115, 138, 139, 172, 178, 181, 188, 196, 206, 229, 241
  Rheingau 181, 196
  Rheinhessen 39, 88, 89, 114, 115, 139, 165, 189, 206, 229, 242
  Württemberg 88, 115, 116, 138, 229, 270
Gesellmann, Albert 42
Ghisolfi, Attilio 163
Giaconda 57, 73, *183*, 185
Giacosa, Bruno 38, 87, *160*, 163
Gibbston Valley 173
Giesen 142, 201, 227
Giesta, Quinto do 42
Gigou, Joël 83
Gilette, Ch. 30, 239
Gillardi 87
Gilliard, Robert 59, 125, 243
Gineste, Dom. de 123
Girardin, Vincent 73, 185
Gitton Père et Fils 227
Giuncheo 203
Giustiniana, La 85
Glatzer, Walter 171, 284
Glen Carlou 73, 241
Glenora *71*
GM vines *see* genetic modification of vines
Godeval 112
Godineau, Père et Fils 83
Gojer, Franz 229
Goldwater Estate *55*, 57, 137, 227
Gonçalves, Faria 39
Gonnet, Charles 115
Gonzalez-Byass 166, *167*
Gorelli 217
Goubert, Dom. les 101, 283
Gouges, Henri 185
Gour de Chaule, Dom. du 101
Grace family 57
Gracia, Viña 61, 137, 185
Graham 269
Graillot, Alain 255
Gramenon, Dom. 101
Grand-Mayne, Ch. 137
Grand Moulas, Ch. du 101
Grand-Puy-Lacoste, Ch. 57
Grand Tinel, Dom. du 101
Grande-Maison, Dom. 239
Grandes Bodegas 265
Grandes Vignes, des 83
Grange des Pères, Dom. de la 141
Grangehurst 187
Granja Fillaboa 37
Grant Burge 255
grapes
  composition of 28
  noble rot 27, 195, *231*, 235, 273
  ripeness 15-16
  *see also* vines
Grasso, Elio 41, 87, 163
Grasso, Silvio 163
Gratien & Meyer 83
Gratien, Alfred 138
Grave, Dom. la 239
Gravner 189, 227
Gray Monk *106*, 111
Graziano, Vittorio 116
Greece 40, 53, 69, 112, 117, 123, 139, 151, 153, 164, 189, 253, 270, 281
  Cephalonia 202
  Crete 115, 122
  Lemnos 117
  Macedonia 285
  Peloponnese 34
  Samos *145*, *151*, 151
  Santorini 34, 38
Green and Red Vineyard 295
green harvesting 24-5
Green Point 73
Gresy, Marchesi di 41, 87, 163
Gries co-op 115, 229
Gries, Muri 115
Grifo, El 121
Gripa, Bernard 125, 205
Grippat, Jean-Louis 125, 255
Grivot, Jean 185
Groot Constantia 187
Gros, Anne, Dom. *180*, 185
Gross, Alois & Ulrike 143, 153, 173, 227
Grosset 45, *70*, 73, *200*, 201
Grotte del Sole, Cantine 169
Grove Mill 227
growth conditions for vines 14-17
Gruaud-Larose, Ch. 57
Grünhaus, Maximin 201
Guardia, La 87
Guelbenzu 265
Guerrieri-Rizzardi 91
Guffens-Heynen 73
Guigal, E 101, 125, 205, *250*, 255, *280*, 283
Guindon, Jacques 122
Guinot, Maison 123
Guiraud, Ch. 239
Gunderloch *196*, 201
Gurgue, Ch. la 57
Gutiérrez de la Vega 153

# H

Haag, Wilhelm *193*
Haag, Weingut Fritz 201
Haas, Franz 45, 111, 153
Hacienda Monasterio 265
Haderburg 169, 227
Haider 164
Haidle, Karl 115
Hainle 270
Hamilton 101
Hamilton Russell 66, 73, 178, 185
Hanging Rock 227
Hardys 101, 255
Hargrave 45, 137
Harlan 57
Harrison 137
Hartenberg 241, 295
Hartford Court 185, 295
harvesting
  green 24-5
  mechanical 27
  picking 26-7
  *tries* 79
Haselgrove 255, 283
Hauner, Carlo 121
Haut-Bailly, Ch. 57
Haut-Bergeron, Ch. 239
Haut-Brion, Ch. 57, 226, 227, 238, 239
Haute Perche, Dom. de 83
Haute-Serre, Ch. de 119
Hautes Vignes, Dom. des 169
Havens 45, 137
Haza, Condado de 265
heat summation scales 15
Hébras, Ch. les 239
Heger, Dr 112, 171, 173, 284
Heidsieck, Charles 138
Heinrich, Gernot 42, 207, 284, 285
Henriques & Henriques 42, 117, *120*, 121, 240, 272
Henry Estate 111
Henry of Pelham *195*
Henschke 111, 119, 201, *251*, 255
Hermitage, Dom. de l' 141
Hertz, Albert 111
Hewitson 141
Heyl zu Herrnsheim, Freiherr 201, 243
Heymann-Löwenstein 201
Hidalgo 166
Hidden Cellars 295
Hiedler 171, 284
Hirtzberger, Franz 114, *199*, 201
history of vines 8-9
  Cabernet Sauvignon 49
  Chardonnay 65
  Chenin Blanc 77
  Garnacha Tinta/Grenache Noir 95
  Gewürztraminer 105
  Merlot 129
  Muscat 147
  Nebbiolo 157
  Pinot Noir 177
  Riesling 193
  Sangiovese 211
  Sauvignon Blanc 221
  Sémillon 233
  Syrah/Shiraz 247
  Tempranillo 259
  Viognier 277
  Zinfandel 289
Hofkeller, Staatliche 243
Hofstätter 111, 169, 185
Hogue Cellars 83, 116, 239
Hohenlohe-Öhringen, Fürst zu 116, 142, 229
Hoodsport 116
Hop Kiln 271, 295
Horta de Nazaré 115
Horton Vineyards 125, 141, 283
Hortus, Dom. de l' 141, 255
Howard Park 57, 73, 201
Huards, Dom. des 203
Hudelot-Noëllat, Dom. 185
Huet 82, 83
Huet, Gaston *79*
Hugel et Fils 111, 171, 173, 201
Humbrecht, Olivier 107, *173*

Hunawihr co-op 111
Hungarovin 115, 116
Hungary 42, 53, 59, 69, 108, 115, 116, 133, 143, 182, 188, 241, 242, 284
  Tokaj 16, 90, 114, 165
Hunter's 73, 111, 227
Huntingdon Estate 239
Hureau, Ch. du 45, 83
Husch 83
hybrid vines 8-9, 240

I

Icardi 153
Icewine (Eiswein) 26, *195*, 195, 199, 273
identification of vines 11
Ifs, Les 119
Ijalba, Viña 112, 265
Illuminati 139
Imesch, Caves 35, 59, 91, 115, 125, 168, 243
Inama 91, 227
India 63
Inniskillin Okanagan 171, 273
Insulares Tenerife 117
integrated management of vineyards 21
irrigation of vineyards 25-6
Irvine, James 137
Isabel Estate 185, 227
Isera co-op 122
Isleta, La 117
Isole e Olena 73, 121, *216*, 217, *253*, 255
Israel 35, 40, 53, 69, 81, 205
Italy 57, 73, 125, 138, 153, 168, 171, 173, 185, 227, 243, 253, 255, 266, 281, 290, 295
  Abruzzo 43, 91
  Basilicata *34*, 34, 122, 150
  Calabria 91, *113*, 122
  Campania 34, 89, 169
  Emilia-Romagna 34, 39, 43, 116, 120, 121, 165, 189, 212, 214, 270, 271
  Friuli-Venezia Giulia 44, 45, 90, 121, 133, 137, 169, 189, 199, 202, 224, 229, 267, 272
  Lazio 35, 43, 58, 120, 121, 137
  Liguria 35, 39, 86, 169, *203*, 203
  Lombardy 43, 68, 122, 163, 169, 182, 203, 271, 273
  Marche 43, 58, 91, 139, 214, *272*, 272
  Piedmont 19, 38, 40, 41, 43, 52, 85, 86, 87, 89, 90, 113, 122, 148, 150, 158, 160, 163, 182, 206, 224, 241, 271, 273, 281, 283
  Puglia 35, *122*, 122, 150, 164, 271, *293*
  Sardinia 58, 99, 101, 122, 139, 150, 164, 267, 273
  Sicily 58, 90, 91, 112, 113, 114, 121, 150, 164, 167
  Trentino-Alto Adige 106, 109, 111, 115, 122, 133, 142, 150, 164, *168*, 169, 170, 182, 199, 203, 224, 229, 242, 273
  Tuscany 24, 35, 43, 52, 58, 68, 120, 121, 122, 133, 137, 139, 188, 207, 212, 213, 214, 217, 273, 283; Elba 35
  Umbria 58, 91, 112, 133, 139, 188, *207*, 207, 214
  Valle d'Aosta 42, 158, 167
  Veneto 91, 122, 130, 133, 137, 169, 188, 267, 272, 273
Ivaldi, Domenico 43, 153
Ivaldi, Giovanni 43

J

Jaboulet, Paul Aîné 101, 125, 153, 204, 205, 252, *254*, 255, 283
Jackson Estate 227
Jacqueson, Henri et Paul 185
Jade Mountain 101, 137, *140*, 141, 255, 283
Jadot, Louis 73, 171, 185
Jamek, Josef 114, 201
Jamet, Jean-Paul & Jean-Luc (France/Rhône) 255, 283
Jamet, Pierre et Fils (France/Loire) 45, 255, 283
Janasse, Dom. de la 101, 205
Japan 85, 281
Jasmin, Dom. Robert & Patrick 255, 283
Jasper Hill 255
Jau, Ch. de 101, 125, 153
Jayer-Gilles, Robert 35
Jekel 168
Jensen, Josh *280*, 281
Jermann 169, 171, 189
Jesús Nazareno co-op 112, 123
Jobard, Dom. François 73
Joguet, Charles *44*, 45
Johner, Karl-Heinz 73, 142, 171, 173, *181*, 185, 241, 284
Jolivet, Pascal 83, 227
Jolys, Ch. *167*
Jordan 57, 187
Josmeyer 111, 171, 173
Jost, Toni 201
J P Vinhos 58, 153, 167
Juliusspital, Weingut 39, 142, *242*, 243
Jung 153
Juris (G Stiegelmar) 207
Justices, Ch. les 239
Justin 45
Justino Henriques, Vinhos 121
Juvé y Camps 166

K

Kaapzicht 187
Kamocsay, Akos 115, 285
Kangarillo Road 295
Kanonkop Estate 57, 187
Kante, Edi 121, 227
Kanu 83, 227
Karina 227
Karthäuserhof 201
Katnook Estate 137, 227
Kazakhstan 35, 151, 199
Keber, Edi 267
Keller, Franz 171, 284
Kendall-Jackson 295
Kenwood 227, 295
KEO 285
Kesselstatt, von 201
Kientzler 59, 111, 153, 173, 201, 243
King Estate *172*, 173
Kiona 83, 153, *199*, 201
Kistler *72*, 73
Klein Constantia 153, 227
Knappstein 111
Knoll, Emmerich 114, 201
Koehler-Ruprecht 114, *197*, 201
Koehly 173
Köfererhof 243
Kollwentz 42
Konzelmann 111
Kourtakis 202, 228
Koutsouyanopoulos 34, 38
Kracher, Alois 43, 153, 207, *229*, 284
Kreydenweiss, Marc 111, 115, 171, 173, 201
Krug 138
Kuentz-Bas 111, 153, 171, 173, 201
Kumeu River 73, 119, 173
Kunde Estate *281*, 283, 295
Künstler, Franz *200*, 201
KWV 241
Kyrgyzstan 117, 199

L

La Jota *44*, 45, 57, 168, 283
La Rosa *134*, 137
La Rural 119
Labarthe, Dom. de 116, 123
Labastide-de-Lévis co-op 116
Labégorce-Zédé, Ch. 57
Labouré-Roi 185
Lackner-Tinnacher 143, 153, 171, 227, 284
Ladoucette, Dom. de 227
Laetitia 171
Lafarge, Dom. Michel *181*, 185
Lafaurie-Peyraguey, Ch. *238*, 239
Laffourcade, Vignobles 83
Lafite-Rothschild, Ch. *49*, 57, *129*
Lafleur, Ch. 45, 137
Lafon, Dom. des Comtes 22, 67, *72*, 73, 178, 185
Lafon-Roc-Épine, Dom. 101
Lafon-Rochet, Ch. 57
Lagar de Fornelos *37*, 37
Lagarosse 239
Lageder 57, 73, 153, 173, 227, 229
Lagoalva de Cima, Quinta de 115
Lagrange, Ch. 57
Lagrezette, Ch. 119
Lagune, Ch. la 57
Laibach 187
Laible, Andreas *108*, 111, 115, 229, 270
Lamartine, Ch. 119
Lambardi 217
Lamberhurst Vineyard Park *142*
Lambrays, Dom. des 185
Lamothe-Guignard, Ch. 239
Lamy-Pillot, Dom. René 73
Lan, Bodegas 37
land area under vines 13
Landiras, Ch. de 239
Landmann, Seppi 111, 243
Landmark 73, 185
Lane Tanner *182*, 185
Langlois-Château 45, 83
Laporte, Dom. 153
Larmande, Ch. 45
Laroche, Dom. 73
Lascaux, Ch. de 125, 205
Lastours, Ch. de 43
Lastra, La 273
latitude, vineyard
  effect on wine flavour 12-13
Latour, Ch. 57
Latour-à-Pomerol, Ch. 137
Latour-Martillac, Ch. 239
Laurel Glen Winery 50, 57
Laurent, Dominique 185
Laurent-Perrier 138, 185
L'Avenir *186*, 187, 241
Laville-Haut-Brion, Ch. 226, *236*, 238, 239
Lawson's Dry Hills 111, 227
Lázaro, Gerardo Méndez 37
Lazy Creek 111
Leacock 42, 117, 121, 240, 272
Leasingham 57, 119, 255
Lebanon 53, 84, 253
Lebreton, Jean-Yves 45, 83
Lebreton, V & V 83
Leccia 153, 164
L'Ecole No 41 137, 239
Leconfield 255
Leeuwin Estate *70*, 73, 201
Leflaive, Dom. *65*, 73
Lehmann, Peter 239, *254*, 255
Leitz, J *193*, 201
Lenswood Vineyards 73, 185, *225*, 227, 295
Leone de Castris 122, 164
Leonetti 57, 137, 217
León-Ville, Bodegas de 188
Léoville-Barton, Ch. *49*, 57
Léoville-Las-Cases, Ch. *52*, 57
Léoville-Poyferré, Ch. 57
Lequin, Louis 171
Leroy, Dom. 73, *178*
Leroy, Maison 185
Letrari 122
Lewis 137
Liards, Dom. des 83
Librandi 91
Lignier, Hubert 185
Lilliano, Castello di 217
Limbardié, Dom. de 137
Limerick Lane 295
Lindemans 57, *237*, 239, 251, 255
Lingenfelder 88, 229
Lionnet, Jean 125, 205
Liot, Ch. 239
Lis Neris-Pecorari 169, 273
Lisini 217
Littorai 185
Livon 189, 273
Loewen, Carl 201
Logis de la Bouchardière, le 45
Lohr, J 271
Loimer, Fred 114
Longridge Winery 84, 187, 206
Loosen, Dr 142, *196*, 201
López de Heredia 265, 284
López, Hermanos 139
Lorenzon/I Feudi di Romans 121
L'Ormarins 173
Los Robles 61
Loubens, Ch. 239
Loupiac-Gaudiet, Ch. 239
Louvière, Ch. la 227
Luisa, Eddi 121
Luneau, Pierre 123, 143
Lungarotti 217
Lupi 87
Luretta 121
Lurton, Hermanos 227
Lusco do Miño 37
Lusenti, Gaetano 121, 165
Lustau 153, 166, 167
Lützkendorf, U 111, 171, 243, 270, 284
Luxembourg 88, 142, 195, 199, 202
Lynch-Bages, Ch. 57
Lytton Springs Vineyard *289*, *294*

M

Macchiole, Le 137, 255, 273
Macchione, Il 217
maceration 29-30
Maculan 57, 137, 273
Madeira Wine Company *121*
Mader, Jean-Luc 111
Madero, Uribes 265
Maffini, Luigi 169
Magdelaine, Ch. 137
Magord, Dom. de 84
Maias, Quinta das 43, 115, 266
Mailly, Producteurs de *181*
Maire, Henri 228, 271
Malabaila 38
Malarctic-Lagravière, Ch. 227
Malle, Ch. de 239
malolactic fermentation 30
Malvirà 38
Malvolti 188
Mann, Albert 111, 153, 173
Männle, Heinrich 111
Mantovana, Mariana 35
Manzaneque, Manuel 73, 265
Manzano, Fattoria di 73, 283
Marcarini *86*, 87
Marcassin 73
Marcoux, Dom. de 101
Marenco, Giuseppe 43, 87
Margaride, Dom. Luis de, Heredos de 115
Margaux, Ch. 26, *49*, *56*, 57, 227
Markezinis 34, 38
Marqués de Cáceres 265, 284
Marqués de Griñón 57, 168, 255, *264*, 265, 272
Marqués de Monistrol 117, 166
Marqués de Murrieta *262*, 265, 284
Marqués de Vargas 265
Marqués de Vizhoja 37
Marribrook 125
Marronniers, Dom. des 111
Martinborough Vineyard *17*, 185
Martinelli *292*, 295
Martinetti, Franco M 41, 85
Martinez (Portugal) 112
Martínez, Faustino 265
Martínez Bujanda 101, 123, 265, 284
Martini, Dom. 229
Martini, Conti 153, 266
Martini, Louis M 295
Marynissen 273
Mas Amiel 101
Mas Blanc, Dom. du 85, 101, 141, 255
Mas de Bressades 205
Mas Bruguière 205
Mas de Daumas Gassac 38, 57, 84, *278*, 283
Mas Gil, Celler 125
Mas Jullien 84, 266
Mas Martinet *97*, *98*, 101
Mas Redorne 141
Mascarello, Bartolo 87, 163
Mascarello, Giuseppe 41, 87, 90, 163
Masciarelli 139
Masi 85, 91, 203
Masía Barril 91, 117
Massa, La 217
massal selection of vines 10
Masseria, Pepe 295
Mastroberardino 34, 89, 112, 169
Mastrojanni 217
Masut de Rive 171

Matanzas Creek Winery 73, *133*, 137, 227, 239
Matarromera 265
Matthew Cellars 239
Matjaz Tercic 189
Matrot 73
Matua Valley 57, 73, 227
maturity charts
  Cabernet Sauvignon 57
  Chardonnay 73
  Chenin Blanc 83
  Garnacha Tinta/Grenache Noir 101
  Gewürztraminer 111
  Merlot 137
  Muscat 153
  Nebbiolo 163
  Pinot Noir 185
  Riesling 201
  Sangiovese 217
  Sauvignon Blanc 227
  Sémillon 239
  Syrah/Shiraz 255
  Tempranillo 265
  Viognier 283
  Zinfandel 295
Maucaillou, Ch. 57
Mauritania 58
Mauro, Bodegas 265, 267
Maxwell 137
Mayr, Josephus 116, 229
Mayr, Thomas 229
Mazzei 139
Mazzolino 43, 202
McCrea Cellars 101, *253*, 255, 283
McDowell Valley 283
McWilliams 43, 153, 239, 255
Mecella, Enzo 272
mechanical harvesting 27
Medalla 119
Meerlust 57, 73
Megyer, Ch. 90, 114
Meín, Viña 267, 270
Meinert 137
Melgaço, Muros de 37
Melgaço, Quinta do 37
Melgares, Julia Roch 141
Melini 273
Mellot, Alphonse 227
Mellot, Joseph 227
Melton, Charles 96, 99, 101, 141, 248, 255
Mendoza, Abel 121
Mendoza, Enrique 153
Méo-Camuzet, Dom. 185
Merlin 73
Meroi, Davino 169
Merryvale 73
Messias, Caves 39, 122
Meste-Duran, Dom. 84, 271
Métaireau, Louis 123, 143
Mexico 55, 81, 161, 163, 168, 215, 225, 229, 263, 270, 271, 293
Meyer, Ernest & Fils 173
Meyer, François 111
Meyer-Fonné, Dom. 111, 171, 173
Mezzacorona 266
Miaudoux, Ch. les 239
Michael, Peter 57, 73
Michel, Dom. Louis 73
Miliarakis Bros. 115
Millérioux, Dom. Paul 227
Millton Vineyard *81*, 83, 201
Mission-Haut-Brion, Ch. la 57
Mission Hill *109*, 111, 171
Mitchell 201
Mitchelton 101, 125, 201, 204, *205*, *255*
Dom. Mittnacht-Klack 111, 173
Moccagatta 87, 163
Moldova 53, 59, 108, 133, 151, 199, 202, 207, 224
Molino, Mauro 41
Molitor, Markus 201
Mollex, Maison 206
Monacesca, La 272
Monbrison, Ch 57
Moncão co-op 37
Monclús 139
Mondavi, Robert *29*, 42, *56*, 121
Mondavi/Chadwick 57, *61*
Monde, Le 173
Monfalletto-Cordero di Montezemolo 41, 163
Mongetto, Il 41
Monin, Philippe 115
Monje 117
Monsanto 217
Mont d'Or, du 115, 168
Mont-Redon, Ch. 101, 113
Mont Tauch co-op 58
Montalcino co-op 217
Montana 73, *223*, 225, 227
Montaribaldi 38
Montbourgeau, Dom. de 228
Monte, Quinta do 37
Monte Bernardi 217
Montecillo 265, 284
Monteillat, Dom. du 255, 283
Montelassi 139
Montenidioli 273
Montepelosio 217
Monterminod, Ch. de 206
Montes 55, 57, 73, 119, 255
Montesissa 41
Montesole 34
Montevertine *212*, 217
Montevetrano 34, 57
Montgilet, Dom. de 83
MontGras *60*, 61, 119, 295
Montille 185
Montmirail, Ch. de 101
Montpertuis 101
Montrachet, Le *65*
Montrose (Australia) 41
Montrose, Ch. (France/Bordeaux) 57
Montus, Ch. 241
Monunt Riley 227
Moondah Brook 272
Moorilla Estate 111
Moorooduc 185
Morandé 119
Mordorée, Dom. de la 101
Morey, Bernard 73
Morgadio 37
Morgante 164
Morgenhof 73, 137, 187, 227
Moris Farms 139, 217
Moro, Emilio 265
Morocco 253
Moroder 139
Morris Wines 43, 88, *143*, *151*, 153
Mortet, Denis 35, 185
Morton Estate 137
Moschioni, Davide 169, 229
Moser, Lenz 43, 114
Moss Wood 57, 239
Mouchão, Herdade de 35
Moueix, Christian *130*, 131
Moulin des Costes 141
Moulin de la Gardette 101
Moulin-de-Launay 239
Moulin Touchais 82
Mount Hope 58
Mount Horrocks *198*, 201, 255
Mount Langi Ghiran 173, 255
Mount Mary 57
Moure, Adegas 37, 123
Mouton-Rothschild, Ch. *56*, 57
Moyer, Dominique 83
Muga, Bodegas 123, *259*, 265
Mugnier, Jacques-Frédéric 185
Mulderbosch 73, 83, *226*, 227
Muller, de 91
Müller, Egon 201
Müller-Catoir 112, 153, 173, *196*, 201, 229, 284
Murana, Salvatore 153, 285
Muré, René 111, *152*, 153, 171, 173, 201, 243
Murphy-Goode 227
Murrieta's Well 239
Murta, Quinta da 38
must concentration 29
Myrat, Ch. de 239


# N


Nada, Fiorenzo 41, 87, 163
Nages, Ch. de 205
Nalles-Magré Niclara 227
Nairac, Ch. 239
Nalle 295
Nalys, Dom. de 101
Nautilus 227
Navarro (USA) 111, 227
Naylor Wine Cellars 58
Nederburg *81*, 83, 171
Neethlingshof 111, 187, 201, 239
Negri, Nino 161
Negro, Angelo 38
Neipperg, von 142, 229
Neipperg, Stephan von 51, 131
Neive, Castello di 38, 163
Nepenthe 73, 239, 295
Nerleux, Dom. de 45
Nerthe, Ch. la 101, 113
Nervi 271
Neudorf 73, 185, 201, 227
Neumeister 139
New Zealand 23, 45, 57, 73, 81, 83, 106, 107, 109, 111, 119, 130, 131, 137, 142, 151, 171, 173, 179, 185, 201, 227, 237, 239, 253, 255, 266, 281, 293
  Auckland 55, 71, 135
  Canterbury 71, 183
  Hawkes Bay 55, 71, 135, 224, 237
  Marlborough 16, 71, *20*, *23*, 135, 183, 199, *219*, 222, *223*, 223, 224, 237
  Nelson 71, 199, 224
  Otago *13*, 183
  Wellington 71, 183, 224, 295
Newlan 295
Newton 57, *69*, 137, 168
Ngatarawa 201
Nicodemi 139
Niebaum-Coppola 295
Niedermayr 116, 229
Niedrist, Ignaz 169, 171
Niellon, Dom. Michel 73
Niepoort 112, 203, 269
Niero, Robert 283
Niero-Pinchon 283
Nieto Senetiner 41, 119
Nigl, Familie 114, 201
Nikolaihof 114, 201
Nino Negri *161*, 163
Nitra Winery 115
Nittnaus 42
Nobilo 142, 227
Noble, Ch. du 239
noble rot 27, 195, *231*, 235, 273
Noblesse, Ch. de la 141
Norman Vineyards 295
Norton 41, 119, 267
Noval, Quinta do 19, 241, *269*, 269
Nuy Wine Cellar 153
Nyetimber 73


# O


oaking 30-1
  Cabernet Sauvignon 51
  Chardonnay *63*, 67
  Riesling 195
  Sémillon 235
  Viognier 279
Oberto, Andrea 41, 163
Ochoa 139, 153, 265
Odderno, 87
Odoardi 34, 91, 113
Ogereau, Dom. Vincent 45, 83
Ojai 255
old vines 19
Oliveira, Manuel Rodrigues de 37
Oliver, Miguel 153
Opitz, Willi 42, 114, 153, 207
Oratoire St-Martin, Dom. de l' 101, 125, 141
Oremus *90*, 114
organic viticulture 21-2
Orlando 201
Ormanni 217
Ornellaia, Tenuta dell' 57, *133*, 137, 227
Orsat, Caves 91
Orsolani 89
Orvalaiz, Bodegas 265
Osborne 166
Ostertag, Dom. 111, 153, 171, 173, *197*, 201, 243
Outeiro de Baixo, Quinta do 39
Overgaauw 241, 243


# P


Pacenti, Siro 217
Paco de Teixeró 39
Pago de Carraovejas 265
Pahlmeyer 73, 137
Paitin 163
Pajzos, Ch. 90, 114
Pakistan 8
Palacio 265
Palacios, Alvaro *97*, 101
Palari 164
Palazzone 112
Palleroudias, les 101
Pallières, Dom. les 101
Palliser Estate 185, 227
Palme, Ch. la 164
Palmer (USA/New York State) 137
Palmer, Ch. (France/Bordeaux) 57, *132*
Palmera, La 117
Paloma 137
Pancas, Quinta da 269
Pancrazi, Marchesi 185
Paneretta, Castello della 217
Panizzi, Giovanni 273
Panther Creek 185
Pantón, Priorato de 123
Paolis, Castel de 121, 153
Papagni 35
Pape-Clément, Ch. 223, 227, 239
Papin, Claude 83
Paradiso, Fattoria (Italy/Emilia-Romagna) 35, 39, 43
Paradiso, Fattoria (Italy/Tuscany) 273
Parducci 59
Paret, Alain 283
Paringer Estate 185
Parodi 169
Paros co-op 122
Parusso, Armando 41, 87, 163
Parxet 166
Pask, C J 137
Pasolini, Mario 122
Passavant, Ch. de 83
Paterna, Doña 37
Paternoster 34, 89
Pato, Luis 38, 39, 42, 122
Paumanok 45
Pavelot, Jean-Marc 185
Pavie-Macquin, Ch. 137
Pavillon, Ch. du 239
Pavy, Gaston 83
Pazo de Barrantes 37
Pazo de Señorans *36*, 37
Pazo de Villarei 37
Peachy Canyon 295
Pecchenino 87
Pech-Céleyran, Ch. *280*, 283
Pech-Redon, Ch. 43, 58, 84
Pecorari, Pierpaolo 267
Pecota, Robert 153
Pedralvites, Quinta de 122
Pedroncelli 295
Pedrosa, Viña 265, 267
Pegasus Bay 239
Pégau, Dom. du *98*, 101
Pegos Claros 58, 167
Pelissero 41, 87, 163
Pellada, Quinta da 269
Pellé, Henry 227
Pellegrino 113, 115, 285
Pelletier, Dom. 91
Peñaflor 119
Peñalba López 267
Penfolds 57, *70*, 73, 101, 141, 251, 252, 255
Pepi, Robert 239
Pepper Tree 137
Peraldi, Dom. Comte 229
Père Caboche 101
Pereira, Manoel Salvador 37
Pereira d'Oliveira 121
Pérez Barquero 167
Pernice 165
Perret, Jean-Claude (France/Savoie) 115
Perret, André (France/Rhône) 125, 283
Perrin, Roger 101
Perrone 153
Pertimali 217
Pervini 164, *293*, 295
Pesos, Quinta dos 38
pests of vines 20
Petrussa 229
*Phylloxera vastatrix* 18
Petaluma *55*, 57, 73, 137, 201
Petersons 239
Petit Metris, Dom. du 83
Petit Val, Dom. du 83
Petits Quarts, Dom. des 83
Petrolo, Fattoria 137, *214*, 217
Pétrus, Ch. *130*, 131, *132*, 137
Peyrade, Ch. de la 153
Peyre-Rose, Dom. 255
Peyrie, Dom. du 119
Peza co-op 115

Pfaffenheim co-op 59, 111, 153, 173, 243
Pfaffenweiler co-op 114
Phelps, Joseph 57, 255
photosynthesis of vines 16
*Phylloxera vastatrix* 18
Pibarnon, Ch. de 125, 141
Pibran, Ch. 57
Pichierri 295
Pichler, Franz Xaver 114, 153, 201
Pichler, Rudi 201
Pichon, Christophe 283
Pichon, Philippe 283
Pichon-Longueville, Ch. 57
Pichon-Longueville-Comtesse de Lalande, Ch. 57
picking grapes 26-7, 79
Pierazzuoli 217
Pieropan 91, 202
Pierre-Bise, Ch. 45, 83, 113
Pierro 73
Pietrafitta 273
Pieve Santa Restituta 121, 217
Pike's Polish Hill River 141, *198*
Pin, Ch. le 137
Pinard, Vincent 227
Pine Ridge 83
Pineraie, Ch. 119
Pinet co-op 169
Pingus, Dom. de *260*, *264*, 265
Pino, Cantina del 90
Pinon, François 83
Pio Cesare 163
Pipers Brook 111, *199*, 201
Pira, Enrico 163
Pira, Luigi 163
Pirineos 117, 139, 265
Pithon, Jo 83
Pittnauer, Hans 171, 207, 284, 285
Pivetta 153
Pizzini *215*, 217
Plageoles, Robert 123, 165
Plaimont, Producteurs 84, 89, 271
Plaisance, Ch. de 83
Planeta 73, 112, 137, 164
Plantagenet 73, 185, 201, 255
planting density of vines 23-4
Pliger, Peter 243
Plumpjack 217
Pochon, E 255
Podere dell'Olivos, Il *40*, 41, *161*, 163
Poggerino 217
Poggio Antico 217
Poggio Salvi 217
Poggio Scalette 217
Poggio al Sole 217
Poggione, Il 139, 217
Poggiopiano 217
Pojer & Sandri 45, 111, 142, 164
Poliziano 57, 122, *214*, 217
Polz, Erich & Walter 139, 143, 227
Ponder Estate 227
Ponsot, Ch. 35, 185
Ponte de Lima co-op 37
Ponzi 173, 185
Portugal 57, 69, 88, 122, 150, 153, 164, 253, 265, 270, 271
Alentejo 35, 38, 58, 167, 203, 260, 261, 263, 266, 268, 270
Bairrada 35, 38, *39*, 39, 42, 43, 58, 59, 89, 268
Beiras 120, 268
Colares 189
Dão 35, 38, 39, 42, 43, 58, 59, 88, 115, 263, 266, 268, 269
Douro Valley 16, 19, 58, 120, 139, 167, 203, 240, 260, 261, 263, 266, 267, 268, 269, 272, 284
Estremadura 35, 58, 112, 120, 188, 266, 268, 284
Madeira 38, 42, 58, 84, 89, 117, 120, 121, *240*, 240, 266, 272
Ribatejo 38, 39, 58, 59, 112, 188, 266
Setúbal Peninsula 35, 58
Vinho Verde 36, 37, 38, 39, 89, 117, 273
Potel, Nicolas 185
Potensac, Ch. 57
Poujeaux, Ch. 57
Poveda, Salvador 138, 141
Pra, Graziano & Sergio 91
Prade, Ch. la 137
Pradeaux, Ch. de 141
Pradelle, Dom. 125, 205
Prager, Franz 201
Pravis 142, 164
pre-phylloxera vines 19
Preston 239, 295
Pride Mountain 45
Prieuré de St-Jean de Bébian 205
Prima & Nuova 111
Primavera, Caves 39, 269
Primitivo Quilés 138, 141
Primo-Palatum 119
Primosic 169, 189
Princic, Alessandro 121, 171, 267
Principiano, Ferdinando 41, 87, 163
Protos 265
Providence 45
Prum, J J 201
pruning of vines 24
Prunotto 38, 41, 87, 163
Puffeney, Jacques 188
Pugliese, Aldo 43
Puiatti 169, 171
Pujol, Castel 241
Pupille, Le 35, 57, 121, 139, 217
Puts, Dom. le 271


# Q


Quady 42, 165, 267
Quail's Gate 122, 165
Quénard, Raymond 205
Querceto 217
Querciabella 57, 171, 217
Quintarelli 45, 85, 138, 203
Quintessa *32-33*
Quivira 227, *293*, 295
Qupé 141, *253*, 255, 283


# R


Rabasse-Charavin, Dom. 101
Rabaud-Promis, Ch. 239
Rabbit Ridge 295
Rabiéga 58
Raccaro, Dario 121
Radford Dale 137
Rafanelli, A *292*, 295
Raffault, Olga 45
Ragose, Le 85, 203
Rahoul, Ch. 239
Raimat 111, 166
rainfall
and vineyards 26
Rainoldi 163
Rame, Ch. la 239
Ramò, Lorenzo 87
Ramonet, Dom. 73
Ramos, João Portugal 35, *263*, 265, 266, 270
Ramos Pinto 266, 269
Ramoser, Georg 229
Rampolla, Castello dei 57, 168, 217
Rapitalà, Adelkam 58
Rasmussen, Kent 87
Raspail-Ay, Dom. 101
Ratti, Renato 87, *159*, 163
Ratto, Giuseppe 87
Rauzan-Ségla, Ch. *52*, 57
Raveneau, Dom. François 73
Ravenswood 295
Ravenswood Lane 227
Ray-Jane, Dom. 141
Rayas, Ch. 96, *100*, 101, 113
Raymond-Lafon, Ch. 239
Rayne-Vigneau, Ch. 239
Raz, Ch. le 239
Rebholz 73, 153, 185, 241, 284
Rectorie, Dom. de la 101
Redbank 255
Redortier, Ch. 101
Reguengos de Monsaraz co-op 266
Reif Estate 201
Reinhartshausen 171
Réméjeanne, Dom. de la 101
Remelluri *259*, 265
Remirez de Ganuza, Fernando 265
Remizières, Dom. des 125, 205, 255
Remoissenet Père et Fils 185
Renou, Joseph 83
Renou, René 83
Renwood 41, 163, 295
Respide-Médeville, Ch. 239
Retiro, Finca El 119, 255, 265
Reverdy, Bernard 227
Revington 111
Rex Hill 173, 185
Reyes, Teófilo 265, 267
Reynolds Yarraman 239
Reynon, Ch. 239
Ribeiro, Vitivinícola del 267, 270
Ricaud, Ch. de 239
Richaud, Marcel 101, 141, 125, 205
Richou, Dom. 45, 83, 113
Richter 201
Ridge Vineyards 57, 58, 122, 137, 141, 168, *290*, 291, 292, *294*, 295
Riecine 217
Rieflé, Dom. Joseph 111
Riegelnegg, Otto 143
Rietine 217
Rieussec, Ch. 239
Rigalets, Ch. les 119
Rigodeira, Quinta da 39
Rijckaert 73
Rillo, Orazio 34
Rinaldi, Giuseppe 90
Rinaldini, Moro Rinaldo 116
Rioja Alta, La 112, 123, 265, 284
Riojanos, Bodegas 265
Rion, Daniel 35, 171, 185
Ripaille, Ch. de 59
ripeness of grapes 15-16
Rippon Vineyard 111, 185
Riscal, Marqués de 123, 227, 265, 272
Rivera 271
Rivetti, Giuseppe see La Spinetta
Rivière Haute, Ch. de la 43
Robaliño 37
Robert-Denogent, Dom. 73
Robertson Winery 84, 187
Roc, Ch. le 164
Roc de Cambes, Ch. 137
Rocailles, Dom. des 139
Rocca, Albino 41, 87, 163
Rocca, Bernarda 169
Rocca, Bruno 41, 87, 163
Rocca di Castagnoli 217
Rocca di Fabri 207
Rocca della Macìe 217
Rocca di Montegrossi 121, 217
Rocche Costamagna 163
Rocche dei Manzoni 41, 163
Rochais, Guy 83
Roche-aux-Moines, Ch. de la *77*, *82*, 83
Roche Redonne, Dom. de 141
Rochelles, Dom. des 45, 83
Rochemorin, Ch. de 227
Roches Neuves, Dom. des 45
Rochioli, J. 185, 227
Rockford 101, *251*, 255
Rockland 168
Roda, Bodegas 265
Rodaro, Paolo 169, 267, 273
Rodero, Bodegas 265
Rodet. Maison Antonin 185
Rodriguez y Berger 34
Roederer Estate *69*, 73, 138
Roger, Jean-Max 227
Roja, Emilio 267, 270
Rolet Père et Fils 271
Rolland, Michel *131*, 131
Rolly Gassmann 111, 153, 171, 173
Romana, Carlo 87
Romanée-Conti, Dom. de la 20, *177*, 179, 185
Romania 35, 40, 53, 59, 89, 90, 108, 112, 115, 133, 151, 164, 182, 188, 202, 224, 271, 284
Cotnari 241
Romanin, Ch. 85
Romassan, Ch. 141
Romeira, Quinta da 38, 189
Romer du Hayot, Ch. 239
Rominger, Eric 111
Ronchi, Umani 139, 272
Ronchi di Cialla 169, 189, 229
Ronchi di Manzano 169, 189, 267, 272
Ronco del Gelso 45, 173
Ronco del Gnemiz 169, 229, 272
Ronco dei Roseti 45
Roques, Quinta dos 35, 43, 88, 115, 265, *269*, 269
Roquetaillade-la-Grange, de 239
Roquette, Dom. la 101
Rosa, Quinta de la 203, 265
Rosemount 73, 99, 101, 141, 255
Rosenblum 295
Rostaing, René 255, 283
Rotaliana, Cantina 266
Rothbury Estate 239, 255, 272
Rotllan Torra 101
Rottensteiner, Hans 116
Rottensteiner, Heinrich 229
Rouget, Dom. Emmanuel 185
Roulerie, Ch. de la 83
Roulot, Dom. Guy 73
Rouquette-sur-Mer, Ch. la 43
Roura 166
Roumier, Georges, Dom. *177*, 185
Rousseau, Armand Père et Fils, Dom. *180*, 185
Routas, Ch. 283
Rouvière, Ch. de la 141
Royal Tokaji Wine Co 90, 114
Ruck, Johann 114
Ruffino 73, 217
Ruggeri, Angelo 188
Ruggeri, L & C 188
Russia 35, 108, 151, 199, 202, 242, 270
Russiz Superiore 45, 137, 169, 173, 267
Russolo 267
Rustenberg 73, 227
Rutherford & Miles 121


# S


Sabathi, Erwin 139
Sablonettes, Dom des 83
Sabon, Roger 101
Saddleback Cellars 171, 217, 295
Saes, Quinta de 43, 89
Saget, Guy 59
Saima, Casa de 39
St-Amand, Ch. 239
St Amant 267
St Antony 201
St Clair 227
St-Cosme, Ch. 101
St-Désirat co-op 255
St Francis 35, 137, 295
St-Gayon, Dom. 101
St George's 189
St Hallett 239, 255, 267
St Helena 171, 173
St-Jean-de-Minervois co-op 153
St-Pourçain co-op 270
Ste-Anne, Ch (France/Provence) 141
Ste-Anne, Dom. (France/ Rhône) 101, 141, 283
Saintsbury 73, 185
Sala, Castello della 73, 112, 185
Saladini Pilastri, Conte Saladino 139
Salaparuta, Duca di 115, 164
Salegg, Schloss 153
Salentein 119, 265
Salette, Le 85
Salettes, Ch. 141
Salnesur, Bodegas 37
Salomon, Fritz 111, 171, 270, 284
Salwey 112, 173, 206
Samos co-op 153
San Felice 217
San Fereolo 87
San Francesco 91
San Giusto a Rentennano 121, 217
San Isidro 265
San Leonardo 45, 57
San Michele Appiano co-op 111, 171, 227, 229
San Pedro 137
San Romano 87
San Vicente, Señorio de 265
Sandeman 269
Sandrone, Luciano 41, 87, 163
Sanford 185
Sansonnière, Dom. de la 45, 83
Sant'Agata, Cantine 113
Santa Barbara 272
Santa Duc, Dom. 101

Santa Lucia 271
Santa Maddalena co-op 229, 273
Santa María dos Remedios co-op 112
Santa Rita 45, 57, 61, 137
Santadi co-op 58, 139, 164, 273
Santar, Casa de 38
Santiago Ruiz 37
Santos Lima, Casa 269
São João, Caves 39, 42
Saracco 153
Sarda-Malet, Dom. 101, 153
Sartarelli 272
Sasbach co-op 171
Sassicaia 57
Sasso 34
Sastre, Hermanos 265
Satta, Michele 217
Sattlerhof 143
Sauer, Horst 201, *242*, 243
Sausal 295
Sauveroy, Dom. du 83
Sauzet, Dom. Etienne 73
Sava co-op 295
Savarines, Dom. des 119
Saxenburg 57, 137, 187, 227, 255
Scala Dei 91, 101
Scarpa 41, 43, 90
Scarsi, Cascina 87
Scavino, Paolo 41, 87, 163
Schaefer, Willi 201
Schaetzel, Martin 111, 153, 171, 173, *242*, 243
Schales, Weingut 114
Schandl, Peter 173, 206
Schellino 87
Schellmann, Gottfried 285
Schenk 265
Scherer, André 111, 171
Scherrer 295
Schiopetto 45, *120*, 121, 169, 171, 173, 202, 267
Schleret, Charles 111, 153
Schlumberger, Dom. 111, 171, 173
Schmitt's Kinder 243
Schneider 114
Schoffit, Dom. 59, 111, 153, 171, 173, 201
Schramsberg 138, 185
Schubert, Max *252*
Schueller, Gérard 111
Scolca, La 85
Scott, Allan 227
Screaming Eagle *54*, 57
Scrimaglio, Franco & Mario 41
seasonal work, vineyard 21
Sebaste, Mauro 163
Sebastiani, Giuseppe 266
Seghesio 41, *215*, 217, 295
Segonzac, Ch. 137
Seifried *109*, 111, 227
Selaks 227, 239
selection of vines 10-11
Seleni Estate 239
Sella & Mosca 99, 101, 267, 273
Selvapiana 121, 217
Semeli Winery 228
Senard, Comte 169
Señorio 112
Señorio del Condestable 138
Seppelt 43, 101, 111, 138, 141, 153, 255
Seresin *71*, 73, 173, 185, 227, 239
Serra, Cantina di Vinchio e Vaglio 41
Sertoli Salis, Conte 163
Setencostas, Quinta das 89, 189
Seuil, Ch. du 239
Seville Estate 255
Sezim, Casa de 37, 270
Shafer *55*, 57, 137, 217
Shaw & Smith 73, 227
Sierra Cantabria 265
Sigalas-Rabaud, Ch. 239
Signorelli 239
Sileni 73, 137
Silver Oak 57
Simi 239
Simoncelli, Armando 122
Simone, Ch. 141
Simonsig Estate 111, 187
Sine Qua Non 101, 141
Sineann 295
Sinfarosa 295
Sipp, Jean 111
Sipp, Louis 111
Skalli/Fortant de France 137, 283
SkillogIee 111
Skouras Winery 34, 228
slope, vineyard 16
Slovakia 108, 114, 182, 284
Slovenia 40, 43, 69, 90, 108, 133, 151, 189, 199, 224, 242, 284
Slovenjvino 116
Smart, Richard 263
Smith-Haut-Lafitte, Ch. *226*, 227
Sobon Estate 138, 205
Sociando-Mallet, Ch. 57
Sogrape 35, 37, 39, 42, 58, 89, 112, 115, 117, 263, 269, 273, 284
soils, vineyard 16-17
  biodynamic/organic comparisons 22
  *for soils affecting specific varieties see under* viticulture, varietal
Sokol Blosser 173
Solar das Bouças 117
Soldera 217
Sorg, Bruno 111, 153, 173, 243
Sorrel, Marc 125, 205, 255
Sottimano 163
Souch, Dom. de 114, 167
Soucherie, Ch. 83
Soulez, Pierre 83
Soulez, Yves 83
Soumade, Dom. la 101
Sours, Ch. de 239
Soutard, Ch. 45
South Africa 43, 45, 57, 73, 78-9, 80, 83, 84, 99, 109, 111, 114, 137, 141, 153, 171, 173, 183, 185, 186, 187, 201, 206, 227, 237, 239, 241, 243, 255, 266, 269, 270, 271, 281, 283, 293, 295
  Constantia 55
  Malmesbury 253
  Olifants River 151
  Paarl 85, 135, 253
  Robertson 71
  Stellenbosch 55, 71, 85, 135, 225, 253
  Walker Bay 66
  Worcester 71, 151
Spadafora 58, 164
Spagnoli, Enrico 142
Spain 53, 57, 73, 91, 111, 117, 125, 137, 138, 141, 150, 153, 165, 182, 199, 227, 249, 255, 260
  Andalucía 167, 267
  Basque Country 114
  Canary Islands 117, 164, 267
  Castilla-León 188, 224, 260, 261, 262, 265, 267, *271*, 271
  Castilla-La Mancha 34, 262, 265
  Cataluña 23, 98, 101, 166, *284*, 285
  Extremadura 167
  Galicia 36, 37, 88, 241, 267, 270
  Madrid 35, 267
  Navarra 69, 98, 120, 121, 263, 265, 284
  La Rioja 69, 98, 112, 120, 123, 253, 261, 262, 281, 284
  Valencia 42, 167
Sparr, Pierre 111
Speri 85, 203
Spice Route 83, *135*, 137, *186*, 187, 255
Spielmann, J-M 111
Spinetta, La 41, 87, *150*, 153, 163
Spottswoode 57, 239
Springfield 73, 227
Stadlmann, Johann 203, 285
Staglin family 217
Stagnari, Hector 241
Stag's Leap Wine Cellars 8, 24, *53*, 57, 168
Stags' Leap Winery 168
Stanko Curin 116
Stanton & Killeen 153
Staple St James 114
Statti, Cantine 91
Steele 73, 171, 295
Steenberg *225*, 227, 239
Stellenzicht 187, 239, 241, 255
Sterling Vineyards 121
Stonechurch Vineyards 39
Stonecroft *110*, 111, 255
Stoney Ridge (Canada) 122
Stonyridge (New Zealand) 57
Stoppa, La 41, 121
Story 138
Storybook Mountain 295
Strofilia 202, 228
Strozzi, Guicciardini 273
Struzziero, Giovanni 34, 89, 112, 169
Suau, Ch. 239
Suavia 91
Suduiraut, Ch. *233*, 239
Suffrène, Dom. de la 141
Sumac Ridge 111, *170*, 171
Sutter Home 292, 295
Swan, Joseph 295
Swanson 137, 217, 239, 255
Swartland Wine Cellar 84, 187
sweet wines
  and botrytized grapes 195, 235, 273
  Viognier 279
Switzerland 69, 130, 137, 182, 202, 243
  Geneva 167
  Ticino 133
  Valais 35, 114, 115, 124, 125, 168, 228, 242, 253
  Vaud *59*, 59


## T


Tachis, Giacomo *211*
Taille aux Loups, Dom. de la 83
Tain l'Hermitage co-op 125, 205
Tajikistan 151, 199
Talbott, Robert 73
Talley 185
Taltarni 119, 138, 255
Taluau, Joël 45
Tamariz, Quinto do 117
Tamás, Iván 270
Tamellini 91
tannin 16, 29, 31
Tanoré 188
Tapada, La *112*, 112
Tapada do Chaves 270
Tardieu-Laurent, Dom. 101, 255
Targe, Ch. de *45*, 45
Tariquet, Ch. du 84, 271
Tasca d'Almerita 57, 73, 115, 164
Tassarola, Castello di 85
Tatachilla 101, 137
Taurino, Cosimo 122, 164
Tavares & Rodrigues 189
Tavignano 272
taxonomy of vines 9
Taylor (Portugal) *268*, 269
Taylors (Australia) 57, 255
Te Mata *55*, 57, 73, 255
Tedeschi 85, 91, 203
Tement, E & M 139, 153, 227
Tempé, Marc 111
Tempier, Dom. *140*, 141
Tenaglia, La 41
Terlano co-op 171, 227
Termeno co-op 111, 229
Terpin, Franco 189
Terrablanca 217
Terraces, The 295
Terranoble *60*, 61
Terras Gauda 37
Terrazas de los Andes 119
Terrazze, Le 139
Terredora di Paolo 34, 89
Terre Rosse 169
Terre da Vino 41, 87
Terrebrune, de (France/Loire) 83
Terrebrune, Dom. de (France/Provence) 141
Terriccio 57
terroir
  and vineyards 14, 22
Tertre-Daugay, Ch. 45
Tertre-Roteboeuf, Ch. le 137
Teruzzi e Puthod *273*, 273
Tessier, Philippe 202
texture of wines 16
Thackrey, Sean 141, 168, 255
Thelema 57, 73, 137, 201
Theulet, Ch. 239
Thévenet, Jean 73
Thiel, Richard 285
Thirty Bench 45, 201
Thivin, Ch. 91
Thomas, Lucien 227
Thomas, Paul 83
Thorin 91
Three Choirs 39, 114, 117
Throwley 165
Thurnhof 112, 116
Tiefenbrunner 112, 116, 142, 153, 169
Tigné, Ch. de 45
Tijou, Pierre-Yves 83
Tinhorn Creek 138
Tirecul-la-Gravière, Ch. *237*, 239
Tissot, Dom. André & Mireille 271
Togni, Philip 42, 153
Tollot-Beaut, Dom. 185
Tommasi, Viticoltori 85
Topolos 35, 295
Torbreck 141
Torelló, Augustí 285
Torii Mor 185
Torino, Michel *16*, 57, 119, 267
Toro Albalá 167
Torraccia, Dom. de 229
Torre Rosazza 273
Torres 57, 73, 111, 137, *140*, 141, 227
Torrevento 271, 295
Tosa, La 41, 121
Touche Noire 83
Tour-Blanche, Ch. la 239
Tour de Bon, Dom. de la 141
Tours des Gendres, Ch. 239
Tours des Verdots, les 137
Tour-Figeac, Ch. la 45
Tour Vielle, Dom. la 101
Tourade, Dom. de la 101
Tours, Ch. des 101
Toutigeac, de 239
Tracy, Ch. de 227
Trapadis, Dom. du 101
Trapiche 119
Travaglini, Giancarlo 271
Tre Monti 35
trellising and training of vines *24-5*
Trentadue 295
Treuil-de-Nailhac, Ch. 239
Trévallon, Dom. de *124*, 125, 205
Trexenta 164
Triacca 163
Triebaumer, Ernst 42
*tries* 79
Trignon, Ch. du 101
Trimbach, F E 111, 153, 173, 201
Trinchero 41
Trollat, Raymond 125
Troplong-Mondot, Ch. 137
Trotanoy, Ch. 137
Trottevielle, Ch. 45
Truchard 137, 255
Tsantalis 285
Tua Rita 57, 137
Tualatin 142
Tucquan 58
Tunisia 151, 253
Turcaud, Ch. 239
Turckheim co-op 111, 171, 173
Turkey 43
Turkmenistan 151, 199, 202
Turley Cellars 168, 295
Txomin Etxaniz 114
Tyee 111, 171
Tyrrell, Bruce 236
Tyrrell's *70*, 73, *238*, 239, 251, 255


## U


Uccellina 35
Uiterwyck Estate *187*
Ukraine 59, 108, 151, 199
Ulrich, Jan 115
Umathum, Joseph 207
Unger, Dr 114
University of California at Davis 289, 290, 293
Uria, Giovanni 87
Uruguay 55, 114, 225, 241, 270, 271, 281
Ürziger Würzgarten 19, *194*
USA 39, 229
  Arizona 54, 292, 295
  California *see under* California
  Colorado 292

Illinois 292
Indiana 292
Iowa 292
Massachusetts 292
Michigan 195
Nevada 292
New Mexico 292
New York State *14*, 45, 58, 70, 85, 88, 122, 137, 164, 199, 273
North Carolina 292
Ohio 292
Oregon 54, 70, 81, 109, 111, 134, 142, 171, 173, 179, 183, 185, 195, 199, 201, 292
Pennsylvania 88
Tennessee 292
Texas 54, 292
Virginia 39, 125, 141, 283
Washington State 42, 45, 50, 54, 57, 70, 73, 81, 83, 101, 109, 116, 130, 134, 137, 142, 153, 163, 199, 201, 217, 225, 237, 239, 253, 255, 283, 295
Usseglio, Pierre 101
Uzbekistan 35, 151, 199

## V

Vacheron, Dom. 227
Vadiaperti 89
Vagnoni 273
Vajra, Giuseppe Domenico 41, *86*, 87, 163
Val Brun, Dom. du 45
Val di Suga 217
Val d'Orbieu, Les Vignerons du 84, 117, *250*, 255
Valandraud, Ch. 131, 132
Valdamor 37
Valdespino 166, 167
Valdicava 217
Valdipiatta 217
Valdivieso 16, 45, 57, 119, *134*, 137, 185
Valdoeiro, Quinta do 38
Valduero 265, 267
Valdumia 37
Vale da Raposa, Quinta do 203, 265, 267, *268*, 269
Valentini 270
Valette 73
Vallarom 122, 171
Valle, Caso do 39
Valle Isarco co-op 243
Valle Pradinhos 266
Vallerosa-Bonci 272
Valley Vineyards 88, 117, 142, 165, 189, 229, 240
Vallis Agri 122
Vallon co-op 89
Vallone 164
Vallouit, Ets L de 283
Valtellina 217
Valtellinese, Enologica 163
Valtravieso 265
Van Loveren 173, 206
Vannières, Ch. 141
Varaldo, Rino 90
Varichon & Clerc 138, 206
Varière, Ch. la 83
varietal differentiation of vines 10
varietal wines 31, 51
Vasse Felix 73, 239, 255
Vatan, André 227
Vavasour 73, 227
Vecchia Cantina, La 169
Vecchie Terre de Montefili 217
Veenwouden 137
Vega de la Reina 272
Vega Sicilia *261*, *262*, 265, 267
Veglio, Mauro 41, 163
Velich, Roland and Heinz 73, *284*, 285
Veramonte 61
Vercesi del Castellazzo 43
Vergelegen 73, 137, 227
Verget, Maison 73
Veritas 101, 141
Vernay, Georges *277*, 283
Versa, La 169
Verset, Dom. Noël 255
Viader 45
Vial-Magnères 101
Viansa 163, 270
Viarengo e Figli, G L 41
Viarte, La 229, 267
Viberti, Eraldo 87
Viberti, Giovanni 163
Vicchiomaggio 217
Vidal, Angel Rodríguez 272
Vidal-Fleury, J 153, 283
Vie de Romans 73, 169, 173, 227, 267
Vietti *38*, 38, 41, 87, 163
Vieux-Château-Certan 45, 137
Vieux Château Gaubert 239
Vieux Donjon, le 101
Vieux Télégraphe, Dom. du 101
Vigna Rionda 87, 163
Vignalta 153, 171
Vigne di San Petro, Le 202
Vigne di Zamò, Le 169, 189, 267
Vignerons Catalans 84, 117, 122
Vignerons de Maury, Les 101
Vigneto delle Terre Rosse 121, *202*, 202
Vilariño-Cambados, Bodegas de 37
Vilerma 267, 270
Villa Bel Air 239
Villa Cafaggio 217
Villa Maria 57, 73, 137, 201, 227
Villa Matilde 34, 89, 169
Villa Mt Eden 295
Villa Pillo 255
Villa Russiz 173, 189, 202, 227, 267
Villa Simone 59
Villa Sparina 85, 87
Villaine, A & P de 35
Villandry, Ch. de *219*
Villano, Gérard 283
Villanova 121, 189
Villard, François 125, 255, 283
Villard Estate 185, 227
Villeneuve, Ch. de (France/Loire) 45
Villeneuve, Dom. de (France/Rhône) 101
Villiera 227, 241
Vin Blanc de Morgex et de la Salle, Cave du 42
Viña Almaviva *50*, 57, 61
Viña Casablanca 57, *71*, 73, *109*, 111, 119, 137, *225*, 227
Viña Tarapacá 57, *253*, 255
Viñas del Vero 111, 137, 265
Vinchio e Veglio Serra co-op 113
Vincor 183
Viñedos y Bodegas 265
Vineland Estate 273
vines
diseases 10, 20
genetics 10
geographical influences on 12-13
growth conditions 14-17
hybrid 8-9, 240
identification of 11
old 19
pests 18, 20
photosynthesis 16
phylloxera-resistant 18
pre-phylloxera 19
taxonomy 9
varietal differences 10
wild 8
yields 22-3
*see also* grapes; history of vines *and* viticulture
vineyards
altitude 16
climate 14-15, 16
land area under vines 13
latitude 12-13
rainfall 26
slope 16
soils 16-17
terroir 14, 22
*see also* vines *and* viticulture
vinification
chaptalization and acidification 29
co-pigmentation 31
fermentation temperatures 28-9
maceration of red wines 29-30
malolactic fermentation 30
must concentration 29
oaking 30-1, 51, 67, 195, 235, 279
yeasts 28, 67, 166, 228
vinification, varietal
Cabernet Sauvignon 51
Chardonnay 67
Chenin Blanc 79
Garnacha Tinta/Grenache Noir 97
Gewürztraminer 107
Merlot 131
Muscat 148-9
Nebbiolo 159
Pinot Noir 179
Riesling 195
Sangiovese 213
Sauvignon Blanc 223
Sémillon 235
Syrah/Shiraz 249
Tempranillo 261
Viognier 279
Zinfandel 291
Vire dos Remedios 123
Virgen de l'Orea 114
viticulture
cloning 10-11, 19
genetic modification 11
green harvesting 24-5
hybridization 8-9, 240
integrated management 21
irrigation 25-6
massal selection 10
picking and mechanical harvesting 26-7, 79
planting densities 23-4
pruning 24
seasonal work 21
trellising and training *24-5*
*see also* vines *and* vineyards
viticulture, biodynamic 22
viticulture, organic 21-2
viticulture, varietal
Cabernet Sauvignon 50-1
Chardonnay 66-7
Chenin Blanc 78-9
Garnacha Tinta/Grenache Noir 96-7
Gewürztraminer 106-7
Merlot 130-1
Muscat 148
Nebbiolo 158-9
Pinot Noir 178-9
Riesling 194-5
Sangiovese 212-13
Sauvignon Blanc 222-3
Sémillon 234-5
Syrah/Shiraz 248-9
Tempranillo 260-1
Viognier 278-9
Zinfandel 290-1
*Vitis* genus 8-9
*V. lambruscana* 58
*V. riparia* 18
Vitis Hincesti 202, 207
Viu Manent 137
Voerzio, Gianni 38, 41, 87, 90, 163
Voerzio, Roberto 41, 87, *157*, 163
Vogüé, Comte Georges de 185
Vojvodina 42, 115, 284
Volpaia, Castello di 122, *211*, 217
Volpe Pasini 189, 267
Vougeot, Clos de 180
Voulte-Gasparets, Ch. la 58
Voyat 153
Vredendal 206

## W

Wachau, Freie Weingärtner 201
Wagner 240
Walch, Elena 111, 173
Warre 269
Warwick Estate 45, *187*
Waterford 227
Weaver, Geoff 201, 227
Weil 201
Weinbach, Dom. *108*, 111, 153, 171, *172*, 173, 201, 243
Weinert 57, 119
Wendouree 119, 141, 153, 255
Wenzel 90
Werheim, Dr 171
Widmann, Baron 227
Wieninger, Fritz 42
wild vines 8
Wildekrans 187
Wilkinson, Audrey 111
WillaKenzie Estate *171*, 171, 173, 185
Willamette Valley Vineyards 185
Williams Selyem 185, 295
Wilson Vineyard 201
Winiarski, Warren 8, *53*
Winkler and Amerine Heat Summation Scale 15
Winner Wines 265
Wirra Wirra 255
Wirsching, Hans 229
Wither Hills 227
Wolf, J L 201
Wolf Blass 201
Wolff-Metternich 111, 229, 270
Wollersheim 122
Woodward Canyon 239
Wootton, 229
Wright, Ken 185
Wynns 57, 251, 255

## Y

Yalumba 31, 84, 101, 141, 239, 278, 280, *281*, 283
Yarra Yering 57, 137, 185, 255
yeasts in vinification 28, 67
*flor* in sherry 166
*voile* in Vin Jaune 228
yields of vines 22-3
*for yields of specific varieties see under* viticulture, varietal
Youngs 41
Yquem, Ch. d' *231*, *233*, *234*, *236*, 239
Yugoslavia, former 35, 53, 116, 133, 224
*see also* Croatia; Slovenia *and* Vojvodina

## Z

Zaca Mesa 141, 255, 283
Zaccagnini, Fratelli 272
Zema 255
Zemmer, Peter 116
Zenato 85
Zeni, A & R 142, 266
Zerbina 35, 217
Zierer, Harald 203
Zimbabwe 186
Zimmerling, Klaus 39, 111, 173, 270
Zind-Humbrecht 106, 107, *110*, 111, 153, 171, 173, *197*, 201, 243
Zündel, Christian 137

# ACKNOWLEDGMENTS


**BIBLIOGRAPHY**

Nicolas Belfrage *Barolo to Valpolicella* (Faber & Faber, London, 1999)

Stephen Brook *The Wines of California* (Faber & Faber, London, 1999)

Bruce Cass (Ed.) *The Oxford Companion to the Wines of North America* (Oxford University Press, Oxford, 2000)

Pierre Galet *Dictionnaire Encyclopédique des Cépages* (Hachette, Paris, 2000) and *Précis d'Ampélographique Pratique* (J F Impression, St-Jean-de-Védas, Montpellier, France, 1998)

Anthony Hanson MW *Burgundy* (Faber & Faber, London, 1995)

Ian Hutton *The Zinfandel Trail* (IGH Publications, Esher, Surrey, England, 1998)

James Halliday *Wine Atlas of Australia and New Zealand* (HarperCollinsPublishers, Sydney, 1998)

Ron S. Jackson *Wine Science, Principles and Applications* (Academic Press, San Diego, California, 1994)

Hugh Johnson *The Story of Wine* (Mitchell Beazley, London, 1989)

Nicolas Joly *Le Vin du ciel à la terre* (Sang de la Terre, Paris, 1997)

Giles MacDonogh *Austria: New Wines from the Old World* (Österreichischer Agrarverlag, Klosterneuburg, Austria, 1997)

Richard Mayson *Portugal's Wines and Wine Makers, 2nd edition* (The Wine Appreciation Guild, San Francisco, 1998)

Alex McKay, Garry Crittenden Peter Dry, Jim Hardie *Italian Winegrape Varieties in Australia* (Winetitles, Adelaide, 1999)

Remington Norman MW *Rhône Renaissance* (Mitchell Beazley, London, 1995)

John Radford *The New Spain* (Mitchell Beazley, London, 1998)

Jancis Robinson (Ed.) *The Oxford Companion to Wine, 2nd edition* (Oxford University Press, Oxford, 1999)



**ACKNOWLEDGMENTS** The authors and publishers would like to thank the countless wineries and individuals all over the world who have given invaluable help and advice with this book, in particular the following:

**ARGENTINA** Susanna Balbo, Vintage S.A.; Pedro Marchevsky, Nicolas Catena
**AUSTRALIA** Cape Mentelle; Louise Helmsley-Smith, D'Arenberg; Robert O'Callaghan, Rockford Wines; Louisa Rose, Yalumba; Adam Wynn, Mountadam
**AUSTRIA** Austrian Wine Marketing Board
**BULGARIA** Domaine Boyar UK
**CANADA** British Columbia Wine Institute; Vintners Quality Alliance, Ontario; Wines of Canada
**CHILE** Wines of Chile
**CZECH REPUBLIC** CMVVU
**CYPRUS** Ministry of Commerce, Industry & Tourism – Cyprus
**FRANCE** Bureau Interprofessionnel des Vins de Bourgogne (BIVB); Conseil Interprofessionnel des Vins d'Alsace (CIVA); Comité Interprofessionnel du Vin de Champagne (CIVC), Comité Interprofessionnel des Vins du Jura; Comité Interprofessionnel des Vins Doux Naturels à Appellations Contrôlées; Comité Intersyndical des Vins de Corse; Conseil Interprofessionnel du Vin de Bordeaux (CIVB); Conseil Interprofessionel des Vins du Roussillon à Appellation d'Origine Contrôlée; Olivier Humbrecht MW; Interprofession des Vins du Val de Loire; Inter Rhône; Maison de la Vigne et du Vin – Gaillac; ONIVINS; Christian Seely, AXA Millésimes; Syndicat des Côtes de Provence; Syndicat Régional des Vins de Savoie; Union Interprofessionnelle des Vins de Beaujolais (UIVB)
**GERMANY** Dr Bernhard Abend; Geheimer Rat Dr von Basserman-Jordan; Jochen Becker-Köhn, Weingut Robert Weil; Weingut Josef Biffar; Reichsrat von Buhl; Ferdinand Erbgraf zu Castell-Castell, Fürstlich Castell'sches Domänenamt; Forstmeister Geltz-Zilliken; Horst Kolesch, Juliusspital; Franz Künstler; Ernst Loosen, Weingut Dr Loosen; Bernd Philippi, Koehler-Ruprecht; Annegret Reh-Gartner, Reichsgraf von Kesselstatt; Stefan Ress, Balthasar Ress; Dr Joachim Schmid, Geisenheim Viticultural Research Institute; Johannes Selbach, Selbach-Oster; Dr Heinrich Wirsching, Hans Wirsching
**GREECE** Boutari
**HUNGARY** Dominique Arangoits, then of Disznókö; Crown Estates (Tokaj Trading House); The Hungarian Food & Wine Bureau; Royal Tokaji Wine Company
**ISRAEL** Golan Heights Winery
**ITALY** Dr Alberto Antononi; ARSSA (Abruzzo); Mario Consorte, Sella & Mosca; Consorzio di Tutela Vini DOC Valtellina; Enoteca Italiana, Siena; Enoteca Regionale Emilia-Romagna; Dott. Giancarlo Montaldo; Regione Autonoma Valle d'Aosta; Regione Campania; Regione Marche; Regione Siciliana (Instituto Regionale della Vite e del Vino); Regione Toscana; Trentino Vini.
**JAPAN** Miyoko Stevenson
**LEBANON** Chateau Musar
**MEXICO** Mexican Association of Grape Growers
**MOLDOVA** Premium Brand Corporation
**NEW ZEALAND** Cloudy Bay; Alan Limmer, Stonecroft; Philip Manson, Winegrowers of New Zealand; Dr Neil McCallum, Dry River; The Wine Institute of New Zealand
**PORTUGAL** Jacques A Faro da Silva, Madeira Wine Company; Francisco Manuel Machado Albuquerque, Madeira Wine Company; ICEP – Portuguese Trade & Tourism Office – Wines of Portugal; Vasco Magalhaes, Sogrape;
**SLOVAKIA** Ministry of Agriculture of the Slovak Republic
**SLOVENIA** Ministry of Agriculture, Forestry & Food – Slovenia
**SOUTH AFRICA** South African Wine Industry Information & Systems; Gyles Webb, Thelema
**SPAIN** Ministerio de Agricultura, Pesca y Alimentacion; Carlos Read, Moreno Wines; Rioja Wine Group; Bodegas Riojanas
**SWITZERLAND** Yvon Roduit, Caves Imesch
**UNITED KINGDOM** Argentine Trade Department, Wine Department; Australian Wine Bureau; Bordeaux Wine Information Service; Ben Campbell-Johnson, J E Fells; Tony Potter, Denbies Wine Estate; English Wine Producers; German Wine Information Service; David Gleave MW; Bill Gunn MW; Margaret Harvey MW; Italian Trade Centre, Wine Department; Geoffrey Kelly; Antony Lacey, Mistral Wines; Paul Lapsley, Southcorp; Della Madison, Madeira Wine Company; Angela Muir MW; Joanna Simon; Westbury Communications; Wines of Chile; Wines from Spain; Wines of South Africa
**USA** Dr Carole Meredith, University of California at Davis (UCD); New York Wine & Grape Federation; Oregon Wine Advisory Board; Lynn Penner-Ash, Rex Hill; Michael Silacci, Warren and Julia Winiarski, Stag's Leap Wine Cellars; Wine Institute of California; Zinfandel Advocates & Producers (ZAP)
**ZIMBABWE** Cairns Wineries

**PHOTO CREDITS**

Photographs of bottles by Steve Marwood.

Photograph on page 8, 'Vintagers and rope makers at work, copy of a wall painting from the tomb of Kha'emwese at Thebes, c. 1450 by Mrs Nina Garis Davies (1881-1965)', from British Museum, London, UK/Bridgeman Art Library.

All other photographs supplied by Cephas Picture Library. All photographs by Mick Rock except: Jerry Alexander 25; Kevin Argue 195; Nigel Blythe 193 (below left and right); Fernando Briones 37 (left); Hervé Champollion 151; David Copeman 121 (left), 240; Andy Christodolo 16, 21 (centre), 24, 26 (below), 50, 51, 59, 61 (left), 99, 119 (left), 131, 141 (left), 148, 158, 198, 206, 225, 278, 281; Juan Espi 81; Bruce Fleming 18 (right), 53; Chinch Gryniewicz 117; Kevin Judd 1, 10 (right), 13, 17 (above), 20 (above), 21 (left), 31 (right), 55, 70, 106, 119 (right), 135, 171 (right), 187 (right), 201, 205 (right), 239, 283; Herbert Lehmann 61 (right), 90; Diana Mewes 4, 259 (below left), 277 (below left); Steven Morris 134; R & K Muschenetz 11, 243 (left); Alain Proust 187 (left); Ted Stefanski 20 (below), 21 (below right), 54, 215, 291; Stephen Wolfenden 31 (left).